OXFORD MEDICAL PUBLICATIONS

AN INTRODUCTION TO HUMAN ANATOMY

AN INTRODUCTION TO HUMAN ANATOMY

J. H. Green

M.A., M.B., B.Chir., Ph.D., F.R.S.C.

*Professor of Physiology, University of London
at The Middlesex Hospital Medical School*

P. H. S. Silver

M.B., B.S., Ph.D.

*S. A. Courtauld Professor of Anatomy, University of London
at The Middlesex Hospital Medical School*

Oxford New York Toronto

OXFORD UNIVERSITY PRESS

1981

Oxford University Press, Walton Street, Oxford OX2 6DP

London Glasgow New York Toronto
Delhi Bombay Calcutta Madras Karachi
Kuala Lumpur Singapore Hong Kong Tokyo
Nairobi Dar es Salaam Cape Town
Melbourne Wellington
and associate companies in
Beirut Berlin Ibadan Mexico City

Published in the United States
by Oxford University Press, New York

British Library Cataloguing in Publication Data
Green, John Herbert
An introduction to human anatomy. –
(Oxford medical publications).
1. Anatomy, Human
I. Title   II. Silver, P. H. S.   III. Series
611   QM23.2   80-40063
ISBN 0-19-261196-8

Set by Western Printing Services Ltd, Bristol
Printed in Great Britain by the
Thetford Press, Norfolk

PREFACE

This book has been written as a companion volume to *An Introduction to Human Physiology*. It has been written primarily for students who are just starting the study of human anatomy. It is hoped that it will prove useful to those involved in physiotherapy, nursing, and other paramedical subjects and, of course, medical students, especially those who have had no grounding in biology or zoology. It is hoped that it will also prove useful to those studying human biology.

The object has been to provide a clear, concise introductory text for the benefit of those who wish to learn more about the structure of the living human body and who wish to relate this structure to function and disease. It is hoped that it will also serve as a brief summary of the salient points for those who wish to revise the subject. To this end, in addition to their mention in the text, a summary of the important muscles, their action and nerve supply and the important nerves and blood vessels has been included.

It is realized that anatomy is best studied in the dissecting room but this is not always possible and even when these facilities are available the amount of time for study is being eroded by the introduction of new subjects into the curriculum. Furthermore, it is the anatomy of the living with which this book is primarily interested.

We have tried to present much of the information in a new way. There is no place for another boring old anatomy book. There are many features of the design of the body which become self-evident and interesting if one considers the different possibilities which could be used to achieve a particular objective. We have employed a great deal of space in trying to explain why things are the way they are. Most students remember easily once they realize how it works, and that very often any rival arrangement would not work. We have tried to suggest this, and to show how you can work much of it out for yourself. Some of the things we have said may appear trite and obvious to experienced anatomists, but we know from our experience of teaching and of examining that simple things are not necessarily so simple to those who are starting a subject.

Nor must only one side of the body be studied. You must be able to apply your knowledge to either side of the body and always remember that, from the front, a subject's left is on your right and that the subject's right is on your left, whereas when viewing the back the subject's right is on your right and the subject's left is on your left. Of course it is the subject's right and left that is important, not your own! The illustrations in this book use either side of the body in a random fashion. Note carefully which side you are looking at.

There are two major problems associated with the study of human structure. The first is where to start. Any arrangement of chapters in a book has its disadvantage. If we start with the brain we soon find that we need to refer to the blood supply from the heart. If we start with the heart we find that we need to refer to the brain and the tenth cranial nerve or vagus. This means that a certain amount of cross-referencing between chapters and a small amount of duplication of information is inevitable.

The second is learning the language used. The majority of the structures, and particularly the muscles, have Latin names although an English translation is sometimes available. In many cases the Latin names have been derived from the Greek or Arabic. Although a knowledge of the derivation of these words is not essential for the study of the subject, such a knowledge explains many of the unusual words encountered. In the opinion of the authors such knowledge makes the subject more interesting and very often makes the words easier to remember. For this reason frequent reference is made in the text to the original meaning of the words used even though this interrupts the flow of the narrative. We start with a brief account of the Latin plural and genitive case endings for the benefit of those who are not familiar with this language.

To reduce the cost of the book to the reader the figures have been reproduced in black and white. The arteries are shown as 'unshaded' tubes, the veins as 'shaded' tubes, and the nerves as tubes with 'sinewaves'. The reader may like to colour these in red, blue, and green respectively to improve the clarity of the illustrations.

We wish to thank the staff of the Technical Graphics Department of Oxford University Press for preparing the illustrations. We are particularly grateful to Mrs Joanna R. Cheney for her invaluable assistance with the production of this book and its index.

The Middlesex Hospital Medical School　　　　　J. H. G.
London　　　　　P. H. S. S.

CONTENTS

CONTENTS

NOTE ON LATIN AND ANATOMICAL LANGUAGE

The first great anatomical book was written by Galen, who died in AD 200. He wrote in Greek. From then until the Renaissance (1450–1500) a mixture of Greek and 'dog' latin was used in anatomical texts. In 1555 order was restored by Vesalius when he published *De Corporis Humanis Fabrica* [*Concerning the Structure of the Human Body*]. This book, with over 800 large pages of text, and numerous accurate and artistic illustrations, is one of the great intellectual achievements of mankind. Vesalius wrote in Latin. However, it was only recently that an internationally agreed terminology has been produced after a protracted effort which first started almost a century ago. The first version, the *Basle Nomina Anatomica* (the BNA) was published in 1895. In 1936 another system emerged in Jena (the JNA). Then in 1950, at the first post-war International Anatomical Congress in Oxford, England, a committee was set up to reach world-wide agreement. Their proposals were accepted at the next meeting in Paris in 1955 (the *Paris Nomina Anatomica*: PNA) and finally ratified in New York in 1960.

There are about 5500 terms in the *Paris Nomina Anatomica* and they are all in Latin—a language without nationalistic or political implications. It was agreed, however, that any of these terms could be translated into the speaker's or writer's mother tongue. In practice, in the English-speaking world, the Latin terms are usually employed, but often there is a curious mixture of English and Latin.

It is interesting to note that many people use a vocabulary of only 2000 words in everyday life (admittedly in many different permutations). This fact puts into perspective the task which a student undertakes when the mastery of anatomy is begun.

It is important to approach your new language in an informed manner and always to find out how the terms have been put together and what they mean.

Most students nowadays have not had the doubtful privilege of learning Latin and it is *essential* therefore to possess a good medical dictionary which explains the etymology of the words. A dictionary which defines 'acetabulum' as 'part of the hip joint' (which you probably already know) is not so good as a dictionary which gives the origin of the word and explains why this particular five syllable word was first used, and why it caught on, as a good nick-name does, and why everyone who knows what it really means instantly recognizes the aptness of the imagery which has been conjured up [see p. 129].

It is still very common practice, especially in clinical circles, to use eponymous terminology, that is, terms which include the names of the alleged discoverer of an anatomical structure or of a disease. For example, the auditory tubes leading to the ear are often referred to as the *Eustachian tubes* after Eustachius, and the uterine tubes are commonly referred to as *Fallopian tubes* after Fallopius. But eponyms in medicine often fail to catch on at the international level because in faraway lands rival local eponyms spring up and suppress the name of the intrusive foreigner. A more serious objection to eponyms is that they constitute a form of terminology which gives no clue to the anatomical structure or disease process whereas a proper descriptive term is often self-explanatory. Eponyms have been eliminated from the *Paris Nomina Anatomica* terminology, but a few, such as those above, are so well established that they will no doubt continue to be used for some time to come.

In English the meaning of a word is often changed by moving its position in the sentence or by using various little words like 'of, by, with, from, for' and so on. In Latin these small words are not used and it is the endings of the words that are of importance in determining their meaning. It is therefore very important for you to notice the endings of Latin words. For example, take the terms 'sella turcica' which means 'the turkish saddle' (the picturesque name for the pituitary fossa in the base of the skull) and 'dorsum sellae' which means 'the back *of the* saddle'. Sell*a* and sell*ae* have different meanings indicated by their different ending '-a' or '-ae'.

Inanimate objects in English are always neuter, but in Latin they may be masculine, feminine or neuter. Quite a number are masculine especially those whose name (nominative case) ends in *-us* (plural *-i*).

The majority of Latin nouns ending in *-a* (plural *-ae*) are feminine. Since adjectives in Latin have to agree with their nouns, the adjectives will also have the feminine ending. Sell*a* is feminine and 'turci*ca*' has the feminine ending to agree with it.

Latin nouns ending in *-um* (plural *-a*) are usually neuter (as dors*um* above) and their adjectives will have a neuter ending.

In anatomy, we are frequently concerned with the Latin genitive (possessive) endings which replace the words 'of the'. Even in English we have an ending change (-'s or -s') as an alternative to using the words 'of the'.

For masculine Latin words ending in *-us* the genitive ending is usually *-i* (plural *-orum*). For feminine words ending in *-a* the genitive ending is *-ae* (plural *-arum*) as with sellae above. For neuter words ending in *-um* the genitive ending is -i (plural *-orum*) as with dors*i* below. With other word endings the genitive form may just have to be remembered. Most Latin dictionaries give both the nominative and genitive form of each word.

Difficulty may be encountered in anatomical terminology because key words are frequently left out because they are 'understood'. Thus when we say this is 'pectoralis major' or here is 'serratus anterior' we have in both cases omitted the word **musculus** (Latin for muscle) in accordance with widely accepted practice. Thus strictly we should say 'musculus pectoralis major'. If we need to put in the missing word it is usual to do so in English! Thus we say 'pectoralis major **muscle**'.

In anatomy we frequently need to use what are called 'comparative adjectives'. This is what we are doing when in English we say 'wide, wider, widest', or 'low, lower, lowest'. Here we have modified the meaning of the words wide and low, by changing their endings, and a similar method is used in Latin.

The regular way of doing this is normally to say for example 'la*tus*' for wide, 'la*tior*' for wider, 'la*tissimus*' for widest. The endings convey the sense of the words.

Take the muscle usually called 'latissimus dorsi': the point here is that *latissimus* is an adjective (meaning 'the widest') and *dorsi* is the genitive case of dorsum (see dorsum sellae above) and means '*of the back*'. Latissimus does not have a noun to which it can attach itself. In fact the word **musculus** again is 'understood', and if we put in the missing word we could say the 'widest *muscle* of the back'. However, it is usual to use Latin adjectives such as 'latissimus dorsi' as if they were one English noun and say this is 'latissimus dorsi'. This process of amalgamation of languages, often in a grammatically illogical way, was one of the principal factors in the evolution of the English language itself from a German Anglo-Saxon dialect, from French and from Latin. There need be no inhibitions in participating today in the same evolutionary process.

'Supra' means above, and the other Latin words we use in comparing this adjective are 'supe*rior*' and 'supre*mus*'—which are almost English words anyway. Notice that on this occasion we do not say suprissimus (to match latissimus) because this word supra is 'irregular'.

The opposite word 'infra' (which means below) is even more irregular and goes via 'infe*rior*' to 'infimus' or most commonly to '*imus*'.

Here are some examples: the *supra* meatal crest, the *superior* vena cava (PNA has vena cava superior!) and linea nuchae *suprema*; *infra* glenoid tubercle, *inferior* vena cava, the thyroidiae *ima* artery (in PNA Latin we would say arteria thyroidiae ima) which means the lowest thyroid artery. Note that **musculus** is masculine in Latin whilst **arteria** is feminine in Latin. This difference in gender of the noun alters the endings of the satellite adjectives.

The Latin for 'and' is 'et' but sometimes instead the suffix -que is added to the end of the previous word. Thus 'levator anguli oris alae*que* nasi' means the 'levator of the angle of the mouth *and* of the nose'. This small but important muscle is an accessory muscle of respiration which prevents the nostril from being sucked in in forced respiration. Its activity is often a useful clinical sign of dyspnoea, for example, in cases of pneumonia.

It is of importance to remember that in the PNA there are no superfluous words. When we speak of 'pectoralis *major*' you will instantly know that there must be another muscle called pectoralis *minor*; if there is a serratus *anterior* there must also be a serratus *posterior*. Do not forget, however, that there may be more than one complementary muscle. In fact, there are two serratus posterior muscles distinguished by a further pair of adjectives *superior* and *inferior*. There are innumerable similar examples.

Finally, quite a number of anatomical words are derived from classical Greek, and usually the Greek words are entirely different from the corresponding Latin words.

Here are a few examples:

Latin	Greek
Mental (chin)	Genial
Navicular (boat-shaped)	Scaphoid
Mala (cheek bone)	Zygoma
Clavicle	Cleido
Fibular	Peroneal
Omental	Epiploic
Lien	Spleen
Jaundice	Icterus
Parietal	Somatic
Visceral	Splanchnic
Brachium (shoulder)	Omos
Umbilicus	Omphalos
Vitreous (like glass)	Hyaline
Digits	Phalanges
Tuba (tube)	Salpinx
Ovum	Oon
Ovarian	Oophoro-

1 Introduction

The human body consists of the trunk and skull together with the two upper limbs (or upper extremities) and the two lower limbs (or lower extremities). The names of the principal bones of the body are shown in FIGURE 1.01.

The use of the words arms and legs for the limbs is slightly different from that in everyday life. The term **arm** refers to the upper arm above the elbow. The part below the elbow is termed the **forearm**. The term **leg** means the part of the lower extremity below the knee. Above the knee it is termed the **thigh**.

The main bone of the arm is the **humerus**. The forearm bones are the **ulna** and **radius**. Note that the adjectival form of these words are *ulnar* = pertaining to the ulna, and *radial* = pertaining to the radius.

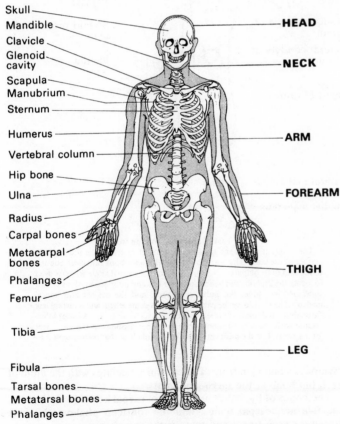

Skull
Mandible
Clavicle
Glenoid cavity
Scapula
Manubrium
Sternum
Humerus
Vertebral column
Hip bone
Ulna
Radius
Carpal bones
Metacarpal bones
Phalanges
Femur
Tibia
Fibula
Tarsal bones
Metatarsal bones
Phalanges

HEAD
NECK
ARM
FOREARM
THIGH
LEG

FIG. 1.01. The principal bones of the body in their anatomical position. Note the anatomical use of the words 'arm' and 'leg'. Arms refers only to the part of the upper limb above the elbow whilst leg means the part of the lower limb below the knee.

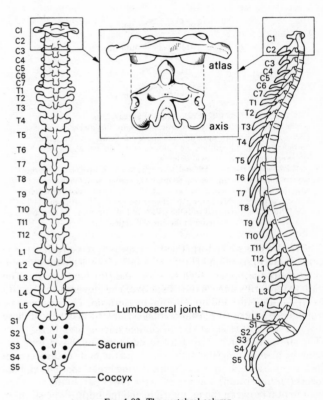

FIG. 1.02. The vertebral column.
The vertebral column consists of 7 cervical, 12 thoracic, and 5 lumbar vertebrae. It joins the sacrum at the lumbosacral joint. The upper two cervical vertebrae are termed the atlas and axis. The atlas articulates with the skull.

The **vertebral column** [FIG. 1.02] extends from the skull down to the **sacrum**. It consists of seven cervical, twelve thoracic, and five lumbar vertebrae. The sacrum itself is formed by five fused sacral vertebrae. Below the sacrum is the coccyx consisting of 1–4 parts.

At the upper end the first cervical vertebra or **atlas** (adjective = *atlanto*) articulates with the **occiput** (adjective = *occipital*) of the skull. A certain amount of flexion and extension of the head on the neck takes place at this joint.

The joint between the first cervical vertebra and second cervical vertebra or **axis** (adjective = *axial*) allows some rotation of the head. Thus one can nod ones head to say 'yes' at the atlanto-occipital joint (C1–skull) and shake ones head to say 'no' at the atlanto-axial joint (C1–C2).

The lower end of the vertebral column, the fifth lumbar vertebra, articulates with the sacrum at the **lumbosacral joint**.

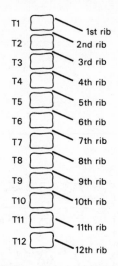

FIG. 1.03. Costovertebral articulations.

The head of each rib usually articulates with the intervertebral disc above its own vertebra and overflows above and below onto the adjacent vertebral bodies. In addition the tubercle of the rib articulates with the transverse process of its own vertebra.

The heads of the 1st, 11th and 12th ribs, however, articulate with the sides of their own vertebrae and do not make contact with the disc above. In addition the two floating ribs (11th and 12th) do not articulate with their transverse processes. Note the gap between the heads of the 10th and 11th ribs. This gap is associated with the failure of the 11th rib to be incorporated into the costal margin.

The twelve **ribs** which form the thoracic cavity are attached to the thoracic vertebrae behind [FIG. 1.03] and to the sternum in front. Usually only the upper seven ribs join the sternum directly. The eighth to tenth ribs join via the ribs above. The eleventh and twelfth ribs are floating ribs and do not join the sternum. The ribs become cartilaginous anteriorly (**costal cartilages**) and the elasticity of these cartilages allows the chest size to change during breathing.

The **sternum**, which consists of fused segments, has a joint between the **manubrium** (the upper part) and the **body of the sternum** (the lower part) which moves slightly during breathing. The **xiphoid process** (xiphisternum) is attached to the lower part of the sternum. The junction between the manubrium and the body of the sternum is termed the **sternal angle**. It is an important landmark for the localization of the second rib [FIG. 1.04].

The main bone of the thigh is the **femur** (adjective = *femoral*), and those of the leg are the **tibia** (adjective = *tibial*) and **fibula** (adjective = *fibular* or *peroneal*, see p. x). The fibula does not reach up to the knee joint, but it forms the lateral malleolus of the ankle joint [FIG. 1.04]

The upper limb is attached to the trunk by the **scapula** and **clavicle**. The head of the humerus forms a joint with the **glenoid cavity** of the scapula. The scapula is attached to the trunk by muscles. The clavicle, which acts as a strut, has joints with both the scapula at its outer end, and with the manubrium of the sternum and first rib at its inner end.

The lower limb is attached to the pelvis. The head of the femur forms a joint with the acetabulum of the **pubis**, **ilium**, and **ischium**. All three pelvic bones form the acetabulum joint socket. Together these three bones form the **hip bone** (previously known as the innominate bone).

The pelvic girdle is completed by the sacrum behind and by the junction between the two pubic bones, termed the **pubic symphysis**

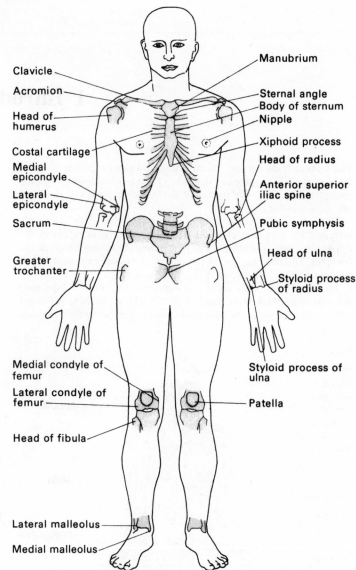

FIG. 1.04. Bony points on the front.

The diagram shows the important points of reference on the anterior surface of the body. These are the places where the bones are close enough to the skin to be palpated. Note particularly the sternal angle which is used to locate the 2nd rib, and hence the other ribs and interspaces; the anterior superior iliac spine; the pubic symphysis; and the medial and lateral malleoli. The clavicle can be palpated along its entire length and a vertical line through its midpoint is termed the 'midclavicular line'. Its point of intersection with the 5th left interspace usually corresponds to the apex beat of the heart. It is the location of the V_4 lead of chest electrocardiography.

(symphysis pubis) in front. The sacrum articulates with the ilium of each hip bone at the **sacro-iliac joints**.

The bones of the hands are the eight **carpal** bones (wrist bones), the five **metacarpal** bones, and the fourteen **phalangeal** bones (three for each finger and two for the thumb).

The bones of the foot are the seven **tarsal** bones, the five **metatarsal** bones, and the fourteen **phalangeal** bones (three for each toe except the big toe which has two).

Since we are dealing with living human beings, and not with skeletons, it is important to try to visualize the bones in position under the skin. To this end, any place when a bone is close enough to the skin to be palpated is of great help. These are termed **bony** **points**. You will find these bony points referred to frequently throughout this book. Some of the more important palpable structures of the front of the body are shown in FIGURE 1.04 and those of the back in FIGURE 1.05.

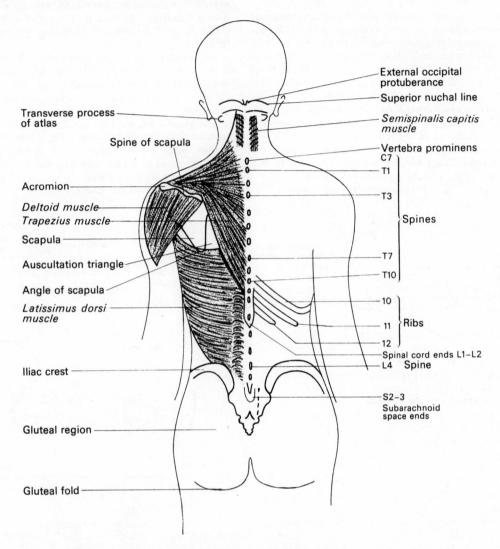

FIG. 1.05. Bony points on the back.

The diagram contains the information concerning the back which every doctor carries round in his head. Trapezius and latissimus dorsi between them cover most of the back. They lie just deep to the skin. The *auscultation triangle* is clear of these muscles and is the place where tradition has it that breath sounds are more easily heard with a stethoscope.

The vertebra prominens (C7) is the first spine which projects unambiguously as the finger passes down the back of the neck although the spine of T1 often projects further than C7. T3 is identified at the level of the spine of the scapula. The oblique fissure of the lung commences behind at this level. T3 is the apex of the lower lobe of the lung; above T3 is the upper lobe; below T3 is the lower lobe. T7 is identified at the level of the inferior angle of the scapula. Count down to T10. At this level the base of the lung finishes. The pleural cavity extends two vertebrae further down just below T12, a distance of 5–6 cm.

The spinal cord terminates at the gap between the spines of L1 and 2. L4 is identified at the level of the highest point of the iliac crests. It is the same level as the umbilicus. L4 marks the top of your trousers or skirt.

The posterior superior iliac spine is about the level of S2 and this identifies the lowest level of the subarachnoid space and spinal theca.

Arch the back and identify the erectores spinae muscles on each side of the midline in the lumbar region (small of your back). The 12th rib may be too short to be palpated at the lateral border of this muscle. The tip of the 11th rib can however always be identified. The tip of the 10th rib and its costal cartilage join the costal margin which leads directly to the sternum.

THE NERVOUS SYSTEM

Introduction

In the nervous system the fundamental cellular unit is the **neurone**. Signals called **nerve impulses** are conveyed to all parts of the nervous system by neurones. A neurone consists of a cell body which has a variable number of processes projecting from it. One of these processes is called the **axon** or dendron, and this conveys nerve impulses away from the cell body. Other processes which convey nerve impulses towards the cell body are **dendrites**. The internal substance of these processes is **axoplasm**. It is continuous with the cytoplasm of the cell body. The same membrane encloses both axoplasm and the cytoplasm of the cell body [FIG. 1.06].

A **nerve fibre** consists of a cylinder of axoplasm surrounded by a protective **myelin sheath** which is interrupted at intervals at **nodes** (of Ranvier) [FIG. 1.06(b)]. This combination is called a **myelinated nerve fibre**. The term **unmyelinated nerve fibre** is used to describe fibres which are also protected and functionally isolated from other adjacent fibres, but in a very much simpler manner [FIG. 1.06(e)]. The insulating cells in both cases in the peripheral nervous system are **Schwann cells**, without which the nerve fibres do not function. The myelin of the myelinated nerve fibre is laid down by the Schwann cell encircling the axon [FIG. 1.06(c) and (d)].

Nerve impulses have chemical and electrical characteristics and it is the special properties of the **cell membrane** covering the whole neurone which allow the conduction of nerve impulses. The electrical changes associated with nerve impulses are termed **action potentials**.

The axon of one neurone may be apposed to another neurone in such a way that the nerve impulses in the first excites nerve impulses in the second. The specialized junction zone is a **synapse**. Synapses conduct in one direction only. The signal is said to **relay** when it is transmitted from one neurone to another. One neurone may have up to a million synaptic connections.

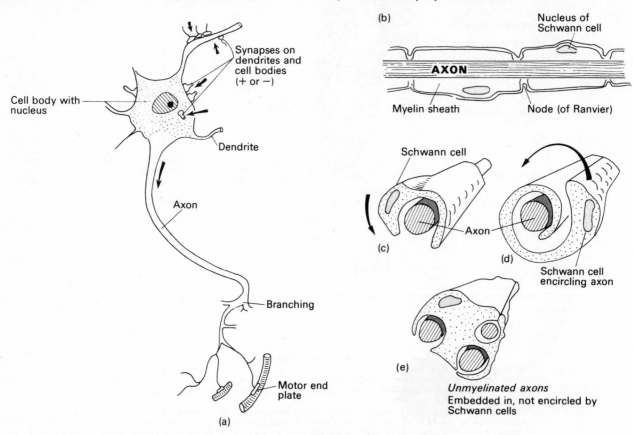

FIG. 1.06. The neurone.

(a) A motor neurone showing cell body with nucleus and nucleolus. There are five dendrites and one axon. Synapses are shown on the dendrites and cell body. The axon is dividing and terminating in multiple motor end plates on voluntary muscle fibres.

(b) The axon is surrounded by a myelin sheath. The nodes (of Ranvier) are gaps in the myelin at intervals of about 1 mm. They enable fast saltatory conduction of nerve impulses to occur.

(c) The start of the myelination process.

(d) The myelination process continuing. The Schwann cell encircles the axon laying down successive layers of myelin. In the mature myelin sheath there is very little cytoplasm left. only the double coils of cell membrane remain which appear to be tightly bound together so that their contiguous surfaces adhere to each other.

(e) Unmyelinated axon embedded in Schwann cells. but there is no Schwann cell rotation.

Axons have different specialized **endings** at their tips depending on their function which may be to excite another neurone (synapse), to cause **secretion** in a gland, or the **contraction** of a muscle (motor end plate). Certain hormones are secreted by neurones [p. 403].

Sensory (afferent) nerve fibres, for example in the skin, are activated at their peripheral extremities by various different physical stimuli such as touch, hot, or cold. At a given moment many fibres will respond only to one type of stimulus. The function of afferent nerve endings, whether they be in the skin, the eye, or the ear, is to respond to the energy conveyed by a particular modality of sensation and to translate it into the 'all or none' language of the nervous system which is the presence or absence of an action potential (rather like the silicon chip microprocessor which responds only to the binary language of 0 and 1).

Motor (efferent) nerve fibres may branch many times so that one axon may ultimately come to supply 200–300 separate muscle fibres. Similarly sensory dendrites may be very complex with numerous branches merging as they approach the cell body.

The **nervous system** is divided into:
1. The Peripheral Nervous System.
    Peripheral somatic nerves
    Autonomic nerves to viscera
2. The Central Nervous System (CNS).
    Brain
    Spinal cord

PERIPHERAL NERVOUS SYSTEM

The branches and cell bodies of the neurones of the central nervous system (CNS) lie entirely within the brain or spinal cord. The neurones of the peripheral nervous system lie partly or entirely outside the CNS. The 'peripheral nervous system' includes the cranial nerves, the thirty-one pairs of spinal nerves, as well as the peripheral parts of the autonomic nervous system. The optic nerve is a special case [p. 270].

Anatomy of typical spinal nerves

The spinal nerves emerge in pairs from each side of the spinal cord via the gaps between the vertebrae (intervertebral foramina) [FIG. 1.07].

By convention the nerve which leaves between the skull and first cervical vertebra is termed the first cervical nerve (C1). The nerve leaving between the first cervical vertebra and the second cervical vertebra is termed the second cervical nerve (C2) and so on. Thus each cervical nerve is named according to the vertebra lying below it. However, the nerve leaving between the seventh cervical vertebra and the first thoracic vertebra is termed the eighth cervical nerve (C8). Thus there are eight cervical nerves and only seven cervical vertebra. The remaining spinal nerves are named according to the vertebrae lying above them. Hence there are eight cervical nerves, twelve thoracic, five lumbar, five sacral, and one coccygeal nerve on each side.

A typical nerve is shown in FIGURE 1.08. It consists of a mixture of efferent and afferent fibres. The motor fibres have their cells of origin in the anterior columns of grey matter in the spinal cord [FIG. 1.08] and leave the cord in the ventral (anterior) spinal nerve roots. The dorsal (posterior) nerve root consists only of afferent sensory fibres. The cells of origin of these fibres are in its spinal (posterior root) ganglion. The formation of the mixed spinal nerve takes place just peripheral to the spinal ganglion.

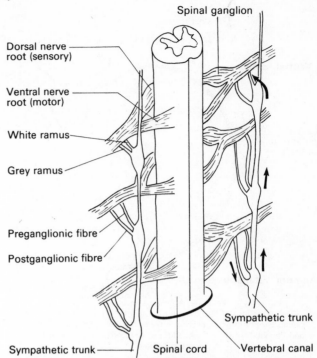

FIG. 1.07. The spinal cord and sympathetic trunk viewed from in front. The segmental arrangement is seen. The dorsal and ventral nerve roots of each segment join to form the mixed spinal nerve root. The swelling associated with the dorsal nerve root (the spinal ganglion) contains the cells of origin of the sensory neurones (but there is no synapse here). The preganglionic sympathetic fibres leave to join the sympathetic trunk and run up, or down, the sympathetic trunk to synapse (relay) in a distant sympathetic ganglion. The postganglionic fibres then rejoin another mixed nerve (see arrows on right).

The ventral and dorsal nerve roots when they reach the theca of the spinal cord are surrounded by individual diverticular sleeves of dura and arachnoid. These continue along the roots as far as the region where they unite with each other. This takes place in the intervertebral foramina. Note that in the cervical and thoracic

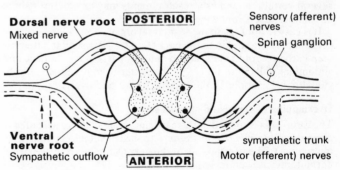

FIG. 1.08. The Bell–Magendie Law. Ventral nerve roots motor. Dorsal nerve roots sensory.

This diagram shows a cross-section of the spinal cord with a central area of grey matter (nerve cells) surrounded by white matter (nerve fibres). The motor nerves have their cells of origin in the anterior (ventral) horns of grey matter. The sympathetic nerves have their cells of origin in the lateral horns of grey matter (thoracic and upper lumbar segments only). The sensory nerves have their cells of origin outside the spinal cord in the spinal ganglion cells.

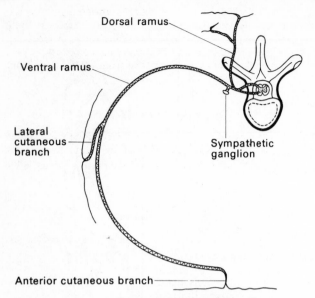

FIG. 1.09. A typical spinal nerve.
The nerve divides into a ventral (anterior) and a dorsal (posterior) ramus. The posterior ramus supplies the erector spinae muscles of the back and the skin overlying them. The rest of the trunk is supplied by the anterior ramus.

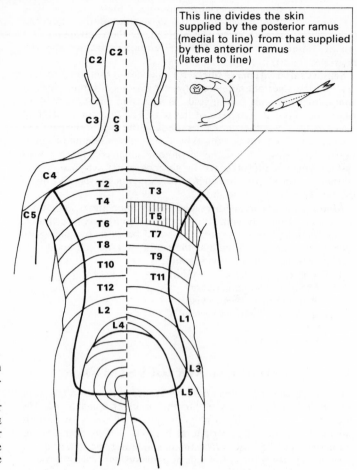

FIG. 1.10. Pattern of skin innervation (dermatomes) of the back.
The lateral line divides the skin medial to the line supplied by the posterior ramus from that lateral to the line supplied by the anterior ramus (insert). It corresponds to the lateral line of the fish (insert).
By superimposing the patterns on the left and right, it will be seen that there is a marked overlap in the skin innervation. Note that C3 and C4 overlap T2.

regions the ventral and dorsal nerve roots are relatively short, but in the lumbar and sacral regions the roots travel a considerable distance forming the **cauda equina** inside the spinal theca.

The branches of the spinal nerves are arranged in a regular pattern on a segmental basis. The posterior (dorsal) ramus [FIGURE 1.09] supplies the erector spinae and its overlying skin. The anterior (ventral) ramus supplies the rest of the trunk lateral to this muscle group (*ramus* (Latin) = a branch, plural = *rami*) as well as the limbs.

Note the rami communicantes which join the sympathetic trunk to the spinal nerves [FIG. 1.07]. Sensory nerves also leave the mixed spinal nerve and re-enter the intervertebral foramina to supply the spinal theca. They are distributed over a distance corresponding to several vertebrae, and as a result patients may experience pain at anomalous levels following back injuries.

It is easy to appreciate the segmental organization of the sensory nerve supply of the skin to the back [FIG. 1.10]. A similar pattern is seen in the rest of the lateral and anterior parts of the thoracic and abdominal walls [FIG. 1.11]. All the posterior rami remain as individual nerves supplying the erector spinae and overlying skin.

Overlapping of segmental spinal nerves

The **skin area** supplied by a particular spinal nerve is called a **dermatome** (*dermis* = skin; *'tome* = to slice or cut into sections as in micro*tome*). The overlapping of the nerve supplies of the adjacent dermatomes is seen in FIGURES 1.10 and 1.11.

As an example, the skin supplied by T10 is the T10 dermatome, whose upper half is overlapped by T9 and whose lower half is overlapped by T11.

This distribution of the sensory fibres of the peripheral nervous system in the skin, establishes what we call 'double assurance' since each area is supplied by two nerve roots.

Spinal nerves supplying the thorax and abdomen do not anastomose with each other. Overlapping occurs because their small peripheral branches find their way into the territory of the adjacent dermatomes.

In the cervical, brachial, and lumbosacral regions the situation is different because the anterior rami of the spinal nerves intermingle soon after they leave the vertebral column so as to form plexuses. In these plexuses the spinal nerves are packaged so as to give rise to composite trunks and cords containing contributions from as many as five different spinal nerves. The branches of the trunks and cord (with one exception in the cervical plexus) all have multiple root values and the congruence of peripheral nerve distribution and dermatome pattern is lost in the neck and limbs.

The cutaneous nerves derived from limb plexuses also demonstrate the principle of overlapping. Diagrams of their distribution usually suggest that a line can be drawn between the territory of one nerve and that of its neighbours. In view of overlapping such a line would be more appropriately represented by a smudge.

Take the back of the hand as an example. If the radial nerve is cut high up, the area of anaesthesia does not cover the whole three and most muscles. This additional complexity is made possible by the connections already established in the limb plexus. Unfortunately the nerve supply of the intrinsic muscles of the hand is an exception to double or triple assurance, as T1 provides an indispensable part of the nerve supply. There is some overlap from C8 but none from T2.

CRANIAL NERVES

In addition to the thirty-one pairs of spinal nerves, there are twelve pairs of cranial nerves which arise from the brain itself. Each has a name, but alternatively they are numbered using roman numerals:

I	Olfactory nerve
II	Optic nerve
III	Oculomotor nerve
IV	Trochlear nerve
V	Trigeminal nerve
VI	Abducent nerve
VII	Facial nerve
VIII	Vestibulocochlear nerve
IX	Glossopharyngeal nerve
X	Vagus nerve
XI	Accessory nerve
XII	Hypoglossal nerve

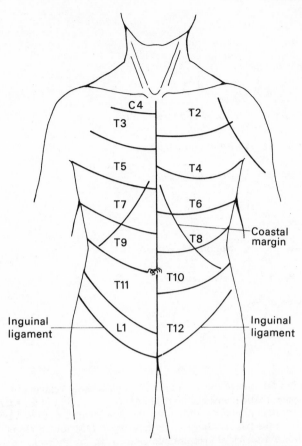

FIG. 1.11. Pattern of skin innervation (dermatomes) of the front. By superimposing the patterns on the left and right the overlapping of the dermatome nerve supply is seen. Such overlapping also occurs in the limbs. Note that C4 overlaps T2.

a half finger area, but is confined to the skin over the first dorsal interosseous and has a diameter of only 2–3 cm. This is because the adjacent ulnar and median nerves share 70–80 per cent of the radial skin nerve area. Similarly if the ulnar nerve is cut the radial side of the ring finger and the back of the hand will not show much abnormality, but the little finger, which is supplied exclusively by the ulnar nerve, will be completely anaesthetic. In diagnosing peripheral nerve injuries the clinician needs to bear overlapping very much in mind.

Dermatome and peripheral nerve patterns of anaesthesia are nevertheless always recognizably different. This means that it is possible to distinguish lesions affecting either segmental spinal nerves, or plexus nerves or peripheral nerves. Anaesthesia of the whole hand (glove anaesthesia) cannot come about from a segmental or peripheral nerve lesion. The diagnosis lies between a hysterical anaesthesia or a tumour which has destroyed the whole hand region of the sensory cortex in the brain.

Similar principles apply to the motor innervation of the trunk musculature. Here we speak of **myotomes** instead of dermatomes. Overlapping occurs in the muscles in the same way as it does in the skin.

In the limbs there is not a double but a triple nerve root supply to

AUTONOMIC NERVOUS SYSTEM

Introduction

The **autonomic** nervous system controls visceral structures such as the salivary glands, the blood vessels and heart, the gut, urinary bladder, and genitalia. The **somatic** nervous system by contrast controls 'voluntary' structures such as the skeletal muscles. Parts of the brain such as the **hypothalamus**, and the vasomotor centre have the ultimate central control over the visceral structures of the body. The autonomic centres of the brain may be profoundly influenced by the highest centres in the cortex.

In the periphery the autonomic is a motor system which acts on its various target organs by means of two separate neurones which **synapse** with each other in a peripheral ganglion. This distinguishes anatomically the autonomic from the somatic nervous system, in which the effector cells in the CNS act directly on their targets and in which therefore only one single neurone is involved.

The autonomic fibres running from the brainstem or spinal cord to these ganglia are called **preganglionic fibres**; the fibres running from the ganglia to the target organs are called **postganglionic fibres**.

Afferent fibres run from the viscera to the CNS but they do not differ in any fundamental respect from somatic afferent fibres. Both sets of fibres involve single neurones whose cells of origin are in the spinal ganglia. Pain fibres from the viscera normally accompany the autonomic fibres. Pain is usually referred to that part of the 'surface' of the body with the same segmental nerve supply as the sympathetic (not parasympathetic) nerves.

The autonomic system is divided into two:
1. The sympathetic nervous system
2. The parasympathetic nervous system

Many organs in the body (not all of them) are supplied by both

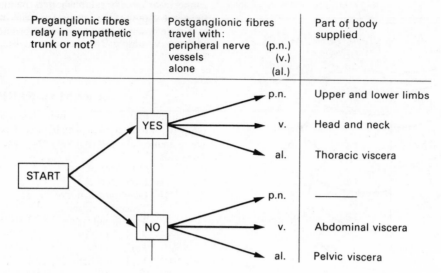

Preganglionic fibres relay in sympathetic trunk or not?	Postganglionic fibres travel with: peripheral nerve (p.n.) vessels (v.) alone (al.)	Part of body supplied
	p.n.	Upper and lower limbs
YES	v.	Head and neck
	al.	Thoracic viscera
	p.n.	———
NO	v.	Abdominal viscera
	al.	Pelvic viscera

FIG. 1.12. Summary of sympathetic pathways.

systems. Their actions are usually antagonistic. If one is excitatory the other is inhibitory. Thus the sympathetic quickens the heart beat while the parasympathetic slows it. The state of activity at a given moment usually depends on the balance struck between these conflicting factors. But the smooth muscle in the blood vessel walls and the sweat glands are examples of structures supplied only by sympathetic nerves; the ciliary muscle in the eyeball (used for focusing) is supplied only by the parasympathetic. In the somatic nervous system the motor nerves are exclusively excitatory.

SYMPATHETIC NERVOUS SYSTEM

The preganglionic fibres all have their cells of origin in the lateral column of grey matter in the spinal cord [FIG. 1.08]. Instead of stretching from one end of the spinal cord to the other, the lateral column exists only from T1 to L2 inclusive. This means that the sympathetic supply of the whole body including the head, has to concertina between these two levels. The two most cranial segments, T1 and 2, supply the head; the two most caudal segments, L1 and 2, supply the pelvic organs. The rest of the body is supplied by the intervening segments.

The preganglionic fibres are motor and therefore leave the spinal cord in the ventral (anterior) spinal nerve roots along with the somatic motor fibres.

The preganglionic fibres conveying impulses to:
    (i)   the head;
    (ii) the upper limb;
    (iii) the thoracic viscera;
    (iv) the lower limb,
all **relay** in the sympathetic trunk.

The preganglionic fibres conveying impulses to:
    (v) the abdominal viscera;
    (vi) the pelvic viscera,

enter but then leave the sympathetic trunk **without relaying** in it.

A **sympathetic trunk** lies on each side of the midline [FIG. 1.07]. Each trunk consists of a rather slender nerve trunk running vertically from the base of the skull to the coccyx. Distributed along its length are a series of **sympathetic ganglia**. In the thoracic region there is one ganglion near each segmental spinal nerve. In the neck, however, although there are eight cervical nerves there are usually only three cervical sympathetic ganglia, the **superior**, the **middle**, and the **inferior cervical ganglia**. In the lumbar and sacral regions the number of ganglia is usually not five but four. In front of the coccyx the trunks unite in the ganglion impar (unpaired).

The preganglionic fibres having left the spinal cord in the ventral roots reach as far as the mixed spinal nerves. Here, within a few centimetres of the spinal cord, the spinal nerves cross immediately behind the sympathetic trunk at right angles. The sympathetic nerve fibres from each spinal nerve form a small bundle called a **white ramus communicans** (often used in plural: rami communicantes) which runs into the sympathetic trunk from the spinal nerve. The preganglionic fibres thus gain access to the sympathetic trunk. The sympathetic pathway then depends upon the part of the body supplied. These pathways will now be considered in more detail. They are summarized in FIGURE 1.12.

Sympathetic supply to the head [FIG. 1.13]

All fibres which are destined to relay in the sympathetic trunk do so in the ganglion or ganglia which are as near to their target organ as possible. We have already mentioned that the sympathetic supply of the head emerges from the spinal cord at the level of T1 and T2. These fibres therefore turn upwards in the sympathetic trunk and relay in the **superior cervical ganglion**. Having done so the postganglionic fibres leave the ganglion as **grey rami communicantes** and travel to their destinations by forming a plexus around the internal or external carotid arteries and their branches.

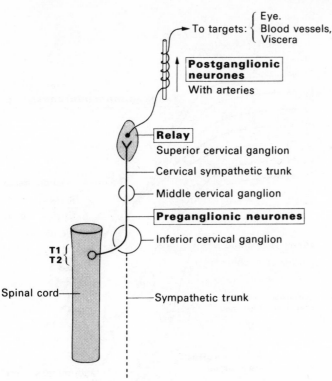

FIG. 1.13. Plan of the sympathetic nerve supply to the head.

Sympathetic supply to the upper limb [FIG. 1.14]

The sympathetic supply of the upper limb emerges from the spinal cord at the levels of **T3, 4, 5, and 6** and enters the sympathetic trunk at these levels. The fibres then turn **upwards** as they are heading for the ganglia closest to the roots of the brachial plexus which are the

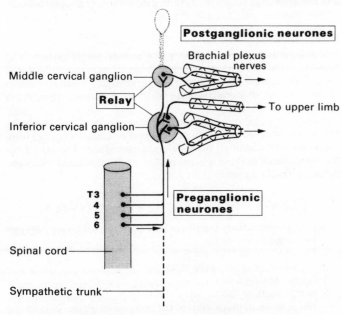

FIG. 1.14. Plan of the sympathetic nerve supply to the upper limb.

middle and inferior cervical ganglia, where they relay. The postganglionic fibres travel in the **grey rami communicantes** from the ganglia to the nerve roots of C5, 6, 7, and 8 and T1. They then become integral components of the branches of the brachial plexus and reach their destinations say via the median, radial, or ulnar nerves, especially their cutaneous branches.

The difference in appearance between the bundles of preganglionic and postganglionic fibres whether they look white or grey is due to the presence of a typical myelin sheath around the former, while the latter are unmyelinated.

Sympathetic supply to lower limb [FIG. 1.15]

The lower limb preganglionic fibres come out of the spinal cord at **T7–12**. They enter the white rami and reach the sympathetic trunk. They then turn **downwards** and are distributed to the roots of the lumbar and sacral nerves having relayed in the ganglia which lie nearby, **lumbar ganglia** for the **femoral nerve** (L2, 3, 4), **lumbar and sacral ganglion** for the **sciatic nerve** (L4, 5; S1, 2, 3) etc.

Notice that the strategy of the course of the preganglionic fibres is the same in these three systems, but the limbs differ from the head in the way the postganglionic fibres reach their destinations.

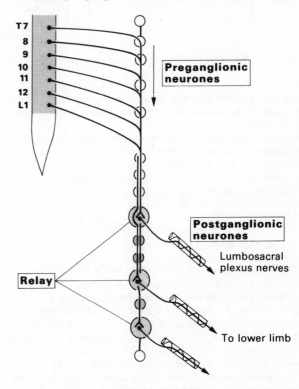

FIG. 1.15. Plan of the sympathetic nerve supply to the lower limb.

Sympathetic supply of the thoracic viscera [FIG. 1.16]

The sympathetic supply of the heart and lungs come from the spinal cord at the level of **T3 and T4**. The fibres surprisingly travel **upwards** in the sympathetic trunk and relay in the cervical ganglia, mostly in the **middle and superior cervical ganglia**. The postganglionic fibres then travel **downwards** to enter the thorax. The bundles travel on their own, and do not accompany other nerves or vessels.

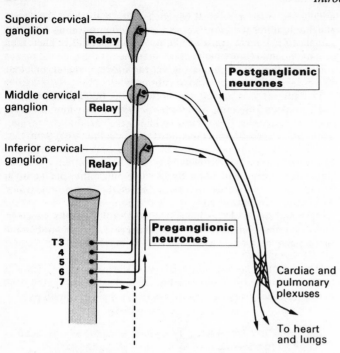

FIG. 1.16. Plan of the sympathetic nerve supply to the thoracic viscera.

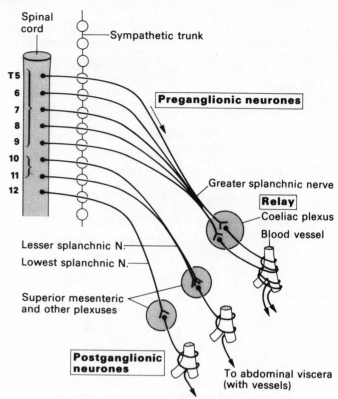

FIG. 1.17. Plan of the sympathetic nerve supply to the abdominal viscera.

They are the **cardiac** and **pulmonary branches** of the cervical ganglia which lie in the cardiac and pulmonary plexuses before reaching their destinations.

The rule regarding the place where the relay takes place in the sympathetic trunk, namely as close to the destination as possible, appears now to have been broken. This is not really so because in its early development the heart lies in what will become the neck. At this stage the heart picks up its nerve supply and then descends to the thorax. The nerve supply of the diaphragm (C3, 4, 5) is also due to its initial development in the neck at the C4 level. The heart, lungs, and diaphragm descend together and drag their nerve supplies down with them.

Sympathetic supply of the abdominal viscera [FIG. 1.17]

In the case of the abdominal viscera the preganglionic supply comes from **T5 to T12**. The fibres enter the sympathetic trunk in the usual way but instead of relaying they leave the chain as three separate nerves, the greater splanchnic (T5–9), the lesser (T10 and 11), and the lowest (T12) which run downwards and pierce the diaphragm. Once in the abdomen they break up into extensive plexuses which surround the main branches of the abdominal aorta, the coeliac trunk, superior mesenteric artery, renal arteries, inferior mesenteric artery, and their branches. Intermingled in the plexuses are scattered groups of nerve cells with which the preganglionic fibres synapse. The postganglionic fibres are guided to their destinations by the smaller branches of the arteries.

Sympathetic supply of the pelvic viscera [FIG. 1.18]

The pelvic viscera are supplied by fibres which leave the spinal cord at **L1 and L2** and enter and immediately leave the sympathetic trunk just like the splanchnic nerves. They mostly pass downwards in front of the common iliac artery and relay in front of the sacrum

in the **presacral nerve** (a misleading name), otherwise known as the **hypogastric plexus**.

The postganglionic fibres find their way to the bladder, uterus, and rectum and genitalia by travelling either free in the pelvic fascia or by accompanying the blood vessels.

Catecholamines

The sympathetic nervous system is a **noradrenergic system**. The activity at the sympathetic postganglionic terminations (with the exception of sweat glands which are cholinergic) is brought about by the release of **noradrenaline** which is an amine of catechol (di-hydroxybenzene).

The activity of the sympathetic nervous system is augmented by the release of catecholamines (noradrenaline and adrenaline) from the suprarenal (adrenal) medulla endocrine gland. The activity of the sympathetic nervous system is blocked by alpha and beta sympathetic blocking agents.

PARASYMPATHETIC NERVOUS SYSTEM

The parasympathetic nerve fibres travels in the following nerves:
1. Oculomotor nerve (III).
2. Facial nerve (VII).
3. Glossopharyngeal nerve (IX).
4. Vagus nerve (X).
5. Sacral outflow (S2, 3, 4).

In the parasympathetic system the relay takes place either in the wall of the viscus itself as for the heart, gut, and bladder, or in

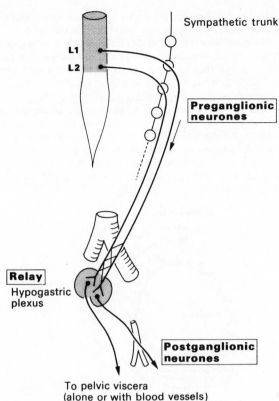

Sympathetic trunk

L1
L2

Preganglionic neurones

Relay
Hypogastric plexus

Postganglionic neurones

To pelvic viscera
(alone or with blood vessels)

Fig. 1.18. Plan of the sympathetic nerve supply to the pelvic viscera.

ganglia which lie close to the target organ as for the salivary glands and eyes.

Thus the vagus fibres to the heart relay in a ganglion close to the sinu-atrial node. The postganglionic fibres lie in the heart and are very short. The plexuses in the bladder and in the wall of the gut between the circular and longitudinal muscle coats and in the submucosa contain many nerve cell relay stations for the parasympathetic system. In the head the relay takes place in one of the four ganglia associated with the outlying branches of the trigeminal nerve. The ganglia are close to but not within the viscera which they supply.

The **oculomotor** preganglionic nerve fibres come from the Edinger–Westphal nucleus, and relay in the **ciliary ganglion**. The postganglionic fibres run in the ciliary nerves and constrict the pupil and cause contraction of the ciliary muscle in accommodation.

The **facial nerve** is secretomotor to the lacrimal gland (relay in **pterygopalatine ganglion**), and the submandibular and sublingual glands (relay in or in the vicinity of the **submandibular ganglion**). The preganglionic fibres come from the lacrimal and superior salivatory nuclei [p. 380]. The postganglionic fibres travel with appropriate branches of the trigeminal nerve which leads them to their destinations [p. 281].

The **glossopharyngeal nerve** supplies the parotid gland from the inferior salivatory nucleus, relaying in the **otic ganglion**. The complex course of the preganglionic fibres is described on page 287. The postganglionic fibres run in the auriculotemporal nerve.

The **vagus** has a very wide distribution to the heart and lungs and to the intestine as far as the distal third of the transverse colon. The

relay in each case is in the viscus itself. The vagus slows the heart, contract the bronchial musculature, and is the excitatory nerve to the intestine. It increases the secretion of digestive juice, increases peristalsis, and relaxes the sphincters.

The **sacral outflow** (S2, 3, 4) supplies the pelvic viscera and external genitalia. Fibres also pass upwards out of the pelvis into the abdominal cavity and supply the distal third of the transverse colon and the descending colon. The sacral outflow takes over in the large intestine where the vagus leaves off.

The parasympathetic nervous system is a **cholinergic system**, that is, the activity at the postganglionic terminations is brought about by the release of **acetylcholine**. Unlike the sympathetic nervous system there is no endocrine gland which augments this activity. The activity of the parasympathetic nervous system is blocked by atropine and hyoscine.

Note that the parasympathetic nervous system is restricted to the trunk and skull. It does not innervate the limbs. All the autonomic nerve fibres in the upper and lower limbs including the hands and feet are sympathetic.

APPLIED ANATOMY OF THE AUTONOMIC NERVOUS SYSTEM

1. **The head**: paralysis of the sympathetic supply of the head is called **Horner's syndrome**. It has the following features:
   (i) drooping (ptosis) of upper eye lid, due to paralysis of the smooth muscle component of the levator palpebrae superioris;
   (ii) constriction of the pupil compared with the opposite side due to paralysis of dilator pupillae muscle;
   (iii) vasodilation of skin on affected side;
   (iv) no sweating possible.
   Any lesion between the spinal nerve roots T1 and 2 and the superior cervical ganglion may produce Horner's syndrome.
2. **Upper limb**: the circulation in the limb can be improved at least temporarily by a sympathectomy (*-ectomy* = removal of). Removing the middle and inferior cervical ganglia produces a postganglionic sympathectomy. Cutting the sympathetic trunk at T2 will produce a preganglionic sympathectomy. The former operation is complicated by Horner's syndrome, while the latter is not. Preganglionic sympathectomy is very successful in treating hyperhydrosis of the hands (excessive sweating).
3. **Thorax**: removal of the sympathetic trunk between T3 and T9 may alleviate the pain of angina. This operation is no longer popular because the patient loses the warning of pain when the heart muscle becomes ischaemic.
4. **Abdomen**: the vagus fibres to the secreting part of the stomach are sectioned to reduce acid secretion in patients with a gastric or duodenal ulcer while not interfering with the vagal supply to other abdominal viscera. This is known as selective vagotomy.
5. **Pelvis**: cutting the pelvic splanchnic nerves in front or behind the common iliac artery may alleviate pain from pelvic viscera.
6. **Lower limb**: short-term improvement in the circulation is achieved usually by removing the lumbar sympathetic trunk—preganglionic sympathectomy of fibres running with the sacral nerves: once popular for patients with ischaemia producing intermittent claudication = intermittent limping (*claudico* Latin = I limp). This operation also paralyses the sympathetic supply to the genitalia and produces sterility (but not impotence) in the male.

BLOOD VESSELS

Arteries

The aorta is the largest artery in the body with a diameter of 2–3 cm. When it emerges from the left ventricle [FIG. 1.19] it is directed upwards, but it soon arches horizontally backwards and to the left (aortic arch). It passes downwards through the thorax close to the front of the bodies of the vertebrae and ends in the abdominal cavity by dividing into the common iliac arteries [FIG. 1.19]. This division takes place at the level of the umbilicus only just to the left of the midline. Before it divides, the aorta gives off branches to the heart, the head and brain, the upper limbs and trunk, and viscera. Its terminal branches supply the pelvis and lower limbs.

The branches (arteries) form **arterioles** which in turn give rise to the smallest blood vessels. These are the **capillaries**. The anatomy and physiology of the circulation revolves around what happens in the capillaries, and the nutrition and function of every tissue in the body depends on the events taking place in them.

The wall of the arteries contains both elastic tissue and smooth muscles fibres. The large arteries have a considerable quantity of elastic tissue in their wall [FIG. 1.20]. They distend with each heartbeat (systole) and spring back again between heartbeats (diastole). This elastic recoil maintains the blood flow through the capillaries during diastole. The smaller arteries, and particularly the arterioles, have muscular coats.

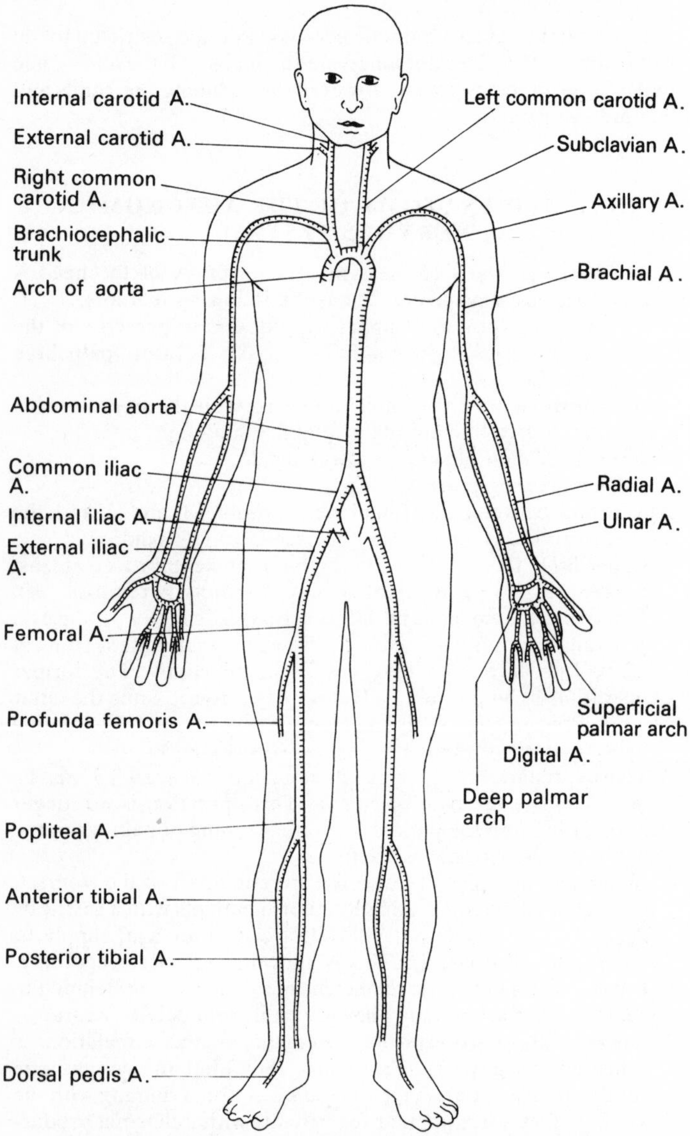

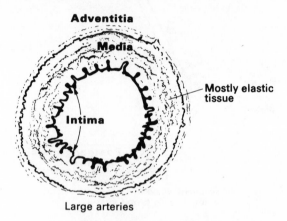

FIG. 1.20. The great arteries.
The three coats are termed intima, media, and adventitia. The media consists mainly of elastic tissue which allows the vessel to expand during ventricular systole and to recoil during ventricular diastole. This allows the capillaries to continue to receive a flow of blood during diastole when there is no blood leaving the heart.

Arterioles

The very small arteries just before the capillaries are termed arterioles. They can be seen with a magnifying lens. Each arteriole supplies a number of capillaries.

The arterioles are the **resistance vessels** of the peripheral circulation, that is, they provide the peripheral resistance that is essential for the maintenance of the arterial blood pressure.

The characteristic feature of the walls of arterioles is that they contain a well-marked encircling coat of smooth muscle which is innervated by the sympathetic nervous system. This sympathetic activity is, in turn, controlled by a collection of cells in the medulla termed the **vasomotor centre**. In addition the arterioles are under the influence of local factors, such as metabolites and temperature, which modify the degree of contraction of the muscle coat. If this muscle contracts, the lumen of the arterioles is reduced and blood flow to the associated capillary network decreases. When the muscle relaxes blood flow increases [FIG. 1.21].

Blood can be distributed to different regions according to functional needs, for example, to the muscles during exercise, or to the intestine after a heavy meal.

Capillaries

The capillaries are thin-walled, one-cell-thick, blood vessels. The lumen is so small that it only allows blood cells to go through in

The figure on the left is labelled with the following:

Internal carotid A.
External carotid A.
Right common carotid A.
Brachiocephalic trunk
Arch of aorta
Abdominal aorta
Common iliac A.
Internal iliac A.
External iliac A.
Femoral A.
Profunda femoris A.
Popliteal A.
Anterior tibial A.
Posterior tibial A.
Dorsal pedis A.

Left common carotid A.
Subclavian A.
Axillary A.
Brachial A.
Radial A.
Ulnar A.
Superficial palmar arch
Digital A.
Deep palmar arch

FIG. 1.19. The principal arteries and arterial 'pulses' of the body.
You should be able to find the radial, ulnar, brachial, common carotid, temporal (external carotid), femoral, popliteal, dorsalis pedis, and posterior tibial pulses.

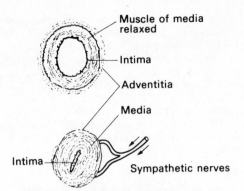

Arterioles: in the lower vessel the media
has contracted and obliterated the lumen

FIG. 1.21. The arterioles.
The media consists of smooth muscle which is innervated by the sympathetic nerve fibres, the majority of which are vasoconstrictor. An increase in sympathetic activity reduces the size of the lumen and increases the resistance to blood flow by the 'resistance blood vessels'.

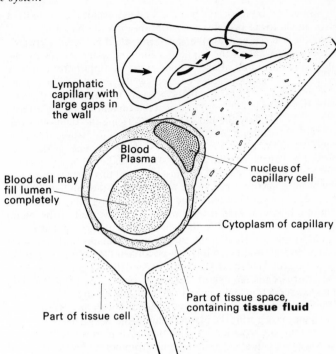

FIG. 1.22. Capillaries, lymphatics, and tissue fluid.
The tissue fluid (interstitial fluid) is formed at the arterial end of the capillaries and reabsorbed at the venous end of the capillary or returned to the blood via the lymphatic channels. The lymphatics also return plasma proteins which have leaked out through the capillary walls. Lymphocytes, in particular, circulate round from blood, tissue spaces, lymphatics, and lymph nodes back to blood again every few hours.

single file. It is through the walls of the capillaries that the nutritional exchanges take place between the blood and tissues.

Although there is no muscle in the capillary walls, as there is in the arterioles, the cells forming the walls are nevertheless able to contract and to obstruct completely the flow of blood. There is present in these cells contractile elements, visible with the electron microscope [FIG. 1.22].

Capillary cells have no nerve supply but they are sensitive to local factors such as the concentrations of oxygen, carbon dioxide, and other metabolites. The accumulation of waste products in the tissues causes relaxation of these cells, and blood now flows freely in the local capillaries, provided, of course, that the arterioles feeding them are also dilated.

Veins

Having seeped through the 'capillary bed', the blood returns to the heart via the venules and veins. Veins are thin-walled structures in comparison with arteries, and the pressure within them is much lower. However, when a vein is grafted by a surgeon into the arterial system to relieve an obstruction (reversed so that the valves stay open) it is able to withstand the higher arterial pressure; in time the wall becomes much thicker.

In man most arteries are accompanied not by one vein but by several (*venae comitantes* (Latin plural) = companion veins), which anastomose freely with each other to form a plexus around the artery. Such an arrangement acts as a **counter-current heat exchanger** which, in the limbs, warms the blood returning to the heart from a cold extremity.

You might have expected half the blood volume to be in the arteries and half in the veins, but in view of what has just been said, it is not surprising that the balance is actually about 20 per cent in the arteries and 80 per cent in the veins. Veins are thus the **main capacitance vessels** of the body.

The veins contain smooth muscle in their walls and the sympathetic activity to this muscle (**venomotor tone**) prevents their overdistension and thus limits the total volume of blood in the veins at any one time. Any increase in the capacity of the veins (such as produced in the lower parts of the body by gravity) will reduce the blood in the heart, and reduce the cardiac output.

Most small veins have valves at intervals along their course, but the big veins do not. In the embryo some veins have valves which subsequently disappear (portal vein, orbital veins). When muscles contract they squeeze on the veins. The presence of valves ensures that the blood is propelled forwards towards the heart and does not flow backwards towards the capillaries which feed them. When the muscles relax the segment of vein fills. This 'muscle pump' reduces the venous pressure in the lower limbs when walking, etc. This, in turn, reduces the capacity of the leg veins, increases the venous return to the heart, and speeds up the circulation of the blood. In the lower limb a break down of the valve mechanism of the subcutaneous veins is a feature of varicose veins.

It is important to note that the muscle pump requires the alternate contraction and relaxation of the muscles. A sustained contraction, such as when standing with the legs as rigid pillars, impedes rather than enhances the return of blood from the legs.

LYMPHATIC SYSTEM

The anatomy of the lymphatic system is associated with certain problems which arise in the tissue spaces. By 'tissue spaces' we mean the narrow gaps between adjacent cells. These spaces contain tissue fluid (or interstitial fluid) which is derived from the blood flowing in the capillaries. The capillary walls are relatively impermeable to the plasma proteins and as a result the tissue fluid has a much lower protein concentration than the blood.

At the arterial ends of the capillaries, fluid is forced outwards through the walls of the capillaries into the tissue spaces by the blood pressure, and at the venous ends the fluid is returned by osmotic (or oncotic) forces exerted by the difference in plasma protein concentrations. The blood pressure falls as the blood passes from the arterial to the venous end of the capillaries. At the arterial end of the capillary the blood pressure driving the fluid outwards exceeds the osmotic pressure sucking it back in, whereas, at the venous end the reverse is the case. At the same time, as fluid has passed out of the capillary at the arterial end, the plasma protein concentration in the capillary must rise, and this further increases the osmotic suction force at the venous end of the capillary. So the balance of power between blood pressure driving fluid out and the osmotic forces sucking it back in again, is reversed as the blood passes from the arterial to the venous ends of the capillaries.

The amount of fluid in the tissue spaces remains fairly constant under normal conditions. It is only possible for the return of fluid at the venous ends of the capillaries to exceed that which passes out at the arterial ends on a short-term basis. Such a movement would lead to a reduction in the interstitial fluid volume and an increase in blood volume as fluid returns to the blood. This return is one of the mechanisms restoring the plasma volume following a haemorrhage.

The opposite situation in which fluid return does not keep pace with the tissue fluid formation is much commoner. However, small such an imbalance may be, if more fluid is formed than reabsorbed then, sooner or later, the tissue spaces will begin to swell and **oedema** will result.

There is also the problem of the removal not only of the plasma protein which has leaked into the tissue spaces, but also of the materials which originates from the tissue cells themselves, which will be unable to pass through the blood capillary walls. Such material, such as cytoplasm escaping from a disintegrating cell, will be osmotically active and will make it more difficult for fluid to get back into the bloodstream. Their presence would lead to oedema.

Under normal conditions oedema does not occur. There must therefore be some form of overflow system which will ensure the escape of surplus tissue fluid, and the return of the leaked plasma proteins to the bloodstream.

In the tissue spaces are to be found blind ending capillary-like vessels. Their walls (unlike blood capillaries) are such that plasma proteins and surplus tissue fluid can readily enter these vessels. They are the **lymphatic capillaries**. They join each other to form larger and larger lymphatic vessels. They contain very numerous valves in their walls, which ensure that when the surrounding tissues compress them, the lymph flows forwards and cannot flow backwards into the tissue spaces. Bacteria, cellular debris, even whole cells (including of course, lymphocytes) can enter the lymphatic vessels along with tissue fluids.

Main lymphatic pathways

Lymph eventually finds its way back into the bloodstream. The **lymph** from the **right side** of the head and neck, the right upper limb, and the right side of the thorax eventually runs by three separate lymph ducts into the venous system at or near the junction of the right internal jugular and subclavian veins, that is, at the root of the neck on the right side [Fig. 1.23]. The **lymph** from the **rest of the body** runs into the **thoracic duct**. This vessel (2–3 mm in diameter) starts in the upper abdomen on the right side of the first and second lumbar vertebrae. The lymph from both lower limbs and the

pelvic and abdominal viscera drain into it. The duct runs upwards close to the bodies of the thoracic vertebrae, shifting over from right to left behind the oesophagus as it does so. It is joined by the main lymph ducts from the left side of the head and neck, the left upper limb, and the left side of the thorax. The thoracic duct empties into the junction of the internal jugular and subclavian veins, that is, at the root of the neck on the left side [Fig. 1.23].

The **lymphatic system** plays a special role in the absorption of **fat** from the gut. During fat absorption the lymph contains so many fat droplets that it looks white like milk. Mesenteric lymph vessels in the mesenteries are often called **lacteals** for this reason.

Lymph nodes

Small lymph vessels (lymphatic capillaries) merge into each other forming vessels of gradually increasing size. They usually run alongside blood vessels. But the lymphatic pathways differ fundamentally from venous pathways because of the succession of **lymph nodes** through which the lymph must percolate.

A number (12–24) of small **afferent** lymph vessels converge on a single lymph node, and the lymph leaves it at the hilum via a single much larger **efferent** duct [Fig. 1.24]. A node consists of a loose meshwork or reticulum (*rete* = net) surrounded by lymph sinuses all enclosed in a capsule. The lymph enters the sinuses first of all, and then works its way through the reticulum towards the efferent duct. Embedded in the reticulum are centres where lymphocytes develop to maturity called germinal centres. There are also other cell types such as individual scavenger cells called phagocytes (*phago* = swallowing; *cyte* = cell). As the lymph passes through the node, bits of cellular debris, whole cells, bacteria, etc. are trapped in the reticulum. The phagocytes engulf this material and then digest and destroy it.

Normal lymph vessels are small, transparent, and invisible to the naked eye. Lymph nodes are bigger (2–3 mm in size). They may be visible at an operation but they are not normally palpable except possibly in the superficial inguinal and submental regions. The **superficial lymphatics** of the limbs and of the neck run with the superficial veins, but the **deep vessels** run with the arteries. Since the lymph vessels, like veins, have valves the pulsation of the arteries may help to propel the lymph along the lymphatic channels.

Lymph nodes are situated in particular sites in the body. For example, in the upper limb they mostly lie just above the elbow on the medial side, and clustered along the various blood vessels in the axilla (arm pit). Their positions will be described in more detail in chapters dealing with particular parts of the body. A good knowledge of the anatomy of the lymphatic system is of fundamental importance in many aspects of medicine.

When **bacteria** invade the tissues, many enter the lymphatic vessels and are swept centrally with the lymph stream. They will be filtered out in the reticulum of the first group of lymph nodes to which they come. In the node, either the bacteria are engulfed and destroyed by the scavenger cells or they succeed in multiplying there and set up infection in the lymph nodes, which become enlarged and painful. If bacteria succeed in overcoming the defence mechanisms of the lymph nodes they may reach the bloodstream and set up **septicaemia**.

Another aspect concerns the spread of **malignant tumours**, especially those derived from epithelial cells. Such tumour cells may enter the lymphatic vessels and travel to the regional lymph nodes where they are trapped in the perinodal space and reticulum. They may then take root and grow, producing a secondary malignant deposit or **metastasis** which is at a distance from the primary

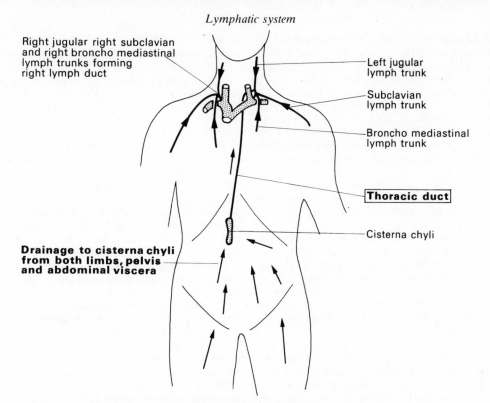

Right jugular right subclavian
and right broncho mediastinal
lymph trunks forming
right lymph duct

Left jugular
lymph trunk

Subclavian
lymph trunk

Broncho mediastinal
lymph trunk

Thoracic duct

Cisterna chyli

**Drainage to cisterna chyli
from both limbs, pelvis
and abdominal viscera**

Fig. 1.23. Principal lymphatic pathways.
The right half of the head and neck. the right upper limb. and most of the right half of the thorax drains via
the three trunks shown into the angle between the internal jugular and subclavian viens. The rest of the body
drains into the thoracic duct which reaches the corresponding angle on the left. The detailed anatomy is subject
to countless variations.

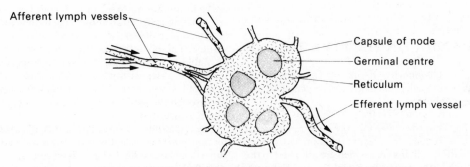

Afferent lymph vessels

Capsule of node

Germinal centre

Reticulum

Efferent lymph vessel

Fig. 1.24. Lymph node.
The lymph passes through the reticulum of the lymph node. having reached it via many afferent vessels. and
leaves via the efferent vessels which commences at the indented hilus of the node. Note the values. Lymphocyte
formation takes place in the germinal centres.

growth. In many instances the treatment of a primary tumour is not difficult but the removal of all the lymph nodes into which a tumour has spread may be impossible. If tumour cells enter the bloodstream they are carried with the venous blood to the lungs where they are filtered out in the capillary bed and may give rise to metastases in the lungs.

In bedside practice two problems frequently arise. In the first a primary infection or tumour is recognized in some part of the body and it is necessary to know which regional lymph nodes are likely also to be affected. You must therefore learn the lymphatic drainage of every region of the body. In the second, the problem is reversed; a lymph node is recognized as being inflamed or the site of a malignant deposit and the question now is 'Where is the primary source of the disorder?'. To answer this question you must learn the catchment area of all groups of lymph nodes.

Lymph nodes are also part of the immunity system of the body. They assist in the production of the white blood cells which distinguish between 'self' and 'non-self', the **lymphocytes**, and their associated **antibodies**.

Lymphatics and the lungs and brain

One organ which needs special mention is the **lung**. The tissues of the walls of the bronchial tree are similar to any other tissues of the body with respect to their blood supply (from the **bronchial**, not the pulmonary. arteries) and lymphatic drainage. but the **alveoli** (singular = *alveolus*) of the lungs are a special case. They consist of an epithelial lining (too thin to be seen with the light microscope) enclosing a small sac filled with alveolar air. Immediately outside this lining, and in contact with it, are many blood capillaries which are the ultimate branches of the pulmonary artery. Now if fluid were to escape through the capillary wall in the way described in other tissues, it would come to lie in the alveoli. Even if it were removed at the venous end of the capillary the presence of this fluid in the alveoli would interfere seriously with blood–gas interchange in the lungs.

This problem, however, does not arise in the normal lung, because the blood pressure in the pulmonary artery is so low that even at the arterial end of the capillaries there is no tendency normally for tissue fluid to be formed.

However, any condition which raises the pulmonary capillary pressure sufficiently (for example, congestive heart failure) will cause what is called 'oedema of the lung', that is fluid in the alveoli. With the patient sitting up in bed, this fluid will collect especially at the base of the lung. Its presence can be heard with a stethoscope. You need to know the surface anatomy of the base of the lung to listen in the right place with a stethoscope to diagnose the first signs of oedema of the lung in such a patient.

The second organ to be mentioned is the **brain**. It is said not to have a lymphatic system at all. But the subarachnoid cerebrospinal fluid [p. 381], in which the brain floats inside the skull, extends for a short distance along perivascular channels inside the brain around the blood vessels. These channels probably subserve the same function for the tissue space of the brain as lymphatics do for tissue spaces in other parts of the body. Oedema of the brain occurs very easily. It raises the intracranial pressure, which raises the venous pressure, which in turn causes more oedema.

EPITHELIUM AND CONNECTIVE TISSUE

The terms 'epithelium' and 'connective tissue' are frequently used in descriptive anatomy. Every cell in the body can be classified as a connective tissue or epithelium.

The cells which line many surfaces on the outside or inside of the body are called **epithelia**. Thus the epidermis, the outside surface of the skin, is an epithelium. It consists of layers piled on top of each other, the deepest consisting of living cells which are attached to the tissues beneath them, while the most superficial die and flake off. Such a stratified (*stratum* = a layer, plural *strata*) epithelium is found in other parts of the body where the surface may be exposed to wear. Many other epithelia are only one cell thick, for example in the stomach and intestine, and throughout most of the respiratory tract where they have no need to withstand physical trauma like the skin.

Other tissues in the body are **connective tissues**. Connective tissues do not usually line spaces and surfaces. This term includes structures such as bone, cartilage, muscle, all kinds of fibrous tissue such as superficial and deep fascia, tendons, and ligaments.

It is difficult to make a clear-cut distinction between epithelia and connective tissue without an elaborate account of development and morphogenesis. As a rule any space in the body which eventually communicates with the outside world is lined by an epithelium. Thus kidney tubule cells, seminiferous tubule cells, pulmonary alveolar cells, etc. are all epithelia. Spaces which do not communicate directly with the outside world such as the knee, peritoneal, or pericardial cavities (*peri* = around; *cardium* = heart) are lined by sheets of connective tissue cells which are usually called mesothelial rather than epithelial cells. The female peritoneal cavity is an exception to this rule because it communicates with the exterior via the genital tract.

SKIN

The skin covers the outer surface of the body. It has an area of about 1.75 square metres. It is the largest 'organ' of the body. It produces vitamin D, stores fat, and is the principal barrier between the body and the external environment. Even quite superficial damage to a substantial proportion of the skin, as occurs in burning, may result in death. The skin suffers from a large number of special diseases; their study is called **dermatology**.

The skin consists essentially of two layers, the dermis which is a vascular connective tissue, and the epidermis which consists of epithelial cells [FIG. 1.25].

The **dermis** is a dense vascular feltwork of collagen and elastic fibres, produced by cells called **fibroblasts**. Leather is made from the dermis of animals. It is pliable, strong, and tough.

The dermis has an extensive sensory nerve supply, and also an autonomic motor supply which regulates the flow of blood in the skin from moment to moment and controls the body temperature.

The thickness of the dermis varies in different regions of the body according to functional needs. Even at birth the dermis of the sole of the foot, anticipating future functional needs, is much thicker than that of the eyelid.

The **epidermis** covers the whole of the dermis on its superficial surface. It is avascular (*a* (Greek) = not) and only contains a few nerve terminations. Whilst the deeper layers of the epidermis consist of living cells, often showing cell division, the more superficial layers consist of dead cells, which flake off when they reach the surface. The epidermis contains a protein called **keratin** which is insoluble in most fluids. The skin is impervious to many noxious fluids and gases with which it comes in contact, as well as being more or less waterproof. A small amount of carbon dioxide is given off through the skin, but the amount is negligible as far as respiration is concerned.

The epidermis is said to be 'stratified' (*stratum* (Latin) = a layer, plural *strata*) because it is multilayered. In addition, because in the superficial layers the dying or dead cells are flattened, it is termed a **stratified squamous** (*squamus* (Latin) = flat) epithelium. Loss of the superficial layer or layers can therefore occur without damage to the underlying living tissues. The skin stands up to everyday abrasions for this reason, and protects the dermis and other tissues. The thickness of the 'dead' layers of the epidermis varies in different parts of the body and becomes much thicker when the skin is exposed to friction or intermittent pressure. Like the dermis then, the epidermis responds readily to functional needs.

The skin is the organ by which body temperature is controlled. In Man (as in all warm-blooded animals) heat must be lost continuously—as it is from the radiator of a motor car—the only questions are how much and how quickly?

Heat loss can be increased by raising the skin temperature by increasing the blood flowing through the dermis, and also dramatically by sweating. The skin temperature and rate at which sweat evaporates on the skin surface are the most important factors in determining the comfort or discomfort in hot places.

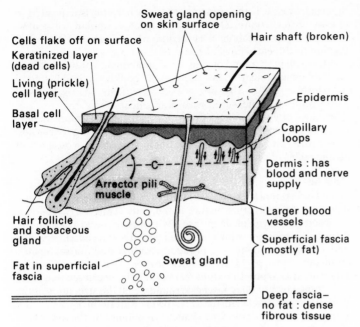

Sweat gland opening on skin surface

Hair shaft (broken)

Cells flake off on surface

Keratinized layer (dead cells)

Living (prickle) cell layer

Basal cell layer

Epidermis

Capillary loops

Arrector pili muscle

Dermis : has blood and nerve supply

Larger blood vessels

Superficial fascia (mostly fat)

Hair follicle and sebaceous gland

Fat in superficial fascia

Sweat gland

Deep fascia — no fat : dense fibrous tissue

FIG. 1.25. The skin.
The diagram shows a section through the epidermis, dermis, superficial fascia, and deep fascia.

The terminal branches of the cutaneous nerves (omitted for simplicity) lie in the dermis close to the epidermis; the epidermis is sufficiently supple so that the nerve endings can respond to physical stimuli transmitted to them through the epidermis. The coiled up secretory part of the sweat glands (like fairies' intestines) lies in the fat deep to the dermis. The tip of the hair follicles and their attendant sebaceous glands also project below the dermis. The capillaries which supply the dermis form loops which are at right angles to the skin surface. The blood vessels, sweat glands, and arrector pili muscles (regulate body temperature) are all supplied by sympathetic motor nerves.

When the skin is 'taken' for a skin graft the plane of the knife is parallel to the skin surface approximately at the level of —c—, removing all the epidermis. The donor area may be very large, say 30 × 10 cm but it will come once more to be completely covered with epithelial cells within 24 hours. This is possible because the cells lining the sweat-gland ducts and hair follicles migrate onto the surface and are transformed into epidermal cells! They re-establish a waterproof, bacteriaproof frontier.

The rate of healing clearly is not affected by the overall size of the donor area. It depends only on the distance between neighbouring follicles and sweat glands and the time it takes for the cells to cover half this distance.

Skin blood flow is modified by variation in the calibre of the blood vessels in the dermis which can change rapidly according to functional needs. The blood vessels are supplied with sympathetic **vasoconstrictor** nerve fibres which are ultimately under the control of heat regulating centres in the brain (in the hypothalamus [p. 402]). These blood vessels are found in the dermis, never in the epidermis. Stimulation of their nerve supply causes the superficial vessels to shut down and hence reduces skin blood flow and heat loss.

The arteries and veins in the skin are connected to each other not only by capillaries, but also, in certain parts of the body directly by arteriovenous anastomoses which short-circuit the capillary bed altogether. This means that skin blood flow can be increased still further even when all available capillaries are fully dilated.

Sweat glands

Sweat glands are downward tube-like extensions from the epidermis many of which run right through the dermis into the underlying fat. They are supplied by *cholinergic* sympathetic nerve fibres (that is, nerves which release **acetylcholine**, not noradrenaline, at their postganglionic nerve terminations). These nerves are also under the control of the heat regulating centre in the hypothalamus. Stimulation of these nerves causes sweating, and an increased heat loss.

Sweating in the **palms of the hands** also occurs in association with emotion and stress. Many teenagers and young adults suffer from this complaint, called **hyperhydrosis**. As has been seen it can be cured completely by cutting the sympathetic nerves supplying the hands. Palmar sweating, as it is termed, varies with the menstrual cycle in the female. It tends to disappear during sleep.

The sympathetic nerves supplying the skin reach it via the ordinary cutaneous nerves. When the sympathetic pathway to a particular skin area is interrupted, the sweat glands are paralysed, and the blood vessels dilate. The skin becomes abnormally warm and dry. The colour of the skin becomes redder. Of special value in diagnosis is the demonstration that when the patient is heated, sweating will occur elsewhere, but the affected area of skin will remain perfectly dry. One way of showing this is to dust the skin with dry starch–iodine powder which turns blue–black when it becomes damp.

Cutaneous nerves

The skin is richly innervated with sensory nerves which form part of the peripheral nervous system. The skin is capable of distinguishing a variety of different 'modalities of sensation'. Thus we can distinguish hot, cold, touch, and pain.

Many different specialized nerve endings have been described in localized parts of the dermis, each one said to be responsible for one particular sensation, but their anatomical distribution is quite inadequate to explain the discrimination of which the skin is capable over the whole surface of the body. Furthermore, many nerves in the skin have no specialized endings and there is an awkward fact that a small area of skin which only responds to 'hot' may not do so for long. At another time it may be insensitive to hot but respond to 'touch' or 'cold'.

Hair follicles

The hair follicles develop from the epidermis and are the source of hair. Hair is one of the tissues of the body which throughout life grows continuously in length. The state of the hair, including the eyebrows, may provide useful clinical data regarding a patient's general health. In hypothyroidism, and in dietary deficiency diseases such as kwashiorkor, the hair virtually stops growing. With advancing age normal hair grows more slowly, becoming thicker and more brittle. Grey hairs can be found in most people by the age of 30. Baldness occurs in normal men given the appropriate hereditory background and a sexually mature endocrine system.

Associated with hair follicles are **sebaceous glands**, whose product called **sebum** (Latin = tallow or grease), consists of fragments of disintegrating gland cells (holocrine secretion). This mode of 'secretion' is quite different from that of sweat gland cells which are themselves not affected by the synthesis of sweat. Sebaceous glands have no nerve supply but are continuously active giving to the hair its sheen and to the epidermis its added suppleness and water resistance. Too much washing removes the protective layer of sebum, and is bad for the skin. Patients with dermatitis are warned against soap and water!

The deepest part of the hair follicles are connected by strands of smooth muscle to the overlying epidermis. When these muscles contract the hairs 'stand on end' at right angles to the skin surface instead of lying flat. At the same time the skin surrounding each hair is raised into a small bump, producing 'goose-flesh'. The **m. arrectores pilorum** (= hair-raising muscles) are supplied by sympathetic nerves like the blood vessels and sweat glands. They contract when it is cold, and at other times, when the sympathetic is active, as in intense fear. The sensation of 'flesh creeping' is brought about by the contraction of these muscles which activate nearby sensory nerve endings. Shakespeare, with amazing insight, likens hair-raising to the quills upon the fretful porpentine (porcupine). Porcupine and hedgehog quills are giant hairs, with correspondingly well developed arrectores pilorum.

Other skin structures

The skin also contains pigment cells called **melanocytes** (melanin-containing cells). They are situated in the superficial parts of the dermis. They put out long spidery processes containing black pigment into the dermis and epidermis. Melanocytes hypertrophy when exposed to the sun, and protect the tissues by darkening it and making it more opaque to damaging ultraviolet rays. They help to avoid sunburn and skin cancer, both caused by ultraviolet rays.

Skin grafting

When skin is taken for a skin graft the plane of section is usually that of —c— [FIGURE 1.25]. It includes the whole epidermis plus part of the dermis but the parts of the sweat glands and hair follicles lying deep remain intact. The donor area (may be very extensive, 30×10 cm) re-epithelializes within 24 hours (!) from cells which come from these sweat gland and hair follicle linings deep from the surface. They are changed from sweat gland cells to epidermal cells. Rapid covering of the apparently large raw surface depends on half the distance separating the hair follicles and sweat glands from each other. It follows from this that the area of the donor site is irrelevant to the time it takes to re-epithelialize. The dermis is first covered completely with a one cell thick epithelium, which then becomes stratified and restores the skin to normal without scarring (consider also the rapid regeneration of the endometrium in the female uterus after menstruation).

FASCIA

Superficial fascia

Just deep to the dermis is a layer called the 'superficial fascia'. It is a loosely arranged meshwork of fibrous tissue whose interstices are filled with fat. This layer plays an important role in determining the contours of the body surface. Its distribution in male and female differs. There is normally more superficial fascia in females in most parts of the body.

On its deep surface the superficial fascia tends to become more fibrous and may form a distinct layer. It does this deep to the female breast, and in the lower abdomen extending into the perineum. Usually this layer blends with the deep fascia.

Fat conducts heat very slowly and the thickness of the superficial fascia is very important in reducing the rate of heat loss under adverse conditions, for example when swimming in cold water. The superficial fascia is the principal region where fat is deposited in obesity. Brown body fat is present in large amounts in a new-born baby. Its metabolism is an important source of heat production in the early days of life.

When you grab a 'fold of skin' on your abdomen or buttock, you are holding not only the skin itself, but in addition a double fold of superficial fascia. In assessing obesity in patients, the thickness of such skin folds is measured with skin calipers.

In most other mammals the superficial fascia is more or less fat-free. The skin is loosely attached to the underlying tissues and can be easily and quickly removed. But not so in Man. You can peel off the 'skin' of an animal but you have to 'flay' a man.

Deep fascia

Deep fascia refers to the layer of fibrous tissue which surrounds and envelops almost every structure in the body. Thus every muscle is enclosed in a layer of deep fascia which is thin and transparent. Groups of muscles which are functionally associated, say, the adductors of the thigh, are grouped together and are segregated from other groups by quite thick layers of deep fascia, called **intermuscular septa**. In FIGURE 7.08 [p. 134] you can see that there is a strong layer of deep fascia surrounding all the structures of the thigh called the **fascia lata** (*latus* (Latin) = wide, as in 'latitude'; *lata* is feminine to agree with *fascia* which is feminine). In the lower limb this deep fascia is extremely strong and even has large muscles, tensor fasciae latae and gluteus maximus, inserted into it.

It has always been traditional to regard deep fascia as mere packing tissue (like saw-dust in a crate filled with china) and to think of it as just connective tissue which has failed to differentiate into something more 'interesting' like muscle or bone. There is now evidence, however, which suggests that in early development the layout and position of muscles is determined first by the way the would-be fascia arranges itself forming the walls of compartments. The muscle cells (myoblasts) then position themselves, like stuffing, in spaces already prepared for them.

The deep fascia allows the various parts of the body to be described in compartments. Thus in the upper arm there is a flexor and extensor compartment [p. 95]. In the thigh there are three, the flexor, extensor, and adductor compartments [p. 134]. When dealing with a complex region such as the neck, it is convenient to be able to subdivide it into four or five compartments [p. 314].

Flexor and extensor retinacula

Retinacula (singular = *retinaculum*) are thickened regions in the deep fascia which develop in the region of many joints. When you extend (dorsiflex) your ankle joint by pointing your foot upwards, the extensor tendons stand out on the upper surface of the ankle joint. The degree to which this happens is limited by the extensor retinacula (superior and inferior) which bind the tendons down and prevent them from standing out like bow strings. Similar retinacula are present behind the medial and lateral malleoli of the ankle joint, on the front and back of the wrist and in other places. In the case of the dorsum of the wrist the tendons stand out much less than they do on the dorsum of the ankle joint.

These thickenings of the deep fascia are produced by transversely running fibres. They represent local thickenings in the deep fascia just as **capsular ligaments** in joints are thickenings in the **joint capsule** and are not independent structures in their own right.

Tendons in such regions are usually surrounded by **synovial sheaths** which allow their often very powerful movements to take place beneath the retinacula with the minimum of friction.

There appears to be a mechanism available to the body, which ensures that the strength of tendons, ligaments, joint capsules, etc. can be adjusted to the mechanical forces to which they are subjected. The basic pattern and strength of these structures is determined long before birth. Because the foetus lives in a more or less weightless state floating in the amniotic fluid, it follows that the general pattern is inherent and genetically determined. This does not deny the critical practical significance of the 'tuning mechanisms' which operate later adjusting the strength of tendons and ligaments (and bone itself) to the stresses to which they are subjected.

Skin flexor and extensor creases

Look at the front of your wrist and fingers. The skin is marked by a succession of flexor creases (by which palmists claim to foretell the future). On the front or back of other joint surfaces similar creases can be seen.

In these regions the skin becomes adherent to the deep fascia, and the intervening superficial fascia is compressed into a thin layer of fat-free fibrous tissue. In the hand these creases often look pinker in colour compared with the rest of the hand.

Flexor creases therefore form a barrier separating the superficial tissues above the crease from those below it (see p. 248, flexor crease of hip joint). Flexor creases are much used by surgeons in choosing sites for making incisions as the resulting scar is not so noticeable.

MUSCLE

Muscle is of three types:
1. Voluntary or striated (striped) muscle.
2. Involuntary or smooth (unstriated, unstriped or plain) muscle.
3. Heart or cardiac muscle.

The words 'voluntary' and 'involuntary' contrast the fact that certain muscles, such as those of the hand, are under direct voluntary control, whilst others, such as those of the intestine, are not. The terms 'striated' and 'smooth' are histological terms which refer to the appearance under the microscope of the two types of muscle. The terms 'striated' and 'smooth' are preferable since the terms voluntary and involuntary must not be taken too literally. The bladder wall consists of involuntary smooth muscle, but micturition is, in a sense, under voluntary control. The diaphragm and intercostals consist of voluntary striated muscle, but breathing takes place most of the time without voluntary intervention. But breathing can be controlled voluntarily, and speech, singing, etc. depend on this fact. The oesophagus has 'voluntary' striated muscle superiorly, but 'involuntary' smooth muscle inferiorly. The muscle of the upper eyelid, the levator palpabrae superioris, is a mixture of both striated and smooth muscle. The iris of the eye has smooth muscle only.

Striated muscle is capable of rapid contraction, whilst smooth muscle only contracts very slowly. If a movement takes place rapidly, for example, the movements of the pharynx in swallowing (which is not under voluntary control in the normal sense) then striated muscle is certain to be present. The pupil of the eye, on the other hand, contracts fairly slowly. Stand in front of a mirror and cover your eyes with your hands. Remove your hands and you will be able to see the speed with which the smooth muscle of the iris of the eye contracts with the light.

Striated muscles have a motor nerve supply which reaches them directly from the brain or spinal cord. These muscles do not contract unless appropriate nerve impulses reach them via these motor nerves from the central nervous system. If the nerves are cut, the muscle fibres are paralysed. They quickly waste, degenerate, and disappear.

The dependence of striated voluntary muscle on its nerve supply, not only for its function, but for its very existence, is in sharp contrast with the basic independance of smooth involuntary muscle and heart muscle to their nerve supplies.

Smooth muscle, as has been seen, is supplied by the autonomic nervous system with its two divisions, the sympathetic and parasympathetic nervous systems. The activity of the sympathetic nerves inhibits the rhythmic wave-like contractions in the gut called peristalsis and will simultaneously bring about contraction in sphincters such as the pylorus of the stomach. In this case it 'stops the works'. The parasympathetic system has the opposite effect, causing increase of peristalsis and relaxation of the sphincters. In the normal gut both sets of nerves are active simultaneously and the final level of contraction depends on the balance established between the two conflicting factors.

Heart (or cardiac) muscle cells are small and short compared with the other muscle types and are striated. The cells often branch and are Y-shaped. The ends of the cells are fixed to each other by the so-called intercalated discs. These are special adhesion areas between adjacent cells. This arrangement allows contraction waves to be conducted from cell to cell.

Heart muscle cells contract rapidly and have the important property of doing so independently of their nerve supplies. Heart cells isolated in a dish continue to beat rhythmically.

Cardiac muscle is also supplied by the autonomic nervous system. The nerve supply of heart muscle does not determine whether it beats or not, only whether it beats more rapidly (sympathetic chronotrophic activity), more powerfully with more complete emptying of the ventricles (sympathetic ionotrophic activity), or more slowly (parasympathetic vagal (X) activity). In practice the actual heart rate depends on the balance between these two divisions of the autonomic nervous system which usually operate simultaneously.

GROSS ANATOMY OF STRIATED MUSCLES

A typical striated muscle consists of a belly with a tendon at each end. The tendons join the belly to the bony skeleton. The belly contains not only contractile fibres but also a non-contractile fibrous tissue 'skeleton' through which the pull of many of the muscle fibres is conveyed to the tendon. It also contains muscle spindles which form part of the proprioceptive system.

Origin and insertion

The term 'origin' and 'insertion' are used to distinguish between the attachments at each end of a muscle. When a muscle contracts it is convenient to pretend that one attachment is fixed (the origin) while the other moves (the insertion). The convention assumes that in the limbs it is the proximal attachments of the muscles which are the **origins** and the distal attachments which are the **insertions**. In the upper limb this convention works reasonably well, but in the lower limb the terms may be misleading if one is thinking functionally. In standing, walking, and running the foot is the only stationary part of the whole limb.

Although the terms 'origin' and 'insertion' are used in their conventional sense, it must be remembered that in many actions the origin and insertions of the muscles are reversed.

The latissimus dorsi and pectoralis major are said to take origin from the rib-cage (and other parts of the trunk) and to be inserted into the humerus near the shoulder joint. The convention assumes that normally the trunk is stationary and it is the upper arm and shoulder which move. But a patient on a pair of crutches uses these muscles with the origin and insertion reversed to swing the body backwards and forwards.

Put your right hand into the left arm pit, palm inwards. Push round backwards as far as possible. Give a cough. You will feel a strong contraction in the posterior wall of the arm pit (axilla) which is the **latissimus dorsi** contracting so as to squeeze on the rib-cage. It is acting as an accessory muscle of **expiration** once again with its origin and insertion reversed.

Another example is the **sternocleidomastoid**. Its name suggests that its fixed origin is the sternum and clavicle and that it is inserted into the mastoid process of the skull and therefore is concerned with head movement. But now, keeping your head still, put your fingers on this muscle just above the medial third of the clavicle. Very slowly take in a long deep breath. Towards the end of inspiration the sternomastoid contracts strongly. It is now acting as an accessory muscle of respiration and it is **inspiration** which is assisted. The head is fixed and the muscle, by reversing origin and insertion, is helping to pull the rib-cage upwards. You will see this muscle contracting in this way in patients with severe dyspnoea, who are 'fighting for breath'.

Whatever the functional circumstances the origin and insertion are unchanged, in rather the same way as the anatomical position is maintained whatever the patient's orientation.

In many instances when muscles contract the origin and insertion do not come closer together, but either maintain a constant distance or even become more widely separated. If you raise this book from the table by flexing your elbow joint, you do so by using the elbow flexors (including the biceps). When you lower it again, you must extend (i.e. straighten) your elbow, but you do not use the extensor muscles (triceps). You use the same muscles as you did when you flexed your elbow. The point is that gravity is the main factor which must be overcome in lifting the book, and in lowering it you use the flexors to put a brake on the force of gravity. Unless you want to lower the book with an acceleration greater than that produced by gravity, you will not use the extensor muscles. Rather surprisingly, the same muscles are used when you go downstairs as when you go up. If the brake put on gravity fails, you fall. Indeed a patient with a wasted quadriceps muscle (knee extensor) has more trouble in walking downstairs than in walking up.

When origin and insertion come closer together the muscle is said to work **concentrically**. When they get further apart, it is said to work **eccentrically**. When there is no change it acts **isometrically**.

In cases where the terms 'origin' and 'insertion' are misleading, it is often convenient to use the term 'attachment' for both.

An important factor concerning muscular contraction is the initial length of the individual fibres in the muscle belly. If a relaxed muscle fibre is say 10 cm in length, it will be capable of shortening to about 50 per cent of this length, that is, to 5 cm. If the origin and insertion are separated by only 5 cm before contraction, this muscle will be unable to have any action, even if it contracts fully.

There are several examples in the body where the muscle fibre lengths are insufficient to exert any mechanical effect. Examples of this are seen in the hamstrings [p. 136] and long flexors of the fingers. It is virtually impossible to grip powerfully with the hand when the wrist joint is fully flexed. This is because the distance between the origin and insertion of the flexor muscles of the fingers has been so reduced that these muscles cannot shorten sufficiently to flex the fingers.

Muscles do not consist purely of contractile fibres. They also contain a non-contractile fibrous 'skeleton' through which the pull of many fibres are transmitted to the tendons. This tissue is an essential ingredient of all muscles. When an athlete goes into training, the contractile elements in the muscles hypertrophy very rapidly but the non-contractile elements hypertrophy rather slowly. There is a danger that an imbalance may occur when, as a result, the muscle is torn.

CLASSIFICATION OF THE NAMES OF MUSCLES

There are nearly three-hundred named muscles in the body. The name given to every muscle is descriptive in one way or another. It is very important to be aware of the meaning behind these names. Unlike the recently adopted SI system of units, where the name of the unit, such as Newton or Joule, tells you nothing about the unit (except possibly that it is in some way connected with the field of study of the person after whom it is named), in anatomy, the name of a muscle often tells you what it does or where it is.

Function names

This group includes about 26 per cent of the muscles. They have names which tell you what they do.

Examples are:

flexor carpi radialis	—*radially flexes wrist*
levator scapulae	—*elevates scapula*
opponens pollicis	—*opposes thumb*
depressor anguli oris	—*depresses angle of mouth*
levator ani	—*elevates anus*
supinator	—*supinates*
tensor tympani	—*tensions ear-drum*

Position names

The commonest group, accounting for about 34 per cent of all muscles, tells you the whereabouts in the body of these muscles, but not what they do.

Examples are:

brachialis	—*in the arm*
infraspinatus	—*below the spine of scapula*
popliteus	—*in popliteal fossa*
subclavius	—*below the clavicle*
tibialis anterior	—*front of tibia*
peroneus longus	—*long muscle attached to fibula*
stapedius	—*attached to stapes in middle ear*

Shape names

The next group accounts for 17 per cent of the muscles.

Examples are:

quadriceps	—*four heads*
triceps	—*three heads*
lumbricals	—*worm-like*
digastric	—*two bellies*
gracilis	—*slim or graceful*
semimembranosus	—*half membranous*

and the **geometric shapes** *such as:*

scalenus	—*like a scalene triangle with all sides different lengths*

trapezius	—*like a trapezium*
rhomboid	—*like a rhomboid*
deltoid	—*like a delta*
quadratus	—*like a square*

Attachment names

The last group of names embracing 23 per cent of the muscles has been formed by simply joining up the origin and insertions of the muscles in that order.

Examples are:

sternocleido-mastoid	—*sternum and clavicle to mastoid process*
brachio-radialis	—*arm to radius*
genio-glossus	—*chin to tongue*
omo-hyoid	—*shoulder to hyoid*

The muscle names in the above examples are not the full names. As has been seen the noun *musculus* (m.) should, strictly speaking, prefix all the above adjectives, but is seldom used. Unless it is rendered superfluous by the context, you must always say biceps **brachii** or biceps **femoris**, because there is a biceps muscle in the thigh as well as the upper arm. There is no need to say triceps *brachii* because there is only one triceps muscle in the whole body. Similarly quadriceps needs no qualification. There are three scalene muscles, and it is important to specify which one, scalenus anterior, scalenus medius, or scalenus posterior.

INTERNAL ARCHITECTURE OF MUSCLE FIBRES

In certain muscles all the fibres run in parallel from one end to the other. Examples are the sartorius, the 'strap muscles' of the neck (sternothyroid and thyrohyoid) and the intercostal muslces [Fig. 1.26(a)].

Fibres may be arranged so that they eventually all unite with a tendon at one or both ends giving to the muscle a **fusiform shape** [Fig. 1.26(b)].

A tendon may run a considerable distance along one edge of the muscle giving it a **unipennate form**. [Fig. 1.26(c)]. The flexor pollicis longus is the best example. The word 'unipennate' really means feathered on one side, like a feather from which half the barbs have been removed. The palmar interosseous muscles are also arranged in this way.

In a **bipennate muscle** [Fig. 1.26(d)] there is a tendon running in the central axis with muscle fibres converging on it from both sides. The dorsal interossei are arranged in this way.

The deltoid is a **multipennate muscle** [Fig. 1.26(e)]. It usually consists of four septa which are attached above to the acromion and clavicle, and three which are attached to the humerus which interdigitate with them from below. The muscle fibres pass obliquely downwards from one set of septa to the others.

It follows from this arrangement that when the shoulder joint is abducted a large number of deltoid muscle fibres are available all in parallel with each other. The more muscle fibres in parallel the greater the strength of movement. The fibres are all rather short and can only therefore produce a limited approximation of the insertion of the deltoid to the tip of the acromion. If you look at your own shoulder first with the arm by the side and then in full abduction, you will realize that the amount of shortening is in fact very slight.

Put your shoulder into extension and lateral rotation. Measure the distance from the clavicle (clavicular end of the deltopectoral groove) to the insertion of the deltoid with your other hand. Now swing your arm fully forwards. You will find the length of the anterior part of the deltoid has been approximately halved. It follows that there must be a group of long uninterrupted fibres on the anterior border of deltoid and also presumably on the posterior border. This is exactly what is found in the dissecting room.

MUSCLE GROUP ACTION

The principal muscles which produce a particular movement are called the **prime movers** or protagonists. Thus when you flex the elbow from the anatomical position with the forearm in supination you feel the biceps brachii contracting; it is one of the prime movers.

At the same time the extensor muscles are the **antagonists**. If they were to contract simultaneously the flexion movement would be very much weaker or would not occur at all. In normal movements the antagonist muscles are automatically relaxed by a physiological reflex mechanism called 'reciprocal inhibition'.

A third group of muscles are called 'synergists'. These are muscles which do not participate directly in the movement, but nevertheless make it possible for the prime movers to act more

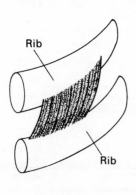

(a) External intercostal muscle

(b) Fusiform

(c) Unipennate

(d) Bipennate

(e) Multipennate

FIG. 1.26. Internal architecture of muscles.

effectively. A good example is the wrist extension which takes place when a fist is made which enables the long finger flexors to contract more powerfully.

The last group of muscles are called **fixators**. These are muscles which anchor the origin of the prime movers. Note in your own hand the contraction of the flexor carpi ulnaris tendon when the little finger is abducted. Now this muscle terminates by insertion into the pisiform bone a considerable distance from the finger and cannot possibly be concerned directly in its movements. The point is that the prime mover, the abductor digiti minimi (abductor of little finger), itself takes origin from the pisiform. If its origin were not fixed by the flexor carpi ulnaris, the movement of the fingers would be much more difficult to control. The muscle would dissipate its force by producing unwanted pisiform movement instead of concentrating its action on the little finger.

The use of the sternomastoid muscle as an accessory muscle of inspiration has been mentioned on page 20. Why doesn't the head flex when this muscle is used in this way? It is because fixator muscles, the extensors in the back of the neck, in particular the semispinalis capitis, contract and prevent such movement from occurring.

A muscle in any particular movement, may act in any one of these four roles. In certain disorders of the central nervous system a muscle may act normally when called upon as a fixator, but fail completely when called upon as a prime mover.

Most muscles in the body act over more than one joint. A good example is the biceps brachii which you should now examine again in yourself. Flex the elbow to a right angle, forearm parallel with (not on) the table and in the supinated position (palm upwards). Feel the belly of the muscle which is contracting strongly. Now without changing the horizontal position of the forearm, turn your hand over strongly into the pronated position (palm downwards). The belly of the biceps is now quite flabby and the tendon slack although the elbow is still flexed to a right angle. If the pronation movement is less powerful you may begin to feel the biceps contracting weakly again.

In this case complications arise because the biceps is a prime mover in elbow flexion and also a prime mover in supination. Hence it is an antagonist to forearm pronation. During flexion and pronation the muscle should contract as it is an elbow flexor but be reciprocally inhibited as it is a supinator. What actually happens depends on the balance of the two conflicting factors, stimulation (+) and inhibition (−) playing simultaneously on the same motor cells in the spinal cord in the segments C5, 6, and 7 which control the biceps muscle. There are four possible combinations between these two movements, all of which you should consider (TABLE 1.1).

TABLE 1.1.

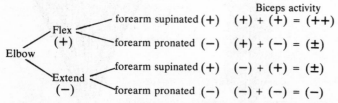

There is obviously another muscle which flexes the elbow which is not influenced by pronation and supination. This is the brachialis, which is inserted into the ulna. It lies deep to the biceps and can easily be felt behind the biceps tendon if you repeat the movements of elbow flexion plus pronation.

The biceps also acts on the shoulder joint and its action is still more complex than we have indicated.

LIGAMENTS

Classification of names of ligaments

The names of ligaments can be classified in the same way as those of muscles. We do not speak of 'origin and insertion' when describing ligaments, but instead simply speak of their 'attachments'. In the same way as certain muscles are named by joining origin and insertion together, so ligaments are named by combining their attachments into a composite word.

There are about two hundred and fifty named ligaments in the body. The word is not only used to describe fibrous bands joining bones together, but it is also used to describe folds of peritoneum which join viscera to each other in the peritoneal and pelvic cavity. For example, 'the gastrosplenic ligament', which is a peritoneal fold, joins the stomach to the spleen. In the pelvis, thickenings of fascia, connecting the uterus or other viscera such as the prostate to the pelvic wall, and thus helping to keep them in place, are also called ligaments.

Function names: 3 per cent

Example:
 suspensory ligament of the penis

Shape names: 20 per cent

Examples:

conoid ligament	—*cone-like*
cruciate ligament	—*crux = cross*
deltoid ligament	—*Greek capital delta Δ*
quadrate ligament	—*square*
ligamentum teres	—*teres = round*
trapezoid	—*like trapezium*
anular	—*anulus = ring*

Attachment names: 55 per cent

Examples:

coraco-acromial ligament	—*coracoid process to acromion*
lieno-renal	—*spleen (= lien) to kidney (= renal)*
pubo-prostatic	—*pubis to prostate*
talo-fibular	—*talus to fibula*

Position names: 20 per cent

Examples:

inguinal ligament	—*ligament of inguinal region*
ligamentum nuchae	—*ligament of neck*

Others

Examples:

ligamentum arteriosum	—*fibrous remnant of ductus arteriosus joining aorta to pulmonary artery*
ligamenta flava	—*yellow ligaments, made of yellow elastic fibres. Other ligaments are white made of collagen.*

Function names are the commonest amongst muscles and the rarest among ligaments; attachment names are the commonest amongst ligaments although they are comparatively rare amongst muscles.

BONES AND BONE FORMATION

Towards the end of the first month of embryonic life the earliest stages in the development of the skeleton can be recognized in the loosely arranged **connective tissue**, called **mesenchyme**. In most parts of the skeleton this is soon converted into **hyaline cartilage**, which in its turn, by the end of the second month, begins to be transformed into bone: that is, a **primary ossification centre** appears in the middle of the shaft of the 'bones' and spread towards, but does not reach the two ends [FIG. 1.27]. This is 'ossification in cartilage', and even at this early stage the shape and proportions of most bones are fairly similar to those seen in the adult.

But in the vault of the skull in most of the bones of the face and in the clavicles, the intermediate stage of cartilage formation is omitted, and these bones are said to 'ossify in membrane'. Indeed, the earliest onset of ossification is seen in the connective tissue of the mandible and clavicle preceding by about two weeks the ossification in cartilage of other bones, such as the humerus, femur, and vertebrae.

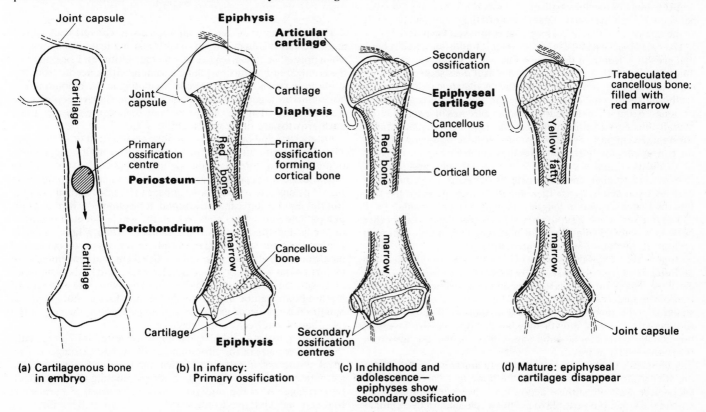

FIG. 1.27. Ossification of a typical long bone (not to scale).

(a) The bone having been sketched out in mesenchyme develops in hyaline cartilage. The shape is similar to that of the mature bone, but the proportions of the different segments of the limbs may not be the same. Thus the carpal bones are as long as the arm or forearm during the stages of development in cartilage. The primary ossification centre appears at about week 8 of foetal life in the middle of the shaft.

(b) Bone formation continues in the shaft beneath the periosteum. and thickens by the laying down of successive layers by the periosteum. The medullary cavity is hollowed out and comes to be filled with red bone marrow. The shaft is now called the diaphysis; the ends are the two epiphyses. The diaphysis next to the epiphysis. is called the metaphysis.

(c) The cartilagenous epiphyses at the ends of the bone now begin to ossify with the appearance of secondary ossification centres. These replace all the cartilage except for the articular cartilage and epiphyseal cartilage. On the medial side of the head of the humerus the articular and epiphyseal cartilages are continuous. On the lateral side they come to be separated by bone.

The epiphysis of the head is formed from three separate secondary ossification centres.

At the lower end of a bone such as the humerus, there are four separate centres. That for the medial epicondyle remains separate from the other three which are for the trochlear capitellum and lateral epicondyle. The figure shows the upper and lower epiphyses after fusion of these multiple centres. In many bones the epiphyses ossify from a single centre.

The periosteum is continuous with the joint capsules, which are lined on their inner surfaces by synovial membrane.

Normal epiphyses appear in response to traction or pressure. At the lower end those for the medial and lateral epicondyles are traction epiphyses. while those for the trochlea and capitellum are pressure epiphyses.

Long bones show more growth in length at one end than at the other. This is called the growing end. Secondary ossification usually begins here first. and the epiphyseal cartilage survives longer (1–2 years) compared with the other end. In the humerus the growing end is proximal. but in the femur the growing end is distal.

(d) In the mature bone both epiphyseal cartilages have disappeared. Cancellous bone develops trabeculae in response to the mechanical stresses and strains to which the bone is subjected. Red marrow is present between the trabeculae at the ends of the bones. but yellow marrow (fat) is found in the hollow shaft.

At birth most long bones have an ossified shaft termed the **diaphysis**, in which the central medullary (marrow) cavity has been hollowed out. This space is filled with red bone marrow, in which red and white blood cells and platelets develop. The ends of the bones however are still (usually) entirely cartilagenous. These regions are called the **epiphyses** (pronounced *epiphyseez*: singular **epiphysis**).

The ossification of epiphyses occurs in infancy, early childhood or later, but not by ossification spreading into them from the shaft. Instead independent or **secondary ossification centres** appear and convert most but not all the remaining cartilage into bone. Some epiphyses have two or three secondary ossification centres which fuse with each other.

A thin plate of hyaline cartilage, 2–3 mm thick, remains however between the epiphysis and metaphysis until the cessation of growth at the age of 15–20 years. This is the **epiphyseal cartilage**.

The epiphyseal cartilage is of the greatest importance as it is here that **growth in length** of the bone occurs. As it grows, the cartilaginous plate tends to become thicker, but as it does so its upper and lower surfaces are converted into bone. A dynamic equilibrium is established between the growth of the cartilage and its replacement by bone, so that for many years the bone increases in length but the overall thickness of the epiphyseal cartilage itself remains fairly constant. Eventually the growth of the cartilage slows down but as the conversion into bone continues the equilibrium shifts in favour of ossification, and the disc of cartilage becomes progressively thinner and thinner and eventually disappears altogether. The epiphysis and diaphysis are then united to each other for the first time by bone. Growth in length is now no longer possible.

The free surface of the epiphysis where it articulates with another bone in a synovial joint does not undergo ossification. Instead it remains throughout life as **articular cartilage**.

Bones are always covered by a fibrous membrane—the **periosteum**, except, of course, where they are covered with articular cartilage. The periosteum is vascular and responsible for the nutrition of the underlying bone cortex. The deep layers of the periosteum contain important cells called **osteoblasts**. These are bone forming cells. During growth they lay down successive layers of bone on the surface in the same way as the bark of a tree lays down successive layers of wood.

A moment's thought suggests that there must also be a mechanism which allows the cavity inside the bone to enlarge so that it keeps pace with the surface deposition. The overall proportions remain the same. It is possible to put the whole shaft of the femur at birth, inside the marrow cavity of an adult bone, which means that all the bone in the infant shaft must have been removed.

The cells which are responsible for removal of bone are called **osteoclasts** (bone destroyers). While new bone is being laid down on the outer surface of the cortex, its inner surface is being eroded by osteoclasts [Fig. 1.28]. Clearly there must be some mechanisms which ensure that the activity of all these cells is co-ordinated, so that the shape of all bones remains unchanged during development. These mechanisms are at present unknown. We only know that the epiphyseal cartilage is responsible for increase in length, while the periosteum is responsible for increase in thickness.

In a child all the bones contain red marrow and blood cell formation takes place in all of them. In the adult most of the bone marrow is converted into fat, and becomes yellow marrow. Red marrow is found in the adult only in the ends of the long bones and in flat bones such as the sternum and pelvis and in the vertebrae and skull.

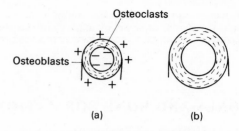

Fig. 1.28. Growth of bone.
Osteoblasts lay down bone on outer surface whilst osteoclasts remove bone from inner surface. Adult bone (b) has medullary cavity larger than bone in childhood (a), hence all childhood bone has been replaced.

The yellow marrow retains the capacity to convert back to red marrow, and this happens for example in pernicious anaemia, or when normal people go to live at very high altitudes. Bone marrow is examined in patients by taking a sample with a needle from the sternum or pelvis. Malignant cells circulating in the blood find red bone marrow a good culture medium and secondary deposits are frequently found at the ends of the long bones or in the vertebrae. Such growths are often very painful.

Osteoclasts and osteoblasts are not only found in growing bone. There is constant 'turn over' of mineral salts in adult bone, deposition of collagen fibres and remodelling of trabeculae, but because the processes of deposition and removal are in perfect equilibrium the shape and size of adult bone appear to be fixed and immutable. But this equilibrium can be disturbed, for example by disuse. Thus, in a patient who simply goes to bed, the whole skeleton begins to decalcify and the bones become thinner. When a limb is put in plaster, the same thing happens locally. A serious problem in space medicine is the absence of gravity. This means that movements can be carried out with very little muscular effort, and both the muscles and skeleton may waste rapidly. The thickness and strength of mature bones can be altered by disuse and disease, but change in length does not occur except in acromegaly where the size of the face, hands, and feet bones increases.

Certain parts of the skeleton are not converted into bone. Thus the anterior parts of the ribs are made of hyaline cartilage and we speak of the 'costal cartilages' which join the true bony ribs to the sternum or to the costal margin. Even the floating ribs are capped by cartilage. Also the intervertebral discs consist of permanent fibrocartilage and give to the vertebral column its flexibility.

The basic plan for bone formation is summarized in Table 1.2.

Sesamoid bones

Certain tendons of the body contain bones which are termed sesamoid bones. They were thought by Galen (AD 180) to resemble a sesame seed. They can be palpated as small round structures in the tendons running to the digits.

The largest 'sesamoid' bone is the patella (knee cap) which is associated with the tendon of quadriceps femoris.

JOINTS

Joints join two or more bones to each other. There are only three basically different types, fibrous, cartilaginous, and synovial joints. These reflect the three stages in the development of bones and

TABLE 1.2.

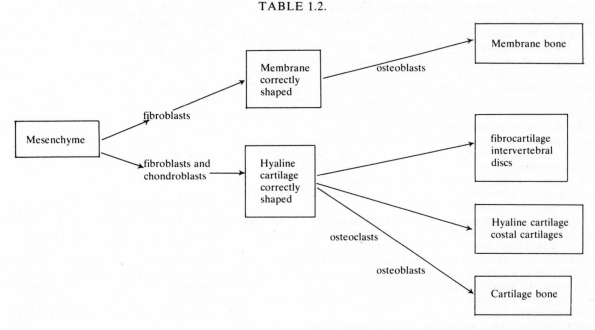

joints from connective tissue in the embryo in which the bones (and joints) are lightly sketched out in connective tissue which then goes on to form hyaline cartilage. The cartilage then begins to form bone towards the end of the second month.

1. Fibrous joints

It may happen that between these new formed bones, the 'connective tissue' cells remain so that the bones are joined together by fibrous tissue with the formation of a **fibrous joint**. An example of such a joint is that between the lower end of the tibia and fibula. This very simple joint is very strong and tearing of the interosseous ligament is virtually unknown. Such a joint is also called a **syndesmosis**.

2. Cartilaginous joints

If the material between the two bones progresses in development to the next step and forms cartilage, then we speak of a cartilaginous joint. These joints are of two types:
(i) **Primary cartilaginous joints**: in which the bones are united by hyaline cartilage. Most of these joints are found between the shaft (diaphysis) and epiphysis of growing bones, and disappear when growth stops or rather growth stops because they disappear.
(ii) **Secondary cartilaginous joints**: these are joints in which the bones are joined by fibrocartilage. Unlike the hyaline cartilage, in primary cartilaginous joints, in which virtually no movement occurs, the fibrocartilaginous joint is thick, pliable, and permanent. The best examples are those between the bodies of the vertebrae. They also occur at the manubriosternal joint, and at the pubic symphysis. They all lie in the sagittal plane.

There is a large group of **disorders** associated with **primary cartilaginous joints** which affect growth. These disorders are widespread and produce deformities in many parts of the skeleton—for example, in the vertebral bodies (producing scoliosis), in the bones in the foot, and in the hip joint.

The epiphyseal cartilage in the neck of the **femur** is of interest.

Looking at a mature bone and visualizing the position of the epiphyseal cartilage it would appear at first sight that the cartilage lies close to the vertical plane and should be subject to considerable shearing strain. In fact if allowance is made for the angle of convergence of the long axes of the femora (so that the knees touch) the situation is improved, but a more important fact is that the cartilage plate is not at right angles to the neck but lies at an oblique angle, so that the epiphyseal plate lies almost in the horizontal plane, and the stability of the joint is greatly increased. Even so 'slipping' of this epiphysis is a common disorder.

The diaphysis of the **humerus** is not cut off square at its upper end. It is cone-shaped. The epiphysis sits on top of it and has a corresponding cone-shaped recess. The epiphyseal cartilage is therefore not a flat disc (as many people suppose). These facts explain the appearances seen on X-ray in this region.

This arrangement obviously increases the strength of this primary cartilaginous joint but even so dislocation sometimes occurs.

3. Synovial joints

Synovial joints differ from the two previous types in that the ends of the two bones are not in direct physical continuity with each other. They are separated by a synovial 'cavity', which contains a small amount of **synovial fluid**, once thought to resemble egg white (*syn* = like; *ovum* = egg). These joints are all '*freely movable*', but the sacro-iliac joint is a special case [p. 59].

The articulating surfaces of bones in synovial joints are covered with a layer of hyaline cartilage, the **articular cartilage**.

In a model synovial joint the articular cartilage would be continuous at its rim with the edge of the epiphyseal cartilage [FIGURE 1.27]. The surfaces which are non-articular, on the inside of the capsule or on any bone inside the joint which is not covered with articular cartilage, are covered with synovial membrane. This provides a smooth lining over the whole of the inside of the joint, and is the source of synovial fluid. Thickenings often develop in particular regions of the capsule, called **capsular ligaments**.

Many joints are also strengthened by non-capsular or **accessory ligaments**. These are found in the three logically possible places:
1. Inside the joint (cruciate ligaments of knee).
2. Incorporated into the capsule itself (patellar ligament of knee).
3. Outside the joint altogether (the sacrotuberous and sacrospinous ligaments of the sacro-iliac joint).

In certain joints such as the inferior radio-ulnar joint, the bones are held together principally by an interarticular cartilage called the triangular articular disc. This is really a powerful ligament with certain additional articular functions. It is extremely strong, and when pulled on with sufficient violence, it fractures the styloid process of the head of the ulna to which it is attached. Obviously nothing would be gained by making it any stronger, as it is already stronger than the bone to which it is attached. This situation is so with many other ligaments in the body [see deltoid ligament of ankle (*below*)]. The strength of ligaments in the body (as well as many muscle tendons) exploits the strength of the skeleton to the maximum.

Synovial joints often contain varying amounts of **fat** situated between the capsule and the synovial membrane. Obviously when a joint moves the shape of the inside of the joint also changes, but the fat can adapt easily as it is liquid at body temperature. Straighten your knee and see the bulges on each side of the patella tendon. This is the infrapatellar pad of fat. A hollow is found here as soon as the knee is bent slightly.

Associated with many joints are **bursae** which are cavities lined with synovial membrane lubricated with synovial fluid. Bursae are found where tendons pass near joints and in other places. They allow those structures which slide over each other when the joint moves, to do so with the minimum of friction. They may or may not communicate with the joint: the suprapatellar bursa of the knee joint usually does whilst the subacromial bursa of the shoulder usually does not. If they do communicate, then excess fluid collecting in a diseased joint will produce a lump near the joint by distending the bursa.

Many joints contain **fibrocartilaginous intra-articular structures** which partly divide (acromioclavicular joint) or completely divide (temporomandibular joint) the joint into two. In the knee the menisci are remnants of partitions which once completely separated the femur from the tibia. Their central part seems to be worn away by joint movement during development. The remaining peripheral rim is non weight-bearing, but plays an important role in joint lubrication.

Synovial fluid is peculiar to the inside of synovial joints and bursae, and while playing a nutritive role, its most important function is to lubricate the contiguous joint surfaces. The slipperiness of synovial fluid varies with the load put on it. Thus when the joint is lightly loaded—flexing non-weight bearing knee joint—the fluid is thin and watery, but under weight-bearing conditions it becomes much move viscous. It is remarkable that this lubricant can change its mechanical properties according to functional needs. It is the best 'multigrade lubricant' in existence.

All **synovial joints** have a good **blood supply** and associated **lymphatics**. They are supplied with **sensory nerve fibres**, and a diseased joint may be very painful. Joint capsules and ligaments are supplied with special nerve endings which provide information to the central nervous system concerning the position of the joint, that is, they provide **proprioceptive information**.

Synovial joints are built to a common basic plan and are subdivided according to the principal structural and functional features which distinguish them. In many cases the movements are much more complex than the names suggest:

(i) Ball and socket joints (polyaxial),
    e.g. shoulder joint, hip joint;
(ii) Hinge joints (one axis),
    e.g. elbow joint, knee joint;
(iii) Pivot joints (one axis),
    e.g. superior radio-ulnar joint, atlanto-axial joint;
(iv) Condyloid joints (two axes),
    e.g. wrist joint;
(v) Saddle joints (two axes),
    e.g. carpometacarpal of thumb, posterior talocalcanean;
(vi) Plane joints (one or two axes),
    e.g. between most carpal and tarsal bones.

Nerve supply of synovial joints

Typically the sensory nerve supply of joints reaches them via the nerve supply of the muscles which produce movements in them. In most parts of the body the same nerves supply the joint, its muscles, and the skin over the joints. For example, the axillary nerve supplies the shoulder joint, the deltoid muscle and the skin over the deltoid. Certain fibres convey information concerning the tension in different parts of the capsule and other ligaments, and this is integrated by the brain with other information especially from the muscle spindles, so that a 'picture' of the position in space of various parts of the body, can be built up. Joints are richly provided with pain fibres.

Blood supply

The periosteum contains many small blood vessels which supply all those parts of the bone surface not covered with avascular articular cartilage. Many blood vessels also reach the bone along the tendons and muscle bellies. Most muscles are attached to the ends rather than the middle of the long bones, and there is usually a 'nutrient artery' which supplies this part of the bone, and the bone marrow. These are small arteries. The nutrient artery of the femur, for example, is smaller than one digital artery.

Immediately above and below most joints there is a ring of anastomosing blood vessels, which supply the epiphysis and metaphysis. All the big blood vessels in the vicinity will contribute to the periarticular anastomosis.

Attachment of joint capsules, ligaments, and tendons to bone

The question naturally arises at this point, as to how capsules, ligaments and tendons are fixed to the bone.

All bones consist of a meshwork of protein collagen fibres. Their arrangement is responsible for shape and strength. In the interstices of the meshwork mineral salts (mainly calcium hydroxyapatite) are deposited which give to bone its hardness. Defects in the collagen fibres result in bones which are very brittle (fragilitas osseum), and defects in the deposition of mineral salts result in soft bones which bend (bow legs in rickets).

Collagen fibres are the main component found in ligaments and tendons but arranged in dense bundles. When they reach the bone surface, the fibres continue into the substance of the bone and become interwoven there indistinguishably with the bone collagen fibres. They are cemented in place by the mineral salts deposited in the bone.

Once this is appreciated it is not so surprising that when a ligament (or tendon) is violently wrenched, it may avulse that part of the bone in which it is embedded. Thus the medial ligaments of the ankle joint may pull away the whole of the medial malleolus of the tibia when the ankle is 'broken'.

2 Exploring the limbs

THE ANATOMICAL POSITION

Stand with your feet together, hands by your sides, and palms facing forwards. This is the anatomical position used when describing the relative position of parts of the body [Fig. 2.01]. **Superior** means towards the head. Alternatively the terms *cephalic* or *upper* may be used. **Inferior** means towards the feet. The alternative terms are *caudal* or *lower*. A position between superior and inferior is termed **medius** or *middle*.

In the front-to-back direction, nearer the front of the body is termed **anterior**, whilst nearer the back of the body is termed **posterior**.

In the side-to-side direction **medial** means nearer to the median plane which divides the body into left and right halves. **Lateral** means farther away from this plane. A position between lateral and medial is termed intermediate.

Nearer to the surface of the body is termed **superficial**. Farther away from the surface is termed **deep**.

Special terms in connection with the upper and lower limbs are used. **Proximal** means nearer to the root of the limb. **Distal** means farther from this root.

The anterior surface of the **hand** is termed its **palmar** surface whilst the posterior surface is termed its **dorsal** surface.

The inferior surface of the **foot** is termed its **plantar** surface. The

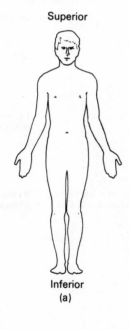

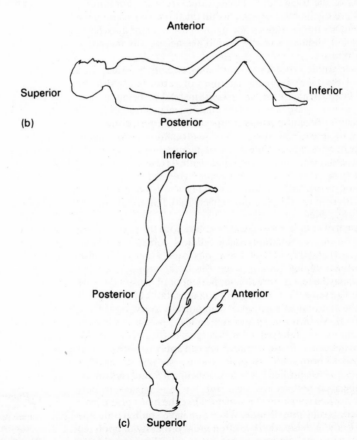

Fig. 2.01. The anatomical position.
The relationship terms used refer to the anatomical position (a). Note. however. that you take your anatomical position around with you. Thus your head is still superior to the trunk when you are lying down (b) and standing on your head (c).

superior part of the foot is termed the **dorsum of the foot** and the superior surface thus becomes the **dorsal** surface.

In the limbs **flexion** of a joint bends the limb whilst **extension** straightens the limb. Movement away from the median plane is termed **abduction**. Movement towards the median plane is termed **adduction**.

In the **fingers** abduction and adduction refer to movements relative to the middle finger, whereas in the **toes** the second toe is taken as the reference toe. The thumb is at right angles to the other fingers and the use of the terms flexion, extension, abduction, and adduction takes this into account [p. 35].

It is important to remember that, when describing the special relationships of organs to each other in anatomical terms, the relationship is always given as if the subject were in, or began in, the anatomical position (even when it is not so). Thus your head is always in a superior position relative to your trunk no matter whether you are standing up, lying down, or standing on your head. You carry your anatomical position axes around with you [Fig. 2.01(b) and (c)].

If you place the side of your hand on this book in front of you, and relax your fingers and thumb, your five digits will come to lie in 'the position of rest', that is, the position in which there is minimal muscular activity as well as minimal tension in the other soft parts. Although there is now practically no muscular activity, the fingers are flexed and the thumb is abducted, because this describes how the position of the fingers and thumb differs from the anatomical position where the fingers are straight and the thumb is touching the base of the index finger. It does not mean that the flexor muscles of the fingers and abductors of this thumb are necessarily in action, quite the opposite.

If you have studied zoology you will, alas, already have acquired the habit of thinking about the anatomy of vertebrates with the trunk and head *parallel with the ground*, instead of being at *right angles to* it, as in the case of man in the anatomical position. It is with words like anterior and posterior, superior and inferior that confusion may then arise because with a quadruped **anterior** means towards the head (cranial) while in man it means towards the front.

It is often convenient to use words like **ventral** (towards the belly) and **dorsal** (towards the back), and **cranial** (towards the head) or **caudal** (towards the tail), because with these words the position of the body with respect of its surroundings, parallel or at right angles to the horizon, does not matter.

A great deal of clinical medicine takes place with the patient lying in bed, or on an examination couch (**clinical** comes from *kline* (Greek) = bed) and the study of dissections usually takes place also with the subject **supine** (*supine* = on the back: *prone* = on the front). You may have a picture of your patient or of a dissected part in your mind's eye in this position, but you must talk (or write) as if it were in the anatomical position. It is important to try, right from the start, to think in terms of the anatomical position no matter what the orientation of the person with respect to the outside world.

It is often necessary to use words which can apply to both sides of the body, say to both limbs. In such cases we say one structure is *medial* or *lateral* to another because words like *left* and *right* must inevitably be incorrect on one side. But 'left' and 'right' are perfectly correct when speaking of a structure like the heart or the liver because there is only one of each. When you have studied only one side of the body in detail, there is often great difficulty in applying this knowledge to the opposite mirror image side. Try to avoid this problem by studying both sides together, and in your memory try to think of them simultaneously.

Remember always that it is **not your** left or right which is meant but that of the subject or patient whom you are studying. Note that when viewing the front of the patient, the patient's left is on your right, but that when viewing from the back, the patient's left is on your left.

ANATOMICAL PLANES

You will see from FIGURE 2.02 that there are also the three planes at right angles to each other: the **transverse**, **sagittal**, and **coronal**.

The word **sagittal** comes from the Latin word for an arrow and the **sagittal plane** gets its name from the sagittal suture which you will see on the vertex of the mature skull, running between the two parietal bones. But the imagery of the term is obscure until you look at a foetal skull. Here [FIG. 2.02(*upper*)] the head of the arrow is

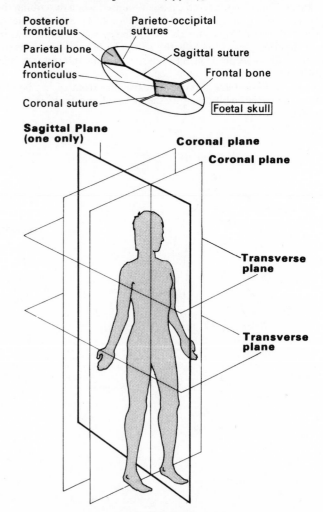

FIG. 2.02. The anatomical planes.

(*Upper*). The fonticuli and sutures of the foetal skull resemble an arrow. The suture corresponding to the shaft of the arrow is termed the sagittal suture (*sagitta* (Latin) = arrow).

(*Lower*). The sagittal plane is a vertical plane which passes through the sagittal suture of the skull. Coronal planes are vertical planes at right angles and pass through or are parallel to the coronal suture of the skull. Transverse planes are horizontal planes at right angles to both the sagittal and coronal planes.

immediately suggested by the shape of the **anterior fonticulus (fontanelle)** (which disappears at about eighteen months of age) and the **shaft** is the suture between the **parietal bones**, and the 'tail' is represented by the parieto-occipital sutures.

The word **coronal** comes from the Latin for a crown, but the term will remain incomprehensible if you think of a modern crown which is worn like a hat with its brim in the horizontal plane. The crown referred to here is the horseshoe-shaped wreath of laurel stretching from one temple to the other, over the top of the head, which crowned the brow of the Roman emperor or victorious athletes in ancient times. The sutures separating the parietal bones from the frontal bone were called coronal because they lie immediately beneath such a crown, and this in its turn defines the direction of the **coronal** or frontal **plane**.

There is only **one sagittal plane**, which splits the body into two symmetrical halves, right and left. If we wish to think of a plane parallel to the sagittal but to one side, we use the term 'parasagittal' (para = alongside). The position of the coronal 'plane' is not fixed in this way. It does not have to go through the coronal suture itself: it only needs to be parallel to it, and at right angles to the sagittal plane. There is no such thing as a paracoronal plane, any more than we would ever use the term paratransverse plane. Any plane which cuts through the body horizontally is a **transverse plane**.

Using these conventions the various movements, and muscle groups involved, of the upper and lower limbs will be considered. Details of the individual muscles involved and their nerve supply will be found in Chapters 5 and 7. We will start with the **upper limb**.

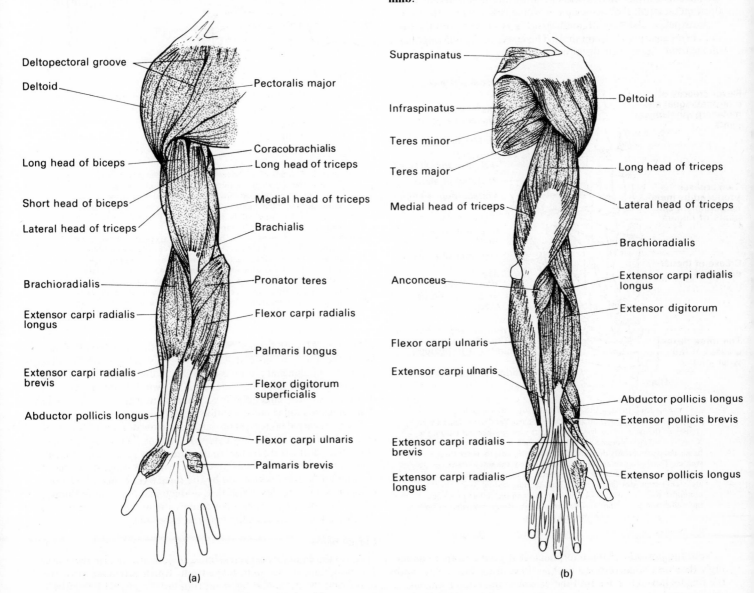

FIG. 2.03. Muscles of the upper limb.

(a) anterior view.   (b) posterior view.

HOW TO FIND YOUR WAY AROUND THE UPPER LIMB

Phalanges and metacarpal bones

Starting first with the 'surface anatomy', or if you prefer, the 'find it on yourself anatomy', the bones of the fingers can be readily palpated. The thumb has two phalanges, a proximal phalanx and a distal phalanx whilst the index, middle, ring, and little fingers have three phalanges, a proximal phalanx, a middle phalanx and a distal phalanx [FIG. 2.04]. Each digit has a metacarpal bone. The metacarpal bone associated with the thumb is termed the first metacarpal; that associated with the index finger is termed the second metacarpal; that associated with the middle finger is the third metacarpal; that associated with the ring finger is the fourth metacarpal, whilst the fifth metacarpal is associated with the little finger (metacarpus digiti minimi). The heads of the phalanges and metacarpals lie at their distal ends. Their bases lie proximally.

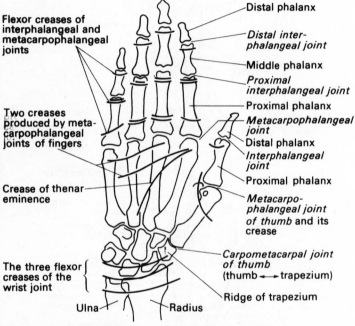

FIG. 2.04. Bones of the hand and wrist and skin creases.
The figure is a tracing from an X-ray of the hand and wrist. The flexure creases of the palm of the hand and fingers were marked out before the X-ray was taken. Compare these lines carefully with those of your own hand, and try to identify the joints. In front of the wrist there are three skin creases. The distal one marks the transition from the soft forearm skin to tough palmar skin and lies near the intercarpal joint, while the intermediate crease coincides fairly closely with the radiocarpal joint. Note the relationship in your own wrist of this crease with the styloid processes of the radius and ulna. The most proximal crease is quite variable in both depth and position.

Note the position of the interphalangeal joints. Rather confusingly they do not correspond to the skin creases [FIG. 2.04]. With the fingers flexed the heads of the proximal and middle phalanges can be easily felt. They each have two condyles with a groove between. The joint is just distal to these condyles since these

condyles articulate with the base of the adjacent phalanx when the finger is extended [FIG. 2.05]. The joint, with the fingers extended, lies about 4 mm distal to the projection formed by the head of the proximal phalanx and 2 mm distal to the head of the middle phalanx. Find these joints on yourself.

The alternate arrangement whereby the base would be proximal and the head distal would be unstable when pressing downwards with the joint flexed [FIG. 2.05(b)].

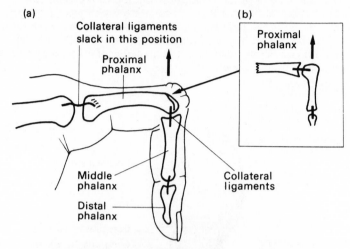

FIG. 2.05. Joints of the fingers. Two possible arrangements.
(a) Correct. The normal arrangement of the three joints of the left index finger, and their relationships to the flexor and extensor creases. When any one of these joints is flexed a bony prominence is produced by the articular surface of the proximal bone. Feel it on yourself. This arrangement is inherently stable.
(b) Wrong. Consider what would happen if this alternative arrangement had been adopted. A force in the direction of the arrow would tend to dislocate the joint. The collateral ligament would be at right angles to such a force and would therefore be unable to prevent the dislocation.
   An exactly similar design problem arises in the knee joint which is like 'a'. Dislocation by a force at right angles to the collateral ligaments in the knee is prevented by having in addition a pair of cruciate ligaments [p. 160].

The knuckles are formed by the heads of the metacarpals. Once again the joint is more distal than is generally realized. This metacarpophalangeal joint lies about 8 mm distal to the projection formed by the head of the metacarpal. (It will be seen on page 48 that the site of the knee joint is very much lower than many people think since a similar consideration applies there.)

The metacarpal bones can be palpated on the dorsum of the hand although they are covered by extensor tendons. The styloid process at the base of the third metacarpal is often very prominent and can be seen when the wrist is flexed. On the palmar surface of the hand the first metacarpal is covered by the **thenar eminence** which contains the short muscles which flex, adduct, and oppose the thumb, whilst the fifth metacarpal is covered by the hypothenar eminence which contains muscles supplying the little finger.

Carpal bones

Seven of the eight carpal bones [FIG. 2.07] are arranged in two rows. The scaphoid, lunate, and triquetrum, which articulate with the radius and the triangular ligament (but not directly with the ulna), form the proximal row. The trapezium, trapezoid, capitate, and hamate form the distal row. Distally, the trapezium articulates

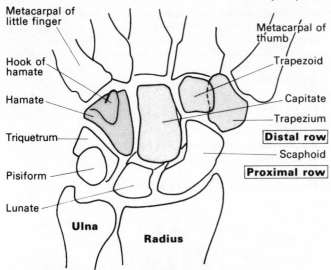

FIG. 2.06. The carpal bones (traced from an X-ray).

In life the gaps between adjacent bones are filled with articular cartilage which does not show up on an X-ray. The triangular cartilage which runs from the styloid process of the ulna to the radius does not show up either.

The distal row of carpal bones (shaded) consists of the trapezius, trapezoid, capitate, and hamate. The proximal row consists of the scaphoid, lunate, and triquetral. The pisiform is a sesamoid bone in the tendon of flexor carpi ulnaris and is only an appendage of the proximal row.

The overall shape of the joint surfaces separating the proximal and distal rows of carpal bones precludes much movement in the plane of the palm of the hand (abduction/adduction) by flexion/extension movements can take place fairly freely. The two rows fit together like pieces of a jig-saw puzzle.

Note that in studying X-rays of the hand it is customary to look at the picture as if you were holding up your own hand in supination in front of your own face.

Usually no carpal bone has started to ossify at birth, but occasionally ossification of the capitate and hamate has started especially in the female. The pisiform is the last to ossify. This ossification begins in about the twelfth year.

Note that the ossification of the true carpal bones proceeds in a regular sequence beginning with the largest (capitate) and finishing with the smallest (trapezoid). In practice a three-year-old child often has three ossified carpal bones, a five-year-old has five, and so on. An X-ray of the wrist showing carpal bone ossification as well as the epiphyses of the radius and ulna is helpful in assessing the **skeletal age** of the child. The pisiform ossifies later at the onset of puberty. It is not a true carpal bone, but only a sesamoid in flexor carpi ulnaris.

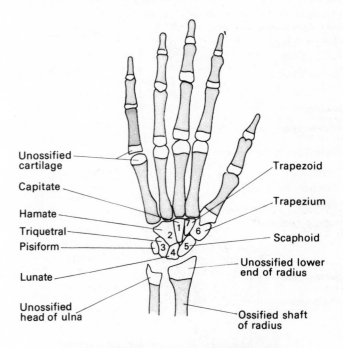

FIG. 2.07. Ossification of the bones of wrist and hand.

The carpal bones ossify in the order indicated by the numbers. Usually ossification in the capitate begins half way through the first year. Approximately one additional carpal bone ossifies each year up to age 6 or 7. There are 29 separate ossification centres shown in the figure which corresponds to the ossification at birth.

The metacarpals and phalanges all begin ossification from their primary ossification centres at the beginning of the 3rd month of intra-uterine development soon after similar primary centres appear in the other long bones of the body.

Unlike most long bones, the bones of the fingers have an epiphysis at one end only. The metacarpal of the thumb differs from those of the fingers in that its epiphysis is at the proximal rather than the distal end.

The secondary ossification centres in the epiphyses appear during the third year. The epiphyseal cartilages eventually disappear at 18–20 years. Up to this time the fingers continue to grow.

Ossification in the lower epiphyses of the radius and ulna commence during the second and fifth years respectively. They are the growing end of these bones, and therefore fuse late (18 years of age).

mainly with the metacarpal of the thumb, the trapezoid† articulates mainly with the metacarpal of the index finger, the capitate† articulates mainly with the metacarpal of the middle finger, whilst the hamate articulates with the metacarpals of the ring and little fingers.

The remaining carpal bone the pisiform is an exception. It lies on the palmar aspect of the triquetrum. It forms a projection at the medial end of the hypothenar eminence and can be moved slightly if the wrist is passively flexed.

The ossification state of the carpal bones acts as an important guide to the developmental age of a child [FIG. 2.07]. The carpal bones ossify in the following order:

1. Capitate	6 months	
2. Hamate	6 months	
3. Triquetrum	2–4 years	
4. Lunate	3–5 years	Age at which ossification normally starts
5. Scaphoid	4–6 years	
6. Trapezium	4–6 years	
7. Trapezoid	4–6 years	
8. Pisiform	12+ years	

† The trapezoid also articulates with the third metacarpal whilst the capitate also articulates with the second and fourth metacarpals.

The triquetrum can be readily palpated as a projection below the lower end of the ulna on the dorsum of the wrist if the wrist is slightly flexed and abducted (hand deviated to the radial side). The capitate can be felt at the base of the third metacarpal if the wrist is fully flexed.

MOVEMENTS OF WRIST AND HAND

The powerful flexors and extensors of the wrist and hand are situated in the forearm [FIGS. 2.03 and 2.09]. Confirm this on yourself. Encircle your right forearm with your left hand, then make a powerful fist with your right hand. You will feel the muscles of the forearm contract. The tendons running to the fingers have to be prevented from bow-stringing when the wrist is flexed or extended. Thus the tendons from the forearm muscles are kept in

points are raised, so that the connecting flexor retinaculum forms the carpal tunnel for the flexor tendons.

The tendons and associated structures proximal to the flexor retinaculum are arranged as follows (from the radial to the ulnar side) [FIG. 2.10]:
1. Brachioradialis tendon.
2. Radial artery (the pulse).
3. Flexor carpi radialis tendon.
4. Median nerve (passing deep to flexor retinaculum but very superficial just proximal to the retinaculum).
5. Palmaris longus tendon (you may not have one).
6. Ulnar artery.
7. Ulnar nerve.
8. Flexor carpi ulnaris tendon (running to pisiform).

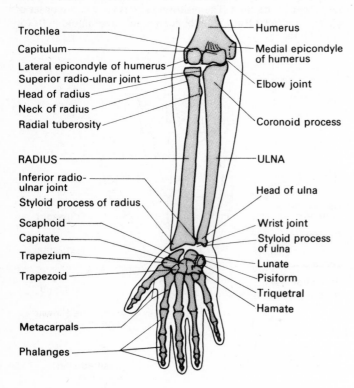

FIG. 2.08. Bones of the forearm, wrist, and hand.

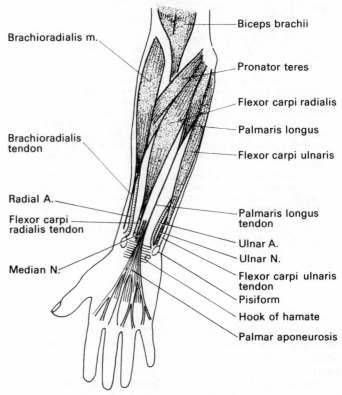

FIG. 2.09. The muscles at the front of the forearm.

place at the wrist by the flexor retinaculum on the palmar surface and the extensor retinaculum on the dorsal surface.

The flexor retinaculum is a strong band of fibrous tissue attached to the pisiform and hook of the hamate on the ulnar side, and to the tubercle of the scaphoid and ridge of the trapezium on the radial side [FIG. 2.10]. These attachment points which are termed the 'four pillars of the carpal arch' can be readily palpated. The hook of the hamate is about 2 cm distal and lateral to the pisiform. The tubercle of the scaphoid can be seen as a prominence if the wrist is fully extended. The ridge of the trapezium is palpable just distal to the tubercle of the scaphoid, and proximal to the base of the first metacarpal, which of course moves with the thumb, while the carpal bone does not. Find all these in yourself.

The flexor retinaculum also gives origin to muscles of the thenar and hypothenar eminence.

If you examine an articulated skeleton you will see that these four

Deeper from the radial to the ulnar side will be the tendons of the flexor pollicis longus, and the four tendons of flexor digitorum superficialis. The four tendons of flexor digitorum profundus lie deeper still. The arrangement of their synovial sheaths are shown in FIGURE 2.11.

At the back of the wrist is the extensor retinaculum which holds down the extensor tendons. This is a thickening of deep fascia which runs from the radius round the back of the wrist to the hamate and pisiform. It is divided by partitions into six compartments which contain the tendons of the following muscles. Find these on yourself. Start with the lower end of the radius and work your way round

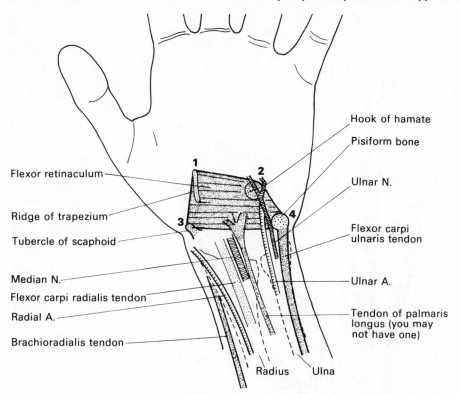

Flexor retinaculum

Ridge of trapezium

Tubercle of scaphoid

Median N.

Flexor carpi radialis tendon

Radial A.

Brachioradialis tendon

Hook of hamate

Pisiform bone

Ulnar N.

Flexor carpi ulnaris tendon

Ulnar A.

Tendon of palmaris longus (you may not have one)

Radius　Ulna

FIG. 2.10. Left wrist: what can be palpated from the front?

Find the pisiform bone. and tracing upwards find the tendon of flexor carpi ulnaris. Immediately adjacent to the tendon find the pulsations of the ulnar artery. Between the tendon and the artery is the ulnar nerve.

Now find 'the pulse' ('the radial artery'). It is easier to feel than the ulnar artery. It lies directly in contact with the radius. Immediately to its lateral side a few centimetres above the wrist is the tendon and belly of brachioradialis which overlies and protects the radial artery. On its medial side is the tendon of flexor carpi radialis which is easy to identify if the wrist is gently flexed against resistance. The palmaris longus lies less than 1 cm on its medial side (you may not have one). Between these two tendons is the median nerve. This nerve is not palpable like a tendon. but it lies very superficially just above the flexor retinaculum: pressure produces a tingling painful sensation.

The flexor retinaculum cannot be palpated as such. but its four corners are attached to bony landmarks which can be identified with the finger:

(1) Ridge (crest) of the trapezium.
(2) Hook of the hamate.
(3) Tubercle of the scaphoid.
(4) Pisiform.

The radius and ulna are shown as dotted lines.

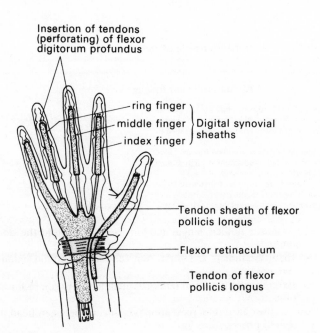

Insertion of tendons (perforating) of flexor digitorum profundus

ring finger

middle finger

index finger

Digital synovial sheaths

Tendon sheath of flexor pollicis longus

Flexor retinaculum

Tendon of flexor pollicis longus

FIG. 2.11. Flexor tendon sheaths.

The tendons to each of the four fingers of flexor digitorum superficialis and profundus lie in the expected relationship (superficial and deep) (*profundus* (Latin) = deep) throughout their course in the forearm, wrist, and hand. In the fingers, however, although enclosed in the fibrous flexor sheath, the tendon of flexor digitorum superficialis splits and allows the profundus tendon to pass through it. The profundus therefore flexes the terminal phalanx, and is the most important flexor muscle of each finger.

The superficialis tendon is inserted into the middle phalanx.

Note that the thumb, which only has two phalanges and therefore only one interphalangeal joint, has only one long flexor tendon, which like the profundus tendons of the four fingers, is inserted into the terminal phalanx.

In both the thumb and fingers the five metacarpophalangeal joints are flexed by short but powerful muscles which lie throughout their course in the palm of the hand. They are the intrinsic muscles of the hand but the flexor digitorum superficialis and profundus are extrinsic muscles of the hand.

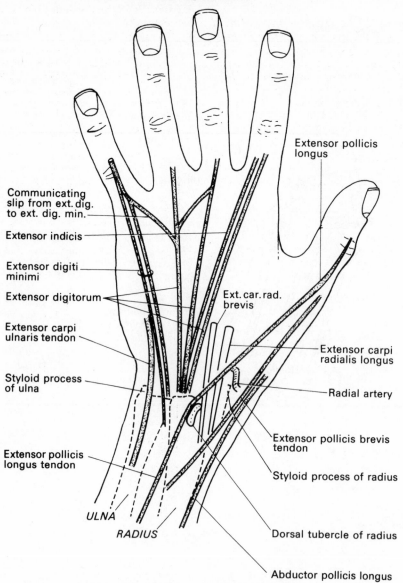

Extensor pollicis
longus

Communicating
slip from ext. dig.
to ext. dig. min.

Extensor indicis

Extensor digiti
minimi

Extensor digitorum

Ext. car. rad.
brevis

Extensor carpi
ulnaris tendon

Extensor carpi
radialis longus

Styloid process
of ulna

Radial artery

Extensor pollicis brevis
tendon

Styloid process of radius

Extensor pollicis
longus tendon

Dorsal tubercle of radius

ULNA

RADIUS

Abductor pollicis longus

FIG. 2.12. Left wrist: what can be palpated from behind?
Find the tendons of extensor carpi radialis longus and brevis and extensor carpi ulnaris.
The first two stand out when you make a firm fist. These three tendons can be traced to
the bases of the metacarpals 2, 3, and 5.
 Find the dorsal tubercle of the radius and the tendon of extensor pollicis longus. Move
your fingers and find the extensor digitorum, extensor indicis, and extensor digiti minimi
tendons. They form a group and cannot be distinguished from each other.
 Note that the reduplicated tendon to the little finger only represents extensor digiti
minimi. The extensor digitorum supplies all the fingers (not the thumb) but the little finger
is only supplied by a cross slip at the last moment from the tendon supplying the ring finger.

to the lower end of the ulna [FIG. 2.12]. The tendons felt, and seen
in a thin person, are:

 1. Abductor pollicis longus and extensor pollicis brevis tendons
    passing over the styloid process of the radius. (Also the inser-
    tion of brachioradialis.)
 2. Extensor carpi radialis longus and extensor carpi radialis
    brevis tendons.
 3. Extensor pollicis longus tendon, running around the dorsal
    tubercle.
 4. Extensor indicis and digitorum tendons (common extensor
    tendons).
 5. Extensor digiti minimi tendon, which lies between radius and
    ulna.
 6. Extensor carpi ulnaris tendon which runs between head and
    styloid process of ulna.

Movements of the thumb

The trapezium and the metacarpal of the thumb form a saddle-shaped joint which on the surface of the trapezium is convex from side to side and concave from front to back. This gives the thumb a wide range of movements since flexion, extension, abduction, adduction, and opposition can occur at this joint as well as circumduction. In addition flexion and extension only can take place at the metacarpal-phalangeal joint and the interphalangeal joint of the thumb [FIG. 2.13]. Try these movements on yourself. Remember that because the metacarpal of the thumb is at right angles to that of the other fingers, all movements take place at right angles to those of the other fingers. Flexion and extension is in the plane of the palm. Abduction and adduction movements are at right angles to this plane. The standard anatomical position is adduction with the thumb closed on the index finger.

1. Abduction. Move the thumb forwards and slightly laterally so that the web between the thumb and index finger is straight and at right angles to the plane of the palm of the hand.
2. Adduction. Bring the thumb back to the index finger.
3. Flexion. Move the thumb medially across the palm, so that the web is twisted forwards.
4. Extension. Move the thumb laterally so that the web is twisted backwards.
5. Opposition. Move the thumb so that it touches the tips of the other fingers in turn. Opposition is a mixture of flexion, adduction, and rotation of the first metacarpal bone.

The bellies of the powerful thumb muscles lie in the forearm. The long flexor of the thumb (flexor pollicis longus) lies deep in the anterior surface of the forearm.

The bellies of the powerful extensor muscles lie deep in the back of the forearm. [FIG. 2.03]. The older names for these extensor muscles gives a clue to their insertion. Abductor pollicis longus (old name extensor ossis metacarpi pollicis) is inserted into the base of the first metacarpal. Due to its point of insertion it acts more as an abductor than an extensor of the thumb [FIG. 2.12 and 2.14].

Extensor pollicis brevis (extensor primi internodii pollicis) is inserted into the base of the proximal phalanx of the thumb. It extends the proximal phalanx of the thumb. It takes origin from the radius.

Extensor pollicis longus (extensor secundi internodii pollicis) is inserted into the base of the terminal phalanx of the thumb. It

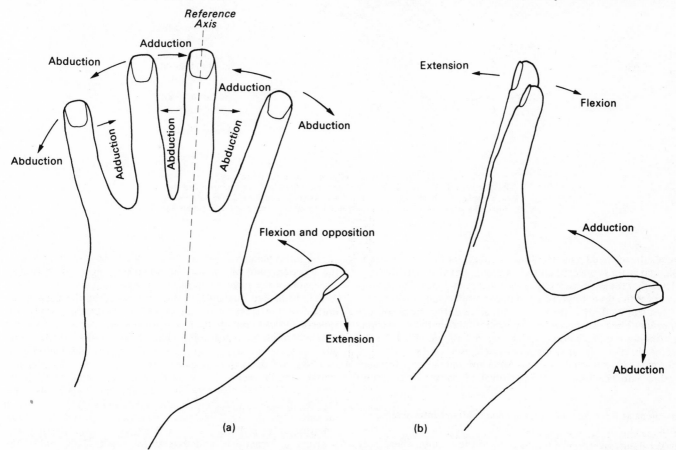

FIG. 2.13. Abduction and adduction of the fingers and thumb.

(a) The movements of abduction and adduction of the fingers take place with respect to an imaginary line which passes through the middle of the middle finger. Note that in the foot the imaginary line passes through the second toe. These conventions, far from being arbitrary, are based on the patterns of the dorsal and palmar (or plantar) interossei which will be described later.

With your hand flat on the table your thumb can only move in the horizontal plane, the plane of flexion and extension of the thumb. Tuck your thumb beneath your palm with the back of your nail against the table: this is opposition.

(b) Abduction of the thumb is in the plane at right angles to the palm of the hand.

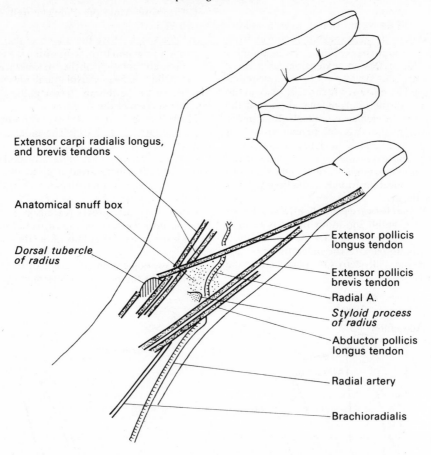

Extensor carpi radialis longus, and brevis tendons

Anatomical snuff box

Dorsal tubercle of radius

Extensor pollicis longus tendon

Extensor pollicis brevis tendon

Radial A.

Styloid process of radius

Abductor pollicis longus tendon

Radial artery

Brachioradialis

FIG. 2.14. Left wrist: what can be palpated from the lateral side?
The view from the lateral side shows the 'anatomical snuff-box'. This lies between the extensor carpi radialis longus and brevis. Find the styloid process of the radius in the floor of the fossa. The snuff-box is most obvious when the thumb is extended—as in 'hitch-hikers thumb'.

extends the terminal phalanx of the thumb. It takes origin from the ulna, and is thus longer than the extensor brevis at both ends.

The other thumb muscles lie in the thenar eminence. These short muscles of the thumb run from the flexor retinaculum, the tubercle of the scaphoid, the third metacarpal and the ridge of the trapezium. Their exact origins are not important but their insertions are. Abductor pollicis brevis is inserted into the outer side of the base of the first phalanx. Opponens pollicis is inserted into the radial side of the first metacarpal. Adductor and flexor pollicis are inserted into the base of the proximal phalanx of the thumb [p. 106].

Movement at the other metacarpophalangeal and interphalangeal joints

Unlike the metacarpophalangeal joint of the thumb, where only flexion and extension occurs, the other metacarpophalangeal joints allow a wide range of movements—flexion, extension, abduction, adduction, and circumduction. Test the range of movements on yourself. Individual movements of the index and little finger are usually easy. Solitary abduction and adduction of either of the other two fingers may require practice.

The interphalangeal joints allow only flexion and extension. If the ligaments are lax, note that a marked degree of backward extension of the fingers is possible at these joints.

The distal interphalangeal joints are flexed by the **deep** flexor muscle of the forearm (flexor digitorum profundus) whilst the **superficial** flexor muscle (flexor digitorum superficialis) flexes the proximal interphalangeal joints. This is the opposite to what common sense would suggest [FIG. 2.15]. The superficial tendon splits to allow the deep one to gain access to the terminal phalanx. Neither tendon acts directly on the metacarpophalangeal joint. Together these muscles close the hand and curl it up by also flexing the metacarpal-phalangeal and wrist joints. Note that when making a fist the wrist extensors (extensor carpi radialis longus, extensor carpi radialis brevis, and extensor carpi ulnaris) also contract as synergists and extend the wrist. This makes the origin–insertion distance of the flexors longer and hence aids their action. Therefore, to reduce a person's grip on you, flex their wrist.

If the wrist is to be flexed against resistance, additional flexor muscles, flexor carpi radialis, flexor carpi ulnaris, and palmaris longus are brought into play.

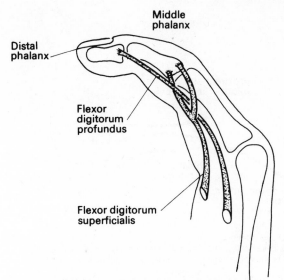

FIG. 2.15. Long flexor tendons of fingers.
Common sense suggests that the superficial tendon should terminate in the distal phalanx. But, in this instance, the body does not do the obvious thing. In each finger the tendon of flexor digitorum superficialis splits as shown in the diagram and allows the profundus tendon to reach the terminal phalanx.

In the foot this pattern is repeated exactly—the superficial muscle being represented by flexor digitorum brevis, and the deep by flexor digitorum longus. Note that in the thumb (big toe) there is only one long flexor tendon. It is as if the middle phalanx is missing.

The fingers are extended again by the extensor muscles in the forearm (extensor digitorum, extensor indicis and extensor digiti minimi). Their action also tends to abduct the fingers giving the spread out appearance. This is the opposite effect to that of flexor digitorum profundus and flexor digitorum superficialis which tend to adduct the fingers.

If you place your hand palm downwards on the table, you will find that you can now abduct and adduct each finger separately. This action is brought about by the interossei (and abductor digiti minimi). By convention these movements of abduction and adduction are related, not to the midline of the body, but to a line running down the centre of the middle finger. Movement away from this line is abduction. Movement towards this line is adduction. It follows that as far as the middle finger is concerned, movement in either direction is considered to be abduction!

The dorsal interosseous muscle which causes abduction of the index finger can be readily palpated between the base of the thumb and the second metacarpal bone.

Movements of the other metacarpal bones

Very little movement of the second and third metacarpals takes place but flexion of the fourth and fifth metacarpals can be seen on making a fist. The heads of the metacarpals cease to be in a straight line and make a convex curve on the dorsum of the hand. The fourth and fifth metacarpal bones can be felt moving when making a fist and when carrying out the thumb opposition manoeuvre. The movement takes place between the base of these metacarpals and the hamate.

The flexion of the fifth metacarpal helps to tighten the grip of the closed hand and thus prevent a cylindrical handle (such as a tennis racket) from slipping [FIG. 2.16].

A blow with a clenched fist is normally struck using the head of the second and third metacarpals (rather than those of the fourth and fifth). This transmits the blow through the capitate and trapezoid to the lunate and scaphoid and thence to the radius. When the fourth and fifth are used, a fracture of the neck of the metacarpal is common.

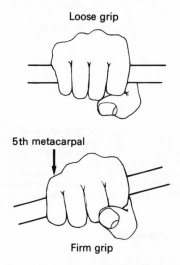

FIG. 2.16. Clenching the fist.
Note the forward movement (remember the anatomical position terminology!) of the head of the 5th and to a lesser extent of the 4th metacarpal bones, when the fist is clenched. Try it yourself and watch the movement.

Movements at the intercarpal and radiocarpal joints

The line of the **intercarpal joint** is S-shaped [FIG. 2.07]. In particular the capitate and hamate fit into a concavity formed for them by the scaphoid and lunate. It follows that in such a joint flexion and extension could occur but not abduction and adduction. In fact the intercarpal joint plays an important part in movements of 'the wrist' which involve flexion and extension. Indeed in wrist flexion the movement in the intercarpal joint is greater than that between the proximal row and the radius [FIGS. 2.17 and 2.18]. In extension the opposite is true.

The joint between the proximal row of carpal bones (scaphoid, lunate, triquetral) and the radius is the **radiocarpal joint**. On the medial side the head of the ulna with its styloid process does not participate directly in this joint. Apart from extension and flexion, abduction (radial deviation) and adduction (ulnar deviation) and circumduction also occur.

The word 'carpus' means 'wrist'. The muscles which act on the wrist (flexor and extensor **carpi** ulnaris or radialis) are however not inserted into the carpal but into the **meta**carpal bones, that is beyond the wrist itself.

Abduction of the wrist is brought about by flexor carpi radialis, extensor carpi radialis longus and extensor carpi radialis brevis working together aided by the extensor and abductor tendon of the thumb [FIG. 2.19(a)].

Adduction of the wrist is brought about by flexor carpi ulnaris and extensor carpi ulnaris working together [FIG. 2.19(b)].

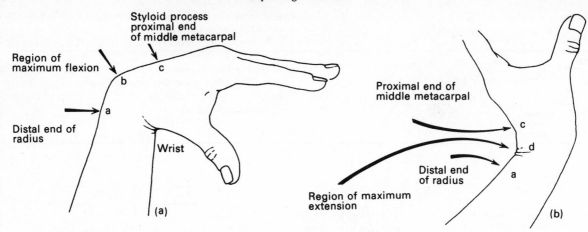

FIG. 2.17. The left wrist in flexion and extension.

Consider the back of your own wrist in full flexion (a), and in full extension (b). Find first the dorsal tubercle at the distal end of the radius. landmark 'a'. It lies just proximal to the radiocarpal joint. Then find the styloid process of the middle (3rd) metacarpal landmark 'c' which defines the level of the carpometacarpal joint.

In flexion these landmarks are easy to find. In extension you need to search for them or alternatively keep a finger of the other hand on each as you move. The region of maximum flexion at 'b' is about half way between 'a' and 'c'. It must be taking place mainly between the proximal and distal rows of carpal bones. Very little flexion occurs at the radiocarpal joint, this joint is important in extension at 'd'.

The distance between the landmarks 'a' and 'c' in extension is less than half that in flexion. Thus the sets of joints at the wrist must slide over each other like three segments of a lobster's tail.

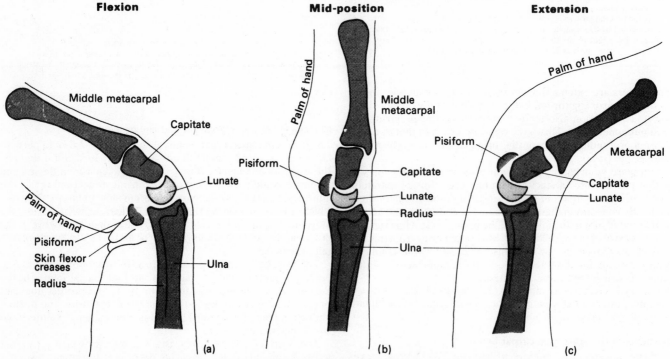

FIG. 2.18. Right wrist joint (X-ray tracing).

The wrist joint is shown in the midposition (b), in full flexion (a) and full extension (c). Some people can produce greater ranges of movement than are shown here. The view is looking through the hand from the side. In such X-rays the carpal bones are superimposed on each other as are the metacarpals of the four fingers, as well as the radius and ulna. In these tracings only the lunate, capitate, and third metacarpal are shown, as well as the pisiform which stands away from the other carpal bones.

Note that the direction of the articular surface of the radius is not square to its long axis, but faces slightly forwards. Little movement between the lunate and radius occurs in flexion. Most of this movement takes place between the lunate and capitate. In extension the reverse situation arises. In hyperextension, which may occur in a fall, the lunate may be squeezed out of the angle between the radius and the capitate, so that it is dislocated in a forward direction. The bone may then impinge on the median nerve in the carpal tunnel.

The outline of the soft parts shown in the figures are clearly visible on an X-ray. This includes the flexor creases on the front of the wrist joint.

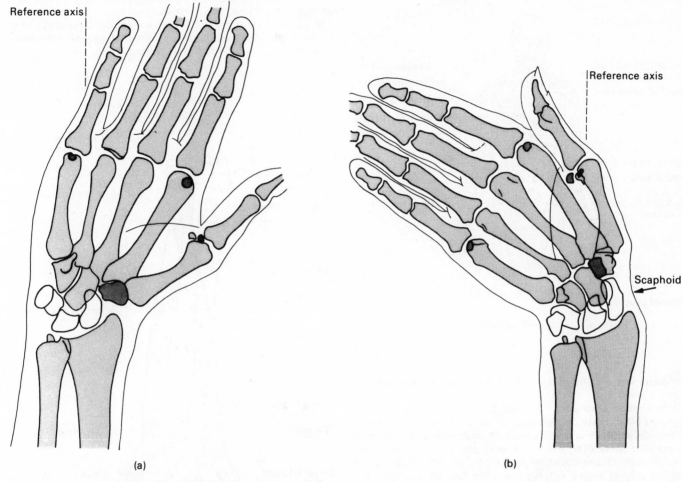

Fig. 2.19. Abduction and adduction of the wrist joint (X-ray tracing).
These movements take place with respect to the hand in the anatomical position. Ulnar deviation (= adduction) (b) usually radial deviation (= abduction) (a). Note that in ulnar deviation the scaphoid is easily palpable (←).
The sesamoid bones often associated with the index and little fingers can be seen. The trapezius and trapezoid bones do not appear as separate entities, because the plane of the X-rays has caused them to be superimposed on each other.
The movements of abduction and adduction take place at the radiocarpal joint. Very little movement takes place at the intercarpal joint, and virtually none takes place at the carpometacarpal joints.

RADIUS AND ULNA

Rather confusingly the head of the radius is its proximal end whilst the head of the ulna is its distal end [Figs. 2.20 and 2.21]. Both bones have a styloid process at their lower ends. Find these on yourself. The styloid process of the radius lies well below the styloid process of the ulna. This is an important fact in the diagnosis and treatment of a 'broken' wrist.

In the anatomical position (supination) the shafts of the radius and ulna are parallel. When pronation takes place the lower end of the radius moves in front of the lower end of the ulna to reach its medial aspect. As a result it is the front surface of the head of the ulna that makes a projection at the back of the wrist in pronation. Palpate it on yourself.

The styloid process of the ulna can be readily felt in the mid-position between supination and pronation. The styloid process of the radius can be palpated on the other side proximal to the base of the first metacarpal between the tendons of abductor pollicis longus and extensor pollicis longus in the upper part of, what is termed, the 'anatomical snuff box' [Fig. 2.14].

The posterior border of the ulna is subcutaneous throughout its length and can be palpated all the way down from the olecranon to the styloid process. The olecranon forms the prominent point of the elbow in flexion, but in extension it tends to disappear into a hollow.

The medial and lateral epicondyles of the humerus are in line with the olecranon in extension. Confirm this on yourself. Fractures of the olecranon disturb this relationship.

THE ELBOW JOINT

The radius and ulna articulate with the humerus at the elbow. It is a hinge joint. In addition the upper end of the radius and ulna

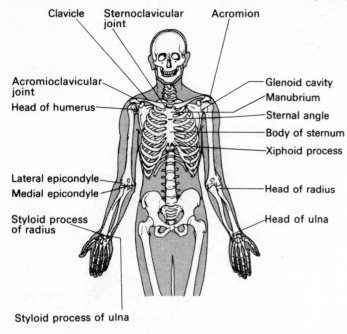

Fig. 2.20. The bones of the upper limb in the anatomical position.

articulate with each other in the superior radio-ulnar joint, which is a pivot joint.

The humero-ulnar joint is between the trochlea of the humerus and trochlear notch of the ulna. It allows mainly flexion and extension although a small degree of abduction and adduction is possible in flexion. Note that since the medial part of the trochlea extends lower than the lateral part, the forearm makes an angle (carrying angle) with the arm when the elbow is extended (p. 43). Since the trochlea is spiral in shape the forearm and arm become parallel when the elbow is bent. Check this point on yourself.

At the superior radio-ulnar joint the head of the radius rotates within the anular ligament during supination and pronation of the forearm and hand [Fig. 2.22]. Note that unlike the lower end of the radius which rotates around the ulna, the head of the radius at its upper end always stays on the lateral side of the ulna [Fig. 2.21]. The rotation of the head of the radius during pronation and supination can be felt by palpating the head of the radius just below the line of the elbow joint, which can be identified below the lateral epicondyle. In pronation the shaft of the radius crosses the shaft of the ulna obliquely. The bones are now in close proximity and cross fusion is likely if a fracture is set in this position. The bones are furthest apart in supination.

The humeroradial joint is between the capitulum (= little head) of the humerus and the head of the radius. The capitulum of the humerus is spherical [Fig. 2.04] and articulates with the concave upper end of the radius. However, if you look at the bones you will see that the capitulum is situated at the front of the humerus and the two surfaces are most closely in contact when the elbow is bent at a right angle and the forearm is mid-way between full pronation and full supination. This is the position usually adopted when using a screw-driver (or cork-screw). Pressure can then be exerted directly

from the humerus to the radius during the process of supination (driving the screw in if right-handed).

When the elbow is straight, force is transmitted from the humerus first to the ulna and then via the interosseous ligament (between the radius and ulna) to the radius and thence to the hand.

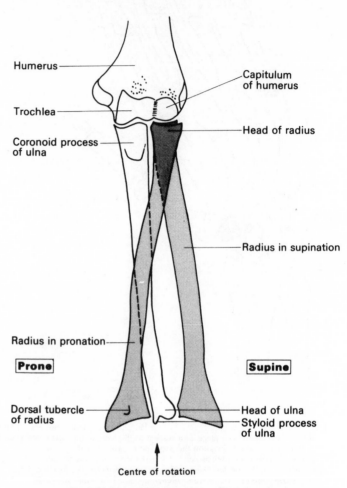

Fig. 2.21. Pronation and supination of left forearm.

The essential component of pronation and supination is the rotation of the lower end of the radius around the head of the ulna, carrying the hand with it.

In the anatomical position the forearm is in supination and the radius is lateral to and roughly parallel to the ulna. In pronation the lower extremity of the radius comes to lie on the medial side of the ulna, the shaft of the radius crossing over in front of the ulna.

At the upper end the head of the radius is held in position by the anular ligament. It is easy to feel the head of the radius rotating in pronation and supination. The two bones are joined at the lower ends by the triangular articular cartilage. The axis around which rotation occurs is indicated by the arrow.

Consider the design of the trochlea of the humerus, which provides a cylindrical system around which the ulna can turn in what is in effect a simple hinge joint. The radius, however, moves in the side-to-side plane in pronation and supination as this figure makes obvious. The articular surface of the radiohumeral joint must be spherical (not cylindrical) in shape, and this accounts for the differences between the ulnar and radial facets at the lower end of the humerus.

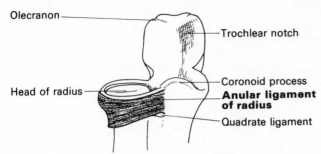

FIG. 2.22. The right anular ligament.

The anular ligament is a weakness especially in a child. A mother pulling sharply on a child's arm may pull the radius out of its socket. It is usually possible to replace it.

Biceps and brachialis flex the elbow (p. 22). Triceps extends it.

THE SHOULDER GIRDLE

The **acromion** [FIG. 2.20] is the highest point of the shoulder and can be easily felt. Find it on yourself or more easily on a model. Palpating backwards and downwards leads to the spine of the scapula. With the arm raised above the head, the spine of the scapula lies at the bottom of a furrow with the deltoid below and trapezius above.

The **clavicle** [FIG. 2.20] can be seen almost as clearly in the living as in a skeleton. If the shoulders are raised and drawn forwards, it stands out clearly. Note that in the living it usually slopes upwards as it passes laterally (articulated skeletons often show it as lying horizontally). Development of the muscles to increase the obliquity and thus raise the clavicle from the chest wall is one method of dealing with nerve compression symptoms due to a cervical rib.

The **acromioclavicular joint** is formed between the lateral end of the clavicle and the facet on the medial aspect of the acromion. The bones are separated by fibrocartilage. The acromioclavicular joint can be easily seen when the clavicle projects above the acromion. When this is not so, palpation round the front reveals a depression corresponding to the junction of the clavicle and the acromion.

The **sternoclavicular joint** is formed between the medial end of the clavicle and both the manubrium and first rib (costal cartilage). The bony surfaces are separated by an articular disc which is attached superiorly to the clavicle and inferiorly to the first costal cartilage. If the shoulders are pulled backwards the sternoclavicular joint can be made to gape at the front.

The clavicle acts as a strut for all movements of the shoulder girdle, and all movements have the sternoclavicular joint as their centre. It is the only place where the bones of the shoulder girdle are attached to the axial skeleton. Animals in which the movements of the shoulder are limited to flexion and extension do not have a clavicle at all, as in the horse and cat. Thus with arm movements, the acromioclavicular joint always lies on the surface of a sphere with the sternoclavicular joint at its centre. Study how the clavicle limits the movement of the scapula on your model, or on yourself standing in front of a mirror.

The shoulder joint is a ball and socket joint between the humerus and the scapula. The hemispherical head of the humerus articulates with the glenoid cavity of the scapula. This glenoid cavity is very shallow and although it is deepened by the glenoid lip, a great deal of movement is possible at this joint, flexion, extension, abduction, adduction, medial and lateral rotation, and circumduction [FIG. 2.23]. However, if you examine the bony skeleton you will see that although flexion (forward movement) can be continued until the arm is vertically above the head, abduction (sideways movement) is

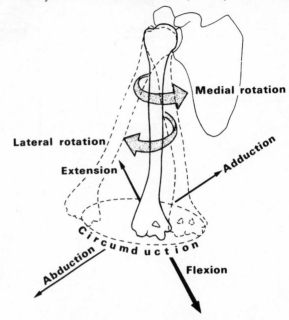

FIG. 2.23. Movements at the shoulder joint.
Movements may be classified as flexion, extension, abduction, adduction, medial and lateral rotation, and circumduction. They are relative to the position of the glenoid cavity which points sideways and slightly forwards.

only possible up to a right angle, then the bony surfaces touch. It is possible to continue this movement until the arm is vertically above the head, but the last part of the movement must be associated with either rotation of the humerus so that the same relationship is reached between the head of the humerus and the scapula as in flexion, or the scapula must rotate on the chest wall. Palpate the scapula on your model and see at which point during abduction scapular movement commences. In most people the scapula moves to a limited degree during abduction to the horizontal position. From then onwards it moves to a considerable degree [FIG. 2.24].

With the arm in the anatomical position, the greater tubercle of the humerus projects beyond the acromion anteriorly. Posteriorly the head of the humerus is tucked under the acromion [FIG. 2.25]

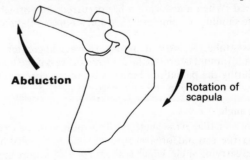

FIG. 2.24. Abduction is usually associated with rotation of the scapula.

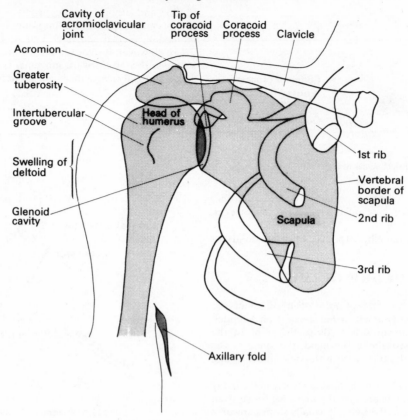

Cavity of acromioclavicular joint — Tip of coracoid process — Coracoid process — Clavicle

Acromion—

Greater tuberosity—

Intertubercular groove—

Head of humerus

Swelling of deltoid

Glenoid cavity—

1st rib

Vertebral border of scapula

Scapula — 2nd rib

3rd rib

Axillary fold

FIG. 2.25. Shoulder joint (X-ray tracing).
The reading of X-rays may be made difficult because complicated distortions occur when the parts of the body are projected simply in one flat plane. The shadows cast by different structures, which may be widely separated from each other in the body, are superimposed on each other.
Note the humerus, whose outline is similar to the normal appearance of this bone. The head fits against the glenoid cavity. The acromion and coracoid process, are somewhat distorted by the angle of view.
The acromioclavicular joint cavity can be seen. The skin lies close to the lateral third of the clavicle and acromion. These bony structures are easy to feel in yourself. The shoulder below the acromion shows a swelling produced by the presence of the deltoid muscle.
In an anteroposterior view of the shoulder joint the coracoid process is foreshortened and its tip may be superimposed on its base.

and does not project. The overlying deltoid muscle which covers the shoulder joint is also thicker in front than behind. This 'rounded' appearance is important to remember when a shoulder dislocation is suspected. When a dislocation has occurred the normal roundness of the shoulder is completely lost and it appears flattened [FIG. 2.26].

The lesser tubercle, intertubercular groove, and greater tubercle can be readily found lateral to the coracoid process if the humerus is rotated during the palpation. Find the tip of the coracoid process just below the clavicle in the deltopectoral groove.

Carrying angle

It has been seen that in the anatomical position, with the forearm in supination, the arm and forearm are not in line with each other [FIG. 2.27]. The carrying angle, which is about $10° \pm 3°$ is more marked in the female, in whom the hips are wider, than in the male. It will be

seen later that the bone of the thigh and leg are also not in line with each other [FIG. 2.29, p. 45].

When the palm faces inwards, the angle disappears. When the palm faces backwards the angle is slightly reversed. The position of the hand moves about 10–12 cm in the medial direction when the forearm is pronated. Carrying a bucket in pronation has the rim knocking against the leg, but in supination with the palm to the front, this does not happen. In addition when the carrying angle is utilized the medial epicondyle on the medial aspect of the elbow rests against the hip bone, and this avoids fatigue in the abductors of the shoulder joint. Everyone suffers fatigue in these muscles when carrying a suitcase. This is because the handle is longitudinal rather than transverse in direction and makes it impossible to exploit the advantages of the carrying angle. How would you design a suitcase handle?

Clinically it is important in the manipulation of fractures round

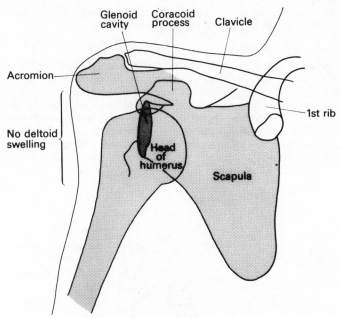

Acromion

Glenoid cavity

Coracoid process

Clavicle

1st rib

No deltoid swelling

Head of humerus

Scapula

FIG. 2.26. Dislocated shoulder joint (X-ray tracing).
The head of the humerus no longer lies in the glenoid cavity. The X-ray shows that the head has come to lie medially. It is impossible to say from such an X-ray whether the head has come to lie in front of or behind the scapula. In practice it is almost always anterior, sandwiched between the body of the scapula and the subscapularis muscle.

FIG. 2.27. The carrying angle.
Note the difference when carrying a bucket with the forearm pronated (left) and supinated (right).

the elbow joint to be aware of the carrying angle. If it is exaggerated pressure will be exerted on the ulnar nerve, and paralysis will result. If this happens the nerve has to be repositioned in front of the elbow joint. In following up a patient, whose fracture you have treated, you would examine ulnar nerve function very carefully

so as to deal with this problem before the nerve is irreparably damaged.

The upper limbs and sport

In many sports the club or bat is held in the two hands. Such games include golf, hockey, cricket, baseball, lacrosse, etc. Although a great deal of speed can be transmitted to the club by swinging the arms, it is the action of the wrists that are all important in achieving the maximum velocity at the head of the club and thus giving the maximum impact to the ball.

In baseball the ball is pitched or thrown using a bent arm whereas in cricket it is bowled with the arm straight. The distance the ball has to travel is similar in these two sports and the velocity achieved by these two balls is very similar although it may be greater in cricket since the bowler is allowed to make a fast run before delivering the ball. In both sports, if the ball is given a rotation by a flick of the wrist it will travel through the air with a swerving pathway rather than in a straight line. The reason for this is that the air in the immediate vicinity of the ball rotates with the ball. On one side this air will be travelling in the same direction as the motion and on the other side it will be travelling in the opposite direction. The velocity of this air relative to the air through which it is passing will be greater on one side than the other. The greater the velocity the lower the pressure (Bernouilli's Principle) which will suck the ball into an arc concave on this side.

Bernouilli's principle can be seen in action when a boat is moored close to the bank in a fast moving stream. The high velocity of the water between the boat and the bank produces a marked fall in the water level which can be clearly seen. It is also the principle of the scent spray and atomizer, where a fast air movement over the top of the tube will suck fluid up the tube and disperse it in the air flow.

Aeroplanes fly because due to the shape of their wings the air moves at a higher velocity over the top surface than over the lower surface and this gives the lift.

If there were no air the ideal trajectory for throwing a ball a long distance would be to throw it at an angle of 45°. Any higher angle would cause the ball to go higher but it would not travel so far; any lower angle and the ball would reach the ground sooner and once again not travel so far. The golf ball is hit so it leaves at a lower angle than 45° but to prevent it hitting the ground sooner it is undercut so it has a spin whereby the top of the ball is travelling in the opposite direction to the direction of travel. The reduced pressure produced at the top of the ball holds the ball in the air in opposition to gravity provided that the golf ball has an indented surface. Such a golf ball travels further than a perfectly smooth one.

A table-tennis ball swerves dramatically when spin is put on it. If side spin is used, it swerves say to the left and then breaks to the right when it hits the table. In baseball the striker only has to cope with the ball in the air, while in cricket the batsman has the added complication of how the ball breaks once it hits the ground. Some people understand these problems intuitively, while others are utterly bamboozled by sidespin, backspin, and top spin. In many games, such as tennis it is obvious from the way the shot is played what kind of spin (or no spin) has been put on the ball. Good players 'disguise' the shot until the last moment. Ordinary players make it obvious what they are going to do long before they actually hit the ball.

The role of the upper limbs in the maintenance of balance during sporting activities is discussed on page 163.

HOW TO FIND YOUR WAY AROUND THE LOWER LIMB

The lower limbs enable us to stand, walk, and run. The ground surface on which we walk may be irregular; we need to be able to walk uphill, downhill, as well as along the flat, and even on occasions to walk along the side of a mountain! In all these cases, if the trunk is to be kept upright, quite complex changes in the position of the large number of joints between the ground and the pelvis are necessary with each step. To do this the brain, and particularly the cerebellum and cerebral cortex, has to be informed of the position of every joint and the degree of contraction of every muscle so that the movements of gait can be co-ordinated.

Maintenance of balance

The maintenance of balance involves three separate mechanisms, proprioception from the legs, eyesight, and the vestibule apparatus in the inner ear. It is possible to maintain balance if one of these is lost. Thus a blind man can maintain his balance using proprioception and his vestibular apparatus only, but he is less steady than a sighted person is when the floor is moving, say, standing in a crowded train. Riding a bicycle in pitch darkness is virtually impossible.

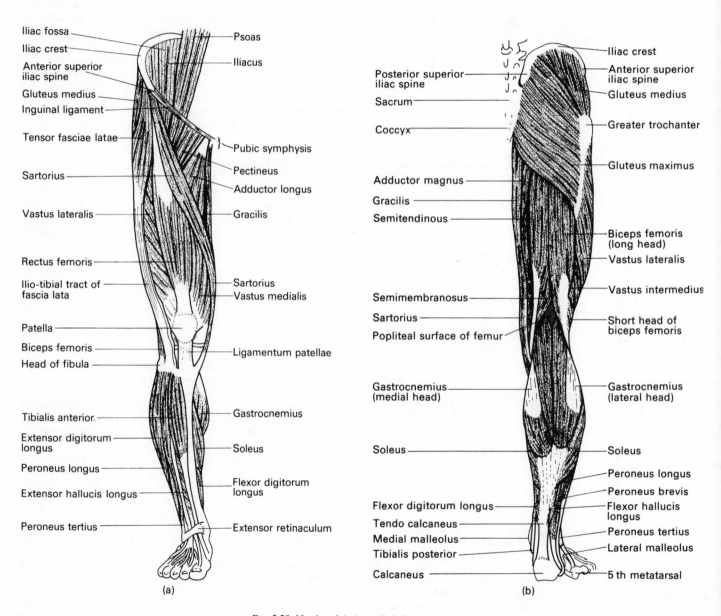

Iliac fossa
Iliac crest
Anterior superior iliac spine
Gluteus medius
Inguinal ligament
Tensor fasciae latae
Sartorius
Vastus lateralis
Rectus femoris
Ilio-tibial tract of fascia lata
Patella
Biceps femoris
Head of fibula
Tibialis anterior
Extensor digitorum longus
Peroneus longus
Extensor hallucis longus
Peroneus tertius

Psoas
Iliacus
Pubic symphysis
Pectineus
Adductor longus
Gracilis
Sartorius
Vastus medialis
Ligamentum patellae
Gastrocnemius
Soleus
Flexor digitorum longus
Extensor retinaculum

(a)

Posterior superior iliac spine
Sacrum
Coccyx
Adductor magnus
Gracilis
Semitendinous
Semimembranosus
Sartorius
Popliteal surface of femur
Gastrocnemius (medial head)
Soleus
Flexor digitorum longus
Tendo calcaneus
Medial malleolus
Tibialis posterior
Calcaneus

Iliac crest
Anterior superior iliac spine
Gluteus medius
Greater trochanter
Gluteus maximus
Biceps femoris (long head)
Vastus lateralis
Vastus intermedius
Short head of biceps femoris
Gastrocnemius (lateral head)
Soleus
Peroneus longus
Peroneus brevis
Flexor hallucis longus
Peroneus tertius
Lateral malleolus
5 th metatarsal

(b)

FIG. 2.28. Muscles of the lower limb (anterior view).
(a) Anterior view.   (b) Posterior view.

Stand on one leg with the eyes open and then with them closed. Most people cannot do this for more than 10–15 seconds with the eyes closed.

Patients with loss of proprioception in their legs, such as occurs in tabes dorsalis (a late manifestation of the venereal disease, syphilis) often say that they do not go out walking on dark nights because they feel unsteady, and stumble and fall. They may give a 'history' of having fallen when soap entered their eye whilst washing the face. These are indications that such patients are using eyesight as the principal mechanism for balance in place of leg proprioception. Unlike normal patients, they are unable to maintain their balance if asked to stand with their arms by their side and their eyes closed. Stand-by to catch them!

The vestibular apparatus in the inner ear [p. 284] gives an indication not only of any change in speed (acceleration) but also of the direction of gravity, that is, which direction is up and down with reference to the centre of the earth. The vestibular mechanism is not highly developed in most people (except possibly in short-sighted or blindfolded tight-rope walkers in circuses) but patients with vestibular disorders may have difficulty in maintaining their balance and often feel dizzy. The same will apply when any drug is given that has side-effects on the vestibular apparatus.

Incidentally, patients with inner-ear disorders should not be allowed to go swimming since they may easily get into difficulties and possibly get drown. Once underwater, if the feet are no longer in contact with the ground, neither proprioception nor eyesight can be used, and the only information as to the direction of the surface of the water comes from the inner ear.

Even with a sighted person it is surprising how little clearance is allowed for the feet when walking. It only needs an unseen pavement stone to be raised a centimetre in the air to trip over it.

When there is a discrepancy between the information coming in from the three mechanisms the result may be nausea and motion sickness. Thus, on board ship the feet may be firmly in contact with the deck, but the eyes may view the horizon moving up and down with the motion of the ship whilst the inner-ear signals that the position of the head relative to the centre of the earth is continually changing.

In view of the complexity of joint movements involved in quite simple movements when using the lower limb, it is convenient to break these movements down into those occurring at the hip joint; those occurring at the knee joint; those occurring at the ankle joint; and those occurring at the intertarsal, metatarsophalangeal, and interphalangeal joints. However, this is an artificial subdivision used to make the study of the subject easier. The human body does not carry out voluntary movement in terms of anatomical muscles, but in terms of motor units [p. 4] in a large number of anatomical muscles at the same time. Some of these motor units will be contracting to bring out the movement (prime movers), others will be relaxing to allow the movement to take place, whilst others will be contracting initially and then slowly relaxing (paying out the slack) to give a smooth graded movement. The co-ordinating centre for this is in the cerebellum and one has only to see a patient with a cerebellar disorder to realize the importance of this co-ordination. Such a patient is not paralysed, but uses the wrong number of motor units to produce the wrong power for the desired movement. As a result their action appears clumsy (particularly on the side of the body of the cerebellar lesion). They tend to fall to this side and may be arrested by the police for appearing drunk.

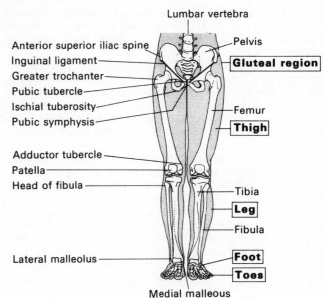

FIG. 2.29. Bony landmarks of the lower limbs.

Labels: Lumbar vertebra; Anterior superior iliac spine; Inguinal ligament; Greater trochanter; Pubic tubercle; Ischial tuberosity; Pubic symphysis; Adductor tubercle; Patella; Head of fibula; Lateral malleolus; Medial malleous; Pelvis; Gluteal region; Femur; Thigh; Tibia; Leg; Fibula; Foot; Toes

GENERAL ARRANGEMENT OF THE LOWER LIMB

The lower limb consists of the gluteal region, the thigh, the leg, the foot, and the toes.

The weight of the upper part of the body is transmitted through the lumbar vertebrae to the **pelvis** which consists of the **sacrum** and **hip bones**. The fifth lumbar vertebra articulates with the sacrum at the **lumbosacral joint**. The sacrum articulates with the hip bones on either side at the **sacro-iliac joints**.

The hip joint is a ball and socket joint formed by the spherical head of the femur (the thigh bone) inserted into the relatively deep acetabulum socket of the pelvis. The femur in man has a comparatively long neck joining the head to the shaft and this greatly increases the range of movement of the hip joint. However, this arrangement is a point of weakness since the weight of the body has to be transferred through almost a right-angle between the neck and shaft. It is not surprising that fractures of the upper end of the femur are common especially in old age.

The **gluteal region** overlies the hip bones at the back and sides. It may be subdivided into the rounded region at the back, the **buttock** (Latin = *natis*) and the region at the side, the **hip** (Latin = *coxa*). The cleft between the buttocks is termed the **natal cleft** (Latin *nato* = I am born). The skin fold below the buttock is termed the **gluteal fold**. The muscles of this region (gluteus maximus, medius, and minimus) are referred to as the **glutei**.

The upper part of the lower limb above the knee is termed the **thigh** (Latin = *femur*). It joins the gluteal region at the **gluteal fold** posteriorly. Medially the thigh joins the perineum. Anteriorly it joins the abdominal wall at the **inguinal ligament**. This junctional inguinal region is also known as the **groin**.

The lower part of the lower limb below the knee is termed the **leg**. The femur articulates with the **tibia** at the **knee joint** (Latin = *genu*) which has the large sesamoid bone (in the tendon of the extensor muscles) called the **patella** in front. The second thinner bone of the leg, the **fibula**, lies laterally. It articulates with the tibia at its upper and lower ends at the tibiofibular joints. Note that the Greek for fibula is **perone** and the adjective *peroneal* is commonly used in place of *fibular*. For example, the **peroneal** muscles and nerves are *fibular* muscles and nerves. The French call the fibula 'le peroné', the peroneal bone.

The hollow at the back of the knee is termed the **popliteal fossa** from *poples* which is Latin for ham, the lower part of the back of the thigh. The muscles at the back of the thigh (biceps femoris, semimembranosis and semitendinosis) are often referred to as the **hamstring** muscles. Their tendons, biceps on the lateral side, semimembranosis and semitendinosis on the medial side form the side boundaries of the popliteal fossa.

The back of the leg is termed the **calf**. The Latin for the calf is *sura* hence the **sural nerve** supplies the skin over this region.

The tibia and fibula articulate with the first of the tarsal bones (*tarsus* = ankle) the **talus** at the **ankle joint**. The prominent lower end of the fibula forms the **lateral malleolus**. The lower end of the tibia forms the **medial malleolus**. The other tarsal bones are the **calcaneus** (*calx* (Latin) = heel) which lies under the talus and forms the heel, the **navicular** (Latin = boat-shaped) and the three **cuneiforms** (*cuneus* (Latin) = wedge; hence these bones are wedge-shaped) and the **cuboid** (the shape of a cube) [FIG. 2.30].

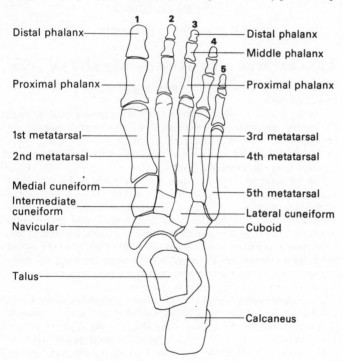

FIG. 2.30. Bones of the foot.

The foot has five metatarsal bones which articulate with the tarsal bones at the **tarsometatarsal joints** and with the proximal phalanges of the toe at the **metatarsophalangeal joints**.

The toes, like the fingers, are referred to as digits (as in flexor *digitorum* longus muscle). The big toe (Latin = *hallux*), like the thumb, has only two toe bones or phalanges (a proximal and a distal phalanx) whilst the other toes, like the fingers, have three (a proximal, middle, and distal phalanx). The proximal end of the metatarsals and phalanges is termed their **base**. The distal end is termed their **head**.

The tarsal and carpal bones are variations on the same plan. For example the pattern of the bones articulating with the bases of the metacarpal and metatarsal bones is the same. The medial, intermediate and lateral cuneiform bones correspond to the trapezium, trapezoid and capitate, while the cuboid which articulates with the fourth and fifth metatarsals obviously represents the hamate which articulates with the fourth and fifth metacarpals.

THE PELVIS AND THIGH

Bony landmarks of the gluteal region

The iliac crest forms the boundary between the gluteal region and the abdomen [FIG. 2.31]. It can be palpated along its entire length from the **anterior superior iliac spine**, which forms a prominence *in front*, to the **posterior superior iliac spine** *behind* which lies in the floor of a dimple a short distance above the buttocks. The posterior superior iliac spine is at the level of the second sacral spine.

The anterior third of the iliac crest corresponds to the well-marked iliac furrow. The highest point of the iliac crest is at the level of the fourth lumbar vertebra and this fact is used as a guide to vertebrae when carrying out a lumbar puncture. Posteriorly the iliac crest is less easy to define especially in a muscular person due to the bulging external oblique abdominal muscle.

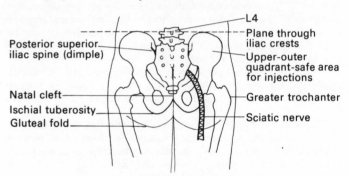

FIG. 2.31. Bony landmarks of the gluteal region.
The sciatic nerve presents a large target when giving injections into the buttocks. Take care not to hit it!

The buttocks are separated by the vertical **natal cleft** which starts at the third sacral spine and leads down to the anus. The lower part of the sacrum and the coccyx lie in the floor of the natal cleft. The transverse **gluteal fold** which limits the buttocks inferiorly is not formed by the lower border of gluteus maximus, the lower border of which runs diagonally downwards, but by a collection of fat distal to the muscle. The fold is formed by the attachment of the skin to deep fascia below the gluteal fold. Above, there is surplus skin to allow for flexion of the hip joint. When bending forwards the gluteal fold disappears.

You sit on your **ischial tuberosities** so it is easy to find. When standing the ischial tuberosity is covered by gluteus maximus, but when sitting the ischial tuberosity appears below the lower border of this muscle. Pressing firmly between the ischial tuberosity and the lower part of the sacrum reveals the strong **sacrotuberous ligament**. The rami of the **ischium** and **pubis** running forward from the ischial tuberosity are also palpable.

The **greater trochanter** of the femur lies close to the surface on the lateral side of the thigh. It may appear as a projection, or due to the bulging gluteus medius above, it may lie in a depression.

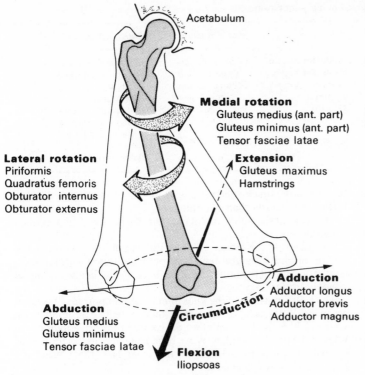

FIG. 2.32. Movements of the hip joint.
The diagram shows the principal movements of the hip joint and the muscles which produce the movements.

The dotted line shows the path taken by the lower end of the femur during the movement of circumduction. If you were to draw a circle in the sand with your heel or toe, the principal movement would be that of circumduction at the hip joint.

Note that because of the length of the femur, quite small movements in the hip joint produce considerable movements at the level of the knee, and even greater movements of the foot.

BONY LANDMARKS OF THE FRONT OF THE THIGH

The **pubic symphysis** lies in the midline between the two pubic bones. The **pubic crest**, which is about 2.5 cm long, runs laterally from the superior margin of the symphysis to the **pubic tubercle**. Below, the **pubic arch** is formed by the pubis and ischium. It runs backwards from the lower part of the pubic symphysis and forms the medial boundary between the thigh and the perineum.

The **inguinal ligament** can be palpated as a resilient band running from the anterior superior iliac spine to the pubic tubercle. It lies in the shallow curved furrow which separates the abdomen from the thigh.

The **head of the femur**, although it lies deep to muscle, can be felt by pressing deeply over the **mid-inguinal point** (mid-way between the anterior superior iliac spine, and the symphysis pubis). If the lower limb is rotated the head of the femur can be felt rolling under the finger. The **shaft of the femur** lies deep to muscle but can be felt if the muscles are relaxed.

MOVEMENTS OF THE FEMUR

The principal function of the hip joint, from an everyday life point of view, is that it allows a person to walk, to sit in a chair, to stand up from the sitting position, to climb stairs, and to run. These are all manoeuvres that a patient with a diseased hip may find difficult.

The site of the muscles controlling the hip joint is not at first sight at all obvious. The chief flexor muscle of the hip, **psoas**, is situated at the back of the abdomen. Its tendon runs deep to the inguinal ligament to reach the lesser trochanter of the femur. The chief extensor muscle, **gluteus maximus**, is situated in the buttocks. The **adductor** muscles (pectineus, adductor longus, adductor brevis, and adductor magnus) are situated in the thigh itself. The chief abductor muscles, **gluteus medius, gluteus minimus**, and **tensor fasciae latae** are found partly under cover of gluteus maximus again in the buttocks.

Put your hand between the greater trochanter and the iliac crest, and stand on one leg. As you do so you will feel these muscles contracting very firmly.

In addition there are a number of small muscles lying deep under cover of the glutei which produce lateral rotation of the hip joint (obturator internus and externus, the gemelli, piriformis, and quadratus femoris). Medial rotation is brought about by the anterior fibres of gluteus medius, gluteus minimus, and tensor fascia lata. Turn your non-weight bearing foot inwards and your other hand will feel this muscle, contracting just below the anterior superior iliac spine.

The range of movements at the hip joint may be confirmed using yourself as a subject, or better still using another person as a model. Lie down on your back (supine position) and place your hands on the anterior superior iliac spines of the pelvis to confirm that the pelvis is remaining stationary in the following manoeuvres.

Raise the right foot in the air with straight leg as far as it will go. This movement is termed **flexion** of the hip joint. With a leg straight, flexion is limited by the muscles at the back of the thigh, the hamstrings, which become taut. These muscles biceps femoris, semitendenosis and semimembranosis run from the pelvis (ischial tuberosity) to the leg (tibia and fibula) and thus act over two joints, the hip joint and the knee joint. If the knee is flexed the tension in these muscles will be relaxed and it is now possible further to flex the hip joint. Confirm this fact.

When a patient is placed in the lithotomy position in the operating theatre (on back, legs apart and raised) knees should be flexed.

Turn onto your front (prone position) and raise one leg backwards at the hip joint. This movement, the opposite of flexion, is termed **extension**. In the female, very little extension from the straight leg position is possible but some is possible in the male. If a male and female of equal stature are studied walking, the male will usually take longer strides than the female even though their legs are the same length. This is because he takes his legs further backward with each step by extending the hip joint. It is possible for the female to keep in step with the male but in order to do so she has to swing her pelvis with each step. This allows a certain amoung of abduction to take place which with a rotated pelvis increases the length of the step.

The principal muscle bringing about hip extension is gluteus maximus supplied by the inferior gluteal nerve. It is particularly important as a muscle extending the limb from the flexed position, such as getting up from a chair and climbing stairs. The hamstrings are also very important.

Returning to the supine position, if the foot is taken out sideways from the midline, the movement is termed **abduction**. It is brought about by gluteus medius and minimus. Abduction is limited by the shape of the head and neck of the femur and the acetabulum. In the baby where the shape is different, it is possible for the limbs to be

Table 2.1. Muscles employed in hip joint movements.

Flexion	Extension	Abduction	Adduction	Medial rotation	Lateral rotation
Iliopsoas	Gluteus maximus	Gluteus medius	Adductor magnus	Gluteus medius	Obturator externus
Rectus femoris	Biceps femoris	Gluteus minimus	Adductor brevis	(anterior part)	Obturator internus
Sartorius	Semitendinosus	Gluteus maximus	Adductor longus	Gluteus minimus	Gemelli
Pectineus	Semimembranosus	(upper part)	Pectineus	(anterior part)	Piriformis
Gluteus medius	Abductor magnus	Tensor fasciae latae	Gracilis	Tensor fasciae latae	Quadratus femoris
(anterior part)	(posterior part)			Iliopsoas	Glutei
Gluteus minimus					(posterior parts)
(anterior part)					Sartorius
					Adductors

abducted to 90°. This is the position used for splinting for the treatment of congenital dislocation of the hip. It is also possible for a young baby to put its toes in its mouth. With continuous practice from childhood it is possible to retain these facilities in adult life and become a contortionist!

If the left leg is first abducted, it is possible to demonstrate an inward movement from the midline of the right leg. This movement is termed **adduction**. It is brought about by the adductor muscles, adductor longus, brevis, magnus, and pectineus supplied by the obturator nerve. These muscles are used to squeeze the legs together for example when horse-riding.

If the foot is turned outward, movement takes place at the hip. It is termed **lateral rotation** of the hip. It is brought about by piriformis, quadratus femoris, and the obturators. This is the position adopted when palpating the femoral pulse since it opens the femoral triangle.

If the foot is turned inwards, the movement at the **hip joint is medial rotation**. It is brought about by the anterior part of the gluteus medius and minimus and tensor fascia lata all supplied by the superior gluteal nerve. As seen above these movements of lateral and medial rotation are used by the female when walking with long strides. They are used by both sexes when turning a corner.

The movements involved in pointing the toe inwards and outwards take place mainly at the hip joint since there is virtually no medial and lateral rotation at either the ankle or the knee joint. Thus if a person who walks with pigeon-toes wishes to walk with their feet straight they have to contract their glutei muscles to rotate laterally the hip joint.

A more complete list of the muscles employed in these movements is given in TABLE 2.1.

The movement of the hip is seldom brought about by a single muscle. It is brought about by groups of muscles working together.

These muscles are considered in more detail in Chapter 7.

THE LEG AND FOOT

Bony landmarks of the knee region

The **condyles** of the femur are subcutaneous and thus can be readily palpated. When the knee is flexed the trochlear surface for the patella at the lower end of the femur is exposed and can be easily felt. The **patella** itself is subcutaneous. When the quadriceps femoris muscle is relaxed, the patella can be moved appreciably. The ligamentum patellae (patellar tendon) can be traced from the patella to the anterior tuberosity of the tibia.

The **adductor tubercle** of the femur can be found by passing the flat of the hand down the medial side of the thigh. The adductor tubercle is the first projection met as the knee is approached. It lies further posteriorly than you think!

The form of the upper end of the tibia can be readily determined when the knee is bent.

The head of the **fibula** can be easily found by following the tendon of biceps (lateral tendon of popliteal fossa) with the knee bent. The head of the fibula lies at the back of, and below the lateral condyle of the tibia.

The actual position of the **knee joint** itself is important. It is often erroneously thought to be higher than it in fact is. Find it on yourself. Flex the knee and find the cavity of the knee joint at the sides of the patellar tendon. Work back from here on each side and you will feel the joint cavity getting narrower. Note that you are at a level well below the patella.

THE KNEE JOINT

The principal movement at the knee joint is one of flexion and extension. The extensor muscles are found at the front of the thigh. The flexor muscles are found at the back of the thigh [FIG. 2.33].

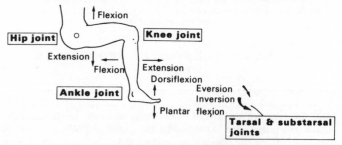

FIG. 2.33. Movements at joints of lower limb.

When the knee is flexed a small amount of medial and lateral rotation is possible. The knee is a modified hinge joint.

SUMMARY OF MOVEMENTS OF THE KNEE JOINT

The muscles which produce the movements at the knee joint are given in TABLE 2.2.

Biceps femoris is the only muscle inserted into the lateral aspect of the leg and hence is the only **lateral rotator**. Note also that the rotational directions (medial rotation and lateral rotation) given above refer to the leg. When the leg is fixed it is the femur which rotates. Rotational movements are limited by the capsule.

Flexion is limited by the back of the leg meeting the thigh.

Table 2.2. Muscles employed in knee joint movement.

Flexion	*Extension*	*Medial rotation*	*Lateral rotation*
Biceps femoris	Rectus femoris	Popliteus	Biceps femoris
Semimembranosus	Vastus medialis	Gracilis	
Semitendinosus	Vastus lateralis	Sartorius	
Sartorius	Vastus intermedius	Semitendinosus	
Gracilis		Semimembranosus	
Popliteus			

Extension is limited by all ligaments except ligamentum patellae and posterior cruciate ligament. **Rotation** (and gliding movements) can only take place freely in flexion.

Locking the knee joint

The articular surface of the medial condyle of the femur is larger than the lateral, and medial meniscus is larger and C-shaped. At the end of full extension, the femur rotates medially around its lateral condyle in its O-shaped meniscus. The medial condyle moves backwards with a screwing movement termed *locking the joint*. Before the knee can be flexed, it has to be unlocked by **laterally** rotating the femur. This is carried out by muscles that **medially** rotate the leg, namely by **popliteus** assisted by sartorius, semitendinosus, and semimembranosus. We are assuming that the tibia is fixed and the femur rotating (rather than the other way round) because this is what happens under weight-bearing conditions.

MUSCLES OF THE LEG

The part of the lower limb below the knee is termed the **leg**.

The muscles which originate from the leg run downwards and act on the ankle joint and the joints of the foot and toes. A muscle, such as gastrocnemius, which originates from the femur, will also have an action on the knee joint.

The investing fascia of the leg sends partitions inwards to the bones which, like those in the thigh, divide the muscles into three groups:

Anterior groups of muscles—dorsiflexors (extensors)—supplied by deep peroneal nerve
Lateral group of muscles—peroneal muscles—supplied by superficial peroneal nerve
Posterior group of muscles—plantar flexors (flexors)—supplied by tibial nerve

The septum running to the fibula which separates the anterior and lateral group of muscles is termed the **anterior septum**. The septum which separates the lateral group from the posterior group is termed the **posterior septum**. The anterior and posterior groups are separated by the **interosseus membrane**.

Note that in the leg, unlike the thigh, there is no medial group of muscles. On the medial aspect of the leg the tibia is, for the most part, subcutaneous and is therefore not giving origin to any muscles. Furthermore, there is no need for a medial septum since the tibia itself is separating the anterior and posterior group of muscles.

Remember also that the fibula lies on the lateral side of the tibia. The fibula forms the lateral malleolus of the ankle joint whilst the medial malleolus is formed by the tibia. Confirm this on yourself. The muscles associated with the fibula are termed peroneal muscles (*fibula* (Latin) = *perone* (Greek) = broach-pin, or skewer).

The term *sural* means pertaining to the calf (*sura* (Latin) = calf).

Remember when comparing the leg with the forearm that, in the anatomical position, the upper limb has the thumb lying laterally to the other fingers, whereas the foot has the big toe lying medially to the other toes. The upper limb needs to be rotated medially at the shoulder joint, the forearm in supination, and the wrist extended at right angles when a comparison is being made. This could be achieved by putting your hands on a table in front of you. Thus you will note that the radial artery now corresponds to the posterior tibial artery.

Because of the possibility of confusion the digits of the hands have been referred to by name as thumb, index finger, middle finger, ring finger and little finger. With the toes, there is less likelihood of confusion, and these are usually numbered from one to five. Toe number one is the big toe and toe number five is the little toe.

Amongst music instrumentalists numbers are used to indicate different fingers but according to the instrument being played the same number is used to indicate different fingers! Thus 'fourth finger' means the ring finger to a pianist, but the little finger to a violinist. Never call the index finger the 'forefinger' as it will be confused with one or both of the rival 'fourth' fingers.

Remember that although the same terms, proximal phalanx, middle phalanx, and distal (terminal) phalanx are used for both the hands and the feet, the small bones are termed **carpal** bones and **metacarpal** bones in the **hands** but **tarsal** bones and **metatarsal** bones in the **feet** (*carpus* = wrist: *tarsus* = ankle).

BONY LANDMARKS OF THE ANKLE REGION

The prominences on the inner and outer sides of the ankle joint are produced by the **medial malleolus** (lower end of tibia) and the **lateral malleolus** (lower end of fibula). The lateral malleolus is lower and more posterior than the medial malleolus. If the lateral malleolus is traced upwards by palpation, it can be shown to be connected to the shaft of the fibula.

At the bottom of the hollow in front of the lateral malleolus the front of the **calcaneus** can be felt. The calcaneus forms the prominence of the heel.

The trochlear surface of the body of the **talus** can be palpated immediately below the tibia when the ankle is plantar flexed. The neck and head of the talus run downward and medially from the body. The head is just proximal to the tubercle of the navicular.

Running the finger along the **lateral border** of the **foot** the following structure can be palpated. The lateral surface of the calcaneus is subcutaneous and just below, and anterior to, the lateral malleolus where its peroneal tubercle can be felt. In front is the **cuboid** and in front again the prominent **styloid process** of the **fifth metatarsal**. This lies half-way along the lateral border of the foot [FIGS. 2.30, 2.35].

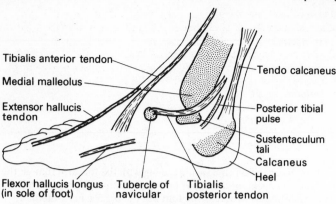

Tibialis anterior tendon

Medial malleolus

Extensor hallucis tendon

Tendo calcaneus

Posterior tibial pulse

Sustentaculum tali

Calcaneus

Heel

Flexor hallucis longus (in sole of foot)

Tubercle of navicular

Tibialis posterior tendon

FIG. 2.34. Medial side of ankle joint showing major features which can be palpated.

The easiest way to find the navicular and sustentaculum tali is to palpate from below upwards. The flexor hallucis tendon. and also those of flexor digitorum longus can easily be felt in the sole of the foot when the toes are flexed and extended. Dorsiflexion with inversion makes tibialis anterior stand out. Plantar flexion with inversion tenses tibialis posterior; find it between the medial malleolus and the tubercle of the navicular.

Find the posterior tibial pulse. It lies behind the medial malleolus.

Running the finger along the **medial border** of the **foot** the **sustentaculum tali** of the calcaneus is felt distal to and two fingers below the medial malleolus, the tubercle of the **navicular**, the **medial cuneiform**, and the **base** of the **first metatarsal**. Note that these bones, unlike those on the lateral aspect of the foot, are clear of the ground (hence the characteristic shape of a wet footprint on a bathroom floor). It is not until the head of the first metatarsal is reached that this part of the foot comes in contact with the ground. In the ball of the big toe the head of the metatarsal does not come in contact with the ground: two sesamoid bones intervene.

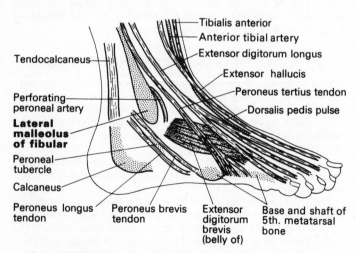

Tendocalcaneus

Perforating peroneal artery

Lateral malleolus of fibular

Peroneal tubercle

Calcaneus

Peroneus longus tendon

Tibialis anterior

Anterior tibial artery

Extensor digitorum longus

Extensor hallucis

Peroneus tertius tendon

Dorsalis pedis pulse

Peroneus brevis tendon

Extensor digitorum brevis (belly of)

Base and shaft of 5th. metatarsal bone

FIG. 2.35. Lateral side of ankle joint showing major features which can be palpated.

You should easily identify in yourself all the structures shown. Plantar flexion tenses the tendocalcaneus. eversion tenses the three peroneal tendons, and extending (dorsiflexing) the toes tenses the extensor digitorum and extensor hallucis tendons. Find the dorsalis pedis pulse. The dorsalis pedis artery runs over the tarsal bones to the cleft between the 1st and 2nd metatarsals.

The arrangement of the tarsal bones in the foot is complex but may be remembered from the following considerations when standing.

The **talus**, which lies on the medial projection of the calcaneus termed the **sustentaculum tali** and therefore lies slightly medial to the main body of the calcaneus, transmits the body's weight ultimately to the **heads** of the three **medial metatarsals** while the **calcaneus** transmits it to the two **lateral metatarsals**.

The **talus** sits on the calcaneus and is therefore at a higher level than the calcaneus. To reach the medial metatarsals (and hence the ground) two more tarsal bones are necessary, the **navicular** and a **cuneiform** bone. There are three cuneiform bones, medial, middle, and lateral. Each is associated with one of the three medial metatarsals.

The **calcaneus** sits on the ground and the lateral side of the foot is in contact with the ground along its entire length. Only a large single bone, the **cuboid**, connects the calcaneus to the two lateral metatarsals.

The arches of the foot formed by this arrangement are discussed on page 148.

ARRANGEMENT OF TENDONS AROUND THE ANKLE JOINT

At the front of the ankle joint the order of tendons, nerves and blood vessels from lateral to medial is:

LATERAL peroneus tertius—extensor digitorum longus—deep peroneal nerve—anterior tibial artery and veins—extensor hallucis longus—tibialis anterior MEDIAL

Behind the ankle joint the order of structures from lateral to medial is:

LATERAL flexor hallucis longus—tibial nerve—posterior tibial artery and veins—flexor digitorum longus—tibialis posterior MEDIAL

Below the medial malleolus the arrangement of tendons is as follows:

Tibialis posterior above the sustentaculum tali leading to the tubercle of the navicular
Flexor digitorum longus below the sustentaculum tali
Flexor hallucis longus in the groove below the sustentaculum tali

Below the lateral malleolus the arrangement of tendons from above downwards is:

Peroneus brevis
Peroneus longus

MOVEMENTS AT THE ANKLE AND TARSAL JOINTS IN THE FOOT

Flexion and extension

The principal movement at both the ankle joint and at the toes is one of flexion (plantar flexion) and extension (dorsiflexion) [FIG. 2.33]. The extensor muscles are found at the front of the leg. The flexor muscles are found at the back of the leg.

Ankle flexion (Plantar flexion)

It is important to note that flexion of the ankle joint can be brought about in two ways. Firstly, it may be brought about by contraction of the **flexor muscles** whose tendons run round the back of the ankle joint to reach the plantar surface of the foot. Secondly, it may be brought about by the contraction of **gastrocnemius–soleus** group which are inserted into the calcaneus. Due to the better leverage gastrocnemius produces a more powerful flexion such as is needed when standing on one's toes. This is because the tendo calcaneus (tendo Achilles) lies some distance behind the ankle joint. However, gastrocnemius is frequently not used when no resistance is encountered. Flex your ankle and knee joints with the foot clear of the ground and see if your gastrocnemius has hardened.

Inversion and eversion

When walking across the side of a hill the foot needs to be able to turn and remain in contact with the ground. This movement is termed **inversion** when the sole of the foot turns medially and **eversion** when it turns laterally. Inversion is easy to remember since inverting means 'turning upside down'. The nearest the foot can get to being turned upside down is being turned so that the sole faces medially. Inversion is always associated with medial rotation of the foot which is corrected by lateral rotation at the hip joint [p. 133].

Although inversion and eversion are often stated to be taking place at the transverse tarsal joints, a moment's consideration will reveal, that since the heel also turns, it must also be taking place at the subtarsal joints. This is discussed further on page 150.

Inversion is brought about by contracting the medial extensor and flexor at the same time. These muscles are **tibialis anterior** and **tibialis posterior**.

Eversion is brought about by muscles on the lateral side of the leg which take origin from the fibula and are termed the peronei muscles. These are **peroneus longus**, **peroneus brevis**, and **peroneus tertius**.

MOVEMENTS AT THE METATARSOPHALANGEAL AND INTERPHALANGEAL JOINTS

The **metatarsophalangeal joints** between the heads of the metatarsal bones and the bases of the proximal phalanges allow flexion, extension, abduction, adduction, and circumduction. However, no rotation is possible at these joints. Abduction and adduction is brought about by the interossei.

The **interphalangeal joints** have tight collateral ligaments and only flexion and extension is possible. In the resting position some of these joints are usually partially flexed. Look at your own feet and see to which toes this applies.

The **distal interphalangeal joints** are **flexed** by the long flexors. The **proximal interphalangeal joints** are **flexed** by both the short flexors and the long flexors. The **metatarsophalangeal joints** are **flexed** by the long and short flexors aided by the interossei and lumbricals.

The **interphalangeal joints** are **extended** by the extensors aided by the interossei which extend the distal interphalangeal joints.

The big toe (*hallucis* (Latin) = of the big toe) has its own long and short flexors and extensors, an abductor and an adductor. The little toe (*digiti minimi* (Latin) = of the little toe) has its own long and short flexors and an abductor. The muscles inserted into the toes closely resemble those inserted into the fingers.

SUMMARY

ANKLE

Dorsiflexion	*Plantar flexion*
Tibialis anterior	Gastrocnemius
Peroneus tertius	Soleus
	Tibialis posterior

Inversion	*Eversion*
Tibialis anterior	Peroneus longus
Tibialis posterior	Peroneus brevis

TOES

Extension	*Flexion*
Extensor hallucis longus	Flexor hallucis longus
Extensor digitorum longus	Flexor digitorum longus

3 The vertebral column and the back

THE POSTERIOR ASPECT OF THE TRUNK
HOW TO FIND YOUR WAY AROUND THE BACK

Bony landmarks

The only parts of the vertebral column that are subcutaneous are the spines [FIG. 3.01]. In the erect posture they lie at the bottom of a longitudinal furrow which runs all the way from the external occipital protuberance of the skull to the sacrum. This furrow is produced by the erector spinae muscles which lies on either side of the spines.

The bodies of adjacent vertebrae are connected by fibro-cartilaginous intervertebral discs with an anterior common ligament in front of the bodies and a posterior common ligament, inside the spinal canal on the back of the bodies. Adjacent spines are joined by interspinous ligaments [FIG. 3.02].

The first cervical vertebra (the atlas) has no spine, but its prominent transverse processes can be palpated. The spine of the second cervical vertebra (the axis) can be readily felt below the nuchal crest

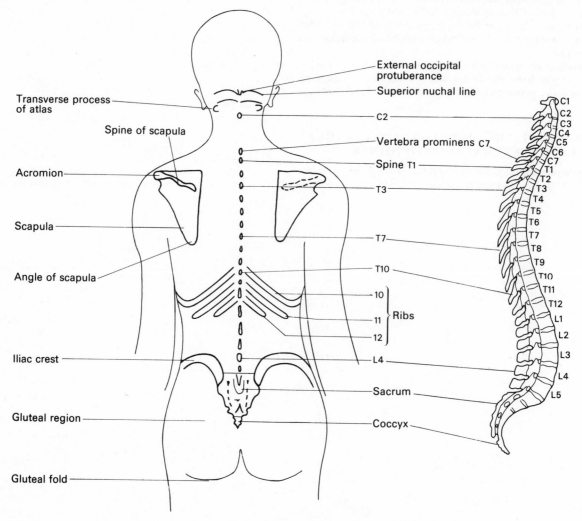

FIG. 3.01. The vertebral column and bony points of the back.

52

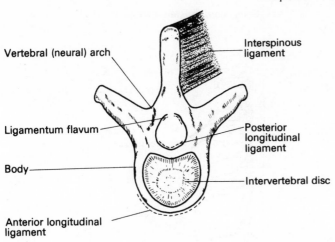

FIG. 3.02. Ligaments uniting the vertebrae in the vertebral column.

of the skull. Normally the spines of the third, fourth, and fifth cervical vertebrae lie too far from the surface to be palpated. Working downwards the next palpable spine is thus C6. The spine of C7 (the vertebra prominens) is taken as the landmark for locating spines in the upper part of the vertebral column. It forms a marked elevation when the neck is flexed. However, the spine of T1 may be more prominent and the two should not be confused. The spine of C7 is the uppermost visible spine, with the palpable (but invisible) spine of C6 above it. In some cases all spines from C2 downwards can be palpated [FIG. 3.03].

Once the spine of C7 has been located, the twelve thoracic spines, five lumbar spines and three sacral spines (the lower two sacral segments have no spines) can be palpated without too much difficulty particularly if the vertebral column is flexed. It is important to remember that the spines of T5–9 are directed downwards. As a result the palpated spine lies over the body of the vertebra one below. In the other regions the spine gives a better indication of the level of the corresponding vertebral body. In the lumbar region where the spines are broader, it is often simpler to feel for the spaces between the spines than the spines themselves.

The upper angle of the scapula is level with the first and second thoracic spines. The lower angle of the scapula lies at a level

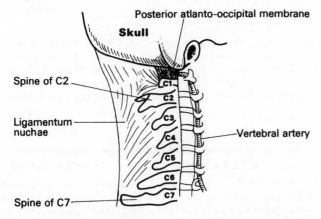

FIG. 3.03. The spines of the cervical vertebrae and the ligamentum nuchae.

between the seventh and eighth thoracic spines. The umbilicus corresponds to the fourth lumbar spine.

The highest point of the iliac crest lies at a level corresponding to the space between the third and fourth lumbar spines. This is an important landmark for a lumbar puncture. Although the spinal cord fills the spinal canal in early embryonic life, at birth the spinal cord only extends down to the third lumbar vertebra that is, about one vertebra below the adult position. The subarachnoid space containing cerebrospinal fluid (C.S.F.) extends down to the second sacral vertebra. A sample of C.S.F. is obtained by passing a needle between the vertebrae into this space. Since puncturing the spinal cord itself would cause permanent damage, the puncture is carried out in the lumbar region below the lower end of the spinal cord usually between the third and fourth lumbar vertebrae.

The posterior superior iliac spine, which is covered by skin and lies in a dimple, is at the level of the second sacral spine.

The coccyx lies at the lower end of the sacrum in a groove which leads to the anus.

Ligamentum nuchae

Ligamentum nuchae [FIG. 3.03] is the name given to the ligament at the back of the neck in the midline between the nuchal crest and the spines of the cervical vertebrae. It gives origin to the muscles of the back in this region. In man it is very inconspicuous.

SPINAL CORD

The spinal cord extends from the **foramen magnum** at the base of the skull to the lower part of the **first lumbar vertebra** [FIG. 3.04]. It lies in a relatively roomy **spinal canal** (vertebral canal) behind the bodies of the vertebrae and in front of the spinal laminae. The spinal cord actually lies nearer the bodies so that there is less movement.

The spinal cord is surrounded by three protecting membranes, pia mater, arachnoid mater, and dura mater. The spinal cord is closely invested with **pia** which sends processes into it. The **arachnoid** is a thin transparent membrane. Between the pia and the arachnoid is the subarachnoid space filled with cerebrospinal fluid which protects the spinal cord from injury. The subarachnoid space communicates in a complex way with the ventricles of the brain and with the central canal of the spinal cord. The outer membrane or **dura** is strong and fibrous. Surrounding the spinal cord, it is a single layer. The dura of the skull is a double layer [p. 331]. That lining the cranial bones ceases at the foramen magnum whilst that lining the brain continues down the spinal cord.

There is a small **subdural space** between the dura and arachnoid of the spinal cord, and a large epidural space, filled with fat, plexus of veins and loose areolar tissue between the dura and the wall of the canal. Local anaesthetics are placed in this space to anaesthetize the nerves in epidural anaesthesia.

The cord shows two swellings. The upper is the **cervical enlargement** which gives rise to the brachial plexus. The second is the **lumbar enlargement** which gives rise to the lumbosacral plexus which innervates the lower limb. If a limb is congenitally absent then the nerve cells which would have innervated it will die and the enlargement will not be present on that side.

The spinal cord terminates as the **conus medullaris**. From here the **filum terminale**, a non-nervous bluish fibrous cord, extends down to the dorsum of the coccyx. The nerves below the conus medullaris form the **cauda equina** (horse's tail) [FIG. 3.04].

The upper cervical nerves run horizontally from the spinal cord to the corresponding intervertebral foramina, but as we descend the

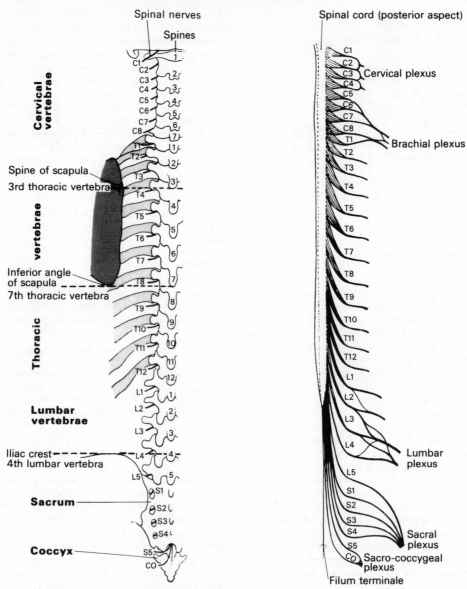

FIG. 3.04. The spinal nerve roots and vertebral column.

nerves pass more and more obliquely until in the cauda equina they run virtually vertically. Thus when considering the effect of an injury involving a spinal transection it is important to remember that spinal cord segments do not correspond to the vertebral segments:

Spinal nerve	Level of spinal cord segment
Upper cervical nerves	Same level
Lower cervical nerves	1 segment up
Upper thoracic nerves	2 segments up
Lower thoracic nerves	3 or 4 segments up
Lumbar nerves	T11–12
Sacral nerves }	T12–L1
Coccygeal nerve }	

For example, the abdominal wall around the umbilicus is supplied by the tenth thoracic nerve (T10). The corresponding segment of the spinal cord is at the level of the body of the sixth thoracic vertebra. Owing to the obliquity of the spines in the region, this point will be indicated by the position of the spine of T5.

VERTEBRAE

THE VERTEBRAL COLUMN AND VERTEBRAE

The vertebral column stretches from the atlas to the coccyx [FIG. 3.01]. Typical vertebrae consist of a **body** which lies in front and a **vertebral (neural) arch** behind [FIG. 3.02]. The bodies of the vertebrae are weight bearing and are connected to their neighbours by intervertebral discs [FIG. 3.02]. The flexibility of the vertebral column as a whole depends on these discs which account for about 20 per cent of its total length.

The function of the neural arches is to protect the spinal cord and cauda equina; they are not weight bearing. Enclosed between the back of the body of each vertebra and its neural arch is the **vertebral foramen**. Its size is such that you can easily get a finger into it. When the vertebrae are joined together in series, the vertebral foramina together form the **vertebral canal**, which therefore runs throughout the whole length of the column.

There are **seven cervical** vertebrae (*cervix* = neck), **twelve thoracic** (all articulating with ribs), **five lumbar** (*lumbus* = the loin or small of the back), **five sacral** fused together to form the sacrum (*sacrum* = sacred bone in Greek mythology), and about four fused together to form the coccyx—all that is left of the human tail.

The cervical part contributes roughly 20 per cent to the total height of the column, the thoracic 40 per cent, the lumbar 25 per cent and the sacrum and coccyx 15 per cent.

The weight supported by the vertebrae increases from above downwards. The size of the weight-bearing bodies also increases from the upper thoracic region downwards to the sacrum. This reduces the increase in pressure which otherwise would be exerted on them and on the intervertebral discs.

Intervertebral discs

The intervertebral disc consists of concentric layers of collagen fibres (**anulus fibrosus**) surrounding a mass of gelatinous material termed the **nucleus pulposus**. The anulus fibrosus joins the cartilage covering the adjacent bodies of the vertebra. The movement allowed by the disc is limited by the surrounding ligaments. The disc acts as a shock absorber when acceleration is applied to the vertebral column. Extrusion of the nucleus pulposus not only interferes with the function of the disc but may press on nerve roots causing pain.

Each disc is separated from the cancellous bone by a layer of hyaline cartilage. This avascular tissue protects the bone from the pressure exerted on it by the disc. In arthritis this cartilage breaks down, as it does in other joints.

Astronauts who have spent many months weightless in space, return to earth 3 cm (1 inch) or more taller presumably because the absence of gravity removes the compression on the intervertebral discs.

THE VERTEBRAE

We will deal first with typical thoracic vertebrae, and then show how the cervical, lumbar and sacral vertebrae are variations on a common pattern.

Typical thoracic vertebra

In FIGURE 3.05 identify:
1. The neural arch.
2. The body.
3. Vertebral foramen.
4. The rib (the rib of course is not part of a thoracic vertebra, but it is present as an integral part of all the other non-thoracic vertebrae).
5. The spine.
6. The lamina.
7. The pedicle.
8. The articular facets for the head and tubercle (angle) of the ribs.

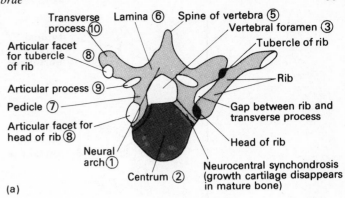

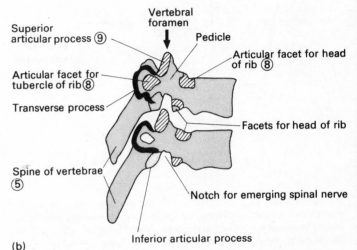

FIG. 3.05. Thoracic vertebrae.

(a) A typical vertebra (seen from above) consists of the neural arch, which covers the spinal cord posteriorly and on each side, and the centrum which lies in front and is weight bearing. The centrum is not co-extensive with the body of the vertebra, but is smaller. Neural arch and centrum are joined at the neurocentral synchondrosis which is just like an epiphyseal cartilage and disappears when maturity is reached. The body of the mature vertebra consists of all the centrum as well as the most anterior part of the neural arch.

(b) Two typical mature thoracic vertebrae (lateral view) are shown. The bodies are normally joined by the intervertebral disc, which is not shown in this figure. The vertebrae have been widely separated for clarity.

The numbers in the figure correspond to those in the text.

9. The synovial joints associated with the two pairs of superior and inferior articular processes.
10. The transverse process.

The gaps between the **laminae** of adjacent vertebrae are filled in on each side of the midline by the **ligamenta flava** (yellow ligaments) and more superficially by muscle (erector spinae).

The gaps at the sides between the **pedicles** of adjacent vertebrae allow the spinal nerves with their covering membranes, as well as blood vessels, to get in and out of the vertebral canal. The pedicles form boundaries of the **intervertebral foramina**, along with the intervertebral disc in front and the synovial intervertebral joint behind. Notice that the pedicles forming the upper boundaries have a deep notch in which the nerves lie, while the inferior boundary has

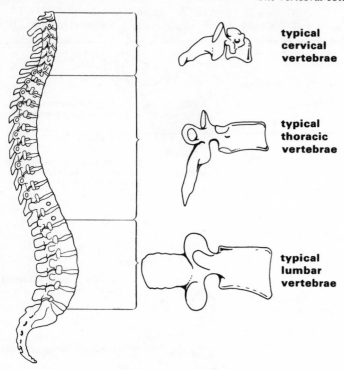

FIG. 3.06. Cervical, thoracic, and lumbar vertebrae.

typical cervical vertebrae

typical thoracic vertebrae

typical lumbar vertebrae

no such notch. This correlates with the general direction of the nerves as they emerge from the spinal cord in a **downward** direction. This is not so marked in the cervical and upper thoracic region, but is of major significance with the other lower spinal nerves.

Note relationship of the intervertebral foramina with their nerves to:

(i) intervertebral disc (anteromedial)
(ii) the synovial joints (dorsal)

Rupture of the disc or arthritic disorders in the synovial joints may press on the nerves giving rise to pain (sciatica, brachial neuralgia, etc.).

The neural arch also gives rise to the **spine** which projects backwards usually in a downward direction. The **transverse process** develops at the side, and points somewhat backwards [FIG. 3.06] as well as laterally. Looked at from behind the vertebra shows two pairs of articular processes projecting at the junction between the lamina and pedicle. They face backwards and slightly outwards, and articulate with the corresponding inferior articular processes which belong to the vertebra immediately above, which face forwards and slightly inwards. They form plane synovial joints, which have thin capsules with enough slack in them so as not to become taut (and painful) at the extremes of movement of the vertebral column.

Each transverse process articulates with the tubercle of its 'own' rib (i.e. T7 transverse process—seventh rib). The head of the rib articulates with the bodies of both T6 and T7, i.e. it is as high up as it can be without losing touch with the body of its own vertebra. The spinal nerves from T1 downwards all lie below their corresponding vertebra, and hence they must lie below their corresponding ribs. In fact the ribs are grooved along their lower borders by the spinal nerves.

Cervical vertebrae

The first cervical vertebrae [FIG. 3.07] is called 'the atlas'. According to Greek mythology the God Atlas supported the weight of the Earth on his back. In the same way the two lateral masses of the atlas vertebra shoulder the weight of the skull, transmitted to them each side of the foramen magnum via the pair of occipital condyles [FIGS. 3.09 and 17.02].

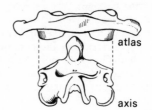

atlas

axis

FIG. 3.07. The first and second cervical vertebrae.

The articular facets on the upper aspect of the lateral masses are concave, and fit the convexity of the **occipital condyles**. The only movement possible at the atlanto-occipital joint is therefore nodding of the head up and down.

The articular facets on the lower aspect of each lateral mass are flat.

The atlas is the only vertebra from which the body appears to be missing. Examination of the atlas and axis soon after birth shows unambiguously that part of the atlas has fused with the axis. It is not really missing, but specially adapted to form the **odontoid process** of the axis.

The base of the skull from the pituitary fossa to the anterior edge of the foramen magnum is formed by the fusion of about four vertebral bodies. This may seem strange but at the other end of the vertebral column five vertebrae fuse together to form the sacrum, while below the sacrum 3–4 vertebrae fuse to form the coccyx while at the opposite end only two centra fuse to form the axis with its odontoid process. Thus the idea that the centra of the first and second cervical vertebrae should fuse to give rise to the odontoid process as part of the axis is not exceptional [FIG. 3.08].

The second cervical vertebra is the **axis**, having its **odontoid process** projecting upwards against the posterior side of the anterior arch of the atlas. It is held in position by the transverse ligament [FIG. 3.18, p. 62]. The joint formed in this way is a pivot joint. When the axis is viewed from above, bilateral flat articular surfaces are seen—the atlanto–axial joints. The weight of the head passes therefore through each occipital condyle and lateral mass of the atlas to the upper surface of the axis on each side of the midline. In typical vertebrae the weight is born by the vertebral bodies in the midline. Look at the axis from below and you will see the normal midline 'body'. Clearly the axis, apart from providing the pivot around which the atlas and head can turn when looking to left or right, also acts as an adaptor reconciling the bilateral weight bearing of the atlas to the midline weight-bearing of C3 [FIG. 3.09].

Note that the sacro-iliac joint [p. 60] and pelvis convert midline to bilateral weight-bearing once more, so that the weight of the body born by the lumbar vertebral bodies, is transferred to the two hip joints.

On each side every cervical vertebra is distinguished from all non-cervical vertebrae by the presence of the **foramen transversarium** through which the vertebral artery runs with its

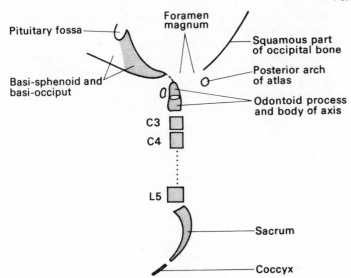

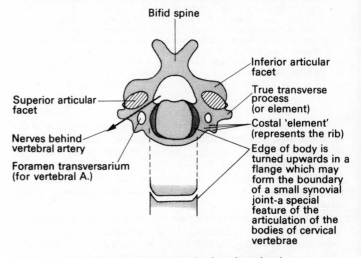

The transverse foramen of a cervical vertebra represents the gap between the transverse process and rib in FIGURE 3.05(a). The anterior boundary of the foramen is the 'costal element' of the cervical transverse process, while its posterior boundary is the 'transverse element'.

A frequent anomaly (1–2 per cent of people) is the existence of a 'cervical rib' in which the 'costal element' of C7 is much larger than it should be. A cervical rib is a quantitative not a qualitative abnormality.

In certain people with a 'cervical rib', the nerve T1 is pressurized as it climbs out of the thorax into the axilla (like a man climbing over a fence which is higher than he thought) not only over the first rib but over the cervical rib as well. Unfortunately this nerve is the principal nerve supply of all the intrinsic muscles of the hand and its paralysis by a cervical rib cripples the hand. The same clinical picture is seen in some patients who do not have a cervical rib but in whom the damage is done by the first rib itself. This is called a 'cervical-rib syndrome'. Treatment is to remove the offending ribs.

FIG. 3.08. Sagittal diagram showing fusion of vertebral components. Fusion occurs between adjacent segments in the tissue which gives rise to the axial skeleton, at both the head and tail ends.
1. The *base of skull* is formed by fusion of four segments between the pituitary fossa and the foramen magnum.
2. The *odontoid process* represents fusion between parts of C1 and C2.
3. The *sacrum* is formed by fusion of five segments.
4. The *coccyx* represents fusion of at least three to four segments.

   The vestige of an intervertebral disc can always be seen between the body of C2 and the odontoid process. Similarly four disc vestiges are visible in a sagittal section of the sacrum (*vestigium* (Latin) = a footprint).

FIG. 3.10. Typical cervical vertebra (seen from above). The nerves, in passing through the intervertebral foramina, are bounded by two synovial joints as well as the intervertebral disc. Inflammatory or arthritic changes in any of these joints may compress the nerves and give rise to pain. Note the flange-like up-turned edge which is a special feature of the bodies of the cervical vertebrae (*lower*).

Lumbar vertebrae

Normally there are **five lumbar vertebrae**. The bodies and intervertebral discs are slightly thicker in front than behind, with the result that the lumbar curvature is convex forward, forming the **lumbar lordosis**.

Because of their position low down in the vertebral column, the lumbar vertebrae are larger and **more robust** than the thoracic vertebrae.

The **spinal cord** usually terminates at the disc between L1 and L2. The conus medullaris lies in the vertebral foramen of L1, but the vertebral foramina of the other lumbar vertebrae only contain the cauda equina enclosed in the '**theca**'. The theca and subarachnoid space continue to the lower border of S2.

Notice the **direction of the facets** for the synovial intervertebral joints. The direction is approximately parasagittal [FIG. 3.11]. These joints allow flexion, extension and side flexion to take place freely, but they must *limit* rotation.

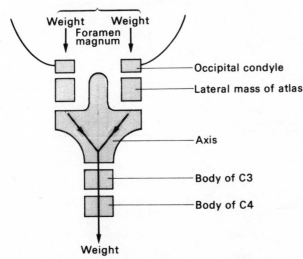

FIG. 3.09. Weight transmission from skull to cervical vertebrae. The weight of the head is transmitted by the two occipital condyles on each side of the foramen magnum to the two lateral masses of the atlas. The weight is borne by the body of C3 and by the other more inferior vertebrae in the midline. The axis is the adaptor which reconciles the bilateral weight bearing of the atlas to the midline weight bearing of C3.

companion veins (venae comitantes) and sympathetic nerves [FIG. 3.10]. The transverse foramen of C7 is exceptional as the artery does not go through it although the other structures mentioned do so [FIG. 3.03].

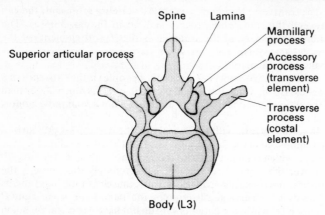

FIG. 3.11. Typical lumbar vertebrae.

Lumbar vertebrae can always be identified because they have neither transverse foramina (found in all cervical vertebrae) nor facets for heads of ribs (found in all thoracic vertebrae).

These vertebrae are thicker and stronger than other vertebrae. The plane of the facets of the articular processes is parasagittal. Note that the upper pair face inwards, and the lower pair look outwards. The transverse process is really a short rib whose head has fused solid with the vertebra. The transverse process of thoracic vertebrae is represented by the accessory process. The mamillary process has no special morphological interest. It gives attachment to the slips of the multifidus muscle (part of erector spinae).

Notice the attachment of the pedicles, more or less square to the posterior aspect of the bodies in all these vertebrae except L5. In this case the pedicles approach the body from the side [FIG. 3.12]. Compare the shape of L4, looked at from above, with the first piece of the sacrum [FIG. 3.15]. It is obvious that, shapewise, L5 is a compromise between L4 and S1. It is another of the body's many adaptors.

The upper surface of the sacrum, like a slipway, slopes downward and forward [FIG. 3.15]. Although the disc between L5 and S1 is the most wedge-shaped in the body, there is a tendency nevertheless for the body of L5 to slide forwards with respect to the first piece of the sacrum.

This danger is guarded against by the direction of the lower pair

of articular facets. In the other lumbar vertebrae they face sideways but here they face *forwards*. They articulate with corresponding S1 facets, which face backwards [FIG. 3.13].

The effectiveness of this arrangement depends on the capacity of the pedicles and laminae of L5 to stand the strain set up between its body and the inferior articular processes. Unfortunately the laminae of L5 on one or both sides may be defective and unable to stand this strain. The resulting downward slipping of the body of L5 is called 'spondylolysthesis'.

The iliolumbar ligaments also help to anchor L5 in place [FIG. 3.15]. Notice the oblique angle (about 60°) which these ligaments make with the direction of the forward thrust of the body weight. This means that the tension in these ligaments has to be twice what it would have been had they been in the direct line of pull.

L5 is the victim of many minor developmental abnormalities. The commonest is **spina bifida occulta**, which is found in 1–2 per cent of the population [FIG. 3.12]. In fact this term is a misnomer, as the spine of the vertebra is missing altogether, as is true in any vertebra showing spina bifida. In mild cases the 'bifid' appearance is produced by an otherwise normal neural arch whose laminae have failed to fuse in the midline. In more advanced cases, e.g. with a meningocoele, the **neural arch** may be represented only by two stunted pedicles, or be missing altogether.

The sacrum

Five vertebral segments fuse with each other to form the sacrum (the sacred bone!). The sacrum tapers slightly as far as the third segment and then in the last two segments tapers rapidly. The narrow tip of the sacrum articulates with the coccyx [FIG. 3.13].

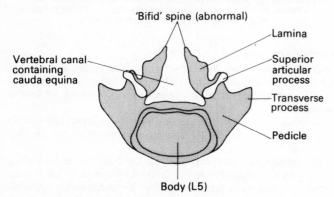

FIG. 3.12. 5th lumbar vertebra.

Compare a typical lumbar vertebra (L2, 3, 4) with the 1st piece of the sacrum, seen from above. It is obvious that L5 is the *adaptor* which reconciles the pedicles of L4 to the alae of the sacrum. This is why the pedicles of L5 join the body obliquely from the side instead of lying in the parasagittal plane. This vertebra is subject to many minor anatomical defects of which the spina bifida shown is the most common (1–2 per cent).

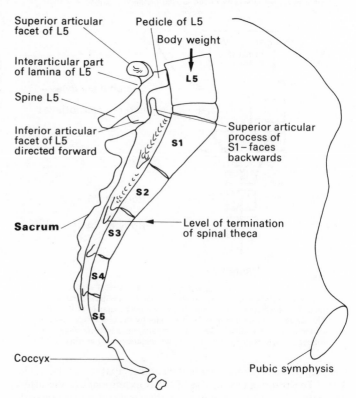

FIG. 3.13. The transmission of the body weight through L5 to the sacrum.

The first three segments articulate on both sides with the iliac part of the innominate bones forming the sacro-iliac joints. The base of the sacrum (S1) of course articulates with L5.

The shape of the sacrum differs in female and male. In the female the sacrum is relatively wider at the level of S1, and this helps to enlarge the inlet of the pelvis.

When viewed from the side, the male sacrum curves forwards and encroaches markedly on the outlet of the pelvis; but in the female the sacrum is flatter carrying [p. 225] the lower segments backwards and enlarging the outlet of the pelvis. These two considerations are very relevant in obstetrics.

Sacral nerves

The vertebral canal contains the membranes which form the theca of the cauda equina as far as the lower border of S2 [FIG. 3.04]. The sacral nerves emerge from the theca [FIG. 3.14] and reach the intervertebral foramina [FIG. 3.14]. Within the foramina the nerves divide. The anterior branches pass out through the anterior sacral foramina, and the posterior through the posterior foramina. The anterior branches are large and contribute to the sacral plexus and supply the lower limb. The posterior branches are very small (in spite of their large foramina) and only supply the **erectores spinae muscles**.

The fifth sacral and coccygeal nerves pass through the sacral hiatus with the filum terminale [FIG. 3.14].

Local anaesthetics can be injected into the sacral canal from below via the sacral hiatus. The lower sacral and coccygeal nerves are then paralysed in the canal as they escape from the spinal theca. This procedure is sometimes employed in obstetrics to render the perineum insensitive and reduce the tribulations of childbirth. The technique is unsatisfactory if the theca extends abnormally low down.

Sacro-iliac joints

One of the principal functions of the sacrum is to distribute sideways towards the hip joints the body weight which comes down in the midline from L5. This is done through the medium of the sacro-iliac joints [FIG. 3.15].

The two sacro-iliac (SI) joints, are modified plane synovial joints. The articular surfaces of the sacrum cover the upper 2½–3 sacral segments [FIG. 12.13, p. 229] (more in male than female) and face sideways and somewhat upwards and backwards.

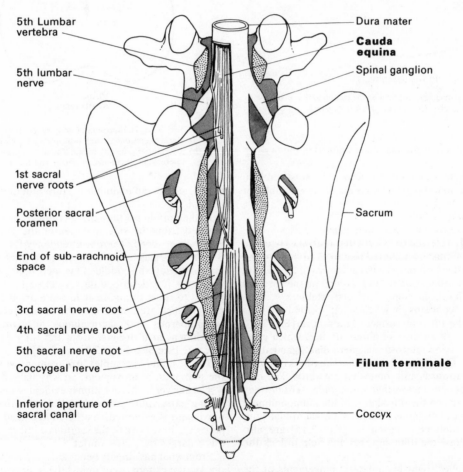

FIG. 3.14. The sacrum and spinal nerve roots.
The sacral nerves leave the sacral canal through a single intervertebral foramen' which leads to both the anterior (pelvic) sacral foramen for the ventral branch and the posterior (dorsal) foramen for the dorsal branch.

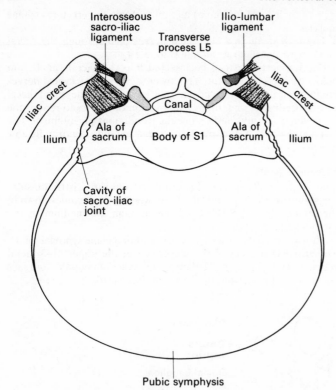

FIG. 3.15. The weight-bearing ligaments joining the sacrum to the pelvis. The inlet of the pelvis (p. 224) is viewed from above.

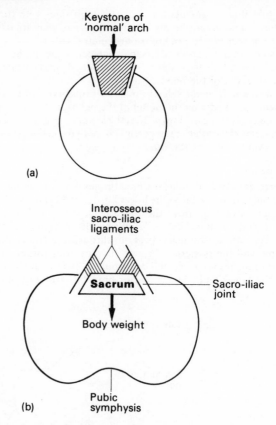

FIG. 3.16. Transmission of body weight from sacrum to pelvis.
(a) Normal keystone of arch principle (not used by body).
(b) Diagram of principle found. The body weight is transmitted by the interosseous sacro-iliac ligaments and joints.

At first sight it appears that the sacrum is wedged in the 'wrong' direction [FIG. 3.16].

In spite of this the sacro-iliac joints are among the most stable in the body. The combination of two factors is mainly responsible for this. The factors are:

(i) The irregular joint surfaces;
(ii) The posterior interosseous sacro-iliac ligaments.

The joint surfaces are referred to as auricular (ear-like) because of their irregularities. It must be admitted that to liken them to the pinna of the ear is an exaggeration. The irregularities of the sacrum and ilium fit into each other exactly, and so long as the surfaces remain in close apposition, the joint is perfectly stable.

The problem now concerns how close apposition can be achieved. Just behind the auricular surface of the sacrum [FIG. 3.15] is a large depression with an area of about 10 cm², which gives attachment to the posterior interosseous **sacro-iliac ligament**. In FIGURE 3.15 this ligament is shown. Notice that the fibres run from the sacrum obliquely upwards and somewhat outwards. The outward directional component means that when under tension they will exert an *inward* force on the dorsal part of the ilium, pulling it medially and locking the contiguous surfaces of the sacro-iliac joint firmly together. The iliolumbar ligaments [FIG. 3.15] are also ideally positioned to draw the ilium inwards and help to lock the sacro-iliac joint.

This mechanism implies slight mediolateral movements of the ilium on each side. This is made possible by the flexibility of the secondary cartilaginous joint at the pubic symphysis.

Unlike many ligaments in other joints, the interosseous sacro-

iliac ligament can maintain posture indefinitely without becoming painful.

In the upright position the sacrum lies in a very oblique position. Viewed from the side it is nearer the horizontal plane than is generally realized. There is a tendency for the sacrum to tilt [FIG. 3.17]. The axis of this movement goes transversely through approximately the middle of the sacro-iliac joints. The interosseous ligaments just described lie very close to this axis and are poorly placed to resist tilting. Instead there are two other accessory ligaments which are attached to the lower pieces of the sacrum—the **sacrotuberous** and **sacrospinous ligaments** [FIG. 3.17]. Their attachment at a distance from the tilt axis gives them excellent leverage [see also FIG. 12.13, p. 229].

The SI joint is a complex arrangement which apparently ensures the minimum of movement. So why have a joint at all?

The answer to this question is evident when the role of the sacrum during parturition is considered. At this time the potentiality for movement is exploited. As the child's head descends through the pelvis the lower end of the sacrum can be displaced backwards and this enlarges the pelvic outlet.

From what has already been said it is clear that such movement of the sacrum cannot occur unless the interosseous ligaments slacken off so as to allow the contiguous auricular surfaces to disengage from each other under non-weight bearing conditions. The sacro-iliac and sacrotuberous ligaments must also relax.

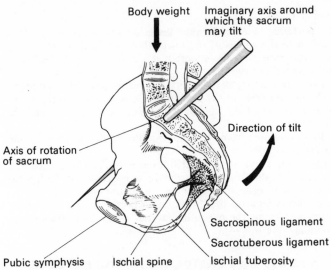

FIG. 3.17. Function of sacrotuberous and sacrospinous ligaments.
The axis of tilting the sacrum passes through the sacro-iliac joint. The
sacrotuberous and sacrospinous ligaments are ideally positioned to resist
the tendency of the lower part of the sacrum to tilt upwards.
(The pelvis is shown cut sagittally – splitting the pubic symphysis and
sacrum.)

This change in the pelvic ligaments is brought about under hormonal influences, and may be noticeable quite early in pregnancy. After the birth the ligaments tighten up again and the auricular surfaces should come to lie once more in correct register with each other.

This joint, not surprisingly, is a common cause of backache in pregnancy and during the puerperium, when things are returning to normal after childbirth. This joint may give trouble at other times in women and also in men.

LIGAMENTS BETWEEN AXIS, ATLAS, AND SKULL

Alar ligaments
The alar (or check) ligaments run from the dens of the axis C2 to the skull (inner sides of condyles of occipital bone) [FIG. 3.18].

Apical ligament
The apical ligament runs from the apex of the dens to the skull at the front border of the foramen magnum [FIG. 3.18]. It is the remains of the notocord.

Cruciform (Cruciate) ligament
The **transverse ligament**, which stretches across the cavity of the atlas, divides it into two compartments, the anterior compartment for the dens and the posterior compartment of the spinal cord.

The superior and inferior longitudinal bundles running vertically fuse with the transverse ligament forming a combined ligament in the shape of a cross termed the **cruciform ligament** [FIG. 3.18]. The superior longitudinal bundle (superior crus) runs to the skull in the foramen magnum. The inferior longitudinal bundle (inferior crus) is attached to the back of the body of the axis.

The cruciform ligament is covered by the **membrana tectoria** which is a continuation of the posterior longitudinal ligament which forms the anterior wall of the spinal canal.

On the posterior aspect of the spinal canal the posterior atlanto-axial and the posterior atlanto-occipital membranes, which extend to the posterior border of the foramen magnum, form the posterior wall.

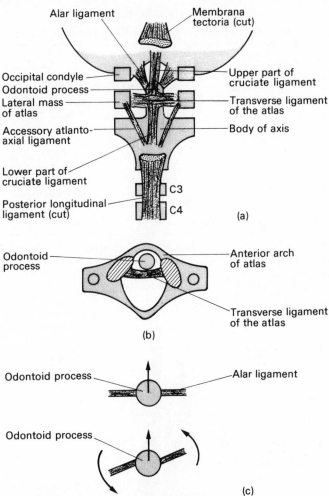

FIG. 3.18. Ligaments of axis, atlas, and skull.
(a) Diagrammatic view from behind of the ligaments, joining the axis, atlas and occipital bone to each other.

Note that the posterior longitudinal ligament covers over this region and finally passes through the foramen magnum and is attached to the basi-occiput. The ligament changes its name to membrana tectoria (*tectum* (Latin) = a roof). In the diagram it has been removed to show what it has been covering.

The transverse ligament of the atlas covers the dorsal surface of odontoid process. The apical ligament, from the tip of the process to the edge of the foramen magnum is of no functional importance and is not shown.
(b) The odontoid process and atlas are shown cut horizontally. The process forms a pivot joint and articulates anteriorly with the anterior arch of the atlas by means of a plain synovial joint. It articulates by means of another synovial joint with the transverse ligament of the atlas.
(c) The odontoid process and its alar (or check) ligaments are shown diagrammatically from above. When the head turns the axis may remain stationary. The alar ligaments join the axis not to the atlas but to the occipital bone of the skull. They are strong ligaments. The diagram shows how these ligaments must be stretched as rotation takes place. The angle of rotation shown represents about the maximum movement possible (20–25 per cent). The chin can turn through 180° in looking from left to right. Only about one-quarter of this movement occurs in the atlanto-axial joint.

These joints are protected by ligaments of great strength. In judicial hanging they remain intact and separation of vertebrae, with rupture of the spinal cord, occurs in the neck at a lower level.

MUSCLES OF THE BACK

The muscles of the back are conveniently divided into four layers.

Muscle	Nerve supply
1st Layer	
Trapezius	X1, C3, 4
Latissimus dorsi	Thoracodorsal N (C6, 7)
2nd Layer	
Rhomboids	Dorsal scapular N (C5, 6)
Levator scapulae	Dorsal scapular N (C5, 6)
Serratus posterior	
Splenius capitis and cervicis	
3rd Layer	
Erector spinae	Dorsal rami of spinal nerves
4th Layer	
Quadratus lumborum	T12, L1, 2, 3, 4

The trapezius rhomboids and levator scapulae are concerned with the movements of the scapula. Details of these muscles will be found on pages 93 and 94. Details of latissimus dorsi which is inserted into the humerus will be found on page 93.

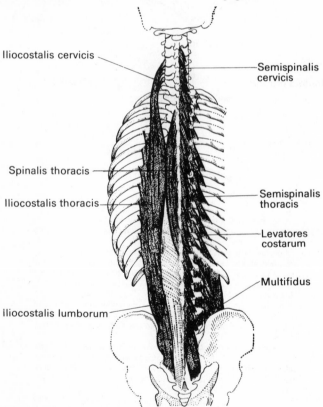

Iliocostalis cervicis

Semispinalis cervicis

Spinalis thoracis

Iliocostalis thoracis

Semispinalis thoracis

Levatores costarum

Multifidus

Iliocostalis lumborum

Erector spinae

Origin: Back of sacrum
Insertion: Base of skull and vertebrae
Nerve supply: Spinal nerves, posterior rami
Action: Extends vertebral column
                  Maintains upright position

The name indicates that it keeps the spine erect. It is the collective name for the group of back muscle (see below).

Lumbar fascia

The posterior layer of **lumbar fascia** covers sacrospinalis. It is attached to the spines of the lower thoracic, lumbar, and upper sacral vertebrae medially. Laterally it is connected to the iliac crest and the angles of the ribs.

It gives origin to latissimus dorsi and serratus posterior inferior.

The middle layer of the **lumbar fascia** is attached to the transverse processes of the lumbar vertebrae, the iliac crest, and the twelfth rib.

Quadratus lumborum lies on the middle layer of the thoracolumbar fascia.

The anterior layer of the lumbar fascia is thin and insignificant. It is attached to the front of the transverse processes of the lumbar vertebrae and forms the iliolumbar ligament below.

The layers join together at the outer border of quadratus lumborum and forms the **origin of internal oblique** and **transversus abdominis** [see p. 194].

A FURTHER CONSIDERATION OF ERECTOR SPINAE (SACROSPINALIS MUSCLE)

The term erector spinae is used for the group of muscles that move the trunk. They are all supplied by the dorsal rami of the spinal nerves.

They may be divided into a medial intermediate and lateral portion, and further subdivided into the region over which they act, *lumborum* which act on the lumbar region, *thoracis* which act on the back, *cervicis* which act on the neck, and *capitis* which act on the head.

Passing upwards from the sacrum the erector spinae divides into three longitudinal sections. The spinalis thoracis, spinalis cervicis, and semispinalis capitis form the first portion closest to the midline. Longissimus thoracis, longissimus cervicis, and longissimus capitis form the second. Iliocostalis lumborum, iliocostalis thoracis, and iliocostorum cervicis form the third and most lateral as far as the angles of the ribs.

Medial portion	*Intermediate portion*	*Lateral portion*
1. Spinalis thoracis	1. Longissimus thoracis	1. Iliocostalis lumborum
2. Spinalis cervicis	2. Longissimus cervicis	2. Iliocostalis thoracis
3. Semispinalis capitis	3. Longissimus capitis	3. Iliocostalis cervicis

The semispinalis (cervicis and thoracis) muscle fibres run obliquely from the transverse processes to the vertebral spines. The spinalis muscles, which run between the vertebral spines, are relatively insignificant.

Medial portion

Semispinalis capitis

Origin: Vertebrae
                  Articular processes C4–7
                  Transverse processes T1–6
Insertion: Skull—occipital bone
                  between superior and
                  inferior nuchal lines
Nerve supply: C1, C2, C3, C4, C5 (dorsal rami)
Action: Extends head
                  Bends head laterally

The name indicates that it is half of a double bellied muscle of the spine connected to the head. Its old name was biventer cervicis (two bellied muscle of the neck). This is an important muscle which stops your head from dropping off when you lean forwards.

Levatores costarum

Origin:	Transverse process of vertebrae, C7–T11
Insertion:	Posterior part of shaft of rib below
Nerve supply:	Intercostal nerves Ventral rami
Action:	Elevates ribs at back

These twelve small muscles run from the transverse process of the vertebra above to posterior part of shaft of the rib. The uppermost muscle runs from the transverse process of C7 to the first rib.

Intermediate portion

Longissimus thoracis

Origin:	Iliac crest. Posterior surface of sacrum, spinal processes of lumbar vertebrae
Insertion:	Vertebrae, thoracic
Nerve supply:	Posterior rami of spinal nerves
Action:	Extends trunk (both sides), Abducts and rotates trunk (one side)

The name indicates it is the longest muscle of the back (*dorsum* (Latin) = back).

Longissimus cervicis (transversalis colli)

Origin:	Vertebrae T1–6
Insertion:	C2–6 vertebrae
Nerve supply:	as above
Action:	Extends vertebrae

The name indicates that it is the longest muscle of the neck (*cervix* (Latin) = neck).

Longissimus capitis

Origin:	Vertebrae C4–7
Insertion:	Mastoid process of skull
Nerve supply:	Posterior rami of C1–5
Action:	Extends head

The name indicates that it is the longest muscle to the head (*caput* (Latin) = head).

Lateral portion

Iliocostalis lumborum

Origin:	Iliac crest. Posterior surface of sacrum. Spinous processor of lumbar vertebrae
Insertion:	Ribs 7–12
Nerve supply:	Posterior rami of lumbar nerves
Action:	Extends trunk (both sides) Abducts and rotates trunk (one side)

The name indicates that it is a flank-rib muscle of the loin (*ilium* (Latin) = flank; *costae* (Latin) = ribs; *lumbus* (Latin) = loin).

Iliocostalis thoracis

Origin:	Ribs 7–12
Insertion:	Ribs 1–6
Nerve supply:	Posterior rami of thoracic nerves
Action:	Extends vertebral column

The name indicates that it is a flank-rib muscle of the back (thoracic region).

Iliocostalis cervicis

Origin:	Ribs 1–6
Insertion:	Vertebrae C4–6
Nerve supply:	Posterior rami of cervical nerves
Action:	Extends vertebral column

The name indicates that it is a flank-rib muscle of the neck.

Splenius muscles

Splenius capitis and cervicis extend the neck.

Splenius capitis

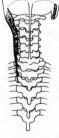

Origin:	Ligamentum nuchae over cervical spines Spinous process of T1–5 vertebrae
Insertion:	Skull—mastoid process of temporal bone
Nerve supply:	C2, C3, C4 (dorsal rami)
Action:	Extends head. Bends and rotates head to same side

The name indicates that it is a muscle of the head that resembles a bandage (*splenion* (Greek) = pad of linen). It covers semispinalis capitis.

Splenius cervicis

Origin:	Ligamentum nuchae Spines of C7–T5
Insertion:	Transverse processes of C1–3
Nerve supply:	C2, C3, C4 (dorsal rami)
Action:	Extends head and neck Bends and rotates head to same side

The name indicates that it is a muscle of the neck that resembles a bandage.

Quadratus lumborum

Origin: Iliac crest and ilio–
                lumbar ligament
Insertion: Twelfth rib, inferior
                border. Transverse
                processes of lumbar
                vertebrae
Nerve supply: T12; L1, 2, 3, 4
Action: Fixes twelfth rib
                Lateral flexor of trunk
                By pulling the last rib
                downwards, it enables
                the dome of the diaphragm
                to descend during inspiration

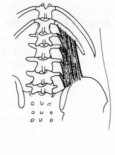

SUBOCCIPITAL TRIANGLE MUSCLES

Four small muscles are found deep to semispinalis capitis in an area known as the suboccipital triangle [FIG. 3.19]. They are all supplied by the suboccipital nerve (C1). These muscles are:

Inferior oblique

Origin: C2 spine
Insertion: Transverse process of C1

Superior oblique

Origin: Transverse process of C1
Insertion: Occipital bone of skull

Insertion is lateral to that of semispinalis capitis.

Rectus capitis posterior major

Origin: C2 spine
Insertion: Occipital bone of skull

Insertion is below inferior muchal line. (lateral impression)

Rectus capitis posterior minor

Origin: Tubercle on back of
             arch of C1
Insertion: Occipital bone of skull
             —medial impression

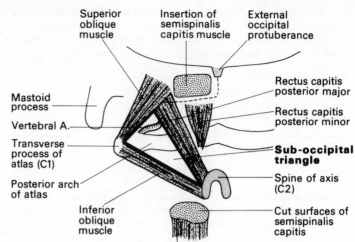

FIG. 3.19. Suboccipital triangle.

The suboccipital triangle is tucked beneath the squamous part of the occipital bone. Its name tells exactly where it lies. To see it the semispinalis capitis must be cut away. The large number of small muscles is probably associated with the need to produce precise movements of the head, and also with the need to provide precise afferent information regarding movements of the head.

Note the shape of the spine of the axis. The inferior oblique muscle runs to the back of the transverse process of the atlas. (Feel both landmarks in yourself.) When this muscle contracts the head turns from side to side. The superior oblique and rectus capitis posterior major form the other two boundaries of the triangle. In its floor is the atlas with the vertebral artery passing behind the lateral mass.

Note that the rectus capitis posterior major is twice as long as rectus capitis posterior minor, because it comes from C2, while the latter comes from C1.

The suboccipital triangle has the arch of the atlas, the posterior atlanto-occipital membrane and the vertebral artery in its floor. The greater occipital nerve (C2) curls round the lower border of inferior oblique before passing through semispinalis capitis [FIG. 3.20].

These muscles are very small and lie very close to the joints where movement takes place and their leverage is correspondingly feeble. They may, however, play an important role in carrying out movements by providing useful proprioceptive information.

Details of the opposing prevertebral muscles which flex the head and neck will be found on page 315, and the abdominal muscles which flex the trunk on page 192.

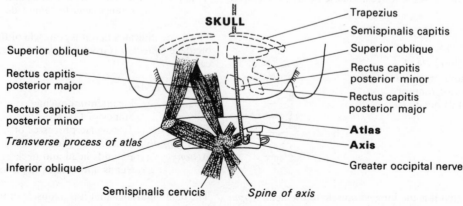

FIG. 3.20. The greater occipital nerve (C2).

INTERVERTEBRAL LIGAMENTS

The bodies of the vertebrae as has been seen are separated by the intervertebral discs forming a slightly movable joint (amphiarthrosis). Compression of these discs through the day results in a slight reduction in height at the end of each day compared with the morning. Old age is associated with a reduction in the thickness of these intervertebral discs and a reduction in stature.

The vertebrae are connected by ligaments on all sides [FIGS 3.02 and 3.22]. These may be subdivided into:

1. Anterior longitudinal ligament.
2. Posterior longitudinal ligament.
3. Interspinous ligaments.
4. Ligamentum flavum.

The anterior longitudinal ligament joins the anterior surface of the bodies of all the vertebrae from the axis (C2) to the sacrum. It continues upwards from the axis (C2) to the atlas (C1) as the anterior atlanto-axial and anterior atlanto-occipital membranes.

The posterior longitudinal ligament joins the bodies of the vertebrae inside the spinal canal.

The interspinous ligaments join adjacent spines. In the neck region, where the spines are deep to the surface, these ligaments extend to the skin in the midline as the **ligamentum nuchae** [FIG. 3.03].

The ligamenta flava (yellow ligaments) join the inner surfaces of the laminae. Their colour is due to the fact that they are composed of yellow elastic tissue.

If available examine the upper part of the vertebral column and study its articulation with the skull. The concave upper articular surfaces of the atlas articulate with the convex condyles of the occipital bone of the skull. The lower articular surfaces of the atlas articulate with the upper facets of the axis. Note how the dens (odontoid process) of the axis extends through the atlas so that its ligaments can be directly connected to the skull. The dens has an articular surface anteriorly with the arch of the atlas. Posteriorly it is held in place by the transverse ligament of the atlas. The dens lies in front of the spinal cord.

MOVEMENTS OF THE SPINAL COLUMN

Bending Forward and Backwards (flexion and extension)

Flexion and extension takes place at the cervical and lumbar vertebrae (TABLE 3.1). The flexion in the thoracic region is minimal. The junction areas between fixed and movable parts of the body are weak spots where fracture-dislocations readily occur.

Table 3.1. Summary of movements of spinal column.

Flexion and extension	Cervical and lumbar (thoracic nil)
Rotation	Thoracic
Lateral flexion	Rotation of vertebrae
Head rotation	30° head and atlas pivoting around dens of axis
Head nodding	Skull and atlas
Head on chest	Skull and atlas plus cervical flexion

As has been seen the atlanto-occipital joint between the skull and the atlas allows nodding of the head. To put the chin down to the chest the neck vertebrae are involved. When the head is extended to obtain a clear airway for artificial resuscitation it is essential to produce extension in the upper cervical region.

When bending forward to touch the toes much of the flexion takes place at the hip joint.

Leaning backwards (hyperextension)

The lumbar spines produce an increased concavity and the motion occurs here.

Twisting

Rotation occurs most easily in the thoracic region. The head can only rotate through an angle of 30° at the joint between the atlas and axis vertebrae. The head and the atlas rotate around the dens of the axis. Further rotation occurs at the other neck vertebrae enabling the head to turn through 90°. Inability to rotate the head makes it difficult to turn round to see if there is any traffic coming before driving off in a car. Even so, rotation of the thorax is needed to look behind you.

Bending to one side (lateral flexion)

In bending to one side a curve convex to the other side is produced. In addition there is a twist.

Spinal column curvatures

The spinal column has marked curvature in the fore and aft direction that is in the sagittal plane.

The normal curvatures are shown in FIGURE 3.05.

Cervical ⎫
Lumbar ⎬ Convex forwards

Thoracic ⎫
Sacral ⎬ Convex backwards

It will be seen that curvature in the cervical and lumbar region is convex whilst in the thoracic and sacral region it is concave in the forward direction.

At birth an infant has only the thoracic and sacral curves. The cervical and lumbar curves appear when the child holds his head up and starts to walk. When bending forward to touch the toes the cervical and lumbar curves disappear giving the infantile shape.

In women the lumbar curve is increased because of the more horizontal direction of the sacrum.

An excessive curvature in the lumbar region is termed **lordosis**. An excessive forward curvature in the thoracic region is termed **kyphosis** (*Kyphos* (Greek) = hump).

Under normal conditions the spinal column is straight in the sagittal plane. Curvature in the lateral direction is known as **scoliosis** (*skolios* (Greek) = crooked).

Standing

In standing upright the weight is fairly evenly distributed between the two feet (usually left greater than right). Looked at from the side the gravity line falls almost half way between the heel and forefoot passing through the tuberosity of the navicular [FIG. 3.21] in front of the ankle joint. In standing upright the muscles in the calf of the leg must contract continuously.

The gravity line passes just in front of the axis of the knee joint, and this joint therefore has a built in 'fail safe' mechanism, which ensures that when the knee extensors cease to contract, the joint nevertheless remains stable. The gravity line passes just behind the axis of the hip joint and this also ensures a 'fail safe' mechanism because in the upright position the hip is very nearly fully extended,

Gravity line

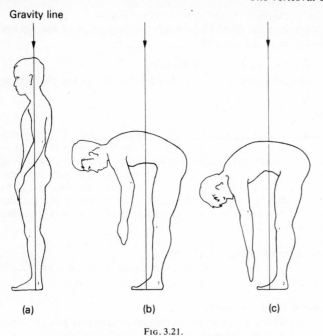

(a)   (b)   (c)

FIG. 3.21.

Bending over: tracings from photographs of the same subject, while erectores spinae muscle activity was recorded.

(a) Bending starts from the upright position. Notice the position of the gravity line passing through the mastoid process, and terminating just in front of the ankle joint.

  The back muscles are contracting gently in normal people. They relax when the subject sways backwards (feel this in yourself).

(b) In bending most of the movement of the trunk takes place at the hip joint. A small additional component is the flexion movement in the lumbar and thoracic vertebrae. Once flexion has started the effect of gravity acts like a prime mover. Gravity is resisted in the vertebral column by the erectores spinae.

(c) Eventually however the intervertebral ligaments become taut and when they do so, quite suddenly these muscles relax, and posture is then maintained by the ligaments. In this subject further bending forward so as to touch the floor can only take place by further hip-joint flexion.

  Note that the gravity line must obviously still pass through the foot. But the forward movement of the head, shoulders and upper limb is exactly counterbalanced by the backward movement of the lower limb and pelvis. Try bending over like this, but begin with your back to a wall.

and stability is ensured by tension in the ligaments at the front of the hip joint like those at the back of the knee.

The upper limbs hang in a position such that they do not unbalance the trunk. The head is delicately balanced on the vertebral column. It is sometimes useful to notice that the mastoid process seen from the side shows where the point of balance of the head lies, and the gravity line of the rest of the body can often be represented by a vertical line produced downwards from the mastoid process.

In Man the upright position can be maintained with the minimum of muscular effort, and in this respect we differ strikingly from the anthropoid apes who stand with the hip and knee flexed.

The centre of gravity of the body lies below the midpoint between the hip joints just in front of S2. This means that the two lower limbs balance the weight of the rest of the body. In infancy the lower limbs are relatively shorter than in the adult, and the centre of gravity lies approximately at the level of the umbilicus and L4.

Back injury, weight lifting, sitting

Back injuries often occur when heavy weights are lifted. Nurses who lift bedridden patients are frequent victims. The lumbar region is most likely to be affected. Pain may come from the intervertebral joints, intervertebral ligaments, or extensor muscles (erector spinae).

Bend over from the standing position and touch the floor [FIG. 3.21(c)]. The principal movement takes place at the hip joint with a small flexion component only from the lumbar spines. When touching the floor the lumbar intervertebral joints are fully flexed and this posture is maintained by tension in the ligaments, in particular the intervertebral discs (posterior part of the anulus fibrosus) the posterior longitudinal ligament, the ligamenta flava, and interspinous and supraspinous ligaments [FIG. 3.22]. The capsules of the synovial zygopophyseal joints are also probably in tension. The intervertebral disc and posterior longitudinal ligament are close to the axis of movement and have poor leverage. The ligamenta flava are exceptional compared with any other ligaments connecting bone to bone in consisting of elastic fibres. All the other 'non-elastic ligaments' will become tight in full flexion and will support the body weight. It is easy to show electromyographically that the erector spinae contraction is ineffective and hence this posture is maintained by the ligaments. You can easily palpate the supraspinous ligament tension in yourself. In this posture the ligaments either support the body weight or break [FIG. 3.22(b)].

It is possible to **sit** upright with the lumbar spines in extension or to adopt the slumped position with the spines in full flexion, the erectores spinae muscles relaxed and the new posture once more maintained by the intervertebral ligaments. Many people regard the first posture as 'good' and the second as 'bad'. They are just alternative postures which avoid prolonged tension in either muscles or ligaments [FIG. 3.23].

So far as the lumbar region is concerned the **weight lifting** position in FIGURE 3.21(b) is very similar to that of the slumped sitting position FIGURE 3.23(a) but with the trunk rotated through 90°. The trunk as a whole is raised in weight lifting by hip extension (gluteus maximus and hamstrings). This puts additional strain on the ligaments until such time as the erectores spinae begin once more to contract and by extending the intervertebral joints, take over the strain previously borne by the ligaments.

It is possible to avoid putting the ligaments under strain by weight lifting as shown in FIGURE 3.24. Here by squatting, it is possible to begin lifting with the lumbar spines in extension in which case the erectores spinae bear the strain throughout. Many people start in the right position and then get into the wrong position, with flexed lumbar spines, as they stand up.

At all times the combined centre of gravity of the body plus the load must pass through the feet otherwise the lifter will fall over. The feet have therefore to be positioned correctly before the movement starts.

The nurse's task of lifting a patient in bed presents an insoluble problem [FIG. 3.24]. Because of the width of the bed it is impossible to put the feet in the right position. The Australian lift is a very useful method which gets round some of the problems of 'normal' weight lifting, but it requires considerable patient cooperation and cannot be used after many surgical operations.

Raising the abdominal and thoracic pressure helps to extend the trunk and also reduces compression forces on the bodies of the vertebrae. A patient with weak abdominal muscles or a paralysed larynx cannot lift heavy weights. A nurse who chats while lifting a

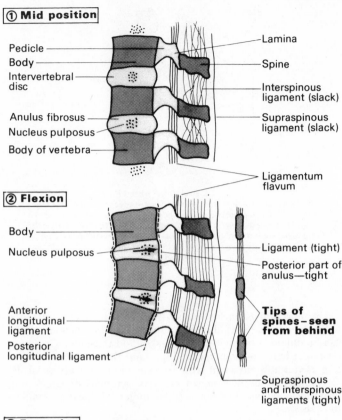

① Mid position

- Pedicle
- Body
- Intervertebral disc
- Anulus fibrosus
- Nucleus pulposus
- Body of vertebra
- Lamina
- Spine
- Interspinous ligament (slack)
- Supraspinous ligament (slack)
- Ligamentum flavum

② Flexion

- Body
- Nucleus pulposus
- Anterior longitudinal ligament
- Posterior longitudinal ligament
- Ligament (tight)
- Posterior part of anulus—tight
- **Tips of spines – seen from behind**
- Supraspinous and interspinous ligaments (tight)

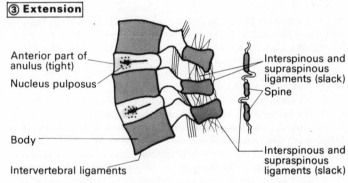

③ Extension

- Anterior part of anulus (tight)
- Nucleus pulposus
- Body
- Intervertebral ligaments
- Interspinous and supraspinous ligaments (slack)
- Spine
- Interspinous and supraspinous ligaments (slack)

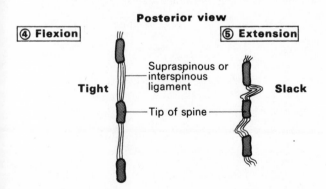

Posterior view

④ Flexion **⑤ Extension**

Tight **Slack**

- Supraspinous or interspinous ligament
- Tip of spine

patient is making her task more difficult. Athletes when weight-lifting raise the abdominal pressure far in excess of the systolic blood pressure for a short time.

Certain back injuries are thought to involve the trapping of the capsule of one of the synovial zygopophyseal joints. This is often associated with twisting the vertebral column while the lifting movement is taking place.

The rupture of the anulus fibrosus allows the nucleus pulposus to escape in a posterior lateral direction, pressing against the spinal theca (which is richly innervated) and nerves of the cauda equina. Sciatica as well as back pain occurs. Such a lesion also causes secondary damage to the synovial joints (osteoarthritis) by disturbing the mutual alignment of adjacent vertebrae. Surgical removal of the herniated nucleus pulposus as well as immobilization of the adjacent vertebrae with a bone graft, will cure the pain without significantly reducing overall mobility.

The ability of the intervertebral ligaments indefinitely to maintain posture without pain contrasts with the pain which develops in many other ligaments when they are under tension. For example the back of the knee joint becomes painful. But the intervertebral discs and the ligamenta flava are unique anatomical components of these joint mechanisms, and it is not surprising that they are functionally unique also.

Middle of the body

Measure the height of yourself or a subject and divide this distance by two. Then determine the part of the body that corresponds to this half-height distance from the floor. It will be readily seen that human beings have lower limbs which are as long as the trunk and skull together.

The middle of the body is usually below the pubic symphysis in the male. In the female the legs are relatively shorter, and the middle of the body lies above the pubic symphysis.

FIG. 3.22. Intervertebral ligaments in midposition, and in flexion and extension.

The anterior and posterior longitudinal ligaments run throughout almost the entire length of the vertebral column. They are anterior and posterior with respect to the bodies of the vertebrae. The anterior longitudinal ligament blends with the anterior parts of the anular ligaments, while the posterior longitudinal ligament, forming the anterior boundary of the vertebral canal, blends with the posterior parts of the anular ligaments. They become tight in extension or flexion respectively.

Because of their closeness to the centre of movement (nucleus pulposus) and poor leverage, these ligaments need to be very powerful to produce a useful effect.

1. Midposition: in intermediate positions the intervertebral ligaments are not in tension, with the exception of the ligamenta flava, which are a special case, as they consist of elastic tissue.
2. Flexion: In flexion the intervertebral disc becomes wedge-shaped forwards. The nucleus pulposus is displaced backwards and the posterior part of the anulus is under direct tension.

   In addition all the slack in the interspinous and supraspinous ligaments is taken up and they are now in tension. The ligamenta flava are stretched further.
3. Extension: In extension the spines of the vertebrae may come in contact with each other. The supraspinous and interspinous ligaments are completely relaxed. The intervertebral disc is now wedged backwards, the nucleus pulposus is displaced forwards, as shown by the arrows, and the anterior part of the anulus is tight.

4 and 5. In flexion the tips of the spines are widely separated. In extension they may rub against each other. Feel the tips of your own lumbar spines in flexion and extension. In flexion the ligaments are in such tension that they feel as hard as bone.

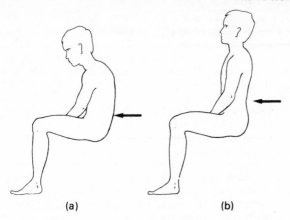

(a)   (b)

Fig. 3.23. Principal sitting postures. Tracing of photographs of the same subject while erectores spinae muscle activity was recorded.

So far as the vertebral column is concerned there are two basically different postures which can be adopted in sitting.

In (b) the erectors spinae muscles contract and the back is 'hollow' in the lumbar region. The posture is maintained by the muscles.

In (a) the muscles have relaxed and the vertebral column is now fully flexed, the hollow has disappeared and posture is now maintained by the intervertebral ligaments.

It is very important to be able to use both these mechanisims alternately, otherwise fatigue and stiffness may occur.

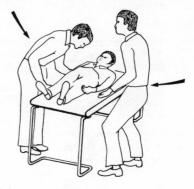

Fig. 3.24. Back strain when lifting.
When lifting a patient from a hospital bed or couch it is impossible to get the feet in the appropriate position because the edge of the bed is in the way. The wider the bed the worse the problem. The lift must come from the hip joint extensors and this puts a great strain on the vertebral column. In the diagram the person on the left is in greater danger of back strain than the person of the right.

In a baby the lower limbs are relatively very short and the mid-point is above the umbilicus at birth. A baby can easily put its big toes in its mouth. Its upper limbs are relatively longer than the lower limb. The mid-point of the body moves downwards with the increase of age. It is at the umbilicus at two years of age.

It is interesting to note that a child is already half its total adult height on its second birthday anniversary.

Body contours

The contours of the body are determined not only by the bony skeleton but also by the muscles covering them and the superficial fat which lies between these muscles and the skin. The muscles themselves are surrounded by a sheath of areolar tissue termed deep fascia. In most parts of the body there is a layer of superficial fat between the deep fascia and the skin. The chief exception is the back of the hands and in the face, the muscles are actually attached to the skin itself [p. 297]. In places where the skin is joined to deep fascia by fibrous strands, the fat is absent and skin creases are formed. The gluteal fold is an example of such a crease.

The fat is firmly attached to the skin, but is only loosely attached to the deep fascia. An indication of the amount of subcutaneous fat can be obtained by measuring the skin fold thickness. Grip your own skin between your index finger and thumb. Between your fingers you now have two layers of skin and two layers of fat. By allowing 3 mm for the skin, you can make a rough estimate of your own subcutaneous fat thickness (don't forget to halve the remaining thickness!).

In the obese the shape and form of the underlying bony skeleton and muscles may be concealed completely. The new-born baby is uniformly covered with a layer of fat which gives it a smooth surface hiding the underlying bones. The only places where the fat is missing is at the joint regions. The absence of the fat here gives rise to deep grooves running round the joints which look as if elastic bands had been placed around these joints.

As the child grows the superficial fat becomes thinner. At puberty, the fat tends to be deposited in specific regions. It is present in the same region in both sexes, but tends to be deposited in large amounts in the female.

The amount of fat in the subcutaneous tissue is a good guide to a patient's nutrition. In babies who are undernourished the limbs are very thin and measuring the circumference of the upper arm is a useful method of assessing nutritional status.

4 The blood supply and the nerve supply to the upper limb

The surface anatomy, muscle groups and important movements of the upper limb have already been considered on pages 27–43. In this and the next chapter the anatomical structures of the upper limb will be considered in more detail including the origin, insertion, and nerve supply of the individual muscles. First we will start with the vascular supply to the upper limb.

BLOOD VESSELS OF THE UPPER THORAX
SUBCLAVIAN ARTERY

The right subclavian artery is one of the two terminal branches of the right brachiocephalic trunk [FIG. 4.01] (the other terminal branch is the common carotid artery). The left subclavian artery arises directly from the aorta.

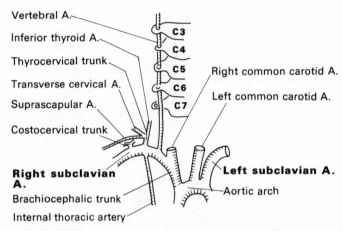

FIG. 4.01. The right subclavian artery and its branches.

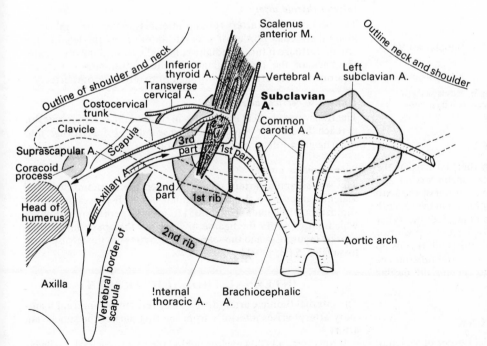

FIG. 4.02. The subclavian artery is divided into three parts by scalenus anterior.

The figure is based on an X-ray. The right and left subclavian arteries differ in their origins but they have the same standard four branches, the vertebral artery, the internal thoracic artery, the thyrocervical trunk and the costocervical trunk.

The costocervical trunk is shown as a dotted line as it lies behind the scalenus anterior muscle. It may come off the first part of the subclavian artery on the left.

The suprascapular artery which is a branch of the thyrocervical trunk accompanies the suprascapular nerve towards the suprascapular notch. The branches of the transverse cervical artery accompany the accessory nerve (superficial branch) and the nerve to the rhomboids (deep branch). These three arteries all supply the muscles of the scapular and take part in the anastomosis around the scapula. Note the association of these vessels with their corresponding nerves in a neurovascular bundle.

The costocervical trunk may come off the first part of the left subclavian artery.

The subclavian artery changes its name to axillary at the outer border of the first rib, where it enters the axilla. 'Subclavian' means 'below the clavicle' (dotted outline). Do you think this is a good name for an artery which has already changed its name when it reaches the clavicle?

Scalenus anterior, which passes in front of the artery to its insertion into the first rib, divides the artery into three parts [FIG. 4.03].

First part of subclavian artery

Branches:
1. Vertebral artery.
2. Thyrocervical trunk.
3. Internal thoracic artery.

This part of the artery in the neck runs behind the internal jugular and brachiocephalic veins, and the sternohyoid and sternothyroid muscles. It is running in front of and above the pleura of the lungs.

Second part of the subclavian artery

Branches: Costocervical trunk.

By the above definition this part of the artery runs **behind** scalenus anterior. The corresponding vein, the subclavian vein, is more superficial and runs **in front of** scalenus anterior.

Third part of the subclavian artery

Branches: Nil.

This part of the artery runs from the lateral margin of scalenus anterior over the first rib to its outer border where it becomes the axillary artery.

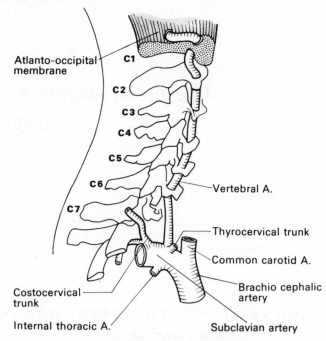

FIG. 4.04. The vertebral artery.
Note that it does not pass through the foramen transversarium of C7.

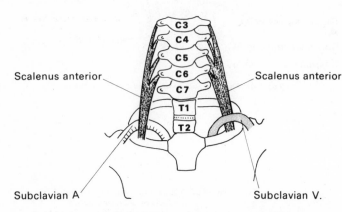

FIG. 4.03. The relationship of scalenus anterior to the subclavian artery and vein.
Subclavian artery passes behind scalenus anterior. Subclavian vein passes in front. Scalenus anterior divides the artery into three parts.

VERTEBRAL ARTERY

The vertebral artery [FIG. 4.04] arises from the first part of the subclavian artery. It passes upwards between longus cervicis and scalenus anterior to enter the foramen transversarium of the sixth cervical vertebra. Note that it by-passes the foramen in the seventh cervical vertebra. It passes through the foramina of all the other cervical vertebrae including that of the atlas. It pierces the posterior atlanto-occipital membrane and dura to enter the skull. It runs up on the anterior surface of the medulla to join the vertebral artery of the other side at the lower border of pons to become the **basilar artery**.

THYROCERVICAL TRUNK

The thyrocervical trunk arises near the medial border of scalenus

anterior between the internal jugular vein and the pleura. Almost immediately it breaks up into:
1. Inferior thyroid artery.
2. Transverse cervical artery.
3. Suprascapular artery.

Inferior thyroid artery

The inferior thyroid artery runs up anteriorly to the vertebral artery along the medial border of scalenus anterior. At the level of the cricoid cartilage it turns medially behind the carotid sheath to enter the back of the thyroid gland. It often anastomoses with the superior thyroid artery (first branch of external carotid artery).

Transverse cervical artery

The transverse cervical artery runs in front of the scalenus anterior to reach the anterior border of levator scapulae. Here it divides into superficial and deep branches [FIG. 4.02].

Suprascapular artery

The suprascapular artery passes between the sternomastoid and scalenus anterior to the suprascapular notch. It passes above the suprascapular ligament [FIG. 4.05] to join the suprascapular nerve which passes under the ligament. Together they pass through the supraspinous fossa and the scapular notch to reach the infraspinous fossa [FIG. 5.03, p. 91].

INTERNAL THORACIC ARTERY

The internal thoracic artery (previously termed the internal mammary artery) arises inferiorly from the first part of the subclavian artery.

It runs vertically downwards behind the first five costal cartilages

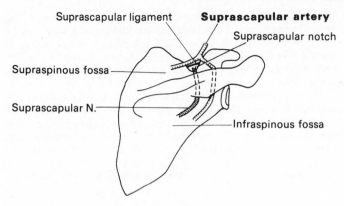

FIG. 4.05. Suprascapular artery.
The suprascapular artery passes above the suprascapular ligament. The suprascapular nerve passes under the ligament. Together they supply supraspinatus and infraspinatus.

about 1 cm from the edge of the sternum. It gives rise to the upper anterior intercostal arteries [p. 173] and perforating arteries which supply the mammary glands in the female [p. 167]. It also sends branches to the mediastinal structures.

Behind the sixth costal cartilage it divides into:

1. **Superior epigastric artery**
   This artery runs down in the rectus sheath to anastomose with the inferior epigastric artery (branch of external iliac artery).
2. **Musculophrenic artery**
   This artery supplies the anterior intercostal arteries to the seventh, eighth, and ninth interspaces.

COSTOCERVICAL TRUNK

The costocervical trunk arises from the second part of the subclavian artery. It divides almost immediately into:

1. Superior intercostal artery.
2. Deep cervical artery.

Superior intercostal artery

The superior intercostal artery passes in front of the neck of the first rib to give rise to the first and second posterior intercostal arteries.

Deep cervical artery

The deep cervical artery supplies the muscle at the back of the neck. It passes between the transverse process of the seventh cervical vertebra and the neck of the first rib. It runs upwards deep to semispinalis capitis. It forms an anastomosis with the cervical branch of the occipital artery.

THE AXILLA

The blood vessels and nerves enter the arm via the **axilla** [FIG. 4.06]. Feel the anterior and posterior walls of the axilla on yourself. The **anterior wall** is in two layers. The superficial layer is the pectoralis major while the deep layer is the pectoralis minor and subclavius surrounded by the clavipectoral fascia. The **posterior wall** is composed of three muscles, teres major, latissimus dorsi, and subscapularis from below upward [FIG. 4.07]. The posterior wall extends lower than the anterior wall.

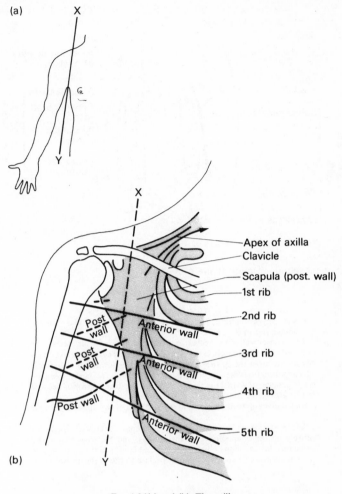

FIG. 4.06(a) and (b). The axilla.
(a) The axilla lies between the upper part of the side of the chest (upper five ribs) and the inner surface of the arm. It is roughly the shape of a three-sided pyramid.
(b) The three sides of the pyramid are:
  1. The upper five ribs (medial).
  2. The muscles in front of the scapula (posterior—dotted lines).
  3. The pectoralis major and the tissues deep to it—clavipectoral fascia and pectoralis minor (anterior-solid lines).
The base of the pyramid, or floor of the axilla. is just the skin of the arm pit. The apex of the axilla is the narrow triangular space bounded by the 1st rib. the clavicle and the scapula leading to the posterior triangle of the neck. The great vessels, the brachial plexus and lymphatics pass through this space.

Rather surprisingly the muscles forming the posterior wall of the axilla are inserted into the front (not the back) of the humerus. Furthermore, the insertion of teres major and latissimus dorsi is very close to that of pectoralis major and all are inserted into the bicipital groove of the humerus. This means that the **lateral wall** of the axilla is only the very narrow intertubercular groove between these insertions. However, the presence of biceps and coracobrachialis between the anterior and posterior wall muscle insertions give the axilla its characteristic appearance.

The **medial wall** of the axilla is made up of the upper five ribs and interspaces covered by digitations of serratus anterior. As will be

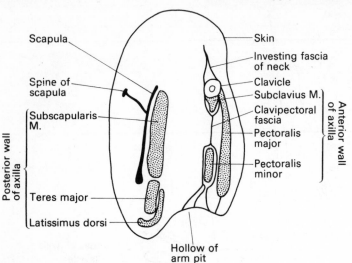

Vertical section through plane of X–Y

(c)

FIG. 4.06(c). The axilla (*contd*).

(c) Vertical section through plane of X–Y showing anterior and posterior boundaries of the axilla.

The anterior wall of the axilla is in two layers. Anteriorly just deep to the skin and superficial fascia is the pectoralis major. The second layer runs downwards from the clavicle and consists of a tough sheet of deep fascia which first encloses the subclavius muscle. It then fills the gap between the clavicle moves upwards; the skin of the arm pit is drawn upwards because of this arrangement.

calvicle moves upwards; the skin of the arm pit is drawn upwards because of this arrangement.

The posterior wall consists of the subscapularis muscle which lies entirely on the deep surface of the scapula, and the teres major and latissimus dorsi tendon, which lie below and lateral to the scapula. The main vessels and nerves are fairly close to the posterior wall.

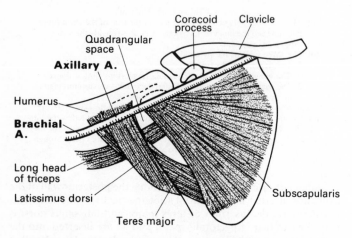

FIG. 4.07. Posterior wall of axilla.

There are three muscles forming the posterior wall of the axilla. The subscapularis covers most of the surface of the scapula. Immediately below this muscle is the teres major. The latissimus dorsi approaches the axilla from behind, and winds around the lower border of teres major and comes to lie in front of the teres major. The axillary artery lies in turn on subscapularis, latissimus dorsi, and teres major. At the lower border of teres major it becomes the brachial artery.

seen in FIGURE 4.07 the insertion of latissimus dorsi forms a sling around teres major. Their insertion into the anterior aspect of the humerus means that these muscles, like pectoralis major, will act as medial rotators of the humerus (not lateral rotators). As will be seen on pages 92–3 pectoralis major adducts and flexes the arm, whilst latissimus dorsi adducts and extends the arm.

The blood vessels, nerves, and lymph vessels enter and leave the arm between the middle third of the clavicle and the first rib. This is the apex of the axilla. It lies in front of the superior border of the scapula and communicates directly with the posterior triangle of the neck. The axilla thus has the shape of a pyramid with the top chopped off (truncated pyramid). The *base* is formed by skin, subcutaneous tissues, and axillary fascia [see also FIG. 4.19].

BLOOD VESSELS OF THE UPPER LIMB

General plan

The **subclavian artery**, which has been seen, arises from the brachiocephalic trunk on the right side and from the aorta on the left side, changes its name at the outer border of the first rib and becomes the **axillary artery**. At the lower border of teres major it changes its name again and becomes the **brachial artery**. The brachial artery divides, opposite the neck of the radius in the forearm, into the **radial artery** and the **ulnar artery**.

AXILLARY ARTERY

The axillary artery [FIG. 4.08], as its name implies, is the artery of the **axilla**, the pyramidal space between the upper part of the arm and the trunk. The artery enters the axilla at its apex. It lies deep to the surface except at its lower end where it becomes superficial.

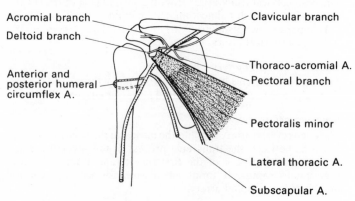

FIG. 4.08. The axillary artery and its branches

Since it is surrounded by many important structures, from a descriptive point of view it is convenient to divide the artery into three parts. The first part of the artery lies above pectoralis minor [FIG. 4.09]. The second part lies behind the muscle. The third part lies below pectoralis minor.

First part of the axillary artery

This part of the artery lies in front of the medial cord of the brachial plexus which in turn lies in front of the first intercostal space, and

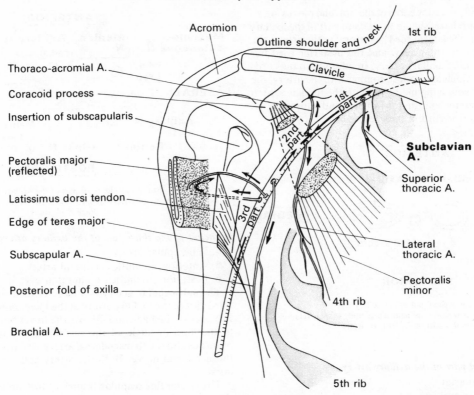

FIG. 4.09. The three parts of the axillary artery.
Like the subclavian artery the axillary artery is divided into three parts by a muscle, in this case the pectoralis minor. The first part runs from the outer border of the 1st rib to the pectoralis minor, the 2nd part lies behind it, and the 3rd part runs from the lateral border of pectoralis minor to the lowest part of the posterior fold of the axilla (teres major muscle with latissimus dorsi twisting round it). There is one branch from the 1st part, two from the 2nd and three from the 3rd. In life the artery here has the cords of the brachial plexus and the axillary vein in close relationship.
The clavipectoral fascia fills in the gap between the lower border of the clavicle and the upper edge of the pectoralis minor. It is anterior to the 1st part of the axillary artery.
In the figure part of pectoralis minor has been removed to expose the artery.

the first digitation of serratus anterior. The lateral and posterior cords lie lateral to this part of the artery. The axillary vein lies on the medial side of the artery. Anterior to the artery are clavipectoral fascia [FIG. 4.10] and pectoralis major.

The **clavipectoral fascia** (previously known as the costocoracoid membrane) runs from the first rib and fascia over the first inter-

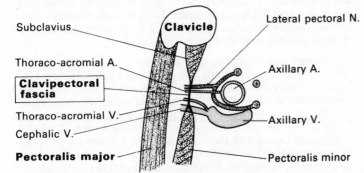

FIG. 4.10. Structures penetrating clavipectoral fascia (costocoracoid membrane).

costal muscle to the coracoid process. It separates into two layers above to enclose subclavius and separates again below to enclose pectoralis minor.

Some fibres continue onwards to be attached to the skin of the floor of the axilla. This connection between the clavicle and the floor of the axilla explains why the skin is drawn upwards in the 'arm pit' when the arms are raised.

The clavipectoral fascia [FIG. 4.10] is pierced by:
(i) cephalic vein;
(ii) thoraco-acromial artery and vein;
(iii) lateral pectoral nerve.

The lateral and medial pectoral nerves form a nerve loop around this part of the axillary artery.

Branches of the first part of the axillary artery:

1. Superior thoracic artery.

This is a small artery which runs along the upper border of pectoralis minor and supplies the upper part of the medial wall of the axilla. It is small and unimportant.

Second part of the axillary artery

The second part of the axillary artery lies behind pectoralis minor.

The lateral, medial, and posterior cords of the brachial plexus are so named because of their relationship to the second part of the axillary artery [FIG. 4.11]. Thus the **lateral cord** lies lateral to the artery, the **medial cord** lies medial to the artery, and the **posterior cord** lies posterior to this part of the axillary artery. In addition to the cords of the brachial plexus which are in close contact with the artery, the artery has the axillary vein lying medially (with the medial cord in between), pectoralis minor and major in front and subscapularis behind (with the posterior cord in between).

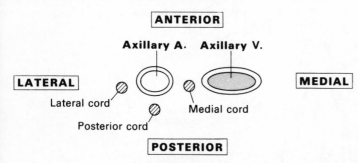

FIG. 4.11. The second part of the axillary artery.
The cords of the brachial plexus are named according to their relationship with this part of the axillary artery.

Branches of the second part of the axillary artery:

2. Thoraco-acromial artery.
3. Lateral thoracic artery.

The **thoraco-acromial artery** arises from the axillary artery under cover of the upper border of pectoralis minor. It pierces the clavipectoral fascia and divides into four divergent branches. The *clavicular branch* runs medially between the clavipectoral fascia and pectoralis major. The *deltoid branch* runs across pectoralis minor to run with the cephalic vein in the groove between deltoid and pectoralis major. The *acromial branch* runs across the coracoid process to reach the acromion where it divides and anastomoses with the suprascapular and posterior circumflex arteries. The *pectoral branches* supply pectoralis major and pectoralis minor.

The **lateral thoracic artery** runs medially and inferiorly in the angle between the chest wall and pectoralis minor. In the female it supplies the lateral part of the mammary gland.

Third part of the axillary artery

By the time the third part of the axillary artery has been reached, the cords of the brachial plexus have become nerve branches. Seven nerves surround the artery [FIG. 4.12]. The *musculo-cutaneous nerve* lies in the lateral side, the *axillary* and *radial nerves* lie posteriorly, the *ulnar nerve* and *medial cutaneous nerve of the forearm* lie medial, and the *two heads* of the *median nerve* lie anteriorly and laterally [see FIG. 4.21, p. 82].

The axillary vein lies on the medial side of the artery and nerves. The medial cutaneous nerve of the arm lies medial to the vein.

The artery in this region is running in front of subscapularis, latissimus dorsi, and teres major (the latter muscles form the posterior fold of the axilla). It is running behind pectoralis major (which forms the anterior fold of the axilla) and has coracobrachialis on its lateral side.

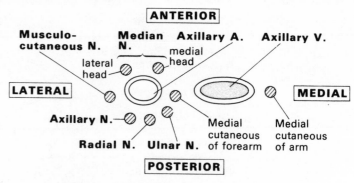

FIG. 4.12. The third part of the axillary artery.
This part of the artery is surrounded by the nerves formed by the brachial plexus.

Branches of the third part of the axillary artery

4. Subscapular artery.
5. Anterior circumflex humeral artery.
6. Posterior circumflex humeral artery.

The **subscapular artery** is a large branch which arises from the third part of the axillary artery at the lower border of subscapularis. It passes to the back and follows the lower border of subscapularis to the inferior angle of the scapula where it gives off muscular branches and the **thoracodorsal artery** which is accompanied by the thoracodorsal nerve. Both the artery and nerve supply latissimus dorsi.

The **circumflex scapular** branch of the subscapular artery passes through the triangular space between subscapularis (superiorly), long head of triceps (medially) and teres major (inferiorly) to reach the infraspinous fossa where it forms important anastomoses around the scapula with the suprascapular and posterior scapular arteries.

The **anterior circumflex humeral artery** is much the smaller of the two circumflex arteries which anastomose and encircle the humerus [FIG. 4.09]. It arises at the lower border of subscapularis and passes laterally under cover of biceps and coracobrachialis to reach the intertubercular groove. Here it sends a branch to the shoulder joint before continuing on round the surgical neck of the humerus to anastomose with the posterior circumflex humeral artery.

The **posterior humeral circumflex artery** is the larger of the two circumflex arteries. It arises at the same level as the anterior artery and reaches the back of the humerus by passing through the quadrangular space with the axillary nerve. The artery continues around the surgical neck of the humerus to anastomose with the anterior circumflex humeral artery. It sends branches to the shoulder joint, and surrounding muscles. In addition, it forms an important anastomosis with the profunda brachii artery.

The **quadrangular space** is bounded from the front by the surgical neck of the humerus laterally, the teres major inferiorly, the long head of triceps medially, and the subscapularis superiorly. Seen from behind the boundaries are the same except that the upper boundary is now teres minor. The point to grasp is that the upper boundary is really the inferior part of the shoulder joint. Think of it as the **cuboid space** with the depth of the third dimension provided by the whole thickness of the shoulder joint. When the shoulder is dislocated the head of the humerus is displaced downwards into the cuboid space, and may paralyse the axillary nerve (about 10 per cent of cases) and cause bleeding from the circumflex vessels.

BRACHIAL ARTERY

The axillary artery becomes the **brachial artery** at the lower border of teres major. If the arm is abducted and an imaginary line drawn from the middle of the clavicle to the middle of the cubital fossa, then the proximal third of this line gives the surface markings of the axillary artery, whilst the distal two-thirds, gives the surface marking of the brachial artery.

The brachial artery lies first on the medial side of the humerus between biceps and triceps. As the cubital fossa is approached the artery comes to lie anterior to the bone. Since the artery is for the most part superficial, its route can easily be established by palpation. It is usually easier to work upward from the cubital fossa, where it lies medial to the biceps tendon, along the medial side of the arm to the axilla. Try this on yourself.

Remember that the nerves of the arm were surrounding the artery when it was the axillary artery and they will therefore be in close relation to the brachial artery at least at its start. The **median nerve** runs down the arm with the brachial artery. Rather confusingly the median nerve at first lies on the lateral side of the artery but at about the middle of the arm it crosses (usually in front of the artery, but occasionally behind) and continues down on the medial side of the artery to the cubital fossa.

The **ulnar nerve** lies at first medial to the brachial artery, but soon leaves it by passing posteriorly through the medial intermuscular septum. Notice that the median nerve occupies the place vacated by the ulnar nerve on the medial side of the artery [see FIG. 4.14, p. 76].

The **radial nerve** lies behind the brachial artery in the axilla but soon leaves it curving laterally between the long and medial heads of triceps to reach the spiral groove of the humerus accompanied by the profunda brachii branch of the brachial artery.

The medial cutaneous nerve of the forearm lies medial to the artery. It pierces deep fascia at the middle of the arm and sends a branch which runs with the medial basilic vein.

Branches of the brachial artery

1. Profunda brachii.
2. Superior ulnar collateral artery.
3. Inferior ulnar collateral artery.
4. Nutrient and muscular arteries.
5. Terminal branches—ulnar and radial arteries.

The **profunda brachii artery** [FIG. 4.13] is given off by the brachial artery shortly after its commencement. It follows the radial nerve into the spiral groove where it divides into the **radial collateral artery** which passes through the lateral intermuscular septum and anastomoses with the radial recurrent artery in front of the lateral epicondyle, and the posterior descending branch (**median collateral artery**) which runs down the back of the humerus and anastomoses with the posterior interosseous recurrent artery behind the lateral epicondyle.

The **superior ulnar collateral artery** which is given off half way down the arm, accompanies the ulnar nerve through the medial intermuscular septum. Behind the medial epicondyle it anastomoses with both the posterior ulnar recurrent artery and with the posterior branch of the inferior ulnar collateral artery.

The **inferior ulnar collateral artery** is an important artery for anastomoses around the elbow joint. For this reason its older name was anastomotica magna artery. It arises 5 cm above the elbow, and divides into anterior and posterior branches. The anterior branch anastomoses with the anterior ulnar recurrent artery in front of the medial epicondyle. The posterior branch pierces the medial inter-

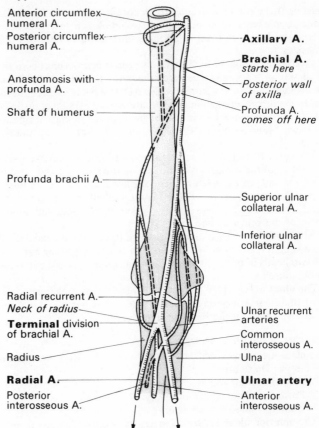

FIG. 4.13. The brachial artery and its terminal branches the radial and ulnar arteries.

The brachial artery is shown throughout its length with respect to the humerus, from which in life it is separated by the brachialis muscle. Confirm in yourself that in the upper part of its course it lies medial to the humerus and that it comes eventually to lie in front of the humerus and elbow joint. In taking blood pressure, the stethoscope is placed over the artery medial to the biceps tendon just above the elbow joint.

The anastomosis round the elbow joint consists of vessels running in parallel to the main artery in front of and behind both medial and lateral epicondyles. The common interosseous artery gives rise to vessels which run down anterior and posterior to the interosseous membrane. The smaller vessels in the figure are subject to endless variations in detail and there is no need for you to remember specific details.

The brachial artery is more liable to variation than any other great vessel in the body. It may be doubled or have a collateral vessel which usually runs to the radial artery; the terminal bifurcation may be high up in the arm, almost in the axilla.

muscular septum and, as has been seen, anastomoses with the superior ulnar collateral and the posterior ulnar recurrent arteries behind the medial epicondyle.

THE ULNAR ARTERY

The ulnar artery [FIG. 4.13] is the larger of the two terminal branches of the brachial artery which bifurcates at the level of the neck of the radius. The ulnar artery usually runs deep to practically all of the muscles which arise from the medial epicondyle (both heads of pronator teres, flexor carpi radialis, palmaris longus, and flexor digitorum superficialis) but it becomes superficial before it reaches the most medial muscle, flexor carpi ulnaris. However, some people have an anomalous superficial ulnar artery, that is, an artery which runs in front of all these muscles just under the skin and deep fascia.

From then onwards, in all cases, the ulnar artery runs down the ulnar side of the forearm between the tendons of flexor digitorum superficialis and flexor carpi ulnaris. At the wrist it passes superficial to the flexor retinaculum on the lateral (radial) side of pisiform but deep to palmaris brevis. At the hook of the hamate it divides into two branches which form the superficial and deep palmar arches.

The surface markings for the latter part of its course in the lower two-thirds of the forearm is a line joining the medial epicondyle to the radial side of the pisiform. The surface markings of the upper part of the artery is a curved line convex medially running from the middle of the cubital fossa to the junction of the upper and middle thirds of the previous line.

At the elbow the **median nerve**, which lies on the medial side of the artery in the cubital fossa, leaves the artery by passing between the two heads of pronator teres (the ulnar artery passes behind the whole muscle).

The **ulnar nerve** which has passed behind the medial epicondyle, joins the ulnar artery on its medial side for its passage down the distal two-thirds of the forearm [Fig. 4.14].

Branches of the ulnar artery

The ulnar artery gives off three branches shortly after its origin at the elbow. These are:

1. Anterior ulnar recurrent artery.
2. Posterior ulnar recurrent artery.
3. Common interosseous artery.

The **anterior ulnar recurrent artery** runs upwards deep to pronator teres and in front of brachialis to reach the anterior surface of the medial epicondyle where, as has been seen above, it anastomoses with the inferior ulnar collateral artery.

The **posterior ulnar recurrent artery** reaches the back of the medial epicondyle by passing upward through the two heads of flexor carpi ulnaris along the course of the ulnar nerve. Here, as has been seen above, it anastomoses with the superior and inferior ulnar collateral arteries.

The **common interosseous artery** is given off from the ulnar artery about 2.5 cm from its commencement. It divides almost immediately into:

(i) Anterior interosseous artery;
(ii) Posterior interosseous artery.

The **anterior interosseous artery** runs down on the anterior surface of the interosseous membrane with the anterior interosseous nerve. It lies at the bottom of the groove between flexor digitorum profundus and flexor pollicis longus.

The anterior interosseous artery gives off a small **median artery** which accompanies the median nerve.

Deep to pronator quadratus it reaches the back of the wrist by piercing the interosseous membrane to join the posterior interosseous artery, which also runs down to the back of the wrist with the posterior interosseous nerve.

The **posterior interosseous artery** reaches the back of the forearm by passing between the radius and ulna (unlike the posterior interosseous nerve, p. 85) before the interosseous membrane has started (it passes superior to the interosseous membrane).

It gives off the posterior interosseous recurrent branch which runs upwards, deep to anconeus, to anastomose with the descend-

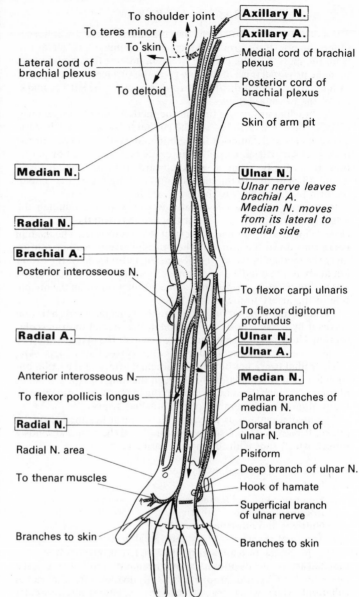

FIG. 4.14. The median. ulnar. and radial nerves.
The position of the three main nerves of the upper limb with respect to the bony skeleton is shown. It will be seen how easily major nerves can be involved when the bones are broken. Nerves may also be paralysed by squashing against the bone. without a fracture.

By the end of the next Chapter. you should be able to work out for yourself the muscles affected by a paralysis of the axillary nerve. the radial nerve. the ulnar nerve. the posterior interosseous nerve. which all lie directly on bone.

Test the major nerves whenever you see a patient with a fracture. Keep on testing them during the next week or two.

In the embryo nerve fibres often grow by climbing along blood vessels. Note that the median nerve runs with the brachial artery on its lateral and then medial side, while the radial nerve evidently followed the profunda artery and the ulnar nerve the superior ulnar collateral artery.

ing posterior branch of the profunda brachii artery behind the lateral epicondyle.

The posterior interosseous artery itself runs down the back of the forearm between the superficial and deep extensor muscles to anastomose with the very much larger perforating anterior interosseous artery which takes over its function in the lower part of the forearm.

At the wrist the ulnar artery gives off:

4. Anterior and posterior carpal branches.

These branches anastomose with similar branches from the radial artery [Fig. 4.16].

In the hand the ulnar artery gives off:

5. Superficial and deep terminal branches.

The superficial terminal branch of the ulnar artery joins the superficial palmar arch. The deep branch runs between flexor digiti minimi brevis and abductor digiti minimi to join the deep palmar arch by twisting round the hook of the hamate.

THE RADIAL ARTERY

The radial artery is the second of the two terminal branches of the brachial artery which, as has been seen, bifurcates at the level of the neck of the radius. Although it is slightly smaller than the ulnar artery it appears at the elbow to be the direct continuation of the brachial artery. The artery is superficial along the entire length of the forearm and hence its route can be confirmed by palpation. It lies in a groove between pronator teres and flexor carpi radialis on its medial side and brachioradialis on its lateral side. The surface markings of the artery is a slightly convex-outwards line running from the middle of the cubital fossa to radial pulse region (in front of the lower end of the radius, 1 cm medial to the styloid process of the radius).

The structures on which the radial artery lies reads like a catalogue of structures on the anterior surface of the radius. These are in succession: tendon of biceps, supinator, pronator teres, flexor digitorum superficialis, flexor pollicis longus, pronator quadratus, and the lower end of radius itself [Fig. 4.15].

The radial nerve joins the artery from the lateral side over the middle third of the forearm. It leaves it again by passing laterally deep to brachioradialis.

When the radial artery is traced distally from the radial pulse, it appears to vanish. It curves backwards just distal to the styloid process of the radius, and can next be found at the back of the hand in the floor of the 'anatomical snuff-box'. See if you can locate a radial pulse here in yourself [see Fig. 2.14, p. 36]. It has passed under the tendons of abductor pollicis longus and extensor pollicis brevis, which form the lateral boundary of the snuff-box, and has come to lie on the scaphoid and the trapezium. The artery passes deep to the tendon of extensor pollicis longus. It leaves the snuff-box and returns to the palmar aspect of the hand by passing between the two heads of the first dorsal interosseous muscles, that is, between the bases of the first and second metacarpal bones. It runs between the transverse and oblique heads of adductor pollicis to reach the deep palmar arch [Fig. 4.16].

Branches of the radial artery

The radial artery **at the elbow** gives off the:

1. Radial recurrent artery.

The **radial recurrent artery** is given off soon after the commencement of the radial artery at the elbow. It runs upwards along the course of the radial nerve to anastomose with the radial

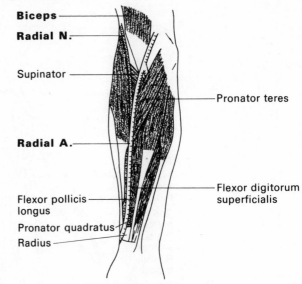

FIG. 4.15. The radial artery in the forearm.
It lies in succession on: tendon of biceps, supinator, pronator teres, radial origin of flexor digitorum superficialis, flexor pollicis longus, pronator quadratus, and the radius itself.

collateral artery (a branch of the profunda brachii artery) in front of the lateral epicondyle.

At the wrist and hand the radial artery [Fig. 4.16] gives off:

2. Anterior carpal artery.

3. Superficial palmar artery.

4. Princeps pollicis artery.

5. Radialis indicis artery.

6. Branches to palmar arches.

The **anterior carpal artery** is a small vessel given off just above the wrist. It anastomoses with the corresponding vessel from the ulnar artery.

The **superficial palmar artery** is given off above the thenar eminence. It runs to the thenar eminence where it usually ends, but it may continue on and join the superficial palmar arch.

The **princeps pollicis artery** (= principal artery of the thumb) is given off by the radial artery as it runs deep to the oblique head of adductor pollicis. The artery runs distally along the first metacarpal deep to flexor pollicis longus. It divides into two branches which become the palmar digital arteries of the thumb.

The **radialis indicis artery** (artery to radial side of index finger) is given off by the radial artery in the palm close to the princeps pollicis artery. It runs between the transverse head of adductor pollicis and the first dorsal interosseous muscle to supply the lateral border of the index finger.

THE SUPERFICIAL PALMAR ARCH

The superficial palmar arch [Fig. 4.16], as its name implies, lies superficial on the palmar side of the hand. It lies in front of the flexor retinaculum, the flexor tendons and the median and ulnar nerves, and is covered only by skin, palmar aponeurosis, and palmaris brevis. The surface markings are an arch convex downwards extending from the radial side of the pisiform to 0.5 cm proximal to upper of the two transverse creases.

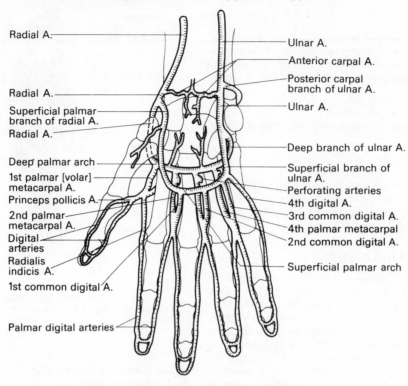

Fig. 4.16. The superficial and deep palmar arches.

The radial and ulnar arteries anastomose with each other on a 50/50 basis when they reach the hand. A radial or ulnar artery cut at the wrist bleeds copiously from both ends, not like an ordinary artery, only from the end nearest the heart. It is usually quite safe to ligate either the radial or ulnar arteries, but not both.

Both the superficial and deep palmar arches give off four main digital branches. The deep arch does not contribute to the ulnar side of the little finger.

Note that the word 'common' digital artery implies that the artery is shared by two digits (at least). The 4th digital artery is not 'common' because it only goes to the little finger.

The principal contribution to the superficial palmar arch usually comes from the superficial branch of the ulnar artery which anastomoses with the radial artery in the thenar eminence. The radial contribution is variable. It may come from the superficial palmar branch of the radial artery itself or from the radialis indicis or princeps pollicis arteries.

Palmar digital arteries

The superficial palmar arch gives off four palmar digital branches which run deep to the nerves (although the superficial palmar arch lies superficial to the nerves).

One digital branch supplies the ulnar side of the little finger. The other three digital branches run towards the three webs between the fingers and then bifurcate and supply adjacent sides of the fingers.

THE DEEP PALMAR ARCH

The deep palmar arch lies deep to the flexor tendons on the bases of the metacarpal bones. It lies about 1 cm proximal to the superficial arch.

The principal contribution to the deep palmar arch usually comes from the radial artery. It anastomoses with the deep branch of the ulnar artery.

The deep palmar arch anastomoses freely with the superficial palmar arch.

The superficial and deep palmar arches form such a good anastomosis between the radial and ulnar arteries, that in most people an adequate blood supply to the hand is still present after either the ulnar artery or the radial artery has become blocked.

Dorsal digital arteries

In addition to the relatively large palmar digital arteries which can be palpated on each side of the fingers (most easily felt as the artery passes over the interphalangeal joints), each digit has two smaller dorsal digital arteries which lie parallel to, but more dorsal than the palmar digital arteries [Fig. 4.17].

The radial artery, as it passes through the snuff-box, gives off a posterior radial carpal branch. This joins a corresponding branch from the ulnar artery to form the **posterior carpal arch**. The posterior carpal arch gives rise to the dorsal arteries.

Perforating branches of the deep palmar arch pass through the interosseous spaces to join the dorsal arteries which supply the small dorsal arteries of the fingers. The dorsal arteries of the thumb arise directly from the radial artery at the base of the first metacarpal bone. The dorsal artery of the index finger also arises from the radial artery. It is given off close to the dorsal arteries of the thumb.

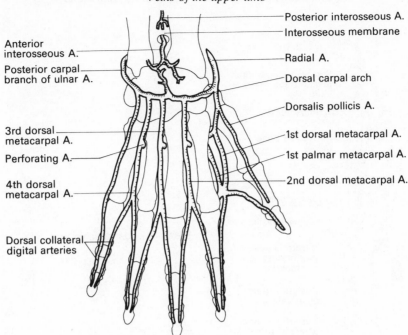

Anterior interosseous A.

Posterior carpal branch of ulnar A.

3rd dorsal metacarpal A.

Perforating A.

4th dorsal metacarpal A.

Dorsal collateral digital arteries

Posterior interosseous A.

Interosseous membrane

Radial A.

Dorsal carpal arch

Dorsalis pollicis A.

1st dorsal metacarpal A.

1st palmar metacarpal A.

2nd dorsal metacarpal A.

FIG. 4.17. The dorsal carpal arch and its branches.
When the posterior interosseous artery fizzles out, the anterior interosseous penetrates the interosseous membrane and takes its place on the back of the wrist. There are not four (as you would expect) but only three perforating arteries joining the palmar and dorsal metacarpal vessels by passing between the metacarpal bones. The radial artery passes through the first intermetacarpal space and can be regarded as the 'missing' fourth perforating arteries.

VEINS OF THE UPPER LIMB

Deep veins

The arteries of the upper limb usually have two or more **venae comitantes** running on either side of them. These vessels are given the same name as the artery and should always be assumed to be present even when only the artery has been mentioned. These venous channels not only provide a pathway for the return of blood to the heart, but their close proximity to the artery allows the counter-current heat exchange mechanism to work in a cold environment. The cold blood returning from the extremities is warmed up by the arterial blood before it reaches the heart. At the same time, the arterial blood is cooled as it passes to the extremity so that the temperature gradient between the limb extremity and the environment is reduced. As a result there will be less heat loss.

Thus the hands may be placed in ice-cold water and yet the blood arriving back at the superior vena cava is still at 37 °C.

Furthermore, the arterial pulsations will tend to improve the venous return by operating the 'muscle pump' action of the valves in veins even without muscular activity.

When an artery is cut the venae comitantes are also severed. It is rare in practice therefore to see pure arterial bleeding.

Superficial veins

The superficial veins are much more variable than the arteries (and the deep veins) and they communicate with each other forming networks. As a result, the venous pattern must be found by inspection. Most people have one or more cephalic, basilic, and median veins. These veins are united by a median basilic and a median cephalic vein [FIG. 4.18].

Cephalic vein

The cephalic vein starts from the venous arch on the dorsum of the hand. It runs across the snuff-box to reach the radial border of the forearm. At the elbow it is joined by the median cephalic vein. It then runs up in lateral groove between biceps and triceps to reach the groove between pectoralis major and deltoid. Here it is accompanied by the deltoid branch of the thoraco-acromial artery. It pierces the clavipectoral fascia and joins the axillary vein just below the clavicle.

Basilic vein

The basilic vein starts from the medial side of the dorsal venous plexus. It runs up the ulnar border of the forearm. At the elbow it is joined by the median basilic vein. It runs up the medial furrow between biceps and triceps separated from the brachial artery by deep fascia. At the middle of the arm, it pierces deep fascia, and at the lower border of teres major joins the venae comitantes and becomes the axillary vein.

Median vein

The median vein runs up the anterior surface of the forearm to the cubital fossa. Here it communicates with the deep veins and divides into the median basilic and median cephalic veins.

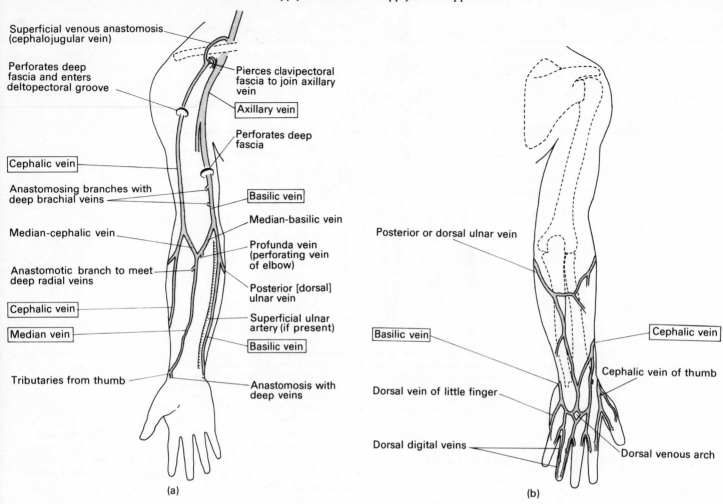

Superficial venous anastomosis
(cephalojugular vein)

Perforates deep
fascia and enters
deltopectoral groove

Pierces clavipectoral
fascia to join axillary
vein

Axillary vein

Perforates deep
fascia

Cephalic vein

Anastomosing branches with
deep brachial veins

Basilic vein

Median-cephalic vein

Median-basilic vein

Profunda vein
(perforating vein
of elbow)

Anastomotic branch to meet
deep radial veins

Posterior [dorsal]
ulnar vein

Cephalic vein

Superficial ulnar
artery (if present)

Median vein

Basilic vein

Tributaries from thumb

Anastomosis with
deep veins

(a)

Posterior or dorsal ulnar vein

Basilic vein

Cephalic vein

Cephalic vein of thumb

Dorsal vein of little finger

Dorsal digital veins

Dorsal venous arch

(b)

FIG. 4.18. Veins of the upper limb.

The pattern of veins on the back of your hand is different from everyone elses. All veins are very variable. The venous return from the hand and fingers travels mainly in the companion veins of the dorsal arteries.

The cephalic vein runs up the radial side of the forearm, enters the deep fascia, in the deltopectoral groove. It eventually goes through the anterior wall of the axilla and drains in the axillary vein above pectoralis minor. It resembles the great saphenous vein. The basilic vein runs up the ulnar side of the forearm and penetrates the deep fascia above the elbow joint. It eventually combines with the (several) venae comitantes of the axillary artery to form the axillary vein. often quite high up in the axilla. The median vein is often absent altogether. or joins the cephalic or basilic veins very early. In this case a vein usually runs upwards and laterally from the basilic to the cephalic vein and is called the cubital vein. Many doctors loosely call any vein in front of the elbow 'the cubital vein'.

Note especially that the anomalous superficial ulnar artery runs alongside the cubital vein and in a pulseless shocked patient an injection may erroneously be given into the artery instead of the vein with disastrous results.

The **median basilic vein** runs superiorly and medially to join the basilic vein. As has been seen, it lies in front of the brachial artery, but is separated from it by the bicipital aponeurosis. It also runs close to the median nerve, and the medial cutaneous nerve of the forearm.

The **median cephalic vein** runs superiorly and laterally to join the cephalic vein.

The superficial veins at the elbow anastomose richly with the deep veins. They usually lie superficial to the cutaneous nerves, but sometimes they lie deep. Be aware of this when you put a needle into these veins.

LYMPHATICS OF THE UPPER LIMB

The lymphatics from the front of the fingers and palm run mainly to the dorsum of the hand. This is where the swelling will be found. In general the lymphatics follow the blood vessels.

The lymphatic vessels draining the lateral side of the hand mostly follow the cephalic vein and run up the lateral side of the forearm and arm. In the deltopectoral groove there is a chain of lymph nodes, which filter the lymph before it reaches the axillary lymph nodes [FIG. 4.19].

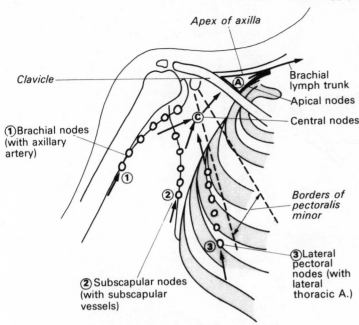

FIG. 4.19. Lymph nodes of axilla.

Axillary lymph nodes become palpable when they are enlarged by disease. They are carefully scrutinized when investigating a tumour of the breast which, if malignant, readily spreads to the axillary lymph nodes. The lymph nodes and lymph vessels run with the blood vessels.

Group 1 nodes. called brachial (not axillary for obvious reasons) run with the axillary vessels, especially the vein. Feel for them against the shaft of the humerus.

Group 2 nodes run with the subscapular vessels. Feel for them along the border of the scapula.

Group 3 runs with the lateral thoracic vessels. Feel for them along the lateral border of pectoralis minor in the angle between the ribs and deep surface of pectoralis major.

All three groups drain into the central lymph nodes (C) which lie in the axillary fat. and then into the apical nodes (A) which lie at the apex or inlet of the axilla. From here the brachial lymph trunk having reached the neck drains into the angle of the internal jugular and subclavian veins.

Tumour cells may find their way into the neck and for this reason it is essential to look carefully for lymph node enlargement in the posterior triangle of the neck as well as in the axilla itself. in suspected cases of malignancy.

The lymphatics draining the medial side of the hand run to the epitrochlear nodes and then to the axillary nodes.

The axillary lymph nodes may be subdivided into three groups of lymph nodes:

1. The brachial group which drain the upper limb consist of a group of lymph nodes and which lie along the axillary blood vessels.

2. The pectoral group which lie in the angle between the anterior and medial wall of the axilla with the lateral thoracic artery drain the mammary gland.

3. The subscapular group along the subscapular artery on the posterior wall of the axilla drain mainly the back.

The axillary lymph nodes lie alongside the blood vessels in the axilla. Thus the brachial group lie along the medial side of the humerus with the axillary artery and veins; the pectoral group lie in the angle between the anterior and medial walls (behind pectoralis minor) with the lateral thoracic artery, and the subscapular group lie along the lower border of the scapula and subscapularis muscle with the subscapular artery.

These three outlying groups of nodes are all connected to each other by nodes which lie in the middle of the axillary fat, the central group, and before leaving the axilla the lymph passes towards the apical group.

In the clinical examination of a suspected case of malignancy of the breast, or in assessing the progress of a patient who has the disease, a meticulous search for palpable axillary lymph nodes is essential. To do this properly you must have a clear picture in your mind as to the places where you can expect to find evidence of enlargement.

Normal axillary lymph nodes are never palpable.

NERVE SUPPLY TO THE UPPER LIMB
THE BRACHIAL PLEXUS

The brachial plexus [FIG. 4.20] is formed by the anterior rami of the fifth, sixth, seventh, and eighth cervical nerves and the first thoracic nerve with contributions for the fourth cervical and the second thoracic nerves.

C5 and C6 join to form the **upper trunk**
C7 forms the **middle trunk**
C8 and T1 form the **lower trunk**

Each trunk divides into an **anterior** and **posterior division** behind the clavicle. The divisions unite to form the **cords** which are named lateral, medial and posterior according to their relation to the second part of the axillary artery, as it passes behind pectoralis minor.

Anterior division of upper trunk and anterior division of middle trunk forms the **lateral cord**.

Anterior division of lower trunk forms the **medial cord**.

All the posterior divisions unite to form the **posterior cord**.

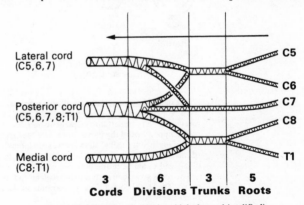

FIG. 4.20. Diagram of right brachial plexus (simplified).

Supraclavicular branches of the brachial plexus

The supraclavicular branches [FIG. 4.21] are given off before the formation of the cords. These are:

(a). **Nerve to subclavius** (C5, 6)

(b). **Dorsal scapular nerve** (C5)

The dorsal scapular nerve (nerve to the rhomboids) passes through scalenus medius to reach levator scapulae which it supplies. It then runs to the rhomboids.

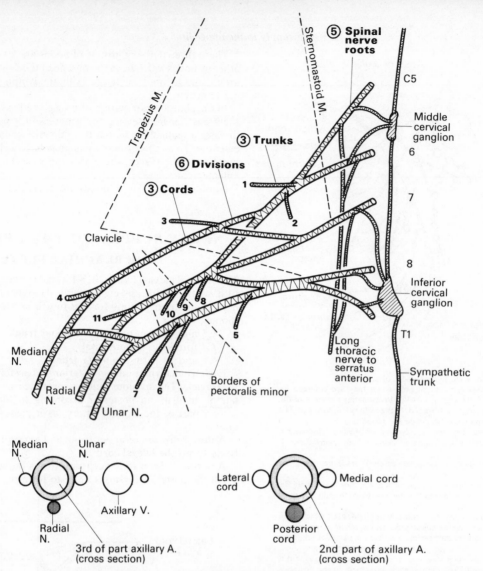

FIG. 4.21. Brachial plexus.

The figure shows diagrammatically the principal features of the brachial plexus. The contributory spinal nerve roots, C5–T1, are all directed towards the apex of the axilla converging like the spokes of a wheel. The roots receive the very important sympathetic autonomic fibres.

The upper part of the plexus lies above the clavicle in the posterior triangle of the neck. Below the clavicle, having reached the axilla, all the nerves are intimately related to the axillary artery. With respect to the 2nd part of this artery, the nerve cords lie lateral, medial, and posterior. Below this level the three main nerves, median, ulnar, and radial lie in the positions exactly corresponding to the cords from which they are derived.

The roots give off branches to the scalene and rhomboid muscles as well as a branch (from C5) to the phrenic nerve. The spinal roots C5, 6, 7 contribute to the nerve to serratus anterior which runs downward behind the other nerve roots.

The numbered branches are:

From the upper trunk:
1. Subscapular nerve (to supra and infraspinatus).
2. Nerve to subclavius.

From the lateral cord:
3. Lateral pectoral nerve to pectoral muscles.
4. Musculocutaneous nerve (to coracobrachialis, biceps, brachialis, and skin).
This cord is joined by a major branch from the medial cord to form the median nerve.

From the medial cord:
5. Medial pectoral nerve (to pectoral muscles).
6 and 7. Medial cutaneous nerves of the arm and forearm. The former is the only brachial plexus nerve which comes to lie medial to the axillary vein.
The medial cord becomes the ulnar nerve.

From the posterior cord:
8, 9, 10. Upper subscapular nerve (to subscapularis), thoracodorsal nerve to latissimus dorsi muscle, and lower subscapular nerve (rest of subscapularis and whole of teres major).
11. The axillary nerve (to teres minor, deltoid, shoulder joint, and skin).
The posterior cord becomes the radial nerve.

(c). Long thoracic nerve (C5, 6, 7)

The long thoracic nerve (nerve to the serratus anterior) arises by three roots. The C5 and C6 roots pass through scalenus medius to join C7 and run to the digitations of serratus anterior behind the mid-axillary

(d). Suprascapular nerve (C5, 6)

This nerve runs with the suprascapular artery in front of scalenus medius to reach the suprascapular notch of the scapula. The nerve passes below the suprascapular ligament whilst the artery passes above it to enter the supraspinous fossa. It supplies the supraspinatus here, and then passing through the great scapular notch and supplies infraspinatus. [Fig. 5.03, p. 91]

(e). Phrenic nerve (C3, 4, 5)

Infraclavicular branches of the brachial plexus

(i). Lateral pectoral nerve (C5, 6, 7) and **medial pectoral nerve** (C8; T1)

The **lateral pectoral nerve** arises from the lateral cord on the lateral side of the axillary artery. It sends a loop around the artery to join the medial pectoral nerve. It then passes through the clavipectoral fascia to supply pectoralis major.

The **medial pectoral nerve** arises from the medial cord as it lies posterior to the axillary artery. The nerve runs between the axillary artery and axillary vein to form a loop with the lateral pectoral nerve which supplies pectoralis minor and pectoralis major.

(ii). Thoracodorsal nerve (C6, 7, 8)

The thorarcodorsal nerve (nerve to latissimus dorsi) arises from the posterior cord. It runs with the subscapular artery along the posterior wall of the axilla to supply the latissimus dorsi.

(iii). Subscapular nerves (C5, 6)

The upper and lower subscapular nerves arise from the posterior cord. They supply subscapularis. The lower scapular nerve also supplies teres major.

(iv). Medial cutaneous nerve of arm (T1)

The medial cutaneous nerve of the arm arises from the medial cord of the brachial plexus. It passes behind the axillary vein and runs down on its medial side to join the intercostobrachial nerve from T2 and supply the skin on the medial side of the arm.

(v). Medial cutaneous nerve of forearm (C8; T1)

The medial cutaneous nerve of the forearm also arises from the medial cord. It runs down the arm on the medial side of the brachial artery, pierces deep fascia at the middle of the arm on the medial side where the basilic vein enters, and divides into two branches. The anterior branch supplies the skin over the anterior and medial aspects of the limb down to the wrist. The posterior branch supplies the skin over the posterior and medial aspects of the forearm down to the wrist.

(vi). Axillary nerve (C5, 6)

The axillary (circumflex) nerve is one of the two terminal branches of the posterior cord, the other being the radial nerve [Fig. 4.22]. The axillary nerve runs between the axillary artery and subscapularis. It passes to the back of the humerus by passing through the quadrangular space [Fig. 5.03, p. 91]. This space is bounded by subscapularis, the capsule of the shoulder joint and teres minor

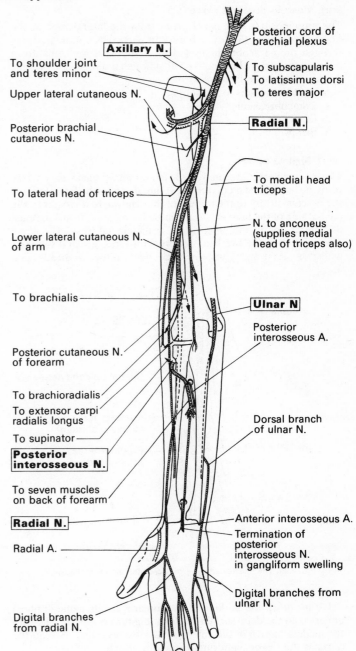

FIG. 4.22. The posterior cord of the brachial plexus, the radial and axillary nerves.

above, the surgical neck of the humerus on the lateral side, the tendon of teres major below, and the long head of triceps on the medial side. The axillary nerve then divides into anterior and posterior branches.

The anterior branch of the axillary nerve passes round the surgical neck of the humerus and supplies deltoid. The posterior branch supplies deltoid and teres minor and then becomes the upper lateral cutaneous nerve of the arm.

(vii). **Musculocutaneous nerve** (C5, 6, 7)

The musculocutaneous nerve arises from the lateral cord of the brachial plexus at the lower border of pectoralis minor. It passes through the coracobrachialis (which it supplies) and runs down between biceps and brachialis (both of which it supplies) to become the lateral cutaneous nerve of the forearm.

Musculocutaneous nerve supplies:
1. Coracobrachialis.
2. Brachialis.
3. Biceps.

(viii). **Median nerve**

The median nerve is formed on the lateral side of the axillary artery from the lateral and medial cords. It runs down on the lateral side of the brachial artery to the middle of the arm where it crosses to the medial side of the brachial artery (usually crosses in front) to reach the lateral side of the ulnar artery in the cubital fossa. The nerve passes between the two heads of pronator teres, which it supplies, whilst the ulnar artery usually passes deep to the two heads [Fig. 4.23].

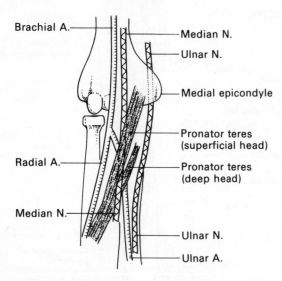

Fig. 4.23. The brachial artery, medial and ulnar nerves at the elbow.

It runs down the midline of the forearm (as its name implies) adherent to the deep surface of flexor digitorum superficialis with the median branch of the anterior interosseous artery. Behind the nerve is the flexor digitorum profundus muscle.

At the wrist the nerve lies between the tendons of flexor carpi radialis and palmaris longus and is very superficial. It then passes deep to the flexor retinaculum in company with the flexor synovial sheaths and breaks up into its medial and lateral divisions.

Branches of median nerve

The median nerve has no branches in the axilla or in the arm. In the forearm it supplies:
1. Pronator teres.
2. Flexor carpi radialis.
3. Flexor digitorum superficialis.
4. Palmaris longus.

The **anterior interosseous nerve** is given off from the median nerve as it passes between the two heads of pronator teres.

The anterior interosseous nerve supplies:
1. Flexor pollicis longus.
2. Flexor digitorum profundus (radial half).
3. Pronator quadratus.

The **median nerve** (and its anterior interosseous branch) thus supplies all the muscles on the front of the **forearm** with the exception of flexor carpi ulnaris and the lateral part of flexor profundus.

The **lateral division** (recurrent division) of the median nerve in the **palm** supplies:
1. Abductor pollicis.
2. Flexor pollicis brevis.
3. Opponens pollicis.

The **medial (recurrent) division** of the median nerve in the **palm** supplies:
1. Lateral two lumbricals.
2. Skin over palmar aspect of hand from thumb to ring finger.
3. Skin over dorsal aspect of distal phalanx of thumb, index, and middle fingers.

The **median nerve** thus supplies all the muscles of the **hand** lateral to the tendon of flexor pollicis longus and the skin of the central part of the palm, the palmar aspect of the lateral $3\frac{1}{2}$ digits and the dorsal aspect of the distal phalanx of the thumb and the two distal phalanges of the index, middle, and lateral half of ring fingers [Fig. 4.24].

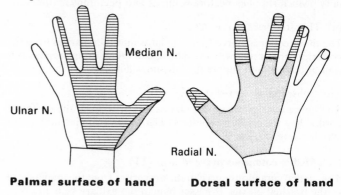

Fig. 4.24. Cutaneous sensory innervation of the hand.

(ix). **Ulnar nerve** (C8; T1)

The ulnar nerve is the continuation of the medial cord. It runs down the upper arm on the medial side of the brachial artery. At the middle of the arm it passes, with the ulnar collateral artery, through the medial intermuscular septum to the back of the arm. It runs down on the medial head of triceps to reach the groove on the lower end of the humerus between the medial epicondyle and the olecranon. The nerve here can be readily rolled under the finger [Fig 4.23].

The ulnar nerve returns to the front of the forearm by passing round the medial side of the elbow joint between the two heads of the flexor carpi ulnaris. In the forearm it passes deep to the flexor carpi ulnaris, on flexor profundus and one-third the way down the forearm meets the ulnar artery. The ulnar nerve lies medial to the ulnar artery for the rest of its course. The ulnar nerve (and ulnar artery) pass superficial to the flexor retinaculum to enter the palm. The nerve runs on the lateral side of the pisiform bone and its superficial branch in the hand passes in front of the hook of the hamate. Pressing the pisiform outwards stimulating the nerve, causes contraction of the palmaris brevis and wrinkling of the skin on the inner aspect of the palm.

Branches of the ulnar nerve

The ulnar nerve supplies the following muscles:

1. Flexor carpi ulnaris.
2. Flexor digitorum profundus (medial half).
3. Palmaris brevis.
4. All eight interossei.
5. Adductor pollicis.
6. Two medial lumbricals.
7. Abductor digiti minimi.
8. Flexor digiti minimi.
9. Opponens digiti minimi.

The palmar cutaneous branch of the ulnar nerve supplies the skin of the inner side of the palm. The dorsal cutaneous branch supplies the skin on the ulnar side of the little finger and adjacent sides of little and ring fingers. The superficial branch in the palm supplies cutaneous branches to the little finger and ulna side of ring finger.

The dorsal cutaneous branches come off 8–10 cm above the wrist. It can be pressed against the medial side of the triquetrum bone. If the ulnar nerve is cut at the wrist the dorsal branch escapes and the skin sensation on the back of the hand is unaffected.

(x). Radial nerve (C5, 6, 7, 8; T1)

The radial nerve is the continuation of the posterior cord of the brachial plexus which, as has been seen, is formed by the union of all three posterior divisions. The radial nerve is thus the largest branch of the brachial plexus [FIG. 4.22].

The radial nerve is formed behind the third part of the axillary artery and runs down at first behind the brachial artery. Then accompanied by the profunda brachii artery it curves laterally between the long and medial heads of triceps to enter the musculospinal groove of the humerus [FIG. 5.03, p. 91]. Here it lies in close contact with bone under cover of the lateral head of triceps. It winds round the back of the humerus to reach the lateral septum which it pierces to enter the front of the arm. It lies between brachioradialis and brachialis then passes in front of the lateral epicondyle. It then gives off the posterior interosseous nerve.

The radial nerve in the arm supplies:

1. Triceps.
2. Anconeus.
3. Brachialis (main innervation is musculocutaneous).
4. Brachioradialis.
5. Extensor carpi radialis longus.

Fracture of the shaft of the humerus is often complicated by paralysis of the radial nerve. The triceps largely escapes, but all the muscles on the back of the forearm are paralysed giving rise to 'wrist drop'.

The posterior interosseous nerve

The posterior interosseous nerve runs from just below the lateral epicondyle round the neck of the radius to the back of the forearm embedded in the supinator. It then breaks up into many branches which makes repair of any injury to this nerve difficult. It is distributed to all the muscles on the back of the forearm (except anconeus).

The posterior interosseous nerve supplies:

1. Extensor carpi radialis brevis.
2. Supinator.
3. Extensor digitorum.
4. Extensor digiti minimi.
5. Extensor carpi ulnaris.
6. Abductor pollicis longus.
7. Extensor pollicis brevis.
8. Extensor pollicis longus.
9. Extensor indicis.

Sensory innervation of radial nerve

The posterior cutaneous nerve of the arm arises from the radial nerve in the lower axilla. It supplies the skin over the medial and posterior aspects of the arm as far down as the olecranon.

The lower lateral cutaneous nerve of the arm arises from the radial nerve in the musculospiral groove. It divides into two branches. The upper branch supplies the skin over the lateral and anterior aspects of the lower half of the arm. The lower branch supplies the skin at the back of the forearm as far down as the wrist.

The radial nerve itself continues down the forearm in front of the head of the radius to meet the radial artery in the middle of the forearm. The radial nerve lies on the outer side of the radial artery until, at about 7 cm above the wrist, the radial nerve passes backwards under the tendon of brachioradialis to become subcutaneous. It divides into two branches. The lateral branch supplies the skin on the outer aspect of the thumb. The medial branch supplies the adjacent sides of the thumb, index, middle, and ring fingers. The radial nerve supplies the back of the fingers with the exception of the distal phalanx of the thumb and the middle and distal phalanges of the fingers which are supplied by the median and ulnar nerves. Two or three branches can be rolled on the tendon of extensor pollicis longus when the thumb is extended (recall dorsal cutaneous branch of ulnar nerve *above*).

SUMMARY

The three main nerves of the upper limb, the ulnar, radial, and median are so named according to their relative positions in the forearm (not the upper arm). The ulnar nerve runs superficially adjacent to the ulnar artery for part of its route on the ulnar side of the forearm. The radial nerve runs superficially adjacent to the radial artery for part of its route on the radial side of the forearm. The median nerve runs down the middle of the forearm between the superficial flexors of the fingers (flexor digitorum superficialis) and the deep flexors (flexor digitorum profundus).

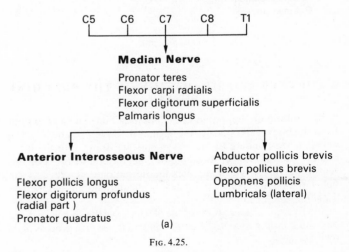

FIG. 4.25.

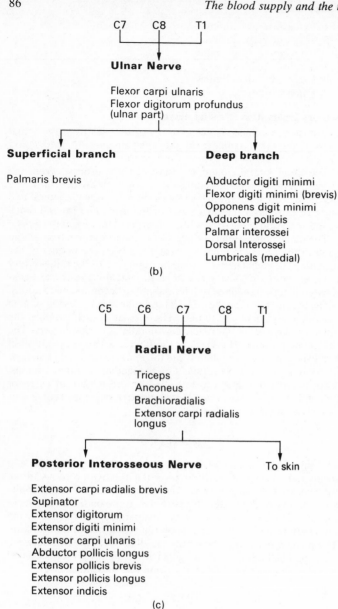

Ulnar Nerve

Flexor carpi ulnaris
Flexor digitorum profundus
(ulnar part)

Superficial branch

Palmaris brevis

Deep branch

Abductor digiti minimi
Flexor digiti minimi (brevis)
Opponens digit minimi
Adductor pollicis
Palmar interossei
Dorsal Interossei
Lumbricals (medial)

(b)

Radial Nerve

Triceps
Anconeus
Brachioradialis
Extensor carpi radialis
longus

Posterior Interosseous Nerve

Extensor carpi radialis brevis
Supinator
Extensor digitorum
Extensor digiti minimi
Extensor carpi ulnaris
Abductor pollicis longus
Extensor pollicis brevis
Extensor pollicis longus
Extensor indicis

To skin

(c)

FIG. 25 (contd).

A FURTHER CONSIDERATION OF THE BRACHIAL PLEXUS

In the embryo the upper limb develops as a 'bud' which grows out from the body wall opposite the segments C5, 6, 7, 8 and T1. The nerves corresponding to these segments are drawn into the limb, and are then no longer available to innervate the trunk. For this reason nerve C4 which supplies the dermatome of skin 3–4 cm below the clavicle comes to lie next to the dermatome supplied by T2 [FIG. 4.26].

The five spinal nerve roots of the plexus leave the spinal cord and pass out laterally each in the gutter-shaped transverse process of its own vertebra.

The upper two nerves (C5 and C6) join each other, as do the lower two (C8 and T1). The middle nerve root (C7) remains on its own. In this way five nerve roots are reduced to three nerve trunks. Notice the symmetry of this pattern.

· Each trunk then divides into an anterior and a posterior division. The three anterior divisions between them supply the anterior or flexor muscles, while the three posterior divisions between them supply the posterior or extensor muscles [FIG. 4.20, p. 81].

All three posterior divisions unite into a single common extensor nerve, the posterior cord, which comes to lie behind the second part of the axillary artery. It supplies directly the three muscles in the posterior wall of the axilla, the subscapularis, teres major, and latissimus dorsi and gives off the axillary nerve which runs *backwards* around the neck of the humerus to supply teres minor and the deltoid. It then changes its name to the radial nerve and supplies the triceps and all the muscles on the back of the forearm via its posterior interosseous branch. Just to spoil the perfection of the rule, the radial nerve also supplies a small part of the brachialis and the brachioradialis muscles which are elbow flexors. Since it contains all three posterior divisions, the posterior cord contains fibres from all the roots of the plexus, that is it contains C5, 6, 7, 8 and T1.

The basic simplicity of the brachial plexus would be more obvious if the three anterior divisions were also all to join to form a single common flexor 'anterior cord', balancing the posterior cord. But instead there are two large nerves, the lateral and medial cords of the brachial plexus. Between them they supply all the anterior or flexor muscles such as the biceps and brachialis, the muscles on the anterior aspect of the forearm and the intrinsic muscles of the hand. The cords are named from their relationship to the second part of the axillary artery.

The upper trunk must remain on the lateral side of the subclavian and axillary arteries and contribute to the lateral cord. It is also evident that T1 and C8 will go to the medial side and contribute to the medial cord. The only hesitation is in working out the destiny of the anterior division of C7. Does it go into the lateral or the medial cord? Recall that the cervical nerves lie above their corresponding vertebrae. A glance at the skeleton shows that nerve C7 lies in the posterior triangle above the subclavian artery (like C5 and 6) and that the simplest thing is for it to join C5 and 6. So the lateral cord has fibres from C5, 6, and 7, the medial cord has C8 and T1. Between them they balance the posterior cord, which has C5, 6, 7 plus C8 and T1.

C5, 6, 7 lateral cord + C8, T1 medial cord	posterior cord C5, 6, 7, 8, T1
Flexor muscles	Extensor muscles

Note that all the branches of the lateral cord have the same root value (C5, 6, 7) as the lateral cord. All branches of the medial cord have the same root value as the medial cord (C8; T1).

The lateral cord or one of its branches often contributes fibres from C7 to the medial cord or to the ulnar nerve. Such nerve fibre interchange does not take place between the lateral or medial cords and the posterior cord.

It is simplest to regard the contribution of the lateral cord to the median nerve and the median nerve itself, as the direct continuation of the lateral cord, because the median nerve lies on the lateral side

Stage 1

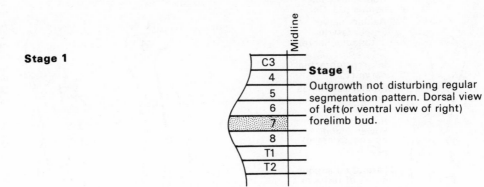

Stage 1

Outgrowth not disturbing regular segmentation pattern. Dorsal view of left (or ventral view of right) forelimb bud.

Stage 2

Outgrowth

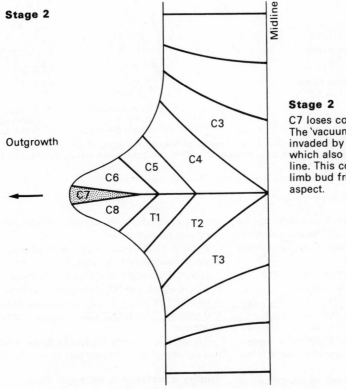

Stage 2

C7 loses contact with midline. The 'vacuum' so formed is then invaded by adjacent segments, which also lose contact with midline. This could also be view of limb bud from dorsal or ventral aspect.

FIG. 4.26. Dermatome formation and limb-bud segmentation.

Stage 1: Outgrowth not disturbing regular segmentation pattern. Dorsal view of left or ventral view of right forelimb bud.

Stage 2: C7 loses contact with midline. The 'vacuum' so formed is then invaded by adjacent segments. which also lose contact with midline. This could also be view of limb bud from dorsal or ventral aspect.

The regular segmentation of the trunk is disturbed by the outgrowth of the limbs. The upper limb outgrowth is pioneered by C7. As a result C6 and C8. and C5 and T1 supply adjacent skin areas. also C4 and T2. The dorsal and ventral axial lines coincide with continuity regions.

The development in the lower limb is similar to that in the upper. but complicated by complex morphogenetic processes not found in the upper limb. The dorsal and ventral axial lines come to lie close to each other in the back of the thigh.

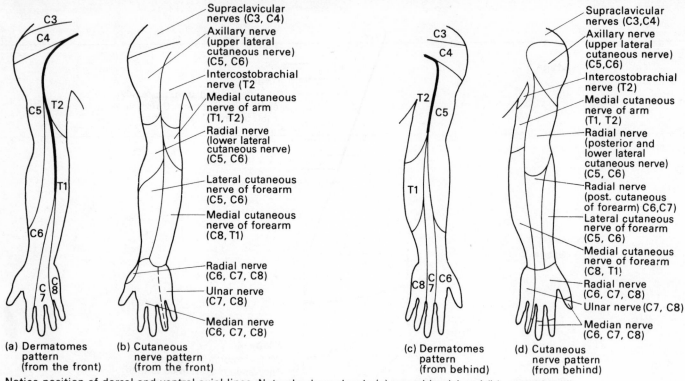

(a) Dermatomes pattern (from the front)

(b) Cutaneous nerve pattern (from the front)

Supraclavicular nerves (C3, C4)
Axillary nerve (upper lateral cutaneous nerve) (C5, C6)
Intercostobrachial nerve (T2
Medial cutaneous nerve of arm (T1, T2)
Radial nerve (lower lateral cutaneous nerve) (C5, C6)
Lateral cutaneous nerve of forearm (C5, C6)
Medial cutaneous nerve of forearm (C8, T1)
Radial nerve (C6, C7, C8)
Ulnar nerve (C7, C8)
Median nerve (C6, C7, C8)

(c) Dermatomes pattern (from behind)

(d) Cutaneous nerve pattern (from behind)

Supraclavicular nerves (C3,C4)
Axillary nerve (upper lateral cutaneous nerve) (C5,C6)
Intercostobrachial nerve (T2)
Medial cutaneous nerve of arm (T1, T2)
Radial nerve (posterior and lower lateral cutaneous nerve) (C5, C6)
Radial nerve (post. cutaneous of forearm) C6,C7
Lateral cutaneous nerve of forearm (C5, C6)
Medial cutaneous nerve of forearm (C8, T1)
Radial nerve (C6, C7, C8)
Ulnar nerve (C7, C8)
Median nerve (C6, C7, C8)

Notice position of dorsal and ventral axial lines. Note also how closely (a) resembles (c) and (b) resembles (d). Figures (a) (c) show segmental nerves supplying skin areas, Figures (b) (d) show cutaneous nerve distribution.

FIG. 4.27. Dermatome pattern and cutaneous nerve supply of upper limb.

of the third part of the axillary artery in a position which exactly corresponds to that of the lateral cord higher up. The ulnar nerve is the direct continuation of the medial cord, while the radial nerve is the direct continuation of the posterior cord.

Thus each of the three cords is represented by one of the three big nerves which supply the upper limb below the axilla.

The brachial plexus also contains sympathetic nerve fibres destined for the upper limb, especially the skin and sweat glands.

Before the five nerve roots have joined to form the three trunks, branches come off to supply the muscle in the vertebral compartment of the neck, the longus capitis and cervicis, the scalene muscles as well as serratus anterior.

Two branches come off from the upper trunk of the plexus at **Erb's point**. One is the suprascapular nerve (to supraspinatus and infraspinatus). The other, the nerve to subclavius, runs downwards across the front of the brachial plexus to the subclavius muscle. This nerve often has a branch which communicates with the phrenic nerve somewhere in the thorax, and this branch is called the accessory phrenic nerve.

A blow at the root of the neck (karate chop) compresses the upper trunk against the vertebral column and produces an Erb's paralysis. Pulling the arm forcibly downwards may also produce an Erb's paralysis. The fifth (and possibly the sixth) cervical nerve root is involved affecting biceps, brachialis, brachioradialis, supinator, and deltoid. The upper limb hangs by the side with the forearm pronated.

The middle and inferior trunks do not give off any side branches, nor do any of the anterior and posterior divisions.

Sensory innervation of the upper limb

FIGURE 4.27 shows the dematomes pattern and sensory nerve pattern of the upper limb. Note how clearly the anterior and posterior patterns resemble each other.

5 The muscles and joints of the upper limb

THE SHOULDER JOINT

The **shoulder joint** is formed between the hemispherical head of the **humerus** and the shallow glenoid cavity of the **scapula** which is deepened by the glenoid lip. It is a ball-and-socket joint which, due to the shallowness of the socket and the laxity of the capsule, allows a very great freedom of movement [FIG. 5.01].

The capsule is attached to the base of the scapula just proximal to the glenoid lip. It is attached to the humerus just distal to the articular cartilage except on the medial side where the attachment runs further down to the level of the surgical neck. This attachment is distal to the epiphyseal line so that a slipped epiphysis of the head of the humerus will involve the shoulder joint.

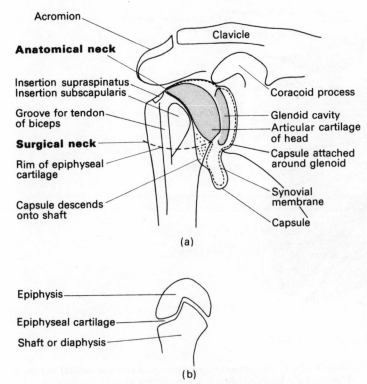

FIG. 5.01. Shoulder joint.
The figure shows the position of the anatomical and surgical necks of the humerus. The latter is a common fracture site.
The epiphyseal cartilage appears to run around the surgical neck in the horizontal plane. But a coronal section through the bone shows that the cartilage is not a flat disc, but is cone-shaped. This is the appearance seen on X-ray in children (b).

An epiphysis is formed by a secondary centre of ossification after the main bone (diaphysis) has ossified. The epiphysis and diaphysis are separated by the epiphyseal cartilage (metaphysis) at the epiphyseal line [see p. 23].

The bone increases in length by new bone being pushed down on the diaphyseal side of the epiphyseal line. A slipped epiphysis in a child is the separation of the epiphysis from the diaphysis. It usually takes place as a fracture through the new bone so that the epiphyseal cartilage goes with the epiphysis itself.

The epiphysis at the upper end of the humerus has three ossification centres (for the head, greater tubercle, and lesser tubercle). These unite to form a single epiphysis by six years of age. The epiphyseal line then runs transversely across at the lowest point of the cartilage of the head.

The epiphysis fuses with the shaft at 18–25 years and then the arm ceases to increase in length. (There are also epiphyses at the lower end which fuse with the shaft at 17 years of age.)

The humerus is unique as a bone in that it has both an anatomical and a surgical neck. The anatomical neck is the constricted portion of the head of the humerus just below the articular cartilage. The surgical neck is the constricted part of the shaft of the humerus just distal to the tubercles (where fractures occur).

The synovial membrane lining the capsule surrounds the biceps tendon with a tube of synovial membrane.

Scapula

The scapula at the back covers the ribs from the second to the eighth. It is very mobile as can be seen by moving the shoulder girdle.

Have a look at a scapula bone. The flat surface is anterior, and is in contact with the chest wall at the back. It is covered and cushioned by the subscapularis muscle [FIG. 5.02]. The posterior surface of the scapula has the prominent spine which leads laterally to the **acromion** (= the highest point of the shoulder). The **supra**spinatus muscle lies above the spine. The **infra**spinatus muscle lies below the spine. The tendons of all three muscles closely envelop the shoulder joint [FIG. 5.03].

The tendon of supraspinatus runs over the top of the shoulder joint and is incorporated into the capsule. Contraction of this muscle coincides with the start of abduction. Because it lies so close to the shoulder joint its leverage is poor. The deltoid is the principal abductor of the shoulder joint with good leverage. The head of the humerus must remain in position in the glenoid cavity when the deltoid first contracts. The upper part of the capsule, strengthened by the supraspinatus tendon, does this during the initial stages. Once abduction too has started, the rotator cuff muscles take over as fixators as soon as the upper capsule slackens. If this part of the capsule, with its associated tendon is damaged, the mechanism of the initial stage of abduction is destroyed. Patients quickly learn to

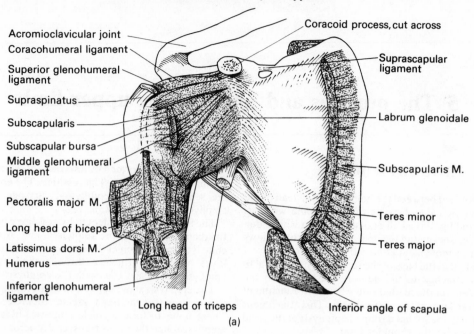

Acromioclavicular joint
Coracohumeral ligament
Superior glenohumeral ligament
Supraspinatus
Subscapularis
Subscapular bursa
Middle glenohumeral ligament
Pectoralis major M.
Long head of biceps
Latissimus dorsi M.
Humerus
Inferior glenohumeral ligament
Long head of triceps

Coracoid process, cut across
Suprascapular ligament
Labrum glenoidale
Subscapularis M.
Teres minor
Teres major
Inferior angle of scapula

(a)

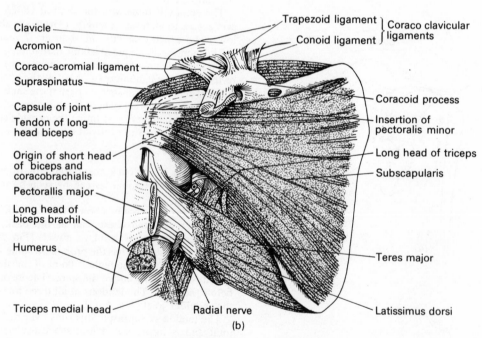

Clavicle
Acromion
Coraco-acromial ligament
Supraspinatus
Capsule of joint
Tendon of long head biceps
Origin of short head of biceps and coracobrachialis
Pectorallis major
Long head of biceps brachil
Humerus
Triceps medial head

Trapezoid ligament ⎫ Coraco clavicular
Conoid ligament ⎬ ligaments
Coracoid process
Insertion of pectoralis minor
Long head of triceps
Subscapularis
Teres major
Latissimus dorsi
Radial nerve

(b)

FIG. 5.02. Right shoulder joint (anterior view).

get the joint moving into abduction by trick movements such as either by leaning sideways or even by using the other hand.

The scapula which is triangular in shape has superior, medial, and inferiolateral borders. Muscles are attached to these borders to move the scapula up and down and forwards and backwards, and rotation upwards so that the glenoid cavity points upward and rotation downwards so that the glenoid cavity points downwards. The important muscles are trapezius and serratus anterior. Serratus anterior which runs between the chest-wall and the medial border of the scapula by many slips is an important muscle for moving the scapula round the chest when lunging, as in fencing. The powerful lower fibres of serratus anterior assisted by trapezius are used to rotate the scapula when raising the arm above the head.

The two scapulae are pulled together when standing to attention by trapezius, levator scapulae, rhomboid minor, and rhomboid major which are also attached to the medial border.

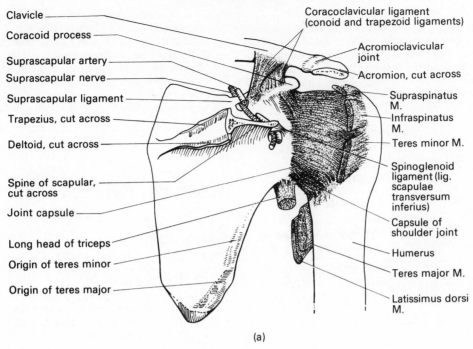

Clavicle

Coracoid process

Suprascapular artery

Suprascapular nerve

Suprascapular ligament

Trapezius, cut across

Deltoid, cut across

Spine of scapular, cut across

Joint capsule

Long head of triceps

Origin of teres minor

Origin of teres major

Coracoclavicular ligament (conoid and trapezoid ligaments)

Acromioclavicular joint

Acromion, cut across

Supraspinatus M.

Infraspinatus M.

Teres minor M.

Spinoglenoid ligament (lig. scapulae transversum inferius)

Capsule of shoulder joint

Humerus

Teres major M.

Latissimus dorsi M.

(a)

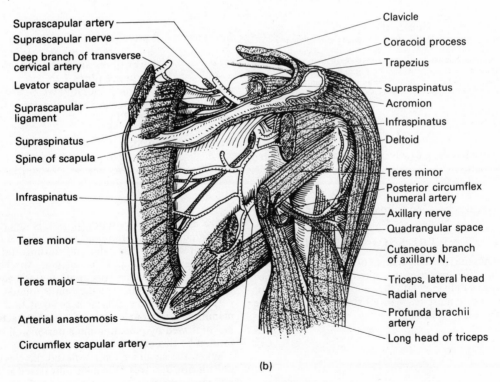

Suprascapular artery

Suprascapular nerve

Deep branch of transverse cervical artery

Levator scapulae

Suprascapular ligament

Supraspinatus

Spine of scapula

Infraspinatus

Teres minor

Teres major

Arterial anastomosis

Circumflex scapular artery

Clavicle

Coracoid process

Trapezius

Supraspinatus

Acromion

Infraspinatus

Deltoid

Teres minor

Posterior circumflex humeral artery

Axillary nerve

Quadrangular space

Cutaneous branch of axillary N.

Triceps, lateral head

Radial nerve

Profunda brachii artery

Long head of triceps

(b)

FIG. 5.03. Right shoulder joint (posterior view).

Ligaments of the shoulder joint [FIGS. 5.02, 5.03]

The coracohumeral ligament runs from the base of the coracoid process to the greater and lesser tubercles and to the transverse ligament between them. The glenohumeral ligaments are thickenings of the capsule in front of the shoulder joint.

The coraco-acromial ligament overhangs the shoulder joint and prevents upward displacement. It may be considered to be forming a secondary socket for the humerus [FIGS. 5.02(b), 5.04].

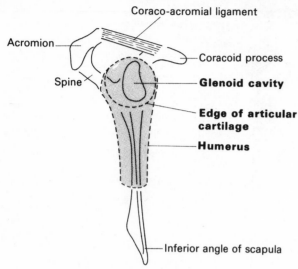

FIG. 5.04. Right shoulder: lateral view.
The right scapula is viewed from the lateral side. Superimposed on it is the outline of the humerus with the articular cartilage of its head shown by the dotted lines. Note the comparatively small size of the glenoid, which is a major contributory factor in the aetiology of dislocations of this joint. The shoulder joint is guarded by the acromion, the coracoid process, and especially by the very powerful coraco-acromial ligament.

Bursae of the shoulder joint

Subscapular bursa

The subscapular bursa lies deep to subscapularis. It communicates with the shoulder joint.

Subdeltoid bursa

The subdeltoid bursa lies between deltoid and the joint capsule.

MOVEMENTS AT THE SHOULDER JOINT

The shoulder joint has the freest movement of any joint in the body with movement in every direction permitted. For convenience these movements are subdivided into the following, but this is for convenience only and it must be realized that all intermediate movements are possible [see p. 41].

Flexion.
Extension.
Abduction.
Adduction.
Circumduction.
Medial rotation.
Lateral rotation.

To study movement at the shoulder joint alone, the scapula must be fixed. Otherwise movements at the shoulder joint are nearly always associated with movements of the scapula on the posterior thoracic wall. Thus when the arm is abducted, the scapula rotates at the same time as the shoulder joint abducts. In raising the arm above the head the humerus rotates laterally.

Rotation around the long axis of the humerus is limited to 90°. Bend your elbow and try the rotational movements of the shoulder joint on yourself.

The individual muscles involved will now be considered in more detail.

MUSCLES UNITING UPPER LIMB TO TRUNK

Pectoralis major

Origin:	(i) Clavicular head: Anterior surface of inner two-thirds of clavicle (ii) Sternocostal head: Anterior surface of sternum. Upper six ribs (costal cartilages) Rectus sheath

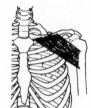

Insertion:	Lateral lip of intertubercular groove (bicipital groove). The lower fibres are inserted deep to the upper fibres
Nerve supply:	Lateral pectoral nerve (C5, 6, 7) and medial pectoral nerve (C8; T1)
Action:	Adducts, flexes or extends, and medially rotates arm. If shoulder girdle is fixed, it can be used as an accessory muscle of respiration drawing the thoracic cavity upwards

In the male pectoralis major [p. 190] is superficial except where it is covered by the anterior border of deltoid. A furrow can usually be seen between the clavicular and sternocostal parts of the muscle when the muscle is put into action. Another furrow, this time between the deltoid and pectoralis major leads to the infraclavicular hollow where the cephalic vein is found.

In the female the muscle is covered to a large extent by the mammary gland, but in both sexes the contraction of the muscle can be confirmed by palpating the anterior wall of the axilla which is formed by this muscle.

When the shoulder joint is flexed the clavicular (upper) fibres pull the arm upwards. When the fully flexed arm is extended against resistance, the sternocostal (lower) head contracts. Confirm this in yourself. Both heads contract simultaneously when the arms are adducted. Put hands on hips, and then squeeze inwards and the whole muscle becomes taught. The latter method is made use of clinically when examining the breast.

Pectoralis minor

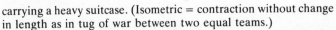

Origin: Third, fourth, and fifth ribs
Insertion: Coracoid process of scapula
Nerve supply: Medial pectoral (C7, 8; T1)
Action: Either draws coracoid process
                   forward and downwards or draws
                   ribs upwards

If the arm is raised above the head pectoralis minor (p. 190) produces a slight swelling below the edge of pectoralis major.

Subclavius

Origin: First rib (junction of first costal
                   cartilage and bony rib)
Insertion: Clavicle on its under surface
                   (subclavian groove)
Nerve supply: Nerve to subclavius (C5, 6)
Action: Draws clavicle towards first rib and sternum
                   Depresses lateral end of clavicle

It is a relatively unimportant muscle since paralysis produces no demonstable effect.

Serratus anterior

Origin: First to eighth ribs. The
                   muscle arises by digitations
                   from the anterior ends of the
                   bony ribs. The lower five
                   interdigitate with external
                   oblique
Insertion: Vertebral border of the
                   scapula
Nerve supply: Nerve to serratus anterior
                   (C5, 6, 7) previously known
                   as the long thoracic nerve
                   of Bell
Action: It draws the scapula round
                   the chest when pushing with
                   the arm flexed. The very
                   powerful lower digitations
                   rotate the scapula upwards

This muscle is important in movements which involve stretching and reaching. It has been termed the fencer's muscle. If it is paralysed the scapula become 'winged', that is it no longer lies flat against the chest-wall. In minor degrees of paralysis winging may not be obvious at rest, but as soon as the arms are raised, especially against resistance, the medial scapular border stands out very obviously.

When the arm is raised above the head the movement takes place not only by abduction of the shoulder joint but also by rotation of the scapula so that the glenoid cavity faces upwards. The serratus anterior and the trapezius muscle work in consort in scapula rotation. The serratus anterior digitations are not evenly distributed along the vertebral border of the scapula as you might expect. Instead five of them (absolute majority out of eight digitations) are focused on the inferior angle. Here they get the best possible leverage.

These fibres are not only of importance for their isotonic action in such manoeuvres as brushing the back of the hair, but also for their isometric action in maintaining the position of the scapula when carrying a heavy suitcase. (Isometric = contraction without change in length as in tug of war between two equal teams.)

The upper digitations of serratus anterior are hidden by pectoralis major, latissimus dorsi, and the scapula. The lower four digitations are visible as characteristic finger-like ridges on the side of the chest-wall.

The serratus anterior forms the medial wall of the axilla, covering the ribs and intercostal muscles.

Latissimus dorsi

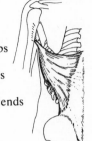

Origin: (i)   Vertebral spines, T7–S2
                   (ii) Iliac crest (posterior part
                         via thoracolumbar fascia
                   (iii) Slips from ninth to twelfth ribs
                   (iv) Slip from angle of scapula
Insertion: Intertubercular groove of humerus
Nerve supply: Thoracodorsal nerve (C6, 7, 8)
Action: Adducts, medially rotates and extends
                   the arm

This muscle which extends from the arm to the pelvis has a very high innervation. If the arm is fixed it pulls the trunk to the arm. Following a spinal transection where the lower part of the body is paralysed, it may be possible to develop this muscle to such an extent to swing the pelvis, that walking with crutches becomes possible. Archery is a sport used to develop this muscle.

The outer border of the muscle forms a ridge on the side of the trunk extending up to the posterior wall of the axilla which is formed by this muscle and teres major. To put the muscle strongly into action, draw the uplifted arm downwards, backwards, and inwards.

This muscle also acts as an accessory muscle of expiration, as acting with origin and insertion reversed it squeezes the rib-cage. In testing this muscle put your hand on the posterior wall of the axilla and ask your patient to cough. The contraction of this muscle should be felt.

The muscle lies behind the origin of teres major at the inferior angle of the scapula, and the tendon twists round it so that it is below the middle third and in front of the lateral third, and is inserted just in front of the insertion of the teres major.

Levator scapulae

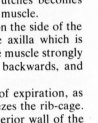

Origin: First to fourth cervical vertebrae
                   Tubercles of the transverse
                   processes of the first and second
                   cervical vertebrae. Posterior
                   tubercles of third and fourth
                   cervical vertebrae
Insertion: Vertebral border of scapula
                   (above base of spine)
Nerve supply: C3 and 4 and dorsal scapular
                   nerve C5
Action: Raises the scapula and hence
                   supports the upper limb

This muscle forms a substantial part of the floor of the posterior triangle of the neck and can be palpated here.

Trapezius

Origin:	(i) SKULL—External occiput protuberance and inner third of superior nuchal line
	(ii) Ligamentum nuchae running from skull to spine of C7
	(iii) Vertebral spines from C7–T12 and supraspinous ligaments between them
Insertion:	(i) Upper border of spine of scapula and tubercle on lower border
	(ii) Acromion (inner border)
	(iii) Clavicle on posterior border of outer third
	All three attachments are continuous with one another
Nerve supply:	XI accessory nerve C3, 4
Action:	Upper fibres pull the scapula and clavicle upwards
	Lower fibres pull the scapula downwards
	Whole muscle pulls scapula to midline as in standing to 'attention'
	Trapezius is also used to rotate the scapula as when raising the arm above the head. The upper fibres working to support the weight of the upper limb
	It is very well developed in oarsmen

Once the lower borders have been made out, the two muscles lying superficially throughout, have the shape of a trapezium, hence the name.

Serratus posterior superior

Origin:	Ligamentum nuchae and spines of C7–T3 vertebrae
Insertion:	Ribs 2–5 (near angles)
Nerve supply:	T2–5 intercostal nerves

This muscle lies deep to the rhomboids.

Serratus posterior inferior

Origin:	Spines of T11–L2 vertebrae via the thoracolumbar fascia
Insertion:	Ribs 9–12
Nerve supply:	T9–11 intercostal nerves

This muscle lies deep to latissimus dorsi.

The use of the term serratus anterior for a muscle implies that there must be a serratus posterior muscle. There are in fact two, serratus posterior superior and serratus posterior inferior. They are often paper thin and are relatively unimportant.

Rhomboid Minor

Origin:	Lower part of ligamentum nuchae Spines of C7 and T1 vertebrae
Insertion:	Vertebral border of scapula (base of spine)
Nerve supply:	Dorsal scapular nerve C5

Rhomboid Major

Origin:	Spines, and supraspinous ligaments of T2–5 vertebrae
Insertion:	Vertebral border of scapula (below base of spine)
Nerve supply:	Dorsal scapular nerve C5
Action:	Retracts the scapula
	Rotates the scapula so that glenoid cavity points downwards

It will be noted that levator scapulae, rhomboid minor, and rhomboid major form a deep layer of muscles attached to the vertebral border of the scapula. They retract and rotate the scapula.

The muscles when contracting show slightly through the overlying trapezius.

MUSCLES OF THE SHOULDER REGION

Deltoid

Origin:	Outer third of anterior surface of clavicle
	Outer border of acromion
	Lower border of spine of scapula
Insertion:	Deltoid tuberosity of humerus
Nerve supply:	Axillary nerve (C5, 6)
Action:	Abducts the humerus
	The anterior fibres acting alone flex the arm, that is, move it forwards
	The posterior fibres extend the arm

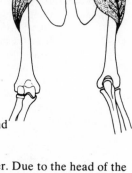

The deltoid forms a cap over the shoulder. Due to the head of the humerus projecting slightly forwards, the anterior part is rounder and the posterior part is flatter. This characteristic roundness in front is completely lost when shoulder dislocation occurs. The deltoid is an extremely important antigravity muscle of the shoulder joint. Without it abduction is virtually impossible.

Supraspinatus

Origin:	Medial third of supraspinous fossa of scapula
Insertion:	Greater tubercle of humerus (uppermost facet)
Nerve supply:	Suprascapular (C5, 6)
Action:	Abducts the humerus

The tendon of supraspinatus is incorporated into the capsule of the shoulder joint. It is said to start the movement of abduction before the deltoid takes over.

The tendon of the supraspinatus and the associated capsule of the shoulder joint act as a ligament which is in tension when the arm is at the side. This prevents upward displacement of the head of the humerus when the deltoid contracts. Once the head begins to rotate the ligament is relaxed, and upward movement of the head is controlled by the rotator cuff muscles especially the lower fibres of the subscapularis and infraspinatus.

The upper capsule and supraspinatus tendon may atrophy in a diseased shoulder joint. This destroys the mechanism of initiating shoulder joint abduction and makes this very important antigravity movement impossible.

Infraspinatus

Origin: Infraspinous fossa of scapula
Insertion: Greater tubercle of humerus (middle facet)
Nerve supply: Suprascapular nerve (C5, 6)
Action: Lateral rotator of the humerus, fixator of head of humerus

Teres Minor

Origin: Infraspinous fossa of scapula (near axillary margin)
Insertion: Greater tubercle of humerus (lowest facet)
Nerve supply: Axillary nerve (C5, 6)
Action: Lateral rotator of the humerus, fixator of head

The tendon of teres minor is closely adherent to the posterior aspect of the capsule of the shoulder joint.

Subscapularis

Origin: Subscapular fossa of scapula
Insertion: Lesser tubercle of humerus
Nerve supply: Subscapular nerves (C5, 6, 7)
Action: Medial rotator of the humerus

The tendon of subscapularis lies close to the anterior aspect of the capsule of the shoulder joint. The muscle may have to be elongated by slow lateral rotation before reducing a dislocation.

Teres major

Origin: Lower part of lateral border and inferior angle of scapula
Insertion: Medial lip of intertubercular groove of humerus
Nerve supply: Lower subscapular nerve
Action: Adducts and medially rotates humerus

Teres major is a thick round muscle which forms part of the posterior axillary wall. The latissimus dorsi tendon twists round it.

MUSCLES OF THE ARM

The muscles of the arm, with the exception of coracobrachialis, run to the forearm and as a result act mainly on the elbow joint. They are divided by the lateral and medial intermuscular septa into an anterior **flexor** and a posterior **extensor compartment**. The anterior group, biceps, coracobrachialis, and brachialis is supplied by the musculocutaneous nerve. The posterior group is composed only of the three heads of triceps supplied by the radial nerve.

Biceps brachii

Origin: Two heads. The long head arises from the scapula just above the glenoid cavity. The short head arises from the tip of the coracoid process of the scapula
Insertion: Posterior part of tuberosity of the radius. There is also an accessory insertion through the bicipital aponeurosis to the deep fascia over pronator teres. This helps to spread out the pull of the muscle in elbow flexion on to the medial aspect of the forearm
Nerve supply: Musculocutaneous (C5, 6, 7)
Action: Flexes the elbow. Supinates the forearm

The tendon of the long head runs inside the shoulder joint and in the intertubercular groove (bicipital groove) of the humerus. Note that the lower tendon runs round the medial side of the radius to be inserted into the radial tuberosity. The action of the muscle will be to rotate the radius and thus bring about supination in addition to flexing the elbow. It has been referred to as the wine-waiter's muscle. It is used to twist the corkscrew in and then to pull the cork out (provided that one is right-handed!).

The **biceps jerk** is a stretch reflex. Since the biceps tendon cannot readily be struck with a patellar hammer the following indirect method is usually employed. With the patient's elbow flexed the examiner presses his thumb onto the tendon of biceps in the cubital fossa. He then hits his own thumb with the patellar hammer which causes the muscle to stretch. Normally the elbow joint flexes momentarily. The smallest invisible twitch can easily be palpated by this technique.

Coracobrachialis

Origin: Tip of coracoid process of scapula
Insertion: Middle of humerus on medial side
Nerve supply: Musculocutaneous nerve (C5, 6, 7)
Action: Flexes and adducts the shoulder. It corresponds to the adductor muscles of the hip joint

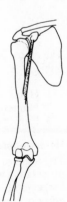

Brachialis

Origin: Lower two-thirds of anterior surface of humerus

Insertion: Ulna, in front of coronoid process

Nerve supply: Musculocutaneous nerve (C5, 6, 7) with anomalous contribution from radial nerve

Action: The flexor of the elbow joint

Because of its insertion only into the ulna it is not involved in pronation and supination like the biceps muscle. Brachialis is in close contact with anterior part of the capsule of elbow joint. It is stripped from the ulna in dislocation of the elbow joint, and myositis ossificans is a common complication in this muscle.

Triceps

Origin: Three heads. Long head: from scapula below glenoid cavity. Lateral head: from ridge on humerus below greater tubercle. Medial head: from posterior surface of humerus below musculo-spiral groove. Like the quadriceps the triceps is in two layers, the medial head corresponding to the vastus intermedius

Insertion: Olecranon of ulna

Nerve supply: Radial nerve (C6, 7) via four branches

Action: Only important extensor of the elbow joint

Loss of the triceps is not as disabling as loss of the elbow flexors, as it is usually not an antigravity muscle.

The **triceps jerk** is a stretch reflex elicited by hitting the muscle just above the olecranon with a patellar hammer causing the elbow joint to extend momentarily. Since gravity also tends to extend the elbow joint, the arm is first abducted horizontally sideways and the forearm allowed to hang loosely downwards.

THE ELBOW JOINT

The elbow joint consists of three joints in a common capsule [FIG. 5.05].

1. Humero-ulnar joint.
2. Humeroradial joint.
3. Proximal radio-ulnar joint.

Humero-ulnar joint

The humero-ulnar joint is formed by the trochlea of the humerus articulating with the trochlear notch of the ulna. Being essentially a hinge joint, it has strong medial and lateral ligaments and the anterior and posterior parts of the joint capsule are relatively weak.

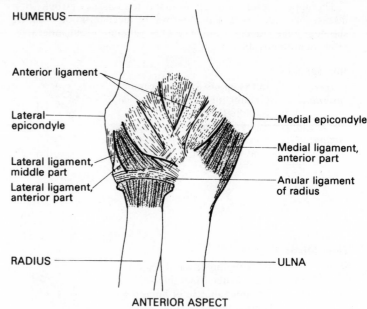

FIG. 5.05. Elbow joint (anterior view).

The trochlear notch of the ulna has a central ridge which separates two concave areas. It is the most lateral area that is in contact with the humerus at full extension.

The trochlea of the humerus has two lips with the medial lower than the lateral. Thus when the elbow joint is extended, the forearm makes an angle with the humerus known as the carrying angle.

When the elbow is flexed, this carrying angle disappears and the forearm is parallel to the arm. This is brought about by the trochlea being spiral not circular. Medial rotation and flexion of the shoulder joint is also needed for the hand to reach the mouth.

Humero-radial joint

The humero-radial joint is formed by the capitulum of the humerus and the upper surface of the head of the radius. The capitulum is spherical when viewed from the front [see FIG. 2.08, p. 32]. Hence the two surfaces are only in contact when the elbow is flexed. Thus this is the position that is adopted when using a screwdriver.

Proximal radio-ulnar joint

The head of the radius rotates under the anular (or annular) ligament attached to the lateral surface of the upper end of the ulna (*anulus* (Latin) = ring; *annulatus* (Latin) = ringed). The ulna has a facet for the head of the radius making a complete circle under which the head of the radius revolves [FIG. 5.06].

Capsule and ligaments of the elbow joint

The fibrous capsule is attached to the humerus, above the coronoid fossa and radial fossa in front, distal to the epicondyles at the side and through the olecranon fossa behind. The origins of the flexor and extensor muscles from the epicondyles lie outside the joint capsule. The capsule is attached to the ulna at the olecranon behind, along the medial side distal to the articular area to the coronoid process in front. On the lateral side it is attached to the annular ligament surrounding the head of the radius.

The capsule is lined by the synovial membrane which is reflected

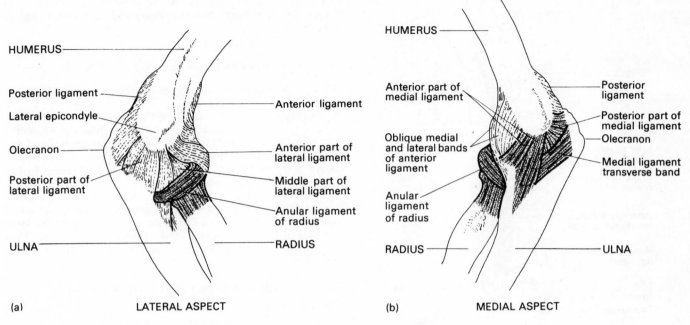

HUMERUS

Posterior ligament

Lateral epicondyle

Olecranon

Posterior part of lateral ligament

ULNA

Anterior ligament

Anterior part of lateral ligament

Middle part of lateral ligament

Anular ligament of radius

RADIUS

(a)   LATERAL ASPECT

HUMERUS

Anterior part of medial ligament

Oblique medial and lateral bands of anterior ligament

Anular ligament of radius

RADIUS

Posterior ligament

Posterior part of medial ligament

Olecranon

Medial ligament transverse band

ULNA

(b)   MEDIAL ASPECT

FIG. 5.06. Elbow joint (medial and lateral view).

on to the non-articular surfaces including the olecranon fossa, radial fossa and coronoid fossa.

The medial ligament [FIGS. 5.05 and 5.06] which strengthens the medial side of the joint capsule consists of three bands. These run from the coronoid process of the ulna to the medial epicondyle of the humerus from the coronoid to the olecranon and from the olecranon to the medial epicondyle. Only the first, from the tubercle of the coronoid to the epicondyle is a true medial ligament that limits abduction at the elbow.

The radial ligament on the lateral side is a strong but narrow band which runs from the lateral epicondyle to the anular ligament. It continues on to be inserted into the lateral aspect of the radius.

The anterior ligament runs from the epicondyles and upper border of the radial and coronoid fossae to the coronoid process and anular ligament. The posterior ligaments run from the epicondyles and floor of the olecranon fossa to the olecranon. These ligaments are weak to allow flexion and extension at the elbow joint.

MOVEMENTS AT THE ELBOW JOINTS

Being a hinge joint, the movements at the elbow are principally flexion and extension. A small amount of abduction and adduction is possible especially when the joint is flexed.

The main muscles about the following movements are:

Flexion	Extension
Biceps	Triceps
Brachialis	
Brachioradialis	
Pronator teres	

In the male extension usually stops before the forearm is in line with the upper arm. In the female hyperextension is often possible.

Pronation and supination

The proximal radio-ulnar joint is concerned with pronation and supination. This is best studied with the elbow bent, since when the arm is extended rotation may be occurring at the shoulder joint. Bend your elbow and turn your palm upwards. The forearm is now supinated. With the other hand palpate the lower end of the ulna at the back of the wrist on the medial side. Now turn your palm downwards to the pronated position and note how this is brought about. The lower end of the radius rotates around the lower end of the ulna [FIG. 5.07].

If you now palpate the upper end of the radius and repeat the supination and pronation, you will feel the head of the radius revolving under the anular ligament.

Note that the ulna itself does not revolve. Palpate the olecranon at the back of the elbow during pronation and supination and confirm this fact.

Note also that the anterior surface of the ulna always points in the same direction. Hence when the forearm is supinated it is the posterior surface of the lower end of the ulna that is superficial whereas when the forearm is pronated, it is the anterior surface of the lower end of the ulna that is prominent. Confirm this on yourself.

If, however, the lower end of the ulna stayed in exactly the same position, then the pronated hand would come to rest in a different position to the supinated hand. Place your ulna along the edge of this book with the forearm supinated. Now pronate your hand without moving the ulna. You will see that the new position of the hand is medial to the previous one.

If the hand is to be rotated around a central axis, then the lower end of the ulna must move in a circle at the same time. This movement takes place at the elbow joint and is a combination of extension and abduction. Try this on yourself, and follow the direction of movement of the lower end of the ulna.

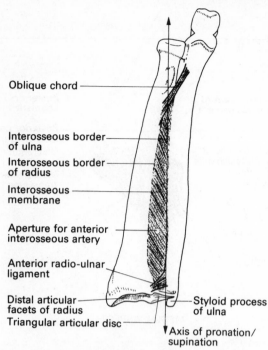

Oblique chord

Interosseous border
of ulna

Interosseous border
of radius

Interosseous
membrane

Aperture for anterior
interosseous artery

Anterior radio-ulnar
ligament

Distal articular
facets of radius

Triangular articular disc

Styloid process
of ulna

Axis of pronation/
supination

FIG. 5.07. Interosseous membrane.
Most of the fibres of the interosseous membrane run downwards and
medially from the radius to the ulna. so as to transmit longitudinal forces
from the radius to the ulna. Recall that the head of the ulna does not
articulate with the carpal bones. The oblique cord has a similar effect in
spreading the load when the radius is pulled downwards. The mechanism
may not work when the hand is suddenly pulled on. Dislocation of the head
of the radius so that it escapes from the anular ligament often occurs in
children.

Distal radio-ulnar joint

A thick triangular plate of fibrocartilage, with its base attached to
the radius and its apex attached to the root of the styloid process of
the ulna, unites the two bones at their lower end, and yet allows the
radius to rotate around the ulna in pronation and supination. This
plate is covered on both surfaces with synovial membrane. Its upper
surface articulates with the ulna. Its lower surface articulates with
the lunate and triquetrum. It thus separates the ulna from the
carpal bones, and prevents the ulna taking part in the wrist joint.

In addition between the two bones at their lower ends is an
articulation between the head of the ulna with the ulnar notch of the
radius. This articulation is surrounded by a lax fibrous capsule lined
with synovial membrane which sends a projection up between the
radius and ulna.

Interosseous membrane

The interosseous membrane joins the radius to the ulna along
almost their entire length. It starts about 2.5 cm below the radial
tuberosity and runs down to blend with the capsule of the distal
radio-ulnar joint.

Starting from the radius the direction of the fibres is downwards
and medial. When the forearm is in supination the membrane is
tight and forces transmitted from the hand and wrist to the radius
may reach the ulna via this membrane. Unfortunately in falling on
the outstretched hand the forearm is in pronation and the membrane is quite slack, because the bones now lie so close together.

The principal function of the interosseous membrane is to
increase the area from which both flexor and extensor muscles can
take origin.

FOREARM MUSCLES

The muscles of the forearm may be conveniently divided into two
groups:
1. The flexor group of muscles lie on the front of the forearm. They
take origin especially from the medial epicondyle of the humerus
('common flexor origin') and from the front of the radius and ulna
and interosseous membrane. They are all supplied by the median
nerve except for the flexor carpi ulnaris and the medial half of
flexor digitorum profundus.
2. The extensor muscles on the back of the forearm are not so
strong as the flexors. They take origin mostly from the lateral
epicondyle of the humerus ('common extensor origin') and from
the back of the radius and ulna and interosseous membrane. They
are supplied by the radial nerve and its branch, the posterior
interosseous.

THE FLEXOR GROUP OF MUSCLES

Pronator teres

Origin:	Medial epicondyle of the humerus
Insertion:	Lateral side of the shaft of the radius
Nerve supply:	Median nerve
Action:	Pronates the forearm. Flexes the elbow joint

Pronator teres forms the medial boundary of the cubital fossa. It is
inserted into the most convex part of the shaft of the radius. This
insertion lies deep to extensor carpi radialis longus and brevis. Since
it is supplied by the median nerve, uniting the muscle with these
tendons would enable the wrist to be extended in cases of permanent radial nerve damage.

If the musculocutaneous and radial nerves are paralysed, pronator teres becomes the only effective flexor of the elbow joint.

Flexor carpi radialis

Origin:	Medial epicondyle of the humerus
Insertion:	Base of second and third metacarpals
Nerve supply:	Median nerve
Action:	Flexes the wrist

Note that the flexors (and extensors) of the wrist are not inserted into the carpal bones, but their tendons cross the carpal bones to be inserted into the metacarpal bones.

The tendon of flexor carpi radialis can be seen standing out at the wrist if the wrist is flexed. It lies to the outer side of the more prominent palmaris longus tendon. The tendon of flexor carpi radialis passes through the **carpal tunnel** in a groove in the trapezium. It lies immediately on the medial side of the radial pulse at the wrist.

Palmaris longus

Origin: Medial epicondyle
Insertion: Palmar fascia
Nerve supply: Median nerve (C7)
Action: Flexes wrist

This is an interesting but unimportant muscle that is absent in about 15 per cent of cases. However, its tendon is very prominent and acts as a good landmark for the other structures at the wrist. Bend your wrist slightly forwards and take your thumb across to touch the little finger. The tendon will stand out as a tense cord.

This muscle represents the 'missing' long flexors which by analogy with flexor digitorum profundus and superficialis, should be inserted into the proximal phalanges. Instead the tendons spread out in the palm of the hand to form the palmar aponeurosis. This forms fibrous strips which split and are attached to the base of the proximal phalanges. In Dupuytren's contracture the metacarpophalangeal joints of the little and ring finger, are principally affected. The business of flexion of the metacarpophalangeal joints has been taken over by the interossei.

Flexor carpi ulnaris

Origin: Two heads:
(i) medial epicondyle
(ii) olecranon and upper two-thirds of posterior border of ulna
Insertion: Pisiform bone
Nerve supply: Ulnar nerve
Action: Flexes and adducts the wrist

The tendon can be seen standing out as it runs to the pisiform by clenching the fist with the wrist slightly bent. The pisiform is regarded as a sesamoid bone in the tendon of this muscle. The pisiform is attached to the base of the fifth metacarpal and to the hook of the hamate, by the pisometacarpal and pisohamate ligaments.

Flexor digitorum superficialis

Origin: Three heads:
(i) medial epicondyle
(ii) tubercle on medial side of coronoid process of ulna
(iii) anterior border of radius
Insertion: Divides into four tendons at wrist. Tendons to ring and middle finger lie in front of tendons to index and little finger as they pass through carpal tunnel. Inserted into sides of middle phalanges
Nerve supply: Median nerve
Action: Flexes proximal interphalangeal joints of the fingers

The tendons split opposite the proximal phalanx of each finger to allow the tendons of flexor digitorum profundus to pass through to their insertion at the base of the terminal phalanx.

Flexor digitorum profundus

Origin: Upper three-quarters of anterior and medial surface of ulna and interosseous membrane
Insertion: Base of terminal phalanges
Nerve supply: Anterior interosseous nerve (branch of median nerve) and ulnar nerve (medial two tendons)
Action: Flexes fingers and wrist

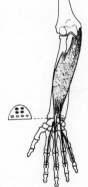

Flexor digitorum profundus as its name implies lies deep to flexor digitorum superficialis. Like this muscle, it also divides into four tendons. The flexor digitorum profundus muscle first flexes the terminal phalanx of each finger, then flexes the middle and proximal phalanges and finally flexes the wrist. It thus rolls up the fingers.

Note that the tendons of this muscle enter the carpal tunnel in line abreast, while those of the flexor digitorum superficialis are in pairs. This allows these eight tendons to fit into the semicircular shape of the tunnel and to make room for the median nerve. When the wrist is slashed the tendons of flexor digitorum superficialis going to the middle and ring finger are in greatest danger, as well as the median nerve.

Flexor pollicis longus

Origin:	Anterior surface of radius and interosseous membrane
Insertion:	Terminal phalanx of thumb
Nerve supply:	Anterior interosseous nerve (branch of median nerve)
Action:	Flexes terminal phalanx of thumb. (This is the only muscle that has this action)

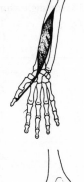

Pronator quadratus

Origin:	Lower quarter of anterior surface of ulna
Insertion:	Anterior surface of radius
Nerve supply:	Anterior interosseous nerve (branch of median nerve)
Action:	Pronates the forearm

THE EXTENSOR GROUP OF MUSCLES

Extensor carpi radialis longus

Origin:	Lower third of lateral epicondylar ridge and lateral intermuscular septum
Insertion:	Base of second metacarpal on dorsum of hand
Nerve supply:	Radial nerve
Action:	Extends and abducts the wrist

Extensor carpi radialis brevis

Origin:	Lateral epicondyle
Insertion:	Base of second and third metacarpals on dorsum of hand
Nerve supply:	Posterior interosseous nerve (branch of radial nerve)
Action:	Extends the wrist

Extensor carpi ulnaris

Origin:	Lateral epicondyle Subcutaneous border of ulna
Insertion:	Base of fifth metacarpal on dorsum of hand
Nerve supply:	Posterior interosseous nerve (branch of radial nerve)
Action:	Extends and adducts the wrist

These three muscles together with flexor carpi radialis and flexor carpi ulnaris allow the wrist to be flexed, extended, abducted, adducted, or moved to any intermediate position.

Try the wrist movements on yourself. It may even be circumducted, that is, the hand may be moved in a circle.

Extensor digitorum

Origin:	Lateral epicondyle
Insertion:	Four tendons with slips to the proximal, middle, and distal phalanges of each finger
Nerve supply:	Posterior interosseous nerve (branch of radial nerve)
Action:	Extends fingers and wrist. The main action is on the metacarpophalangeal joint of each finger. The tendons to the index, ring, and little finger tend to abduct these fingers away from the middle finger

Extensor indicis

Origin:	Ulna
Insertion:	Own tendon to index finger
Nerve supply:	Posterior interosseous nerve
Action:	Extends index finger

Extensor digiti minimi

Origin: Lateral epicondyle
Insertion: Common extensor tendon of little finger
Nerve supply: Posterior interosseous nerve (branch of radial nerve)
Action: Extends little finger

Extensor pollicis longus

Origin: Ulna and interosseous membrane
Insertion: Base of terminal phalanx of thumb
Nerve supply: Posterior interosseous nerve
Action: Extends phalanges and metacarpal of thumb

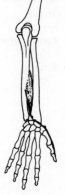

The direction of the tendon of extensor pollicis longus changes when it reaches the dorsal tubercle of the radius (Lister's tubercle). The tendon may be frayed at this point as a late complication of a fracture of the radius a centimetre or two above the wrist joint. The patient cannot extend the top joint of the thumb [FIG. 5.08].

The older name for these three thumb muscles gives a better clue to their insertion and action. These names are ext. ossis. metacarpi pollicis (abductor pollicis longus), ext. primi. internodii pollicis (extensor pollicis brevis), and ext. secundi internodii pollicis (extensor pollicis longus) [FIG. 5.08].

Anconeus

Origin: Posterior aspect of lateral epicondyle
Insertion: Upper quarter of posterior surface of ulna
Nerve supply: Radial nerve
Action: Weak extensor of elbow

The anatomical snuff-box

The depression seen at the base of the thumb when the thumb is extended is termed the anatomical snuff-box [FIG. 5.08].

Anterior boundary—Tendon of extensor pollicis brevis with the tendon of abductor pollicis longus
Posterior boundary—Tendon of extensor pollicis longus
Proximal boundary—Styloid process of radius
Distal boundary—Base of metacarpal of thumb and part of trapezium

The floor is made up of the scaphoid and trapezium. The pulsations of the radial artery can be felt as it runs through this space.

The styloid process of the radius is easy to find in the snuff box. Just distal to it the scaphoid bone (often fractured) can be palpated when the wrist is put into ulnar deviation. The tubercle of the scaphoid can be felt on the front of the wrist.

Abductor pollicis longus

Origin: Ulna, interosseous membrane, and radius
Insertion: Base of metacarpal of thumb
Nerve supply: Posterior interosseous nerve
Action: Abducts and also extends the thumb

Extensor pollicis brevis

Origin: Radius and interosseous membrane
Insertion: Base of proximal phalanx of thumb
Nerve supply: Posterior interosseous nerve
Action: Extends proximal phalanx of thumb

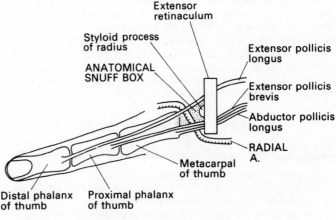

FIG. 5.08. Anatomical snuff-box.

Supinator

Origin: Ulna
Insertion: Radius
Nerve supply: Posterior interosseous
                 nerve (C6)
Action: Supinates the forearm
                 Note that the muscle has been cut
                 through and lifted away from the
                 radius. The arrow shows the direction
                 of pull.

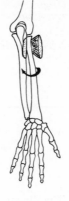

The supinator muscle winds round the neck of the radius to its insertion. As a result contraction of the muscle rotates the radius and hence supinates the forearm. Note that the tendon of biceps also winds around the radius and the mode of action of the two muscles is basically similar.

Brachioradialis

Origin: Upper two-thirds of lateral
                 supracondylar ridge and
                 lateral intermuscular
                 septum
Insertion: Just above styloid
                 process of radius
Nerve supply: Radial nerve
Action: Rotates the forearm from
                 either prone position or
                 supine position to mid-
                 prone–supine position.
                 Flexes the elbow

This muscle is the exception of the rule that the muscles originating from the lateral epicondylar region are extensors. Although the old name for this muscle was 'supinator longus', it only supinates as far as the mid-prone–supine position. It is a powerful flexor of the elbow in this position.

The muscle twists shortly after its origin so that its anterior surface becomes posterior. The muscle is superficial throughout except at its insertion where its tendon is covered by those of abductor pollicis longus and extensor pollicis brevis.

The **brachioradialis jerk** is demonstrated by tapping the tendon with the elbow at right angles and the forearm in the mid-prone–supine position, that is with the thumb pointing upwards. It is often called the supinator jerk.

STRUCTURE OF THE WRIST JOINT

The proximal row of carpal bones at the wrist, from lateral to medial, are:

Scaphoid—Lunate—Triquetrum

The distal row of carpal bones, again from lateral to medial are:

Trapezium—Trapezoid—Capitate—Hamate

The **pisiform** lies on the palmar surface of the triquetrum.

The term wrist joint refers to the radiocarpal joint and the midcarpal joint. The movements of the midcarpal joint are only significant when the wrist is flexed. In wrist flexion the midcarpal joint makes a greater contribution to the movement than the radiocarpal joint. In extension their importance is reversed.

The radiocarpal joint is formed by the concave lower end of the radius and the triangular fibrocartilage articulating with the convex surfaces of the scaphoid, lunate, and triquetrum. As has been seen the ulna does not form part of the radiocarpal joint.

The fibrous capsule which surrounds the joint is attached to the margins of the articular surfaces. The capsule is lined with synovial membrane, which may occasionally herniate through the fibrous capsule. This capsule may communicate with that of the inferior radio-ulnar joint on the outer side of the triangular ligament.

The capsule is strengthened in front by the anterior ligament [FIG. 5.09] and behind by the posterior ligament [FIG. 5.10]. The anterior ligament runs from the lower end and styloid process of the radius and the head of the ulna to the scaphoid, lunate, triquetrum, and capitate.

On the lateral side, the radial lateral ligament runs from the styloid process of the radius by the tubercle of the scaphoid and to the trapezium.

The ulnar collateral ligament, on the medial side, is a rounded cord running from the styloid process of the ulna to the triquetrum.

MOVEMENTS AT THE WRIST JOINT

Since the wrist can be moved in four different directions, flexion, extension, abduction, and adduction, there must be at least four different muscles involved. In fact there are five.

If the cross-section of the wrist is considered to be approximately a rectangle, these muscles could have been connected to the mid-point of each side of the rectangle. Alternatively these can be connected to the four corners. In the first case the movement would have been brought about by the muscles acting singly, the flexor muscle producing flexion, the abductor muscle producing abduction, etc. In the second case, movements would be brought about by the muscles acting in pairs. The two anterior muscles would produce flexion, the two lateral muscles would produce abduction and so on.

The arrangement of the wrist joint muscles is similar to the second case. The muscles of the radial side are termed *carpi radialis* muscles. The muscles on the ulnar side are termed *carpi ulnaris* muscles. The muscles in front are termed flexors. Those at the back are termed extensors. The total of five is due to the fact that there are two extensor carpi radialis muscles, a longus, and a brevis.

The movements at the wrist joint and the muscles which produce these are:

Flexion	Extension	Abduction	Adduction
Flexor carpi radialis	Extensor carpi radialis longus	Flexor carpi radialis	Flexor carpi ulnaris
Flexor carpi ulnaris	Extensor carpi radialis brevis	Extensor carpi radialis longus	Extensor carpi ulnaris
	Extensor carpi ulnaris	Extensor carpi radialis brevis	

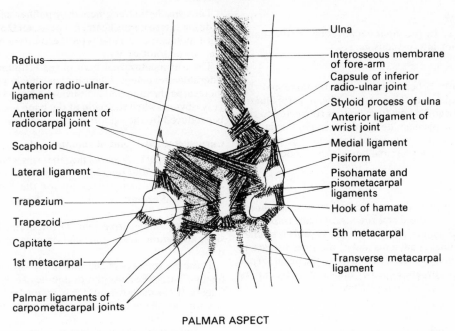

Radius —
Anterior radio-ulnar ligament —
Anterior ligament of radiocarpal joint —
Scaphoid —
Lateral ligament —
Trapezium —
Trapezoid —
Capitate —
1st metacarpal —
Palmar ligaments of carpometacarpal joints —

Ulna —
Interosseous membrane of fore-arm —
Capsule of inferior radio-ulnar joint —
Styloid process of ulna —
Anterior ligament of wrist joint —
Medial ligament —
Pisiform —
Pisohamate and pisometacarpal ligaments —
Hook of hamate —
5th metacarpal —
Transverse metacarpal ligament —

PALMAR ASPECT

FIG. 5.09. Ligaments of the wrist joint (anterior view).

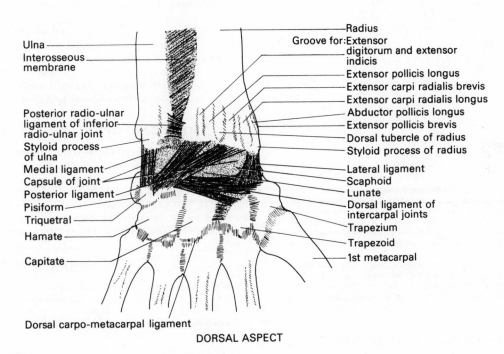

Ulna —
Interosseous membrane —
Posterior radio-ulnar ligament of inferior radio-ulnar joint —
Styloid process of ulna —
Medial ligament —
Capsule of joint —
Posterior ligament —
Pisiform —
Triquetral —
Hamate —
Capitate —

Radius —
Groove for: Extensor digitorum and extensor indicis —
Extensor pollicis longus —
Extensor carpi radialis brevis —
Extensor carpi radialis longus —
Abductor pollicis longus —
Extensor pollicis brevis —
Dorsal tubercle of radius —
Styloid process of radius —
Lateral ligament —
Scaphoid —
Lunate —
Dorsal ligament of intercarpal joints —
Trapezium —
Trapezoid —
1st metacarpal —

Dorsal carpo-metacarpal ligament

DORSAL ASPECT

FIG. 5.10. Ligaments of the wrist joint (posterior view).

In addition all the tendons passing anterior to the wrist joint such as palmaris longus and the flexor of the fingers will tend to flex the wrist joint. The tendons on the back of the joint, such as the extensors of the fingers will tend to extend the wrist joint. Abduction will be aided by abductor pollicis longus and extensor pollicis brevis.

As has been seen, with the possible exception of flexor carpi ulnaris which is inserted into the pisiform, none of these muscles are inserted into the carpal bones. They are all inserted into the bones immediately distal, namely, the base of the metacarpals. This avoids straining the carpometacarpal joints.

Nerve supply to wrist joint

The wrist joint is supplied by the ulnar nerve (dorsal branch) and the anterior and posterior interosseous nerves.

Pisiform joint

The pisiform joint is formed between the pisiform bone and the palmar surface of the triquetrum. The tendon of flexor carpi ulnaris is inserted into the pisiform. The pisiform bone is connected to the hook of the hamate by the pisohamate ligament, and to the base of the fifth metacarpal by the pisometacarpal ligament. The pisiform can thus be considered to be a sesamoid bone in the tendon of the ulnar flexor of the wrist.

Midcarpal joint

The joint between the proximal row of carpal bones, scaphoid, lunate, and triquetrum, and the distal row, trapezium, trapezoid, capitate, and hamate is S-shaped [FIG. 5.11]. The capsule is strengthened by palmar and dorsal ligaments and at the sides by extensions of the radial and ulnar collateral ligaments to the distal carpal bones. The bones of each row are joined by interosseous ligaments which form a complex series of joint cavities.

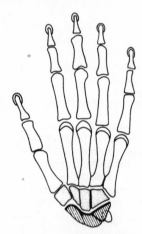

FIG. 5.11. Midcarpal joint.

This joint lies between the proximal (scaphoid. lunate. triquetral) and distal rows of carpal bones (trapezium. trapezoid. capitate. hamate). The figure shows that the midcarpal joint cavity is 'S' shaped: abduction/ adduction movements are therefore not possible. Flexion (especially) and extension take place in conjunction with the same movements in the radiocarpal joint.

Fracture of the scaphoid is a very common injury. The bone breaks into two. The distal fragment has a normal blood supply, but the proximal usually becomes ischaemic and undergoes avascular necrosis because the blood vessels cannot bridge the fracture site [see p. 163].

CARPOMETACARPAL JOINTS

The carpometacarpal joints are between the trapezium, trapezoid, capitate, and hamate, and the five metacarpals. The trapezium articulates only with the metacarpal of the thumb. The trapezoid articulates mainly with the second metacarpal, the capitate mainly with the third metacarpal and the hamate with both the fourth and fifth metacarpals. More accurately all the other carpal bones articulate with two metacarpals except the capitate which articulates with three.

The capsule is strengthened by palmar and dorsal ligaments. In addition interosseous ligaments between the hamate, capitate, and third metacarpal usually separate the two medial joints from the rest of the joint cavity.

The carpometacarpal joint of the little finger is capable of considerable movement. Demonstrate this in yourself by moving the metacarpal head to and fro through about 3 cm. This movement is very obvious when the palm of the hand is cupped. The movement is produced by the opponens digiti minimi.

Carpometacarpal joint of thumb

The metacarpal bone of the thumb forms a saddle-shaped joint with the trapezium. This joint has its own capsule and allows a great deal of movement which makes up for the limitation of movement (flexion and extension only) at the thumb's metacarpophalangeal joint.

This carpometacarpal joint of the thumb allows flexion, extension, abduction, adduction, and opposition, as well as circumduction.

These movements of the thumb are different from those of the fingers. This is because the thumb lies at right angles to the other fingers. It has its flexor surface lying medially, and the thumb itself lies anterior to the fingers. Thus starting from the anatomical position:

Extension—the thumb moves laterally

Flexion—the thumb moves medially

Abduction—the thumb moves anteriorly (away from the palm)

Adduction—the thumb moves posteriorly (towards the palm, returning to original position)

Opposition—the thumb is abducted, medially rotated, and flexed so that the palmar surface of its terminal phalanx can be brought into contact not only with the tips but with any part of any of the four fingers

METACARPOPHALANGEAL JOINTS

At the metacarpophalangeal joints the rounded distal heads of the metacarpals articulate with the concave proximal ends of the proximal phalanges [FIG. 5.12].

The fibrous capsule is strengthened by strong lateral and medial ligaments. It is strengthened in front by a fibro–cartilage which is attached to the proximal phalanx and which is grooved for the flexor tendons. The margins of these fibro-cartilages are united by

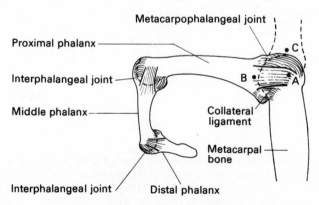

FIG. 5.12. Metacarpophalangeal and interphalangeal joints.

the transverse metacarpal ligament. The lumbrical tendon runs in front of this transverse ligament whereas the interossei tendons run behind it [FIG. 5.13].

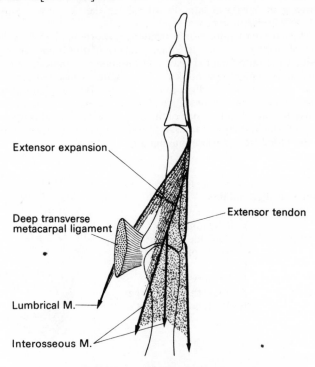

FIG. 5.13. The movements of the metacarpophalangeal and inter-phalangeal joints are shown.
The interossei lie in the plane behind the deep transverse metacarpal ligaments, which are attached to the anterior aspects of the capsules of the metacarpophalangeal joints. The lumbrical muscles are attached to the deep flexor tendons and are therefore in a plane in front of these ligaments. Both sets of muscles reach the extensor tendon on the back of the fingers, via the extensor expansion.

Movements at the metacarpophalangeal joints

The metacarpophalangeal joints of the fingers allow flexion, extension, abduction, and adduction. The metacarpophalangeal joint of the thumb allows only flexion and extension. When the joints are flexed abduction and adduction are impossible.

INTERPHALANGEAL JOINTS

The trochlear distal ends of the proximal phalanges articulate with the proximal end of the more distal phalanges forming the interphalangeal joints. These joints have strong medial and lateral ligaments so that only flexion and extension is possible. In full extension, the fingers may be at an angle of less than 180° with the back of the hand. If the ligaments are lax, the fingers may in some people be bent back to 90°.

Flexor retinaculum

The flexor retinaculum is a strong band of deep fascia which prevents the flexor tendon bow-stringing at the wrist. It runs from the tubercle of the scaphoid and ridge of the trapezium on the lateral side to the hook of the hamate and the pisiform on the medial side.

It forms a tunnel with the underlying carpal bone, through which the flexor tendon runs. Identify the four bony landmarks.

Extensor retinaculum

The extensor retinaculum is a band of deep fascia which holds down the extensor tendons at the back of the wrist. It runs from the radius diagonally downwards and inwards to the hamate and pisiform [p. 34].

THE MUSCLES OF THE HAND

The muscles of the hand are divided into:
   (i) muscles of the thenar eminence acting on the thumb.
   (ii) muscles of the hypothenar eminence acting on the little finger.
   (iii) interossei and lumbrical muscles.

MUSCLES OF THE THENAR EMINENCE

The muscles of the thenar eminence on the thumb side of the palm are abductor pollicis brevis, flexor pollicis brevis, opponens pollicis, and adductor pollicis. Abductor pollicis brevis, flexor pollicis brevis, and opponens pollicis are supplied by the median nerve (recurrent branch). Adductor pollicis is supplied by the ulnar nerve.

Abductor pollicis brevis runs from the flexor retinaculum, the tubercle of the scaphoid, and the ridge of the trapezium to the radial side of the base of the proximal phalanx of the thumb. It carries the first metacarpal anteriorly so that the web between the index finger and the thumb is straight. This motion of the thumb is termed abduction.

Flexor pollicis brevis runs from the flexor retinaculum and the ridge of the trapezium to be inserted with abductor pollicis brevis on the radial side of the base of the proximal phalanx of the thumb.

Opponens pollicis lies deep to abductor pollicis and flexor pollicis brevis. It runs from the flexor retinaculum and the ridge of the trapezium to the whole length of the radial side of the first metacarpal bone. Its action is to oppose the thumb which is a combination of flexion, abduction, and rotation of the first metacarpal to bring the thumb over to the little finger.

Adductor pollicis arises from two heads. The transverse head arises from the distal part of the palmar surface of the third metacarpal. The oblique head arises from the capitate, trapezoid, and trapezium as well as the second and third metacarpal bones. Both heads are usually inserted into the inner side of the proximal phalanx of the thumb.

Abductor pollicis brevis

Origin:	Tubercle of scaphoid, ridge of trapezium, and flexor retinaculum
Insertion:	Outer side of base of proximal phalanx of thumb
Nerve supply:	Median nerve
Action:	Abducts the thumb

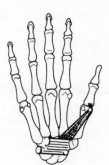

Flexor pollicis brevis

Origin: Ridge of trapezium and flexor retinaculum
Insertion: Outer side of base of proximal phalanx of thumb
Nerve supply: Median nerve
Action: Flexes proximal phalanx of thumb

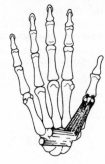

The tendon of flexor pollicis brevis runs lateral to that of flexor pollicis longus. There are usually sesamoid bones in the tendon.

Opponens pollicis

Origin: Ridge of trapezium and flexor retinaculum
Insertion: Radial side of metacarpal of thumb along whole length
Nerve supply: Median nerve
Action: Opposes thumb, that is, it brings the thumb over to the little finger. This movement is a combination of flexion, adduction, and rotation of the thumb metacarpal

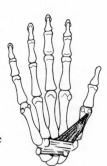

Adductor pollicis

Origin: Transverse head: distal part of palmar surface of meta-carpal of middle finger (third metacarpal)
Oblique head: capitate. trapezoid, trapezium, and metacarpals of index and middle fingers (second and third metacarpals)
Insertion: Inner side of proximal phalanx of thumb
Nerve supp: Ulnar nerve (deep branch)
Action: Adducts the metacarpal of the thumb

All these muscles principally move the carpometacarpal joint of the thumb. The only muscle acting directly on this joint is the oppo-nens. The others are inserted into the base of the proximal phalanx and produce their action on the carpometacarpal joint indirectly through the metacarpophalangeal joint.

MUSCLES OF THE HYPOTHENAR EMINENCE

The muscles of the hypothenar eminence on the little finger side of the hand are palmaris brevis, abductor digiti minimi, flexor digiti minimi, and opponens digiti minimi.

Palmaris brevis is a thin muscle which forms a vertical furrow in the skin. It arises from the palmar aponeurosis and flexor

retinaculum. It runs superficial to the other muscles of the hypothe-nar eminence to be inserted into the skin on the ulnar margin of the hand. It is supplied by the superficial branch of the ulnar nerve. Pressing the pisiform laterally stimulates the nerve and causes palmaris brevis to contract.

Abductor digiti minimi runs from the pisiform bone to the ulnar side of the base of the proximal phalanx of the little finger. It abducts the little finger. Flexor digiti minimi, originates from the flexor retinaculum and hook of the hamate, is inserted into the same place. Opponens digiti minimi lies in a deeper plane. It too originates from the flexor retinaculum and hook of the hamate but is inserted into the ulnar side of the fifth metacarpal (metacarpal of little finger). Its action is to draw the fifth metacarpal forwards and to the radial side as when making a fist.

Abductor digiti minimi

Origin: Pisiform and flexor retinaculum
Insertion: Medial side of base of proximal phalanx of little finger
Nerve supply: Ulnar nerve
Action: Abducts the little finger It also tends to flex it

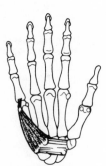

Flexor digiti minimi

Origin: Hook of the hamate and flexor retinaculum
Insertion: Inserted with abductor digiti minimi into medial side of base of proximal phalanx of little finger
Nerve supply: Ulnar nerve
Action: Flexes and abducts the little finger

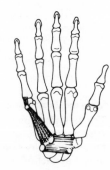

Opponens digiti minimi

Origin: Hook of hamate and flexor retinaculum
Insertion: Ulnar side of palmar aspect of metacarpal of little finger
Nerve supply: Ulnar nerve
Action: Opposes little finger

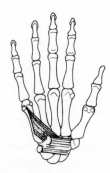

The metacarpal of the little finger is drawn forward and moved to the radial side. Make a fist and you will see this action.

Palmaris brevis

Origin: Palmar aponeurosis and flexor retinaculum
Insertion: Skin on medial border of the hand
Nerve supply: Ulnar nerve (superficial branch)
Action: Causes vertical furrows in the skin on medial edge of palm which deepens the hollow in the palm of the hand

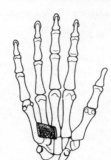

As has been seen the muscle may be seen to contract when the ulnar nerve is stimulated by pressing the pisiform laterally.

THE INTEROSSEI AND LUMBRICALS

The interossei run from the metacarpal bones to the base of the proximal phalanges. Taking the middle finger as the axis of movement the four palmar interossei are adductors, while the four dorsal interossei are abductors. Each palmar interosseous arises from a single head from the metacarpal bone of the finger on which it acts. Each dorsal interosseous arise by two heads from the bones between which it lies.

The middle finger has no palmar interossei since it is already in the midline and therefore cannot be adducted to this line. However, movement in either the radial or the ulnar direction will constitute abduction. It therefore has two dorsal interossei.

The first dorsal interosseous is the most prominent and enables you to remember that dorsal interossei abduct. Find it on yourself. The two heads arise from adjacent sides of the first and second metacarpal bones and it is inserted into the radial side of the base of the proximal phalanx of the index finger. The first dorsal interosseous contracts when the index finger is abducted. The radial artery passes between the two heads as it passes from the back of the hand to enter the palm.

The four lumbricals run from the tendons of the flexor digitorum profundus in the palm to the radial side of the extensor tendons of each of the four fingers.

They are said to flex the metacarpophalangeal and extend the interphalangeal joints. The interossei, in addition to abduction and adduction, have a similar but more powerful action [p. 111].

Four palmar interossei

Origin: Metacarpals of thumb, index, ring, and little fingers
Insertion: Side of base of proximal phalanx and extensor expansion. Thumb and index finger tendons run to ulnar side. Ring and little finger tendons run to radial side
Nerve supply: Ulnar nerve (deep branch)
Action: Adduct the fingers and thumb. In addition they flex the metacarpophalangeal joints and extend the interphalangeal joints

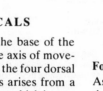

It follows that the middle finger has no palmar interosseous since it cannot be adducted towards itself. However, it has two dorsal interossei, one on each side, since whichever way it moves it will be abducted from the reference line.

Four dorsal interossei

The thumb and little finger have their own abductors and therefore do not have dorsal interossei.
Origin: Two heads from adjacent metacarpals, thumb-index metacarpals
Insertion: Side of base of proximal phalanx and extensor expansion. Index and one middle finger tendon run to radial side. Other middle finger tendon and ring finger tendon run to ulnar side

Four lumbricals

As has been seen the four lumbrical muscles arise in the palm from the tendons of flexor digitorum profundus.
Origin: From radial side of index tendon
From radial side of middle finger tendon
From adjacent sides of ring finger tendon
From adjacent sides of little finger tendon
Insertion: Radial side of extensor expansion of each finger
Nerve supply: Median nerve—outer two lumbricals
ulnar nerve—inner two lumbricals
Action: Flex the metacarpophalangeal joints and extend the interphalangeal joints [see p. 112]

COMPARTMENTS OF THE PALM

The midpalmar space is a large potential space deep to the tendons of flexor digitorum superficialis, flexor digitorum profundus and their lumbricals. It is bounded posteriorly by the fascia over the interossei and laterally by the fibrous septum which runs from the third metacarpal bone [Fig. 5.14].

The thenar space is a similar space which lies deep to the tendons of the index finger. This space is bounded posteriorly by the fascia over adductor pollicis. The surface marking of this space is the crease on the skin termed the *life-line* (flexion line of the thumb).

SYNOVIAL SHEATHS OF FLEXOR TENDONS

A synovial sheath surrounds the flexor tendons. The sheaths for the index finger, middle finger and ring finger stop at the distal transverse crease [Fig. 5.15]. The sheath of the little finger joins the common flexor sheath and extends to the proximal transverse crease. Proximally it extends 2.5 cm above the flexor retinaculum.

Flexor pollicis longus has its own synovial sheath which also extends 2.5 cm above the flexor retinaculum.

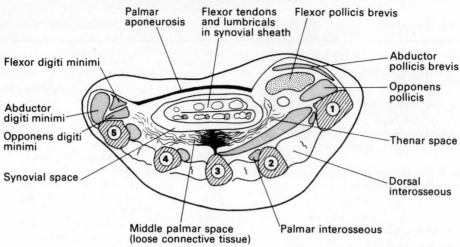

FIG. 5.14. Cross-section of the palm of the hand to show the middle palmar and thenar spaces.

The middle palmar and thenar 'spaces' in the palm of the hand lie in the plane of cleavage between the metacarpal bones and interosseous muscles (behind) and the long flexor tendons enclosed in their common synovial sheath at this level (in front). The term space may be misleading: it is filled in life with a small amount of delicate connective tissue which is somewhat thicker in front of the 3rd metacarpal where it may form a partition separating the two spaces.

Note especially that the adductor pollicis, because of the way the transverse head is attached to the 3rd metacarpal, forms the posterior boundary of the thenar space. Infection may be set up in the 'spaces', which may become very distended. Only then is the word 'space' justified.

Drainage of the thenar space may be effected by an incision through the web of the thumb in front of the adductor pollicis. The middle palmar space can be approached through the webs on one or both sides of the ring finger.

These spaces extend upwards through the carpal tunnel into the forearm in the plane between the radius, ulna and pronator quadratus and the posterior surface of the flexor digitorum profundus.

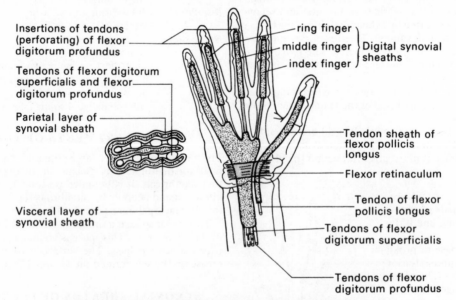

FIG. 5.15. Synovial tendon sheaths in wrist and palm.

The synovial membrane has two layers. One is attached to the outer surface of the tendons and the other to the tissues amongst which the tendons are lying. By analogy with the body cavities, this layer covering the tendon is called the visceral layer, and the layer covering the adjacent tissues is the parietal layer. Synovial fluid lubricates the synovial membrane and allows them to slide over each other without friction. If the synovial membrane becomes inflamed due to injury or infection (tendosynovitis) excessive synovial fluid collects and produces a swelling whose shape and distribution follows that of this figure. Friction may be heard and felt by the observer and pain is usual.

If bacteria enter the synovial space (finger prick), infection will spread immediately throughout the synovial space. The tendon may be destroyed by an infection in its own synovial sheath.

THE HAND AS A FUNCTIONAL UNIT

Infections and the hand

The hand is a common site of injury and infection. Since swelling in a confined space can cut off the arterial blood supply and cause necrosis, it is important to know the compartments of the hand.

The bone of the distal phalanx of each digit is particularly vulnerable. It is surrounded by dense fibrous tissue forming the pulp of the finger. An infection here (whitlow) can create sufficient pressure to cause bone necrosis unless the pressure is relieved by an incision. Fortunately some blood can reach the bone via blood vessels in the attached tendon.

Infections can spread along the synovial sheaths surrounding the tendons.

The dense palmar aponeurosis prevents the pus from an abscess in the palmar space reaching the surface. Incisions are necessary to allow it to escape. These incisions must be made without damaging the arterial palmar arch. There are two main compartments in the palm in which abscesses form.

A further consideration of the intrinsic muscles of the hand

The term 'intrinsic' implies that the muscle begins and ends in the hand. The extrinsic muscles (the term is rarely used in practice) have their bellies in the forearm and only the tendons are to be found in the hand.

In your own hand identify the thenar and hypothenar eminences. The former is the fleshy mass in front of the metacarpal of the thumb, while the latter is in front of and on the medial side of the fifth metacarpal. The term 'hypothenar' is a misnomer. It is a relic from the days before the anatomical position. In the anatomical position these muscles lie medial to, rather than below, the thenar mass, as the name seems to imply. If you are holding this book in your hands then the ulnar border of the hand is 'below', and it is easy to understand how the term hypothenar first came into use. It is now too firmly ingrained in the language to be displaced.

Notice the hollow of the palm of the hand which is present even when the fingers and thumb are flattened. The tendons running through the palm of the hand to the thumb and fingers cannot usually be seen here but they can be palpated when the fingers are flexed.

Movements of the hand and digits

It is very important to understand the terminology used to describe the movements of the hand and its five digits. When we speak of **adduction** or **abduction** of the hand we mean movement at the wrist joint with the limb in the anatomical position with the forearm in supination. The hand moves nearer the midline of the body in adduction (ulnar deviation of the wrist) or away from it in abduction (radial deviation of the wrist). Note that when the forearm is pronated, so that the palm faces backwards, it is not longer in the anatomical position, but the words are still used to describe what the wrist movements would be if it were in the anatomical position. In this case adduction of the wrist is still ulnar deviation (although it now carries the hand away from the body) and abduction is still radial deviation (although it carries it towards the body). To avoid confusion many people speak only of radial or ulnar deviation of the wrist and do not use adduction and abduction.

The fingers are capable of **flexion** and **extension** in all three joints. In addition the metacarpophalangeal joints are capable of **abduction** and **adduction**.

Now **abduction** and **adduction of the digits** has a different meaning from that in the rest of the body including the wrist. The hand (like the foot) is a microcosm of its own. Place your hand palm down flat on the table with the fingers together and in extension. Spread your fingers out like a fan. This movement in any of the fingers is called abduction. It is defined as movement away from an imaginary line running through the middle of the middle finger. The middle finger can therefore be abducted in the radial or ulnar direction. The arrangement of the interosseous muscles which mainly produce these movements, are attached in a pattern around the middle finger which makes this convention logical. Adduction implies the return of the fingers towards this imaginary line. Note that it is not the middle finger itself, but the position it occupies in the resting state that is the reference line. If it were the position of the finger itself, it would be impossible to talk about its movements of abduction and adduction.

Movements of the thumb

The thumb as has been seen has its own rules. In the anatomical position the thumb lies at the side of the hand with the thumb nail at right-angles to the finger nails. Flex and extend the interphalangeal (or top) joint of the thumb. The movement takes place in the plane of the palm of the hand, whilst, of course, flexion of the fingers takes place in a plane at right-angles to the palm. So the plane of flexion of the thumb is different from the plane of flexion of the fingers. Having said this, the next question concerns the plane of abduction and adduction of the thumb—it must be at right-angles to its own plane of flexion and extension, that is, not in the plane of the palm of the hand, but in the plane at right-angles to it.

When you placed your palm flat on the table and abducted and adducted your fingers you may, without thinking, have pointed your thumb sideways. This movement was not abduction but extension of the thumb. Turn your hand over so that it is palm upwards and direct your thumb straight up at the ceiling. This is abduction. Adduction returns the thumb to the original position.

Finally extend your thumb again with the hand still palm upwards; the opposite movement, flexion, not only returns the thumb to its original position at the side of the palm, but also carries it right across the palm, so that the tip touches the base of the little finger. While this movement is taking place the whole thumb rotates in its own long axis so that the pulp faces towards the palm or the tips of the other fingers. This rotation movement is called **opposition** of the thumb. Flexion and opposition are inextricably combined with each other; you cannot have one without the other. The opposite of 'opposition' is therefore called 'extension', because extension is the opposite of flexion.

Note that the metacarpophalangeal joint of the thumb, capable only of flexion and extension, is a pure hinge joint. It is therefore a much simpler joint than the corresponding metacarpal-phalangeal joints of the fingers.

The complex movement of circumduction which takes place at the carpometacarpal joint of the thumb renders superfluous any movements other than flexion and extension in the other thumb joints. In many people the flexion movement in the metacarpophalangeal joint is restricted to only 20–30°. Even the interphalangeal joint cannot flex to 90°.

This carpometacarpal joint of the thumb is the most important joint in the hand. Injury to this joint cripples the hand because so many hand functions involve the thumb in opposition—either the tip as in holding a pen, or the whole length of the thumb when a handle is gripped.

The **muscles in the thenar eminence** play a major role in thumb movements. They are the abductor pollicis brevis, the flexor pollicis brevis, and first palmar interosseous muscles, the adductor pollicis and the opponens pollicis.

It has been seen that the abductor pollicis brevis, the flexor pollicis brevis, the first palmar interosseous, and adductor pollicis are all inserted into the base of the proximal phalanx, although the movements which the abductor and adductor produce take place at the carpometacarpal joint of the thumb, that is between the first metacarpal and the trapezium.

The movements of flexion of the thumb as a whole, as well as abduction and adduction, take place between the trapezium and the first metacarpal. One would expect the muscles mainly responsible for initiating these movements to be inserted directly into this metacarpal. In fact the opponens pollicis is the only muscle which is inserted in this way. All the others are inserted into the base of the proximal phalanx and therefore exert their effect on the carpometacarpal joint via the metacarpophalangeal joint especially via its collateral ligaments. The absence of abduction and adduction in this joint has already been commented on [p. 104], and it is obviously associated with the need for the collateral ligaments to be in tension at all times so that the contraction of the abductor and adductor pollicis muscles can be transmitted directly to the metacarpal bone and thence to the carpometacarpal joint.

When the thumb is abducted against resistance, the abductor pollicis brevis can be seen on the front of the thenar eminence. It is in the ideal position from a mechanical point of view. On the medial side of this muscle is the flexor pollicis brevis which covers the front of the tendon of flexor pollicis longus. This can be felt moving to and fro when the interphalangeal joint is flexed and extended, but indistinctly because the flexor brevis intervenes between your fingers and the tendon.

95 per cent of the opponens pollicis is deep to the abductor pollicis brevis and flexor pollicis brevis. There is a thin strip about 2–3 mm wide where the muscle is subcutaneous (and therefore available for direct palpation) between the subcutaneous part of the thumb metacarpal and the lateral border of the abductor brevis.

The adductor pollicis can be palpated in the web of the thumb, and should be approached from the dorsal not from the palmar surface of the web.

The abductor and flexor pollicis brevis muscles reach their insertion via a sesamoid bone which lies near the head of the metacarpal. It is connected to the proximal phalanx by a short tendon. The first palmar interosseous (regarded by some authorities as the deep head of the flexor brevis and not a separate muscle in its own right) and the adductor pollicis are inserted into the phalanx via another sesamoid bone which is medial to the first one. These two sesamoids are often called the radial and ulnar sesamoid bones of the thumb. The tendon of flexor pollicis longus runs between the sesamoids in exactly the same way as the flexor hallucis longus tendon runs between the pair of sesamoids in the flexor hallucis brevis muscle in the ball of the big toe.

The origins of the thenar muscles are somewhat variable and here the muscles often fuse and intermingle with each other. Their insertions, upon which their movements mainly depend, are more consistent. Note the two heads of origin of the adductor pollicis, the transverse head coming from the front of the shaft of the third metacarpal, and the oblique coming from the bases of the second

and third metacarpals and from their adjacent carpal bones, the capitate and trapezoid. Some of the fibres of the oblique head run parallel with flexor pollicis brevis and their action is similar to that of the flexor. They pass between the tendon of the flexor pollicis longus and the metacarpal bone and are attached to the lateral or radial sesamoid with the flexor brevis. The direction of these 'adductor' fibres makes it possible for it to participate in the movement of opposition. If the median nerve is cut and the opponens is paralysed, the adductor pollicis, which is supplied by the ulnar nerve and therefore would escape, will continue to function. Some patients, after median nerve section, produce a remarkable degree of opposition of the thumb, which may trick the observer into underestimating the damage done to the median nerve, because of the action of the oblique fibres of the adductor pollicis.

The function names of these muscles describes the principal movements in which they take part. It is wrong (as well as illogical) to think that other muscles cannot also produce these various movements.

Feel the tendon of flexor pollicis longus on the palmar surface of the proximal phalanx. It is inserted into the base of the distal phalanx. Its action is to flex the interphalangeal joint; it is the only muscle which can do this. It will also tend to flex the two other more proximal joints of the thumb.

The **extension** movements of the thumb are controlled by the extensor pollicis longus and brevis muscles. The abductor pollicis longus also participates (feel it in yourself) in extension of the carpometacarpal joint, while extensor pollicis brevis extends the proximal phalanx and the extensor pollicis longus extends the distal phalanx.

Extensor pollicis longus is the only muscle which extends the interphalangeal joint, and is the antagonist to flexor pollicis longus. The tendon of this muscle may fray after a fracture of the lower end of the radius (Colles' fracture) and the top joint of the thumb is then in permanent flexion—although of course it can be extended passively.

Movements of the little finger

The **hypothenar muscles** are a modified mirror image of three of the thenar muscles. The abductor digiti minimi runs along the medial side of the fifth metacarpal and is inserted into the medial side of the base of the proximal phalanx. Note that unlike the thumb the movement of abduction takes place in the metacarpophalangeal joint. The muscle takes origin from the pisiform and from the ligaments and tendon attached to this bone. The fixator action of the flexor carpi ulnaris during abduction of the little finger was referred to on page 22.

The flexor digiti minimi takes origin lateral to abductor digiti minimi but is inserted with the abductor into the medial side of the proximal phalanx. These muscles sometimes share a sesamoid bone just like the two corresponding muscles in the thenar eminence.

The opponens digiti minimi lies deep to these two muscles and is inserted into the shaft of the fifth metatarsal in a manner similar to that of opponens pollicis. This muscle can only act on the carpometacarpal joint of the little finger, that is, between the metacarpal and the hamate bone.

The contraction of this muscle is one of the factors responsible for producing the hollow of the palm of the hand, and for giving firmness to the grip on the ulnar side, when pressure is exerted against the tool or handle.

Very few people appreciate the mobility of the fifth metacarpal bone. The head can be moved backwards and forwards, actively or passively, through a distance of about 3 cm. The metacarpal of the neighbouring ring finger is also mobile. The second and third metacarpals, however, are much more firmly fixed presumably as they support the fingers which are so frequently used in gripping objects between the 'fingers and thumb'.

Hold the little finger of your left hand between the fingers and thumb of your right hand. Keeping the finger in the same position, move the hand to and fro. If you relax the intrinsic muscles, especially opponens digiti minimi, you will be surprised by the movement which takes place.

Movements produced by the interossei

The terms 'interossei' or interosseous muscles is used to name those muscles which lie between the metacarpal bones of the hand.

All the interossei are inserted, or should we say attached, to the extensor tendons on the backs of the proximal phalanges of the fingers by means of the so called 'extensor expansions'. This unfortunate term gives quite the wrong idea. In fact the extensor expansion is in no sense an expansion of the extensor tendon itself. It happens that the insertion of the interossei is via an aponeurosis which *merges* with the extensor tendon. The distinction between expansion and merging is critical in understanding the anatomy of the interossei.

Identify the edge of the extensor expansion in your own hand in the following manner. Put the middle or ring finger of the left hand into a circle as shown in FIGURE 5.16 and with the finger and thumb of your right hand palpate the sides of the proximal phalanx. If you squeeze your left finger and thumb together, and even better, at the same time move the proximal interphalangeal joint from side to side (by abducting or adducting the metacarpophalangeal joint) you will feel the edge of the extensor expansion as it runs from palmar to dorsal along the whole length of the proximal phalanx. In many people with rather long slender fingers the extensor expansion can easily be seen; even in people with short 'meaty fingers' the edge of the expansions can always be palpated.

It is immediately obvious from its position, that tension in the so-called extensor expansion must flex the metacarpophalangeal joints.

If you look at the back of the hand in old people, there are usually to be seen three deep gutter-like grooves running between the metacarpal bones. In younger people the back of the hand is more or less flat. The strength and bulk of the interossei is not appreciated because they are only conspicuous by their absence (as in the old) when they begin to waste.

The interossei are in two groups the palmar and dorsal. Since there are five metacarpal bones, including that of the thumb, there are four interosseous spaces. Each space contains two interosseous muscles, the dorsal which lies immediately deep to the skin of the back of the hand and web of the thumb, and the palmar which lies immediately anterior to the dorsal interossei. These cannot be palpated.

The four **dorsal interossei** [p. 107] are all bipennate muscles which take origin from most of the shafts of the two adjacent metacarpal bones.

Study the first dorsal interosseous muscle in your own hand. Place the ulnar border of your left hand on the table with the palm facing to the right and thumb abducted so that it is parallel to the table. Palpate the belly of this muscle between the first and second metacarpals and feel it harden as you point your index finger upwards. This movement of the index finger was abduction [see p. 35]. It is clear that the first dorsal interosseous muscle is an abductor of the index finger.

Now flex the metacarpophalangeal joint against resistance, while extending the two interphalangeal joints. Once more the muscle can be felt contracting strongly. It is not only an abductor but it is also a flexor of the metacarpophalangeal joint.

The second and third dorsal interossei lie in the interosseous spaces between the second and third, and third and fourth metacarpals, and both muscles are inserted into the extensor expansions on each side of the middle finger.

The fourth dorsal interosseous lies between the fourth and fifth metacarpals and is inserted into the extensor expansion on the ulnar side of the ring finger. All the dorsal interossei abduct the three middle fingers. It is obvious that the muscles which abduct the two 'outside' digits (the thumb and little finger) cannot be interosseous in position.

The **palmar interossei** [p. 107] are unipennate muscles which take origin from the medial sides of the metacarpals of the thumb and index finger, and on the lateral side of the metacarpals of the ring and little fingers. They are inserted via the extensor expansions to the extensor tendon of the corresponding finger. This statement does not apply to the palmar interosseous muscle of the thumb, which as we have already mentioned [p. 110] is inserted into the base of the proximal phalanx via the ulnar sesamoid bone.

The action of the palmar interossei is to adduct the fingers, that is to move them towards the imaginary axis passing through the middle of the middle finger.

The pattern of the interossei in the hand is symmetrical around the middle finger, which is moved by a pair of dorsal interossei. The index and ring fingers are moved by a dorsal and palmar interosseous, while the thumb and little finger only have palmar interossei attached to them.

The two sets of interossei differ from each other in that the dorsal muscles abduct the digits, while the palmar ones adduct them at the metacarpophalangeal joints. When both muscles contract simultaneously the abduction and adduction components neutralise each other, and their common action, namely flexion of the metacarpophalangeal joint and extension of the interphalangeal joints takes place.

Since the extensor 'expansion' merges with the extensor tendon just proximal to the proximal interphalangeal joint, the interossei are also able to pull on this tendon and to extend the two interphalangeal joints, as well as flexing the metacarpophalangeal joint. This is useful because when the metacarpophalangeal joint is flexed (say to 90°) any tension in the belly of extensor digitorum cannot be transmitted to the two terminal phalanges because its tendon is

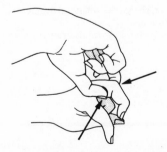

FIG. 5.16. Feel your own extensor expansion.
Put the middle finger of the left hand in a circle with the thumb. Palpate the sides of the proximal phalanx with the other hand and feel the extensor expansion as you abduct and adduct the metacarpophalangeal joint.

fixed to the base of the proximal phalanx. The extensor 'expansion' joins the tendon beyond this point of attachment so that the interphalangeal joints can still be flexed and extended independently of the position of the metacarpophalangeal joint.

Now flex the proximal interphalangeal joint to 90°. You will now be quite unable to extend the terminal phalanx, because the extensor tendon—the only structure capable of extending the top joint—is attached to the base of the middle phalanx and any tension either in extensor digitorum communis or in the interossei cannot extend beyond this point of attachment.

Note that in the thumb this problem does not arise because there are two separate extensor muscles, extensor pollicis longus and extensor pollicis brevis which are attached independently into the terminal and proximal phalanges. Consequently in the thumb the terminal phalanx can always be flexed or extended, no matter what the position of the other joints may be.

Movements produced by the lumbricals

The lumbrical muscles (*lumbricus* = a worm, and these small muscles deserve this contemptuous name) are said to take origin from the radial side of the four tendons of flexor digitorum profundus. They also become entangled in the synovial sheath which surrounds these tendons in the palm of the hand. The insertion is said to be into the extensor expansion on the radial side of the fingers, that is, the first lumbrical joins the first dorsal interosseous, the second joins the second dorsal interosseous, the third joins the third palmar interosseous, and the fourth joins the fourth palmar interosseous.

These muscles have a bulk equal to about 10 per cent of that of the interosseous muscle with which they share the extensor expansion. In theory they can bring about radial deviation of all the fingers at the metacarpophalangeal joints (provided the flexor digitorum profundus tendon is taut enough for this to happen) as well as flexion at the metacarpophalangeal joints. In view of their small size it is hard to see how these muscles can be of significant assistance to the interossei producing flexion of the metacarpophalangeal joint, an extremely powerful movement.

The problem of the function of the lumbricals is prejudged by most people, who are misled by the words 'origin' and 'insertion'. The 'origin' of the lumbricals from the tendons of another muscle, is very curious and interesting. It is reasonable to look here for their special function. It seems likely that these little muscles act synergically with the interossei in the following way. When the fingers are moved by the interossei, as in the upstroke in writing, bellies of the antagonist muscles in the forearm are reciprocally inhibited. This is a purely negative event (inhibition) which does nothing to overcome the drag of the long flexor tendons which run in the wrist and hand and fingers for a distance of about 20 cm. The drag produced by these tendons probably exceeds the friction in the joints of the fingers. But the lumbricals pull the longest flexor tendons towards the fingers so that the drag is largely eliminated. This increases the precision of movement of the fingers.

Ulnar nerve lesion

Following an ulnar nerve lesion, all the interossei are paralysed as well as the two ulnar side lumbricals. This means that the patient has lost the principal flexors of the metacarpophalangeal joints and the principal extensors of the interphalangeal joints. For this reason the fingers are held with the metacarpophalangeal joints extended and the interphalangeal joints flexed. The deformity is less obvious in the index and middle fingers because the two radial side lumbricals will escape, being supplied by the median nerve. As they are attached to the extensor expansions in the manner previously described, they can exert some influence on the posture of the fingers to a degree which is noticeable when all the interossei are paralysed.

Joints of hand and 'cracking'

Examine the metacarpophalangeal joints in your own hand. Note that abduction and adduction can take place in addition to flexion and extension. In the index and little fingers some degree of rotation can also occur when the fingers grip an object such as a cricket ball.

Abduction and adduction occurs freely when the fingers are extended but when the metacarpophalangeal joints are flexed abduction and adduction no longer takes place. Try to produce this movement passively in yourself. You will fail to do so. It is obvious that it is the joints themselves and their ligaments which prevent these movements from occurring. In the extended position the collateral ligaments are slack, but in the flexed position they become tight. Examine the bones of the hand and you will see that the shape of the head of the metacarpals of the fingers is such that in extension the distance [Fig. 5.12, p. 104] from A → B is much less than the distance A → C. What prediction would you make about the shape of the head of the metacarpal of the thumb when viewed from the the side?†

The stability of the joints in extension depends to a marked degree on atmospheric pressure and the molecular cohesion of the synovial fluid as well as the longitudinal pull exerted by the tone in all the muscles. If the finger in extension is pulled distally, the joint 'cracks'. This is due to nitrogen and the other gases suddenly coming out of solution as the sudden reduction in pressure breaks the cohesion of the synovial fluid. A delay is needed for the gases to go back into solution before another 'crack' can occur.

During this interval the instability of the joint is very obvious. Would you expect to be able to crack the metacarpophalangeal joint of the thumb?‡

---

† It has to be circular, so that the collateral ligaments can remain in tension irrespective of flexion and extension.
‡ No, because the collateral ligaments are tight in the extended position.

6 The blood supply and nerve supply to the lower limb

The surface anatomy, muscle groups, and important movements of the lower limb have already been considered on page 44. In this and the next chapter the anatomical structures of the lower limb will be considered in more detail including the origin, insertion, and nerve supply of the individual muscles. First we will start with the vascular supply to the lower limb.

BLOOD SUPPLY TO THE LOWER LIMB
FEMORAL ARTERY

The **femoral artery** is the continuation of the **external iliac artery** which changes its name to the **femoral artery** as it passes deep to the inguinal ligament half-way between the anterior superior iliac spine and the pubic symphysis. The femoral artery runs superficially through the **femoral triangle**, and then under cover of sartorius through the **adductor canal**. It becomes the **popliteal artery** as it passes through an opening in adductor magnus en route to the popliteal fossa at the back of the knee. This opening is about two-thirds the way down the thigh. Note that due to the femur having a long neck, the artery is able to take an almost direct vertical path downwards without encountering the shaft of the femur [FIG. 6.01].

The **femoral vein** lies medial to the artery at the inguinal ligament. Further down the femoral vein becomes posterior to the femoral artery [FIG. 6.02].

The **femoral nerve** lies lateral to the femoral artery at the inguinal ligament, usually separated from it by a part of psoas. The vessels, including the lymphatic vessels are enclosed in the **femoral sheath**. The nerve is not.

The **femoral triangle** [FIG. 6.03] lies on the front of the upper part of the thigh. Its boundaries are the inguinal ligament [p. 191], the medial boundary of sartorius and the medial boundary of the adductor longus. It is roofed over by skin, superficial fascia, and deep fascia. Its floor is muscular: the psoas, pectineus, and adductor longus. It contains the femoral nerve and the femoral artery and vein with their branches or tributaries. At the apex of the triangle medial to the shaft of the femur, about half-way down the thigh, the sartorius muscle overlaps superficially the adductor longus and in doing so gives rise to the **subsartorial canal** [FIGS. 6.01 and 6.04]. The canal is triangular in cross-section, bounded by the vastus medialis and adductor longus and roofed over by the sartorius.

The **femoral vessels** pass through the femoral triangle to the subsartorial canal which they traverse to its lower end. At this point the lower edge of adductor longus has been reached, and the vessels then slip through a large opening in adductor magnus [FIG. 6.05] and reach the popliteal fossa behind the knee joint. On reaching the

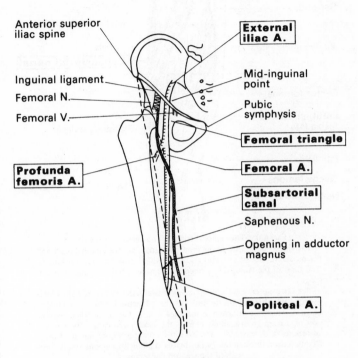

FIG. 6.01. The femoral artery.
Owing to the shape of the femur, the femoral artery is able to take a direct route vertically downwards from the mid-inguinal point where it lies close to the flexor aspect of the hip joint to the popliteal fossa where it lies close to the flexor aspect of the knee joint.

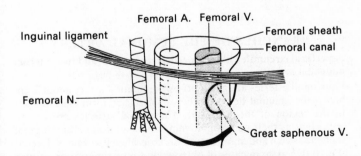

FIG. 6.02. The right femoral artery, vein, and nerve.
The femoral pulse is found at the mid-inguinal point (midway between the anterior superior iliac spine and the pubic symphysis). The femoral artery and the femoral vein lie in a common tubular fibrous sheath termed the femoral sheath. The femoral vein lies medial to the artery. The femoral nerve lies outside the femoral sheath and lateral to it.

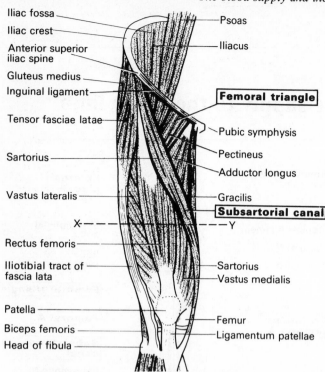

FIG. 6.03. The femoral triangle and subartorial canal.

The three boundaries of the femoral triangle are the inguinal ligament, the medial border of sartorius and the medial border of adductor longus. The floor of the triangle is formed by iliopsoas, pectineus, and adductor longus.

The sartorius covers the deep recess between the vastus medialis which overlies the shaft of the femur and the adductor longus, or adductor magnus, which are attached to the linea aspera. The result is the conversion of a deep groove into a tunnel—the subsartorial canal. The femoral vessels traverse this canal and eventually emerge in the popliteal fossa having passed through the gap or hiatus in adductor magnus about one-third the way up the femur.

popliteal fossa the artery again changes its name to popliteal artery [FIG. 6.01].

During its passage through the adductor canal the femoral artery is crossed from lateral to medial by the **saphenous nerve**; it has the nerve to vastus medialis on its lateral side.

BRANCHES OF THE FEMORAL ARTERY

Superficial circumflex iliac, superficial epigastric, and two external pudendal arteries

Four small arteries are given off by the femoral artery about 1 cm below the inguinal ligament [FIG. 6.04]. They form anastomoses in the region of the hip joint. These small arteries are not so important as their companion veins which all drain into the great saphenous vein and may become varicose when that vein is affected [FIG. 6.07]. Also in cases of obstruction to the superior or inferior vena cava the superficial epigastric vein may become very distended. Its anastomosis with the superior epigastric [p. 173] vein in the rectus sheath provides an alternative route for the venous return following obstruction to either vena cava.

The **superficial circumflex iliac artery** runs laterally below the

inguinal ligament to the anterior superior iliac spine where it anastomoses with the superior gluteal artery, the deep circumflex iliac artery, and the ascending branch of the lateral circumflex artery.

The **superficial epigastric artery** runs upwards over the inguinal ligament to reach the anterior abdominal wall.

The **(superficial) external pudendal artery** crosses the femoral vein to reach the penis and scrotum. or clitoris and labia.

The **deep external pudendal artery** runs medially to reach the scrotum or labia.

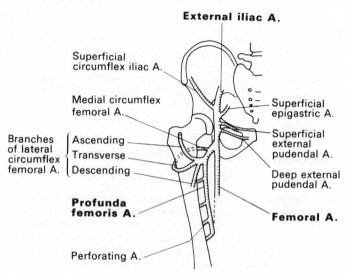

FIG. 6.04. The branches of the femoral artery (simplified).

PROFUNDA FEMORIS ARTERY

The profunda femoris artery arises from the femoral artery about 3 cm distal to the inguinal ligament. It curves round behind the femoral artery and leaves the femoral triangle by passing behind adductor longus close to the femur. It gives off the lateral and medial circumflex femoral arteries and four perforating arteries [FIGS. 6.04 and 6.06]. (In older terminology the femoral artery proximal to the profunda femoris was termed the *common femoral artery*, the continuation after the profunda femoris was termed the *superficial femoral artery*. Profunda femoris is Latin for *deep femoral artery*.)

It is important to realize that the profunda is the **principal blood vessel** supplying the **thigh**. It anastomoses via its branches with other non-femoral-artery vessels in the anastomosis round the hip joint and in the cruciate anastomosis. The femoral artery below the profunda gives off very few branches in the thigh but is distributed to the lower limb from the knee joint downwards. It has no anastomosis to speak of, and if the main artery is obstructed anywhere below the profunda, gangrene of the foot and tissue death from ischaemia is likely. But obstruction above the profunda, instead of making things worse as most students assume, is not so serious. This is because non-femoral-artery blood enters the profunda artery from the anastomosis round the hip joint and from the cruciate anastomosis, and by flowing in the 'wrong' direction can enter the femoral artery and supply the leg and foot.

A penetrating wound which grazes the shaft of the femur on its medial side may damage both the main artery in the adductor canal

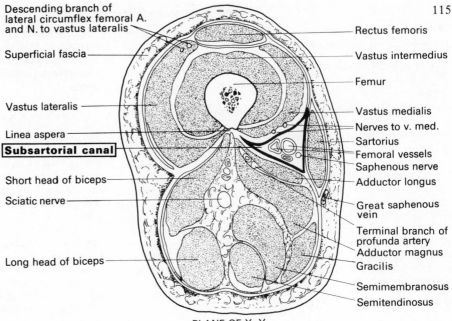

Descending branch of lateral circumflex femoral A. and N. to vastus lateralis

Superficial fascia

Vastus lateralis

Linea aspera

Subsartorial canal

Short head of biceps

Sciatic nerve

Long head of biceps

Rectus femoris

Vastus intermedius

Femur

Vastus medialis

Nerves to v. med.

Sartorius

Femoral vessels

Saphenous nerve

Adductor longus

Great saphenous vein

Terminal branch of profunda artery

Adductor magnus

Gracilis

Semimembranosus

Semitendinosus

PLANE OF X–Y

Fig. 6.05. Transverse section through middle of thigh. The Section is at the level X–Y in Figure 6.03.
Note the femoral artery in the subsartorial canal.

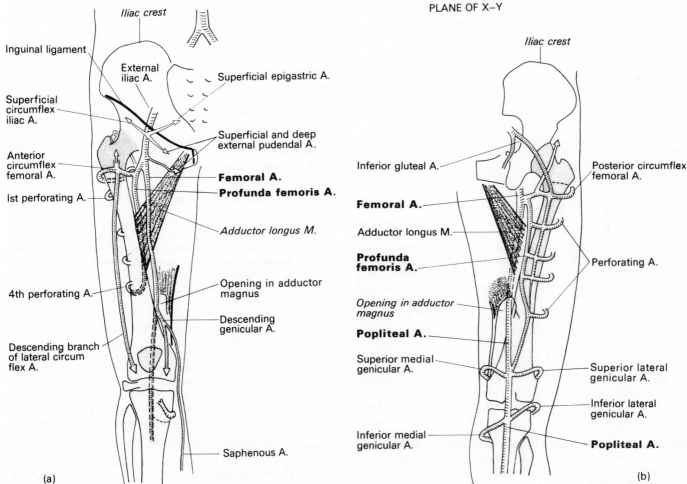

Iliac crest

Inguinal ligament

External iliac A.

Superficial epigastric A.

Superficial circumflex iliac A.

Superficial and deep external pudendal A.

Anterior circumflex femoral A.

Femoral A.

Profunda femoris A.

1st perforating A.

Adductor longus M.

Opening in adductor magnus

4th perforating A.

Descending genicular A.

Descending branch of lateral circum flex A.

Saphenous A.

(a)

Iliac crest

Inferior gluteal A.

Posterior circumflex femoral A.

Femoral A.

Adductor longus M.

Profunda femoris A.

Perforating A.

Opening in adductor magnus

Popliteal A.

Superior medial genicular A.

Superior lateral genicular A.

Inferior lateral genicular A.

Inferior medial genicular A.

Popliteal A.

(b)

Fig. 6.06. Blood supply of the thigh.

(a) Anterior view. (b) Posterior view.

as well as the profunda which is just behind the femoral artery only separated from it by the insertion of adductor longus.

Medial and lateral circumflex femoral arteries

The circumflex arteries are given off near the origin of profunda femoris.

The **medial circumflex artery** runs backwards between psoas and pectineus, and then between adductor brevis and obturator externus to anastomose with the inferior gluteal artery, the first perforating artery and the transverse branch of the lateral circumflex artery. Together these form the cruciate anastomosis (*crux* = a cross).

The **lateral circumflex artery** runs laterally between the branches of the femoral nerve to reach sartorius. It passes deep to sartorius and rectus femoris and divides into transverse, ascending, and descending branches. The transverse branch passes through vastus lateralis to join the cruciate anastomosis. The ascending branch passes upwards deep to tensor fasciae latae to the hip joint. The descending branch runs downwards along the border of vastus lateralis to take part in anastomoses round the knee. It runs with the femoral nerve branch to vastus lateralis.

Perforating arteries

The four perforating arteries are so called because they pierce the adductor muscles [FIG. 6.08] to reach the hamstrings. The lower two pass through adductor magnus only, to reach the hamstrings. The uppermost perforating artery contributes to the cruciate anastomosis, via the longitudinal anastomotic vessel [FIGS. 6.05 and 6.08].

Descending genicular artery

The descending genicular artery arises [FIG. 6.06] from the femoral artery just above the opening in adductor magnus. It divides into two branches. The superficial branch, the saphenous artery, follows the saphenous nerve to the medial side of the leg. The deep branch supplies the knee joint.

Note the pattern of the branches of the medial and lateral circumflex vessels. The **lateral** has three branches, the *horizontal* branch which takes part in the cruciate anastomosis on the back of the thigh, an *ascending* branch to the hip joint and a *descending* branch to the knee.

The **medial circumflex** has *horizontal* and *ascending* branches just like the lateral circumflex. The missing descending branch of the medial circumflex is represented by (part of) the **descending genicular artery**, which in its distribution to the anastomosis around the knee joint exactly matches on the medial side, the distribution of the descending branch of the lateral circumflex, on the lateral side.

POPLITEAL ARTERY

The **popliteal artery** [FIGS. 6.06 and 6.09] starts at the opening in adductor magnus in continuation with the femoral artery. It passes down on the popliteal surface of the femur, on the posterior ligament of the knee joint and on popliteus to the lower border of popliteus where it divides into the **anterior tibial** and the **posterior tibial arteries**. The bifurcation is at a level at the back which corresponds to the tuberosity of the tibia in front. The popliteal artery lies deep to the surface, and hence the popliteal pulse can be difficult to palpate. To find it flex the knee and press forwards against the tibia. When one leg is crossed over the other, the foot often moves up and down in time with the pulse.

The popliteal artery gives off a number of branches (superior medial and lateral geniculate, middle geniculate, inferior medial, and lateral geniculate) which form anastomoses around the knee joint (*Genu* (Latin) = knee, hence geniculate = pertaining to the knee).

ANTERIOR TIBIAL ARTERY

The anterior tibial artery arises from the popliteal artery at the lower border of popliteus. It passes to the anterior compartment of the leg by passing between the tibia and fibula above the interosseous membrane. It runs down the front of the leg on this interosseous membrane in the anterior tibial compartment to become the dorsalis pedis artery (artery of the dorsum of the foot) in front of the ankle joint.

The anterior tibial artery lies on the interosseous membrane and has tibialis anterior on its medial side and extensor digitorum longus and extensor hallucis on its lateral side. It runs between the muscles attached to the tibia and those attached to the fibula. Near the ankle joint, it is crossed by the tendon of extensor hallucis longus. It passes deep to the two extensor retinacula.

The anterior tibial artery as it passes down the leg is accompanied by the **deep peroneal nerve** which lies laterally, then in front, and then laterally again with respect to the artery [p. 126].

The deep peroneal nerve is a terminal branch of the **common peroneal nerve** which reaches the front of the leg by passing laterally round the neck of the fibula.

The surface marking of the anterior tibial artery is a line running from the mid-point between the tuberosity of the tibia and the head of the fibula, to the mid-point between the two malleoli.

The anterior tibial artery gives off anastomotic branches around the knee joint and the ankle joint. The branches in the region of the knee joint are the **anterior tibial recurrent**, and the **posterior tibial recurrent arteries**. The branches around the ankle joint are **anterior lateral** and **medial malleolar arteries**.

DORSALIS PEDIS ARTERY

The anterior tibial artery becomes the dorsalis pedis artery in front of the ankle joint. Accompanied by the deep peroneal nerve, the dorsalis pedis artery runs down the dorsum of the foot over the talus, navicular, and medial cuneiform (where the dorsalis pedis pulse may be readily felt). The nerve may be on either side. The artery runs deep to the most medial tendon of extensor digitorum brevis. The dorsalis pedis artery then turns through a right-angle and passes into the sole of the foot between the bases of the first and second metatarsals where it unites with the lateral plantar artery to form the plantar arch.

A branch of the dorsalis pedis artery, the first dorsal metatarsal artery, continues down the dorsum of the foot between the first and second toes and supplies the medial one and a half toes.

The other branches of the dorsalis pedis artery are the **tarsal artery** which runs laterally under extensor digitorum brevis and the **arcuate artery** which forms the **dorsal metatarsal arteries** supplying the clefts between the four lateral toes.

Clinically it may be important to find the pulse of the dorsalis pedis. Look for it between the extensor tendons going to the first and second toes. Place your ring finger at the proximal end of the first intermetatarsal space, and put your middle and index finger on the dorsum of the foot immediately proximal to this point. You are palpating the last 2–3 cm of the artery.

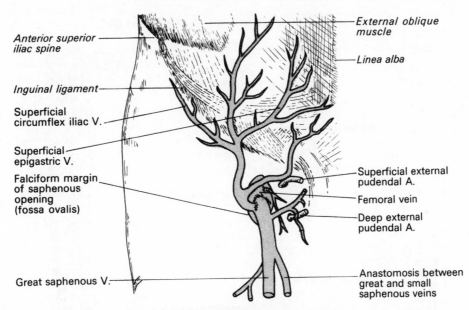

FIG. 6.07. Fossa ovalis of thigh (saphenous opening).
The great saphenous vein passes through the saphenous opening and penetrates the femoral sheath to join the femoral vein. The saphenous opening (fossa ovalis) lies about 4 cm below and lateral to the pubic tubercle. It is covered by a thin layer of investing fascia termed the *cribriform fascia* (cribrum (L) = sieve) since it is perforated by many structures. The saphenous opening has a well-marked crescentic superior and lateral edge.

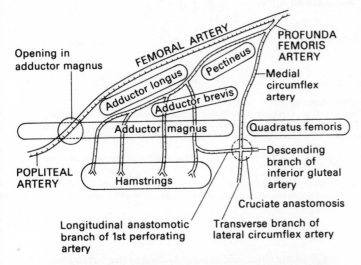

FIG. 6.08. Diagrammatic representation of the profunda femoris artery and its branches in the thigh. Viewed from side.
The four perforating arteries reach the hamstring muscles by perforating adductor magnus. The upper two also perforate adductor brevis. The 1st perforating artery gives off a longitudinal branch which forms the cruciate anastomosis with the circumflex arteries and the descending branch of the inferior gluteal artery.

THE POSTERIOR TIBIAL ARTERY

The posterior tibial artery [FIG. 6.09] starts at the lower border of popliteus as the other terminal branch of the popliteal artery. It runs down the back of the leg accompanied by the tibial nerve. It passes deep to the tendinous arch of the soleus muscle and runs down the leg in the plane of cleavage between the superficial and deep muscles. The former are all inserted into the tendo calcaneus.

It does not lie on the interosseous membrane like the anterior tibial artery but on the posterior surface of tibialis posterior. At the ankle it becomes superficial and passes behind the medial malleolus to end at the mid-point between the medial malleolus and the tuberosity of the calcaneus by dividing into the medial and lateral plantar arteries. The pulse can be felt here.

Peroneal artery

The peroneal artery [FIG. 6.09(b)] arises from the posterior tibial about 3 cm below its origin. It runs laterally between soleus and tibialis posterior to reach a fibrous canal close to the fibula and deep to flexor hallucis longus. It passes behind the ankle joint medial to the peroneal tendons and divides into lateral calcanean branches. It sometimes gives off quite a large perforating peroneal branch which runs in front of the lateral malleolus where it can be palpated. Always look for this pulse. Sometimes the dorsalis pedis is a small artery, in which case this branch is usually large.

Medial plantar artery

The medial plantar artery [FIG. 6.09(b)] is the smaller of the two terminal branches of the posterior tibial artery. It runs with the medial plantar nerve on the medial side of the sole of the foot to supply the medial 1½ toes.

Lateral plantar artery

The lateral plantar artery [FIG. 6.09(b)] is the larger of the two terminal branches of the posterior tibial artery. It starts between abductor hallucis and flexor accessorius and then runs forwards and laterally with the lateral plantar nerve between flexor digitorum brevis and flexor accessorius to reach the base of the fifth metatarsal. Here it turns medially to form the plantar arch which joins the dorsalis pedis artery at the first interspace. The plantar arch lies deep on the second, third, and fourth metatarsals. It supplies the outer 3½ toes.

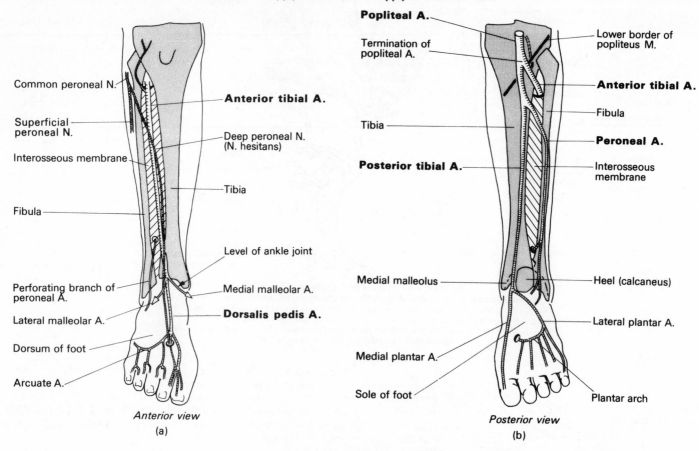

FIG. 6.09. Blood supply of the leg.
　　The blood supply of the leg is distributed via three arteries. the anterior tibial and the posterior tibial and peroneal arteries (synonymous for fibular artery). It is important to note that they are all branches of the popliteal artery which is virtually an 'end artery'. Apart from a small contribution from the longitudinal anastomosis joining the perforating arteries the blood supply comes therefore almost entirely from the femoral artery. Obstruction of the femoral or popliteal artery therefore has disastrous effects on the foot and leg.
　　The blood vessel pattern has certain resemblances to that in the forearm and hand. Note the dorsalis pedis behaving like the termination of the radial artery as it passes through the two heads of the 1st dorsal interosseous to the plantar arch.

VENOUS DRAINAGE OF LOWER LIMB

DEEP VEINS

The small arteries are accompanied by venae comitantes (usually two, one on either side of the artery) which act as heat exchangers, so that cold blood from the skin of the feet is warmed up by the arteries as it runs towards the heart. These companion veins have the same name as the adjacent artery. The venae comitantes of the anterior and posterior tibial arteries unite in the popliteal fossa to form the single popliteal vein which becomes the femoral vein at the aperture in adductor magnus. It lies on the medial side of the femoral artery in the femoral triangle. Passing deep to the inguinal ligament on the medial side of the artery it becomes the external iliac vein which leads to the common iliac vein, inferior vena cava and right side of the heart.

The alternate contraction and relaxation of the surrounding muscles, such as when walking, aids the venous return along these veins by a muscle pump action. The valves in the veins allow segments of the veins to be alternatively filled and emptied of blood and due to the direction the valves are pointing, this blood is directed towards the heart.

When there is no muscular activity, the blood returns along the veins by the *vis a tergo*, the force from behind, from the blood pressure in the arteries which is transmitted through the arterioles and capillaries to the veins. When standing still the pressure in the veins of the feet has to be over 100 mmHg, to overcome the effect of gravity on the venous column between this point and the heart. Under these conditions, the valves play no part, since they must be open to allow a venous return. When walking, with the muscle pump in operation, the venous pressure in the dependent veins falls markedly. Because of the high venous pressure it is not surprising that the feet swell in normal people during the day. When walking all day it is wise to put on another pair of larger sized boots at least once, preferably twice.

The effect of gravity, by increasing the venous pressure when standing still, distends the veins and by so doing increases the capacity of the circulation (the veins are the main capacitance vessels). As a result less blood is available to fill the ventricles of the heart during diastole and the cardiac output falls. There is thus a greater cardiac output when lying (say, 6 litres/minute) than when standing still (say, 4 litres/minute). When walking about, the venous return and hence the cardiac output increases again. A patient who has lost the tone in veins (venomotor tone), as a result of long bed rest, will lose consciousness if suddenly changed from the lying to the standing position.

Since stasis can lead to blood clotting, deep vein thrombosis may occur in the veins of the lower limbs following periods when no muscular activity has occurred, such as during a long plane flight.

Note that it is the alternate contraction and relaxation of the surrounding muscles that operates the 'muscle pump'. A sustained contraction or spasm closes the vein and impedes the venous return. Thus a patient who is shuffling round the wards with legs as rigid columns with both flexor and extensors contracted, aided on either side by nurses and physiotherapists, is unlikely to be increasing the venous return from the legs. What is needed for a good venous return is springy walking steps where first the flexors then the extensors contract. Each group of muscles is then alternately contracting and relaxing.

Deep vein thromboses are dangerous because apart from any local effects, there is also the possibility of the blood clot becoming detached and passing up through the right side of the heart to the pulmonary artery causing a pulmonary embolus, which may be fatal. Superficial vein thromboses are usually more adherent to the vessel wall and, as a result, less likely to cause pulmonary embolisms.

The deep veins in the calves of the legs between the superficial and deep muscle groups form a rich plexus in which the blood stagnates and may thrombose. This may happen during an operation in which the muscles are paralysed by the anaesthetic and give rise later (7–10 days) to a post-operative pulmonary embolus.

SUPERFICIAL VEINS

Great saphenous vein

The presence of a superficial vein running the whole length of the medial side of the lower limb is a feature of the anatomy of man. The great saphenous vein [Fig. 6.10] is just beneath the skin in the superficial fascia and saphenous means 'easy to see'. There is a small **saphenous artery** (not at all easy to see) which runs with it at the knee joint but this artery usually peters out before it reaches the ankle. The **saphenous nerve** is a branch of the femoral nerve [p. 122] and also joins the vein at the knee. It eventually supplies the skin on the medial side of the foot. The great saphenous vein and the cephalic veins of the upper limb are homologous.

The **great saphenous vein** starts on the medial side of the dorsal venous arch on the dorsum of the foot. It passes upwards in front of the medial malleolus where it can always be rolled against the bone. It continues upwards over the subcutaneous part of the tibia

and comes to lie about a hand's breadth behind the medial border of the patella. Having reached the thigh it runs up over the adductor muscles to the **saphenous opening** 2–3 cm below and lateral to the pubic tubercle. Here it pierces the **cribriform fascia** and femoral sheath to terminate in the femoral vein.

The great saphenous vein has numerous tributaries and communicates frequently along its route with the deep veins. Being a superficial vein it is especially liable to dilatation because it has no support from the leg muscles. When the vein dilates, the valves become faulty and varicosities form (varicose veins).

The great saphenous vein is the longest vein in the body. Although, like all veins, it is thin-walled, the vein is strong enough

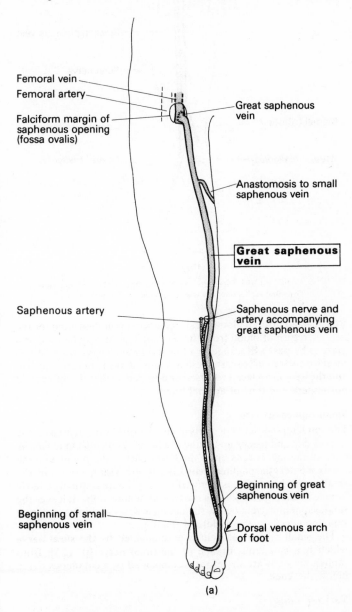

(a)

FIG. 6.10(a). Superficial veins of the lower limb. Anterior view. The great saphenous vein is accompanied by the saphenous nerve and artery.

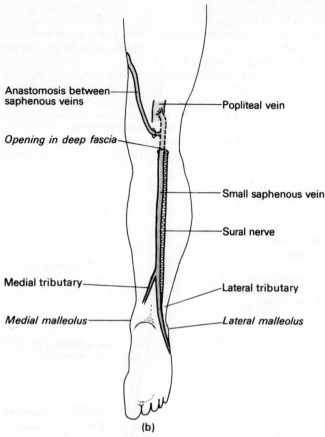

FIG. 6.10(b). Superficial veins at the lower limb. Posterior view. The small saphenous vein is accompanied by the sural nerve.

to withstand arterial blood pressure. Since it is not essential for the venous return of blood from the feet and legs, it may be used as a graft to by-pass a blockage in a femoral or coronary artery. When used for such a purpose, the vein is removed, the tributaries tied off and the vein is replaced reversed end-to-end so that the valves will not impede the flow of arterial blood.

Small saphenous vein

The small saphenous vein starts from the dorsal venous arch on the lateral side and passes up behind the lateral malleolus to reach the back of the leg. It runs up to the lower part of the popliteal fossa where it enters the **popliteal vein**. Like the great saphenous vein, the small saphenous vein has many tributaries. There is usually a fairly large connecting vein joining the two saphenous veins. It leaves the small saphenous vein below the popliteal fossa and joins the great saphenous vein at the middle of the thigh.

The small saphenous vein is accompanied by the **sural nerve** which is a cutaneous branch of the tibial nerve [p. 125]. Both saphenous veins are therefore accompanied by a cutaneous nerve below the knee.

Varicose veins

Tortuous and often as fat as a finger, varicose veins are caused in many cases by blood flowing in the communicating veins under pressure from the deep into the superficial veins. This does not

happen normally because the direction of the valves in the communicating veins prevents it. It is very important, before varicose veins are treated by thrombosing or removing them, to establish that the deep veins are patent. Otherwise treatment may remove the one remaining venous pathway.

LYMPHATIC DRAINAGE OF THE LOWER LIMB

The lymphatics draining the medial aspect of the foot and the medial side of the leg run up with the great saphenous vein to the **superficial inguinal lymph nodes** which are situated along the upper part of the great saphenous vein and along the inguinal ligament. In addition to draining the lower limb, these lymph nodes also drain the lower abdominal wall, the external genitalia, the perineum and the anal region, and the fundus of the uterus [p. 251].

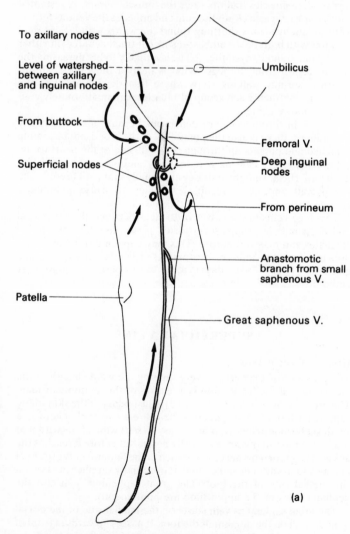

FIG. 6.11(a). Lymphatic drainage of lower limb following great saphenous vein.

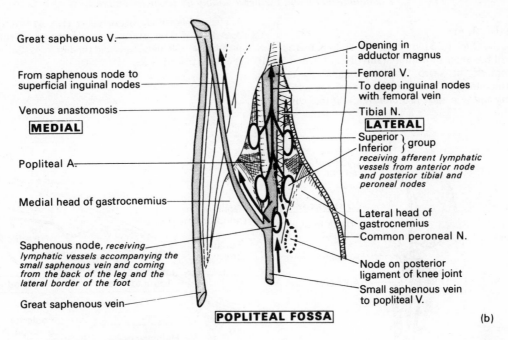

Great saphenous V.

From saphenous node to
superficial inguinal nodes

Venous anastomosis

MEDIAL

Popliteal A.

Medial head of gastrocnemius

Saphenous node, *receiving
lymphatic vessels accompanying the
small saphenous vein and coming
from the back of the leg and the
lateral border of the foot*

Great saphenous vein

Opening in
adductor magnus

Femoral V.

To deep inguinal nodes
with femoral vein

Tibial N.

LATERAL

Superior ⎱
Inferior ⎰ group
*receiving afferent lymphatic
vessels from anterior node
and posterior tibial and
peroneal nodes*

Lateral head of
gastrocnemius

Common peroneal N.

Node on posterior
ligament of knee joint

Small saphenous vein
to popliteal V.

POPLITEAL FOSSA

(b)

FIG. 6.11(b). Lymphatic drainage of lower limb following small saphenous
veins.

The lymph nodes are found at the termination of the great and small
saphenous veins. Other deep nodes are present in the popliteal fossa
alongside the main vessels, and in the inguinal region in the femoral canal
alongside to the femoral vein.

The catchment area of the inguinal glands is very extensive. The nodes
are arranged along the terminal parts of the great saphenous vein and the
four veins which drain into it.

It is helpful to note the resemblance of the lymph node patterns in the
two limbs. The popliteal and supracondylar lymph nodes resemble each
other. The superficial inguinal nodes running vertically along the great
saphenous vein closely resemble the nodes alongside the terminal centimetres of the cephalic vein. A 'bubo' is a lump in the groin from any
cause. Bubonic plague was often due to a flea bite at the ankle, which sets
up infection in the regional nodes in the groin.

The lymphatics draining the lateral aspect of the foot run up with
the small saphenous vein to the **lymph nodes** in the **popliteal fossa**.
From here lymphatics pass up with the popliteal and femoral vessels
to the **deep inguinal lymph nodes**. These surround the upper part of
the femoral vein [FIG. 6.11].

The inguinal lymph nodes drain into the **external iliac nodes**.

NERVES OF THE LOWER LIMB

The lower limb is supplied by the lumbosacral plexus of the spinal
cord which corresponds to the brachial plexus supplying the upper
limb. This plexus may be conveniently subdivided into a lumbar
plexus and a sacral plexus.

THE LUMBAR PLEXUS

The lumbar plexus [FIG. 6.12] is formed by the anterior rami of the
first, second, third, and fourth lumbar nerves (L1–4) with components from T12 and L5. It lies in front of the transverse processes of
the lumbar vertebrae in the psoas muscle.

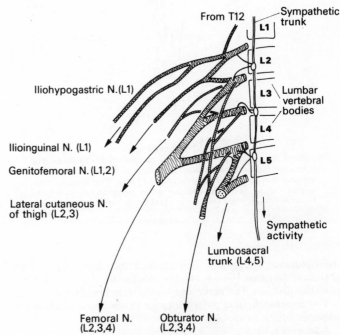

From T12

Sympathetic
trunk

L1

L2

Lumbar
vertebral
bodies

L3

L4

L5

Iliohypogastric N.(L1)

Ilioinguinal N. (L1)

Genitofemoral N. (L1,2)

Lateral cutaneous N.
of thigh (L2,3)

Sympathetic
activity

Lumbosacral
trunk (L4,5)

Femoral N.
(L2,3,4)

Obturator N.
(L2,3,4)

FIG. 6.12. Lumbar plexus.

The branches are shown in the Figure. The obturator nerve and lumbosacral trunk run downwards into the pelvis and emerge from the psoas
muscle on its medial border. The other nerves appear at various levels
along its lateral border, with the exception of the genitofemoral nerve
which appears at the middle of the anterior surface.

The sympathetic fibres joining the nerve roots supply the blood vessels
and sweat glands in the lower limb.

L4 is sometimes called the nervus furcalis because it forks, one component going to the lumbar plexus, the other to the sacral plexus via the
lumbosacral trunk.

The psoas muscle, in which the lumber plexus is embedded, gets its
nerve supply directly from L2 and L3 (not shown in diagram).

Obturator nerve (L2, 3, 4)

The obturator nerve [FIG. 6.13] arises in the abdomen from the ventral branches of the anterior primary rami of L2, L3, and L4. It is formed in front of the sacro-iliac joint behind psoas. The obturator nerve runs along the lateral wall of the pelvis lateral to the internal iliac artery and vein and the ureter. It leaves the pelvis with

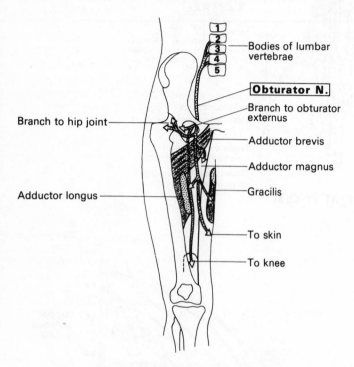

FIG. 6.13. Obturator nerve.
The obturator nerve passes through the obturator canal and divides into its anterior and posterior divisions as it does so. These two nerves lie respectively in front of or behind the adductor brevis. The obturator nerve supplies the muscles in the adductor compartment as well as sensory branches to the hip and knee joints and skin over the adductor muscles.

the obturator artery and vein through the upper part of the obturator foramen [FIG. 6.14] and divides into an anterior and posterior division. The nerve and vessels lie in the obturator canal.

The **anterior division** passes anterior to obturator externus and adductor brevis to supply:

Adductor longus
Adductor brevis
Gracilis
Hip joint

A branch joins the medial anterior cutaneous nerve and saphenous nerve to form the subsartorial plexus which supplies the skin over the inner side of the thigh and knee joint.

The **posterior division** supplies:

Obturator externus
Adductor brevis
Adductor magnus (part of)
Knee joint

The **accessory obturator nerve** (L3, 4) is only present in about one-third of cases. It is a slender nerve which runs down in front of psoas to supply pectineus and the hip joint. Occasionally it supplies the adductor muscles.

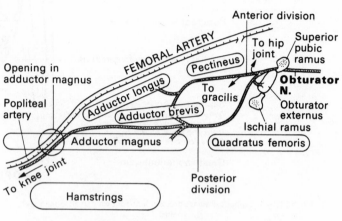

FIG. 6.14. Diagrammatic representation of the obturator nerve in the thigh and the muscles it supplies.
The terminal branch of the posterior division runs with popliteal artery and supplies the knee joint.

Femoral nerve (L2, 3, 4)

The femoral nerve [FIG. 6.15] has the same nerve roots as the obturator nerve but is derived from the dorsal branches of the anterior rami of L2, L3, and L4. The femoral nerve appears at the lateral border of psoas and runs down in the groove between psoas and iliacus. It lies behind the fascia of the abdomen and thus lies outside the femoral sheath as it enters the thigh by passing deep to the inguinal ligament lateral to the femoral artery. It ends almost immediately by dividing into its branches.

The **femoral nerve** supplies the following **muscles**:
Iliacus
Rectus femoris
Vastus lateralis   } Quadriceps femoris
Vastus medialis
Vastus intermedius
Sartorius
Pectineus

The **cutaneous branches** of the **femoral nerve** are:
Medial anterior cutaneous of thigh
Lateral anterior cutaneous of thigh
Saphenous nerve

The saphenous nerve is the only branch of the femoral nerve to extend below the knee. It follows the femoral artery and only becomes superficial below the medial side of the knee. It then runs

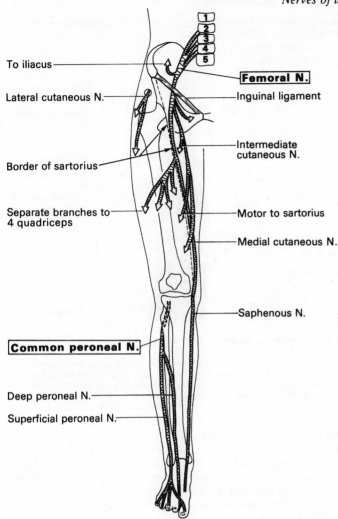

To iliacus

Lateral cutaneous N.

Border of sartorius

Separate branches to 4 quadriceps

Femoral N.

Inguinal ligament

Intermediate cutaneous N.

Motor to sartorius

Medial cutaneous N.

Saphenous N.

Common peroneal N.

Deep peroneal N.

Superficial peroneal N.

FIG. 6.15. Femoral nerve.

Whilst still in the abdominal cavity the femoral nerve supplies the iliacus muscle and gives off the branch to pectineus.

As soon as it enters the thigh, like the obturator nerve it separates into two divisions. The anterior division supplies the skin and sartorius muscle. The posterior division supplies the four heads of the quadriceps muscle by separate nerves which run with the blood supply of these muscles. It also gives off the saphenous nerve which supplies the medial side of the foot as far as the ball of the big toe.

As the cutaneous nerves come to the surface along the sartorius muscle and supply the skin below this muscle. there must be another 'non-femoral-nerve' to supply the skin over the femoral triangle.

This is the function of the branches of the genitofemoral nerve.

down the leg with, and anterior to, the great saphenous vein. It passes in front of medial malleolus and supplies the skin as far as metatarsophalangeal joint of the first toe (big toe).

Note that all these cutaneous branches, as well as the lateral cutaneous nerve of the thigh, reach the surface along the line of the sartorius muscle.

The femoral nerve is tested by eliciting the 'knee jerk' [p. 135].

Iliohypogastric nerve (L1)

The iliohypogastric nerve appears at the lateral border of psoas

[FIG. 6.12]. It runs down between the kidney in front and quadratus lumborum behind to the iliac crest. It pierces the aponeurosis of transversus abdominis and runs in the plane between transversus and the internal oblique. Here it divides into its two branches. The iliac branch passes through internal oblique and external oblique and runs over the iliac crest to supply the skin over gluteus medius. The hypogastric branch also pierces the internus oblique and becomes cutaneous by piercing the external oblique about 2 cm above the superficial inguinal ring. It does not reach the midline.

Ilio-inguinal nerve (L1)

The ilio-inguinal nerve appears at the lateral border of psoas and runs down below and parallel to the iliohypogastric nerve in front of quadratus lumborum and behind the kidney. It pierces transversus abdominis and internal oblique to reach the inguinal canal. It leaves the inguinal canal via the superficial inguinal ring and supplies the skin of the upper and medial part of the thigh and the scrotum or labium.

Genitofemoral nerve (L1, 2)

The genitofemoral nerve emerges from the anterior border of psoas at the level of L3. It runs under the fascia of psoas behind the peritoneum and ureter. At the level of the inguinal ligament it divides into a genital (external spermatic) branch and a femoral (crural) branch. The genital branch enters the inguinal canal and supplies the cremaster muscle, skin of the scrotum and skin on the inside of the thigh. The femoral branch enters the thigh in the femoral sheath on the lateral side of the femoral artery and supplies the skin on the upper part of the front of the thigh, above the sartorius muscle.

In the male this nerve as well as spinal cord segments L1 and L2 can be tested by eliciting the **cremasteric reflex**. The skin on the thigh is scratched and the testis is drawn upwards by the cremaster muscle.

Lateral cutaneous nerve of thigh (L2, 3)

The lateral cutaneous nerve appears at the outer border of psoas and runs across iliacus in the direction of the anterior superior iliac spine. It passes deep to the inguinal ligament to reach the groove between sartorius and tensor fasciae latae. It divides about 5 cm below the anterior superior iliac spine into an **anterior branch** which supplies the skin of the anterolateral aspect of the thigh down to the knee, and the **posterior branch** which supplies the skin over the greater trochanter and buttocks. The nerve pierces the fascia lata to reach the skin, and may be kinked at this point, giving rise to pain.

THE SACRAL PLEXUS

The sacral plexus is formed by the fourth and fifth lumbar nerves and the anterior rami of the first, second, and third sacral nerves which are passing out through the anterior foramina of the sacrum. The fourth lumbar nerve is unusual in that it divides into two. As has been seen, its upper branch contributes to the lumbar plexus, but it also has a lower branch which unites with the fifth lumbar nerve to form the **lumbosacral trunk** which runs over the sacro-iliac joint to reach the sacral plexus [FIG. 6.16].

The sacral plexus lies in the pelvis, behind the pelvic colon, ureter and hypogastric plexus and in front of piriformis and the pelvic fascia. The superior gluteal artery, as it leaves the pelvis, passes between the lumbosacral trunk and the first sacral nerve. The

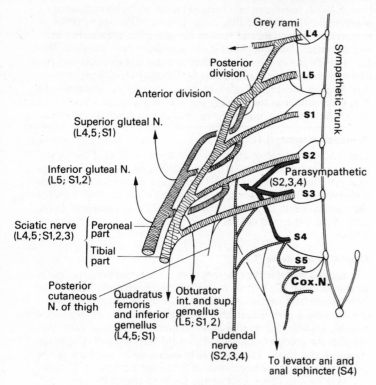

FIG. 6.16. Sacral plexus.

The principal component of the sacral plexus is the sciatic nerve. Each nerve root separates into an anterior and posterior division. The anterior divisions supply the morphological flexors (hamstrings and calf muscle) while the posterior divisions supply the morphological extensors (anterior tibial group). The basic scheme is like that in the upper limb but there is nothing here corresponding to the trunks of the brachial plexus.

The pudendal nerve is important as it supplies the genitalia. The sacral parasympathetic fibres from S2, 3 and 4 leave the nerve roots and join the pudendal nerve in the side wall of the pelvis some distance away from the sacrum. Paralysis of these nerves results in impotence, and retention of urine.

The sympathetic ganglia lie beside the sacral nerve roots as they leave the sacral foramina and send grey rami (only) to the plexus. They are concerned only with the sympathetic supply of the lower limb, as the pelvic viscera get their sympathetic supply from the hypogastric plexus.

inferior gluteal artery runs between the first and second sacral nerves [FIG. 6.17].

Superior gluteal nerve (L4, 5; S1)

The superior gluteal nerve leaves the pelvis through the greater sciatic foramen above piriformis. It is accompanied by the superior gluteal artery. The superior gluteal nerve supplies:

Gluteus medius
Gluteus minimus
Tensor fasciae latae

Inferior gluteal nerve (L5; S1, 2)

The inferior gluteal nerve leaves the pelvis through the greater sciatic foramen below piriformis. It supplies:

Gluteus maximus

Nerve to quadratus femoris (L4, 5; S1)

The nerve to quadratus femoris leaves the pelvis through the

greater sciatic foramen below piriformis. It runs deep to superior gemellus and obturator internus to supply:

Quadratus femoris
Inferior gemellus

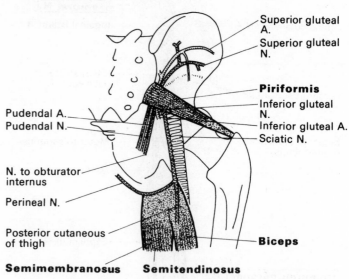

FIG. 6.17. The nerves and arteries of the gluteal region (simplified). Note the key position of piriformis in separating the superior gluteal vessels and nerves from the inferior gluteal vessels and nerve. Note also the large target the sciatic nerve makes if an injection into the buttocks is not given in the upper and outer quadrant.

Nerve to obturator internus (L5; S1, 2)

The nerve to obturator internus leaves the pelvis through the greater sciatic foramen with the pudendal nerve and internal pudendal artery. They cross behind the ischial spine and sacrospinous ligament and then re-enter the pelvis through the lesser sciatic foramen. The nerve supplies the superior gemellus as well as the obturator internus.

Nerve to piriformis (S1, 2)

This nerve runs straight to piriformis which arises from within the pelvis.

Posterior cutaneous nerve of thigh (S1, 2, 3)

The posterior cutaneous nerve of the thigh leaves the pelvis through the lower part of the greater sciatic foramen (below piriformis) on the medial or posterior aspect of the sciatic nerve. For this reason it was previously known as the *small sciatic nerve*. The nerve gives off numerous small branches to the skin over the back and medial side of the thigh and gluteal branches to the gluteal region. The perineal branch passes across the origin of the hamstrings to reach the skin of the scrotum or labia and root of penis or clitoris. The nerve itself continues down immediately under deep fascia to the popliteal fossa where it pierces deep fascia to supply the skin on the posterior surface of the upper part of the leg.

SCIATIC NERVE (L4, 5; S1, 2, 3)

The sciatic nerve [FIG. 6.18] is the largest nerve in the body. It is a very broad nerve with a transverse diameter of about 2 cm. Hence the large target for a needle inserted into the buttocks!

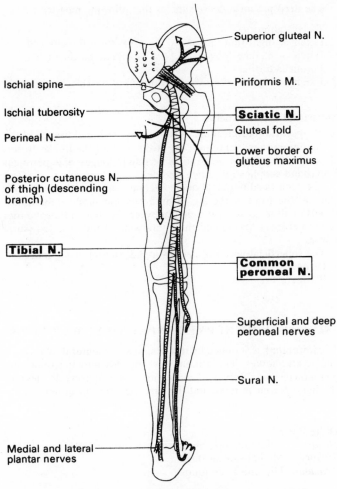

Superior gluteal N.

Ischial spine

Piriformis M.

Ischial tuberosity

Sciatic N.

Perineal N.

Gluteal fold

Lower border of
gluteus maximus

Posterior cutaneous N.
of thigh (descending
branch)

Tibial N.

**Common
peroneal N.**

Superficial and deep
peroneal nerves

Sural N.

Medial and lateral
plantar nerves

FIG. 6.18. The sciatic nerve.

The sciatic nerve is covered from behind first by gluteus maximus and then by the long head of biceps. It lies directly on bone as it leaves the pelvis. It is then cushioned by the gemelli and by quadratus femoris before it comes to lie on the posterior surface of adductor magnus.

Notice that it lies at the junction of the middle and medial one-thirds of the line joining the ischial tuberosity to the greater trochanter. In a patient lying face-down awaiting an injection into the buttock. the hip joint will be externally rotated. the great trochanter displaced medially. and the nerve will now lie about half way along this line.

The hamstring muscles are normally supplied directly by the sciatic nerve. The muscles below the knee are supplied by its terminal branches. the tibial and common peroneal nerves. while the sole of the foot is supplied by the two terminal branches of the tibial nerve.

Notice that the common peroneal nerve resembles the posterior interosseous nerve of the forearm. The medial and lateral plantar nerves resemble the median and ulnar nerves (respectively) once the latter pair have entered the hand.

The path of the superficial and deep peroneal nerves are shown in FIGURE 6.15. The nerve supply of the skin of the dorsum of the foot is complex. The superficial peroneal supplies all the toes except for the cleft between the 1st and 2nd toes. The medial and lateral borders of the foot have their own nerves, the saphenous nerve and sural nerves respectively.

The sciatic nerve, at its commencement, lies deep to gluteus maximus. It separates from the posterior cutaneous nerve of the thigh by passing deep to the long head of biceps (the posterior cutaneous nerve of the thigh runs superficial to this muscle). It then

runs down the back of the thigh between adductor magnus and the hamstrings.

About two-thirds of the way down the thigh it divides into the **common peroneal nerve**, which runs laterally, and the **tibial nerve**.

The **sciatic nerve** supplies the following **muscles:**
Biceps (long head) (S1, 2, 3)
Semitendinosus (L5; S1, 2)
Semimembranosus (L4, 5; S1)
Adductor magnus (L4, 5)

The branches to the hamstrings come off in the buttock before the nerve reaches the thigh. The hamstrings usually escape following a knife wound severing the sciatic nerve. The tibial and common peroneal nerves are often separate right from the beginning, in which case the common peroneal may pass through the piriformis muscle, instead of passing below it.

Tibial nerve

The tibial nerve (older names *medial popliteal nerve* and more distally *posterior tibial nerve*) appears to be the direct continuation of the sciatic nerve. It runs down vertically through the middle of the popliteal fossa, and then down the back of the leg accompanied by the posterior tibial artery. Together they pass behind the medial malleolus. The tibial nerve ends, like the artery between the medial malleolus and the tuberosity of the calcaneus, by dividing into the **medial** and **lateral plantar nerves.**

During its passage down the leg, the tibial nerve lies between soleus and gastrocnemius *behind*, and, in succession, tibialis posterior flexor digitorum longus and the tibia *in front*.

The **tibial nerve** supplies the following **muscles:**

Gastrocnemius (S1, 2) ⎫ Triceps surae
Soleus (L5; S1, 2) ⎭
Plantaris (L4, 5; S1)
Popliteus (L4, 5; S1)
Tibialis posterior (L3; S1, 2) ⎫ Muscles of deep
Flexor hallucis longus (L5; S1, 2) ⎬ compartment of the
Flexor digitorum longus (L5; S1, 2) ⎭ calf

The tibial nerve is tested by eliciting the **ankle jerk** [p. 141].

Sural nerve

The sural nerve is given off by the tibial nerve in the popliteal fossa and runs down between the two heads of gastrocnemius to become cutaneous half-way down the leg. In company with the small saphenous vein, it passes behind the lateral malleolus and runs along the lateral border of the foot to the fifth toe (little toe).

Medial calcanean nerve

The medial calcanean nerve is given off by the tibial nerve in the lower part of the leg. It supplies the skin over the medial part of the heel.

Medial plantar nerve (L4, 5; S1)

The medial plantar nerve corresponds to the median nerve of the hand [p. 84]. It runs between abductor hallucis and flexor digitorum brevis in company with the medial plantar artery.

The **medial plantar nerve** supplies the following **muscles:**
Abductor hallucis
Flexor hallucis brevis
Flexor digitorum brevis
Two medial lumbricals

The medial plantar nerve also supplies the skin over the medial side of the sole. the plantar aspect of the medial 3½ toes and the dorsal aspect of the middle and terminal phalanges of these toes.

Lateral plantar nerve (S1, 2)

The lateral plantar nerve corresponds to the ulnar nerve. It runs diagonally across the plantar aspect of the foot from the medial to the lateral side with the lateral plantar artery between flexor digitorum brevis and flexor digitorum accessorius, and divides into superficial and deep branches.

The **lateral plantar nerve** supplies the following **muscles**:
Abductor digiti minimi
Adductor hallucis
Flexor digiti minimi
Interossei
Three lateral lumbricals

The lateral plantar nerve also supplies the skin on the plantar aspect of the lateral 1½ toes.

Common peroneal nerve

The common peroneal nerve is the smaller of the two terminal branches of the sciatic nerve. It inclines laterally along the inner border of biceps (which is running laterally to its insertion into the head of the fibula) and winds round the neck of the fibula. Here it lies deep to peroneus longus. The common peroneal nerve almost immediately divides into its **deep peroneal** and **superficial peroneal** terminal branches.

The common peroneal nerve can be rolled under the finger at the back of the neck of the fibula. See if you can find it on yourself. It may be paralysed here following a blow or kick on the leg, or by the upper edge of a too tight leg plaster [FIG. 6.19].

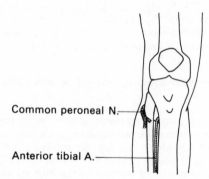

Common peroneal N.

Anterior tibial A.

FIG. 6.19. The common peroneal nerve may be rolled under the finger as it runs round the neck of the fibula on the lateral side of the knee.

Deep peroneal nerve

The deep peroneal nerve (previously known as the *anterior tibial nerve*) passes through extensor digitorum longus to reach the plane between this muscle and extensor hallucis longus and tibialis anterior where it joins the anterior tibial artery. Together they run down on the interosseous septum to the front of the ankle where they enter the dorsum of the foot mid-way between the two malleoli. Here the anterior tibial artery becomes the dorsalis pedis artery. The deep peroneal nerve and the dorsalis pedis artery run forward over the dorsum of the foot. The nerve only supplies a small area of skin in the cleft between the first and second toes.

The **deep peroneal nerve** suppies the following **muscles**:

Tibialis anterior ⎫
Extensor digitorum longus ⎬ Muscles of the anterior
Extensor hallucis longus ⎪ tibial compartment
Peroneus tertius ⎭
Extensor digitorum brevis (on dorsum of foot)

Superficial peroneal nerve

The superficial peroneal nerve is the other terminal branch of the common peroneal nerve in front of the neck of the fibula. It runs down in the septum separating peroneus longus and peroneus brevis and supplies these two muscles.

The superficial peroneal nerve becomes superficial in the lower third of the front of the leg. It branches and supplies the dorsal aspect of all toes except the cleft between the first and second toes (deep peroneal nerve) and the lateral side of the little toe (sural nerve).

The **superficial peroneal nerve** supplies the following **muscles**:
Peroneus longus
Peroneus brevis

SUMMARY OF NERVE SUPPLY TO LOWER LIMB

The lower limb is supplied by three nerves, the femoral, obturator, and sciatic nerves. The sciatic nerve divides into the tibial and common peroneal nerves. The common peroneal nerve divides into the deep peroneal nerve and the superficial peroneal nerve.

Obturator nerve (L2, 3, 4)
Motor: All adductors, obturator externus, gracilis
Sensory: Medial side of thigh
Articular: Hip and knee joints

Femoral nerve (L2, 3, 4)
Motor: Iliacus, quadriceps femoris, sartorius, pectineus
Sensory: Medial side of thigh. Below knee saphenous nerve
              branch supplies medial side of leg and foot
Articular: Hip and knee joints

Sciatic nerve and its branches (tibial and common peroneal nerves) (L4, 5; S1, 2, 3)
Motor: Hamstrings. Adductor magnus (part of)
              All muscles below knee
Sensory: Nil in thigh. Whole of leg and foot (except anterolateral
              surface of leg and medial border of foot)
Articular: All joints

Tibial nerve (branch of sciatic nerve)
Motor: Muscles of posterior compartment of leg. Muscles of
              sole of the foot
Sensory: Strip down back of leg. Sole of foot. Plantar aspect of
              toes. Dorsal aspect of terminal phalanges
              Lateral side of fifth toe
Articular: Ankle joint

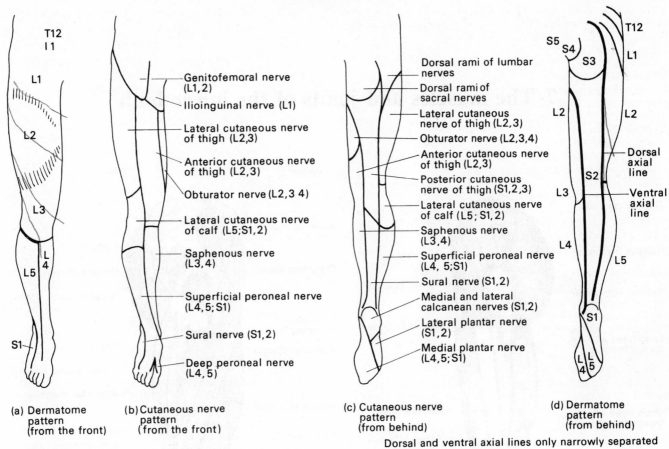

FIG. 6.20. The dermatome and cutaneous nerve pattern of the lower limb.

Common peroneal nerve (branch of sciatic nerve)

The common peroneal nerve divides into the superficial and deep peroneal nerves.

Motor: Muscles which produce dorsiflexion and eversion of the foot (tibialis anterior and peronei). Extensors of toes

Sensory: Lateral surface of leg. Dorsum of foot. All toes (except lateral side of fifth toe). Cleft between first and second toes supplied by deep peroneal branch. Rest of foot supplied by superficial peroneal branch

Articular: Ankle joint

Sensory innervation of the lower limb

FIGURE 6.20 shows the dermatomes pattern and sensory nerve pattern of the lower limb. Note that the dorsal and ventral axial lines are only narrowly separated.

7 The muscles and joints of the lower limb

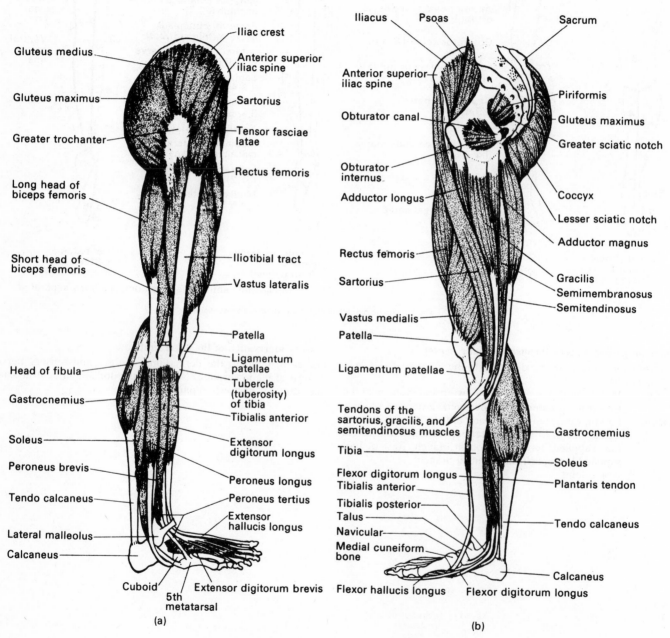

FIG. 7.01. The muscles of the lower limb.

(a) Lateral view.   (b) Medial view.

THE HIP JOINT

Structure of the hip joint [FIGS. 7.02 and 7.03]

The **head of the femur** is covered with cartilage except at the **fovea** where the **ligament of the head of the femur** (ligamentum teres) is attached. This is a mesentery carrying an artery to the head of the femur. It has little physical strength in itself. The cartilage of the head is thicker above and this part takes the weight of the body. The head of the femur is roughly two-thirds of a sphere.

The socket of the pelvis, the **acetabulum**, is deepened by a fibro-cartilaginous rim which is termed the **acetabular lip** (labrum acetabulare). This lip has a slightly smaller circumference than the underlying acetabulum and this, the atmospheric pressure and the molecular cohesion of the synovial fluid tend to hold the head of the femur in place.

The term **acetabulum** was used to describe the cup and ball with which clowns entertained their masters in bygone years. The ball was connected to the cup with a piece of string, which clearly represents the ligamentum teres connecting the head of the femur to the hip bone.

The acetabulum is deficient inferiorly. Here the **acetabular notch** is bridged over by the acetabular lip and by the **transverse ligament**. The articular cartilage of the acetabulum is horse-shoe shaped [FIG. 7.03].

The depth of the acetabulum gives stability and, thanks to the length of the neck of the femur, a wide range of movements is possible. Fracture of the neck of the femur often results in its shortening with reduction in movement at the hip joint.

The most important property of the hip joint is that it should be stable. It is often better from a clinical point of view to have the hip fixed in a good position than to have a mobile one which is unstable and causes the patient to fall over.

Articular capsule of the hip joint [FIG. 7.03]

The articular capsule of the hip joint is very strong. It is attached to the acetabular lip except superiorly and posteriorly where it is attached to bone. The capsule is attached to the femur along the anterior intertrochanteric line. It turns upwards and backwards in front of the lesser trochanter and then runs along the back of the femoral neck. The deeper fibres are reflected back along the neck of the femur forming the **retinacula**. The capsule is more lax in the flexed position and hence movements are freer in this position.

Thickened parts of the joint capsule constitute the iliofemoral, ischiofemoral, and pubofemoral ligaments.

Ligaments of the hip joint [FIG. 7.02]

Iliofemoral ligament

The iliofemoral ligament, which has the shape of an inverted Y, is a very strong ligament which limits extension of the hip in the erect posture. Since the centre of gravity of the body when standing falls behind the hip joint, this ligament becomes taut and enables the posture to be maintained without the need for hip flexor or extensor muscle activity.

The ligament arises from the anterior inferior iliac spine and runs to the upper and lower part of the intertrochanteric line [FIG. 7.02].

Ischiofemoral and pubofemoral ligaments

The other ligament of the hip joint are the ischiofemoral ligament which runs from the ischium below the acetabulum to blend with the posterior part of the capsule of the hip joint, and the pubofemoral ligament which runs from the pubis and obturator membrane to join the anterior and inferior part of the capsule.

Nerve supply to hip joint

The nerve supply to the hip joint comes from the femoral nerve, the obturator nerve, the sciatic nerve and the nerve to quadratus femoris.

Blood supply of hip joint

The blood supply to the hip joint is derived from the superior and inferior gluteal arteries, the obturator artery and the medial and lateral circumflex arteries. See page 163 for special discussion of this important topic.

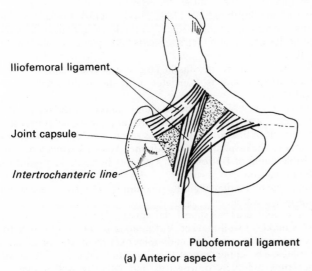

Iliofemoral ligament

Joint capsule

Intertrochanteric line

Pubofemoral ligament

(a) Anterior aspect

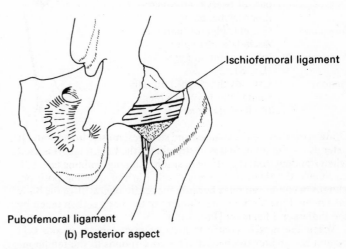

Ischiofemoral ligament

Pubofemoral ligament

(b) Posterior aspect

FIG. 7.02. Ligaments of the hip joint (simplified).

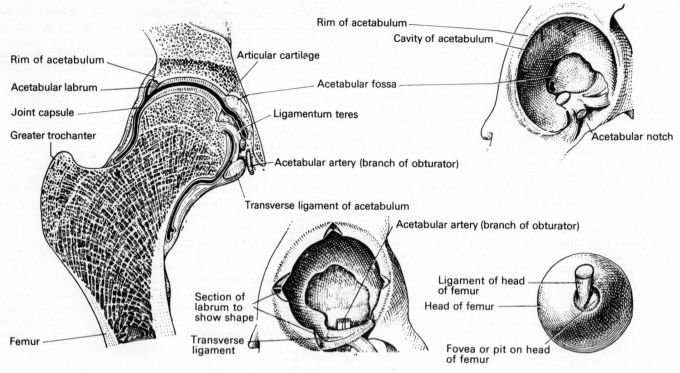

FIG. 7.03. The hip joint.

MUSCLES OF THE GLUTEAL REGION

EXTENSORS, ABDUCTORS AND LATERAL ROTATORS OF THIGH

Gluteus maximus [FIGS. 7.04 and 7.05]

Origin: Ilium between iliac crest and the posterior curved line
Lower part of sacrum, upper part of coccyx, and sacro-tuberous ligament

Insertion: Gluteal ridge of femur and fascia lata of thigh

Nerve supply: Inferior gluteal nerve (L5; S1, 2)

Action: Extends the thigh. Laterally rotates the thigh. Tightens the iliotibial tract

Gluteus maximus is the coarsest and strongest muscle in the body. It extends the hip joint, that is, it straightens the bent thigh. It is used when standing from the sitting position, jumping, walking up a hill, etc. Once the thigh has been extended in the standing position, gluteus maximus can relax because the centre of gravity of the body falls behind the hip joint. The weight of the body is then taken by the iliofemoral ligament [FIG. 7.02].

When the muscle contracts a depression appears in the skin behind the greater trochanter. This corresponds to the tendinous part of the muscle.

As has been seen, the lower border of the muscle does not correspond to the gluteal furrow. This transverse furrow is formed by fat.

In the standing position, but not in the sitting position, the **ischial tuberosity** is covered by the muscle. On the other hand, the **greater tuberosity** of the femur is covered by the muscle in the sitting position, but when standing is only covered by the tendon of gluteus maximus.

The upper two-thirds of the gluteus maximus is inserted into the fascia lata and iliotibial tract, along with the tensor fascia lata. The gluteus maximus can therefore act on the knee joint as well as the hip joint. Extension of both these joints takes place simultaneously in walking.

This attachment to the fascia lata is a special feature of the anatomy of man and is associated with the achievement of an upright stance.

Bursae may be present between gluteus maximus and the greater trochanter, the origin of vastus lateralis, and the ischial tuberosity. Enlargement of this latter bursa gives rise to weaver's bottom. The character in Shakespeare's *A Midsummer Night's Dream*, who bore the risible name of Bottom, was a weaver by trade and good for another laugh on that account.

Due to the relaxation of gluteus maximus, the buttocks are soft in the upright position.

Gluteus maximus is a superficial muscle. The origins of the two muscles makes a well-marked V, the apex of which lies over the coccyx and whose 'arms' end in dimples over the posterior superior iliac spines. The upper border forms a thin faint ridge. The lower border runs obliquely downwards and laterally well above the gluteal furrow.

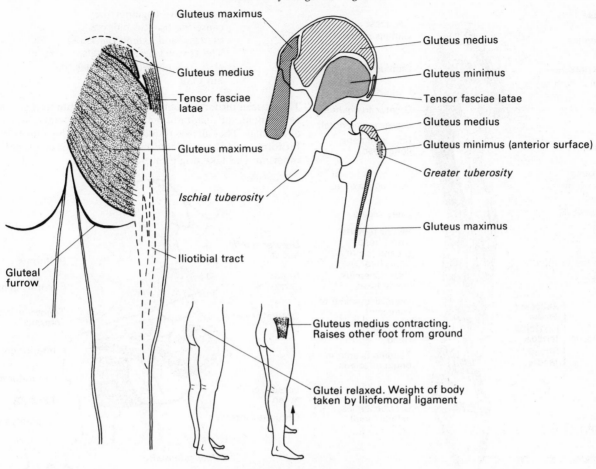

FIG. 7.04. The glutei muscles.
Note their origins and insertions.
In the standing position the centre of gravity falls behind the hip joint. The Y-shaped iliofemoral ligament [FIG. 7.02] is taut. The glutei muscles are relaxed. Confirm this on yourself. Gluteus maximum is used to straighten the bent thigh. Gluteus medius and minimus are used to raise the other foot from the ground when walking.

Gluteus medius

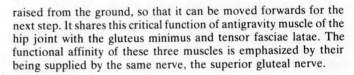

Origin: Ilium—between the posterior and middle curved lines

Insertion: Greater trochanter of femur —oblique ridge on lateral surface

Nerve supply: Superior gluteal nerve (L4, 5; S1)

Action: Abducts the thigh. The anterior fibres medially rotate the thigh. The posterior fibres laterally rotate the thigh

Gluteus medius is a very important muscle in walking. By abducting the pelvis on the thigh, it pulls the trunk over to the side of the lower limb which is in contact with the ground and brings the centre of gravity of the body over this foot. This allows the other foot to be raised from the ground, so that it can be moved forwards for the next step. It shares this critical function of antigravity muscle of the hip joint with the gluteus minimus and tensor fasciae latae. The functional affinity of these three muscles is emphasized by their being supplied by the same nerve, the superior gluteal nerve.

Gluteus minimus

Origin: Ilium between the anterior and middle curved lines

Insertion: Greater trochanter of femur *at front*. The insertion is close to the capsule of the hip joint, and there is usually a bursa deep to the tendon

Nerve supply: Superior gluteal nerve (L4, 5; S1)

Action: Abducts the thigh

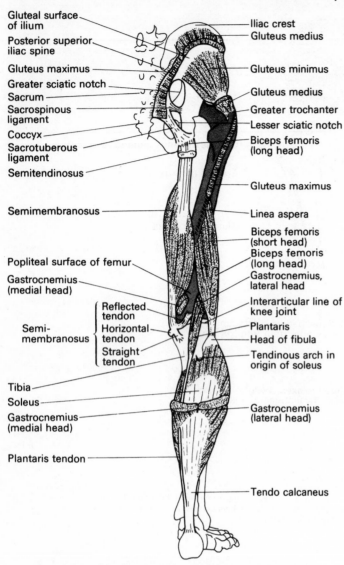

Gluteal surface of ilium
Posterior superior iliac spine
Gluteus maximus
Greater sciatic notch
Sacrum
Sacrospinous ligament
Coccyx
Sacrotuberous ligament
Semitendinosus
Semimembranosus
Popliteal surface of femur
Gastrocnemius (medial head)
Semi-membranosus { Reflected tendon / Horizontal tendon / Straight tendon }
Tibia
Soleus
Gastrocnemius (medial head)
Plantaris tendon

Iliac crest
Gluteus medius
Gluteus minimus
Gluteus medius
Greater trochanter
Lesser sciatic notch
Biceps femoris (long head)
Gluteus maximus
Linea aspera
Biceps femoris (short head)
Biceps femoris (long head)
Gastrocnemius, lateral head
Interarticular line of knee joint
Plantaris
Head of fibula
Tendinous arch in origin of soleus
Gastrocnemius (lateral head)
Tendo calcaneus

Fig. 7.05. The deep muscles at the back of the lower limb.
Note the angle the shaft of the femur makes with the vertical (and the tibia) and the direction of pull of the muscles acting on the hip joint.

posteriorly. Together they cause the tract to pull on the tibia and extend the knee. The tensor fasciae latae also abducts and medially rotates the hip joint

The muscles piriformis, obturator externus, obturator internus, the two gemelli, and quadratus femoris are small lateral rotator muscles of the thigh. They all pass round the back of the hip joint [FIG. 7.06]. Piriformis and obturator internus take origin inside the pelvis. The other muscles take origin outside the pelvis.

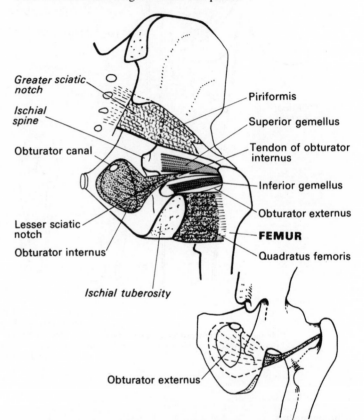

Greater sciatic notch
Ischial spine
Obturator canal
Lesser sciatic notch
Obturator internus
Ischial tuberosity
Obturator externus

Piriformis
Superior gemellus
Tendon of obturator internus
Inferior gemellus
Obturator externus
FEMUR
Quadratus femoris

Fig. 7.06. The small lateral rotator muscles of the thigh viewed from behind.
These muscles lie deep to the glutei. Obturator internus and obturator externus lie on opposite sides of the obturator membrane. The two gemelli lie on either side of the obturator internus tendon. Quadratus femoris lies between the inferior gemellus and adductor magnus. Piriformis which originates from the sacrum in the pelvis is an important landmark for blood vessels and nerves leaving the pelvis through the greater sciatic notch.

Tensor fasciae latae

Origin: Ilium. Anterior one-fifth of iliac crest and anterior superior iliac spine

Insertion: Iliotibial tract. The insertion is about 7.5 cm below the greater trochanter

Nerve supply: Superior gluteal nerve (L4, 5; S1)

Action: Tenses the iliotibial tract by pulling it superiorly and anteriorly. The gluteus maximus tenses the tract by pulling it superiorly and

Piriformis

Origin: Sacrum—second, third, and fourth segments and upper border of greater sciatic notch

Insertion: Greater trochanter of femur (highest point)

Nerve supply: Sacral 1 and 2

Action: Laterally rotates the thigh. It acts as padding inside the birth canal

Piriformis is an important landmark for nerves and blood vessels leaving the pelvis through the great sciatic foramen. Above—the superior gluteal artery and nerve pass above it. Below—the sciatic nerve, the posterior cutaneous nerve of thigh, the inferior gluteal nerve, the pudendal nerve, the nerve to obturator internus, and the nerve to quadratus femoris pass below it. So also do the inferior gluteal artery and the internal pudendal artery.

It should be noted that the pudendal artery and nerve and the nerve to obturator internus return to the pelvis by passing through the lesser sciatic foramen.

Obturator externus [see also p. 163]

Origin:	Obturator membrane (medial half of *outer* surface) and from the adjacent pubis (body and inferior ramus) and ischium (ramus)
Insertion:	Femur (trochanteric fossa)
Nerve supply:	Obturator nerve (L2, 3, 4)
Action:	Laterally rotates the thigh

Obturator internus

Origin:	Obturator membrane (lateral half of *inner* surface)
Insertion:	Femur (inner surface of greater trochanter)
Nerve supply:	Nerve to obturator internus (L5; S1, 2)
Action:	Laterally rotates the thigh

The tendon passes out of the pelvis through the lesser sciatic foramen and runs round the pulley-shaped surface of the ischium with the superior and inferior gemelli to be inserted into the front of the inner surface of the greater trochanter.

Superior gemellus [Fig. 7.06]

Origin:	Ischial spine (upper margin of lesser sciatic notch)
Insertion:	Femur (greater trochanter) with obturator internus
Nerve supply:	Nerve to obturator internus (L5; S1, 2)
Action:	Laterally rotates the thigh

This muscle runs along the upper border of the tendon of obturator internus. The two gemelli hide this tendon and cushion it where the sciatic nerve crosses posteriorly.

Inferior gemellus [Fig. 7.06]

Origin:	Ischial tuberosity (lower margin of lesser sciatic notch)
Insertion:	Femur (greater trochanter) with obturator internus
Nerve supply:	Nerve to quadratus femoris (L4, 5; S1)
Action:	Laterally rotates the thigh

This muscle runs along the lower border of the tendon of obturator internus.

Quadratus femoris

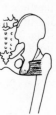

Origin:	Ischial tuberosity (lateral border)
Insertion:	Femur (quadrate tubercle on back of greater trochanter)
Nerve supply:	Nerve to quadratus femoris (L4, 5; S1)
Action:	Laterally rotates the thigh

This muscle lies between the inferior gemellus and adductor magnus. Do not confuse the name of this muscle with that of quadriceps femoris.

MAIN FLEXORS OF THIGH

The main flexors of the thigh are **psoas** and **iliacus** which arise within the abdomen. They enter the thigh by passing deep to the inguinal ligament to be inserted into the lesser trochanter of the femur.

Iliacus, which arises from the anterior surface of the iliac fossa, lies on the lateral side of psoas which arises from the lumbar vertebrae forming the posterior abdominal wall. The fibres of iliacus are inserted into the tendon of psoas. Together they are referred to as **iliopsoas**. Their combined tendon runs over the front of the capsule of the hip joint before passing backwards to reach the lesser trochanter of the femur.

Psoas

Origin:	Vertebrae T12–L5
Insertion:	Lesser trochanter of femur
Nerve supply:	L2 and 3
Action:	Flexes the thigh. It is also a medial rotator

In psoas spasm the hip joint is flexed and the thigh is medially rotated.

Iliacus

Origin:	Iliac fossa and sacrum
Insertion:	Into the tendon of psoas which runs to lesser trochanter of femur
Nerve supply:	L2 and 3.
Action:	Flexes the thigh

Iliopsoas is the main flexor of the thigh. If the lower limb is fixed, iliopsoas flexes the trunk on the thigh, as in sitting up from the supine position.

MUSCLES OF THE THIGH

The muscles of the thigh are enclosed in deep fascia like a stocking. It is this fascia, thickened on the lateral side, that constitutes the iliotibial tract. As its name implies this tract runs from the ilium to the tibia. It runs from the iliac crest to the lateral tuberosity of the tibia. Three partitions run from the encircling deep fascia to the bony ridge which runs down the posterior surface of the femur termed the **linea aspera** [Fig. 7.07]. These partitions divide the

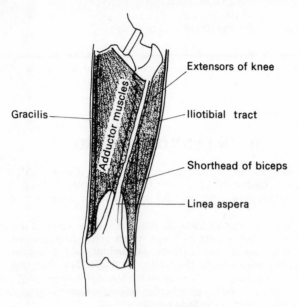

FIG. 7.07. The linea aspera.
This bony ridge running down the posterior surface of the femur is an important attachment line for many of the thigh muscles and their enveloping fascia.

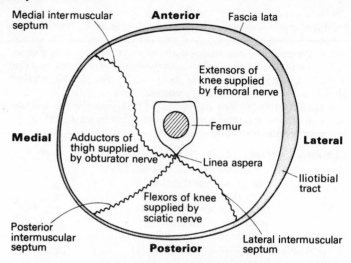

FIG. 7.08. Fascia and muscle compartments of the thigh.
The fascia lata surrounds the muscle of the thigh like a stocking. It is thickened on the lateral side forming the iliotibial tract. The fascia lata is attached above to the iliac crest, the inguinal ligament, the pubic and ischial bones and sacrotuberous ligament. Below it is attached in front and at the sides to the capsule of the knee joint, the lateral and medial borders of the patella, the tuberosity of the tibia and the head of the fibula. Behind it provides a roof over the popliteal fossa and is continuous with the fascia of the back of the leg.

The muscle compartments are formed by the intermuscular septa which run inwards from the fascia lata to the linea aspera of the femur.

The subsartorial canal really lies in the cleft corresponding to the position of the medial intermuscular septum.

thigh muscles into three compartments each with its own special nerve supply [Fig. 7.08]. These are:

Medial muscles—adductors—supplied by obturator nerve.

Anteriolateral muscles—extensors of the knee—supplied by the femoral nerve.

Posterior muscles—flexors of the knee—supplied by sciatic nerve.

In man the shape of the upper end of the femur with its wide neck takes the shaft of the femur out of line with that of the tibia (see p. 42—carrying angle). As a consequence the extensors of the knee will tend to displace the patella laterally. The lower lateral end of the femur is buttressed on its anterior surface to prevent this displacement.

In the chimpanzee the femur and tibia are in the same straight line. The lower end of the femur has no lateral buttress. Since its feet are farther apart the chimpanzee has to walk in a different way to man. It has to swing the pelvis forwards through a considerable angle with each step and transfer the body weight from foot to foot.

In man although the heads of the femur are far apart, the legs and feet are close together. As a result the centre of gravity of the body is very nearly vertically over either foot, and only a slight tilting of the pelvis (by gluteus medius) is necessary to transfer the centre of gravity so that it is exactly over one foot. Contact of the other foot with the ground is then no longer necessary to maintain the balance. Try this on yourself. See how little pelvic tilt is necessary to enable you to stand first on one leg, then on the other.

ADDUCTORS OF THIGH

The medial group of muscles adduct the thigh. They are used, for example, to keep the thighs closely against the saddle when horse riding. They can not only bring the abducted lower limb back to the non-abducted position, but can continue this movement to cross one thigh over the other. The adductor group of muscles are:

Adductor longus

Origin:	Pubis immediately below pubic crest
Insertion:	Linea aspera of femur
Nerve supply:	Obturator nerve (L2, 3, 4) anterior division
Action:	Adducts thigh. It is also a lateral rotator and can flex the extended thigh

This is a triangular shaped muscle, narrow at its origin and broad at its insertion.

Pectineus

Origin: Pubis
Insertion: Upper half of line running from lesser trochanter to linea aspera of femur
Nerve supply: Femoral nerve (occasionally obturator nerve)
Action: Adducts the thigh. It also has a flexor action

Pectineus, as its name implies, arises from the pecten (Latin = comb) which is the old name for the pubic bone. Pectineus lies in the same plane as adductor longus. Although it is an adductor, it is usually supplied by the femoral, not the obturator, nerve. As it is inserted near the hip joint its leverage is weak.

Adductor magnus

Origin: Ischial tuberosity and inferior pubic ramus
Insertion: Femur—line running from greater trochanter, along linea aspera to the adductor tubercle of the femur (with gap for femoral artery and vein)
Nerve supply: Obturator nerve (posterior division) and sciatic nerve
Action: Adducts thigh. The posterior part of the muscle extends the thigh

Gracilis

Origin: Pubis (body and inferior ramus)
Insertion: Tibia (upper part of medial surface)
Nerve supply: Obturator nerve (anterior division)
Action: Adducts thigh. Flexes the knee joint. Medially rotates the leg

Gracilis is a long slender muscle which runs along the medial border of the thigh and knee. Two intercommunicating bursae separate the tendon at its insertion from the tibial collateral ligament of the knee and from the overlapping tendon of sartorius.

Adductor brevis

Origin: Pubis—outer surface of body and inferior ramus
Insertion: Femur—line running from lesser trochanter to linea aspera, and upper part of linea aspera
Nerve supply: Obturator nerve
Action: Adducts the thigh

Adductor brevis lie in a plane behind that of pectineus and adductor longus and in front of adductor magnus.

EXTENSORS OF KNEE, FLEXORS OF HIP

Quadriceps femoris

The quadriceps femoris is composed of four muscles—rectus femoris, vastus lateralis, vastus medialis, and vastus intermedius [FIG. 7.09]. All four muscles are inserted into the upper border of the patella which connects them via the patellar ligament to the tibia. The **knee jerk** is a stretch reflex. The muscle spindles in quadriceps are stimulated by hitting the patellar tendon with a patellar hammer just above the insertion into the tibia.

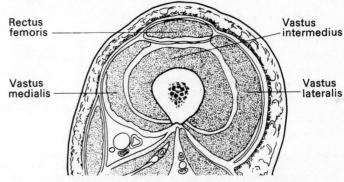

FIG. 7.09. Quadriceps.
A cross-section of the thigh showing vastus intermedius forming a deep layer around the femur with vastus medius. rectus femoris. and vastus lateralis forming the superficial layer of quadriceps femoris.

Rectus femoris

Origin: Straight head: anterior inferior iliac spine
Reflected head: long groove above acetabulum
Insertion: Patella: via tendon 7.5 cm long inserted into upper border of patella
Nerve supply: Nerves (2) to rectus femoris. These are branches of the femoral nerve (L2, 3, 4)
Action: Extends the leg. Flexes the thigh

This muscle acts on both the hip and the knee joint.

Vastus lateralis

Origin: Femur: anterior intertrochanteric
                   line (upper half), greater
                   trochanter (lower margin),
                   linea aspera (outer lip)
                   and lateral epicondylar
                   line (upper part)
Insertion: Patella: via flattened
                   tendon 6 cm long into
                   upper quarter of
                   border of patella
Nerve supply: Nerve to vastus lateralis
                   branch of femoral nerve
                   (L2, 3, 4)
Action: Extends the leg

This muscle acts on the knee joint only.

Vastus medialis

Origin: Femur: anterior intertrochanteric
                   line (lower part), linea
                   aspera (inner lip) and medial
                   epicondylar line
Insertion: Patella: via very short
                   tendon into upper half of
                   medial border of patella
Nerve supply: Nerve to vastus intermedius.
                   Branch of femoral nerve
                   (L2, 3, 4)
Action: Extends the leg

This muscle acts on the knee joint only.

Vastus intermedius

Origin: Femur: anterior and
                   lateral surfaces
Insertion: Patella (upper border)
Nerve supply: Nerve to vastus intermedius.
                   Branch of femoral nerve
                   (L2, 3)
Action: Extends the leg

This muscle acts on the knee joint only.

Note that the anterior and lateral surface of the femur gives origin to vastus intermedius, but the medial surface is virtually free of muscle attachments. The majority of muscles are attached to the femur along the linea aspera.

Quadriceps femoris is the powerful extensor of the knee, that is, it straightens the flexed knee. It does so through its insertion into the patella which is connected to the tibia via the ligamentum patellae. The patella is thus acting as a sesamoid bone in the tendon of this muscle. In addition, the rectus femoris component is a flexor of the hip. Feel this muscle contracting as you sit in your chair and flex your hip joint.

Although the origin of **rectus femoris** is covered by other muscles, it is superficial for most of its extent and gives the fullness to the front of the thigh. The anterior convexity of the thigh is, however, partly due to the anterior convexity of the femur. The **vastus lateralis** gives the convex surface on the outer side of the thigh. The fleshy part of **vastus medialis** extends further down on the medial side and hence the fullness of the lower half of the thigh is greater on the medial side.

Sartorius

Origin: Anterior superior iliac
                   spine
Insertion: Upper part of medial
                   surface of tibia
Nerve supply: Femoral (L2, 3)
Action: Flexes the thigh.
                   Flexes the leg. Laterally
                   rotates the thigh

Sartorius is the tailor's muscle (*sartor* (Latin) = tailor) so called after the cross-legged sitting position adopted by tailors. It is a long slender muscle which runs diagonally across the front of the thigh from the outer aspect of the pelvis to the medial side of the tibia. It flexes both the hip and knee joints, and by bringing the origin and insertion closer together laterally, rotates the thigh as in the crossed-leg position.

FLEXORS OF KNEE, EXTENSORS OF HIP

THE POSTERIOR OR FLEXOR GROUP OF MUSCLES

Hamstrings

The posterior group of muscles biceps femoris, semitendinosus, and semimembranosus, collectively known as the hamstrings, are **extensors of the hip joint** and **flexors of the knee**. These muscles work over the two joints, so that without changing their contracted length, when the hip joint is flexed by other muscles, the hamstrings will flex the knee. Conversely when the knee is actively extended, the tendons of the hamstrings will tend to extend the hip joint.

It is the shortness of these muscles which run from the ischial tuberosity of the pelvis to the tibia and fibula, that makes it difficult for some people to bend forwards and touch their toes with the knees extended. The tautness of the tendons prevents full flexion of the hip. As soon as the knees are bent, this manoeuvre becomes easier.

The origin of the hamstrings from the ischial tuberosity is covered by gluteus maximus. They become superficial below the inferior border of this muscle and give the fullness to the back of the thigh.

Biceps femoris

Origin: Two heads long and short.
                   Long from ischial tuberosity of
                   pelvis. Short from lower part
                   of linea aspera and upper
                   quarter of lateral epicondylar
                   line
Insertion: Head of fibula
Nerve supply: Sciatic nerve, or in 'high
                   bifurcation' the common peroneal
                   nerve
Action: Flexes the knee.
                   Extends the hip

When the knee is bent, the prominent tendon of biceps can be felt marking the lateral border of the popliteal fossa. It acts as a good guide to the head of the fibula. Try this on yourself.

Semitendinosus

Origin:	Ischial tuberosity
Insertion:	Medial surface of shaft of tibia superiorly
Nerve supply:	Sciatic nerve
Action:	Flexes the knee. Extends the hip. Medial rotator of flexed knee

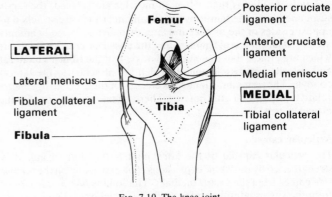

FIG. 7.10. The knee joint.
The joint is viewed from the front in the flexed position. Note the four ligaments. the fibular collateral ligament which is clear of the joint capsule. the tibial collateral ligament which is part of the joint capsule. and the two internal cruciate ligaments.

Semitendinosus is tendinous *below* and fleshy *above* (as its name implies). It is inserted with sartorius and gracilis behind the tubercle of the tibia. Its tendon can be readily palpated running with the tendon of semimembranosus on the medial side of the popliteal fossa when the knee is bent. The tendon of semitendinosus is the more lateral and more superficial of the two tendons. Feel them on yourself.

Semimembranosus

Origin:	Ischial tuberosity
Insertion:	Tibia. Groove at back of medial condyle
Nerve supply:	Sciatic nerve
Action:	Flexes the knee. Extends the hip. Medial rotator of flexed knee

Semimembranosus is fleshy *below* and membranous *above* (as its name implies). Just below the ischial tuberosity it lies deep to biceps and semitendinosus. At the knee the tendon of semimembranosus lies deep to that of semitendinosus, and can be easily located standing out as a prominent cord when the knee is flexed.

Since semitendinosus has its main muscle bulk near its origin where semimembranosus is membranous, whilst semimembranosus has its muscle fibres near the knee joint where semitendinosus is tendinous, the two muscles are able to lie side by side without causing a marked bulge at the back of the thigh.

STRUCTURE OF THE KNEE JOINT

The knee joint is essentially a hinge (ginglymus) joint. However, some rotation and gliding movement is possible in flexion. The joint is formed by the femur, tibia and the patella. The fibula lies outside the knee joint [FIG. 7.10].

Although the shaft of the **femur** makes an angle with that of the tibia. the lower borders of the two condyles are at the same horizontal level. Thus, if you stand a femur on the table. you can see the angle the shaft makes with the vertical when it is in the body.

The groove at the lower end between the two condyles for the patella is termed the **trochlear surface** (*Latin* = pulley). The lateral convex area is larger, extends higher and projects much further forward than the medial area to prevent lateral displacement of the patella during extension.

The articular areas of the two condyles at the upper end of the tibia are shallow. They appear to be deepened by the two crescentric fibrocartilaginous structures termed the menisci, or semilunar cartilages [p. 138].

The **patella** has a large lateral articular surface and a small medial articular surface with a ridge between the two. Thus if you place a patella bone on the table with the apex away from you, it will fall to the side to which it belongs [FIG. 7.11].

Anterior view   *Posterior view*

FIG. 7.11. The patella.

Due to the curvature of the lower end of the femur and the relative flatness of the articular surfaces of the patella, only a small area of the articular surfaces of the patella is in contact with the femur at any given time. As a result the patella may be fractured at its point of contact with the femur by an indirect blow. It may also be broken by the sudden pull of the quadriceps muscle.

The arrangement of articular surfaces at the knee joint should theoretically lead to a weak and unstable joint. In practice, however, the joint is very strong and dislocations are rare. This is because the joint is strengthened by very strong ligaments both inside and outside the joint and by the muscles which act upon it.

The basic plan is that, when all muscles are relaxed, the leg is suspended from the femur by four ligaments which are attached to opposite sides of the medial and lateral condyles. These ligaments are the **tibial collateral ligament** and the **posterior cruciate ligament** which are attached to the medial condyle and the **fibular collateral** and **anterior cruciate** ligaments which are attached to the lateral condyle. The terms anterior and posterior for the cruciate ligaments relate to their tibial attachments. These ligaments are aided by the articular capsule.

Articular capsule

The articular capsule of the knee joint is thin but strong. It is strengthened by muscular expansions. The expansion on the *medial side* comes from the vastus medialis. That on the *lateral side* comes from the vastus lateralis and from the iliotibial tract. These expansions are also attached to the sides of the patella.

The capsule is absent *in front*. Instead its place is taken by the patellar tendon, the patella and the lower part of quadriceps femoris. The patellar tendon is about 5 cm long. It runs from the lower border and lower part of the posterior surface of the patella to the lower part of the tubercle of the tibia.

The capsule *at the back* is strengthened by an expansion from semimembranosis.

Tibial collateral ligament

The tibial collateral ligament [FIG. 7.10] is a flat, broad ligament on the medial side of the knee joint. It is attached *above* to the medial epicondyle of the femur and *below* to the medial surface of the tibia about 8 cm distal to the joint line. The deeper fibres of the ligament are firmly adherent to the medial meniscus.

The ligament covers part of the insertion of semimembranosus. The tendons of semitendinosus, gracilis and sartorious are separated by a bursa as they are inserted into the tibia anterior to the tibial collateral ligament.

Since the ligament lies behind the axis of movement of the knee joint, it will be taut in full extension. This ligament together with a taut fibular collateral ligament and a taut anterior cruciate ligament limits extension when the leg comes in line with the thigh. When the knee is flexed the ligaments are relaxed slightly.

Fibular collateral ligament

The fibular collateral ligament [FIG. 7.10] is a round and cord-like ligament about 5 cm long lying on the lateral side of the knee joint. It is attached *above* to the lateral epicondyle of the femur and *below* to the head of the fibula at the insertion of the tendon of biceps femoris (which splits to allow the ligament through).

The fibular collateral ligament is separated from the capsule by fatty tissues and is not attached to the lateral meniscus.

This ligament is taut in full extension.

Anterior cruciate ligament

The anterior cruciate ligament lies inside the knee joint. It originates from the anterior part of the intercondylar area of the upper surface of the tibia. It runs upwards posteriorly and laterally to be attached to the posterior part of the medial surface of the lateral condyle of the femur. This ligament tightens in extension, but its main function is to withstand forces displacing the femur backwards with respect to the tibia [FIG. 7.12].

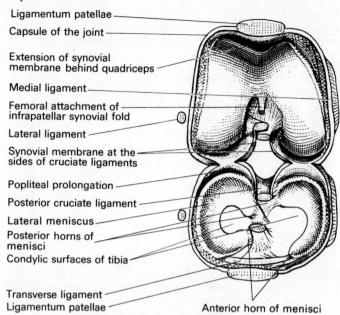

Ligamentum patellae
Capsule of the joint
Extension of synovial membrane behind quadriceps
Medial ligament
Femoral attachment of infrapatellar synovial fold
Lateral ligament
Synovial membrane at the sides of cruciate ligaments
Popliteal prolongation
Posterior cruciate ligament
Lateral meniscus
Posterior horns of menisci
Condylic surfaces of tibia
Transverse ligament
Ligamentum patellae
Anterior horn of menisci

FIG. 7.12. The knee joint opened up.
Note the two menisci (semilunar cartilages) and the points of attachment of the cruciate ligaments.

Posterior cruciate ligament

The posterior cruciate ligament also lies inside the knee joint. Because the ligaments cross like an **X** they are termed the cruciate ligaments.

The posterior cruciate ligament starts, as its name implies, from the posterior part of the intercondylar area of the upper surface of the tibia. It runs upwards, anteriorly, and medially to be attached to the anterior part of the lateral surface of the medial condyle. This ligament tightens in flexion, but its main function is to resist forces displacing the femur forwards.

BURSAE AROUND THE KNEE

There are a number of bursae around the knee which may become inflamed (bursitis) leading to swelling [FIG. 7.13].

Prepatellar bursa

The prepatellar bursa lies subcutaneously between skin and superficial fascia in front of the patella and the ligamentum patellae. Enlargement due to friction is commonly known as 'housemaid's knee'.

Infrapatellar bursa

The infrapatellar bursa lies subcutaneous in front of the lower part of the ligamentum patellae and the tuberosity of the tibia. Its enlargement has been termed 'parson's knee'.

Other bursae around the knee

The **suprapatellar bursa** lies deep proximal to the patella between the tendon of quadriceps femoris and the femur. It is usually connected to the cavity of the knee joint. The deep **infrapatellar bursa** lies deep between the lower end of the patellar tendon and the tibia. The **semimembranosus bursa** lies between semimembranosus and

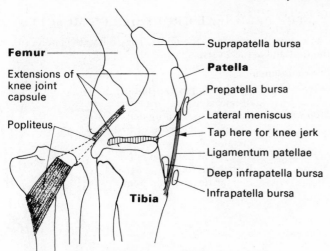

FIG. 7.13. Bursae around the knee and popliteus muscle. Posterior view of popliteus (left). Lateral view of knee joint (right) showing position of bursae. The origin of popliteus from the lateral condyle of the femur lies inside the knee joint.

the inner head of gastrocnemius. This bursa may communicate with the knee joint. The tendon of the popliteus is surrounded by a bursa-like extension of the synovial lining of the knee joint as it penetrates the capsule on the lateral side. The bursa between sartorius, gracilis, and semimembranosus and the tibial ligament has already been referred to.

Popliteus

Origin: Femur—lateral condyle inside capsule of knee joint
Insertion: Tibia—popliteal line on posterior surface
Nerve supply: Tibial nerve (L4, 5; S1)
Action: Unlocks knee by laterally rotating the femur. Flexes knee joint

TIBIOFIBULAR JOINTS

If you look at the talus you will see that the anterior part of the trochlear surface of the talus is wider than the posterior part. Since this bone lies between the lower ends of the tibia and fibula at the ankle joint, it follows that dorsiflexion of the ankle joint will increase the separation between these two bones. The bones come closer together again in plantar flexion. For these movements to occur, there must be joints between these two bones. These joints are termed the tibiofibular joints.

Superior tibiofibular joint

The superior tibiofibular joint below the knee joint is formed between the facet on the back of the lateral tuberosity of the tibia and the facet on the front of the head of the fibula. It is a synovial joint. The capsule of this joint is strengthened by the surrounding ligaments. Only a very small amount of movement takes place at

this joint when the lower ends of the tibia and fibula separate with dorsiflexion.

Interosseous membrane

The fibres of the interosseous membrane between the tibia and fibula run diagonally downwards and laterally from the tibia to the fibula [p. 118]. Thus, forces transmitted up the tibia are transferred to the fibula. The interosseous membrane gives origin to extensor digitorum longus on its anterior surface and tibialis posterior over most its posterior surface.

Inferior tibiofibular joint

The tibia and fibula come in contact again just above the ankle joint. The convex medial surface of the fibula articulates with the concave lateral surface of the tibia. These surfaces are covered with cartilage which is continuous with the ankle joint.

The inferior tibiofibular joint has an anterior ligament in front of the joint, a posterior ligament behind the joint, a strong interosseous ligament, and a horizontal yellow elastic transverse ligament under cover of the posterior ligament. The important structure joining the bones is the interosseous ligament which is a thickening of the lower end of the interosseous membrane. This joint is a syndesmosis.

THE ANTERIOR GROUP OF LEG MUSCLES

Tibialis anterior

Origin: Tibia—upper half of lateral surface
Insertion: Medial side of medial cuneiform and first metatarsal (metatarsal of big toe)
Nerve supply: Deep peroneal nerve
Action: Dorsiflexes the ankle. Inverts the foot

Although extension of a limb joint usually means the movement that moves the limb into a straight line, this is not so in the case of the ankle joint. It is flexion that brings the foot into line with the leg. To prevent confusion **extension** of the ankle joint is usually referred to as **dorsiflexion** as a reminder that the dorsum of the foot is moving towards the front of the leg. The muscles which bring about dorsiflexion are, however, referred to as extensor muscles.

Flexion of the ankle joint is termed plantar flexion.

Inversion means turning the foot so that the sole looks medially. Such an action also raises the medial border of the foot and increases the longitudinal arch of the foot and causes the toes to point inwards.

As its name indicates, tibialis anterior originates from the tibia. The other extensor muscles originate from the fibula. The tibia lies on the medial side of the leg. It is no surprise therefore that the tendon runs to the medial side of the foot. The tendon of tibialis anterior stands out further than any other tendon when the foot is dorsiflexed and inverted and can be followed very easily to its insertion.

Since the muscle is termed tibialis *anterior* there must be a tibialis *posterior*. Otherwise, it would have been called simply *tibialis*. Tibialis posterior is a plantar flexor muscle and will be considered later. Tibialis anterior and tibialis posterior acting together are the principal inverters of the foot.

Extensor digitorum longus

Origin: Fibula—upper three-quarters of anterior surface and intermuscular septum
Insertion: Middle and distal phalanges of lateral four toes
Nerve supply: Deep peroneal nerve
Action: Extends the interphalangeal and metatarsophalangeal joints of toes 2, 3, 4, and 5. Dorsiflexes the foot

The tendon of extensor digitorum longus passes in front of the ankle joint under the superior and inferior extensor retinacula and divides into four slips. Each slip is joined on its lateral side by a tendon of extensor digitorum brevis to form a dorsal expansion. This expansion is joined by the tendon of a lumbrical and two interossei, so that a total of five muscles are inserted into the middle and distal phalanges of toes 2, 3, and 4.

The little toe has only one extensor, one lumbrical and one interosseous tendon making a total of three tendons inserted into its middle and distal phalanx.

The interossei and lumbricals flex the metacarpophalangeal joints and extend the interphalangeal joints. The pattern of extensor tendons and interossei is very similar to that in the hand.

Extensor hallucis

Origin: Fibula—middle of anterior surface
Insertion: Big toe—terminal phalanx
Nerve supply: Deep peroneal nerve
Action: Extends the big toe. Dorsiflexes the ankle

The muscle forms a ridge between tibialis anterior and extensor digitorum longus in the lower part of the leg. The tendon can be easily seen on the dorsum of the foot if the big toe is actively extended.

THE PERONEAL GROUP OF LEG MUSCLES

These muscles arise from the fibula. They run to the lateral side of the foot and act as everters.

Peroneus longus

Origin: Fibula—upper two-thirds of lateral aspect
Insertion: Lateral aspect of medial cuneiform and base of first metatarsal
Nerve supply: Superficial peroneal nerve
Action: Everts the foot. Plantar flexes ankle joint

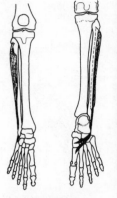

At first it is difficult to see how a muscle which is passing behind the lateral malleolus on the lateral side of the foot could be inserted into the first metatarsal on the medial side of the foot. To do so it has to pass across the sole of the foot from the lateral to the medial side. It does so by running over the lateral surface of calcaneus and enters a groove on the plantar surface of the cuboid before running across the sole of the foot to the base of the first metatarsal and the adjoining part of the medial cuneiform. It is this pathway that makes it an **everter** of the foot, that is, the sole is turned laterally and the lateral border of the foot is raised and the medial side is depressed. This muscle supports the transverse arch of the foot. Inversion and eversion takes place mainly at the talocalcanean and talocalcaneonavicular joints, that is in the foot itself not in the ankle joint.

Peroneus brevis

Origin: Fibula—lower two-thirds of lateral aspect
Insertion: Fifth metatarsal—styloid process
Nerve supply: Superficial peroneal nerve
Action: Everts the foot. Plantar flexes the ankle joint

Peroneus brevis, like peroneus longus, passes behind the lateral malleolus. The two tendons lie here in the same synovial sheath with the peroneus brevis lying deep. They then separate. Peroneus brevis passes above the peroneal tubercle on the calcaneus whilst the peroneus longus tendon passes below. The tendon of peroneus brevis can be located running to the projecting styloid process of the fifth metatarsal on the lateral side of the foot if the foot is strongly everted. The styloid process may be avulsed when this muscle contracts violently when the ankle is twisted.

Peroneus tertius

Origin: Fibula—lower quarter
              and anterior surface
Insertion: Shaft of fifth metatarsal
Nerve supply: Deep peroneal nerve
Action: Everts the foot.
              Dorsiflexes the ankle

This muscle is not always present. It is probably a separated part of extensor digitorum brevis.

EXTENSOR RETINACULA OF THE ANKLE

The **superior extensor retinaculum** is a strong band running across the anterior surface of the leg from the tibia to the fibula just above the ankle joint. It is a thickening of the deep fascia not a separate structure in its own right. The **inferior extensor retinaculum** runs from the calcaneus laterally to the medial malleolus and to the medial deep fascia of the sole. It lies just below the ankle joint.

The extensor retinacula hold down the tendons of tibialis anterior, extensor digitorum longus, extensor hallucis longus, and peroneus tertius.

Peroneal retinacula of the ankle

The tendons of peroneus longus and brevis pass behind the lateral malleolus. They are held in place by the **superior** and **inferior peroneal retinacula**.

THE POSTERIOR GROUP OF LEG MUSCLES

The posterior group of leg muscles are plantar flexors of the ankle joint and flexors of the toes. They are conveniently divided into the superficial muscles, gastrocnemius, soleus and plantaris, and the deep muscles, popliteus, tibialis posterior, flexor digitorum longus, and flexor hallucis longus. The former group are all inserted in the tendocalcaneus.

SUPERFICIAL MUSCLES

Gastrocnemius and soleus together form what was previously known as the *triceps surae*. This muscle complex is a powerful plantar flexor of the ankle. The common tendon by which the muscles are inserted into the calcaneus is often referred to as the tendo achillis. Achilles, in Greek mythology, was held by this tendon when he was immersed by his mother in the River Styx to obtain immunity from injury. This was the only part of his body not covered by the water and hence his only vulnerable part. The modern name for this tendon is the **tendo calcaneus**. It is the strongest tendon in the body but rupture occasionally occurs [p. 154]. A pseudobursa separates it from the upper part of the calcaneus.

Gastrocnemius

Origin: *Lateral head*: Femur—
              lateral epicondyle
              and lateral
              epicondylar line.
              Medial head: Femur—
              popliteal surface
              above medial
              epicondyle
Insertion: By tendo calcaneus into
              calcaneus
Nerve supply: Tibial nerve
              (L5; S1, 2)
Action: Plantar flexes ankle
              joint. Flexes knee
              joint

A sesamoid bone is often present in the lateral head, and is visible on X-ray.

Soleus

Origin: Fibula—back of head and
              upper third of shaft
              Tibia—middle third of
              medial border. A fibrous arch
              joins these two origins
Insertion: With gastrocnemius by
              tendo calcaneus into
              calcaneus
Nerve supply: Tibial nerve (L5; S1, 2)
Action: Plantar flexes ankle joint

Unlike gastrocnemius, soleus has no action on the knee joint. Being a red muscle it is thought to be important for stabilizing the ankle joint when standing whilst gastrocnemius, a white muscle, is thought to be used for rapid movement. Both muscles can be seen in action by standing on tip-toe.

The *ankle jerk* is a stretch reflex. With the subject lying or kneeling on a chair, the tendo calcaneus is struck with a patellar hammer. The ankle plantar flexes.

Plantaris

Plantaris is a small unimportant muscle which runs from the lateral epicondyle between gastrocnemius and soleus to be inserted into the medial side of the tendo calcaneus. Its rupture occasionally occurs. Being unimportant, it may be removed and used as a source of suture material. It is supplied by the tibial nerve. This muscle is not always present [see also p. 154].

DEEP MUSCLES

Tibialis posterior

Origin:　　　Interosseous membrane
　　　　　　　Fibula—back of head and
　　　　　　　　upper two-thirds of
　　　　　　　　medial surface of shaft
　　　　　　　Tibia—lateral to
　　　　　　　　vertical line
Insertion:　　Tuberosity of navicular.
　　　　　　　Medial cuneiform with
　　　　　　　slips to other tarsal
　　　　　　　bones (except talus), second,
　　　　　　　third, and fourth metatarsals
Nerve supply: Tibial nerve
Action:　　　Plantar flexes ankle
　　　　　　　joint. Inverts foot

The tendon of tibialis posterior passes behind the medial malleolus deep to the flexor retinaculum and passes superficial to the talus above the sustentaculum tali to reach many of the bones of the foot. It is easy to palpate.

The flexor retinaculum, running from the medial malleolus backwards and downwards to the medial tubercle of the calcaneus, holds down the tendon of tibialis posterior as well as those of flexor hallucis longus and flexor digitorum longus as they pass behind the medial malleolus. It also covers the posterior tibial artery and the tibial nerve.

Flexor digitorum longus

Origin:　　　Tibia—posterior surface
　　　　　　　medial to vertical line
Insertion:　　Each tendon passes through
　　　　　　　the tendon of flexor
　　　　　　　digitorum brevis to be
　　　　　　　inserted into the distal
　　　　　　　phalanx of toes 2–5
Nerve supply: Tibial nerve (L5; S1, 2)
Action:　　　Flexes phalanges of toes.
　　　　　　　Draws longitudinal arches
　　　　　　　of foot together

In the upper limb both the flexors of the distal phalanges (flexor digitorum profundus) and of the middle phalanges (flexor digitorum superficialis) are in the forearm. In the lower limb the flexor of the distal phalanges (flexor digitorum longus) lies in the leg whilst the flexor of the middle phalanges is in the sole of the foot. The origin of this muscle (flexor digitorum brevis) is the medial process of the calcaneus.

Flexor hallucis longus

Origin:　　　Fibula—lower two-thirds
　　　　　　　of posterior surface
Insertion:　　Base of distal phalanx
　　　　　　　of toe 1 (big toe)
Nerve supply: Tibial nerve (L5; S1, 2)
Action:　　　Flexes phalanges of big toe.
　　　　　　　Important support to the medial arch
　　　　　　　in walking, not in standing

The tendon of flexor hallucis longus grooves the back of the lower end of the tibia, the back of the talus and the under surface of the sustentaculum tali. It crosses the sole of the foot above the tendon of flexor digitorum longus to reach the big toe.

Sustentaculum (Latin = support) tali is the process of the calcaneus which is supporting the talus. It can be palpated below the medial malleolus.

THE ANKLE JOINT

The ankle joint is a hinge joint, or more appropriately a *mortise and tenon* joint, where the distal ends of the tibia and fibula (the mortise) embrace the talus (the tenon).

The **capsule** of the joint is tight at the two sides, but lax at the front and back. It is attached around the articular cartilage. The synovial membrane lines the capsule. It may run up between the tibia and fibula for about 5 mm.

The capsule is strengthened on the medial side by the triangular-shaped **medial (deltoid) ligament**. The apex is attached to the medial malleolus. The base is attached, from anterior to posterior, to the tubercle of the navicular, the inferior calcaneonavicular ligament, the neck of the talus, the sustentaculum tali and the body of the talus. The ligament is in two layers superficial and deep [FIG. 7.14].

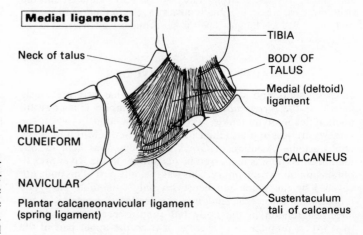

FIG. 7.14(a). Medial ligaments of the ankle joint (simplified)

The **lateral ligament** has three bands. These are the anterior talofibular ligament, the calcaneofibular ligament and the posterior talofibular ligament [Fig. 7.14].

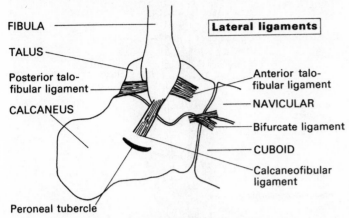

FIBULA
TALUS
Posterior talo-fibular ligament
CALCANEUS
Peroneal tubercle

Lateral ligaments

Anterior talo-fibular ligament
NAVICULAR
Bifurcate ligament
CUBOID
Calcaneofibular ligament

Fig. 7.14(b). Lateral ligaments of the ankle joint (simplified).

Nerve supply

The ankle joint is supplied by the deep peroneal, tibial, and saphenous nerves.

Blood supply

The ankle joint receives its blood supply from the anterior tibial artery, the posterior tibial artery, and the peroneal artery.

THE SOLE OF THE FOOT

The structure of the sole of the foot is complex and for simplification is best considered in layers. These are:

Skin
Cutaneous nerves and blood vessels running in fatty subcutaneous tissue
Plantar aponeurosis
First layer of muscles—
abductor hallucis, flexor digitorum brevis, and abductor digiti minimi
Second layer of muscles—
tendons of flexor hallucis longus and flexor digitorum longus with all four lumbricals and flexor digitorum accessorius which are all attached to the tendons of the latter.
Third layer of muscles—
flexor hallucis brevis, flexor digiti minimi brevis, and adductor hallucis
Fourth layer of muscles—
tendons of tibialis posterior and peroneus longus, three plantar interossei and four dorsal interossei

Nerve supply

Abductor hallucis, flexor hallucis brevis, flexor digitorum brevis, and medial lumbrical supplied by medial plantar nerve.

All other muscles supplied by lateral plantar nerve.

The **medial plantar nerve**, which corresponds to the median nerve in the hand, supplies the skin over the medial 3½ toes on the plantar surface of the foot and the corresponding distal phalanges on the dorsum of the foot.

The **lateral plantar nerve**, which corresponds to the ulnar nerve in the hand, supplies the skin over the lateral 1½ toes and the corresponding distal phalanges on the dorsum. It runs eventually between the heads of the adductor hallucis in the same way as the ulnar nerve runs between the two heads of adductor pollicis.

Blood supply

The three arteries of the leg are the anterior tibial artery, the posterior tibial artery, and the peroneal artery.

The **anterior tibial artery** is the smaller of the two terminal branches of the popliteal artery. It passes through the interosseous membrane and runs down on the anterior surface of this membrane with the deep peroneal nerve to reach the front of the ankle joint midway between the medial and lateral malleoli. Here the artery changes its name and becomes the **dorsalis pedis artery**. The dorsalis pedis artery runs across the dorsum of the foot to the first interosseous space where it passes deep between the two heads of the first dorsal interosseous muscle to form the **plantar arch** with the **lateral plantar artery**. The **first dorsal metatarsal artery** is a branch of the dorsalis pedis and continues in the dorsum of the foot to supply adjacent sides of the first and second toes.

The **posterior tibial artery** is the larger of the two terminal branches of the popliteal artery. It passes down behind the tibia between flexor digitorum longus and soleus accompanied by the tibial nerve. It passes behind the medial malleolus and divides, under cover of the flexor retinaculum, into the **lateral plantar artery** and the **medial plantar artery**.

The **peroneal artery** is given off by the posterior tibial artery shortly after its commencement. It descends close to the fibula between flexor hallucis longus and tibialis posterior. It passes behind the ankle joint and breaks up into a number of lateral calcanean branches.

In the sole of the foot the **medial plantar artery** and the **medial plantar nerve** run between abductor hallucis and flexor digitorum brevis. The **lateral plantar artery** and **lateral plantar nerve** run between flexor digitorum brevis and abductor digiti minimi. They pass to the lateral side of the sole of the foot between flexor digitorum brevis and flexor digitorum accessorius, that is, between the first and second layers of muscles.

The plantar arterial arch is formed by the lateral plantar artery which, having crossed to the base of the fifth metatarsal, crosses back to the base of the first interosseous space between the third and fourth layers of the sole to join the perforating dorsalis pedis artery. Branches from the arch supply the medial and lateral sides of the toes.

FIRST LAYER OF MUSCLES

Flexor digitorum brevis

Origin:	Calcaneus—medial process
Insertion:	Divides into four slips to be inserted into middle phalanges of toes 2–5
Nerve supply:	Medial plantar nerve
Action:	Flex the proximal interphalangeal joint and metatarsophalangeal joints of lateral four toes

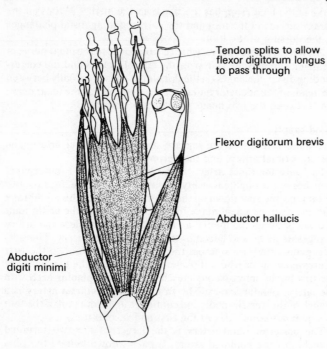

FIG. 7.15. 1st layer of sole of foot—3 superficial muscles.

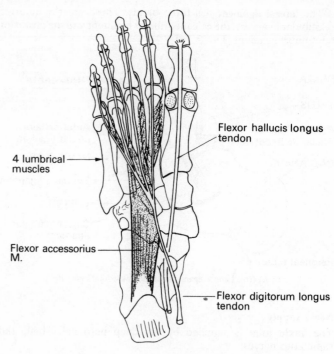

FIG. 7.16. 2nd layer of sole of foot—2 tendons and 2 associated muscles.

Abductor hallucis

Origin: Calcaneus—medial process
                 and flexor retinaculum
Insertion: Medial side of base of
                 proximal phalanx of toe 1
                 (big toe)
Nerve supply: Medial plantar nerve
Action: Abducts and flexes the big toe

Abductor digiti minimi

Origin: Calcaneus—both processes
Insertion: Lateral side of proximal
                 phalanx of toe 5 (little
                 toe)
Nerve supply: Lateral plantar nerve
Action: Abducts the little toe

SECOND LAYER OF MUSCLES

Tendon of flexor digitorum longus

Tendon of hallucis longus

Lumbricals (4)

Origin: Tendons of flexor digitorum
                 longus (medial side of tendon
                 to toe 2, adjacent sides of
                 other tendons)
Insertion: Base of proximal phalanx and
                 extensor expansion of toes 2–5
Nerve supply: First lumbrical = medial plantar
                 nerve. Second, third, and fourth

lumbricals = lateral plantar nerve
Action: Flex the metatarsophalangeal
                 joints. Extend the
                 interphalangeal joints

Flexor digitorum accessorius

Origin: Calcaneus (*two heads*—
                 concave inner surface and
                 in front of lateral
                 tubercle)
Insertion: Lateral margin of flexor
                 digitorum longus
Nerve supply: Lateral plantar nerve
Action: Pulls tendon of flexor
                 digitorum longus so it
                 acts on the toes in
                 a more direct line

THIRD LAYER OF MUSCLES

Flexor hallucis brevis

Origin: Cuboid
Insertion: Two tendons:
                 (i) Medial side of base
                      of proximal phalanx
                      of toe 1 (big toe)
                      with abductor hallucis
                 (ii) Lateral side of base
                      of proximal phalanx of
                      toe 1 with adductor hallucis
Nerve supply: Medial plantar nerve
Action: Flexes the proximal
                 phalanx of the big toe

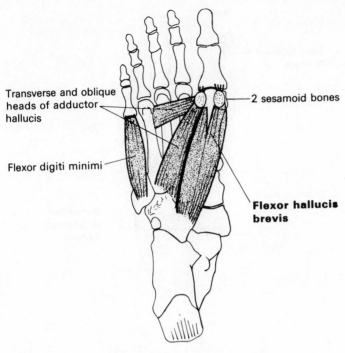

FIG. 7.17. 3rd layer of sole of foot—3 muscles.

FOURTH LAYER OF MUSCLES

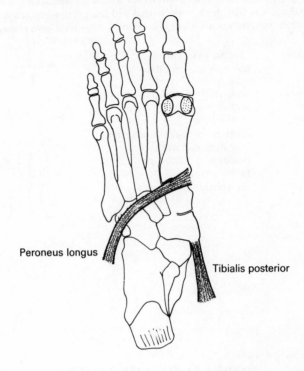

FIG. 7.18. 4th layer of sole of foot—2 tendons.

There is a sesamoid bone in each of the two tendons. The tendon of flexor hallucis longus passes between the two sesamoid bones to reach the base of the distal phalanx of the big toe.

Flexor digiti minimi brevis

Origin: Base of fifth metatarsal
Insertion: Lateral side of base of
                  proximal phalanx of toe 5
                  (little toe)
Nerve supply: Lateral plantar nerve
                  (superficial branch)
Action: Flexes metatarsophalangeal
                  joint of little toe

Adductor hallucis

Origin: *Transverse head*: plantar
                  metatarsophalangeal
                  ligaments of toes 2–5
                  Oblique head: base of second, third, and
                  fourth metatarsals
Insertion: Lateral side of base of
                  proximal phalanx of toe 1
                  (big toe)
Nerve supply: Lateral plantar nerve
                  (deep branch)
Action: Adducts big toe

The transverse head increases the transverse curvature by drawing the heads of the metatarsals closer together. The oblique head flexes as well as adducts the big toe.

Tendon of tibialis posterior

Tendon of peroneus longus

Plantar interossei (3)

Toes 3, 4, and 5 have plantar interossei which are **adductors** and move these toes towards the second toe which is the reference line
  [It will be remembered that the reference finger for the interossei of the hand is the middle finger.]
  Toe 1, the big toe, has its own adductor:

Origin: Single origins from the
                  plantar and medial aspects
                  of the third, fourth, and
                  fifth metatarsal bones
Insertion: Medial side of base of
                  proximal phalanx and
                  extensor expansion of
                  same toe
Nerve supply: Lateral plantar nerve
Action: Adduct lateral three toes
                  towards second toe. Flex
                  metatarsophalangeal
                  joints and feebly extend
                  interphalangeal joints

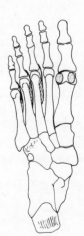

Dorsal interossei (4)

Toes 2, 3, and 4 have dorsal interossei which are **abductors**. The second toe has two since both medial and lateral movements are abduction from reference line. Toes 1 and 5 have their own abductors:

Origin:	Double origin from metacarpals on either side
Insertion:	Base of proximal phalanx and extensor expansion
Action:	Abduct toes 2, 3, and 4 away from midline of second toe. Flex metatarsophalangeal joints and feebly extend interphalangeal joints

Reference axis

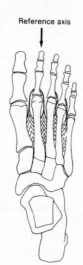

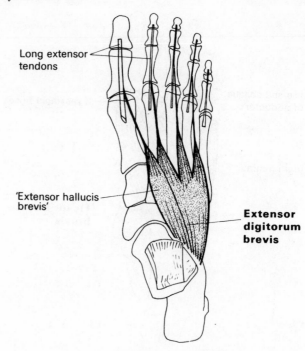

Long extensor tendons

'Extensor hallucis brevis'

Extensor digitorum brevis

FIG. 7.19. Dorsum of foot—extensor digitorum brevis. extensor hallucis brevis. and long extensor tendons.

DORSUM OF THE FOOT

Apart from tendons, there is only one muscle on the dorsum of the foot, extensor digitorum brevis. It lies on the lateral part of the dorsum of the foot in front of the lateral malleolus where it can be readily felt.

Extensor digitorum brevis

Origin:	Calcaneus (anterior part of dorsal surface) and inferior extensor retinaculum. The muscle forms four segments
Insertion:	The outer three tendons join the tendons of extensor digitorum longus running to toes 2, 3, and 4. The inner tendon (extensor hallucis brevis) is inserted into the base of proximal phalanx of toe 1
Nerve supply:	Deep peroneal nerve (lateral terminal branch)
Action:	Extension of toes

The slip going to the big toe, is often called extensor hallucis brevis. There is no slip to toe 5. [It may be represented in peroneus tertius.]

Hallux valgus

Hallux valgus is a permanent lateral displacement of the big toe. It is commoner in women than men. Once it starts, the direction of pull of the extensors of the big toe tend to make it worse. The head of the first metatarsal becomes exposed to pressure on the medial side of the foot. The proximal phalanx of the big toe is no longer in line with the first metatarsal. This leads to a painful swelling termed a bunion. It can be treated surgically and a new straight metatarsophalangeal joint fashioned.

Hammer toe

Hammer toe is the name given to the condition when, usually the second, metatarsophalangeal joint is permanently dorsiflexed and the proximal interphalangeal joint plantar flexed. As a result this interphalangeal joint sticks up in the air.

8 The lower limb as a functional unit

THE FOOT AS A FUNCTIONAL UNIT

The foot is a highly specialized structure and has many features which are peculiar to man. The hand is not so specialized in terms of its gross anatomy; it owes its precision of function to the richness of its sensory and motor nerve supplies, and to the way in which correspondingly large sections of the central nervous system—especially in the cerebral cortex—are devoted to its control. Almost all the anatomical features of the hand are represented in some way in the foot, but the foot contains a number of structures which are not present at all in the hand.

When standing the weight of the body is fairly evenly divided between the feet although in many people more weight bears on the left foot. In each foot the weight is evenly divided between the heel and the forefoot. Between the heel and the forefoot the sole of the foot only comes in contact with the ground on the lateral side. This is because the middle part of the sole of the foot is arched upwards to a considerable degree on the medial side and to only a slight degree on the lateral side.

The gravity line, for balance to be maintained, must fall between the points of contact with the ground, namely the heel and the forefoot. In fact it usually passes through a line joining the tubercle of the navicular on the medial side and the base of the fifth metatarsal on the lateral side [FIG. 8.01].

Movement of the foot on the leg takes place when the ankle joint is dorsi- or plantar-flexed. This, as has been seen, is brought about by the movement of the talus (the tenon) in the mortice of the ankle joint, the lower ends of the tibia and fibula. In addition the foot is capable of its own complex intrinsic movements, that is, movements between the tarsal bones themselves.

The **talus** has a body, neck and head. It is the body of the talus [FIG. 8.02] that articulates at its upper articular surface with the *weight-transmitting lower surface* of the tibia, and at its sides with the medial and lateral malleoli. All the weight borne by the foot has to be transmitted through the talus. The direction of the two weight-pathways, one to the heel, and one to the forefoot, are shown in FIGURE 8.01. Notice that they both run obliquely, and each may be resolved into a vertical component and a horizontal component. In both cases the vertical component represents the pressure of the foot on the ground. The horizontal components will have a tendency to make the longitudinal arch of the foot spread. The longitudinal arch would collapse if it were not for the presence of structures which act like the tie-beam of the arch.

The head of the talus is directed forwards and articulates with the navicular on the **medial side** of the foot. From here the weight pathway passes through the three cuneiform bones and reaches the head of the medial three metatarsal bones.

The bone in the heel is the calcaneus (*calx* = heel). The anterior part of this bone is directed laterally where it articulates with the

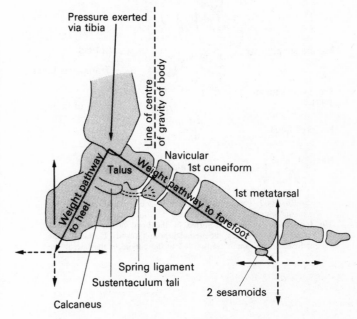

FIG. 8.01. Weight pathways.
The body weight at the foot can be resolved into two oblique weight pathways. These in turn can be resolved into the pairs of dotted arrows at the front of the foot and at the heel which are in contact with the ground. These resultant forces are balanced under static conditions by the forces indicated by the vertical and horizontal arrows. The vertical solid arrows represent the pressure on the ground. The forces represented by the horizontal solid arrows is generated by friction (if any) and by tension in the ligaments and muscles in the sole of the foot.
Note the position of the line of centre of gravity. The 'pressure exerted via tibia' force line is not vertical but slightly oblique since it is running from the centre of gravity of the body.

cuboid on the **lateral side** of the foot. The cuboid articulates directly with the two remaining metatarsals.

The weight pathway to the forefoot reaches metatarsals 1, 2, and 3 through the talus, navicular, and cuneiforms, and reaches metatarsals 4 and 5 through the talus, calcaneus, and cuboid.

Between the navicular and the first, second, and third metatarsals are the **medial**, **intermediate**, and **lateral cuneiform** bones (*cuneus* = a wedge: cuneiform writing, etc.). The intermediate and lateral cuneiforms are very wedge-shaped, the thin end of their wedges being directed downwards. They give to this part of the foot a very well marked **transverse arch** [FIG. 8.03]. The strength of this arch depends primarily on the interosseous ligaments which are very powerful and placed so that they exert the best leverage. These bones articulate with each other by means of plane synovial joints

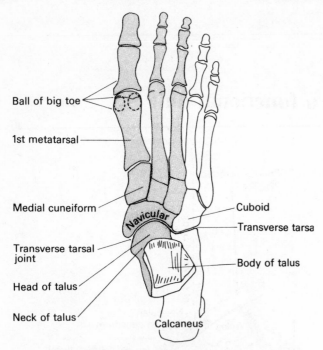

FIG. 8.02. The bones of the foot as seen from above.
The medial longitudinal arch is shaded. The lateral arch is not shaded. The former has weight transmitted by the head of the talus to the navicular. cuneiform bones. and medial three metatarsals. The latter bears weight from the calcaneus transmitted to the cuboid to the lateral two metatarsals.

which lie as close to the dorsum of the foot as possible. The joint cavities are guarded dorsally only by a very thin capsule. On the plantar surface the whole non-articular area is taken up with interosseous ligaments. This arrangement helps to neutralize the tendency of the arch to flatten downwards. The transverse arch is also supported by the tendon of peroneus longus—when the muscle is contracting.

The **joints** and **ligaments** at the **bases** of the **metatarsals** are arranged in a manner which repeats the arrangement between the cuneiforms. There are plane synovial joints dorsally; on the plantar surface very strong interosseous ligaments cover the whole of the extensive area available. A blow on the dorsum of the foot here breaks the bones but does not rupture the ligaments.

The five **metatarsal** bones fan out slightly as they run forwards. They are connected to their tarsal bones [FIG. 8.04] by strong ligaments on their plantar aspects. They are also connected to each other anteriorly by strong interosseous ligaments attached to their necks.

The pressure on the forefoot is evenly spread and in the normal foot all five metatarsal heads bear their share of the body weight. The head of the first metatarsal is separated from the ground by the pair of sesamoid bones in the two tendons of the flexor hallucis brevis (just like the sesamoids in flexor pollicis brevis). This means that there are really not five but six structures which transmit the body weight from the forefoot to the floor. The weight is borne more or less equally by all six of them. Thus about one-third (i.e. two-sixths) of the forefoot body weight is transmitted by the very stout metatarsal of the big toe, and one-sixth via each of the other metatarsals.

Longitudinal arches of the foot

It is customary to describe the medial and lateral components of the longitudinal arch as separate arches. They, of course, merge with each other.

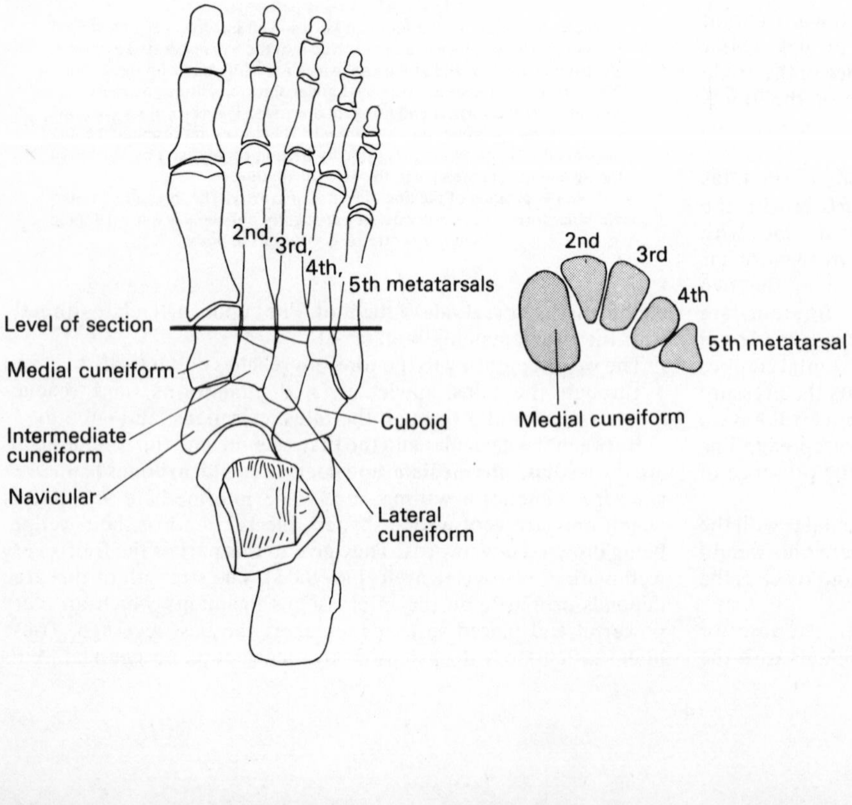

FIG. 8.03. Transverse arch of foot.
The transverse arch is the product of the wedge shape of the intermediate and lateral cuneiform bones. The shapes of the bases of the 2nd and 3rd metatarsal bones agree closely with the wedge-shape of the cuneiforms as you would expect.
As one proceeds towards the forefoot the transverse arch is less evident. The heads of the metatarsals lie in the same horizontal plane. and the transverse arch has disappeared.

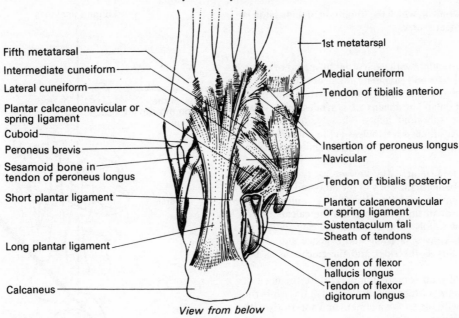

Fifth metatarsal

Intermediate cuneiform

Lateral cuneiform

Plantar calcaneonavicular or
spring ligament

Cuboid

Peroneus brevis

Sesamoid bone in
tendon of peroneus longus

Short plantar ligament

Long plantar ligament

Calcaneus

1st metatarsal

Medial cuneiform

Tendon of tibialis anterior

Insertion of peroneus longus

Navicular

Tendon of tibialis posterior

Plantar calcaneonavicular
or spring ligament

Sustentaculum tali

Sheath of tendons

Tendon of flexor
hallucis longus

Tendon of flexor
digitorum longus

View from below

Fig. 8.04. Ligaments on plantar aspect of foot.

Consider the **medial longitudinal arch** by looking at the foot skeleton from the medial side, and at your own foot. [Fig. 8.02]

The arch runs from the heel to the ball of the big toe and heads of the second and third metatarsals. It is composed by:

1. The calcaneus.
2. The head of the talus.
3. The navicular.
4. The cuneiform bones.
5. The first three metatarsals which between them bear four-sixths of the forefoot body weight.

The bones of the foot are bound tightly to each other by powerful interosseous ligaments on their plantar surfaces. The exception to this is the head of the talus which is not bound directly to either the calcaneus or to the navicular [Fig. 8.01].

The head of the **talus** is often likened to the keystone of an arch, but who has heard of a keystone that is not rigidly fixed to the stones on each side of it? In the foot the neck of the talus lies on, and articulates with, part of the **calcaneus** called the **sustentaculum tali** (supporter of the talus). The sustentaculum tali projects medially like a shelf, and can be palpated in yourself [p. 50]. The anterior border of the sustentaculum is joined to the navicular by a strong ligament. This is the **plantar calcaneonavicular ligament** or **spring ligament**. The head of the talus lies on the upper surface of this ligament which shows cartilaginous features on its upper articular surface. It is obvious looking at Figure 8.05 that this ligament is an integral and absolutely essential feature of the medial longitudinal arch. If it were to be cut, the head of the talus would be driven between the calcaneus and navicular by the body weight, and the arch would collapse completely. This is another example of a ligament essential in the maintenance of posture, in this case the posture of the foot.

The **lateral longitudinal arch** runs from the heel to the heads of the fourth and fifth metatarsals. It consists of the calcaneus, the cuboid, and the lengths of these two metatarsal bones. Under weight-bearing conditions the base of the fifth metatarsal may lightly touch the ground [Figs. 8.01 and 8.05].

It is important not to think of the two longitudinal and transverse arches as separate structures each in its own plane at right angles. When arches intersect at right angles, as happens in the foot, only a three-dimension word is satisfactory to describe the whole picture.

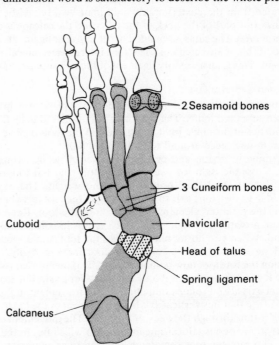

2 Sesamoid bones

3 Cuneiform bones

Cuboid

Navicular

Head of talus

Spring ligament

Calcaneus

Fig. 8.05. Bones of the foot from below.
The bones forming the medial longitudinal arch are shaded. Note that the spring ligament is an integral part of this arch.

The sole of the foot is a dome in which the longitudinal dimension is about twice that of the transverse.

Joints of the foot

Most bones in the foot articulate with each other by means of plane synovial joints. The **navicular** and **cuboid** lie adjacent to each other, but usually there is no synovial joint here, only a strong interosseous ligament. They articulate by means of a **syndesmosis**. Also exceptional are the two subtaloid joints. These joints although quite separate from each other, nevertheless act in concert in the movements of the foot around the talus called inversion and eversion.

The **posterior subtaloid joint**, also known as the **posterior talocalcanean joint**, lies beneath the body of the talus and interrupts the posterior weight-bearing pathway [Fig. 8.01], which passes obliquely backwards through the talus to reach the calcaneus.

This joint is a shallow saddle joint in which the calcaneus provides a surface which is convex laterally and concave anteroposteriorly (think of the talus as the rider and the calcaneus as the horse).

The **anterior subtaloid joint** is more complex. Here the neck of the talus articulates with the sustentaculum tali [Fig. 8.05] while the head articulates below with the spring ligament and anteriorly with the concave proximal surface of the navicular.

This complex joint is a synovial ball and socket. The head of the talus is the 'ball' and the socket consists of the calcaneus and navicular jointed together by the spring ligament into which [p. 142] the superficial part of the deltoid ligament also runs. This is the **talocalcaneonavicular joint**. A better name would be the **talocalcaneo-springligamento-navicular joint** which draws attention to a unique articulation in which the brunt of the weight-bearing falls on a ligament.

The rest of the capsule of this joint is thin and weak. It is separated from the posterior subtaloid joint by a tunnel between the posterior part of the neck of the talus and the calcaneus called the **sinus tarsi**. The sinus is bounded by the capsular ligaments of the two joints but it also contains the strong **interosseous talocalcanean ligament**. This is an accessory ligament of the two subtaloid joints.

Inversion and eversion

The movements of inversion and eversion take place principally at the two subtaloid joints. The talus may be considered to be fixed in the mortice of the ankle joint and the rest of the foot to move more or less in one piece around the talus [Fig. 8.06].

Examine inversion and eversion in yourself while sitting in a chair. Cross the right leg over the left. Roll your left foot on the floor so that the lateral border bears the weight. The foot has rotated in its own long axis so that the sole of the foot faces inwards. This is the essential element in inversion of the foot. Repeat this movement with your right foot which is not in contact with the ground. When you do so, not only does the foot rotate around its own long axis so that the sole of the foot faces medially, but in addition the forefoot now points inwards. This means that rotation of the foot is taking place not only around its long axis, but around a vertical axis also. From examination of this movement in yourself you will see that this vertical axis lies just in front of the ankle joint. In fact it runs through the neck of the talus. The movement swings on the interosseous talocalcanean ligament which lies in the sinus tarsi. This movement is sometimes referred to as *adduction of the forefoot*, but note that as the forefoot is directed inwards the heel points slightly outwards.

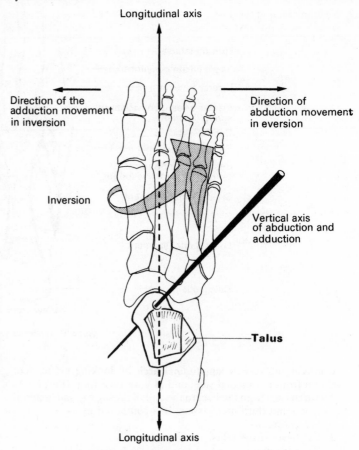

Fig. 8.06. Movements in inversion and eversion.

Thus inversion and eversion are complex three dimensional movements which may be resolved into two components. Inversion of the sole of the foot is always attended by adduction of the forefoot; eversion of the sole is associated with abduction of the forefoot.

How is it possible then to change the angle of the sole of the foot, when walking on rough ground, without alteration in the direction in which the foot points? The answer lies at the hip joint.

Let us first consider the reverse situation. It is possible to change the direction in which the toes are pointing without inversion or eversion of the sole. Stand up with your heels close together and your feet parallel. Now point your toes to each side so that each foot rotates laterally through 90°. As you repeat this movement rapidly, put your hand over the hip joint. You will find that it is the medial and lateral rotation of the hip joint which is mainly responsible for the direction in which the foot points.

To return to the question of inversion, it follows that when the foot is inverted or everted to bring the sole of the foot to a required angle, any unwanted adduction or abduction can be neutralized by rotation at the hip joint.

The interplay between the hip joint and knee joint also occurs during knee locking, but here is an even more impressive example of the interdependence of the weight-bearing joints in the lower limb.

Now look again at your own foot during inversion and eversion.

You will see that the shape of the medial longitudinal arch changes considerably during these movements. In inversion the arch is very obvious, in eversion much less so.

Try this under weight-bearing conditions. Stand with your feet parallel and separated by a few centimetres. Now try to make your patellae touch by twisting your knees towards each other (medial rotation of hip joint), and then twist in the opposite direction so that the patellae are directed as far laterally as possible (lateral rotation of the hip joint). Your feet should remain in contact with the floor in the same place. As you carry out these movements you will see that the soles of your feet on the medial side are moving up and down through a distance of about 3 cm.

The postures of the foot are very numerous and depend on the degree of inversion or eversion, and on rotation of the hip joint. Also it is possible for a certain amount of unwanted adduction or abduction of the forefoot to be 'taken up' in the knee joint under weight-bearing conditions. This may have happened when you inverted your left foot which was in contact with the ground at the beginning of this secion.

The **transverse tarsal joint** is often stated to be the site of inversion and eversion. This is not so in normal people. If it were so, the heel would not move at all!

The transverse tarsal joint itself is a misnomer as it is two quite separate articulations. The first is that between the head of the talus and the navicular (one of the components of the talocalcaneoligamentonavicular joint); the second is that between the calcaneus and the cuboid on the lateral side of the foot. These two joints lie almost in line with each other. Attention was drawn to them first by Chopart as a convenient method of amputating the forefoot with a knife in an emergency. He was not concerned for one moment with function, and the operation is seldom performed nowadays.

In normal people all the **intertarsal joints** are capable of sufficient movements to give to the foot its suppleness. There is no doubt that movement freely takes place between the head of the talus and the navicular. If correspondingly free movement took place between the calcaneus and cuboid then it should be possible to fix the heel, say, by placing the heel in a vice, without interfering with inversion and eversion of the forefoot. However, such fixation does have a very limiting effect on movement.

In young people when the talus is fused to the calcaneus (after an accident or operation), an extraordinary mobility develops with time in the calcaneocuboid joint, and the movements of the forefoot then appear to be similar to the normal in spite of the immobility of the heel.

To summarize, we have the remarkable situation in which the head of the talus, the 'keystone' of the medial longitudinal arch, is apparently not in physical continuity with the 'stones' on each side of it. As the talus remains stationary in inversion and eversion this movement of the whole arch around the head of the talus is responsible for the change in appearance of the whole foot which takes place during these movements.

A FURTHER CONSIDERATION OF THE MUSCLES AND LIGAMENTS OF THE FOOT

So far the following ligaments have been considered:
1. Calcaneonavicular, or spring ligament.
2. The interosseous talocalcanean ligament in the sinus tarsi.
3. The short interosseous ligaments joining adjacent bones to each other especially on their plantar surfaces.

4. The deep transverse interosseous ligaments of the metatarsal bones.

There are a number of other ligaments which must be mentioned. First over the dorsum of the foot is the bifurcate (forked) ligament [Fig. 8.07] whose stem is attached to the calcaneus and which runs forwards and splits as it does so and is then attached both to the navicular and to the cuboid.

On the plantar surface the calcaneus and cuboid which forms parts of the lateral arch are joined by the short plantar (or calcaneocuboid) ligament. The arch is also supported by another ligament, the long plantar, which lies superficial to the short plantar ligament. It is attached behind over a considerable area of the calcaneus and in front it runs forwards and is attached to the cuboid and the bases of the metatarsal bones especially on the lateral side of the foot. Note the groove on the plantar surface of the cuboid in which the peroneus longus tendon lies. This groove separates the attachments of the short and long plantar ligaments. The latter converts the groove into a tunnel.

These ligaments withstand the tendency of the longitudinal arches to spread and flatten under the strain of the body weight [Fig. 8.04]. This tendency is also resisted by the plantar aponeurosis, which is attached behind to the calcaneus and in front over a wide area to the forefoot and the base of the toes. As it lies in a very superficial position it gets better leverage than the long and short plantar ligament. The tendency is also resisted by any of the intrinsic or extrinsic muscles which run longitudinally in the sole of the foot, provided that they are contracting vigorously.

The arches of the foot and their disorders have been the objects of many misunderstandings and misconceptions. The basic shapes of the arches are present in the foot almost as soon as the skeletal elements can be recognized in the embryo. In a plump well nourished baby the sole of the foot seems flat—but this is because of the copious soft tissues in the sole of the foot and has nothing to do with the foot skeleton.

The maintenance of the arches depends on the ligaments under static conditions. Muscles which run longitudinally such as flexor digitorum longus, flexor hallucis longus, the short flexors and abductors of the toes (the superficial layer of muscles in the sole of the foot) are without any effect on the foot unless the tips of the toes dig into the ground—which they do not do in ordinary standing. If you stand up, even with shoes on, you will realize that the pressure exerted by flexor hallucis longus on the terminal phalanx of the big toe, is so slight that it barely blanches the skin. If you take your shoes off and stand on the edge of a table with the toes projecting beyond the edge, it is obvious that none of the muscles inserted into the toes can be exerting any effect on foot posture.

Consider now the muscles concerned with control of the ankle and subtaloid joints. Note the remarkable fact that in spite of its key role in the ankle joint, there is not a single muscle attached to the talus. The tibialis anterior seems to be well sited to sustain the medial longitudinal arch but it cannot do this unless the slack has first been taken out of the extensor retinaculum. The tendon of this muscle is very easy to feel. Dorsiflex your ankle and palpate it. In normal standing the tendon of tibialis anterior is quite slack. So are the tendons of tibialis posterior, peroneus longus and peroneus brevis. The rotation of the hip joint is crucial in determining foot posture under static conditions.

Under active conditions the situation is that the foot is now subject to greatly increased strain, and is at the same time supported to a considerable degree by the extrinsic and intrinsic muscles of the foot.

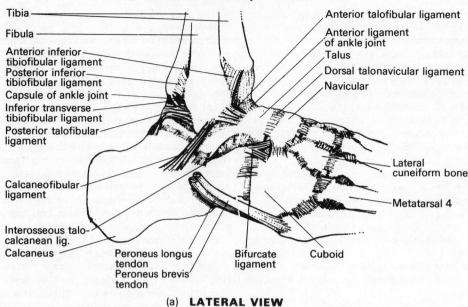

Fig. 8.07(a). Ligaments of the foot (lateral view).

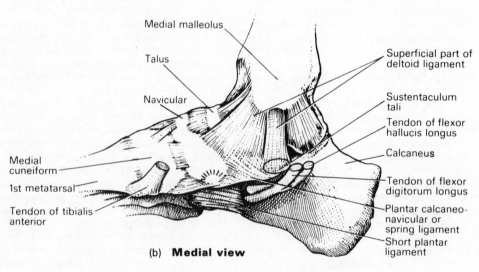

(b) **Medial view**

Fig. 8.07(b). Ligaments of the foot (medial view).

If you stand upright with the foot flat on the ground, and then stand on your toes, many of the extrinsic and intrinsic muscles of the foot come into play, and the depth of the longitudinal arch becomes more evident.

THE INTRINSIC MUSCLES OF THE FOOT

As has been seen the **superficial layer** of foot musculature consists of three muscles—flexor digitorum brevis, abductor digiti minimi, and abductor hallucis.

Flexor digitorum brevis

This muscle takes origin from the calcaneus just in front of the subcutaneous part of this bone and behind the attachment of the long plantar ligament [Fig. 7.15, p. 144].

It has four tendons which are inserted into the second phalanges of the toes other than the big toe. Each tendon splits to allow the corresponding flexor digitorum longus tendon to pass through on its way to the terminal phalanx. The pattern is exactly like that in the hand, where the muscles are called flexor digitorum profundus and superficialis, but whereas flexor digitorum brevis is an intrinsic muscle of the foot, flexor digitorum superficialis lies mostly in the forearm. Can you see why such an arrangement in the hand would not work?† You can easily feel flexor digitorum brevis in yourself in the sole of the foot.

Abductor digiti minimi

This muscle runs from the lateral side of the calcaneus to the lateral border of the fifth toe. It produces considerable movement in the little toe. [FIG. 7.15, p. 144].

Abductor hallucis

This muscle takes origin from the medial side of the calcaneus and runs along the medial edge of the sole of the foot into the medial side of the base of the proximal phalanx of the big toe. The big toe can be abducted and adducted so that its tip moves about 1–1.5 cm, but the real function of this muscle is to act as a flexor of the metatarsophalangeal joint. With the foot on the ground press the big toe downwards and you will see and feel the belly of this muscle contracting [FIG. 7.15, p. 144]

Note that the big toe is fixed to the second metatarsal by a strong transverse ligament and that no movements, such as occurs at the carpometacarpal joint of the thumb, are to be seen in the corresponding joint in the foot. The opponens pollicis is not represented in the foot for this reason.

The **third layer** of musculature in the sole of the foot consists of flexor hallucis brevis, adductor hallucis and flexor digiti minimi which are all represented in the hand. [FIG. 7.17, p. 145].

Flexor hallucis brevis

Flexor hallucis brevis [FIG. 7.17, p. 145] takes origin from the sole of the foot in front of the calcaneus and quickly divides into two parts. Each has a sesamoid bone in its tendon which constitute the two weight-bearing parts of the ball of the big toe. The tendons are attached to each side of the base of the proximal phalanx. The tendon of flexor hallucis longus slides between them on its way to its insertion into the distal phalanx. This arrangement avoids pressure on this tendon in standing. Sometimes these sesamoids are fractured.

Adductor hallucis

Adductor hallucis [FIG. 7.17, p. 145] has two heads of origin, one from the cuneiform bones adjacent to the lateral head of flexor brevis, and the other from the soft tissues on the plantar aspect of the heads of the metatarsals as far as the fifth. The first head is called the oblique and the second the transverse. The muscle is inserted into the lateral sesamoid bone of the flexor hallucis brevis, just as the abductor hallucis brevis is attached to the medial sesamoid. These two muscles strikingly resemble the corresponding muscles in the thenar eminence of the hand in the way they utilize the sesamoid bones which really belong to the flexor brevis.

Note that the big toe, unlike the thumb, has no opponens muscle.

---

†Because the fingers are much longer than the toes their movement is greater. You would predict therefore much longer fibres in flexor digitorum superficialis than in flexor digitorum brevis. This is taken care of in the body by this muscle taking origin, not in the hand, but high up in the forearm. The fact that it now has to run across the wrist is a further complication requiring the muscle fibres to be longer still. This complication has not been entirely solved by the finger flexors since a tight grip is impossible when the wrist is flexed [p. 36].

Flexor digiti minimi

Flexor digiti minimi [FIG. 7.17, p. 145] runs forwards from its origin from the sole of the foot in front of the calcaneus and shares its attachment with the abductor digiti minimi. This arrangement is a mirror image of the two corresponding muscles on the medial side of the foot, and closely resembles the arrangement in the hand.

Interossei

It has been seen that there are three plantar interossei and four dorsal interossei. The pattern of the dorsal interossei is organized around the second metatarsal [FIG. 8.08] in the same way as in the hand it is organized around the middle (third) metacarpal. Abduction and adduction of the toes is defined as movement with respect to the axis of the second toe, and not to the third. In practice movement of the intermediate toes is so rarely described that the difference between hand and foot is not important [pp. 146, 147].

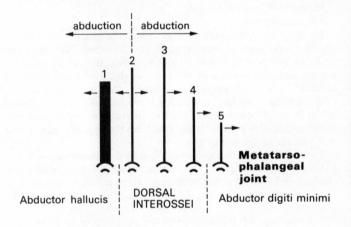

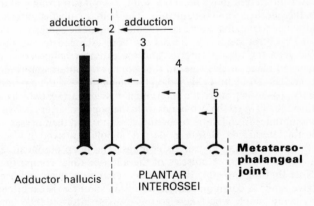

FIG. 8.08. Abduction and adduction of the toes.
Abduction is movement away from the line of the second toe. Adduction is movement towards the line of the second toe.

Extensor digitorum brevis [FIG. 7.19, p. 146]

You can easily feel the belly of this muscle on the dorsum of your own foot just in front of the lateral malleolus. Note that it has four (not five) tendons. The missing one is that which might have gone to supply the little toe. Some people believe that peroneus tertius represents the 'missing' part of this muscle.

Flexor accessorius and lumbrical muscles

These muscles, which are also intrinsic muscles of the foot, are all associated with the flexor digitorum longus muscle and are described with that muscle on page 144.

THE EXTRINSIC MUSCLES OF THE FOOT

Anterior tibial compartment

The extrinsic extensor muscles of the foot lie in the anterior tibial compartment. This compartment is the space between the tibia and fibula seen from the front. Its deep boundary is the interosseous membrane.

It contains four muscles:

Tibialis anterior
Extensor hallucis
Extensor digitorum longus
Peroneus tertius

Look at the tendons of these muscles at the ankle joint. Tibialis anterior is the most prominent of all when the ankle is dorsi-flexed and slightly inverted. Follow it down to its insertion, which overlaps the medial cuneiform and the base of the first metatarsal and straddles the intervening joint. Invert strongly and you will appreciate that the tendon extends a considerable distance round towards the plantar aspect of the foot. It is almost continuous with the insertion of peroneus longus which has the opposite action, that is it plantar flexes and everts.

Extensor hallucis is easily identified by deliberate extension of the big toe. Follow it upwards from the big toe, along the dorsum of the foot and in front of the ankle joint. Repeat the identification of the four extensor digitorum tendons, which are connected to the other four toes.

Now having noted the order of the insertion of these muscles from medial to lateral and having traced their tendons upwards a few centimetres above the ankle joint, it is not surprising that their bellies lie in a similar sequence. The belly of tibialis anterior is attached to the tibia over a considerable area. The extensor hallucis and extensor digitorum longus are both attached to the fibula. Between the tibialis anterior and the extensor hallucis is a well marked plane of cleavage. In this gap lies the interosseous membrane [Fig. 6.09, p. 118], and on the membrane the deep peroneal nerve passes in front of the anterior interosseous artery (which came through between the two bones from the popliteal artery above the membrane) from lateral to medial, hesitates and then passes back to the lateral side. Before modern terminology arrived, it went by the amusing name of nervus hesitans! The deep peroneal nerve supplies these three muscles of the anterior tibial compartment. Note that the investing deep fascia is extremely thick and has no 'give' in it. A common condition of 'shin soreness' is experienced by many people who take vigorous exercise intermittently, and is due to swelling of the muscles subjected to unwonted activity. In this confined space this may seriously reduce their blood supply, giving rise to ischaemic pain and even degeneration of muscle.

The peroneus tertius muscle completes the contents of the anterior tibial compartment. It runs down on the lateral side to be attached to the shaft (usually) of the fifth metatarsal bone. From its position it must be a dorsiflexor and everter, and by producing these movements together you can easily identify it in yourself. The insertion of this muscle varies considerably.

Calf muscles

The extrinsic flexor muscles of the foot lie in the calf on the back of the leg between the knee and the heel. There are two groups of muscles, the superficial and deep.

The **superficial group of muscles** consist of the gastrocnemius and soleus and the plantaris which sometimes may be absent. As the gastrocnemius has two heads this muscle with the soleus is sometimes called the triceps surae (three-headed muscle of the leg). It is inserted into the calcaneus via the tendo calcaneus (also called Achilles tendon). The plantaris muscle arises with the lateral head of gastrocnemius, and its long tendon lies between gastrocnemius and soleus and is attached to the *medial* side of the tendo calcaneus. This slender tendon is said sometimes to rupture, but in view of the tiny muscle to which it is attached this seems dubious. The tendon is sometimes made use of as suture material so that the patient can be repaired with biologically speaking perfectly compatible material.

Note the origin of the gastrocnemius above the knee joint—it is a typical 'two-joint muscle' in which the movements in the two joints are in 'opposite directions' so far as the muscle is concerned.

In walking and running, extension of the knee joint lengthens the muscle while plantar flexion of the ankle joint shortens it. The overall result is that the net length of the muscle changes little. The soleus on the other hand is a typical one joint muscle as it takes origin below the knee joint and therefore acts principally on the ankle joint. What other joints would it influence? This muscle is important in controlling the ankle joint under *static* conditions. If the gastrocnemius were to do this it would tend simultaneously to flex the knee joint. Under static circumstances the knee is in extension with the gravity line teetering just in front of the joint axis, and this action on the knee, albeit a weak one, would disturb this equilibrium.

Stand up on your toes and explore the calf of the leg with your hands. Note the medial head of gastrocnemius is slightly longer than the lateral. Pinch the Achilles tendon between finger and thumb. It is surprisingly painful. A patient in deep coma does not respond to the most painful compression of this tendon.

Rupture of this tendon occurs typically in middle-aged people who undertake some unwonted violently energetic manoeuvre. It can also occur in trained athletes.

The **deep group of muscles** consist of the tibialis posterior, the flexor hallucis and flexor digitorum longus. These are the three muscles whose action is antagonistic (in most respects) to those of the three principal muscles in the anterior tibial compartment. Thus tibialis posterior balances tibialis anterior, flexor hallucis balances extensor hallucis, and flexor digitorum longus balances extensor digitorum longus.

Tibialis posterior can be palpated as the tendon which runs from just behind the medial malleolus (of the tibia) to the tubercle of the navicular when the ankle joint is gently plantar flexed and inverted against resistance. Its main insertion is into the tubercle of the navicular on the medial side of the foot about 1.5 cm behind the insertion of the tibialis anterior. Flexor hallucis is inserted into the terminal phalanx of the big toe and the tendon can be seen and felt as it passes through the sole of the foot. In the same way the tendons of the flexor digitorum longus can be felt.

It is evident that the sequence of these tendons is the same from medial to lateral on the plantar as it is on the dorsal surface of the foot. By analogy with the anterior tibial compartment you would very reasonably expect that the bellies of the three muscles in the deep compartment of the calf would be in the same sequence from medial to lateral as their own tendons. This is not what happens in the body!

Note that the muscle belly of the flexor digitorum longus is the most medial in the deep compartment, although its tendons have the most lateral insertion when they reach the foot. You can palpate this belly in yourself. It lies sandwiched between the soleus muscle and the middle third of the subcutaneous surface of the tibia. Flexion of the small toes is accompanied by contraction of the muscle belly and this makes it easy to identify in yourself. No contraction is felt here when the big toe is flexed. In fact the bellies of flexor hallucis longus and tibialis posterior cannot be palpated at all as they are completely concealed by the superficial calf muscles.

The tibialis posterior covers the whole of the interosseous membrane and 'overflow' on to the adjacent surfaces of the fibula and tibia, which are at right angles to the membrane. The flexor hallucis belly, which is attached to the back of the fibula, is lateral to that of the tibialis posterior, and these two muscles, including their tendons, run side by side in the relationship you would expect. Flexor digitorum longus takes origin from the posterior surface of the tibia, and thus must cross both tibialis posterior and flexor hallucis longus before reaching its insertion. It crosses the former about 5–6 cm above the ankle joint, that is in the leg, and the flexor hallucis longus 5–6 cm distal to the ankle joint, that is in the foot. In both cases it crosses superficial to the other muscle.

We need to comment on the insertion of all three muscles. The tibialis posterior is mainly inserted into the tubercle of the navicular, but it gives slips of various sizes to all the carpal bones except the talus and to the four lesser metatarsals. It is antagonist to tibialis anterior only in that it is a weak ankle plantar flexor while the tibialis anterior is a dorsiflexor. In inversion of the foot they are both protagonists, and their dorsal and plantar flexion components will cancel each other out.

Thus $(I + DF) + (I + PF) = 2 \times I$
where $(I + DF)$ is the **in**version and **d**orsiflexion produced by tibialis anterior and $(I + PF)$ is the **in**version and **p**lantar **f**lexion produced by tibialis posterior and $(DF) = - (PF)$.

The flexor hallucis is a powerful muscle with muscle fibres reaching down almost as far as the ankle joint. The tendon runs very close to the capsule of the posterior talocalcanean joint and it makes a groove in the talus as it does so [FIG. 8.07]. It then runs below the sustentaculum which it also grooves. When the toe digs into the ground in walking and running, this muscle is of major importance in protecting the ligaments of the foot from strain. Removal of the big toe may rob the medial arch of one of its principal supports.

The flexor digitorum longus is the principal long flexor of the other toes. It passes behind the medial malleolus just lateral to the tendon of tibialis posterior. The direction of its tendons is diagonally across the foot from medial to lateral and the pull is oblique. Attached to the tendons on their lateral side is a muscle called the flexor accessorius. You will realize from the direction of this muscle that its combined effect plus the flexor digitorum longus itself, is a pull in the long axis of the foot. In addition there are four lumbrical muscles connecting the four long flexor tendons to the extensor expansions on the medial side of the four small toes. The arrangement is like that in the hand. In the foot the lumbricals are very slender.

Interossei

There are dorsal and plantar interossei in the foot and their arrangement is similar to that in the hand except that the movements of abduction and adduction take place in relation to a plane running through the second toe (that next to hallux), not the third, as you might expect. There are four dorsal interossei and three plantar [pp. 145, 146]. Note that the big toe has no plantar interosseous muscle to compare with the first palmar interosseous of the hand which goes to the thumb.

Peroneal muscles

In the lateral compartment at the lateral side of the leg are to be found the peroneus longus and brevis muscles.

The tendons pass behind the lateral malleolus and the peroneus brevis tendon can easily be seen and felt as it passes forwards to its insertion to the styloid process of the fifth metatarsal. Gently plantar flex against resistance and then evert and you will find the tendon easily (if you are in the right place!).

Now identify the peroneus longus tendon which lies just below the peroneus brevis. There is often a prominent tubercle of bone which projects between the tendons as they pass over the surface of the calcaneus. The peroneus longus then runs in the groove on the plantar surface of the cuboid in the tunnel formed by the long plantar ligament. Look at an articulated foot. Put a pencil in the groove—it points straight at the insertion (it is bound to do so) of the muscle which straddles the joint between the first metatarsal of the big toe and the medial cuneiform bone. Note that its insertion almost joins that of the tibialis anterior muscle [p. 139]. When it is contracting, this muscle is ideally placed to support the transverse arch of the foot [FIG. 7.18, p. 145].

This tendon turns through about 90° as it enters the sole of the foot and you can see on the cuboid bone an articular facet here which marks the place where a sesamoid bone in the peroneus longus tendon articulates with it. The hyaline cartilage (which is avascular) on the articular surface of the sesamoid protects the tendon from pressure as it turns this sharp corner.

The muscle not only raises the lateral border of the foot in eversion, like the peroneus brevis, but it also depresses the medial border. This is one of the mechanisms which is responsible for changing the shape of the sole of the foot during inversion and eversion.

In man the sole of the foot lies flat on the ground. In this respect we differ from the anthropoids which walk on the outer border of the foot, as it were, in inversion. The migration across the sole of the foot of the insertion of peroneus longus and the development of the peroneus tertius are two special features of human anatomy associated with this fact.

Peroneus tertius differs from peroneus longus and brevis in that it is a dorsiflexor of the ankle while they are both plantar flexors. Its attachment to the distal part of the shaft of the fifth metatarsal gives it good leverage compared with the other muscles, which pass so close to the axis of the ankle joint, and this compensates to a considerable degree for the small size of the muscle. The dorsal and plantar flexion components more or less cancel themselves out when all three muscles act as everters.

OSSIFICATION OF THE BONES OF THE FOOT

The bones of the foot are all ossified in cartilage. The shafts of the metatarsals and most of the phalanges begin to ossify by the eighth week of foetal life, like the other long bones in the body.

The talus and calcaneus (in the posterior weight-line) are both well ossified at birth as is the cuboid (on the lateral weight-bearing side of the foot). Roughly one bone per annum commences to ossify after birth starting with the lateral cuneiform (the bone next to the cuboid), then the middle and medial cuneiforms and lastly at three or four years of age the navicular [FIG. 8.09].

In some patients the navicular bone is mishapen and distorted. This is primarily due to intrinsic developmental causes but two other factors are contributory. First the navicular is sandwiched between the three cuneiform bones and the head of the talus, all of which ossify ahead of the navicular, and secondly about two-thirds of the weight transmitted to the forefoot is transmitted via this bone.

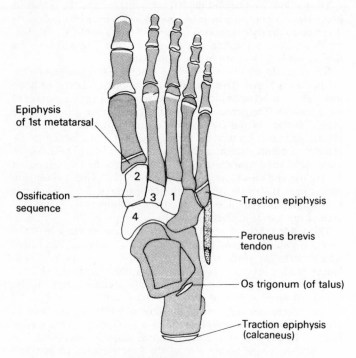

Epiphysis
of 1st metatarsal

Ossification
sequence

Traction epiphysis

Peroneus brevis
tendon

Os trigonum (of talus)

Traction epiphysis
(calcaneus)

FIG. 8.09. Ossification of bones of foot.
The unshaded bones are not ossified at birth. Their sequence of ossification is shown by the numbers (= years).
The traction epiphysis of the calcaneum, which is invariably present, is associated with the pull of the tendo calcaneus. The posterior talofibular ligament may wrench off the posterior tubercle of the talus. and an os trigonum, occasionally seen, may be mistaken for such a fracture. The separate ossification of the styloid process of the 5th metatarsal may be mistaken for a fracture which may be due to direct injury or to violent pull of the peroneus brevis. Vesalius described this anomaly.

The pattern of ossification in the phalanges and metatarsals is similar to that in the hand. Note the epiphysis of the metatarsal of the big toe and compare with the thumb.

The calcaneus is remarkable for the presence of a traction epiphysis into which the tendo calcaneus is inserted. It appears at 8–9 years and fuses with the rest of the calcaneus 4–5 years later. This epiphysis is a common source of pain in the heel between the ages of 8–13.

The posterior tubercle of the talus, into which the posterior talofibular ligament is inserted, may appear as an independent centre, and so may the styloid process of the fifth metatarsal into which the peroneus longus is inserted. Both these epiphyses may remain permanently separate from the rest of the bone. Such anomalies are not merely of academic interest as they may be mistaken for fractures on an X-ray.

A FURTHER CONSIDERATION OF THE ANKLE JOINT

In the ankle joint, the body of the talus lies beneath the lower articular surface of the tibia. The medial malleolus of the tibia projects further downwards and articulates with a comma-shaped facet on the medial side of the body of the talus. The medial malleolus prevents medial displacement of the talus with respect to the tibia. On the lateral side, the lateral malleolus of the fibula projects downwards, like the medial malleolus but to a greater degree. This malleolus prevents lateral displacement of the talus. The general shape of the ankle joint is often likened to a mortice and tenon joint, with which carpenters are familiar.

The lower ends of the tibia and fibula must be firmly attached to each other to prevent any sideways play in the joint. In fact they are joined by a very considerable thickening of the lower 3–4 cm of the interosseous membrane. Rupture of this ligament is unknown. In many animals the tibia and fibula are united here by bone. In man the arrangement is more complex. Look at the body of the talus from above. Note that it is slightly wedge-shaped, being wider in front than behind. If the joint is to be stable in spite of this fact, there needs to be a mechanism which allows slight variation in the distance separating the two malleoli so that in dorsiflexion they can be wider apart than is the case in plantar flexion. The attachment of the tibia and fibula by means of this ligament allows slight outward movement of the lateral malleolus to take place when the wide part of the talus lies in the ankle joint. There is some 'give' in the system. The small movements of the lateral malleolus are transmitted to the rest of the fibula and because of the great length of the shaft and the position of the interosseous ligament which acts like a fulcrum, greater movement occurs in the head (upper end) of the fibula. This is accommodated by means of the plane synovial superior tibiofibular joint.

The important ligaments of the ankle joint lie at the sides. On the medial side is the deltoid ligament [FIG. 8.07 and FIG. 27.14, p. 142]. The deep part of this ligament joins the medial malleolus to the whole of the non-articular surface on the medial side of the body of the talus and is concerned only with the ankle joint itself. The superficial part of the ligament is attached principally to what would otherwise be the free edge of the plantar calcaneonavicular (the spring) ligament. This attachment 'overflows' at each end of the spring ligament and is also attached to the navicular in front and the sustentaculum tali (supporter of the talus) behind. Now the head of the talus lies on the spring ligament, and its body lies on the central part of the calcaneus behind the sinus tarsi. The superficial part of the deltoid ligament helps to keep the talus and tibia in contact, but it does so indirectly by its attachments to the structures on which the talus is lying. Obviously this ligament will be involved in any movements between the talus and calcaneus which may take place. It is a 'two-joint ligament' while the deep part of the deltoid is a 'one-joint ligament'.

The lateral side of the ankle concerns the talus and the fibula. One might expect a ligament corresponding to the deep part of the deltoid ligament joining these two bones together. When you look at the lateral side of the body of the talus you will see that nearly all of it is an articular surface for the deep surface of the lateral malleolus. The expected ligament is in fact in two parts separated by the articular surface. The anterior talofibular ligament runs from the neck of the talus, while the posterior talofibular runs from the posterior tubercle towards the groove on the deep surface of the malleolus. The anterior ligament is tight in plantar flexion, while the posterior ligament is tight in dorsiflexion [FIG. 7.14, p. 142 and FIG. 8.07(c)].

On the lateral side is another ligament running downwards and backwards from the malleolus to the calcaneus. It is the calcaneofibular ligament. It clearly resembles the superficial part of the deltoid ligament as its support for the talus is indirect. It is also involved in movements of the talocalcanean joints, and you can easily feel it in yourself deep to the peroneal tendons when the foot is passively inverted.

THE KNEE JOINT AS A FUNCTIONAL UNIT

The knee joint [FIG. 8.10] is a modified synovial joint of the hinge variety. The two convex condyles of the lower end of the femur articulate with the two slightly concave condyles on the upper surface of the tibia. The fibula does not participate directly in the knee joint. The knee joint is really a (non-identical) twin articulation. This is evident as soon as the upper surface of the tibia is examined: the two condyles, which are completely separated from each other by the intermediate non-articular area, have a shape and size that differ slightly.

The hip joints are separated by the whole width of the pelvis but the knees touch each other. The long axis of the shaft of the femur must therefore incline inwards at an angle of about 10–15° to the vertical. As the legs are parallel, it is obvious that the femur joins the tibia at an angle. This means that the body's weight will be transmitted through the lateral rather than the medial side of the knee. In FIGURE 7.10, p. 137 the femoral condyles are shown from below: note the width of the lateral condyle compared with the medial. This increases the area of the weight-bearing surface and reduces the pressure which these joint surfaces must withstand.

In front of the knee joint is the patella. This is a sesamoid bone in the tendon of the quadriceps muscle. Notice that it does *not* lie at the level of the line of the knee joint, but above it by almost the whole of the length of the palpable part of the patellar tendon (4.5 cm). The patella tendon joins the bluntly pointed lower end of the patella to the tibial tuberosity. On each side of the tendon the joint capsule passes laterally. The patella tendon is incorporated into the joint capsule.

Stand with your feet parallel and fairly close together. Place your hands over the front of each knee joint so that your thumbs are in contact with the medial femoral condyles near the adductor tubercles. Flex the joints by 15–20° and straighten them again. Do this several times. You will realize from the backward twisting of your thumbs that towards the end of extension the femur is rotating medially in its own long axis. The use of both hands makes this rotation or locking much more obvious. It follows that this movement is not confined to the knee joint, but must also take place in the hip joint. Indeed some disorders in the hip joint may interfere with the normal mechanisms of the knee and give rise to pain etc. in the knee.

The knee joint capsule is strengthened on the medial side by the medial (collateral) ligament, which runs in continuity with the tendon of the adductor magnus at the adductor tubercle to the medial side of the tibia. Its attachment to the tibia is a very extensive one, about 1 cm in width and only terminating 7–8 cm below the knee. This ligament is a flattened band, which cannot be palpated as a separate structure, but of course if you know where it is situated you can put your fingers on it.

The corresponding ligament on the lateral side is a rounded cord. It runs from the lateral epicondyle of the femur to the head of the fibula. It is separated by a considerable space from the joint capsule and can easily be felt in yourself in the following way.

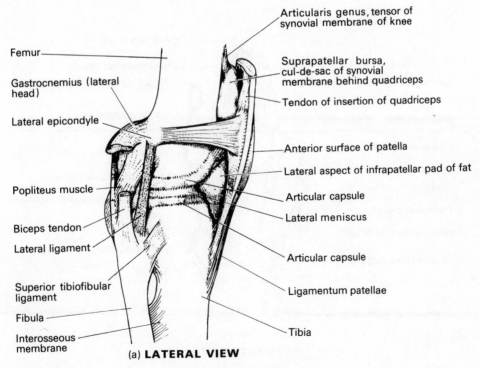

Femur

Gastrocnemius (lateral head)

Lateral epicondyle

Popliteus muscle

Biceps tendon

Lateral ligament

Superior tibiofibular ligament

Fibula

Interosseous membrane

Articularis genus, tensor of synovial membrane of knee

Suprapatellar bursa, cul-de-sac of synovial membrane behind quadriceps

Tendon of insertion of quadriceps

Anterior surface of patella

Lateral aspect of infrapatellar pad of fat

Articular capsule

Lateral meniscus

Articular capsule

Ligamentum patellae

Tibia

(a) **LATERAL VIEW**

FIG. 8.10(a). Lateral view of knee joint.

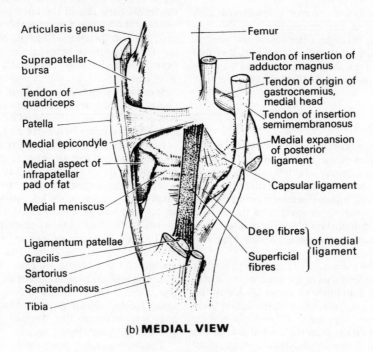

Articularis genus

Suprapatellar bursa

Tendon of quadriceps

Patella

Medial epicondyle

Medial aspect of infrapatellar pad of fat

Medial meniscus

Ligamentum patellae

Gracilis

Sartorius

Semitendinosus

Tibia

Femur

Tendon of insertion of adductor magnus

Tendon of origin of gastrocnemius, medial head

Tendon of insertion semimembranosus

Medial expansion of posterior ligament

Capsular ligament

Deep fibres

Superficial fibres

} of medial ligament

(b) **MEDIAL VIEW**

FIG. 8.10(b). Medial view of knee joint.

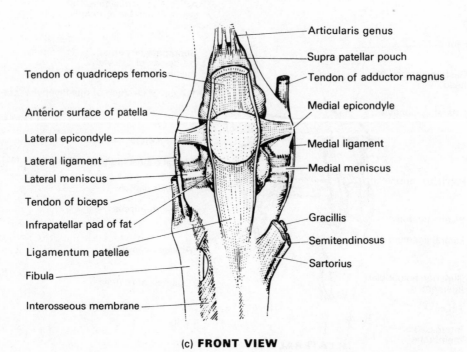

Tendon of quadriceps femoris

Anterior surface of patella

Lateral epicondyle

Lateral ligament

Lateral meniscus

Tendon of biceps

Infrapatellar pad of fat

Ligamentum patellae

Fibula

Interosseous membrane

Articularis genus

Supra patellar pouch

Tendon of adductor magnus

Medial epicondyle

Medial ligament

Medial meniscus

Gracillis

Semitendinosus

Sartorius

(c) **FRONT VIEW**

FIG. 8.10(c). Anterior view of knee joint.

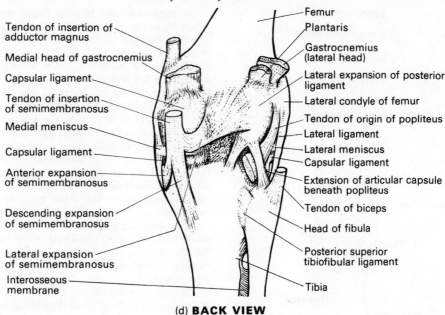

Tendon of insertion of adductor magnus
Medial head of gastrocnemius
Capsular ligament
Tendon of insertion of semimembranosus
Medial meniscus
Capsular ligament
Anterior expansion of semimembranosus
Descending expansion of semimembranosus
Lateral expansion of semimembranosus
Interosseous membrane

Femur
Plantaris
Gastrocnemius (lateral head)
Lateral expansion of posterior ligament
Lateral condyle of femur
Tendon of origin of popliteus
Lateral ligament
Lateral meniscus
Capsular ligament
Extension of articular capsule beneath popliteus
Tendon of biceps
Head of fibula
Posterior superior tibiofibular ligament
Tibia

(d) BACK VIEW

FIG. 8.10(d). Posterior view of knee joint.

Sit as in FIGURE 8.11 with the right ankle crossed over the left knee. The right thigh and leg should be more or less in the horizontal plane. With your right hand find the lateral border of the patellar tendon. By sliding the skin upwards and downwards identify the edges of the lateral condyles of the tibia and femur. In fact your finger lies in a triangular depression, whose base is the tendon and whose sides (one flat, one curved) are the tibia and femur. Work your way backwards until the apex of the triangle is reached. You will be able to feel the ligament running longitudinally [FIG. 8.11]. Because of the position of your right leg, the ligament is easy to feel as it is under tension.

With your other hand push on the medial side of the joint and the ligament will stand out more clearly on its own. Trace it downwards to the head of the fibula. Pressing on the inside of the joint obviously tends to separate the lateral condyles and would do so if it were not that the lateral ligament resists it. In many men this ligament can be seen.

Obviously the collateral ligaments prevent separation of the condyles on either side of the joint. As they run vertically they cannot resist horizontal forces tending to displace the femur backwards and forwards with respect to the tibia. Ideally ligaments to resist horizontal displacement should lie in the horizontal plane. In the event there are two ligaments inside the knee joint, called the anterior and posterior cruciate (*crux* = a cross; *cruciate* = like a cross) ligaments.

In FIGURE 8.12 you see that they do not cross each other at right angles, but both lie close to the horizontal plane. They are named according to their attachments to the intercondylar area of the upper surface of the tibia.

When the tibia is displaced forwards, or the femur displaced backwards, the anterior cruciate ligament will resist the movement. In the opposite displacement the posterior cruciate ligament becomes tight. When the knee is flexed the femoral condyles do not roll backwards like a roller across the upper surface of the tibia. They are prevented from doing this by the cruciate ligaments. The condyles do not roll, they slip or skid on the tibia.

When the knee is extended the reverse happens. The two condyles do not roll forwards but skid on the tibia. Eventually the anterior cruciate becomes tight. As it is attached to the lateral condyle of the femur movement of this condyle is then arrested, but movement of the medial condyle continues for a further 10–15° and it is for this reason that the femur rotates medially around its own long axis. In fact the rotation takes place around the point where the anterior cruciate ligament is attached to the lateral condyle [FIG. 8.13]. Notice in this figure the curve described by the medial condyle. The centre of this curve is the attachment of the anterior cruciate ligament.

The rotation mechanism has other attendant anatomical features. The lateral condyle of the tibia is almost circular in shape while the medial one is elongated in the antero-posterior direction to allow for the wheeling movement of the medial condyle. In the same way the medial and lateral menisci (or semilunar cartilages) differ when looked at from above.

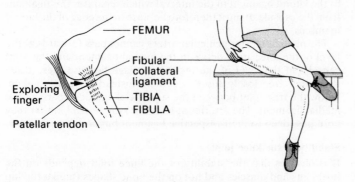

FEMUR
Fibular collateral ligament
TIBIA
FIBULA
Exploring finger
Patellar tendon

FIG. 8.11. A method of finding the fibular collateral ligament.

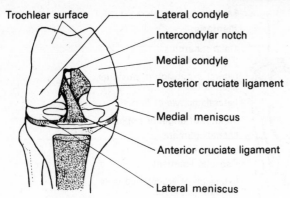

Trochlear surface — Lateral condyle

Intercondylar notch

Medial condyle

Posterior cruciate ligament

Medial meniscus

Anterior cruciate ligament

Lateral meniscus

(a) *Cruciate ligaments from front (knee flexed)*

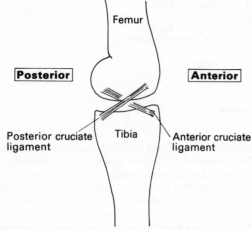

Femur

Posterior — Anterior

Posterior cruciate ligament — Tibia — Anterior cruciate ligament

(b) *Cruciate ligaments from medial side*

FIG. 8.12. Cruciate ligaments.

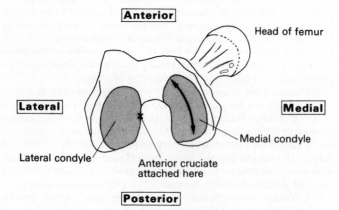

Anterior

Head of femur

Lateral

Medial

Lateral condyle

Medial condyle

Anterior cruciate attached here

Posterior

FIG. 8.13. Distal end of femur.
The lateral condyle is wider than the medial condyle. The curvature of the medial condyle, centred on the point of attachment of the anterior cruciate ligament, is shown by the curved arrow. The weight bearing areas of the condyles are shaded. The head of the femur is included to help in identifying the medial side of the bone.
Note the marked forward projection (upwards in this diagram) of the lateral condyle when viewed from the angle. This feature helps to prevent lateral displacement of the patella when the quadriceps contracts.

The **menisci** are two fibrocartilagenous intra-articular structures which fill in the angle between the flat tibial and curved femoral condylar articular surfaces. The peripheral edge fuses with the capsule of the knee joint, so that in effect the medial meniscus is attached to the medial ligament but the lateral meniscus is not attached to the lateral ligament because as we have already seen, this ligament is not incorporated into the capsule.

The menisci are non-weight-bearing structures which are probably important in improving the lubrication of the inside of the joint. Their shapes when looked at from above differ in accordance with the functional differences in the condyles during locking. Thus the lateral meniscus is an almost complete circle whose horns reach the anterior and posterior surfaces of the tibial tubercle, and are attached to the bone very close to each other. The horns of the medial meniscus on the other hand are widely separated from each other, and the general shape, instead of being more or less circular, (viewed from above) is very much elongated in the anteroposterior direction. These differences reflect the rotation of the lateral condyle around the same axis, and the anteroposterior twisting of the medial condyle which takes place when the knee joint is locked.

The popliteus muscle has a fleshy attachment to the back of the upper quarter of the shaft of the tibia. It is a fan-shaped muscle and its tendon is attached to a characteristic depression on the lateral side of the lateral condyle of the femur *inside* the knee joint. This tendon is covered by synovial membrane which projects through a gap in the capsule of the joint forming a synovial sheath around the tendon. The popliteus is a weak flexor of the knee joint, and because of its oblique direction it helps to 'unlock' the knee joint—that is, under weight-bearing conditions, it rotates the femur laterally, and at the same time to a small extent rotates the tibia medially. The closeness of the pull of this muscle both to the axes of flexion and locking or unlocking means that its effect is weak. It is extension which locks the knee joint, and flexion which unlocks it. The mechanics of the joint make it impossible to have one movement without the other. Similar complex movements occur in the carpometacarpal joint of the thumb.

Blood supply to knee joint

The blood supply of the knee joint is derived from the extensive 'anastomosis round the knee joint'. There are five principal vessels, a pair of superior and inferior genicular arteries, and the middle genicular artery which enters the joint capsule from behind. Notice the symmetry of the medial and lateral superior genicular arteries which both pass forwards from the popliteal artery above the femoral condyles. The inferior lateral genicular artery passes deep to the lateral ligament in the interval which separates this ligament from the capsule. It must therefore lie close to the edge of the lateral meniscus.

The medial inferior genicular artery cannot pass forwards at the level of the knee joint between the medial ligament and the joint capsule, at the level of the medial cartilage because these three structures are closely fused together. Instead it passes 3–4 cm below the true joint between the superficial and deep parts of the medial ligament. The arteries at least are symmetrical in that they both pass deep to their respective ligament but at different levels.

Stability of the knee joint

It is obvious that the stability of the knee joint depends on the ligaments and muscles and not on the bone shapes (unlike the hip joint). The knee is especially susceptible to forces which twist it in its own long axis, because most of the tendons and ligaments run vertically and therefore can have only minimal influence on forces which twist in the horizontal plane. Usually the foot is on the ground and the body is twisting and turning as when playing football or skiing. The weight of the body acts on the knee joint via a very long lever which is the femur. Excessive movement of this kind commonly damages the medial meniscus. The relative mobility of the medial femoral condyle has already been seen. The fact that the medial cartilage is fixed to the medial ligament means that it is unable to escape being damaged and torn. The lateral meniscus is only attached to the capsule of the knee joint, not the lateral ligament, and in addition there are a few fibres of the popliteus muscle which are attached directly to the posterior horn of the cartilage and this may help to pull it out of harms way under certain circumstances.

Rupture of one of the collateral ligaments occurs when the knee is subject to violence on the medial or lateral side. Violence in the anteroposterior direction damages the cruciate ligaments. Not infrequently collateral and cruciate ligaments are ruptured simultaneously. Such injuries, as well as cartilage injuries, are common amongst 'body contact' sports, but are much less common in sports like tennis or squash although they depend to a high degree on the ability to twist and turn.

A further consideration of bursae around the knee joint
There are many bursae around the knee joint. The suprapatellar bursa lies in the interval between the front of the shaft of the lower end of the femur and the quadriceps muscle as it approaches the patella. The bursa extends 3–4 fingers breadths above the patella. When the knee is examined clinically a hand is always placed above the patella to displace fluid from the bursa back into the joint. The presence of excessive fluid in the joint may not declare itself until this is done. The suprapatella bursa occasionally does not communicate with the knee joint.

A second bursa is that separating the lower end of the semimembranosus tendon from the back of the knee joint capsule on the medial side. This communicates with the knee in about 50 per cent of cases, and may give rise to a lump in the popliteal fossa. Compression will make the lump disappear temporarily as the fluid is squeezed back into the joint.

There are bursae between the skin and the tibial tuberosity and also the surface of the patella. The latter may become enlarged if the skin in the vicinity is infected by abrasions or scratches.

On the lateral side the synovial membrane which surrounds the popliteus tendon as it passes through the joint capsule, may give rise to a swelling.

A further consideration of the muscles acting on the knee joint
The angulation of the thigh and leg means that the pull of the extensor muscles, the quadriceps femoris, is correspondingly oblique to the patellar tendon which lies in the long axis of the leg. There must be a tendency for the patella to be displaced laterally. This is guarded against by the prominence of the lateral condyle of the femur [FIG. 8.13] and also by the way in which the lowest fibres of the vastus medialis extend right down to the patella itself and run in the horizontal plane, i.e. they pull the patella medially. Lateral dislocation of the patella can occur, and this and other related problems may be treated by repositioning the attachment of the patella tendon in a more medial site on the tibia.

Note that the 'oblique-pull problem', which applies to the quadriceps extensor muscles, does not apply to the hamstring flexor muscles, because they are attached to the ischial tuberosity, which is in the same parasagittal plane as the knee joint. The general direction of the pull of the hamstrings is therefore in line with the leg from the start.

When the knee is flexed and extended under non-weight-bearing conditions, the rotation of the knee joint, takes place in a slightly different manner from that which we have described. In this case the femur remains 'stationary' and it is the tibia (taking the foot with it) which now rotates laterally. In describing a movement such as this it is essential to say which bone is regarded as fixed, and which as moving. A statement such as 'the knee rotates medially' is without meaning unless the movement of the individual bones is identified.

The knee is **extended** by the quadriceps femoris muscle, that is, by the vastus medialis, rectus femoris, vastus lateralis, and vastus intermedius. Note that the rectus femoris is the only one which takes origin above the hip joint, from the anterior inferior iliac spine and from the bone lateral to the spine. Note also that quadriceps femoris lies in two planes. The vastus intermedius clothes the shaft of the femur (except for the linea aspera) and lies deep. Its superficial surface is completely overlaid by the three other muscles.

The rectus femoris is a weak flexor of the hip joint. During walking and running, as the knee joint extends, the hip joint also is straightened. The hip joint movement requires an increase in the length of the fibres, while of course the knee movement requires a decrease. In practice these factors more or less neutralize each other, so that the rectus fibres maintain a fairly uniform length. This argument cannot apply to the other components of the quadriceps which are strictly 'one joint' muscles.

The hamstring muscles, the semitendinosus, semimembranosus, and biceps femoris, all take origin from the ischial tuberosity. The two former lie on the medial side of the back of the thigh. Semimembranosus is inserted close to the knee joint, to the back of the medial condyle of the tibia, where its attachment is marked by a deep groove. The semitendinosus tendon is almost half the length of the whole muscle (hence the name) and it is inserted into the tibia on its medial subcutaneous surface with the sartorius and gracilis muscles, about 5–6 cm below the joint. The superficial muscle, the semitendinosus, whose thin tendon can be seen and felt on the inner border of the popliteal fossa gets a much better leverage than the semimembranosus because it is attached at a distance from the knee joint. Palpate deep to the semitendinosus tendon and you will find just above the knee joint the short but thick tendon of semimembranosus. Note that the muscle belly lies in the lower half. The upper half of semimembranosus is thin and wide. The bellies of these two muscles fit together 'head-to-tail' like sardines in a tin [see p. 137].

On the lateral side of the back of the thigh is the biceps muscle. The long head is attached to the ischial tuberosity, but the short head comes from the linea aspera and joins the long head. The muscle is inserted into the head of the fibula closely associated with the lateral ligament of the knee. The short head is a 'one-joint muscle' unlike the long head.

When the knee is flexed by about 90° you will find, if you attempt to point your toe inwards or outwards that there is about 10–15° of rotation in the knee joint. Palpate the hamstring tendons as you do this, and you will see that they are producing this movement, with the assistance, no doubt from the popliteus muscle. The leverage exerted by the hamstrings is much more powerful when the knee is flexed to a right-angle. In body-contact sports a player's knee is very vulnerable if the knee is in this position and the player is the object of rough play.

THE HIP JOINT AS A FUNCTIONAL UNIT

The hip joint [FIG. 8.14] is a ball-and-socket synovial joint. The head of the femur fits deeply into the acetabulum and is very stable, except when the joint is flexed and adducted (as when one leg is crossed over the other). If a force is now directed longitudinally along the femur the head can come out of the socket.

The hip has a wide range of movement in the sagittal plane (flexion and extension) and in the coronal plane (abduction and adduction). In addition rotation (medial and lateral) takes place around an axis running through the middle of the head of the femur and extending down to the heel. If you stand up and point your foot first to the right and turn to the left—the principal movement is

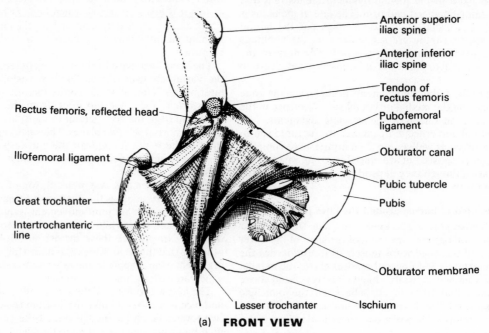

Anterior superior iliac spine

Anterior inferior iliac spine

Tendon of rectus femoris

Pubofemoral ligament

Obturator canal

Pubic tubercle

Pubis

Rectus femoris, reflected head

Iliofemoral ligament

Great trochanter

Intertrochanteric line

Obturator membrane

Ischium

Lesser trochanter

(a) FRONT VIEW

FIG. 8.14(a). Hip joint (from the front).

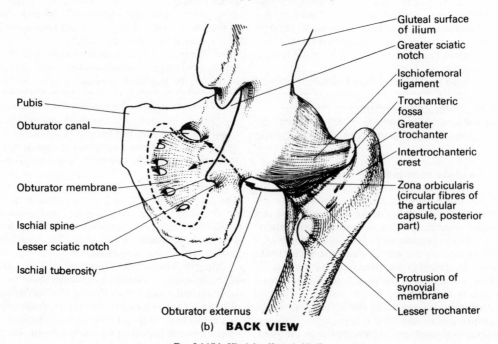

Gluteal surface of ilium

Greater sciatic notch

Ischiofemoral ligament

Trochanteric fossa

Greater trochanter

Intertrochanteric crest

Zona orbicularis (circular fibres of the articular capsule, posterior part)

Pubis

Obturator canal

Obturator membrane

Ischial spine

Lesser sciatic notch

Ischial tuberosity

Protrusion of synovial membrane

Lesser trochanter

Obturator externus

(b) BACK VIEW

FIG. 8.14(b). Hip joint (from behind).

taking place in the hip joint not in the foot itself. Confirm this on yourself by putting your hand on your greater tuberosity.

The combined movement of circumduction is also possible. In this case the whole foot can be steered around the circumference of an imaginary circle.

Certain movements of this joint are restricted by secondary non-joint factors. For example, when the knee is flexed, flexion at the hip (even in thin people) is limited by the contact of the thigh with the abdominal wall. When the knee is extended, tension in the hamstring muscles limits flexion of the hip in many people. Abduction is limited in the same way by tension in the adductor muscles on the medial side of the thigh. Extension is the movement that is limited only by tension in the hip-joint ligaments.

The acetabulum and weight-bearing

The deep recess in the hip bone in which the head of the femur lies is the acetabulum. The weight of the body reaches the upper part of the acetabulum from the sacro-iliac joint and is transmitted via the ilium. The bone connecting these two joints is extremely thick whereas other parts of the ilium are so thin as to be transparent. As has been seen, up to the age of about sixteen, the acetabulum is separated in three parts by a trifoil cartilage. The ilium, ischium, and pubis all contribute to the acetabulum, but the body weight is only transmitted through the ilium.

The acetabulum has a large non-articular area (about 25 per cent of the total) called the acetabular fossa. It is filled with fat and fibrous tissue which is continuous with the ligament of the head of the femur. This ligament is covered by synovial membrane and is attached to the fovea (Latin = pit) near the middle of the head of the femur. The ligament may help to limit certain movements but its position so close to the axes of movements, must render it very ineffective. In childhood the ligament provides the main route for the blood supply to the head of the femur.

The acetabular fossa is approached by the acetabular notch. This notch is bridged by the transverse acetabular ligament. The whole of the edge of the acetabulum and this ligament have a fibrocartilagenous rim or lip—2–3 mm wide—attached to them which is the acetabular labrum. The labrum adds to the depth of the acetabulum, and gives to its edge a flexibility which the bony lip itself does not have [FIG. 7.03, p. 130].

The fibrous capsule of the hip joint is attached directly to the hip bone, to the outer surface of the labrum and to the transverse acetabular ligament. At the femoral end the capsule is attached anteriorly to the intertrochanteric line. This means that seen from the front the whole of the neck of the femur is inside the hip joint. At the back the capsule covers about two-thirds of the neck of the femur, so that the lateral one-third is extracapsular. The edge of this part of the capsule is not attached to the bone at all, and synovial membrane may be seen pushing out between the capsule and the neck of the femur.

The capsule is thickened in three places forming the iliofemoral, pubofemoral, and ischiofemoral ligaments. These ligaments twist around the neck of the femur and are all in tension when the hip joint is extended in the anatomical position. The iliofemoral ligament [FIG. 8.14] is one of the strongest in the body and is attached over the whole length and width of the intertrochanteric line. It is often compared to an inverted Y i.e. λ.

Notice that the obturator externus muscle takes origin from the outer surface of the obturator membrane, and that it runs outwards and backwards below the hip joint, and then winds round behind the neck into a shallow groove which runs over the top of the lateral

part of the neck into the digital fossa. This relationship seems very complex but the complexity is entirely artificial and is the product of the convention of the anatomical position. The joint develops in the 'foetal position' when it is almost completely flexed [FIG. 8.15]. Put

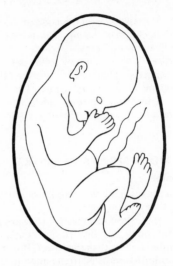

FIG. 8.15. Foetal position: foetus *in utero*.
4½-month human foetus with all body members perfectly formed. The hip joint is 'flexed' by more than 90°. The fibres in the hip joint capsule, in so far as they can be recognized at this stage. run straight between their attachments, and so does obturator externus. The twisted features of these structures appear when the hip joint is extended after birth.

the hip bone and the femur into this position and you will at once see that this muscle now runs straight from origin to insertion. When the joint is extended the muscle and its tendon necessarily becomes twisted around the femoral neck. Exactly the same consideration applies to the capsule with its three thickened ligaments which also become twisted around the neck when the joint is extended.

Avascular necrosis of head of femur

The blood supply of the head and neck of the femur is of great clinical importance. It is derived from blood vessels found in the ligamentum teres and in the retinacula of the neck. During growth the head is joined to the neck by the avascular barrier of the epiphyseal cartilage. During this period the head is nourished principally by the vessels running in the ligamentum teres. Once this cartilage has disappeared with the cessation of growth, these vessels tend to degenerate and the blood vessels in the retinacula then become all important.

Fracture of the neck of the femur, which is a common injury in the elderly, will tear the retinacula and cut off the principal blood to the head, which undergoes avascular necrosis. Healing is then very slow and may never take place. The degree to which the blood supply of the head is reduced depends on the position of the fracture. If there is still intact retinaculum in continuity with the head and neck above the fracture site, then the blood supply may be adequate. Thus if the fracture is close to the intertrochanteric crest, the whole retinaculum on the back of the neck will survive intact. If it is very close to the head (subcapital fracture) then the whole retinaculum will be torn through and only the few inadequate small vessels in the ligamentum teres remain to nourish the head.

On X-ray, bone undergoing avascular necrosis appears denser than surrounding normal bone. This is because disuse atrophy with decalcification occurs in the normal bone, whereas in the avascular bone the calcium loss takes place very slowly because of the poor blood supply.

Avascular necrosis is also seen in the head of the femur in young people following dislocation of the hip. This is because the dislocation ruptures the ligamentum teres.

Avascular necrosis is also common following fractures of the scaphoid bone in the wrist. The bone is nourished by blood vessels which mostly enter at its lateral extremity. The blood supply to the bone medial to the fracture line which articulates with the lunate is therefore largely cut off. Healing is very slow and may never take place.

MOVEMENTS IN SPORT

Walking and running

What is the reason for swinging the arms in walking and why is this done in a different way when running?

The leg which is off the ground imparts torque around the vertical axis which is neutralized by the forward movement of the opposite arm. The upper limb is lighter than the lower. But the torque effect of the moving lower limb is reduced to the minimum because of the position of its centre of gravity which is very close to the midline. Knees are very close together. The upper limb lies further away from the same vertical axis because of the width of the shoulder, and if we compare the wide separation of the elbows with the close proximity of the knees, this point is obvious. So in walking the small upper limb can balance the twisting movement which is exerted by the larger lower limb.

In running the effect of the speed of movement becomes more important. In fact, the leverage produced by a moving part of the body depends on the distance of its centre of gravity from the axis (with respect to which we are considering the movement), its weight and on the square of the velocity. In running, the speed with which the arms move obviously increases, but we are interested also in their position. Obviously by raising the arms, abducting the shoulders, the centre of gravity of the limb is displaced away from the central vertical axis and the leverage is correspondingly increased. No such displacement of the centre of gravity takes place in the lower limb.

Some runners think of their upper limbs as so much luggage, but there must be an optimum weight for the limb in a particular individual, so that he can achieve this essential balance in the most economical way. It is instructive to try to run without arm movements.

In many sports such as tennis, in which a racquet is used, the movements of the other arm is vital in maintaining poise and balance, by counteracting the torque effects not only of the other arm but the racket as well.

The effect of the position of the arms is dramatically shown by the pirouetting figure skater. Here the skater approaches the pirouette spot at speed. The energy of the forward movement is now translated into the energy which spins the skater in one place. The rotation is slow (comparatively) when the arms are outstretched but as the arms are folded across the chest their centre of gravity comes closer and closer to the axis, and their leverage is reduced, but as the total energy in the system remains the same (except for that wasted in churning up the ice) the velocity of rotation increases. The skater can slow the spin by extending the arms again.

In games like golf, cricket, baseball, and hockey the player stands sideways to the ball and would seem to have the option of playing either to his left or to his right. But when you know that in most right-handed people the left shoulder is normally slightly higher than the right, it can at once be seen that the posture most people adopt implies only a slight amplification of an asymmetry which is already present. Obviously to produce a mirror image posture with the right shoulder higher than the left would require a considerable displacement of the natural posture of the trunk and shoulder. Even to stand with shoulders level is a departure from most people's accustomed stance.

Once the shoulder positions and slight twist of the thoracic spines which goes with it is established, then the left hand must grip the handle above the right. Interesting problems arise if you consider the consequences of double hand grips in tennis. A right-handed person first grips the racket near the top of the grip. This means that the left hand must go below it (producing the hand positions of a left-handed golfer) and the appropriate shot must be the backhand, not the forehand. Many of the world's leading tennis players now use a double-handed backhand. It has the advantage of increasing considerably the power available at or just before the moment of impact so that the direction can be disguised much later than in a one handed backhand, or alternatively the ball can be hit much harder. It has the disadvantage of restricting the reach and this in its turn puts a greater premium on mobility and speed about the court. Also the 'other' hand is no longer available to adjust the balance during the time when the player is committed to the shot. It is interesting that only one top class player uses a double handed grip from time to time for the forehand stroke. Here the hand positions are 'wrong' but it is as a doubles player that he excels and here it frequently happens that the ball is aimed straight at a player, who is standing close to the net, and in countering such a stroke the players' reach is not relevant.

9 The thorax

THE THORACIC WALL

HOW TO FIND YOUR WAY AROUND THE FRONT

In the midline the **sternum** (= sword) is subcutaneous. It lies in a depression produced by the pectoralis major on the two sides. The sternum is made up of the **manubrium**, (*manubrium* = a handle; *manus* = a hand) the body of the sternum and its xiphoid process.

The notch at the top of the manubrium sterni between the two medial ends of the clavicle is termed the **jugular notch**. An alternative name is the suprasternal notch. Place your finger in this jugular notch and run it down the midline for a distance of 5 cm. A horizontal ridge will be felt running across the sternum. This is the **manubriosternal junction**. It is termed the sternal angle since the manubrium makes an angle with the body of the sternum at the point. The older name was angle of Louis. The sternal angle may appear as a visible ridge, or it may be felt as a slight depression. It is an important landmark since the costal cartilage of the second rib joins the sternum at this point. Even in very fat people it is easy to feel [FIG. 9.01].

The **xiphoid process** or xiphisternum can usually be felt below the sternum in the midline in the subcostal angle between the seventh costal cartilages.

Counting ribs

The ribs are counted downwards from the second which is located using the sternal angle. The spaces between the ribs are known as **interspaces**. They are named according to the rib above [FIG. 9.02]. Count from above. Notice how narrow is the gap between the ribs, in the midaxillary line.

Interspaces are best located by placing the finger in the gap between two ribs and leaving it there whilst the next finger searches for the next space. Using all the fingers it usually is possible to locate spaces 2–5 simultaneously in this way at a distance of about 7 cm from the midline. The spaces are further apart in this region and if the hand is removed from the chest, the finger may be replaced in the same space in mistake for the one below.

SURFACE MARKING OF HEART

The surface marking of the heart [FIG. 9.02] is represented by a quadrilateral ABCD formed by joining the following four points.

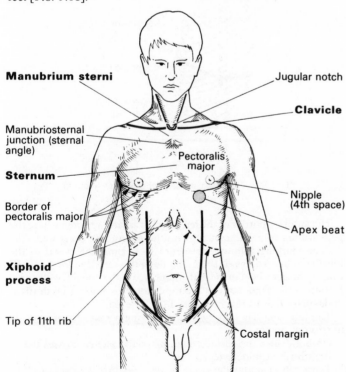

FIG. 9.01. Important landmarks of the thorax.

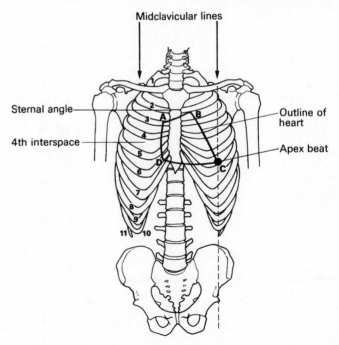

FIG. 9.02. The surface marking of the heart.

A: Upper border of third right costal cartilage 1 cm to right of sternum.

B: Lower border of second left costal cartilage 2 cm to left of sternum.

C: Apex beat area in the fifth intercostal space and midclavicular line about 8.5 cm from midline in an adult.

D: Sixth right costal cartilage 1 cm to right of sternum.

1. The upper border is formed by the left atrium.
2. The left border is formed by the left ventricle.
3. The lower border is formed by the right ventricle.
4. The right border is formed by the right atrium.

The position of the apex beat is very important clinically. The downmost and outermost point where the impulse of the heart can be felt distinctly indicates the position of the apex of the heart. If it is not present in its normal position then the heart must have enlarged, say in high blood pressure, or been displaced, say by a lung on one side collapsing. When the heart is enlarged the impulse may be felt over a large area.

SURFACE MARKING OF LUNGS AND PLEURA

The surface markings of the apices of the **lungs** follow the pleura and extends up to 2.5 cm above the medial third of the clavicle protected by the sternocleidomastoid muscle [FIG. 9.03].

The anterior border of the **right lung** follows the pleura down to the sixth costal cartilage in the midline. The surface marking then crosses:

Sixth rib at midclavicular line

Eighth rib at midaxillary line to reach side of T10 vertebra.

The **main (oblique) fissure** runs from the T2 spine across the angle of the scapula to the junction of the sixth costal cartilage with the sternum. It is a straight line. The **horizontal (accessory) fissure** runs from the fourth costal cartilage at the sternal border straight and horizontal to the main fissure. These fissures divide the right lung into three lobes.

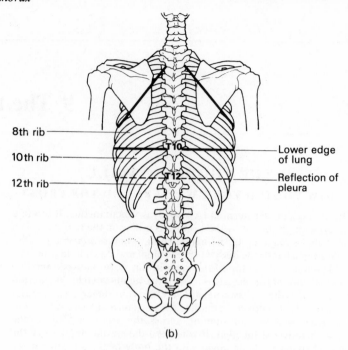

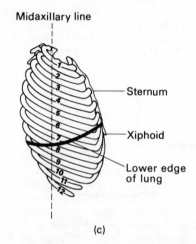

FIG. 9.03. The surface markings of the lungs and pleura.

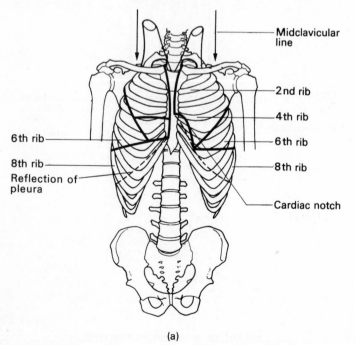

The anterior border of the **left lung** has a cardiac notch where the pericardium comes in contact with the pleura and chest wall. The surface marking follows the pleura in the midline down to the fourth costal cartilage. It then runs along this cartilage for 4 cm and runs to the apex beat in the fifth interspace in the midclavicular line. It follows the same path as the right side crossing: Eighth rib at midaxillary line to reach side of T10 vertebra.

The left lung has only two lobes having a main fissure but no horizontal fissure.

On both sides the surface marking of the **pleura** crosses the:

Eighth rib at midclavicular line

Tenth rib at midaxillary line below twelfth rib to side of T12 vertebra.

Because it is so low at the back, care must be taken not to injure

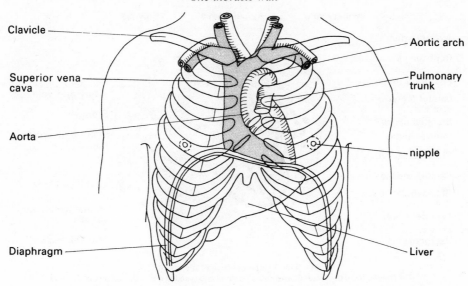

FIG. 9.04. The relations of the heart and diaphragm to the chest wall.

the pleura when operations are carried out on the kidney [see FIG. 11.16, p. 212].

Note that the lower border of the lung having reached the sixth rib in the midclavicular line runs horizontally round the chest wall. It happens because of the oblique direction of the ribs, that this horizontal line cuts the eighth rib in the midaxillary line, and the tenth rib posteriorly. The costodiaphragmatic recess of the pleural cavity extends about two ribs below the lung margin, when this is possible. That means that in the midaxillary line it extends down to the tenth rib, and at the back it reaches just below the twelfth rib. In inspiration the lungs move downwards into this recess; in expiration they withdraw from it.

Due to the domed-shape nature of the diaphragm, the bulk of the lungs is very much higher than is generally realized. A moment's thought will confirm this fact. The hilus of the lung is at the level of the sternal angle. Find this point on yourself. Next find your lower costal margin on the right side. The liver is tucked under this costal margin. It is unusual for the liver to appear below the costal margin unless it is pathologically enlarged. The diaphragm lies on top of the liver which is a large organ and reaches up to the fourth interspace on the right side, that is, to the nipple level [FIG. 9.04]. The bulk of the right lung is above this point as may be confirmed by percussion. On the left side the pericardial sac depresses the diaphragm slightly. The apex beat gives a rough guide to the level of the diaphragm since the pericardium is attached to the diaphragm at this point. In the midline the diaphragm is attached to the xiphisternum. Thus with quiet breathing much of the pleural cavity associated with the lower ribs is obliterated by the two layers of pleura, visceral, and parietal being in contact with one another without any intervening lung tissue.

SURFACE MARKING OF THE MAMMARY GLANDS

The mammary glands extend from the second rib to the fifth rib vertically and from the edge of the sternum at the level of the fourth rib to the fifth rib in the midaxillary line. The axillary tail runs up along the lower border of pectoralis major to the third rib, and thus

comes very close to the lowest axillary lymph nodes. The nipple normally corresponds to the fourth interspace about 10 cm from the midline, that is, just lateral to the midclavicular line.

The mammary glands lie in superficial fascia overlying pectoralis major, serratus anterior and, to a small extent, rectus abdominis, and external oblique [FIG. 9.05].

Development of the breast

The secretory components of the mammary gland are developed from downgrowths of surface ectoderm which send columns of cells into the underlying mesoderm. These columns develop a lumen and become ducts and milk alveoli (acini). The ducts are lined with

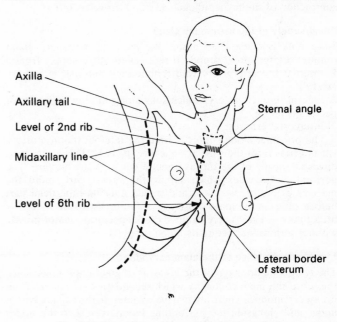

FIG. 9.05. The surface markings of the mammary gland.

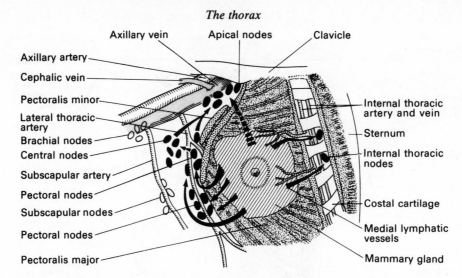

Axillary vein   Apical nodes   Clavicle

Axillary artery
Cephalic vein
Pectoralis minor
Lateral thoracic artery
Brachial nodes
Central nodes
Subscapular artery
Pectoral nodes
Subscapular nodes
Pectoral nodes
Pectoralis major

Internal thoracic artery and vein
Sternum
Internal thoracic nodes
Costal cartilage
Medial lymphatic vessels
Mammary gland

FIG. 9.06. Lymph drainage of the breast.
The lymphatics of the breast drain into the pectoral nodes and hence to the axillary nodes. and
into the internal thoracic nodes. The breast also has lymphatic connections with the posterior
triangle of the neck and downwards to the upper abdomen and liver. which may be of importance
in malignant disease.

columnar epithelium whilst the acini have cuboidal epithelium. The nipple is a depressed area at first but later becomes an elevation which is surrounded by a pigmented areolar area. The nipple may remain depressed making breast feeding difficult.

The milk-producing system of the breast consists of 15 to 20 lobes arranged in a sector basis around the nipple (like slices of a round cake) each with its own opening at the nipple. Shortly before its opening each duct has a diluted ampulla, the lactiferous sinus, to which milk is transferred by the action of the hormone oxytocin on the myoepithelial cells surrounding the milk alveoli during suckling. Milk is ejected from each of these ducts during suckling by the contraction of the myoepithelial cells and active secretion.

Blood supply of the mammary gland

Since milk is made from blood, the lactating mammary gland requires a large blood supply. It receives its blood supply from:
1. Internal thoracic artery (old terminology 'internal mammary artery').
2. Lateral thoracic artery, which runs along the lower border of pectoralis minor.
3. Intercostal arteries.

The principal blood supply is from the internal thoracic artery which arises from the subclavian artery [p. 173]. This artery runs down the inside of the thorax about 3 cm from the midline. The perforating branches which supply the breast curl round the margins of the breast tissue and spread over its superficial part before entering the main substance of the breast tissue. Thus the breast may be extensively stripped off the pectoralis major muscle without seriously affecting its blood supply.

Lymphatic drainage of the mammary gland

The lymphatic drainage of the breast is of great importance since these are the main routes by which secondaries are spread from breast carcinomas. There are two main routes, to the axillary lymph nodes and alongside the perforating blood vessels to the nodes along the internal thoracic artery. In addition lymphatics may pass

through pectoralis major and reach nodes deep to the clavipectoral fascia below the clavicle on the inner side of the axillary vein [FIG. 9.06].

Lymphatic from the inner lower quadrant of the breast may communicate with the abdominal lymphatics by passing between the lateral border of rectus abdominis and the ninth rib.

In the gland itself there is a subareolar plexus under the skin of the areola and a lymphatic plexus on the fascia over pectoralis major.

THE THORACIC CAVITY

The thoracic cavity [FIG. 9.04] provides an air-tight container for the lungs that makes breathing possible. It also contains the heart in its own pericardiac sac, blood vessels, the thymus gland, and structures passing between the neck to the abdomen such as the oesophagus, thoracic duct, sympathetic trunks, and vagus nerves.

Due to the elastic recoil of the lungs the pressure in the thoracic cavity is subatmospheric. The pressure is usually about -2 mmHg but it becomes -8 mmHg during inspiration. As a result of this negative pressure, air will be sucked in if there is a hole in the chest or in the lungs (for example a burst bulla). Any air entering the thoracic cavity allows the lungs to collapse and impairs breathing. This condition is termed a **pneumothorax** (*pneuma* (Greek) = air).

Although the oesophagus passes through the thorax, it is air-tight as far as the thoracic cavity is concerned. However, an inflated balloon in the oesophagus connected to a pressure recorder provides a convenient way of monitoring intrathoracic pressure changes.

THE THORACIC CAGE

The thoracic cage is formed by the twelve thoracic vertebrae, the twelve ribs with their costal cartilages, the intercostal muscle in between and the sternum [FIG. 9.07].

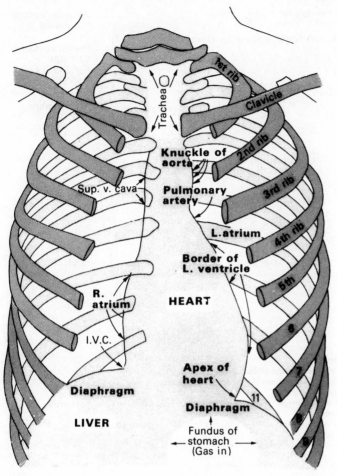

FIG. 9.07. Chest X-ray.

The tissues of the body vary in their translucency to X-rays. Bone which contains calcium atoms is relatively opaque to X-rays. For this reason the ribs in this X-ray show up well, but the costal cartilages which contain no calcium do not. Cavities containing air, such as the trachea and the fundus of the stomach will 'show up' because they contain one of the most transparent substances in existence, namely air. The lungs contain air, and the lung markings are produced by the blood in the blood vessels, which is relatively opaque to X-rays compared with the air in the lungs. The gas in the fundus of the stomach shows up because it is surrounded by soft tissues, which are relatively opaque compared with gas. The trachea can be recognized because the gas in its lumen shows up against the soft tissues of the neck. The shadow of the sternum is lost against the opacity of the heart and vertebral bodies. It is all a question of contrast.

When looking at an X-ray of the chest, first look at the skeleton to see whether the X-ray is an exact postero-anterior (p.a.) view, or whether it is oblique. Look at the medial ends of the clavicles with respect to the vertebral column. They should be exactly symmetrical. If they are not either the patient has been incorrectly positioned, or there may be a scoliosis. The left half of the diaphragm is lower than the right.

When such an X-ray is taken the patient stands with a photographic plate (shielded against the light) against the front of his chest. The X-ray source is behind him. The shoulders are positioned so as to bring the scapulae round to the sides so that they do not obscure the lung fields. The patient takes a deep breath so as to separate the ribs as far as possible. The X-ray source is as far away from the patient as possible (usually 3 metres). The final picture is life size. It is the 'negative' which is then examined. Prints are not made use of in radiology, because a great deal of detail present in the negative is inevitably lost in making a print.

Look at:
1. The trachea.
2. The silhouette of the heart and great vessels:

superior vena cava
right atrium
inferior vena cava } right side

knuckle of aorta
pulmonary artery
left atrium
left ventricle } left side

3. The lung markings (most marked at the hilus and radiate from there).

The upper thoracic inlet is very small being only about 10 cm wide and 5 cm deep. Measure this on yourself and see how small this is compared with the distance between the shoulders. It is bounded by the first thoracic vertebra, the first ribs running downwards and forwards, and the manubrium sterni.

The lower thoracic boundary has led to much confusion. It is not generally realized where the diaphragm is in a living person. Although it takes origin from the lower six costal cartilages and from the upper lumbar vertebrae, it has a very high dome and as a result much of what is commonly thought to be the thoracic region of the body is in fact below the diaphragm and occupied by the abdominal contents. Although theoretically the thoracic outlet is bounded by the twelfth and eleventh ribs posteriorly and laterally and by the costal cartilages of the tenth, ninth, eighth, and seventh ribs anteriorly, the liver is tucked under the costal margin on the right side. The stomach and spleen are on the left side. The diaphragm is above these structures [FIG. 9.05].

Thus the dome of the diaphragm is at the level of the fourth interspace (level of the nipples) and most of the thoracic contents are above this level. The apex beat of the heart represents a slightly lower level of the diaphragm on the left side. The diaphragm is attached to the xiphoid process of the sternum in the midline.

STERNUM—SUMMARY

The sternum is a flat bone consisting of three parts [FIG. 9.08]:
1. Manubrium.
2. Body of sternum.
3. Xiphoid process (xiphisternum).

The **body** of the sternum is formed by the fusion of four parts, and transverse ridges can sometimes be felt marking the junctions.

The **manubrium** is the upper part of the sternum. It is about 5 cm long and articulates with both the clavicle and the first rib on each side. The notch in the midline at the top of the manubrium is termed the **jugular notch** (old name suprasternal notch).

The junction between the manubrium and the body of the sternum is termed the **sternal angle** (angle of Louis). As has been seen it is an important landmark for finding the position of the second rib (costal cartilage). It is felt as a ridge or slight depression when palpating the sternum in the midline. This **manubriosternal joint** allows a slight hinge-movement during breathing.

The manubrium gives origin to sternomastoid. The manubrium and the body of the sternum give origin to pectoralis major.

The **xiphoid process** (xiphisternum) is joined by the xiphisternal joint to the body of the sternum. It gives attachment to the rectus abdominis (anteriorly) and the diaphragm (posteriorly).

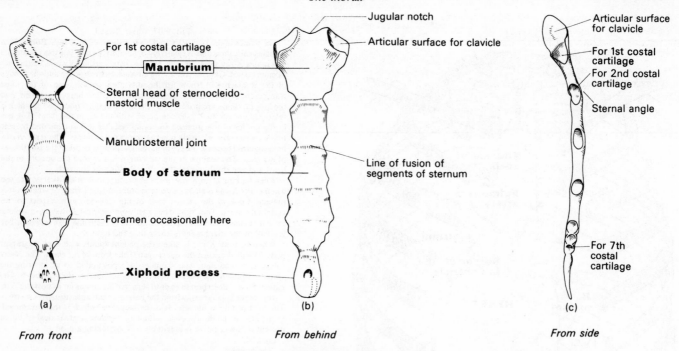

FIG. 9.08. The sternum.

The sternum consists of the manubrium (manubrium sterni), the body of the sternum, and the xiphoid process (xiphisternum).

The 'lines of fusion' represent the places where the sternebri have fused with each other. The sternum, like the spinal vertebrae is bilateral in origin. A foramen may occur as shown when the two halves fail to unite.

The sternum is a 'flat bone' and contains red bone marrow in the adult. A sample of red marrow may be withdrawn (somewhat painfully) from the sternum with a special needle.

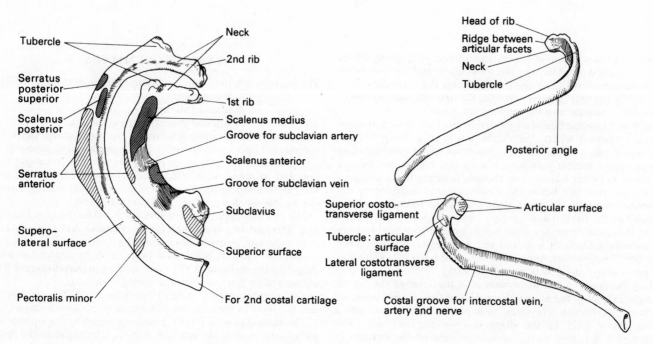

FIG. 9.09. The ribs.

The body of the sternum is formed by the fusion of four segments called *sternebrae*. The segments are connected with each other by discs (like intervertebral discs) but they disappear completely. The costal cartilages typically articulate with the sternum in the same way as the head of the rib articulates with the vertebral bodies and discs. So long as the sternal discs are present the costal cartilage is attached by an internal ligament to the disc and articulates above and below with the bony segments by means of two synovial joints. When the sternaebrae fuse, the internal ligament disappears and the two small joints run together.

RIBS—SUMMARY

A typical rib [FIG. 9.09] consists of:
1. Head.
2. Neck (attachment for costotransverse ligament).
3. Tubercle (facet for articulation with transverse process).
4. Shaft.
5. Costal cartilage.

The bend in the shaft at the **angle** marks the lateral limit of the erector spinae muscle.

The costal cartilages extend forward from the anterior ends of the bony ribs.

Rib joints

Rather surprisingly the heads of typical ribs do not join the vertebral column at the middle of the body of their corresponding vertebra but at the **intervertebral disc above** to which they are attached by an internal ligament [FIG. 9.10]. In addition there are usually two

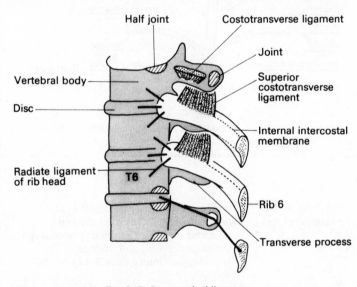

FIG. 9.10. Costovertebral ligaments.

facets on the head. The lower facet articulates with its own vertebra. The upper facet articulates with the vertebra above. The heads of the first, eleventh, and twelfth ribs are exceptions. They have only one facet and articulate only with their own vertebra [FIG. 1.03, p.2]. These joints are termed **costovertebral joints**.

In addition the rib articulates with the transverse process of its vertebra with a synovial joint with the exception of ribs 11 and 12 where the attachment is by ligaments only. These joints are termed

costotransverse joints. This joint is strengthened by the costotransverse ligaments [FIG. 9.11].

The neck of the rib is joined to the transverse process of its own vertebra by the **costotransverse ligament** [FIG. 9.11] and lateral costotransverse ligament. The neck is also connected by the **superior costotransverse ligament** to the transverse process of the vertebra above.

At the front end the rib joins its own costal cartilage at its costochondral joint.† The costal cartilage in turn joins the sternum. The first costal cartilage joins the manubrial part of the sternum directly. The second to seventh costal cartilages have synovial joints with the sternum. The eighth, ninth, and tenth costal cartilages fuse with the seventh to form the costal margin. The eleventh and twelfth ribs are floating ribs and do not join the costal margin.

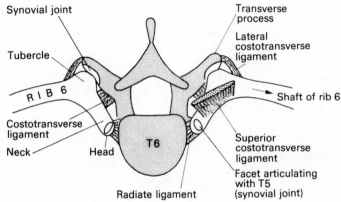

FIG. 9.11. Typical costovertebral articulations.

INTERCOSTAL SPACES

The intercostal spaces are named according to the rib above. Thus the second interspace lies between rib 2 and rib 3. There are eleven interspaces and eleven intercostal (but 12 thoracic) nerves. The space below the twelfth rib is termed the subcostal space and T12 may be referred to as the subcostal nerve.

The interspace contains the intercostal muscles, and two sets (main and collateral) of blood vessels and nerves. The usual arrangement of the main neurovascular bundle, which is tucked under the lower border of the rib, is for the vein to be uppermost, the artery next and the nerve lowest [FIG. 9.12]. The collateral vessels and nerves run just above the next lower rib.

INTERCOSTAL ARTERIES

Each has a posterior intercostal artery and two anterior intercostal arteries with the exception of spaces 10 and 11, the floating ribs spaces, which have only posterior intercostal arteries. The anterior and posterior intercostal vessels anastomose freely within the intercostal spaces. When the aorta is obstructed this provides an alternative pathway.

† The Latin for a rib is *costa* hence *costal* means pertaining to a rib. The Greek for a cartilage is *chondros* hence *chondral* means pertaining to a cartilage.

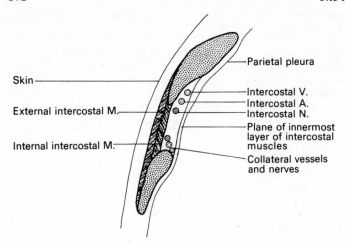

FIG. 9.12. Blood vessels and nerves in the intercostal space.

Posterior intercostal arteries

Interspaces 1 and 2   Branches from superior intercostal artery (from costocervical trunk [FIG. 9.13]).

Interspaces 3–11   Branches from aorta.

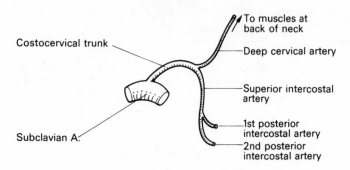

FIG. 9.13. The costocervical trunk.

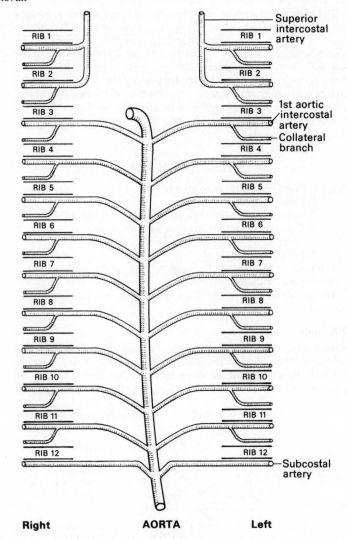

FIG. 9.14. Posterior intercostal arteries.

These are shown diagrammatically in FIGURE 9.14. Each artery at the angle of the rib gives off a collateral branch which runs at a lower level in the interspace. These posterior intercostal arteries anastomose with the anterior intercostal arteries.

Anterior intercostal arteries [FIG. 9.15]

Interspaces 1–6   Branches from internal thoracic artery, which runs beside the sternum.

Interspaces 7–9   Branches from musculophenic artery (terminal branch of internal thoracic artery), which runs along the costal margin.

Interspaces 10–11   Nil.

INTERCOSTAL VEINS

Each space is drained by a posterior intercostal vein and two anterior intercostal veins.

Posterior intercostal veins

These are shown diagramatically in FIGURE 9.16.

Interspace 1   Drains into brachiocephalic vein.

Interspaces 2–3   Drains into superior intercostal vein.

Interspaces 4–11   Drains into azygos veins on the right side and superior (accessory) hemiazygos (interspaces 4–8) and inferior hemiazygos (interspaces 9–11) on the left side.

Subcostal vein   Drains into azygos on right side and
(below rib 12)   inferior hemiazygos on the left side.

The right superior intercostal vein and the hemiazygos veins usually drain into the azygos vein but the arrangement is very variable. The left superior intercostal vein may join the brachiocephalic vein.

Anterior intercostal veins

Interspaces 1–6   Drain into internal thoracic vein.

Interspaces 7–9   Drain into musculophenic vein.

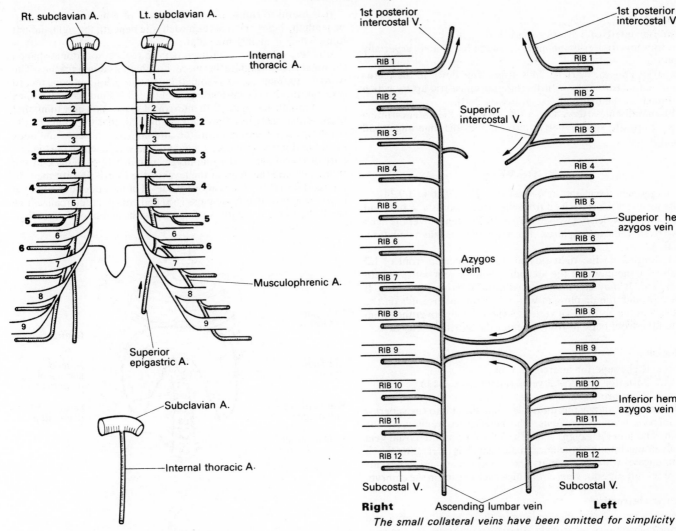

Fig. 9.15. Anterior intercostal arteries.

Fig. 9.16. The azygos system of veins draining the posterior thoracic wall.

INTERCOSTAL NERVES

The intercostal nerves are formed by the anterior primary rami of the thoracic nerves which run round in the subcostal groove deep to the internal intercostal muscle.

Branches:
1. Lateral cutaneous branch.
2. Anterior cutaneous branch.
3. Muscular branches.
4. Collateral branch. This arises at the level of the angle of the rib and runs in the interspace parallel to the main nerve but at a lower level.

First intercostal nerve

The first thoracic nerve gives a large branch to the brachial plexus. It has no lateral cutaneous branch.

Tenth intercostal nerve

This supplies the skin over the umbilicus.

Subcostal nerve (T12)

The twelfth thoracic nerve (subcostal nerve) gives a lateral branch supplying the skin over the anterior superior iliac spine. The main nerve trunk continues forwards to supply the muscles of the anterior abdominal wall.

In putting a needle through the intercostal space it is usual to keep away from the rib above, so as not to damage the intercostal vessels or nerves. The collaterals may be injured if the needle scrapes the rib below.

INTERNAL THORACIC ARTERY

The internal thoracic artery (previously known as the internal mammary artery since it supplies the mammary gland in the female) arises from the subclavian artery. It runs down the anterior thoracic wall deep to the costal cartilages. At the sixth interspace it ends by dividing into the superior epigastric and musculophrenic arteries. [FIG. 9.15].

Branches:

1. Anterior intercostal arteries space 1–6.
2. Perforating arteries to mammary gland in the female, especially in spaces 2, 3, and 4.
3. Superior epigastric artery. This artery runs down in the rectus sheath and anastomoses with the inferior epigastric artery which is a branch of the external iliac artery [p. 250].
4. Musculophrenic artery. This artery gives off the anterior intercostal arteries to spaces 7–9. It also supplies the diaphragm and pericardium.

AZYGOS VEIN

The azygos vein drains the right chest wall [FIGS. 9.16 and 9.24].

The term azygos is derived from the Greek words *a* = not, and *zygon* = yoke, which is a wooden crosspiece fastened over the neck of two oxen. It means unpaired. The azygos vein is thus a unilateral structure.

It is formed on the right side in the abdomen in front of L2 vertebra by the union of the ascending lumbar vein with the subcostal vein. It may communicate with the inferior vena cava. It passes through the diaphragm with the aorta and ascends on the right side of the aorta to T4 vertebra where it loops over the hilum of the lung from back to front to enter the superior vena cava.

Tributaries:

1. Right superior intercostal vein.
2. Posterior intercostal veins (interspaces 4–11).
3. Right bronchial veins.

The corresponding veins on the left side drain into the superior and inferior hemiazygos veins. The hemiazygos veins cross the midline and join the azygos vein. The left superior intercostal vein usually drains into the brachiocephalic vein. It is interesting that it communicates with the coronary sinus of the heart and represents part of the left superior vena cava which once existed in the embryo.

Vascular obstruction

The azygos vein system and the internal thoracic and superior and inferior epigastric veins provide alternative pathways for the venous return to the heart when the great veins, the superior or inferior venae cavae, are obstructed. The two vein systems anastomose freely with each other via the intercostal spaces. If the superior vena cava is obstructed in the superior mediastinum the blood may find its way into the internal thoracic veins, and flow in the 'wrong' direction into the inferior epigastric veins. The blood then runs into the external iliac vein and reaches the heart via the inferior vena cava.

In **coarctation of the aorta** the lumen may be completely obliterated. The collateral circulation travels via the internal thoracic artery and its anastomotic communications with the posterior intercostal arteries, in which it flows in the wrong direction back into the aorta below the obstruction.

Such cases are often diagnosed 'by accident' in adults who do not realize that they have an anomaly.

DIAPHRAGM

The diaphragm forms the partition between the thoracic and abdominal cavities.

It is useful to think of it as consisting of a mainly fibrous part beneath the heart and a part on each side beneath the right and left lungs which is mainly muscular.

It is very difficult to get beginners to realize how dome-shaped the diaphragm is and furthermore, how high its dome reaches. The word 'diaphragm' suggests something flat like a long-playing record disc, but this is very misleading see FIG. 9.17. Find your own apex beat. It usually lies in the fifth intercostal space. You may think that because the heart lies above the diaphragm that the dome of the diaphragm will project on to the abdominal wall below the apex beat, but this does not take account of the way in which the highest part of the dome of the diaphragm is much further back in the thorax, behind the heart at the level of the fourth rib (not the fifth intercostal space), that is, above the level of the apex beat. Find the fourth rib on yourself [see page 165] and you will then realize how small the thoracic cavity is compared with the abdominal cavity.

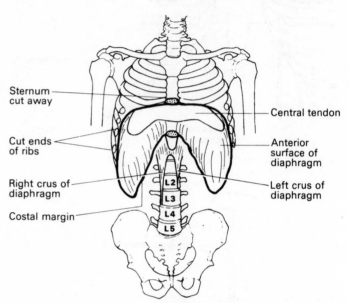

FIG. 9.17. The diaphragm.

The diaphragm [FIG. 9.18] is attached to the xiphoid process in front, the costal margins at the sides, and to the lateral, medial, and median arcuate ligaments. The **lateral arcuate ligament** is a thickening of the anterior layer of the lumbar fascia running in front of the quadratus lumborum from the tip of the twelfth rib to the transverse process of L1. The **medial arcuate ligament** is continuous in front of the psoas muscle and is attached to the body of T12 [FIG. 9.19].

Note that the **aorta** enters the abdominal cavity almost exactly in the midline. It really goes behind rather than through the diaphragm. It is bounded on each side as it does so by two strong muscular downward extensions called the right and the left crus of the diaphragm. The right crus may reach down as far as L3, the left as far as L2. There is a fibrous strap running in front of the aorta joining the crura to each other which is the **median arcuate ligament**. (Do not confuse this midline ligament with the *medial* arcuate ligament above.)

The thickening of the fascia covering the quadratus lumborum and the psoas so as to form the medial and lateral arcuate ligaments is very like the thickening of the obturator internum fascia (the white line) which gives origin to the levator ani.

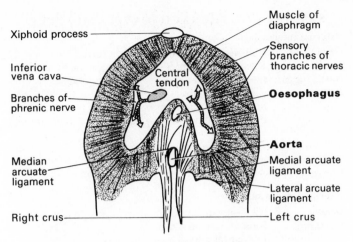

Fig. 9.18. The diaphragm from below.

The central tendon of the diaphragm is joined to the costal margin by radiating striated muscle fibres. Contraction causes downward movement of the central tendon and upward and outward movement of the costal margin. Both movements enlarge the chest and cause inspiration.

The opening for the inferior vena cava also transmits the right phrenic nerve. which ramifies (surprisingly) on the abdominal surface. The left phrenic penetrates the central tendon on its own.

The splanchnic nerves and azygos and hemiazygos veins and thoracic duct pass through the aortic opening or the adjacent crura.

The vagi. now called anterior and posterior gastric nerves. pass through the oesophageal opening.

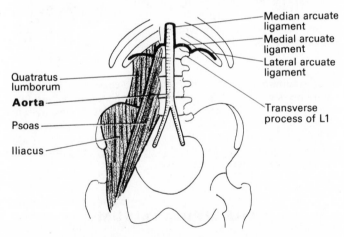

Fig. 9.19. The median, medial, and lateral arcuate ligaments of the diaphragm.

The muscle fibres of the diaphragm all run towards the central 'tendon' to which they are attached. This membrane is trifoil in shape, and roughly follows the shape of the peripheral attachments of the diaphragm.

The principal structures which go through the diaphragm are the inferior vena cava and the oesophagus.

The **oesophagus** passes through the muscular part of the diaphragm posteriorly 2 cm to the left of the midline, at the level of T10. It has a variable relationship to the fibres of the two crura. When the diaphragm contracts it pinches the lower end of the oesophagus and helps to prevent regurgitation upwards from the stomach.

The **inferior vena cava** on the other hand goes through the central 'tendon' and this means that in inspiration (when the venous return increases) the opening of the inferior vena cava will, if anything, be enlarged. The opening is at the level of T8 or the tip of the eighth costal cartilage from the front.

The **thoracic duct** accompanies the aorta behind the diaphragm. The **vagi**, now called anterior and posterior gastric nerves, pass through anterior and posterior to the oesophagus. The inferior vena cava is accompanied by the **right phrenic nerve**. The **left phrenic nerve** pierces the central tendon on its own. Note that the phrenic nerves perhaps surprisingly, go through the diaphragm and then branch out and supply it on its abdominal surface.

The **greater**, **lesser**, and **lowest splanchnic nerves** and the **azygos** and **hemiazygos veins** pass through the crura. The **sympathetic trunk** slips between the psoas muscle and the medial arcuate ligament, en route from the thorax to the abdomen.

The musculature of the diaphragm may be deficient at the xiphoid process where the superior epigastric vessels enter the abdomen, or in the region of the junction of the medial and lateral arcuate ligaments on the left side. The oesophagus may be a very loose fit in the oesophageal hiatus, in which case a hiatus hernia may occur. Usually part of the stomach passes up through the hiatus and comes to lie in the thorax. Hiatus hernia can be difficult to diagnose as they frequently come and go.

It cannot be too strongly emphasized that, although breathing takes place most of the time without conscious thought, the diaphragm consists of striated muscle. The same is true of all other respiratory muscles.

The motor nerve supply of the diaphragm comes from the **phrenic nerves** (C3, 4, 5). The origin of these nerves from the neck is due to the fact that the diaphragm was initially formed in this region and then migrated downwards with the heart, taking its motor nerve supply with it. (Remember that the heart in like manner, gets its nerve supply from the vagus and from the cervical sympathetic ganglia.) After reaching its final position the periphery of the diaphragm is also invaded by sensory branches from those thoracic nerves which cross the costal margin (T7–12). Thus the diaphragm has a single motor supply from the phrenic nerve but has sensory supplies partly from the phrenic and partly from the thoracic nerves. The periphery of the diaphragm therefore gets its motor supply from the phrenic nerve and its sensory supply from the six thoracic nerves which cross the costal margin. The separate origin of motor and sensory nerve supplies is unusual but not unique. The tongue and the face show the same phenomenon in an even more striking way.

The **sensory fibres** supplying the diaphragm also supply the pleural, pericardial and peritoneal membranes which line it. Indeed these membranes are morphologically intrinsic parts of the diaphragm, not membranes stuck on afterwards.

If a lesion affects the phrenic nerve part of the diaphragm then pain may be referred to the shoulder (C4), but if it affects the periphery, then pain will be referred to somewhere in the midline of the abdominal cavity. For example if the T10 part of the diaphragm is affected say by pleurisy, then pain will be at the umbilicus. The nerve supply of the diaphragm contains the seeds of some terrible diagnostic blunders, because of the way pain from a thoracic lesion may be referred to the abdomen.

The diaphragm is the principal inspiratory muscle. In coma the intercostal muscles are paralysed before the diaphragm, leaving the diaphragm as the only respiratory muscle. Expiration takes place by virtue of the elastic recoil of the lungs. It is only when breathing out against resistance that the expiratory muscles are needed. Patients with a broken neck, in whom all the expiratory muscles are paralysed, can speak with remarkable force and emphasis using only the elastic recoil.

The terms origin and insertion are very unsatisfactory when thinking of the diaphragm, because it is influential not only in producing descent of the central tendon but also in raising and causing expansion of the lower part of the rib cage and costal margin, from which the diaphragm is usually said to take origin.

Demonstrate the movement of the diaphragm in the following way. Lie supine. Breathe quietly. Place your left hand about 5 cm in front of your umbilicus. Try to touch your hand with your abdominal wall. The only way it is possible to do this is by contracting the diaphragm without movement of the thoracic cage so that the abdominal wall bulges forwards. Next take a tape measure and measure your minimum waist line with maximum contraction of the transversus abdominis and then with maximum abdominal expansion. Compare this with the maximum and minimum measurements of the thoracic cage at the level of T4 when standing up. Can you produce your maximum chest expansion without breathing in? You should be able to; you need to be able to displace your diaphragm upwards (by contracting the abdominal muscles) so as to reduce the volume of the thorax by a volume similar to the expansion of the rib cage. If you can do this, there is no need to breathe in.

The circumference of the thoracic cage can be varied by 7.5–10 cm in a normal male at the level of the nipples (chest 'expansion'). The waist line circumference can be varied to about the same degree if your abdominal muscles are in good trim.

Bare area of the diaphragm

As has already been seen the serous membranes covering the surfaces of the diaphragm, pericardium, pleura and peritoneum, always have the same sensory nerve supply as the diaphragm which they are lining. But note that part of the diaphragm [p. 208] is not covered with peritoneum, that part which is in contact with the triangular bare area of the liver. We might call this region 'the bare area of the diaphragm'.

Breathing

In order to allow breathing to take place, the volume of the thoracic cavity must be alternatively increased and decreased. This is brought about in two ways.

Firstly, the dome-shaped diaphragm muscle descends as it contracts. This increases the size of the thoracic cavity and at the same time compresses the abdominal contents causing the abdominal wall to bulge forwards.

Secondly, the ribs articulate with the vertebral column such that, as they are elevated, both the anterioposterior and the lateral dimensions of the thoracic cavity are increased. The ribs are set at an oblique angle hence elevation of the sternum increases the anterioposterior distance. The lower ribs can also rotate around an axis joining the front and back of the rib, thus increasing the transverse diameter. Elevation of the ribs is brought about by the contraction of the external intercostal muscles aided by sternomastoid and the scalenes which act as accessory muscles of respiration.

SUMMARY OF THORACIC CAVITY MUSCLES

External intercostals

Origin:	Ribs—lower border
Insertion:	Ribs—upper border
Nerve supply:	Intercostal nerves
Action:	Elevates the ribs
	Muscle of inspiration
	Fibres run downwards
	and forwards

The name indicates that it is the outside muscles between the ribs (*inter* (Latin) = between).

Internal intercostals

Origin:	Ribs—upper border
Insertion:	Ribs—lower border
Nerve supply:	Intercostal nerves
Action:	Elevates and possibly
	depresses ribs. Fibres
	run upwards and forwards

The name indicates that it is the inside muscle between the ribs.

Diaphragm

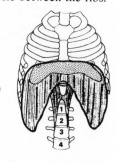

Origin:	Xiphisternum. Lower six
	costal cartilages
	Lumbar vertebrae
Insertion:	Central tendon
Nerve supply:	Phrenic nerve (C3, 4, and 5)
	and T7–12
Action:	Muscle of inspiration

When respiratory muscles are relaxed the chest is at its resting respiratory level. The chest volume then is the functional residual capacity. The respiratory muscles are muscles of inspiration. Although expiration is mainly brought about by the elastic recoil of the lungs, the intercostals continue to contract during early expiration and by 'paying out the slack' allow the chest to return to its resting respiratory level more slowly.

THE MEDIASTINUM

The superior, anterior, middle, and posterior mediastina [Fig. 9.24, p. 181] lie between the lungs. There is no inferior mediastinum.

The important structures in the superior mediastinum are:

1. Oesophagus.
2. Thymus.
3. Trachea.
4. Thoracic duct.
5. Vagus nerve
6. Phrenic nerve.
7. Superior vena cava and its tributaries the right and left innominate veins.
8. Aortic arch and its branches to the brachiocephalic trunk, the left common carotid artery and the left subclavian artery.

THE OESOPHAGUS

The term oesophagus (esophagus) is associated with the Greek *phagein* to eat. It is the 'eating tube' or gullet [FIG. 9.20].

The pharynx changes into the oesophagus at the level of the cricoid cartilage. The oesophagus is about 25 cm long. It passes down the superior and posterior mediastinum in front of the vertebral column following the anterior–posterior curvatures of the spine and slightly to the left of the midline. At the level of the body of the tenth thoracic vertebra it passes through the diaphragm and enters the stomach at the oesophageal-cardiac sphincter at the level of the eleventh thoracic vertebra. This corresponds to a point posterior to the seventh left costal cartilage in front.

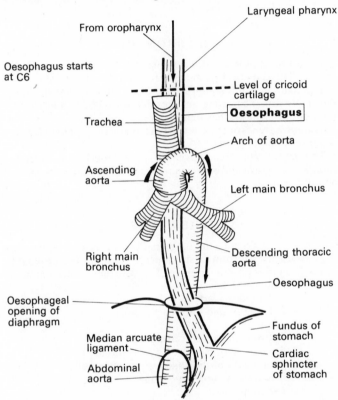

FIG. 9.20. The oesophagus.

The oesophagus commences at the level of C6 behind the cricoid cartilage. It is directly continuous with the laryngeal pharynx. The cricopharyngeal sphincter guards this junction.

The oesophagus lies in the neck for a few centimetres and enters the thorax via the thoracic inlet. It lies just in front of the vertebral bodies in the superior mediastinum above the tracheal bifurcation; below this level the descending thoracic aorta is its immediate posterior relation and the heart and pericardium lie anteriorly.

The oesophagus terminates as it enters the stomach at the cardiac sphincter having passed through the muscular part of the diaphragm.

Note that in the upper thorax the oesophagus lies behind the trachea. Below the bifurcation it lies in front of the aorta, with the heart enclosed within the pericardial cavity in front.

Structure of the oesophagus

The oesophagus has four layers from within outwards:
1. Mucous membrane lined by stratified squamous epithelium.
2. Submucosa with mucous secreting glands.
3. Muscle layer:
   (i) inner circular.
   (ii) outer longitudinal.

In the upper third of the oesophagus it is striated muscle. In the lower two-third it is smooth muscle. The transition is a gradual one.

4. Outer fibrous layer.

It will be noted that there is no serous coat except for the peritoneum over the front and left side of the abdominal part of the oesophagus. This absence makes water-tight anastomoses of the oesophagus difficult.

Relations

As it passes through the superior and posterior mediastinum the oesophagus comes in contact with many important structures.

The right and left vagus nerves form a plexus around the lower part of the oesophagus. From this plexus the right vagus emerges behind the oesphagus whilst the left vagus emerges in front of the oesophagus. These branches then pass through the diaphragm to supply the abdominal contents with parasympathetic nerve activity [p. 215]. They are now called gastric nerves [p. 288].

The **thoracic duct** as it runs upwards from the cisterna chyle [p. 222] to the neck, starts to the right of the oesophagus. Between T8 and T4 it gradually crosses behind and runs up on the left side of the oesophagus.

The **trachea** lies immediately in front but very slightly to the right of the oesophagus. Thus after the bifurcation of the trachea, the oesophagus passes behind the left bronchus. On either side the oesophagus is close to the pleura of the lungs.

The **recurrent laryngeal nerve** as it passes upwards in the last part of its journey lies in the groove between the oesophagus and the trachea. On the right side where the nerve loops round the subclavian artery, this will apply only to the neck region. On the left side, the nerve passes around the aortic arch and the ligamentum arteriosum. It will therefore be found in both the left superior mediastinum as well as the neck [FIG. 9.24].

The relations of the **blood vessels** to the oesophagus are complex. In the superior mediastinum the trachea is immediately in front of the oesophagus. Hence the blood vessels can only be in direct relation with the oesophagus on the two sides and behind. In the lower part of the mediastinum there is no trachea and the oesophagus then lies directly behind the pericardium and left atrium of the heart. Thus a sword swallower, if he is not careful, can puncture his left atrium!

The **descending thoracic aorta** lies behind the oesophagus. The **azygos vein** lies behind the oesophagus to the right.

The **thyroid gland** in the neck is separated from the oesophagus by the trachea except posterolaterally.

The oesophagus has **three constrictions**, at its start, where the left bronchus crosses it and as it enters the stomach. These are common sites for carcinoma to develop. The oesophagus is seen on X-ray when the patient sips and swallows radiopaque 'barium'. Just above the left main bronchus it is narrowed a little by the arch of the aorta. Below this bronchus it is indented by the left atrium of the heart. A barium swallow may be very helpful in demonstrating an aortic aneurysm, or enlargement of the left atrium in mitral stenosis. On X-ray of a patient who is standing up a considerable gap intervenes between the oesophagus and the thoracic vertebrae.

Swallowing is considered further on page 325.

THE THYMUS

The thymus extends from the lower pole of the thyroid gland to the fourth costal cartilage in the thorax. It lies in front of the trachea in the neck, behind the sternohyoid and the sternothyroid, and extends down into the anterior mediastinum in front of the great vessels and pericardium. It fills the superior mediastinum in infancy and childhood, but in the adult it is very inconspicuous.

The thymus is an essential part of the 'lymphatic system' as the source of thymus-dependent lymptocytes (T-lymphocytes).

THE TRACHEA

The trachea is about 10 cm long. It starts at the cricoid cartilage at the level of C6 vertebra and ends opposite T5 vertebra where it divides into the left and right bronchi. It starts superficially in the neck and becomes deeper as it descends into the thorax. It is lined with ciliated columnar epithelium [FIG. 9.21].

The trachea has 16–20 incomplete 'rings' of cartilage which are completed behind by smooth muscle. The interval between the rings is filled by a fibro-elastic membrane. These rings, numbered from above downwards, are used as landmarks. The isthmus of the thyroid gland usually lies in front of the second, third, and fourth rings. The inferior thyroid veins leave the thyroid at the lower border of the isthmus and run downwards over the trachea. The presence of these veins makes a low tracheostomy more difficult than a high one.

Immediately behind the trachea is the oesophagus. On the lateral sides of the trachea is the carotid sheath with the common carotid artery, and also the recurrent laryngeal nerve and the lobes of the thyroid gland.

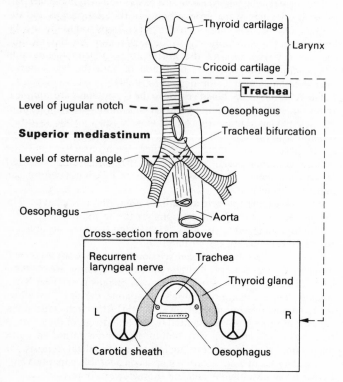

FIG. 9.21. The trachea.

In the thorax the trachea lies behind the thymus, the left brachiocephalic vein, the aortic arch and its branches, the brachiocephalic trunk and the left common carotid artery.

Since the left recurrent laryngeal nerve loops round the ductus arteriosus, it will be on the left side of the trachea in the thorax. This does not apply to the right recurrent laryngeal nerve since this loops round the right subclavian artery higher up in the neck.

The recurrent laryngeal nerve supplies all the muscles of the larynx (except the cricothyroid which is supplied by the superior laryngeal nerve). It is therefore a very important nerve for speech.

Also on the left side of the trachea in the thorax will be the aortic arch, the left common carotid artery and the left subclavian artery. On the right side will be pleura of the right lung, the right vagus and the brachiocephalic trunk.

THE BRONCHI

The left bronchus is about 5 cm long and it enters the lung at the level of T6 vertebra. It is more horizontal than the right.

The right bronchus is shorter, about 2.5 cm long, and in a more direct line with the trachea. As a result foreign bodies tend to enter the right bronchus rather than the left.

The aorta arches over the left bronchus and the pulmonary artery is above it and in front.

The azygos vein arches over the right bronchus to enter the superior vena cava and once again the pulmonary artery is above it and in front.

THE LUNGS

The surface marking of the lungs and pleura were considered on page 166. The **hilus** (previously known as the hilum) of the lungs lie at the level of the sternum angle. The structures which enter and leave the lung at the hilus are:

1. Pulmonary artery.
2. Superior and inferior pulmonary veins.
3. Bronchus.
4. Bronchial arteries and veins.
5. Pulmonary plexus of nerves formed by branches of the vagus (parasympathetic) and the sympathetic trunk.
6. Lymphatics.

The lymphatics remove inhaled carbon and other minute solid particles to the hilus lymph nodes, but as there is no excretory pathway for insoluble matter, they remain there permanently.

The **left lung** has two lobes termed the **superior** and **inferior** lobes [FIG. 9.22]. They are separated by an **oblique fissure** which extends deep into the lung tissue, to the hilus. The surface markings of this fissure run from the spine of T3 to the junction of the sixth rib with its costal cartilage.

The **right lung** has three lobes termed **superior, middle**, and **inferior lobes**. In addition to having an oblique fissure, the right lung has a **horizontal fissure** which extends backwards horizontally from the junction of the fourth rib with the sternum to meet the oblique fissure in the mid-axillary line.

Owing to the liver, the right dome of the diaphragm is higher than the left and hence the right lung is shorter than the left. On the other hand the space available to the left lung is reduced by the presence of the heart and pericardium on the left side of the thoracic cavity. The left lung therefore has a cardiac notch. The part of the superior

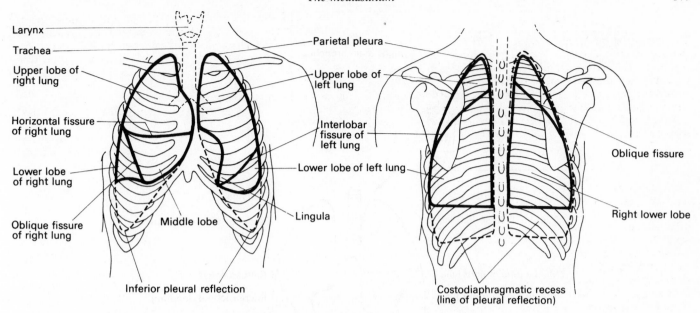

FIG. 9.22. Relations of the fissures of the lungs to thoracic wall.
The lobe of the lungs are separated from each other by the fissures. These are inpushings of the visceral pleura which extend right down to the hilus of the lung. The lobes demarcated by the fissures. two lobes on the left and three on the right are functionally separate units as they are anatomically separate.

lobe of the left lung projecting below this notch is termed the **lingula** [FIG. 9.23].

Each lung is covered with visceral pleura. Parietal pleura also lines the pleural cavity. Inflammation of these layers is termed pleurisy. Normally there are no adhesions between the two layers of pleura. They are in intimate contact separated by only a very thin film of pleural fluid. There is therefore virtually no interpleural space. However, in the case of a pneumothorax (air between lung and chest wall) or a pleural effusion (fluid between lung and chest wall) the pleura of the lung will be separated from the pleura of the chest wall. This fact may be detected clinically by the alteration in the breath sounds and percussion note.

The lungs are highly elastic and if the thoracic cavity is opened, they collapse expelling most of their air. In open-chest surgery respiration has to be maintained by intubating the trachea with a cuffed tube and using a positive pressure pump alternately to inflate and deflate the lungs.

This elasticity enables expiration to take place passively, that is, without muscular activity. When the elastic tissue is lost as in emphysema this facility is lost. The chest becomes bigger (barrel-shaped) and respiration is impaired. Unfortunately, there is at present no known way of replacing the elastic tissue. The patients can, however, be given bronchodilator drugs to make respiration easier.

Each lung has an apex and base. Note that both the lungs and the heart are thought anatomically to resemble a pyramid and hence to have a base and an apex. The lung pyramid is upright. The apex of the lung extends into the root of the neck. The base of the lung sits on the diaphragm. The heart pyramid, however, is inverted and points slightly to the left. The base of the heart is thus close to the hilus of the lungs, whilst the apex of the heart [p. 165] lies inferior to the base of the diaphragm in the fifth left interspace. It is marked by the apex beat.

Bronchopulmonary segments

The trachea bifurcates into two bronchi, the right principal bronchus and the left principal bronchus. The dividing into two is repeated several times until the segmental bronchi which supply a segment of the lung are reached. Blockage of this segmental bronchus will prevent air reaching the segment. The air in the segment will be absorbed into the blood and the segment will collapse.

The detailed arrangement of the bronchopulmonary segments is of importance to thoracic surgeons when they are removing diseased parts of the lungs. The general arrangement is of importance to those carrying out postural drainage of the lungs. The names of bronchopulmonary segments commonly employed are:

Right lung		*Left lung*	
Upper lobe	Apical ⎫ Posterior ⎭ Anterior	Apicoposterior Anterior	Upper lobe
Middle lobe	Lateral Medial	Superior lingular ⎫ Inferior lingular ⎭	corresponds to middle lobe of right lung
Lower lobe	Apical (inferior lobe) Anterior basal Medial basal Lateral basal Posterior basal	Apical (inferior lobe) Medial basal Lateral basal Posterior basal	Lower lobe

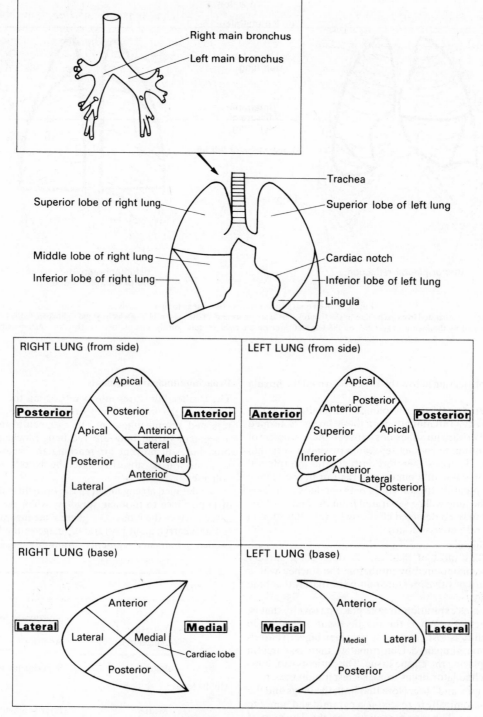

FIG. 9.23. The lung segments.

The left lung is smaller than the right because of the encroachment of the heart on the left side. This explains the cardiac notch and the diminutive size of the left medial basal lobe bronchopulmonary segment as well as the orientation of the two segments in the lingula which lie anterior and posterior with respect to each other instead of medial and lateral as they do in the middle lobe of the right lung.

The overall difference in size is reduced, however, because the diaphragm is higher on the right side than on the left. The final result is that the right lung is only 10 per cent bigger than the left.

The right main bronchus has a greater calibre than the left for this reason. Because of the position of the heart the left main bronchus has to be longer than the right. Since the hili of the lungs are at the same level, simple geometry explains the difference in angles made by the main bronchi with respect to the trachea.

HEART AND PERICARDIUM

The heart lies in the middle mediastinum [Fig. 9.24]. It is a muscular pump consisting largely of specialized heart muscle fibres, which have built into the cytoplasm the extraordinary property of contracting rhythmically. This property shows itself in the embryo very early on in the presumptive heart muscle cells before the 'anatomical' heart has appeared and long before any of the control mechanisms (nerves, hormones, etc.) have had time to develop.

The heart is unique in that its function throughout its own complex morphogenesis, is vital to survival. Other organs in the embryo, the brain, the kidneys, the liver, etc., are not called upon to perform vital functions while they are developing. For this reason heart disorders are a frequent cause of miscarriage and infant morbidity and mortality.

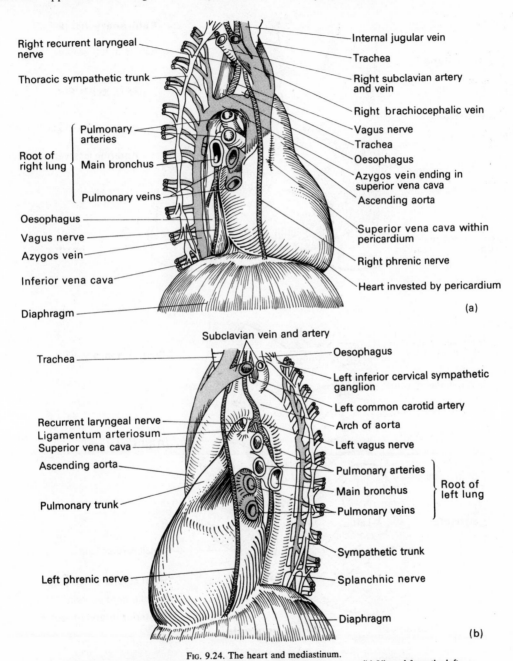

FIG. 9.24. The heart and mediastinum.

(a) Viewed from the right.   (b) Viewed from the left.

The pericardium and heart constitute the middle mediastinum. Above the line joining the sternal angle to T4 lies the superior mediastinum (roughly above the base of the heart); behind the heart is the posterior mediastinum, and in front (deep to the sternum) is the anterior mediastinum.

The heart consists of four separate chambers [Fig. 9.25]. Blood from the whole body (apart from the lungs) including the heart itself flows first:

1. Into the right atrium. }
2. Into the right ventricle. } right heart.
3. Via the pulmonary trunk and lungs.
4. Into the left atrium. }
5. Into the left ventricle. } left heart.

The right atrium and right ventricle are often referred to as 'the right heart'. The left atrium and ventricle are called 'the left heart'. Clinicians speak of 'right-sided' and 'left-sided heart failure'.

The heart is really a midline organ, with a bilaterally symmetrical nerve supply. The simple basic plan is modified in two ways:

1. the apex of the heart is diverted to the left and thus gives rise to the familiar fact that the heart beat is to the left of the midline:

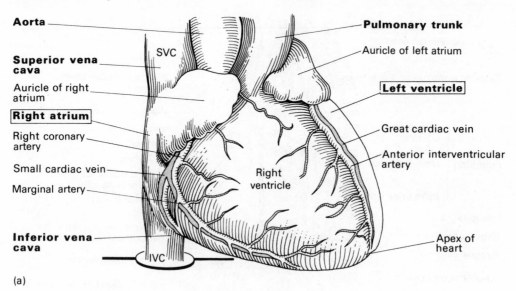

(a)

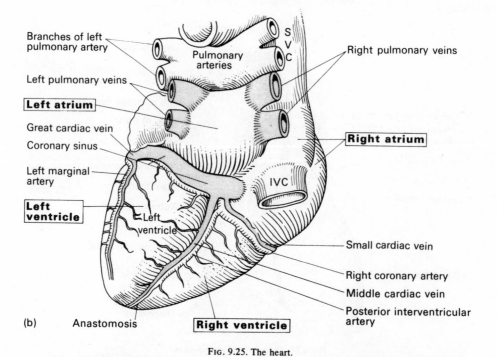

(b)

Fig. 9.25. The heart.

(a) Viewed from the front.   (b) Viewed from behind.

From the front the right atrium and right ventricle are well seen and only a small part of the left side of the heart is visible. From behind the position is reversed. The left atrium is in contact with the oesophagus (sword swallowers beware!).

2. the heart turns around its own longitudinal axis, so that when looked at from in front the view of the right heart is increased at the expense of the left.

When seen from behind the opposite is true, and the left heart is increased at the expense of the right. Turn your face to the left about 45°, without moving your shoulders. The rotation is about the same as that of the heart.

Pericardium

The heart is totally enclosed in the pericardium. This is a fibrous sac which is continuous with the central tendon of the diaphragm below and extends 2–3 cm along the superior vena cava, the aorta and pulmonary trunk above. The inferior vena cava goes through the central tendon as it is about to enter the right atrium and the four pulmonary veins go through the posterior aspect of the pericardium, two on each side, on their way to the left atrium. The fibrous pericardium blends with the adventitia of all these great vessels as they approach or leave the heart. Note that the lower part of the pericardium is part of the central tendon of the diaphragm. We are not dealing with two separate structures secondarily fused together.

The pericardial cavity, that is the space between the outer surface of the heart and the inner aspect of the fibrous pericardial membrane is lined by a continuous serous membrane. The serous membrane covering the heart itself is the visceral layer of serous pericardium, while that part lining the fibrous membrane is the parietal layer of the serous pericardium. This lining renders smooth and slippery the two surfaces which face each other across the pericardial cavity and is exactly like the visceral and parietal serious lining of the peritoneal or pleural cavities. It also 'secretes' a small amount of serous fluid, which allows the movements of the beating heart to take place with the minimum of friction or disturbance of adjacent tissues.

It may be surprising but the fibrous pericardium is really a split off part of the body wall. It becomes separated in the embryo from the ribs and intercostal muscles by the developing lungs [FIG. 9.26].

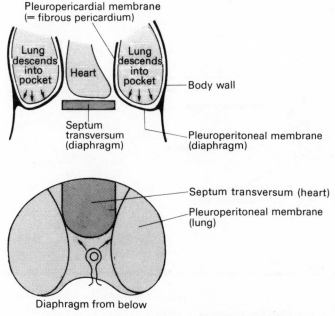

FIG. 9.26. Development of the fibrous pericardium and the diaphragm. The different components must fuse with each other.

The fibrous pericardium with its parietal serous lining can be the source of pain just as much as the rib cage and parietal pleura. Indeed unless you realize the morphological significance of the fibrous pericardium, it is possible neither to make sense of the presence of the phrenic nerves (somatic not a visceral nerve) between the heart and lungs, nor to explain the pain which accompanies pericarditis.

When the pericardium is opened, the pericardial cavity can be examined. The heart is free in the cavity except where it is anchored by the great vessels. A finger pushed upwards immediately to the left of the inferior vena cava comes to lie behind the left atrium in a blind pocket bounded *at each side* by the pairs of pulmonary veins (which must cross the pericardial cavity if they are to reach the left atrium), and *distally* by the reflection of the visceral layer of serous pericardium back onto the fibrous pericardium. This space at the back of the heart is called the **oblique sinus**. It would have been called the vertical sinus if the apex of the heart had not been directed obliquely to the left [FIG. 9.28, p. 184].

The pericardial cavity also contains the **transverse sinus** [FIG. 9.27]. You can push your finger from left to right behind the pulmonary

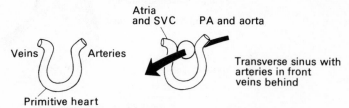

FIG. 9.27. Development of the transverse sinus between the arteries and veins in the primitive heart.

trunk and ascending aorta and it will emerge in front of the superior vena cava. This channel is lined by serous membrane. Its 'floor' is the left atrium, its 'roof' the right pulmonary artery [FIG. 9.29].

The pericardium apart from its confluence with the diaphragm [FIG. 9.28], has fibrous attachments to the deep surface of the body of the sternum, and to the pretracheal fascia.

Blood supply

The blood vessels supplying the heart can be seen on its outer surface, where the large arteries and veins run just beneath the transparent layer of the visceral layer of serous pericardium, cushioned in fat. There are usually two coronary arteries [FIG. 9.30]. The blood mostly returns to the right atrium via the coronary sinus.

The **coronary arteries** both arise from aortic sinuses. There are three aortic sinuses associated with each of the cusps of the aortic 'semilunar' valves. They are slight bulges in the aortic wall immediately above each of the three flaps of a size comfortably fitting the tip of the index finger [FIG. 9.30].

From the anterior sinus, the right coronary artery arises, and from the left posterior the left coronary arises. The apparent asymmetry is explained by the 'rotation' of the heart. The arteries run laterally behind the pulmonary trunk in the atrioventricular groove. As the name 'coronary' suggests they meet on the back of the heart to complete 'the crown'. Unfortunately the anastomosis is not sufficient for heart muscle survival when one or other of the two arteries is obstructed by a blood clot.

The left coronary (bigger than the right) runs to the left and soon gives off the anterior interventricular artery. This runs down the

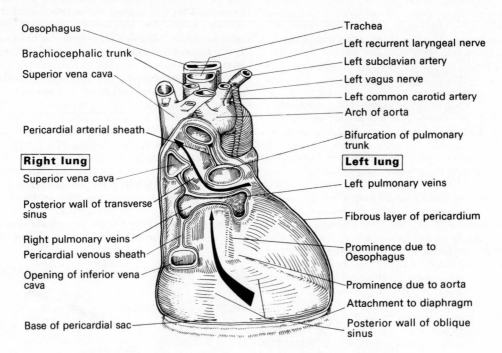

Oesophagus

Brachiocephalic trunk

Superior vena cava

Pericardial arterial sheath

Right lung

Superior vena cava

Posterior wall of transverse sinus

Right pulmonary veins

Pericardial venous sheath

Opening of inferior vena cava

Base of pericardial sac

Trachea

Left recurrent laryngeal nerve

Left subclavian artery

Left vagus nerve

Left common carotid artery

Arch of aorta

Bifurcation of pulmonary trunk

Left lung

Left pulmonary veins

Fibrous layer of pericardium

Prominence due to Oesophagus

Prominence due to aorta

Attachment to diaphragm

Posterior wall of oblique sinus

FIG. 9.28. The pericardium viewed from the front after removal of the heart.
The pericardium is completely blended with the central tendon of the diaphragm. It extends upwards over
the great vessels for a distance of 2–3 cm.

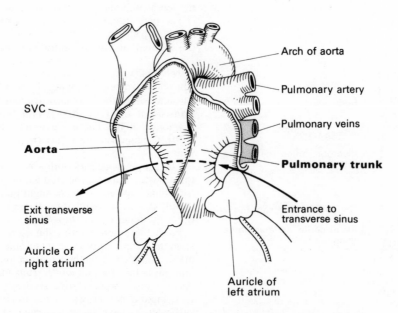

SVC

Aorta

Exit transverse sinus

Auricle of right atrium

Arch of aorta

Pulmonary artery

Pulmonary veins

Pulmonary trunk

Entrance to transverse sinus

Auricle of left atrium

FIG. 9.29. The transverse sinus.
The arrows show the entrance and exit. It runs between the aorta and
pulmonary artery in front and the veins behind inside the pericardial
cavity.

anterior edge of the interventricular septum to the apex of the heart in the anterior interventricular groove. It often continues around the apex into the posterior interventricular artery. The left coronary artery supplies two other large brances, one which runs along the left margin of the heart and another which supplies the posterior surface of the left ventricle.

The right coronary artery first gives off a large marginal branch which runs along the inferior border of the heart, and then the posterior interventricular branch which runs down the posterior margin of the septum and anastomoses with the corresponding branch of the left coronary, usually before it gets to the apex.

There are numerous intracardiac anastomoses between adjacent

blood vessels. Either of the two arteries may supply the sinu-atrial node or the atrioventricular node or the atrioventricular bundle (bundle of His). Evidence of abnormality in the site of initiation of the heart beat or in the conducting system does not therefore provide evidence as to which coronary artery is affected.

The **veins of the heart** run with the branches of the coronary arteries. Their names, however, do not correspond to those of their companion arteries [FIG. 9.31]. They are not of much importance in clinical practice.

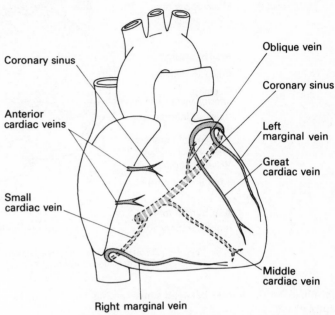

FIG. 9.31. The coronary veins.
The coronary veins run mainly into the coronary sinus which empties into the right atrium.

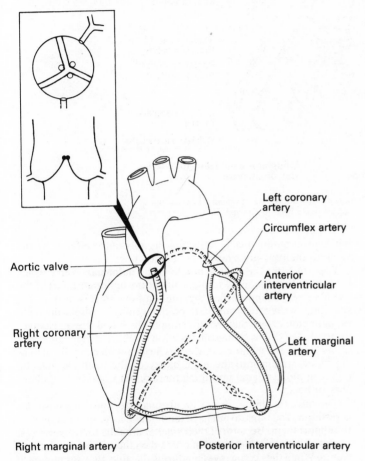

FIG. 9.30. The coronary arteries supplying the heart.
The right and left coronary arteries are the first branches of the aorta. They run forwards around each side of the pulmonary trunk.
Note that although the sinu-atrial node lies in the right atrium, it very frequently receives a substantial part of its blood supply from the left coronary artery. The A-V bundle as it lies in the septum can obtain its blood supply from either coronary artery.
These vessels are only of interest when they become partially or totally blocked leading to the pain of angina pectoris or sudden cessation of the circulation. In suitable cases the blockage can be by-passed with a vein graft. The number of coronaries coming off the aorta varies. A single common stem may be present. there may be more than two. or the blood supply may come off the pulmonary trunk.

INSET: The aortic valve: its cusps and the origin of the coronary arteries.

The veins mostly drain into the **coronary sinus**. This lies in the atrioventricular groove on the back of the heart, and has first the left and then the right coronary artery as its companion. It drains into the right atrium by means of an opening 3–4 mm in diameter which is just above and to the left of the opening of the inferior vena cava. The coronary veins have no valves, nor are they compressed by the contracting cardiac muscle during systole, because they lie on the surface of the heart.

When the inside of the right atrium is inspected the points to note are that the superior and inferior venae cavae (and the coronary sinus) empty into that part of the chamber which has a smoother wall, while anteriorly in front of the crista terminalis the wall is ridged by the musculi pectinati (like the teeth of a comb) [FIG. 9.32]. The crista extends over the front of the opening of the superior vena cava. We mention these 'academic points' so that we can explain the term 'sinu-atrial node'.

The right atrium has a dual origin. The smooth walled part, into which all the important veins drain, was once part of the sinus venosus of the embryo; the rough walled part was derived from the atrium of the embryo. The atrium of the embryo assumed into itself part of the sinus venosus so as to form the composite chamber which we know as the right atrium.

The sinu-atrial (S-A) node is situated just in front of the opening of the superior vena cava close to the upper extremity of the crista

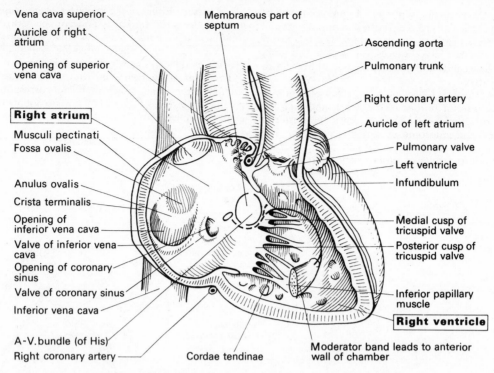

FIG. 9.32. The interior of the right side of the heart showing the right atrium and right ventricle opened up.
The orientation is the same as that of the outside of the heart in FIGURE 9.25(a).

terminalis in the border between sinus venosus and atrium. It is aptly named.

The interatrial septum separates the right atrium from the left. Looked at from the right the septum shows the **fossa ovalis** bounded by the **anulus ovalis**. Shining a light through the septum from the left reveals at once that the tissue in the floor of the fossa ovalis is thin and transparent while the rest is thick and muscular. The interatrial septum in the foetus allowed blood to pass from the right to the left atrium through a patent **foramen ovale**. The transparent part of the septum just mentioned provided a flap-like valve, which closes only once for the first and last time when the baby takes its first breath.

Tricuspid valve and right ventricle

Below and to its left the atrium leads via the tricuspid opening into the right ventricle. This opening is the largest in the heart and is guarded by three large cusp-like flaps which form the **tricuspid valve**. Blood can flow without impedance from atrium to ventricle, but blood regurgitation during ventricular systole instantaneously causes the flaps to shut.

FIGURE 9.32 shows the orientation of the three flaps.

(i) Anterior.
(ii) Medial–septal.
(iii) Posterior.

The flaps each have attached to their free edges and ventricular surfaces a series of fibrous 'strings', the chordae tendinae. The chordae in their turn are attached to nipple-like muscular projections, which are part of the ventricular muscle, called papillary muscles. The more vigorously the heart muscle contracts the more vigorously the papillary muscles pull on the flaps, so there is a

mechanism automatically adjusting the resistance offered by the valve to the pressure exerted by the heart.

The exit from the right ventricle is the **pulmonary valve** which leads into the pulmonary trunk. It lies at the uppermost part of the chamber. It is approached by the outflow tract which is the smoother walled tapering part of the ventricle, also called the infundibulum. Note that the direction of the blood stream changes by nearly 180°. This change of direction is reflected in the presence of the supraventricular crest which is a ridge deflecting the incoming blood stream into the main cavity of the ventricle and the outgoing blood stream into the infundibulum towards the pulmonary orifice.

The valve has three cusps of about equal size, one being posterior in position. The flaps do not have tendinous chords or other devices to support them. Because of their smaller area the force exerted on these valves is less than half that exerted on the tricuspid and mitral valves. When they come together during diastole, they support each other. The edge of each flap is thickened over the area of contact, forming the lunule of the cusp. The tip of the cusp is considerably thickened to form the nodule. The pulmonary valve is a semilunar valve, as the flaps bear a fanciful resemblance to a half moon.

Having entered the pulmonary trunk, the blood is distributed to the two lungs by means of the right and left pulmonary arteries.

Trabeculae carniae

The walls of both ventricles are markedly trabeculated. Some muscle bundles simply form ridges, others form bridges which may jump gaps of 1–2 cm. The moderator band runs from the interventricular septum to the base of the anterior papillary muscle of the right atrioventricular valve. It carries with it the right branch of the

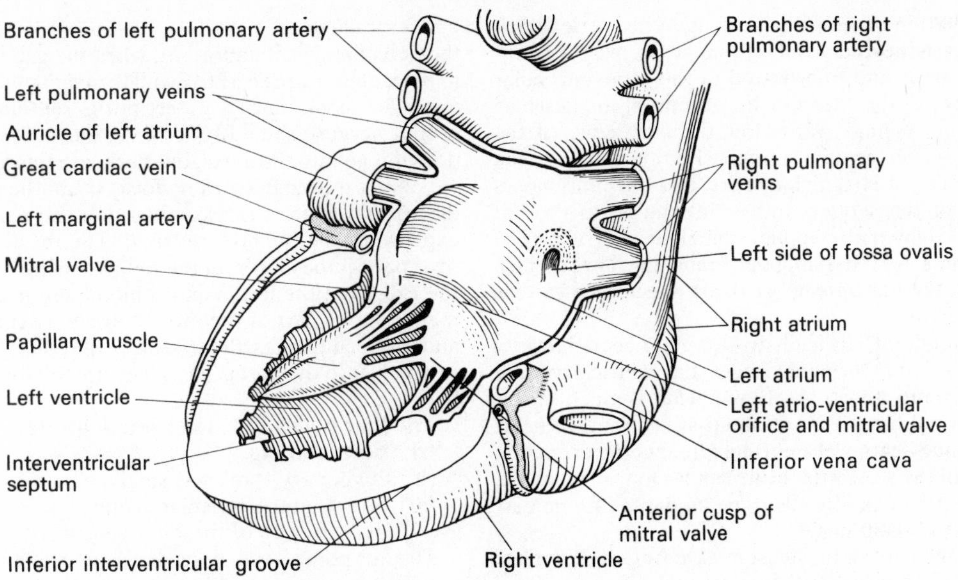

FIG. 9.33. The interior of the left side of the heart showing the left atrium and the left ventricle opened up. The orientation is the same as that in FIGURE 9.25(b)

atrioventricular bundle (of His). This structure is said to reduce or moderate the tendency of the right ventricle to bulge outwards during systole. In the human heart it is very variable in its form and position.

The left atrium

Blood returns from the lungs into the left atrium. There are usually two veins from each lung which open separately into the heart [FIG. 9.33]. Initially in the embryo there was only one common pulmonary vein just as there is a single pulmonary trunk. The wall of this vein and four of its tributories was then assumed into the wall of the atrium in the same way as the sinus venosus was assumed into the right atrium. The pulmonary vein part of the left atrium is smooth walled. The auricular appendage with its trabeculated wall is all that remains in the adult of the embryo's left atrium.

The interatrial septum viewed from this side always shows vestiges of the foramen ovale, even when the foramen has been obliterated. In about 25 per cent of normal people a narrow tortuous channel remains through which the atria communicate with each other.

Mitral valve and left ventricle

Blood flows into the left ventricle from the left atrium through the left atrioventricular orifice. The flaps which guard the opening and prevent regurgitation into the atrium during ventricular systole are arranged on the same plan as those of the tricuspid valve, but it is a bicuspid valve having only two flaps and is usually referred to as the **mitral valve**.

The two flaps are anterior and posterior in position, and with their chordae tendinae and papillary muscles, are more heavily built than those on the right side. This enables them to stand up to the higher pressure exerted by the left ventricle.

The exit from the left ventricle is via the **aortic valve** into the ascending thoracic aorta. The valve is approached by a narrow non-muscular segment in the uppermost part of the left ventricle, called the aortic vestibule. The direction of the blood stream in this ventricle has to turn almost through two right angles. It makes a hair-pin bend, and it is the anterior flap of the mitral valve around which the reverse of direction must take place.

The **interventricular septum** is as thick as the wall of the left ventricle—that is about three times as thick as the wall of the right. In the upper posterior part of the septum there is an elliptical area about 1 cm in diameter which is a thin transparent membrane—the **membranous part of the interventricular septum**. On the right side the septal cusp of the right atrioventricular valve runs across the middle of this membrane. This means that the lower part of the membrane divides the right ventricle (below the septal cusp) from the left ventricle, but the upper part (above the cusp) separates the right atrium from the left ventricle!

The two atrioventricular openings are encircled completely by rings of fibrous tissue. For this reason the calibre of each opening remains constant whatever the state of contraction or relaxation of the heart itself. This means that the flaps contact each other in the same way at each heart beat. The same principle, of non variation of the calibre of openings guarded by flaps also applies to the aortic and pulmonary semilunar valves. The openings of the venae cavae on the other hand do change, and indeed heart muscle fibres may be found in their walls a considerable distance away from the heart. There is a quasi-sphincter like mechanism here.

We have already indicated the position of the sinu-atrial node where the excitation process leading to the heart beat is initiated. This spreads through both atria apparently without any special conducting system.

However, the presence of fibrous rings surrounding both atrioventricular valves should provide a barrier impervious to the spread to the ventricles of the heart beats initiated in the atrium, were it not for the presence of special conducting tissue, the atrioventricular bundle (of His), which crosses this obstacle and provides a functional link between atria and ventricles. With this link the four chambers of the heart beat in a co-ordinated manner; without it the two atria contract synchronously as do the two ventricles, but the pairs of chambers do so entirely independently of each other.

Normally the excitation process is picked up by the atrioventricular node, which lies in the interatrial septum just above the opening of the coronary sinus and transmitted to the atrioventricular bundle. This runs towards the interventricular septum and lies in its muscular part skirting behind and below the very edge of the membranous part of the septum previously referred to. The atrioventricular bundle (of His) divides here, and one limb passes down the left side of the septum and is distributed to the left ventricle, while the right limb supplies the right ventricle. The atrioventricular bundle (of His) and the branching fibres which stem from it to form the Purkinje network are specialized cardiac muscle (not nerve) fibres.

Damage to the bundle of His leads to complete heart block in which (if the patient survives the initial stages) the ventricle beats at its own inherent rhythm—about 25–30/min. That is much slower than the slowest normal heart beat of about 40–45/min. In heart block the atria continue beating at a normal rate, but the excitation process originating in the sinu-atrial node can no longer reach the ventricle. A cardiac pacemaker needs to be implanted to increase the ventricular rate to 70–80/min.

The intrinsic arrangement of the heart muscle fibres is complex. Many fibres are arranged in a spiral and pass through the walls of both ventricles as well as the septum. When the ventricles contract, the apex of the heart twists forwards and medially towards the midline. If the heart simply became smaller when it contracted without twisting, there would be no apex beat.

Notice the relationship of the ascending aorta to the pulmonary trunk. The pulmonary orifice is slightly to the left of the aortic, and the trunk appears to spiral around the left side of the aorta. This relationship confuses many people. Note that if the rotation of the heart itself [p. 182] is neutralized so that the right and left ventricles are symmetrically arranged then the openings come to lie in the expected relationship to each other with the pulmonary artery to the right and the aorta to the left.

The twisting movement of the ventricles during systole combined with their heavily trabeculated walls gives to the blood stream a spiral impetus. It is likely that the relationship of these two great arteries is markedly influenced by this dynamic factor, and that they rotate around each other because the blood streams are ejected from the two ventricles in just such a way. The blood vessels as we see them in the adult have been shaped and moulded in the embryo by the blood stream.

The ligamentum arteriosum connects the pulmonary trunk to the arch of the aorta on its inner concave surface. The recurrent laryngeal nerve 'hooks' round it. It represents the remains of the ductus arteriosus, which in the foetus up to the time of birth allowed the pulmonary artery blood to reach the aorta instead of going into the lungs. At birth the lungs expand and the branches of the two pulmonary arteries expand also. The pulmonary artery pressure slowly falls to a level very much lower than that in the aorta. Within 24 hours the lumen of the ductus closes because of the reflex contraction of the smooth muscle in its wall. If the ductus remains open then the blood will flow from the aorta to the pulmonary artery (patent ductus arteriosus).

Other abnormalities of the heart are worth mentioning: dextrocardia may occur, with or without transposition of the other viscera.

The interatrial septum may be defective and the foramen ovale may persist. This is not necessarily of much functional significance.

The membranous part of the interventricular septum may also be defective in which case blood will flow through the defect from the left ventricle to the right. The juxtaposition of the bundle of His to the defect makes it important, when the gap is repaired surgically, to place the suture so as to avoid damaging the bundle and producing heart block. Such a defect in the septum reduces the cardiac output because some of the blood ejected from the left ventricle does not go into the aorta, but back into the right ventricle instead.

Fallot's tetralogy (*tetra* = four) is another common congenital disorder in which the ascending thoracic aorta is enlarged at the expense of the pulmonary trunk. The result is that the aorta lies over part of the exit from the right ventricle as well as the whole of the exit from the left. Venous blood enters the aorta. Part of the membranous part of the interventricular septum is also defective, and this facilitates further the entry of reduced blood from the right ventricle into the aorta. This gives rise to cyanosis, the typical 'blue baby'.

The four points in Fallot's tetralogy are:
(i)   Aorta too big.
(ii) Pulmonary trunk too small.
(iii) Patent interventricular septum.
(iv) Hypertrophy of the right ventricle.
The last point is the direct consequence of the raised pressure in the right ventricle resulting from point (iii).

Nerve supply to heart

The nerve supply of the heart comes from the vagus (parasympathetic) and sympathetic nerves. The vagi have in general an inhibitory effect on the heart and when they are cut the heart rate rises per minute to about 140. When they are stimulated the heart slows. The sympathetic nerves speed up the heart rate, and when the vagi are inactive the heart rate may be raised to 160–180 beats per minute.

Branches of the vagi reach the sinu-atrial node region where they relay. Very short postganglionic fibres then travel to the sinu-atrial node itself. Nerve fibres also reach the atrioventricular node and are found in association with the atrioventricular bundle. The preganglionic fibres come from the nucleus ambiguus in the medulla oblongata.

The sympathetic nerves emerge from the spinal cord at the level of T3, 4, 5, with the anterior (motor) roots. They reach the sympathetic chain and without relaying they travel upwards as far as the three cervical sympathetic chain ganglia, especially the superior and middle. Recall that the heart develops on the upper side of the diaphragm, high up in the neck. The diaphragm is supplied mainly by C4, and the heart from the nearest sympathetic ganglia. When the heart and diaphragm (and lungs) 'descend into the thorax' they take their nerve supply with them.

Many of these autonomic nerves form plexuses in which both types of nerve become inextricably mixed together.

Pain from the heart is due to ischaemia. The pain fibres travel with the sympathetic nerves. Pain which has its origin in the heart is called angina pectoris (*angina* = strangling; *pectoris* = of the chest). The chest seems to have been gripped between the hands of a malevolent giant. It is a referred pain.

About half the medical profession die from heart disease. For the sake of your colleagues, study this subject carefully.

Referred pain

The relief of pain used to be said to be the first business of a doctor. Pain is a subjective phenomena and is impossible to measure objectively. For this reason it hardly lends itself to scientific study, and has been to a great degree neglected for this reason.

Pain is conveyed to the conscious brain from the surface of the body by nerve fibres which lie in the dermis, and relay in the posterior horn of grey matter. The impulses are then conveyed to the higher centres via the lateral spinothalamic tract on the opposite side in a complex manner through a series of 'gates' which may be opened or closed by other events occurring in the central nervous system.

Evidently pain is experienced in 'the body image', that is in the picture or image of self which is built up during the early months of infancy and childhood. Pain happens in this body image, even when the corresponding part of the body has been removed. After an amputation the phantom limb remains and pain can be experienced in it.

A problem arises in the interpretation by the brain of pain messages which originate in parts of the body which do not exist in the body image, say from the intestine, or diaphragm, or heart. In such cases the pain is 'referred'. That is, it is experienced in another part of the body remote from the lesion itself.

It is well known that a lesion affecting the central parts of the diaphragm gives rise to a pain near the point of the shoulder. If it is due to diaphragmatic pleurisy it will be exacerbated by coughing or deep breathing, not by moving the shoulder joint. The usual explanation starts with the fact that C4 not only is the principal nerve supply of the diaphragm but is also the nerve which supplies this part of the shoulder region. When the pain impulses reach the brain, they are interpreted as coming from that part of the body image which is supplied by C4, that is from the shoulder, not from the diaphragm. You cannot experience pain in the diaphragm because it is not part of your body image.

Pain from viscera such as the heart and intestine travel in fibres which run with the sympathetic nerves. They enter the spinal cord at the same level as the motor sympathetic nerves leave it. Pain is felt in part of the corresponding segment of the surface of the body, that is in the corresponding dermatome. For some reason the pain is not experienced in the whole dermatome, only in that part which lies furthest removed from the spinal cord. Thus in the example just quoted, pain is not felt in the whole shoulder region supplied by C4, only near the tip of the shoulder. If you know the sympathetic nerve supply of a viscus you can usually work out the likely site of referred pain.

Thus the heart is supplied by T3, 4, 5. The parts of the T3, 4, 5 dermatomes furthest removed from the spinal cord lie at the midline in front of the sternum. Pain may radiate in either direction from here, either downwards into the epigastrium (where it may be thought to have originated in an abdominal lesion) or upwards to the ulnar border of the hand (most distal part of the T1 dermatome) or even up into the neck.

The peripheral part of the diaphragm gets its sensory nerve supply from the lower six thoracic nerves. A lesion in this territory will be referred to the most distal parts of dermatomes (T7–12), that is somewhere on the anterior abdominal wall in the midline. Pleurisy in the T10 part of the diaphragm will give pain at the umbilicus.

The principal sympathetic supply of the small intestine and ascending colon comes from T10. The T10 dermatome terminates at the umbilicus. This is where pain is experienced when a lesion affects these parts of the intestine. The classical story of appendicitis illustrates referred pain and localized pain very clearly.

First, when the infection is confined to the appendix itself, pain is referred to the umbilicus. It is intermittent or colicy in nature. When the hand presses on the umbilicus, the pain is unaffected. When the hand presses on the right iliac fossa, however, the pain at the umbilicus gets worse. Later the infection spreads through the wall of the appendix and peristalsis, the producer of colicy pain, ceases.

As soon as the infection spreads to the parietal peritoneum, which lines and is part of the body wall with an ordinary somatic nerve supply, the pain becomes localized in 'the right place'. Pressure of the hand in the right iliac fossa, now increases the pain in this region. Notice that the pain was first referred and then localized.

Thus in considering say a pain in the shoulder, consider local causes first, then consider referred pain. It is essential to know that the phrenic nerve supplies not only both surfaces of the diaphragm, the parietal pleura and parietal peritoneum, but also the fibrous pericardium, and below the diaphragm, the liver, gall-bladder, and pancreas. Anatomy enables you to think of the logical possibilities, and to start making a differential diagnosis.

A pain at the umbilicus, could be due to diaphragmatic pleurisy or peritonitis, or a lesion in the intestine. The nature of the pain, and the factors which make it better or worse, will often give clues as to the source of the pain.

Some people do not feel pain—it may be 'congenitally absent'. Children with this disability often acquire great reputations for daring and bravery and are frequent victims of fractures and other injuries. These fascinating people are always in great danger as they have never undergone the conditioning in care and prudence which pain instils remorselessly in all of us.

It is interesting that many patients who believe that they suffer from a heart lesion experience pain in the left mammary region. Their knowledge and experience of the apex beat clearly persuades them that this is where cardiac pain should be. But left mammary pain is very rarely associated with a heart lesion. On the other hand, many patients do not recognize the significance of 'retrosternal pain', and because of their preconceived ideas about where heart pain should be, attribute it to 'indigestion'.

10 The abdominal wall

Introduction

By 'the abdomen' we mean the abdominal cavity, its lining membranes, and the abdominal wall. The alimentary canal from the lower end of the oesophagus to the rectum lies in the abdominal cavity. In addition the liver, kidneys, and suprarenals are amongst other abdominal contents.

The abdominal cavity is bounded behind by the lumbar vertebrae, above by the diaphragm, laterally and anteriorly by the muscles of the anterior abdominal wall, while below it is continuous with the pelvic cavity.

The coils of the intestine, and other viscera, are covered by a smooth gleaming membrane which is the peritoneal membrane (visceral layer). In addition the deep surface of the abdominal wall is lined by the same membrane (parietal layer). These visceral and parietal layers of the peritoneum enclose the peritoneal cavity. Normally this is only a *potential* cavity as the two layers are in close contact separated by a minimum of tissue fluid. It allows the intestinal coils in their peristaltic writhings to slither over each other, and over the abdominal wall. Between the abdominal wall and the peritoneal membrane there is a strong intervening layer of fascia and also fat.

Anterior abdominal wall muscles

The 'anterior' abdominal or belly wall runs downwards from the rib cage and costal margin to the pelvis, where it is attached along the iliac crest at the side, and to the body of the pubis and pubic symphysis at or near the midline. The gap between the anterior superior iliac spine and the pubic tubercle (identify in yourself) is bridged by the inguinal ligament.

The anterior abdominal wall is constituted by certain muscles and their associated aponeuroses. It is conventional to regard the flank as part of the anterior abdominal wall, although it is really lateral in position.

Notice that in yourself the interval between the iliac crest and the costal margin is only 3–4 cm. In many articulated skeletons this interval is very exaggerated. Notice that the umbilicus is level with the highest points of the two iliac crests, that is, they are all level with the top of your trousers or skirt. In the small of the back, this level cuts the tip of the fourth lumbar vertebra. Identify this in yourself while sitting in the slumped position with the lumbar vertebrae in flexion.

The principal muscles in the anterior abdominal wall [FIG. 10.01] are:

1. Rectus abdominis.
2. External oblique.
3. Internal oblique.
4. Transversus abdominis.

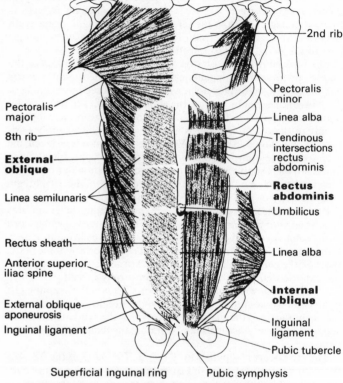

FIG. 10.01. Muscles of the anterior trunk wall. On the left side the external oblique muscle has been removed.

Rectus abdominis

The rectus abdominis [FIG. 10.01] runs from the pubic symphysis and pubic crest vertically upwards to the fifth, sixth, and seventh ribs (costal cartilages). The left and right muscles are close together at their origin but they diverge from each other and become wider and thinner as they ascend.

There are usually three transverse tendinous intersections in the muscles:

1. At the umbilicus.
2. Between the umbilicus and xiphoid process.
3. At the xiphoid process.

These intersections are seen or palpated as depressions on the skin. They are usually more prominent in the male, because most men are more muscular than women, and most women have more superficial fascia (and subcutaneous fat) than men.

Identify the muscles in yourself by lying down in the supine position (= on your back) and raise your head or your legs. The rectus muscles are acting as trunk flexors. Identify the lateral border on each side as it crosses the costal margin at the tip of the ninth costal cartilage. The lateral border of the rectus gives rise to the groove called the **linea semilunaris**.

Note that at the level of the umbilicus the muscle is interrupted by a well marked groove. Even if you are not as slim as you might be, this groove produced by the tendinous intersection is always palpable. See if you can find the two other intersections between the umbilicus and the costal margin.

The rectus muscles are enclosed in the rectus sheath which represents the aponeuroses of the other abdominal wall muscles. Between the recti in the midline is the **linea alba**.

The rectus sheath

The aponeuroses (pleural of 'aponeurosis') of the external oblique, internal oblique and transversus abdominis muscles envelop the rectus abdominis muscle in a compex manner before joining together in the midline to form the linea alba. The linea alba stretches from the pubic symphysis to the xiphoid process and forms a median abdominal furrow between the recti.

The arrangement whereby the aponeuroses of the three abdominal muscles form the rectus sheath by enveloping the rectus abdominis muscle is as follows [FIG. 10.02]:

1. **Between xiphoid process and midpoint between umbilicus and pubic symphysis**
   (i) External oblique aponeurosis passes in front.
   (ii) Internal oblique aponeurosis splits part in front, part behind.
   (iii) Transversus aponeurosis passes behind.

This part of the rectus sheath is entirely fibrous except for about 3–4 cm below the costal margin where it also contains muscle fibres which are derived from transversus abdominis.

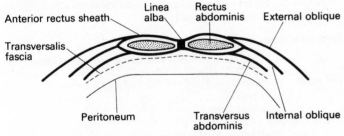

Anterior rectus sheath — Linea alba — Rectus abdominis — External oblique

Transversalis fascia

Peritoneum — Transversus abdominis — Internal oblique

(a) *Between costal margin and arcuate line*

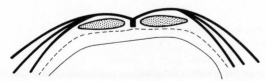

(b) *Between arcuate line and pubic symphysis*

FIG. 10.02. The rectus sheath.
The anterior rectus sheath extends above the costal margin. The posterior rectus sheath commences at the costal margin and terminates at the arcuate 'line'. Below this level the rectus muscle is in contact with the transversalis fascia.

2. **Between midpoint and pubic symphysis**
   (i) External oblique aponeurosis passes in front.
   (ii) Internal oblique aponeurosis passes in front.
   (iii) Transversus aponeurosis passes in front.

The rectus abdominis thus lies on the transversalis fascia in the lower abdomen. The lower free edge of the posterior rectus sheath at the midpoint between umbilicus and pubic symphysis forms the **arcuate line** (semilunar fold of Douglas). In many people the posterior rectus sheath gradually fades over a distance of 2–3 cm instead of doing so abruptly at the arcuate line.

External oblique

The external oblique [FIG. 10.01] originates from ribs 5–12. The muscle fibres originating from ribs 5, 6, 7, 8, and 9 interdigitate with the muscle fibres of serratus anterior which runs to the scapula [p. 93]. The fibres originating from ribs 10, 11, and 12 interdigitate with the fibres of latissimus dorsi which runs to the humerus [p. 93].

The external oblique is inserted into the linea alba, the whole of the inguinal ligament and the anterior half of the iliac crest.

The muscle is supplied by T7–12 nerves.

Although the posterior fibres of this muscle are vertical, the majority of the muscle fibres run downwards and forwards to form the aponeurosis. There are no muscle fibres below the line joining the umbilicus and the anterior superior iliac spine. The aponeurosis of the muscle begins below and medial to this line.

The lower border of the aponeurosis runs from the outer superior iliac spine to the pubic tubercle. Between the tubercle and the midline it is attached to the pubic crest. Between the spine and the pubic tubercle the aponeurosis has a 'free' border in which the last centimetre or so is turned posterior forming the inguinal ligament. The inguinal ligament has important relationships to the inguinal and femoral canals [p. 199].

Internal oblique

The internal oblique [FIG. 10.01] arises from the lumbar fascia, the anterior two-thirds of the iliac crest, and the outer half of the inguinal ligament.

The fibres run upwards and forwards to be inserted into the linea alba and ribs 7–12 (costal cartilages). The lowermost fibres run in front of the spermatic cord, join transversus to form the conjoint tendon which curves down behind the cord to be inserted in the iliopectineal line (a bony ridge on the brim of the true pelvis situated partly on the ilium and partly on the pubis).

The nerve supply is T7 to T12 and L1.

Transversus abdominis

The transversus abdominis [FIG. 10.03], as its name implies, runs transversely across the abdomen with horizontal muscle fibres except where it folds downwards to form the conjoint tendon.

The transversus originates from the thoracolumbar fascia, from the lower six ribs (costal cartilages), from the anterior two-thirds of the iliac crest and from the outer third of the inguinal ligament.

It is inserted into the linea alba and via the conjoint tendon into the iliopectineal line.

The nerve supply is T7–T12 and L1.

The inguinal ligament

The inguinal ligament (Poupart's ligament) is the lower border of the aponeurosis of external oblique which turns inwards and gives

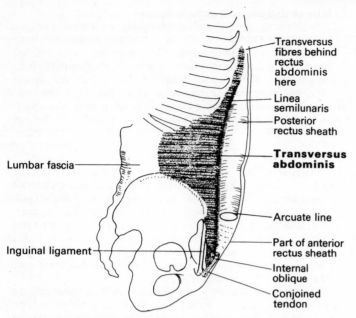

Lumbar fascia

Inguinal ligament

Transversus fibres behind rectus abdominis here

Linea semilunaris

Posterior rectus sheath

Transversus abdominis

Arcuate line

Part of anterior rectus sheath

Internal oblique

Conjoined tendon

Posterior rectus sheath terminates at arcuate line

FIG. 10.03. Transversus abdominis muscle and rectus sheath.

ABDOMINAL MUSCLES SUMMARY

Rectus abdominis

Origin:	Pelvis—pubic bone and pubic symphysis
Insertion:	Costal cartilages of ribs 5, 6, and 7. Xiphoid process
Nerve supply:	T7–12 (intercostal nerves and subcostal nerve)
Action:	Pulls pelvis towards rib cage—flexes the trunk

This muscle is used when raising the shoulder or legs off a couch from the supine position.

Three tendinous intersections are present in the muscle and can usually be seen readily in the male, one at the xiphoid, one between the xiphoid and the umbilicus, and one at the umbilicus. In palpating the abdomen in a patient thought to have an enlarged liver, the tendinous intersections of the rectus may be mistaken for the liver. The edge of an enlarged liver moves up and down with respiration whereas the tendinous intersections do not. When the patient raises his head off the pillow the rectus contracts and the intersections become more obvious, but the edge of the liver is no longer palpable.

External oblique

Origin:	Ribs 5–12
Insertion:	Linea alba Iliac crest—pubic tubercle Inguinal ligament
Nerve supply:	T6, 7, 8, 9, 10, 11, 12 Iliohypogastric (L1)
Action:	Side flexes and/or rotates trunk Increases intra-abdominal pressure

The origin from ribs 5, 6, 7, 8, and 9 interdigitate with serratus anterior which is attached to the upper eight ribs, whilst the origin from ribs 10, 11, and 12 interdigitate with latissimus dorsi, which takes origin from the lower six ribs.

Internal oblique (obliquus abdominis internus)

Origin:	Iliac crest and inguinal ligament. Lumbar fascia
Insertion:	Ribs 10, 11, 12 (costal margin) Linea alba Pubic bone
Nerve supply:	T7–12 Iliohypogastric nerve, Ilioinguinal nerve
Action:	Flexes and rotates trunk Increases intra-abdominal pressure

Transversus abdominis

Origin:	Ribs 7–12 (costal margin) Iliac crest. Inguinal ligament. Lumbar fascia
Insertion:	Pubic bone Linea alba
Nerve supply:	T7–12. Iliohypogastric nerve. Ilioinguinal nerve (L1)
Action:	Increases intra-abdominal pressure

origin in its outer part to the internal oblique and transversus abdominis muscles. The inguinal ligament extends from the anterior superior iliac spine to the pubic tubercle. The inner 2 cm is expanded and runs to the pectineal line of the pubis as the pectineal part of the inguinal ligament (also known as lacunar or Gimbernat's ligament). Pecten† (Latin = comb) is the old name for the pubic bone, hence pectineal which means pertaining to the pubis. Pectineus is a muscle that arises from the pubis and is inserted into the femur.

The inguinal ligament in its medial quarter separates the inguinal canal from the femoral canal. They will be described in more detail later. The external inguinal ring is a 'gap' in the external oblique aponeurosis through which pass the spermatic cord in the male and the round ligament in the female [p. 198].

Muscles attached to the iliac crest

Outer lip
  anterior half   external oblique (insertion)
Middle lip
  anterior two-thirds   internal oblique (origin)
Inner lip
  anterior two-thirds   transversus abdominis (origin)

Muscles attached to inguinal ligament

Entire length
  external oblique   (insertion)
Outer half
  internal oblique   (origin)
Outer third
  transversus abdominis   (origin)

† Pecten should not be confused with *pectus* which means the chest in Latin; hence pectoral muscles (pectoralis major and minor) which are chest muscles.

Action of abdominal muscles

The abdominal muscles flatten the lumbar curvation of the spine and increase the intra-abdominal pressure. They are used in:

1. Forced expiration, coughing etc.
2. Defaecation (straining at stool).
3. Second stage of childbirth.
4. Valsalva manoeuvre.

Loss of muscle tone gives a protruding abdomen.

The action of touching the left foot with the right hand involves twisting the trunk. This action is brought about by the external oblique of the right side and the internal oblique of the left side.

Linea alba

The linea alba (Latin = white line) runs from the pubic symphysis to xiphoid process of the sternum. It is formed by aponeurosis fibres of right and left abdominal muscles [FIG. 10.01]. The linea alba is as narrow as the pubic symphysis below, and as wide as the xiphoid process above.

Just above the pubic symphysis the **pyramidalis muscles** (supplied by T12) are attached to the linea alba on each side of the midline. One or both may be absent.

The **umbilicus** is a scar in the linea alba which once gave origin to the umbilical cord. It is a site of weakness. Umbilical hernia is found in the newborn and in obese adults. Sometimes the upper part of the linea alba is defective and herniae may also occur there.

A further consideration of the abdominal muscles and their attachment to the rectus sheath and lumbar fascia

The **rectus abdominis** muscles lie one on either side of the midline. The muscle fibres run vertically up and down between the costal margin above, and the body of the pubis below. Above, each muscle is about 7 cm wide but narrows down to about half this width as it approaches its attachment to the pubis.

The general direction of the **external oblique** muscle fibres is downwards and medially. It is attached to the lower eight ribs some distance away from the costal margin. The slips of attachment interdigitate with those of the serratus anterior (from upper eight ribs) and of the latissimus dorsi (lower six ribs).

The most posterior fibres of this muscle are not oblique. They run vertically downwards to the iliac crest. This part of the muscle has a free posterior border and it is thicker here than in the rest of the muscle. The action of these fibres is to produce lateral flexion of the trunk.

Note that the fleshy part of the external oblique approaches, but does not overlap, the lateral border of the rectus abdominis. When it reaches the rectus, the external oblique becomes aponeurotic and then contributes to the anterior rectus sheath. Also in the lower part of the abdominal wall the muscle fibres cease and become aponeurotic long before they reach the rectus along a line passing upwards and medially from the anterior superior iliac spine towards the umbilicus.

The lower edge of this aponeurosis forms the inguinal ligament. It runs from the anterior superior iliac spine downwards and medially to the pubic tubercle. Not only is it considerably thickened here, but in addition it is turned back on itself so that the inguinal ligament lies in the plane at right angles to the aponeurosis.

The inguinal ligament in its medial quarter separates the inguinal canal from the femoral canal [p. 200]. The external inguinal ring is a 'gap' in the external oblique aponeurosis through which passes the spermatic cord in the male and the round ligament of the uterus in the female [p. 242].

The **internal oblique** muscle lies deep to the external oblique. The general direction of its fibres is at right angles to that of the external oblique, that is, they run upwards and medially instead of downwards and medially. Note that the direction of the external oblique fibres on one side of the body is in line with that of the internal oblique fibres on the opposite side.

The upper attachment of the internal oblique is to the costal margin where the slips are intimately related to those of the transversus abdominis and the diaphragm both of which are also attached to the costal margin.

The muscle fibres reach almost as far as the lateral border of the rectus, like the external oblique. The muscle then becomes aponeurotic.

Behind, instead of having a free border like the external oblique, the internal oblique is attached via the lumbar fascia to the spines and transverse processes of the lumbar vertebrae.

Below it is attached to the curled back edge of the outer two-thirds of the inguinal ligament and to the 'conjoint tendon' further medially. The internal oblique in this region has an important relationship to the inguinal canal which will be described in more detail below.

The internal oblique is the key to understanding the anatomy of the abdominal wall.

The **transversus abdominis** lies deep to the internal oblique and has attachments which are similar in most respects to those of the internal oblique. The nerves and arteries of the abdominal wall run in the plane of cleavage between the internal oblique and transversus abdominis.

The direction of the fibres of the transversus is transverse. They are ideally situated to bring about compression of the abdominal contents.

It is important to realize that these three muscles are really **inserted** across the midline into the **linea alba** which runs between the two recti from the lower border of the xiphoid process down to the pubic symphysis.

A question will immediately occur to you: 'How do the three muscles at the side get past the rectus abdominis which apparently bars their passage to the midline?' There are three logical possibilities at any individual level of the rectus muscle:

1. To go in front.
2. Behind.
3. Both in front and behind.

The last is the method used typically.

The three muscles reach the midline via their aponeuroses which form the rectus sheath.

Between the costal margin and about half way between the umbilicus and pubic symphysis, the rectus sheath is formed in the following manner. The internal oblique aponeurosis splits, one half running in front of the rectus and the other running behind. They then reunite with each other at the medial border of the rectus and blend into the linea alba.

The aponeurosis of the muscle superficial to the internal oblique, the external oblique, will obviously run in front of the rectus while the aponeurosis of the transversus must pass deep. These two aponeuroses do not remain as separate structures in their own right. Instead they fuse with those of the internal oblique, so that we say that the anterior rectus sheath consists of half the internal oblique aponeurosis plus the external oblique aponeurosis, while the posterior rectus sheath consists of the rest of the internal oblique aponeurosis fused with that of the transversus abdominis.

However, in the lower third of the way between the umbilicus and the pubic symphysis, the posterior rectus sheath is apparently not present at all [FIG. 10.04]. The posterior sheath either ends abruptly or it may gradually fade out over a distance of a centimetre or two. The edge where the posterior sheath stops short is called the **arcuate line**. It is often stated that 'below the arcuate line, all three muscles fuse together to form the anterior rectus sheath'. It is possible, however, to regard the conjoint tendon as part of the 'missing' posterior rectus sheath. This will be referred to again when the inguinal canal is considered.

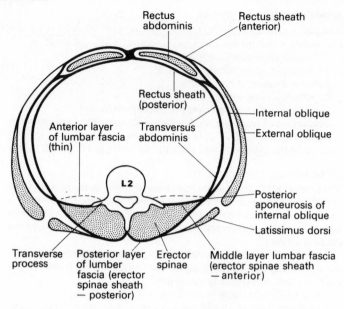

FIG. 10.05. Rectus sheath and lumbar fascia.

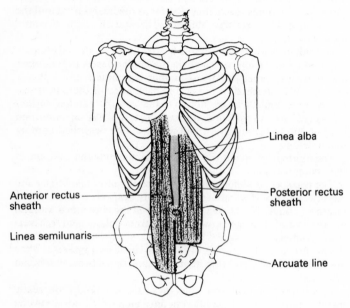

FIG. 10.04. The arcuate line.

Lumbar fascia

Having considered the attachments of the abdominal muscles at the front, it is equally important to consider their attachments at the back.

The internal oblique and transversus abdominis muscles take origin from the lumbar part of the thoracolumbar fascia termed the **lumbar fascia**.

FIGURE 10.05 shows a cross-section through the trunk. The rectus muscle in front, apart from any action it may have in holding the abdominal viscera in place, is one of the principle flexor muscles of the trunk. Its antagonist, the principal extensor muscle of the trunk, is the erector spinae [p. 62]. This muscle lies behind the laminae and transverse processes of the lumbar vertebrae in this region, and is separated from its neighbour on the other side of the midline by the lumbar spinous processes and the ligaments which join these spines together.

Notice in FIGURE 10.05 the lumbar fascia. It is conventionally described as being in three layers, the posterior layer running superficial to the erector spinae, the middle layer running to the tips of the transverse processes, and the anterior layer running in front of the quadratus lumbrum to be attached to the front of the transverse processes. Compared with the other layers the anterior layer is thin and transparent.

The erector spinae is enclosed by two layers of the lumbar fascia. The posterior layer covers its superficial surface in exactly the same way as the anterior rectus sheath covers the superficial surface of

the rectus muscle. This layer is, if anything, thicker than the rectus sheath. It is attached to the tips of the spines and to the supraspinous ligament when it reaches the midline. It is confluent in the midline with the fascia from the other side.

The deep surface of the erector spinae projects a considerable distance laterally beyond the tips of the transverse processes. This part of the muscle is covered on its deep surface by the middle layer of the lumbar fascia, just as the deep surface of the rectus muscle is covered by the posterior rectus sheath. This layer of the lumbar fascia is attached to the tips of the transverse processes and to the intertransverse ligaments which join the tips together. It is about as thick as the rectus sheath.

The posterior and middle layers of the lumbar fascia meet as shown in FIGURE 10.05. Continuing laterally you can see that they are attached eventually to the internal oblique muscle. In this region the transversus abdominis muscle fuses with it on its deep surface.

This thick layer of fascia to which the middle and posterior layers of the lumbar fascia are attached medially, and to which the internal oblique and transversus abdominis muscles are attached laterally, has no official name. As it bridges the gap between these muscles and the named layers of fascia we will refer to it as the **posterior aponeurosis of the internal oblique**.

The whole arrangement bears a striking resemblance to the rectus sheath. The middle and posterior layers of the lumbar fascia really form the 'erector spinae sheath'.

You will have noticed that the external oblique muscle has not been mentioned as a contributor to the erector spinae sheath. This is because, as has already been seen, the external oblique has a free posterior border and does not extend round towards the back far enough to reach the erector spinae. In fact, its place is taken by the latissimus dorsi which develops from the same sheet of muscle as that which gave rise to external oblique. The aponeurosis of this muscle is attached to the posterior layer of lumbar fascia, that is, to the superficial layer of the erector spinae sheath in the same way as the external oblique is attached to the superficial layer of the rectus sheath.

THE ABDOMINAL WALL NERVES

The nerve supply of the anterior abdominal wall comes from the lower six thoracic nerves and the first lumbar nerve (T7–L1 inclusive).

At the anterior ends of the ribs the intercostal nerves (T7–11) enter the abdominal wall in the plane between the transversus abdominis and the internal oblique. They supply the abdominal muscles, cross the linea semilunaris, and enter the rectus sheath deep to the rectus muscle. They then pierce the rectus abdominis to become cutaneous nerves supplying the skin over the abdominal wall.

The subcostal nerve (T12) runs behind the kidney. It pierces the aponeurosis of transversus and is then in the same plane as the others (between transversus and internal oblique). It supplies the muscles surrounding the inguinal canal and it is important therefore that it is not damaged in an appendix operation, since such damage may lead to an inguinal hernia [p. 199].

The anterior primary rami of L1, 2, 3, and 4 form the **lumbar plexus**. This plexus gives rise to:

L1	Iliohypogastric nerve
	Ilio-inguinal nerve
L1 and L2	Genitofemoral nerve
L2 and L3	Lateral cutaneous nerve of thigh
L2, L3 and L4	Femoral nerve
	Obturator nerve

The last three nerves have already been considered on page 120.

Iliohypogastric nerve (L1)

This nerve passes behind the kidney and runs to the iliac crest where it pierces the aponeurosis of transversus to run between transversus and internal oblique. It divides into iliac and hypogastric branches.

The iliac branch supplies the skin over gluteus medius. The hypogastric branch becomes cutaneous above the superficial ring.

Ilio-inguinal nerve (L1)

This nerve passes behind the kidney at a lower level than the iliohypogastric nerve. In front of the iliac crest it pierces the transversus and runs between this muscle and the internal oblique to the inguinal canal. It leaves the canal at the superficial inguinal ring to supply the scrotum or labia and upper and medial part of the thigh.

It supplies the internal oblique and transversus.

Genitofemoral nerve (L1 and 2)

This nerve runs behind the peritoneum at the back of the abdomen and divides into the genital and femoral branches.

The genital branch runs down on the lateral side of the external iliac artery and enters the inguinal canal at the deep inguinal ring. It supplies the cremaster muscle and skin of scrotum and medial aspect of the thigh.

The femoral (or crural branch) passes under the inguinal ligament to supply the skin on the upper part of the front of the thigh.

Nerve supply to anterior abdominal wall

The seventh rib with its costal cartilage is the last one to reach the sternum. The sixth thoracic nerve, which must lie above the seventh rib and costal cartilage, therefore terminates in front of the sternum. The seventh thoracic nerve will therefore be the first to reach the abdominal wall. Note that T10 passes downwards and medially and supplies the region of the umbilicus. T12 reaches the pubic symphysis (the last nerve to reach the midline). L1 supplies the abdominal musculature in the inguinal region to the side of the midline but does not reach the midline.

Many people have the deeply rooted idea that all the abdominal nerves run in a downward and forward direction, as is the case with T10, 11, 12 and L1. This is not so above the umbilicus.

The point is that the direction of the nerves before they reach the abdominal wall is materially affected by the direction of the interchondral part of the space in which they have just been running. It is true that the bony ribs are directed downwards and forwards, but with the exception of the 'floating ribs', they all have costal cartilages which curve upwards to a greater or lesser degree to join the adjacent cartilage in the formation of the costal margin. In doing so they deflect the intercostal nerves T7, 8, and 9 in an upward direction.

The result is that the nerve T7 [Fig. 10.06] runs *upwards* more or less parallel to the costal margin, T8 runs upwards and medially, and T9 runs horizontally crossing the costal margin at the transpyloric plane [p. 222].

The lower six thoracic nerves describe a fan-shaped pattern on the abdominal wall. The position of these nerves is fundamental in understanding the strategy of many abdominal incisions.

Adjacent nerves overlap each others territory both in the motor nerve supply of the muscles and in the sensory supply of the skin. Thus the upper half of the T10 dermatome is also supplied by T9, and the lower half is also supplied by T11. T9 and T11 supply adjacent areas of skin and muscle. If T10 is cut no obvious defect is evident clinically. A surgeon can afford to cut one of the segmental nerves supplying the abdominal wall without any obvious weakness developing. This means that the motor units which remain after nerve section will hypertrophy rapidly so that the weakness in the affected muscle is quickly made good.

Note that L1 is a special case. This nerve [Fig. 10.06] passes to the inguinal region and supplies the internal oblique and transversus abdominis muscles as they pass over the inguinal canal. It is the principal nerve supplying the muscles which constitute the main mechanism preventing an inguinal hernia. If this nerve is cut, say during an appendicectomy, then an inguinal hernia may develop later. The point is that L1 is the lowest and therefore the last nerve

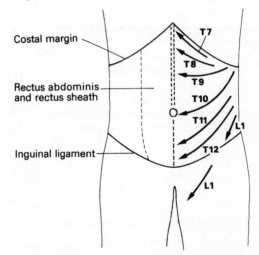

Fig. 10.06. Nerve supply to anterior abdominal wall.
Six thoracic nerves supply the abdominal wall in the midline as far down as the pubic symphysis. Note T10 supplies the umbilicus—T9 is horizontal: above T9 the nerves fan upwards: below they fan downwards.

to supply the abdominal wall. It is overlapped above by T12, but it is not overlapped from below by L2. Section of L1 should always be carefully avoided.

A very similar pattern occurs in the upper limb, when you consider T1, the last nerve contributing materially to the brachial plexus. This nerve is overlapped by C8 but not by T2. Section of T1 produces permanent wasting of all the intrinsic muscles of the hand, and section of L1 can be expected to have a corresponding effect on the masculature of the abdominal wall around the inguinal canal.

ABDOMINAL WALL BLOOD VESSELS

Arterial supply to anterior abdominal wall [FIG. 10.07]
The **internal thoracic** artery gives rise to:
    1. **Superior epigastric artery.**
    2. **Musculophrenic artery.**

The **external iliac** artery gives rise to:
    1. **Inferior epigastric artery.**

This artery, given off just above the inguinal ligament, passes medial to the vas at the deep inguinal ring before running on the deep surface of the rectus abdominis to anastomose in the rectus sheath with the superior epigastric artery which comes down from the internal thoracic artery.

2. **Deep circumflex iliac artery.**
This artery also arises from the external iliac artery just above the inguinal ligament. It runs towards the anterior superior iliac spine.

The **femoral artery** gives rise to:
    1. **Superficial epigastric artery.**
    2. **Superficial external pudendal artery.**
    3. **Superficial circumflex iliac artery.**
The superficial epigastric artery runs upwards across the inguinal ligament to supply the lower part of the anterior abdominal wall. The superficial circumflex iliac supplies the region of the anterior superior iliac spine.

This artery also arises from the external iliac artery just above the inguinal. It runs towards the anterior superior iliac spine.

The abdominal wall is also supplied by branches of the lower intercostal arteries and the four lumbar arteries.

Venous drainage of anterior abdominal wall
The superficial veins below the umbilicus drain downwards into the femoral vein and thus into the inferior vena cava. The superficial veins above the umbilicus drain upwards via the internal thoracic vein into the subclavian vein and thus into the superior vena cava. With obstruction to the portal vein a plexus of distended veins around the umbilicus may give rise to a *caput medusae* appearance

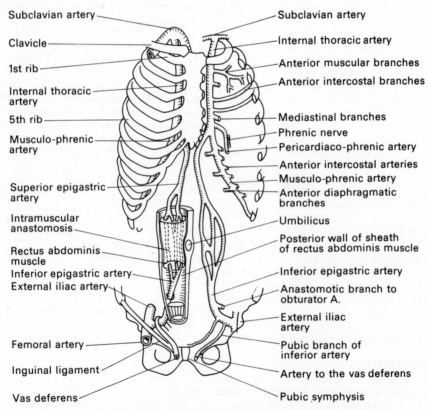

Subclavian artery
Clavicle
1st rib
Internal thoracic artery
5th rib
Musculo-phrenic artery
Superior epigastric artery
Intramuscular anastomosis
Rectus abdominis muscle
Inferior epigastric artery
External iliac artery
Femoral artery
Inguinal ligament
Vas deferens

Subclavian artery
Internal thoracic artery
Anterior muscular branches
Anterior intercostal branches
Mediastinal branches
Phrenic nerve
Pericardiaco-phrenic artery
Anterior intercostal arteries
Musculo-phrenic artery
Anterior diaphragmatic branches
Umbilicus
Posterior wall of sheath of rectus abdominis muscle
Inferior epigastric artery
Anastomotic branch to obturator A.
External iliac artery
Pubic branch of inferior artery
Artery to the vas deferens
Pubic symphysis

FIG. 10.07. The blood supply to the anterior abdominal wall.
The abdominal wall is supplied in addition to the vessels shown, by the four lumbar branches of the abdominal aorta, and iliolumbar artery.
The anastomosis of the accompanying veins is of greater clinical importance than that of the arteries, as they provide a bypass of an obstruction in either the superior or inferior vena cava.

(head of Medusa with snakes as hair), since this is a point of communication between the portal and systemic venous systems [p. 221]. The epigastric veins are also dilated when the inferior vena cava is obstructed, as they provide an alternative route for the blood so as to by-pass the obstruction. Certain liver diseases may obstruct not only the portal vein but also the inferior vena cava which is deeply embedded in the liver.

ABDOMINAL FASCIA

The muscles of the anterior abdominal wall are lined on their deep surface by a layer or layers of fibrous membrane, called the **transversalis fascia**. This in turn is separated from the peritoneal membrane itself by a layer called 'the extraperitoneal fat'. Even in emaciated patients this layer is evident, but there is very little fat left in it.

As the transversalis fascia is traced backwards at the level of the kidneys it splits into two layers, one of which runs in front of the kidney and one of which runs behind. These two layers are called the **renal fascia**. Traced further medially they lie in front of or behind the renal vessels and reach the front or back of the aorta and inferior vena cava. They can be traced to the midline.

Between the renal fascia and the kidney itself is a layer of very delicate fat—the **perirenal fat**. This layer is well marked and quite thick in normal people. In the kitchen the perirenal fat of an animal is called 'suet'.

The transversalis fascia is continuous above with a layer of connective tissue (unnamed) between the diaphragm and peritoneum, and below it is continuous with the well marked layer in front of the psoas and iliacus, called the **fascia iliaca**.

THE INGUINAL CANAL

The inguinal canal is present in both male and female. The inguinal canal **in the male** transmits the **spermatic cord** from inside the abdominal cavity to the scrotum. It passes right through the whole thickness of the abdominal wall just above the medial half of the inguinal canal, and has a length of about 4 cm. It is a potential weakness of the abdominal wall, and the site of **inguinal hernia**. The canal is present in the female but it only transmits the round ligament of the uterus to the labium majus. The **superficial inguinal ring** which is the distal end of the canal can always be identified clinically in the normal male, but not in the female. The **deep inguinal ring**, where the canal begins, cannot be identified clinically, but it lies above the inguinal ligament just lateral to the pulsation of the femoral pulse.

The **spermatic cord** includes the ductus deferens, more commonly referred to as the vas deferens and the blood supply, venous drainage, and lymphatic drainage of the testis (testicular artery, pampiniform plexus of veins); the epididymis; and the vas itself (artery to the vas from the inferior vesicle artery). In addition a third artery runs from the deep epigastric artery to supply the cremaster muscle. All three arteries anastomose with each other.

The spermatic cord can be located through the skin of the scrotum above the upper pole of the testis. It is smaller in diameter than the little finger. Following the cord upwards by invaginating the scrotum leads the finger tip to the superficial inguinal ring. The little finger tip is just too big to enter the superficial ring.

The **inguinal canal** is described as having:

(i) a floor;
(ii) posterior wall;
(iii) anterior wall;
(iv) roof.

The **floor** is the gutter-like upper surface of the inguinal ligament. The medial one-third is the upper surface of the pectinal part of the inguinal ligament.

The **posterior wall** consists of the transversalis fascia throughout its whole length, but this is deficient at the deep inguinal ring. It is strengthened medially by the conjoint tendon.

The **anterior wall** consists throughout its length of the aponeurosis of the external oblique muscle, which is deficient at the superficial ring. It is strengthened in its lateral third by the muscular attachments of the internal oblique to the inguinal ligaments.

The **roof** is formed by the 'arching fibres of the internal oblique', which run from the attachment just mentioned to the conjoint tendon [FIG. 10.08].

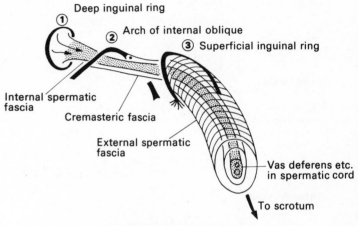

FIG. 10.08. Right spermatic cord.
The spermatic cord commences at the deep inguinal ring and terminates as it reaches the epididymis and testis in the scrotum. The testis is an organ displaced from the abdominal cavity. The spermatic cord is the umbilical cord of the testis, and provides the life support and sole communication systems of the testis and epididymis.
The three coverings of the cord are accumulated one by one as it traverses the canal and passes through the three 'rings'. The arching fibres of the internal oblique contribute muscle fibres to the cremasteric fascia, and the presence of muscle distinguishes it from the internal and external spermatic fascia.

The key to understanding the inguinal canal is the internal oblique muscle. All the other adjacent parts of the abdominal wall, such as the transversalis fascia and the external oblique muscle, maintain a constant relationship to the whole length of the canal, but the relationship of the internal oblique to the canal changes.

The general direction of the fibres of the internal oblique, as has been seen, is upwards and medially. This is not true of its lowest fibres which are closely related to the inguinal canal. They run in an arch more or less parallel to the inguinal ligament [FIG. 10.09]. The lowest fibres of this muscle take origin from the curled up edge of the inguinal ligament and running first upwards then horizontally then downwards, they are attached via the 'conjoint tendon' to the pectineal line of the pubis. They are reinforced by transversus abdominis fibres and arch over the canal obliquely from anterior to posterior. They reinforce the anterior wall in its lateral third, they form the roof in the middle third, and reinforce the posterior wall in its medial third by giving rise to the conjoint tendon.

One wonders how many students in professional examinations

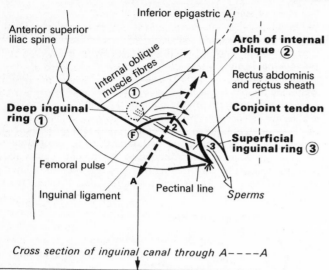

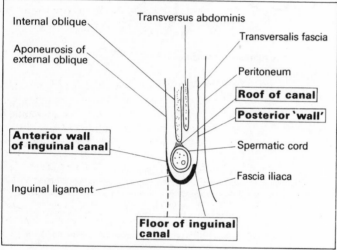

FIG. 10.09. The inguinal canal.
(a) Diagram of right inguinal canal.
(b) Cross-section of inguinal canal through A– – – –A.

drawing attention with this witty word to the long hazardous voyage undertaken by the male gonad, the testis, down the posterior abdominal wall to the pelvis and through the narrows of the inguinal canal until it reaches the safety of the scrotum. Unless you are aware of all this, the presence of the inguinal canal in the female makes no sense at all. Although the voyage of the female gonad, the ovary, stops at the pelvis, the continuation pathway is still there.

The gubernaculum blazes the trail for the inguinal canal through the three layers of the abdominal wall—they are the transversalis fascia, the internal oblique, and external oblique muscles. Each layer produces a diverticulum of fascia, so that the round ligament is ultimately surrounded by three diverticular sleeves. In the male they form the covering of the cord. The innermost is continuous with the transversalis fascia at the deep inguinal ring and is the **internal spermatic fascia**; the second is continuous with the lower border of the internal oblique, where the canal passes beneath its arching fibres and is the **cremasteric fascia**; the third and most superficial, is continuous with the external oblique aponeurosis at the edge of the external ring and is the **external spermatic fascia** (even in the female!).

The cremasteric fascia contains striated muscle fibres which are attached to the inguinal ligament or the conjoint tendon and are inserted to the coverings of the testis. When these fibres contract the testis is drawn upwards.

SUMMARY OF INGUINAL CANAL

	Lateral third	*Middle third*	*Medial third*
Posterior wall	Transversalis fascia Deep ring	Transversalis fascia	Transversalis fascia Conjoint tendon
Roof		Arching fibres of internal oblique muscle and transversus abdominis muscle	
Anterior wall	Internal oblique muscle External oblique aponeurosis	External oblique aponeurosis	External oblique aponeurosis Superficial ring
Floor	Inguinal ligament	Inguinal ligament	Inguinal ligament (pectineal part of)

Transmits in female: Round ligament of uterus
Three coverings
Funiculus vaginalis (obliterated processus vaginalis)

Transmits in male: Vas deferens (ductus deferens)
Testicular artery
Artery to the vas
Artery to cremaster muscle
Pampiniform plexus of veins
Lymphatics
Genital branch of genitofemoral nerve

have had their careers cut short because they thought the conjoint tendon looked like a tendon? It is a flat sheet of connective tissue, bearing no resemblance whatever to a tendon. It is an aponeurosis. Perhaps it actually represents the part of the posterior rectus sheath which seems to be missing below the arcuate line.

The relationship of the internal oblique muscle and conjoint tendon, crossing obliquely over the top of the inguinal canal from front to back, is the key anatomical fact in understanding this region.

In the female the round ligament of the uterus which the canal transmits is a vestigeal remnant of the gubernaculum. The gubernaculum runs from the ovary to the fundus of the uterus (ligament of ovary) and then in the broad ligament of the uterus as the round ligament of the uterus to the deep inguinal ring. It then traverses the length of the inguinal canal and terminates in fibrous strands in the labium majus.

The term **gubernaculum** was coined by John Hunter. The word means either the helm of a ship or the helmsman. Hunter was

Ilio-inguinal nerve through external ring
only

Sympathetic nerves to epidiymis, vas, and
testis

The genital branch of the genitofemoral and the ilioinguinal nerve supply the skin of the upper part of the scrotum or labium majus. The former is the motor nerve to the cremaster. These nerves are tested when the cremasteric reflex is elicited clinically.

The descent of the testis is considered further on page 236.

INGUINAL HERNIA

Inguinal hernia is a very common condition. It is more frequent in the male than the female. There are two principal types called oblique and direct. An oblique (or indirect) hernia enters the canal at the deep inguinal ring just lateral to the inferior epigastric artery. The contents of the hernia may consist of the great omentum with its contained fat, or it may include a loop of small or large intestine.

When the hernia develops, the peritoneum lining the abdominal cavity is stretched out and caused to protrude like a diverticulum down the canal. This is referred to as the sac of the hernia. The diverticulum in the region of the deep inguinal ring, is called the neck of the sac. This is the usual place for strangulation to occur. Strangulation of a hernia containing the alimentary canal will first obstruct the lumen, and also cut off the blood supply. Such a strangulated hernia, if untreated, may be fatal.

In certain congenital and other herniae the processus vaginalis persists and a new diverticulum does not then need to be developed.

An **oblique hernia** passes down the whole length of the canal from the deep ring to the scrotum. To reach the scrotum the hernia must force its way between the inguinal ligament and the arching fibres of the internal oblique and conjoint tendon, and then pass through the superficial inguinal ring. This always comes to be stretched out and enlarged. Some chronic inguinal herniae are very large in size. When a patient lies down and relaxes, a hernia which has reached the scrotum may return to the abdomen either on its own, or with a little judicious assistance. It usually recurs when the patient stands up again.

An alternative variety of inguinal hernia is called a **direct hernia**. It is the hernia of old age. In these cases the abnormality is due to weakness and thinning which takes place in the conjoint 'tendon' which forms the posterior wall of the canal in its medial third. Remember that the superficial ring is a gap in the external oblique aponeurosis which forms the anterior wall of the canal in this region. Obviously if the conjoint tendon gives way, the hernia will bulge via the superficial ring straight or directly through the medial third of the canal into the scrotum. There is nothing to stop it. Such herniae return to the abdomen much more readily than oblique herniae when the patient lies down.

In many oblique herniae the inguinal canal is essentially normal. The strategy of most hernia operations is then to 'darn' the gap separating the arch of the internal oblique and conjoint tendon from the inguinal ligament. The problem in the male is to do this without obstructing the pampiniform plexus in the spermatic cord. In the female the gap can be obliterated completely.

In a direct hernia the problem is more difficult because of the inherent weakness of all the aging connective tissues in the region. The giving way of the conjoint tendon is simply a particular example of this general weakness.

Summary of mechanism of the inguinal canal

The mechanism which normally prevents the development of an inguinal hernia depends on two main factors.

The first is that the deep and superficial inguinal rings are placed so that they do not overlap. Each ring represents an area of absolute weakness in the posterior and anterior walls. Where the posterior wall is weak the anterior wall is reinforced by the internal oblique muscle fibres. Where the anterior wall is weak the posterior wall is reinforced by the conjoint tendon. Furthermore, the obliquity of the canal ensures that when the abdominal pressure rises the posterior wall is pushed against the anterior wall. The pressure exerted is the same as that tending to push the abdominal contents down the canal, so, in theory, hernia should not occur.

The second factor concerns the arching fibres of the internal oblique which form a 'constriction' through which an oblique hernia must pass. Here we are dealing with striated muscle which can stand up to stresses and strain over a long period without giving way as many fibrous structures such as transversalis fascia and the external oblique aponeurosis tend to do. The gap between the internal oblique and conjoint tendon and the inguinal ligament is also blocked to some extent by the cremaster muscle whose contraction causes elevation of the testis.

FEMORAL HERNIA

Femoral canal

As has been seen [p. 113] the femoral nerve (lying lateral), the femoral artery, and the femoral vein (lying medial) pass down into the thigh behind the inguinal ligament. On the medial side of the femoral vein lies the **femoral canal**. In the embryo there is a large lymph node greater in cross-section than the vein and artery. This gland (Clocquet's gland) comes to be relatively much smaller in later development but the space in which the gland originally lay grows in proportion with the rest of the body and comes to be of such a size that in the adult you can put your finger tip into it. This space (the femoral canal) remains permanently on the medial side of the femoral vein, behind the inguinal ligament, lateral to the edge of the pectineal part of the inguinal ligament, and in front of the pectineal line and ligament.

The femoral blood and lymphatic vessels are all enclosed for 3–4 cm in the upper part of the thigh in a sleeve of fibrous tissue called the **femoral sheath**. The sheath is continuous at its commencement below the inguinal ligament with the fascia lining the surface of the anterior abdominal wall muscles, that is, with the transversalis fascia and fascia iliaca. The former is continuous with the anterior half (from 9 o'clock to 3 o'clock; FIG. 10.10) of the femoral sheath, while the latter, lining the front of the muscles immediately behind the canal, passes downwards to be continuous with the posterior half of the femoral sheath. Note that the femoral nerve which in the abdomen is 'outside' the plane of the transversalis fascia and fascia iliaca is therefore excluded from the femoral sheath.

The sheath funnels down as it goes and eventually fuses with the adventitia lining the outer surfaces of the blood vessels. The femoral canal runs downward for a distance of about 3 cm on the medial side of the vein to a blind end deep to the cribriform fascia of the thigh. A femoral hernia in the femoral canal stretches the femoral sheath deep to the cribriform fascia and produces a lump in this region below and lateral to the pubic tubercle. Quite small femoral herniae may threaten life by causing intestinal obstruction or strangulation.

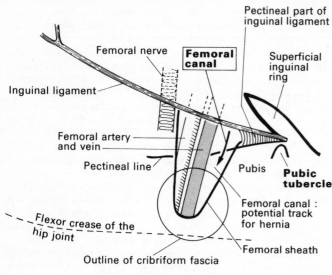

FIG. 10.10. The femoral canal.

There is no 'mechanism' to prevent femoral hernia in the way there is a 'mechanism' preventing inguinal hernia. The peritoneum and extraperitoneal connective tissue are thickened and tough in this region. Because of the differences in the shape of the pelvis in the two sexes, the femoral canal is larger and femoral hernia more common in the female.

Note that the medial part of the inguinal ligament separates the superficial inguinal ring from the femoral canal. It is always possible to push an inguinal hernia upwards and medially and to palpate the pubic tubercle. It is not possible to do this with a femoral hernia, because 'the lump' is at the cribriform fascia 2–3 cm below and lateral to the pubic tubercle. This is how a surgeon distinguishes these two types of hernia.

MOVEMENTS INVOLVING THE ABDOMINAL MUSCLES

The most important function of the abdominal muscles is to protect the contents of the abdominal cavity, and to maintain the viscera in position. This is especially true of the intestine, as it is the 'bowels' which gush out when the abdominal wall is slashed with a knife or the stitches come out of a surgical wound.

The rectus abdominis muscle is one of the principal flexor muscles of the trunk, so that because of its attachment to the pelvis, it brings about flexion of the lumbar vertebrae. When the pelvis is fixed it will tend to pull down the thoracic cage in expiration.

Because of their direction of pull the external and internal oblique muscles produce rotation of the trunk. Thus when sitting in your chair, you turn your shoulders to the left, the external oblique muscle on the right and the internal oblique on the left will contract simultaneously to produce this movement.

The vertically running fibres in the posterior part of the external oblique are involved in side flexion. Note that in many gymnastic movements such as side flexion to the left, the initial movement is brought about by the muscles in the left flank, but once the movement has started gravity takes over as the 'prime mover' and it is then the muscles on the right which contract vigorously so as to counterbalance the effect of gravity.

The transversus abdominis is especially concerned with compression of the whole abdominal cavity associated with a high intra-abdominal pressure. This occurs whenever expulsive efforts are made as in defaecation, coughing, in parturition, and in lifting a heavy weight. In cases such as weight lifting the larynx is closed and a strong expiratory effort is made. This involves the muscles of the thoracic cage as well as those of the abdomen. In such cases the diaphragm is relaxed and the high intra-abdominal pressure produces a high intrathoracic pressure. This is also what happens in coughing. Such a manoeuvre impedes the entry of blood into the thorax and the distended neck veins stand out.

It is also possible to raise the abdominal pressure by using the abdominal muscles especially the transversus abdominis, while continuing to breath normally. For this to happen it is obviously necessary for the diaphragm to contract.

Note that the tension felt over the rectus abdominis muscle in expiration does not necessarily indicate strong contraction of the muscle itself. It is often tension in the rectus sheath which you are feeling. If you contract your belly wall so that the rectus and rectus sheath are concave, it follows that contraction of the rectus would tend to pull the abdominal wall out again. This is the shape of the abdominal wall at the end of expiration (inspiratory muscles relaxed) in non-fat people where the elastic recoil of the lungs is drawing the diaphragm upwards and the abdominal wall inwards.

In movements of the lower limbs while sitting or lying in the recumbent position, the abdominal muscles play an important role as fixator muscles of the pelvis. If you raise both feet off the floor as you sit in your chair, you will instantly notice the contraction of your abdominal muscles.

These facts are of importance in patients who have had an abdominal operation and whose wound has not yet healed. Coughing and deep breathing are inhibited by pain, and this may predispose to the onset of pneumonia.

Movements of the legs may also give rise to abdominal pain so normal leg movements are also inhibited. The muscles will waste rapidly, and in addition the blood in the leg veins will not be propelled vigorously by the 'muscle pump'. The resultant venous stagnation will increase the likelihood of a deep venous thrombosis in the leg veins.

Abdominal reflex

If the skin of the abdominal wall is stroked (with, for example, the pointed handle of a patellar hammer) there is normally a reflex contraction of the abdominal musculature. This looks like a simple protective spinal reflex, but complex neural pathways involving the brain are involved. In a patient who has had a cerebral vascular accident (a 'stroke') the knee jerk, and other tendon reflexes, are very exaggerated but the abdominal reflex disappears on the affected side.

11 The abdominal contents

ABDOMINAL VISCERA

Embryological considerations

The layout of the viscera in the abdominal cavity appears to be very complex and difficult to think of 'all in one piece'. This is because the fundamentally simple plan has become obscure and complicated during development. The tortuosities of the gut need to be unravelled before the simple pattern emerges [FIG. 11.01].

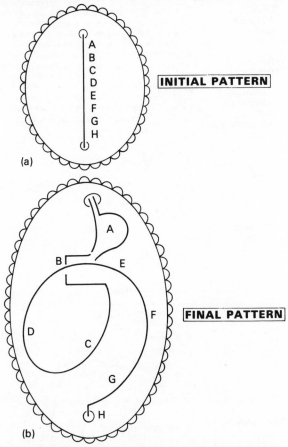

FIG. 11.01. Development of gut in peritoneal cavity (seen from the front).
(a) In the earliest stages of development the gut traverses the abdominal cavity in the midline. The letters A–H represent the presumptive regions shown in (b).
(b) After stomach and midgut rotation the straight line is converted into an 'open loop'. Note the cross-over of transverse colon and 2nd part of duodenum.

A Stomach.	E Transverse colon.
B Duodenum.	F Descending colon.
C Jejunum and ileum.	G Pelvic colon.
D Ascending colon.	H Rectum.

Initially the intestine traverses the abdominal cavity in the midline straight from the diaphragm to the floor of the pelvis [FIG. 11.02]. It is attached throughout its length by a dorsal mesentery to the midline. There is also a ventral mesentery (in midline) which joins the anterior abdominal wall above the umbilicus to the stomach and first part of the duodenum. The umbilical opening is very large.

The stomach then has a ventral, as well as a dorsal mesentery. The lesser curve faces the ventral mesentery and the greater curve

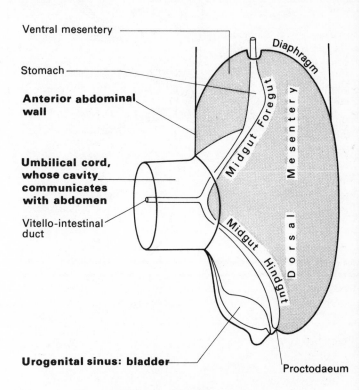

FIG. 11.02. Diagram of abdominal cavity seen from the left at an early stage of development (4–5 weeks).

The gut is joined to the posterior abdominal wall by the dorsal mesentery (shaded) and to the diaphragm and anterior abdominal wall above the umbilicus by the ventral mesentery (shaded). These structures all lie in the sagittal plane. The swelling shown represents the stomach.

The foregut becomes the stomach and the proximal half of the second part of the duodenum. It is supplied by the coeliac trunk. The midgut includes the gut from the distal half of the second part of the duodenum to the junction of the middle and distal thirds of the transverse colon. It is supplied by the superior mesenteric artery. The hindgut continues to the level of the anal valves in anal canal. It is supplied by the inferior mesenteric artery. (The arteries are not shown on this diagram.)

faces the dorsal mesentery. It has two flat sides one facing to the right and the other facing to the left.

The ventral mesentery comes to contain the liver. The dorsal mesentery comes to contain the spleen, derived from a number of splenunculi which usually fuse with each other.

The ventral mesentery is no longer a single structure joining the stomach directly to the diaphragm and anterior abdominal wall: it is divided by the liver into two parts, the first running from the stomach to the liver, the second from the liver to the abdominal wall [Fig. 11.03].

The dorsal mesentery of the stomach now has three zones, the part above the spleen, the part at the level of the spleen, and the part below the spleen.

The pancreas develops from two separate pancreatic buds. One develops in the ventral mesentery close to the liver. The other develops in the dorsal mesentery growing in such a way that its tail lies in contact with the hilus (hilum) of the spleen.

Rotation of the stomach

At this point the stomach 'rotates'. Use your left hand (hand flat,

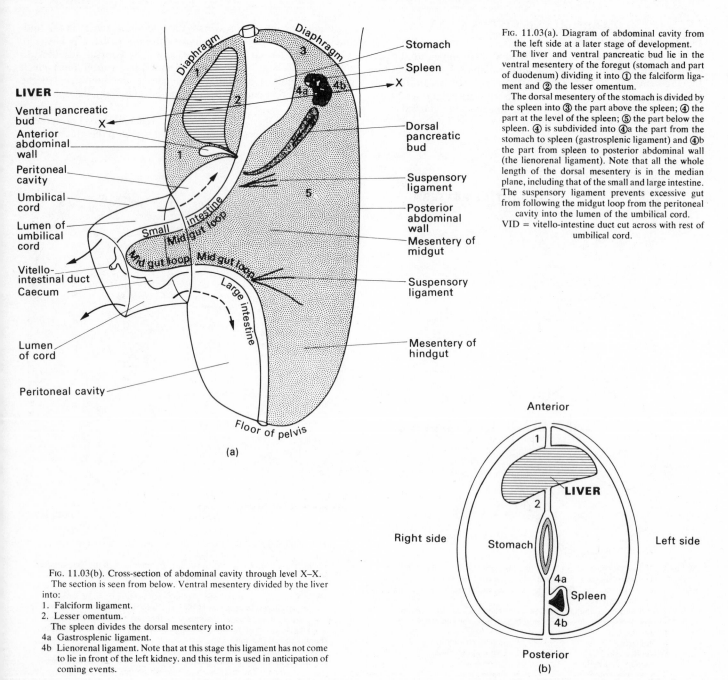

Fig. 11.03(a). Diagram of abdominal cavity from the left side at a later stage of development.

The liver and ventral pancreatic bud lie in the ventral mesentery of the foregut (stomach and part of duodenum) dividing it into ① the falciform ligament and ② the lesser omentum.

The dorsal mesentery of the stomach is divided by the spleen into ③ the part above the spleen; ④ the part at the level of the spleen; ⑤ the part below the spleen. ④ is subdivided into ④a the part from the stomach to spleen (gastrosplenic ligament) and ④b the part from spleen to posterior abdominal wall (the lienorenal ligament). Note that all the whole length of the dorsal mesentery is in the median plane, including that of the small and large intestine. The suspensory ligament prevents excessive gut from following the midgut loop from the peritoneal cavity into the lumen of the umbilical cord.

VID = vitello-intestine duct cut across with rest of umbilical cord.

(a)

Fig. 11.03(b). Cross-section of abdominal cavity through level X–X.
    The section is seen from below. Ventral mesentery divided by the liver into:
1. Falciform ligament.
2. Lesser omentum.
    The spleen divides the dorsal mesentery into:
4a Gastrosplenic ligament.
4b Lienorenal ligament. Note that at this stage this ligament has not come to lie in front of the left kidney. and this term is used in anticipation of coming events.

(b)

fingers pointing forwards, with the thumb upwards) as a model of your own stomach before rotation. Your thumb is the oesophagus, the web of the thumb and index finger are the lesser curvature (true ventral border of stomach), while the edge of the palm and ulnar border and little finger represent the greater curvature (true dorsal border of stomach). The palm faces to the right and represents the right surface of the stomach; the back of your hand faces to the left and represents the left surface. Flex your wrist so that the palm now faces backwards (if you use your *left* hand you cannot turn your hand in the wrong direction). The orientation is now similar to that of the stomach in the adult with the lesser curve to the right and the greater to the left and with anterior and posterior surfaces.

The rotation of the stomach and the new position of the greater curve on the left, affect the position of the spleen which is also carried out to the left. The rotation takes place to such a degree that the pylorus and duodenum are carried to the right.

The simple attachment of the dorsal mesentery to the midline is modified by the stomach 'rotation'. The attachment of the stomach mesentery appears to shift across the posterior abdominal wall so as roughly to follow the shape of the greater curvature, so that instead of being attached vertically in front of the aorta in the midline it comes to lie in front of the left kidney [FIG. 11.04] and then runs transversely across the abdominal wall at about the level of L3 [FIG. 11.07].

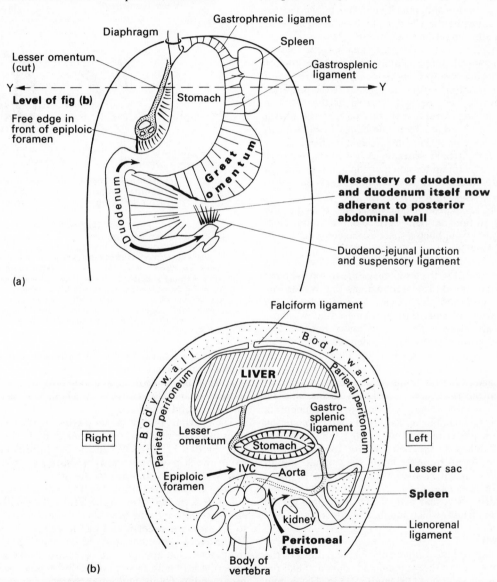

FIG. 11.04. The consequences of stomach rotation.

(a) View from the front, liver removed, to show position of stomach now 'rotated' and duodenum adherent to posterior abdominal wall, and spleen displaced to left. The suspensory ligament (of Treitz) anchors duodenojejunal junction, passing upwards to right and blending with right crus of the diaphragm.

(b) Cross-section of abdominal cavity level Y – – – Y. Lesser sac is the space between posterior (old right) surface of stomach and posterior abdominal wall. Note the peritoneal fusion behind lesser sac so that dorsal mesentery attachment shifts from the midline to the front of the left kidney.

At the same time to the right of the midline, the duodenum and its mesentery become adherent to the posterior abdominal wall. So we say 'the duodenum is retroperitoneal'. It appears to be part of the posterior abdominal wall.

The fusion of the mesentery of the duodenum with the posterior abdominal wall, means that the pancreas which was situated in this mesentery must also come to be retroperitoneal in position. The body of the pancreas continues across the posterior abdominal wall, and its tail remains in contact with the spleen.

The upper part of the dorsal mesentery of the stomach attaches it to the diaphragm. It is the gastrophrenic ligament. At the level of the spleen the mesentery is in two sections, from stomach to spleen, and from spleen to posterior abdominal wall, that is, the gastrosplenic and lienorenal ligaments. The dorsal mesentery below the spleen forms the greater omentum [FIG. 11.04].

Notice that the attachment of the greater omentum runs across the posterior abdominal wall approximately level with the greater curvature of the stomach, *when the stomach is empty*. If there were a short mesentery running straight from the stomach to the posterior abdominal wall it would become taut as soon as the stomach became distended. During development this part of the dorsal mesentery is greatly increased in length, and hangs down as far as the pelvis before retracing its steps to be attached to the posterior abdominal wall. This is the **greater omentum**. As a result the stomach can be greatly distended without tension developing in the mesentery. The greater curvature of the stomach may reach the pelvic cavity in a practised glutton [FIG. 11.05].

The **omental bursa**, also known as the **lesser sac**, is simply the space between the posterior (old right) surface of the stomach and the posterior abdominal wall. Such a space is the consequence of the 'rotation' of the stomach.

The ventral mesentery of the stomach is divided into two regions by the liver. The first runs from the lesser curvature to the liver and is termed the **lesser omentum**. The second joins the liver to the anterior abdominal wall and diaphragm. This is termed the **falciform ligament**. Note that since the ventral mesentery was only present above the umbilicus it has a free lower border. This gives rise to the free border of the lesser omentum and the free border of the falciform ligament in which the ligamentum teres lies.

The **epiploic foramen**† (additus to the lesser sac or foramen of Winslow) leads from the main peritoneal cavity to the omental bursa. It is bounded in front by the free edge of the lesser omentum which contains the portal vein, the hepatic artery, and the common bile-duct. Behind is the posterior abdominal wall in the form of the inferior vena cava. Below is the point where the duodenum becomes retroperitoneal. Above is the tail of the caudate lobe of the liver. This is a thin strip of liver tissue which separates the portal vein from the inferior vena cava.

The stomach, spleen, liver, and part of the duodenum are all supplied by the **coeliac trunk**, which supplies 'the foregut'.

Rotation of the midgut

Another major complication of the basically simple pattern of the gut is caused by the rotation of the jejunum and ileum and half the large intestine. This is the rotation of the 'midgut' loop. The midgut is supplied by the superior mesenteric artery.

Between the sixth and tenth weeks of intra-uterine development, most of the gut lies not in the abdominal cavity but in the extension

† *Epiploic* (Greek) = *omental* (Latin).

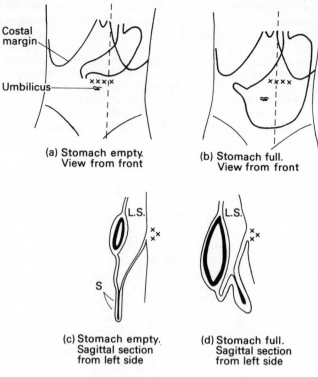

(a) Stomach empty.
View from front

(b) Stomach full.
View from front

(c) Stomach empty.
Sagittal section from left side

(d) Stomach full.
Sagittal section from left side

FIG. 11.05. The stomach empty and full.
(a) and (b) view from the front.
(c) and (d) sagittal section from left side.
When the stomach is empty the greater curvature does not reach the umbilicus. When it is full the greater curvature reaches the pelvis. The great omentum is attached to the posterior abdominal wall along the lower border of the body of the pancreas (XXXX). When empty a short omentum would suffice, but when it is full a long omentum is necessary. The length of the greater omentum is more than sufficient to allow distension.
The surplus (s) becomes adherent to itself.
L.S. = lesser sac.

of the abdominal cavity into the umbilical cord, and it is here that rotation takes place [FIG. 11.06].

At first the gut projects as a loop into the umbilical cord with its mesentery and blood vessels attached. The caecum soon appears and is a useful landmark.

The proximal part of the loop grows rapidly while the distal part does not. The proximal part comes to lie to the right of the distal loop FIGURE 11.07(b). When the gut returns into the abdominal cavity the disposition is shown in FIGURE 11.07(c). It is easy to see how you can get from one to the other.

The first part to return to the abdominal cavity is the descending colon. It is pushed over to the left by the small intestine which is crowding back into the central part of the cavity. The last structure to return is the caecum and the appendix.

The liver is immediately above the opening of the umbilicus, and the caecum and appendix lie very close to the gall-bladder at this time. Occasionally this relationship becomes permanent; there is no ascending colon, and an appendicitis in such a patient could easily be mistaken for a lesion in the gall-bladder, stomach, duodenum, or right kidney.

Occasionally a loop of intestine remains in the umbilical cord at birth. Precautions must always be taken in tying the cord in a new-born baby to avoid including such a loop in the ligature.

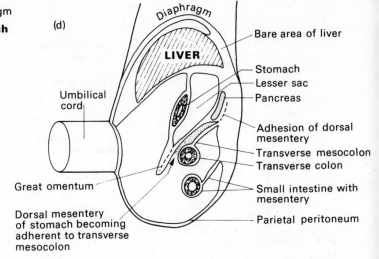

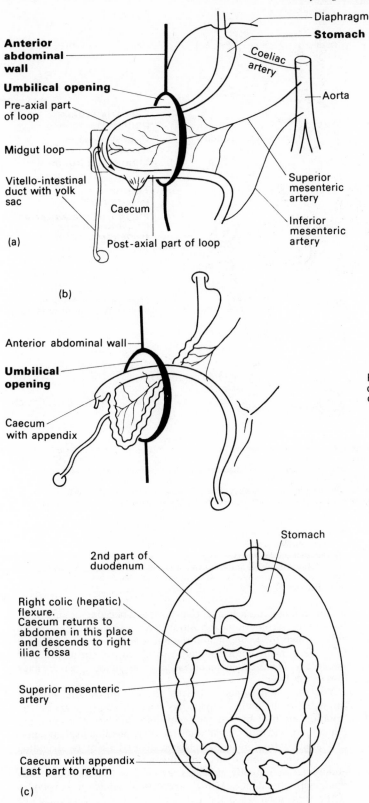

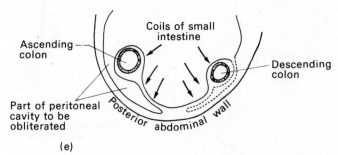

FIG. 11.06. Rotation of midgut.

The midgut herniates into the umbilical cord during weeks 6–10 of intra-uterine life. Then gut returns to abdominal cavity. The midgut extends from the middle of 2nd part of duodenum to last one-third of transverse colon.

(a) 1st stage: the herniated loop rotates anticlockwise around an 'axis' formed by its blood supply the superior mesenteric artery.

(b) 2nd stage: pre-axial part of the loop elongates considerably to the right of the superior mesenteric artery as midgut rotates.
3rd stage: gut returns to abdomen. probably very rapidly.

(c) 4th stage: fixation after returning to the abdominal cavity (week 10) involves adherence of mesentery of the colon to the posterior abdominal wall or to great omentum. It takes considerable time and may never be completed. Vitello-intestinal duct has disappeared.
The cross-over point produced by rotation is between the first third of the transverse colon and the second part of the duodenum.
Note the superior mesenteric artery lies in front of the duodenum. In the event of contrarotation (clockwise) the superior mesentric artery and the transverse colon will be behind the duodenum.

(d) Sagittal section of abdominal cavity from left side to show the fusion taking place between the anterior surface of the transverse colon and mesocolon and the posterior aspect of the greater omentum (stippled area). Note that the pancreas, originally lying in the dorsal mesentery of the stomach has been secondarily incorporated into the posterior abdominal wall.

(e) Diagrammatic cross-section of posterior abdominal wall seen from below. The descending colon and its mesentery, with contained blood vessels. lymphatics and nerves adherent to posterior abdominal wall. The adherent layers are being absorbed (dotted lines) and will disappear completely.
The ascending colon has not yet been fixed in position. When it has been, part of the peritoneal cavity shown, will be obliterated.
In mobilizing the colon, the surgeon opens up again the plane enclosed by the dotted lines. This can be done without interfering with the blood supply of the gut.

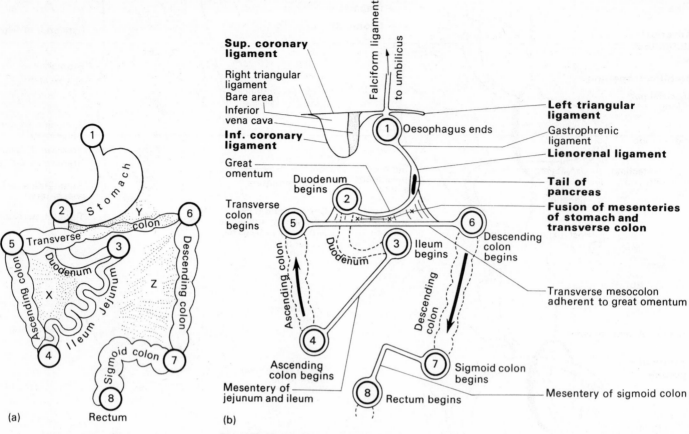

Fig. 11.07. Final position of alimentary canal and mesenteries.

(a) General layout of alimentary canal in the abdominal cavity. Compare with (b).

The stippled areas show where the mesentery of the ascending colon (**X**), transverse colon (**Y**), and descending colon (**Z**) are adherent to the posterior abdominal wall (**X–Z**) or to the dorsal mesentery of the stomach (**Y**). The nos. 1–8 correspond to (b).

(b) General layout of peritoneum seen after removal of those parts of gut which have mesenteries. The alimentary canal, following the rotation of the stomach (with consequential adhesion of the duodenum and its mesentery with contained head and body of the pancreas to the posterior abdominal wall) and of the midgut loop, enters or leaves the greater sac of the peritoneal cavity in eight places.

① Oesophagus enters peritoneal cavity.
② Duodenum leaves peritoneal cavity and becomes retroperitoneal.
   Between ① and ② is the stomach.
③ Duodenum enters at duodenojejunal junction.
④ Caecum leaves. Between ③ and ④ is small intestines.
⑤ Right colic flexure enters.
   Between ④ and ⑤ is the ascending colon.

⑥ Left colic flexure leaves.
   Between ⑤ and ⑥ is the transverse colon.
⑦ Descending colon enters.
   Between ⑥ and ⑦ is the descending colon
⑧ Pelvic colon leaves and becomes continuous with the rectum.
   Between ⑦ and ⑧ is the pelvic colon.

The small intestine pushes against the mesenteries of the ascending and descending colons with the result that they and their mesenteries adhere to the posterior abdominal wall. They become retroperitoneal in the same way as the duodenum. The mesentery of the transverse colon does not adhere to the posterior abdominal wall but fuses instead with that part of the dorsal mesentery of the stomach which has formed the greater omentum.

When looking at the abdominal contents the simple pattern in FIGURE 11.01 is hidden by the convolutions of the jejunum and ileum. But these do not affect the simplicity of the mesentery which runs in a straight line across the posterior abdominal wall. The basic pattern of the attachment of the mesentery to the posterior abdominal wall is also concealed by the disappearance of the mesentery of the ascending and descending colons. This means that there are apparently gaps which the eye must bridge.

You would think that operating on the duodenum or ascending and descending colon would be very difficult because they are so far back in the retroperitoneal position. This is so unless the surgeon cuts through the tissue behind the gut where secondary adhesion took place in the embryo, thus restoring to the gut its primeval mesentery. This is called '**mobilization**'. The gut must be fixed back in position later.

You may well wonder why the straight run of the intestine through the abdominal cavity has become complicated by the 'rotation' both of the stomach with its secondary effects on the duodenum, and of the small intestine. One possible explanation is that failure of rotation to take place at all or failure of the ascending or descending colons to become adequately fixed to the posterior abdominal wall greatly increases the chance of a volvulus developing with attendant intestinal obstruction and death. A volvulus is the term used when the gut twists on itself and occludes its lumen.

THE PERITONEUM

The inner walls and organs of the abdominal cavity are covered by a membrane of flattened endothelial cells on fibro-elastic connective tissue termed the **peritoneum**. It has a light shiny gloss. The peritoneum covering the inside of the abdominal cavity is termed the **parietal layer**. That covering the organs (viscera) is termed the **visceral layer**.

It is the covering of peritoneum that makes water-tight end-to-end anastomoses of the digestive tract possible. There is no peritoneum covering the oesophagus in the thorax and water-tight anastomoses here are more difficult to achieve.

The reflexions of the peritoneum in the abdomen are at first sight very confusing. Every organ must have an area not covered by peritoneum to allow the nerves and blood vessels to enter and these nerves and blood vessels have to reach the posterior abdominal wall without passing through the peritoneum. As has been seen this is achieved by either having part of the organ adherent to the posterior abdominal wall or by having it suspended via a mesentery, that is, two layers of peritoneum between which the blood vessels are running.

The jejunum and ileum of the small intestine are suspended by a mesentery and so is the pelvic colon. The duodenum and ascending and descending colons, however, are attached to the posterior abdominal wall and therefore do not have a mesentery. The transverse colon is suspended by its mesentery, the transverse mesocolon.

SUMMARY

The omentum or epiploon

The term **omentum** is derived from the Latin for an apron. The corresponding Greek word for apron is **epiploon**. The adjective *epiploic* thus means *pertaining to the omentum*. Rather confusingly, it is usual to use the noun **omentum** derived from the Latin and the adjective **epiploic** derived from the Greek although the adjective **omental** is also used.

The **greater omentum** or gastrocolic omentum is a fold of peritoneum attached to the greater curvature of the stomach above which, after dipping down over the small intestine like an apron, returns to fuse with the transverse colon and hence with the transverse mesocolon. The cavity between the descending and ascending folds is the cavity of the greater omentum.

The **lesser omentum** or gastrohepatic omentum is a fold of peritoneum passing from the lesser curvature of the stomach to the transverse fissure of the liver. The right free edge encloses all structures entering or leaving the transverse fissure of the liver. These include the portal vein, hepatic artery, and bile-duct together with nerves and lymphatics.

The entry to the **omental bursa** or **lesser sac** behind the stomach formed by the greater and lesser omenta is termed the **epiploic foramen** [FIG. 11.04].

THE LIVER

Developmental considerations

It is helpful in understanding the liver [FIG. 11.08] and its peritoneal connections to remember its development. The liver can be

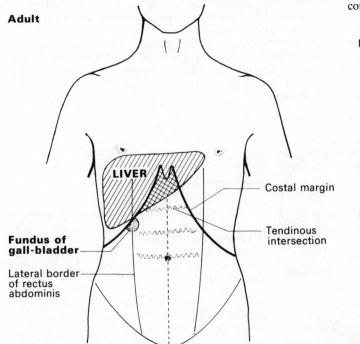

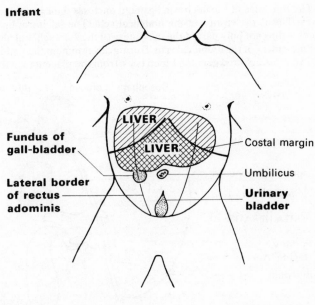

FIG. 11.08. Surface anatomy of the liver in adult and infant.
   In the adult the liver is concealed and protected by the ribs except in the epigastrium where it is exposed from the tip of the 9th right costal cartilage (where lateral border of rectus and costal margin cross) to the tip of the 8th left. The diaphragm and liver are in close apposition, and on inspiration the liver moves downwards with respect to the costal margin. The tendinous intersections of the rectus (stippled) may be mistaken by beginners for the liver, but do not move downwards, and when the rectus contracts (raise head off pillow) the intersections become more obvious, but the liver cannot then be felt at all.
   In the infant the liver almost reaches the umbilicus. Note that a considerable part of the empty urinary bladder lies in the abdominal cavity, unlike the adult.

considered to have developed in the ventral mesentery of the foregut [p. 202]. These two parallel sheets of peritoneum join the lesser curvature of the stomach directly to the diaphragm and to the anterior abdominal wall (via the deep surface of the linea alba) as far down as the umbilicus.

The liver as it develops between these two sheets expands on the right into the right sheet and on the left into the left [FIGS. 11.03 and 11.04].

The ventral mesentery is now sub-divided into two sections, one running from the lesser curvature of the stomach to the liver, the other running from the liver to the superior and anterior abdominal walls. The first section becomes the **lesser omentum** and the second becomes the **falciform ligament** [FIGS. 11.04 and 11.09].

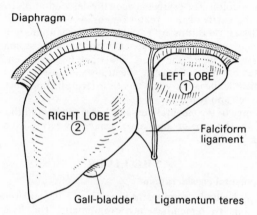

FIG. 11.09. The liver viewed from the front.

The free edge of the falciform ligament encloses the **ligamentum teres**. This is the vestige of the **umbilical vein**. The falciform ligament is thus not only part of the mesentery of the stomach, but also the **mesentery of the umbilical vein**. During development the umbilical vein brought oxygenated blood back from the placenta via the umbilical cord and umbilicus and was itself connected to the left branch of the portal vein. From here the blood did not pass into the liver, as you might expect, but instead into a special vessel, the **ductus venosus**, which short-circuited the liver and ran directly into the inferior vena cava. Just as the obliterated umbilical vein is represented by the ligamentum teres, so the obliterated ductus venosus is represented by the **ligamentum venosum**. The ligamentum venosum lies in the deep groove between the left lobe and caudate lobe of the liver. Note that this is exactly in-line with the groove for the falciform ligament and ligamentum teres. The ligamentum venosum swings to the right and skirts round the caudate lobe on its way to the inferior vena cava.

The final shape of the liver is determined by the surrounding structures. The upper surface of the liver is convex and fits exactly into the concavity of the diaphragm mainly on the right side. This is the **diaphragmatic surface**.

The **visceral surface** is more complex. Notice the position of the gall-bladder with its fundus projecting slightly beyond the edge of the liver. The neck of the gall-bladder is directed towards the porta hepatis. The porta hepatis (gateway to the liver) contains the right and left hepatic ducts as well as the hepatic artery and portal vein. The blood in both the hepatic artery and the portal vein runs in the same direction, that is, into the liver.

The groove for the ligamentum teres, the edge of the liver, the gall-bladder, and the porta hepatis form the boundaries of a rectangular area called the quadrate lobe [FIG. 11.10].

Further back, in line with the quadrate lobe, bounded by the groove for the ligamentum venosum, the upper posterior surface of the liver, the inferior vena cava, and the other side of the porta hepatis, is the caudate lobe. The tail of the caudate lobe (*cauda* (Latin) = a tail) is the narrow spit of liver pinched up between the inferior vena cava and portal vein.

When the hepatic artery and portal vein both divide for the first time, their right branches go to the right lobe, and their left branches supply the quadrate and caudate lobes as well as the left lobe. From the position of the porta hepatis this is just what you would expect.

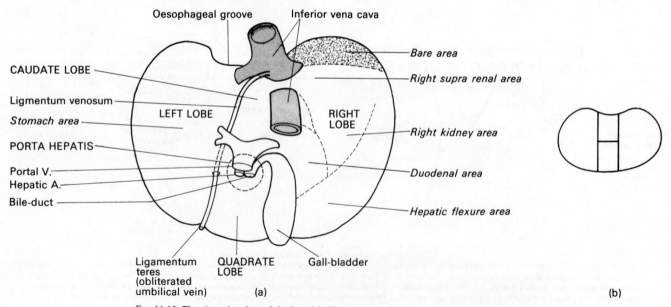

FIG. 11.10. The visceral surface of the liver (a). The structures shown form an H-shaped pattern (b).

When considering the peritoneal relations of the liver the term **right component** of the liver is used to describe that part which expanded into the right peritoneal sheet, that is, the quadrate lobe, caudate lobe, and right lobe. The **left component** expands into the left peritoneal sheet and coincides with the left lobe of the adult anatomy.

On the left side, from what has been said, you would expect the reflection of the left sheet of the peritoneum to encircle the liver in a simple continuous line. This is exactly what happens except at the **left triangular ligament**. Here the peritoneum strikes out sideways in the coronal plane and then returns to the original digression point [Figs. 11.11 and 11.12].

On the right side of the falciform ligament the peritoneum, when it is level with the left triangular ligament, also strikes out to the

right in the coronal plane, forming the **upper part of the coronary ligament** of the liver. Eventually the peritoneum will return to the original digression point, as it did on the left side.

The returning layer is adherent to the first layer and forms the right triangular ligament for a few centimetres. So far the right side mirrors the left. But further progress is complicated by the presence of the inferior vena cava which is both retroperitoneal and is embedded in the liver. This means that that part of the liver in which the inferior vena cava is embedded must also be retroperitoneal. (If it were not for this fact the return of the right leaf to the digression point could be a mirror image of the return of the left leaf.)

The returning layer parts company with the right triangular ligament and takes the course shown in FIGURE 11.12. It passes in front of the inferior vena cava and then returns to its digression point. In

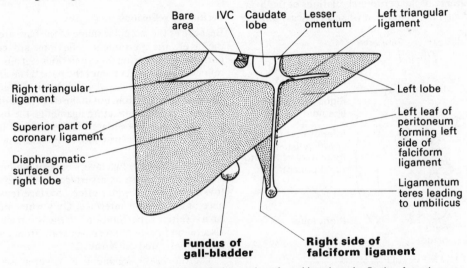

FIG. 11.11. Diagram of liver to show the diaphragmatic surface with peritoneal reflections from above.

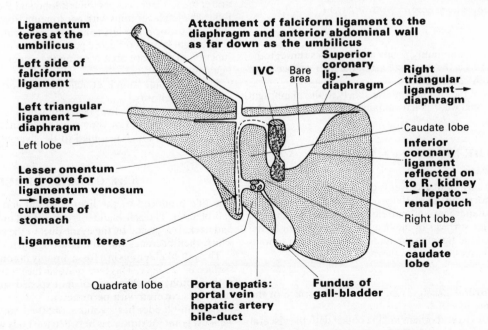

FIG. 11.12. Diagram of liver to show visceral and posterior surfaces with peritoneal reflections.

company once more with the left leaf, the two layers of peritoneum run into the lesser omentum. The right leaf runs down in the groove, for the ligamentum venosum encloses the vessels in the porta hepatis, and reaching the groove for the ligamentum teres, it gets back to the right side of the falciform ligament once more.

The **bare area** of the liver is a triangular area whose apex is the right triangular ligament, whose base is the inferior vena cava, and whose sides are the two separate components of the coronary ligament. The upper part has already been mentioned. The lower part of the coronary ligament runs from the liver to the posterior abdominal wall, that is mainly on to the right kidney. It forms the upper boundary of the hepatorenal pouch. This pouch is the commonest site for the formation of a subphrenic abscess.

The diagnosis and treatment of subphrenic abscess depends on understanding the anatomy of the peritoneal relations of the liver. They are shown in FIGURE 11.13. Note that when an abscess starts in

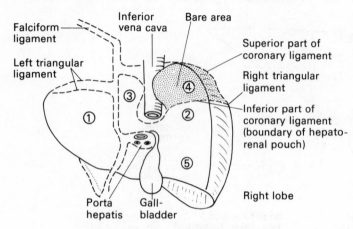

FIG. 11.13. Liver viewed from behind.
Regions in which subphrenic abscesses may occur:
① Left of mesenteries.   ④ Bare area.
② Hepatorenal pouch (commonest). ⑤ Right lobe.
③ Lesser sac.

the liver itself, it most commonly points and ruptures through the bare area, because this is the weakest part of the liver surface since it has no peritoneum covering it.

The attachments of the two mesenteric sheets to the diaphragm and anterior abdominal wall, exactly mimic the attachments to the liver itself.

THE LIVER—SUMMARY

Diaphragmatic surface

The diaphragmatic surface of the liver is divided into right and left lobes by the falciform ligament (falciform = sickle shaped). It extends from the diaphragm and anterior abdominal wall to the anterior and superior surfaces of the liver. It extends down to the umbilical region and the ligamentum teres (remains of left umbilical vein) is found in its inferior free border.

Visceral surface

The fissures and visceral structures on the visceral surface of the liver form the letter I.

A large number of visceral organs are in contact with the visceral surface of the liver [FIG. 11.10]. These are:

1. Stomach.
2. Oesophagus.
3. Duodenum.
4. Right kidney.
5. Suprarenal.
6. Hepatic flexure of large intestine.

Bare area

The liver is covered on all sides with peritoneum except in the bare area where it is in direct contact with the diaphragm, the inferior vena cava and the right suprarenal gland. The connective tissue connecting the liver and diaphragm in the bare area contains the veins of Retzius which form part of the portal–systemic anastomosis system [p. 220].

Peritoneal reflexions

Because of the irregular shape of the bare area and the ligaments of the liver, the peritoneal reflexions are complex. The visceral peritoneum covering an organ does not just come to an end. It is reflected back and becomes the parietal layer covering the abdominal wall. The presence of a thin layer of fluid between these layers allows a gliding action, but in the case of the diaphragm and the liver the presence of connecting ligaments, the bare area and the cohesion of water between their peritoneal layers, the two move together. Thus the liver moves up and down with the diaphragm during respiration.

The fact that the two layers of peritoneum tend to stick together is often referred to as a surface tension effect. This is incorrect. It is a cohesion of water effect. Surface tension is only of importance at a water–air interface. Only when air enters the peritoneal cavity (after a perforation of the stomach or intestine) do the visceral and parietal layers separate from each other as can easily be demonstrated on X-ray.

The falciform ligament is covered on both surfaces with peritoneum. At the upper surface of the liver the two layers separate; the left layer covers the left lobe and the right layer the right lobe. The left layer joins with peritoneum from the under surface of the liver to form the left triangular ligament.

The right layer is reflected posteriorly on to the diaphragm at the junction of the bare area forming the upper layer of the coronary ligament.

The lesser omentum is attached to the porta hepatis and fissure for ductus venosus.

The bare area of the liver lies mainly behind the right lobe. It is in contact with the inferior vena cava, the right suprarenal gland, and the diaphragm which is joined to the liver by connective tissue.

THE GALL-BLADDER

The bile produced by the liver is stored and concentrated in the gall-bladder. The gall-bladder is considered to have a fundus, body, and neck. It is joined by the cystic duct to the common hepatic duct which then becomes the common bile-duct.

The gall-bladder lies in a fossa directly in contact with the visceral surface of the liver. This part of its surface is therefore not covered with peritoneum. The rest of its exposed surface, including the fundus, is covered with peritoneum.

The gall-bladder has a volume of 50 ml but it is distensible. The weakest point for rupture is between the body and neck. The neck is narrow and S-shaped.

The bile-duct system

The right and left hepatic ducts join to form the common hepatic duct but the length of this duct and the point of union is very variable. It lies in the lesser omentum.

The cystic duct is a side cul-de-sac from this common hepatic duct which continues on as the common bile-duct.

The common bile-duct is 7.5 cm long. It is conveniently divided into:

1. Supraduodenal part.
2. Retroduodenal part.
3. Infraduodenal part.

The supraduodenal part of the common bile lies in the free border of the lesser omentum with the hepatic artery and portal vein. The retroduodenal part runs behind the first part of the duodenum in front of the portal vein. The infraduodenal part runs behind the pancreas to join the pancreatic duct at the hepatopancreatic ampulla (ampulla of Vater) to open into the duodenum at the greater duodenal papilla. A gall-stone impacted in the ampulla may cause bile to enter the pancreatic duct giving rise to pancreatitis. The exit into the duodenum is protected by the sphincter of Oddi which has to be relaxed to allow bile to enter the duodenum.

Abnormalities

Since cholecystectomy is a common operation the possibility of an anatomical abnormality is of great importance. Such abnormalities include:

(i) The two hepatic ducts not joining until just above the duodenum.

(ii) The possible presence of an hepatocystic duct which runs from the liver directly to the gall-bladder.

(iii) An accessory hepatic duct which joins the common hepatic duct at some point along its course.

(iv) An abnormal cystic artery or accessory cystic artery which arises from the left hepatic artery or gastroduodenal artery instead of the more usual right hepatic artery.

(v) Abnormal right hepatic artery arising from superior mesenteric artery instead of from the common hepatic artery.

(vi) Accessory hepatic arteries from the left gastric or superior mesenteric arteries are quite common.

Nerve supply

The gall-bladder is supplied by the vagus (X) nerve. Gall-bladder contraction is also mediated by the gastro-intestinal hormone **cholecystokinin**.

THE SPLEEN

The spleen is a single organ found only on the left side of the body deep to ribs 9–11. It is usually remembered as being 1 inch thick, 3 inches wide, and 5 inches long, weighing 7 ounces. In the metric system it is not so easy to remember being 2.5 cm thick, 7.5 cm wide, 12.5 cm long and weighing 200 g.

As has been seen the spleen lies in the dorsal mesentery of the stomach. It lies between that organ and the left kidney. It lies on the left side of the body in the long axis of the tenth rib. It is about three-fifths as long as this rib. It is separated from the ribs by the diaphragm and the costodiaphragmatic recess, but not by the lungs (which do not get down so far). Fractured ribs often tear through the diaphragm and rupture the spleen, which is very soft and friable.

The convex outer surface of the spleen is in contact with the diaphragm. The notched border is in front. The visceral surface of the spleen is Y-shaped being indented for:

1. Stomach (in front).
2. Kidney (behind).
3. Splenic flexure of large intestine (below).

The lienorenal ligament (*lien* (Latin) = spleen; *renes* (Latin) = kidneys) connects the spleen to the posterior abdominal wall in front of the left kidney, and contains the blood vessels which enter the spleen at the hilus as well as the tail of the pancreas which extends to the hilus of the spleen. The hilus is found at the posterior part of the stomach area. It is joined to the stomach by the gastrosplenic ligament which contains the short gastric arteries which run to the stomach from the splenic artery.

The diaphragmatic surface is convex and smooth. On the visceral surface the hilus is seen with the splenic vessels. Behind the hilus is a concave zone [FIG. 11.14] which fits on to the left kidney and in front is the gastric impression. The lower pole sits on top of the phrenicocolic ligament, and is indented by the left (or splenic) flexure of the colon. The edge of the spleen is seen at the junction of the diaphragmatic and gastric surfaces. The notches are always evident.

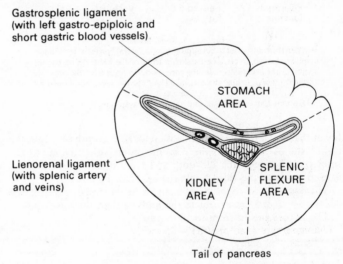

Gastrosplenic ligament
(with left gastro-epiploic and
short gastric blood vessels)

STOMACH
AREA

Lienorenal ligament
(with splenic artery
and veins)

KIDNEY
AREA

SPLENIC
FLEXURE
AREA

Tail of pancreas

FIG. 11.14. The spleen: visceral surfaces.

The tail of the pancreas reaches the hilus via the lienorenal ligament.

The splenic artery is a branch of the coeliac trunk. The splenic vein runs across the posterior abdominal wall just above the pancreas. It is joined by the inferior mesenteric vein and then forms the portal vein by its junction with the superior mesenteric vein to the right of the midline.

Palpating the spleen

The spleen is not normally palpable. The hand is placed under the left costal margin, while the other presses forwards against ribs 9, 10, and 11. The patient takes a deep breath. If the spleen is about three times as big as normal it may become palpable at the end of inspiration. Sometimes the spleen is so much enlarged that its edge may reach beyond the umbilicus. The notches in the edge are then very easily recognized. It is possible to 'miss' such an enlarged spleen by palpating too near the costal margin.

THE KIDNEY

The kidney produces urine. It is also an endocrine gland producing hormones erythropoietin, renin, and dihydroxy-vitamin D.

The two kidneys lie at the back of the abdomen behind the peritoneum and immediately in front of the posterior wall [FIG. 11.15]. Due to the presence of the liver, the right kidney usually lies

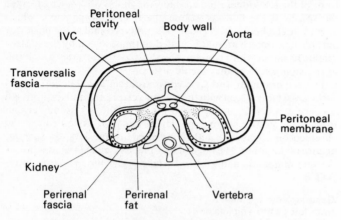

FIG. 11.15. Renal fascia.
When the transversal fascia approaches the lateral border of the kidney, it splits into two layers, one passing in front of the kidney, one passing behind. They eventually meet the corresponding layer from the opposite side in front of or behind the aorta and IVC. The two layers form the renal fascia.
Between the renal fascia and the renal capsule is a thick layer of perirenal fat.

lower than the left. The left extends up to the eleventh rib whilst the right extends only up to the eleventh interspace, that is, to the interspace between the eleventh and twelfth ribs. Both kidneys extend down to the third lumbar spine [FIG. 11.16].

The kidneys lie slightly obliquely with the upper pole of the kidney about 2.5 cm nearer to the midline than the lower pole.

The kidney lies in front of four muscles:

Diaphragm (and pleura)—superiorly
Psoas—medially
Quadratus lumborum—posteriorly
Transversus abdominis—laterally

Since the kidney lies in front of the twelfth rib, the subcostal nerve (T12), the iliohypogastric nerve and the ilioinguinal nerve cross its posterior surface.

The **right kidney** lies behind the liver and hepatic flexure of large intestine. The hilus lies behind the second part of the duodenum.

The **left kidney** lies behind the stomach, pancreas, spleen, and splenic flexure of large intestine.

The suprarenal (adrenal) glands lie at the upper poles of the kidneys.

The kidney lies in the paravertebral gutter which tends to fix its position. It is enclosed in a fascial envelope containing fat.

The renal vessels and ureter join the kidney at the hilus. The renal artery usually divides into four branches, three of which go in front of the ureter and one behind [FIG. 11.17].

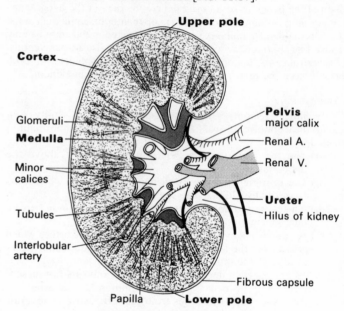

FIG. 11.17. Vertical section of right kidney.

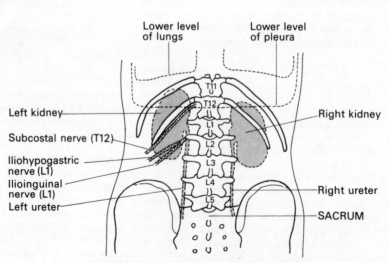

FIG. 11.16. The kidneys and the ureters from behind.

The ureter

The ureter is joined by its pelvis on the medial side of the kidney. The ureter is a muscular tube about 25 cm long. It runs down on psoas deep to and adherent to the peritoneum. It crosses in front of the iliac artery and runs down the side wall of the pelvis in the direction of the ischial spine. It reaches the bladder by crossing the floor of the pelvis accompanied by a plexus of veins. In the male it enters the bladder under cover of the seminal vesicles [p. 239]. It passes obliquely through the bladder wall to enter the bladder at the corners of the trigone. The two ureteric entry points are only 2.5 cm apart [p. 230].

There are three constrictions where a kidney stone may lodge. These are:

(i) at the beginning of the ureter;
(ii) as it crosses the iliac artery; and
(iii) as it enters the bladder.

When viewed from the front the right ureter lies behind the second part of the duodenum. It runs behind the right colic artery and the testicular or ovarian artery and lies behind the root of the mesentery and the terminal ileum.

The left ureter lies behind the left colic artery, the testicular (or ovarian) artery, and the pelvic colon and mesentery.

The surface marking of the ureter from the back is a line joining a point 5 cm lateral to the L1 spine and the posterior superior iliac spine.

THE PANCREAS

The pancreas is both an exocrine gland producing pancreatic juice for the digestion of food and an endocrine gland producing the hormones insulin and glucagon. Insulin is produced by the ß-islet cells. Glucagon is produced by the *a*-islet cells.

The pancreas consists of:

1. Head.
2. Uncinate process.
3. Neck.
4. Body.
5. Tail.

The head of the pancreas lies in the C-shaped concavity of the duodenum. Its extension, the uncinate process, passes behind the superior mesenteric artery (*uncinatus* (Latin) = hooked). Since the rest of the pancreas lies in front of this artery, the artery gives the impression that it is passing through the pancreas.

The head of the pancreas, its body and tail all develop from the dorsal pancreatic bud while the uncinate lobe develops independently from the ventral bud. The different relationships of the different parts of the pancreas to superior mesenteric vessels is due to their different developmental histories.

The tail of the pancreas lies in the lienorenal ligament and reaches the hilus of the spleen. It may be damaged during a splenectomy operation (surgical removal of spleen).

The pancreas is triangular in cross-section (apex upwards) and has an anterior, an inferior, and a posterior surface.

Relations

Posterior to pancreas

The head and neck of the pancreas lies in front of important structures. The bile-duct, which passes behind the first part of the duodenum, then lies in a groove behind the head of the pancreas and enters the second part of the duodenum with the main pancreatic duct.

The portal vein is formed by the union of the splenic vein with the portal vein behind the neck of the pancreas.

The inferior vena cava and abdominal aorta lie behind the head and neck [FIG. 11.18]. The superior mesenteric artery is given off by the abdominal aorta in this region and, as has been seen above,

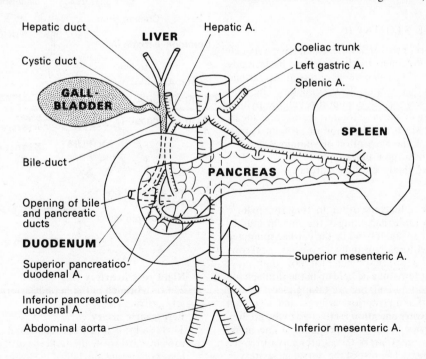

FIG. 11.18. The blood supply to the pancreas (uncinate process not shown).

passes between the body and uncinate process of the head of the pancreas.

The tail of the pancreas lies in front of the left suprarenal gland and the left kidney.

Anterior to pancreas

Since the pancreas lies behind the stomach, it forms part of the stomach bed (see omental bursa, p. 207).

The transverse mesocolon is attached to the lower anterior part of the body of the pancreas and to the head.

The gastroduodenal artery runs in front of the pancreas between its head and neck.

Pancreatic duct

The main pancreatic duct runs through the whole length of the pancreas. It enters the duodenum with the bile-duct in the second part of the duodenum. An accessory pancreatic duct may enter the duodenum 2 cm proximal to the main duct.

Blood supply to pancreas

1. Superior pancreaticoduodenal artery.
2. Inferior pancreaticoduodenal artery.

Lymphatic drainage

The lymphatic channels follow the splenic artery. They also follow the bile-duct to the liver.

Nerve supply

The secreting glands of the pancreas are supplied by parasympathetic secretomotor fibres in the vagus (X) nerve. Pancreatic secretion is also mediated by the gastro-intestinal hormones **secretin** and **cholecystokinin** (previously known as and identical to pancreozymin).

THE STOMACH

The stomach is a dilated part of the alimentary canal lying between the oesophagus and the duodenum. Its development has already been discussed on page 201. The oesophagus enters at the cardiac sphincter which prevents reflux back up the oesophagus. The food leaves the stomach and enters the duodenum through the pyloric sphincter. It is a storage organ allowing food to be taken as meals instead of continuously. It starts the digestion of protein. It produces the intrinsic factor for the absorption of vitamin B_{12}

The stomach is divided into three parts:
1. The fundus (above the level of the cardia).
2. The body.
3. The antrum.

The cardiac and pyloric sphincters differ in that the pyloric sphincter contains a considerable thickening of the circular muscle coat so that it can be both seen and felt, while the cardiac sphincter shows no such specialized features. Normally the oesophageal contents are never held up by the cardiac sphincter for more than a few moments, but the gastric contents may be held up in the stomach for several hours while digestion and mixing are taking place.

The stomach has anterior and posterior surfaces, and a greater and lesser curvature. The **greater omentum** is attached to the greater curvature and to the posterior abdominal wall along a line more or less parallel with the horizontal part of the greater curvature; i.e. along the body of the pancreas. The rest of the dorsal mesentery is the gastrosplenic and lienorenal ligaments (at the level of the

spleen) and the gastrophrenic ligament above the spleen. These are just the names of different parts of the same dorsal mesentery.

The **lesser omentum** is attached to the lesser curvature of the stomach and is connected to the liver along the groove in which the ligamentum venosum runs. Unlike the great omentum the lesser omentum has a free border to the right. The peritoneum of the two omenta continues over the anterior and posterior surfaces of the stomach and are continuous with each other.

The 'rotation of the stomach' has been described on page 202. It is inevitable that once the stomach has rotated there will be a space between the posterior surface of the stomach and the posterior abdominal wall. This is the omental bursa. The structures in the posterior (abdominal) wall of the bursa form the 'stomach bed'.

Relations of the stomach

Anterior—In front of the stomach lies, from right to left, the liver (gastric area), the abdominal wall, and the diaphragm.
Posterior—The structures behind the stomach constitute the stomach bed. These structures are:
1. Left crus of diaphragm.
2. Pancreas.
3. Left kidney and suprarenal gland.
4. Transverse colon.
5. Spleen.

Blood supply of stomach

The upper concave curvature of the stomach is the lesser curvature. The lesser omentum is attached here and brings with it the right and left gastric blood vessels. The greater omentum is attached to the greater curvature and contains the right and left gastro-epiploic arteries, and the short gastric arteries [Fig. 11.19].

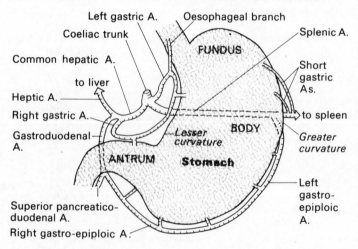

FIG. 11.19. Blood supply to stomach.

1. **Right gastric artery**
   This is a branch of the hepatic artery. It anastomoses with the left gastric artery.
2. **Left gastric artery**
   This is a branch of the coeliac trunk (the smallest branch). It runs upwards over the left crus of the diaphragm to reach the lesser omentum and hence the lesser curvature of the stomach. It gives off oesophageal branches.

3. **Right gastro-epiploic artery**
   This is a branch of the gastroduodenal artery which arises from the hepatic artery.
4. **Left gastro-epiploic artery**
   This is a branch of the splenic artery.
5. **Short gastric arteries**
   These are branches of the splenic artery.

Details of the blood supply of the stomach are only of critical importance to the operating surgeon. It will be noted that it all comes from the coeliac trunk. Originally most of it entered the dorsal mesentery from the coeliac trunk and supplied the greater curvature, but when the spleen developed there most of this was pirated by the spleen and only a few small blood vessels, the short gastric arteries, remain to supply the stomach. In the adult they are branches of the splenic artery.

The alternative route to the stomach is for an artery to travel either to the oesophagus (where it is extraperitoneal) or to the retroperitoneal part of the duodenum and then in each case to run in the appropriate direction towards the stomach. This is the route taken by the left gastric artery and by the right gastric and the right gastro-epiploic arteries. The left gastro-epiploic artery is a branch of the splenic artery. All these vessels anastomose freely with each other along the lesser and greater curvatures, and also in the stomach wall.

Venous drainage of the stomach

The gastric veins drain into the portal vein. In the region of the oesophagus they anastomose with the oesophageal veins (see portal–venous anastomoses, p. 220).

Lymphatic drainage of the stomach

The upper area of the stomach drains towards the cardiac sphincter. The lower area drains towards the pyloric sphincter. The lateral area drains towards the spleen. The lymph glands lie along the route of the corresponding arteries and they may be named accordingly.

1. Coeliac trunk group. All drain into this group which leads to the thoracic duct.
2. Splenic group.
3. Hepatic group.
4. Left gastric group. These lie in lesser omentum and drain lesser curvature of stomach.
5. Right gastro-epiploic group and pyloric group. These lie in greater omentum and drain lower area of stomach.
6. Virchow's gland in the neck near termination of thoracic duct.

Nerve supply to the stomach

The stomach is supplied both by sympathetic and parasympathetic nerves. The former reach it from the greater splanchnic nerves which come from the T5–9 segments of the spinal cord, and pass down from the thorax to the abdomen through the diaphragm without relaying. Relay takes place in the coeliac ganglia (solar plexus in the boxing world) and the postganglionic fibres run with the appropriate branches of the coeliac trunk to the stomach. These nerves inhibit peristalsis and close the pyloric sphincter.

Of greater practical importance is the parasympathetic supply from the tenth cranial nerve, the vagus. The two vagi run in the posterior mediastinum in close company with the oesophagus. In the lower one-third of the oesophagus (smooth muscle only in its wall) there is a nerve plexus to whose anterior part the left vagus, and to whose posterior part the right vagus, make the greater contribution. At the cardia the plexuses form the anterior and posterior gastric nerves. Branches from these nerves especially the anterior, pass to the fundus, body and pyloric sphincter of the stomach. They cause an increase in peristalsis, relaxation of the pylorus, and secretion from the fundus of the stomach.

Notice that the term 'gastric nerves' is misleading. These nerves do not only supply the stomach as the name suggests, but the intestine as far almost as the left colic flexure, as well as branches to the viscera such as the spleen, liver, kidneys, and pancreas.

The fibres pass to the stomach either closely associated with the gastric arteries and their branches, or, they gain access to the lesser omentum and pass to the pylorus direct. They may also reach the hepatic artery and turn downwards to the pylorus.

It is possible for the surgeon to section the branches of the vagus which are exclusively destined for the stomach and hence to prevent the secretion of acid in patients with gastric ulcer without interfering with the nerve supply of other abdominal viscera. Modern operations are called selective or highly selective vagotomy.

It will be remembered that the stomach secretions are also mediated by the gastro-intestinal hormone **gastrin** and that gastric motility is depressed by the hormone **enterogastrone**. Histamine H_2 blocking agents also reduce the acid secretions by the stomach.

Surface markings of the stomach

The cardia of the stomach lies just to the left of the midline behind the point where the sixth costal cartilage joins the sternum. The pylorus lies just to the right of the midline at the level of the transpyloric plane (lower when standing up or full). The pylorus is therefore about a hand's breadth below the cardia. The cardia may reach as high as the fourth intercostal space. Having fixed the position of the two sphincters it is easy to visualize the greater and lesser curvatures. The stomach is much higher than most people imagine. Pain which is gastric in origin is referred to the epigastrium.

The wall of the stomach consists of at least three layers of smooth muscle, longitudinal, circular, and oblique. The mucous membrane is very thick and in the non-distended stomach they form thick folds most of which run longitudinally. They are called **rugae** and are visible in an X-ray after a barium meal.

THE SMALL INTESTINE

The small intestine consists of:
1. The duodenum.
2. The jejunum.
3. The ileum.

THE DUODENUM

The duodenum extends from the pyloric sphincter of the stomach to the duodenojejunal junction. It is about 25 cm long and runs in two-thirds of a circle around the head of the pancreas.

It is useful from a descriptive point of view to divide the duodenum into four parts.

First part: Superior part (5 cm in length)
Second part: Descending part (7.5 cm in length)
Third part: Horizontal part (10 cm in length)
Fourth part: Ascending part (2.5 cm in length)

With the exception of the first part which resembles the stomach in having the lesser omentum attached to its upper border and the greater omentum attached to its lower border, the posterior surface of the duodenum and its mesentery, as we have seen, are attached firmly to the posterior abdominal wall. As a result the duodenum is immobile.

The bile-duct and the pancreatic duct enter at the greater duodenal papilla in the second part of the duodenum about 10 cm from the pylorus. This entry point may be viewed using an endoscope, and instruments are made to enter and explore the bile and pancreatic ducts. It is now possible to remove gall-stones impacted here in the hepatopancreatic ampulla (ampulla of Vater) by this route.

An accessory pancreatic duct may enter nearer to the pylorus at the lesser duodenal papilla. It has no bile-duct companion.

In the duodenum the secreting gland penetrates below the sub-mucosa (Brunner's glands) and this enables the duodenum to be distinguished from the rest of the small intestine histologically.

The secretions entering the duodenum from the liver, pancreas, and from its own mucous membrane are all alkaline and neutralize the acid stomach contents which should reach the duodenum in small amounts, if the pylorus is functioning normally. Duodenal ulcers are confined to the first part of the duodenum presumably because the contents here are still acid. Pain from the duodenum is felt in the epigastrium.

Surface markings

The duodenum begins at the level of the transpyloric plane just to the right of the midline and is directed upwards and to the right. The most useful fact to remember is that the third part of the duodenum runs horizontally at the level of L3, i.e. above the umbilicus. It commences on the right at this level just medial to the linea semi-lunaris. The fourth part runs upwards for 2 cm and runs into the duodenojejunal junction which is only slightly below the trans-pyloric plane at the level of L2.

Blood supply

Superior pancreaticoduodenal artery.
Inferior pancreaticoduodenal artery.

The duodenum lies in the territory of the coeliac trunk and also of the superior mesenteric artery. These two major blood vessels anastomose with each other in the middle of the second part. The 'last' branch of the coeliac axis is the superior pancreaticoduodenal while the 'first' branch of the superior mesenteric is the inferior pancreaticoduodenal.

The veins drain into the portal vein.

JEJUNUM AND ILEUM

The mucous membrane of the jejunum is much thicker than that of the ileum and is thrown up into a succession of transversely running folds, called valvulae conniventes (small winking valves—evidently someone thought these folds looked like eyelids!). Between finger and thumb the normal jejunum feels much 'meatier' than the ileum.

The **jejunum** is so named because it was thought to be empty (*jejunus* (Latin) = empty). **Ileum** is the Latin word for intestine which in turn is derived from the Greek *eilein* = to roll. There is no hard and fast rule as to where the jejunum becomes the ileum. The upper two-fifths is jejunum, the lower three-fifths the ileum, but there is no obvious dividing line. The diameter of the small intestine becomes smaller as the food passes along it. However, the blood vessels, which also become smaller, form more and more loops but the presence of fat in the mesentery makes these loops more dif-ficult to see. The branches from these loops to the intestine are long in the jejunum and short in the ileum.

The collections of lymphoid tissue known as **Peyer's patches** are only found in the ileum and not in the jejunum. They are usually found opposite the mesentery. They become inflamed and may perforate in certain diseases.

The mesentery which attaches the jejunum and ileum to the posterior abdominal wall extends from the left side of L2, across the third part of the duodenum to the right iliac fossa. At laparotomy the proximal part of the small intestine usually lies to the left.

Meckel's diverticulum

When this is present, it represents a remnant of the vitello-intestinal duct which, in the embryo, extended from the small intestine via the umbilical cord to the placenta. It is a blind pouch found about 0.5 m proximal to the ileo-caecal valve on the opposite side to the mesen-tery. It is usually indistinguishable under the microscope from the ileum, but sometimes it contains gastric or pancreatic tissue. Meckel's diverticulitis is indistinguishable at the bedside from appendicitis [FIG. 11.20].

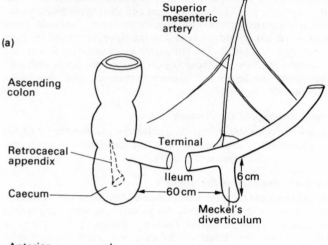

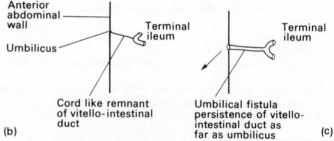

FIG. 11.20. Meckel's diverticulum.
(a) Meckel's diverticulum is due to abnormal peristance of the gut end of the vitello-intestinal duct, which once joined the yolk sac (found at the placental end of the umbilical cord) to the terminal ileum. It is com-mon (2 per cent of 'normal' people). It must lie on the antemesenteric border of the gut at the termination of the superior mesenteric artery itself, not one of its branches. It has the same calibre as the ileum and is histologically identical to it, but it may contain pancreatic tissue and gastric mucosa. It may become inflamed and perforate, and give rise to a clinical picture indistinguishable from appendicitis. The appendix is shown in its commonest retrocaecal position.

Other abnormalities derived from the vitello-intestinal duct include an 'adhesion' (b) or a fistula (c) or other abnormalities. A fistula will be obvious at birth as intestinal contents will discharge at the umbilicus. An adhesion will remain silent unless it gives rise to an 'acute abdomen' by causing intestinal obstruction.
(b) Cord like remnant of vitello-intestinal duct.
(c) Umbilical fistula: persistance of vitello-intestinal duct as far as umbilicus.

THE LARGE INTESTINE

The large intestine which extends from the ileocaecal valve to the anal canal consists of:

1. Caecum.
2. Ascending colon.
3. Transverse colon.
4. Descending colon.
5. Pelvic colon.
6. Rectum.

The outer longitudinal muscle coat in the large intestine is replaced by three longitudinal muscular bands known as taeniae coli (taenia = a tape; taeniae (pl.)). Appendices epiploicae are small sacs of fat attached to the large intestine with blood vessels which penetrate the walls of the intestine. These two facts constitute a weakness of the wall, and pouches of mucous membrane may herniate through the wall producing the clinical condition termed diverticulitis. The presence of taeniae coli and appendices epiploicae are useful in distinguishing large from small intestine in the operating theatre.

The caecum

As its name implies this is a blind sac or cul-de-sac at the start of the large intestine (*caecus* (Latin) = blind). It lies in the right iliac fossa although it may be high up under the liver [p. 204].

The **appendix** is attached to the posteromedial aspect of the caecum. It has a mesentery with the appendix artery in its free border. Since the caecum may also have a mesentery making it highly mobile, the position of the appendix is very variable. It may point towards the anterior abdominal wall, or it may be found behind the caecum (most common), behind the ileum or in the pelvis close to the uterine tube, the right ovary or the bladder. For this reason the symptoms of appendicitis may be misleadingly associated with those of other organs.

Ascending, transverse, and descending colons

The ascending colon is attached to the posterior abdominal wall. It runs upwards on the muscles iliacus, quadratus lumborum, and transversalis. At the right colic flexure, beneath the liver, it becomes the transverse colon which is mobile being attached to the transverse mesocolon and greater omentum [p. 207]. The left colic flexure is usually higher than the right flexure. Here it becomes the descending colon which is again attached to the posterior abdominal wall.

Pelvic colon

The pelvic colon is about 40 cm long. It is mobile since it has a surprisingly long mesentery. The attachment of the mesentery to the abdominal wall is shaped like an inverted **V**. One limb runs along the brim of the pelvis and the other runs down to the midline.

The rectum

The rectum (*rectus* (Latin) = straight), is only about 12 cm long. It starts at S3 and follows the curve of the sacrum. At the level of the tip of the coccyx it turns backwards through 90° to become the anal canal. It lies in the pelvis [see p. 233].

In the male the prostate gland lies in front. In the female the vagina and uterus lie in front.

The anal canal

The anal canal is 4 cm long. The thickened part of the circular muscle over 2.5 cm constitutes the internal sphincter controlled by the autonomic nervous system. The striated muscle external sphincter is a subcutaneous muscle which is more superficial than the internal sphincter. It runs from the coccyx around the anal canal to the perineum. It is supplied by the pudendal nerve (S2, 3, 4).

The levator ani muscle is inserted between the two sphincters.

In front of the anal canal in the male is the membranous urethra and bulbo-urethral (Cowper's) glands. In front of the anal canal in the female is the perineal body and the vagina. At the sides of the anal canal in both sexes is the ischiorectal fossa [p. 234] and the levator ani muscles [FIG. 11.21].

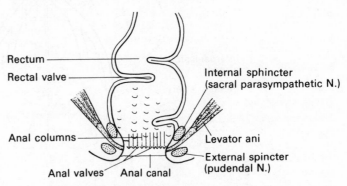

FIG. 11.21. Rectum and anal canal.

The columnar mucous membrane epithelium of the intestine changes to the stratified squamous epithelium of the skin at a zig-zag line in the region of the small valves between the vertical folds (columns of Morgagni) of the anal canal.

The spaces between them just above the anal valves are referred to as the sinuses of Morgagni. The anal columns join to form the anal valves (valves of Morgagni). Tearing of these valves gives rise to a fissure in ano.

THE ABDOMINAL AORTA

The aorta enters the abdomen through the diaphragm at the level of T12 vertebra. At the level of the umbilicus (L4 vertebra) it bifurcates into the two common iliac arteries [FIG. 11.22].

Below the diaphragm it gives off three groups of vessels. The first group supplies the gut, and constitutes:

  (i) the coeliac trunk;
  (ii) the superior mesenteric artery;
  (iii) the inferior mesenteric artery.

The second supplies the genito-urinary system and constitutes:

  (i) the gonadal (i.e., testicular or ovarian) arteries;
  (ii) the suprarenal arteries;
  (iii) renal arteries.

The third group of branches supply the body wall and are the lumbar pairs of branches (how many would you expect to find? Answer on p. 220), the inferior phrenic arteries, and the direct 'body wall' continuity of the aorta which is the median sacral artery.

The inferior phrenic artery is said to give off 'a branch' to the suprarenal gland. Earlier in development the phrenic artery supplies the suprarenal and only a small branch goes to the diaphragm. As

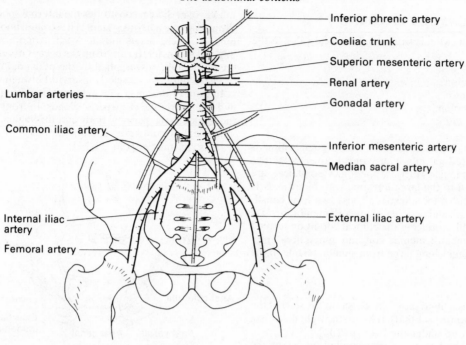

Fig. 11.22. Branches of the abdominal aorta.

growth continues the branch to the diaphragm becomes larger than the parent stem.

Thus the branches in the abdomen are:
1. Inferior phrenic arteries (2).
2. Coeliac trunk.
3. Superior mesenteric artery.
4. Middle suprarenal artery.
5. Renal arteries (2).
6. Inferior mesenteric artery.
7. Gonad arteries (testicular or ovarian) (2).
8. Lumbar arteries.
9. Median sacral artery.

1. INFERIOR PHRENIC ARTERIES

There are two inferior phrenic arteries, the right and the left. The right inferior phrenic artery passes behind the inferior vena cava to reach the diaphragm on the right side. The left inferior phrenic artery passes behind the oesophagus to reach the diaphragm on the left side.

Branches:
Right superior suprarenal artery.
Left superior suprarenal artery.

2. COELIAC TRUNK

The coeliac trunk is a short trunk. It is surrounded by the coeliac sympathetic plexus (the coeliac ganglia). The coeliac trunk lies immediately behind the lesser sac (omental bursa). The trunk splits into three arteries.
1. Common hepatic artery.
2. Splenic artery.
3. Left gastric artery.

1. Common hepatic artery

The common hepatic artery runs to the right behind the lesser sac and then runs upwards in the right free border of the lesser omentum with the bile-duct to reach the liver [Fig. 11.19].
Branches:
(a) Gastroduodenal artery
This is a short artery which runs downwards behind the junction of the stomach and duodenum and divides into:
(i) Right gastro-epiploic artery
This artery runs along the greater curvature of the stomach and anastomoses with the left gastro-epiploic artery. It sends branches to the stomach.
(ii) Superior pancreaticoduodenal artery
This artery runs to the duodenum and head of the pancreas. It anastomoses with the inferior pancreaticoduodenal artery which is a branch of the superior mesenteric artery.
(b) Right gastric artery
This artery runs along the lesser curvature of the stomach and anastomoses with the left gastric artery.
(c) Hepatic artery
Close to the liver this artery divides into right and left hepatic arteries. Branch:
Cystic artery
This artery arises from the right branch of the hepatic artery. It supplies the gall-bladder. It is very variable.

2. Splenic artery

This is the largest branch of the coeliac trunk. It runs along the upper border of the pancreas behind the lesser sac, crosses in front of the left kidney and reaches the spleen via the lienorenal ligament. It enters the spleen at its hilus.

Branches:
  (a) **Left gastro-epiploic artery**
      This artery runs along the greater curvature of the stomach and anastomoses with the right gastro-epiploic artery.
  (b) **Short gastric arteries**
      These arteries reach the stomach close to the oesophagus.
  (c) **Pancreatic branches**
      These arteries run to the pancreas.

3. **Left gastric artery**

This artery arises directly from the coeliac trunk. It sends branches to the oesophagus and then runs via the lesser omentum to the lesser curvature of the stomach. It anastomoses with the right gastric artery.

3. SUPERIOR MESENTERIC ARTERY

The superior mesenteric artery supplies the digestive tract from the second part of the duodenum to the middle of the transverse colon of the large intestine.

The artery is given off from the abdominal aorta at the level of L1 vertebra. It appears between the pancreas and the third part of the duodenum to enter the mesentery.

Branches:
1. **Inferior pancreaticoduodenal**
   This artery supplies the pancreas and duodenum. It anastomoses with the superior pancreaticoduodenal artery which is a branch of the gastroduodenal artery which in turn is a branch of the common hepatic artery.
2. **Intestinal branches**
   The superior mesenteric artery gives off numerous branches in the mesentery to supply the small intestine. They all arise from the left side. These arteries anastomose with one another forming a series of arches. These arches have a single tier at the start of the jejunum but increase to five tiers at the end of the ileum. The outermost arch gives off vasa recta to supply the jejunum and ileum.
3. **Ileocolic artery**
   This artery supplies the last part of the ileum, the caecum, the appendix, and the first part of the ascending colon. It anastomoses with the right colic artery.
4. **Right colic artery**
   This artery divides into ascending and descending branches which anastomose with middle colic and ileocolic arteries respectively. It supplies the ascending colon and right flexure.
5. **Middle colic artery**
   This artery enters the mesocolon, divides into the right and left branches which anastomose with the right and left colic arteries respectively. It supplies the large intestine in the region of the transverse colon [FIG. 11.24].

4. MIDDLE SUPRARENAL ARTERIES

These small arteries run across the crus of the diaphragm to reach the suprarenal gland on each side.

5. RENAL ARTERIES

The renal arteries arise from the abdominal aorta at the level of L2 vertebra, that is, slightly below the origin of the superior mesenteric artery. Since the abdominal aorta lies to the left of the midline the right renal artery is longer than the left.

The **right renal artery** runs behind the inferior vena cava. It runs behind the head of the pancreas, behind the second part of the duodenum and in front of the right crus of the diaphragm.

The **left renal artery** runs in front of the left crus of the diaphragm and behind the pancreas.

Branches:
1. **Inferior suprarenal artery**
   This artery supplies the suprarenal gland.
2. **Terminal branches of renal artery**
   The renal artery usually divides into four branches at the hilus of the kidney. Three of these branches usually pass in front of the ureter, whereas the fourth usually passes behind.

Blood supply to the suprarenal gland [FIG. 11.23]

It has now been seen that the suprarenal or adrenal gland is supplied by three arteries:
  1. Superior suprarenal artery.
  2. Middle suprarenal artery.
  3. Inferior suprarenal artery.

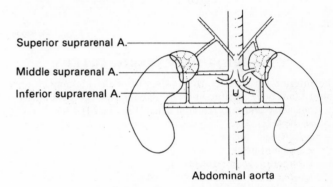

Superior suprarenal A.
Middle suprarenal A.
Inferior suprarenal A.
Abdominal aorta

FIG. 11.23. Blood supply to suprarenal glands.

The superior suprarenal artery is a branch of the inferior phrenic artery (the uppermost branch of the abdominal aorta). The middle suprarenal artery arises directly from the abdominal aorta. The inferior suprarenal artery is a branch of the renal artery.

6. ARTERIES TO THE GONADS

These are:
Testicular arteries in the male.
Ovarian arteries in the female.

The blood supply to the gonads arises from high up in the abdominal cavity because the gonads first developed in this region. The testes descend from this region to the scrotum taking their blood supply with them. The ovaries descend only as far as the pelvis.

The **right gonadal artery** arises behind the pancreas and passes down behind the peritoneum on psoas. It crosses **in front** of the inferior vena cava but behind the right colic and ileocolic arteries. It crosses in front of the ureter and external iliac artery.

The **left gonadal artery** follows a similar course but passes behind the left colic artery and sigmoid artery.

In the male the **testicular arteries** pass through the internal ring and inguinal canal to reach the testis and epididymis in the scrotum.

In the female the **ovarian arteries** reach the ovaries via the suspensory ligament.

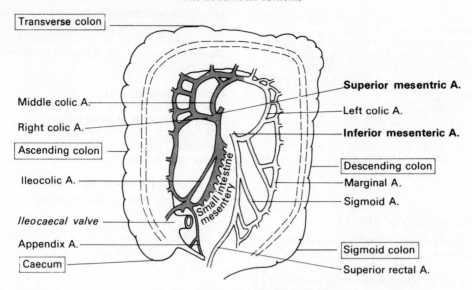

FIG. 11.24. Blood supply to large intestine.

7. INFERIOR MESENTERIC ARTERY

The inferior mesenteric artery arises from the abdominal aorta about 4 cm above its bifurcation. It supplies the large intestine from the transverse colon to the anal canal [FIG. 11.24].
    Branches:
      1. Left colic artery.
      2. Sigmoid arteries.
      3. Superior rectal arteries.

8. LUMBAR ARTERIES

There are four arteries on each side corresponding to L1 to L4 vertebrae which pass round the bodies of the vertebrae to supply the muscles of the back. The artery corresponding to L5 arises from the median sacral artery. The aorta has already divided by the time it gets to L5.

9. MEDIAN SACRAL ARTERY

This artery arises from the back of the abdominal aorta near its bifurcation. It runs down the middle of the sacrum to the coccyx.

BLOOD SUPPLY TO THE LARGE INTESTINE

The large intestine receives its blood supply from the superior and inferior mesenteric arteries [FIG. 11.24].
    Superior mesenteric artery branches to the large intestine:
    1. Ileocolic artery.
    2. Right colic artery.
    3. Middle colic artery.
    The superior mesenteric artery supplies the large intestine as far as about two-thirds of the transverse colon.
    Inferior mesenteric artery branches to large intestine:
    1. Left colic artery.
    2. Sigmoid arteries (two or more).
    3. Superior rectal artery (continuation of inferior mesenteric artery).

The inferior mesenteric artery supplies the large intestine from the splenic flexure to the anal canal.
    There are very good anastomoses between adjacent arteries except between the lowest sigmoid artery and the superior rectal artery. An anastomotic vessel runs in the mesenteric border of the large intestine from the ileocaecal junction to the beginning of the rectum which is the **marginal artery**.

THE PORTAL VEIN

The superior mesenteric vein and the splenic vein join behind the pancreas to form the portal vein [FIG. 11.25]. It passes behind the first part of the duodenum and runs upward in the free border of the lesser omentum with the bile-duct and hepatic artery to the liver. At the liver it divides into two branches. The ligamentum teres (the remains of the umbilical vein) and the obliterated ductus venosus are both connected to the left branch. Ligamentum teres is so called because it is round or cylindrical in shape (*teres* (Latin) = cylindrical).
    The portal vein is the venous drainage of the whole of the gastro-intestinal tract and brings the products of digestion to the liver, but note that it is also the venous drainage of the spleen.
    In foetal life, the portal vein is joined to the inferior vena cava by the **ductus venosus**. Thus blood arriving at the portal vein along the umbilical vein bypasses the liver. The ductus venosus closes at birth but other small anastomoses between the portal and systemic venous systems persist into adult life. Blood may flow through these anastomoses in cases of high blood pressure in the portal vein (portal hypertension) due to portal venous obstruction in the liver or elsewhere [FIG. 11.26].

Portal–systemic anastomoses

1. *Rectum*
Superior rectal (portal)→Inferior rectal (systemic) veins. The superior rectal vein drains into the splenic vein. The inferior rectal vein drains into the internal iliac vein. Their distension leads to haemorrhoids (piles).

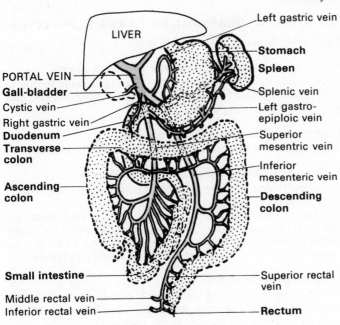

FIG. 11.25. Portal vein and its tributaries.

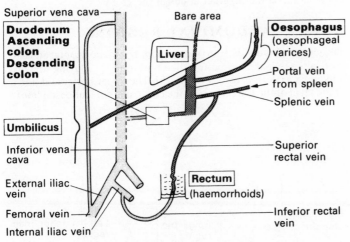

FIG. 11.26. Portal-systemic anastomoses.

2. *Lower oesophagus*

The lower oesophageal veins are partly portal and partly systemic. Their distension and rupture may lead to haematemesis in portal obstruction.

3. *Umbilicus*

The anastomosis between the veins along the ligamentum teres and the veins of the abdominal wall at the umbilicus leads to a plexus of distended veins around the umbilicus known as the caput medusae (Latin = head of Medusa. Medusa had snakes for hair).

4. *Behind the peritoneum*

There are anastomoses between the retroperitoneal veins; anywhere the gut is retroperitoneal: duodenum, ascending and descending colon.

5. *Bare area of liver*

There is an anastomosis between the portal and systemic veins over the bare area of the liver—the principle is the same as in 4.

Portal–caval shunt

The formation of a portal–caval shunt is a surgical procedure to reduce the portal venous pressure by establishing a low resistance anastomosis between the two circulations. However, it will reduce liver blood flow and impair the liver's metabolic activity. For example, the liver converts the ammonia from the deamination of amino acids to urea. If the liver fails to do this ammonia intoxication will develop and the patient will die.

THE INFERIOR VENA CAVA

The **femoral vein** lies medial to the femoral artery. The femoral vein becomes the **external iliac vein** and having received the **internal iliac vein** becomes the **common iliac vein**. The two common iliac veins join **behind** their corresponding arteries to form the **inferior vena cava** (IVC) [FIG. 11.27]. The IVC runs up on the right side of the abdominal aorta. It will be remembered, however, that the veins corresponding to the branches of the **abdominal aorta** leading to the **digestive tract** run to the **portal vein** and not the inferior vena cava.

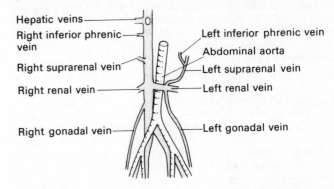

FIG. 11.27. Tributaries of inferior vena cava (lumbar veins not shown).

The inferior vena cava passes behind the third part of the duodenum, the head of the pancreas, and the first part of the duodenum and it forms a fossa on the visceral surface of the liver.

The inferior vena cava passes through an opening in the diaphragm at T8. The abdominal aorta, lying deeper, passes through an opening in the diaphragm with the thoracic duct at T12. Since the inferior vena cava in the upper part of the abdomen lies in a plane anterior to the aorta it is easy to remember that the renal veins (which run to the inferior vena cava) lie in front of the renal arteries [see also p. 249].

After an intrathoracic course of only 1 cm the inferior vena cava enters the right atrium.

Tributaries:
1. Lumbar veins.
2. Right testicular (or ovarian) vein. The left testicular vein joins the left renal vein.
3. Right and left renal veins.
4. Right suprarenal vein.
5. Hepatic veins.
6. Inferior phrenic veins.

Note that the hepatic veins join the IVC while it is embedded in the posterior surface of the liver, and that they have no existence beyond the confines of the liver itself.

LYMPHATIC DRAINAGE OF ABDOMEN

Axillary lymph nodes
The lymph vessels of the upper abdominal wall rather surprisingly run to the axillary lymph nodes. These nodes drain into the subclavian lymph trunk which opens into the subclavian vein (or internal jugular vein) on the right side and joins the thoracic duct on the left side.

Anterior mediastinal lymph nodes
The lymph vessels of the abdominal wall drain into anterior mediastinal lymph nodes. These lymph nodes also drain the anterior pericardium and thymus. These lymph vessels join the tracheobronchial lymph node vessels to form the bronchomediastinal lymph trunk. On the right side it joins the right lymph duct. On the left it joins the thoracic duct.

Inguinal lymph nodes
The lymph vessels of the lower abdominal wall run to the inguinal lymph nodes. The lymphatic vessels from these nodes run deep to the inguinal ligament to join the external iliac lymph nodes.

Aortic lymph nodes
The lymph vessels from the abdominal viscera turn to the aortic lymph nodes. These are sub-divided into two groups.

(i) **Pre-aortic lymph nodes**
The pre-aortic lymph nodes lie in front of the abdominal aorta. They drain the digestive tract and these lymph vessels are the main route for the absorption of fat. The lymph vessels unite to form the intestinal lymph trunk which joins the **cisterna chyli** and hence the thoracic duct. The thoracic duct passes up in front of the vertebral column crossing to the left side at T5 vertebra to empty its contents into the left brachiocephalic vein.

(ii) **Para-aortic lymph nodes**
The para-aortic lymph nodes lie alongside the abdominal aorta. As has been seen they drain the lower part of the body. The lymph vessels form the right and left lumbar lymph trunks which join the cisterna chyli and hence the thoracic duct. They also drain the kidneys and gonads.

Cisterna chyli
The cisterna chyli is the sac-like reservoir at the beginning of the thoracic duct. It is thin-walled, about 6 cm long and lies in front of the L1 vertebra between the right crus of the diaphragm and the aorta.

Afferent lymph vessels
Right and left lumbar lymph trunks.
Intestinal lymph trunk.
Lower posterior intercostal lymph vessels.

Efferent lymph vessel
Thoracic trunk which leads to the left brachiocephalic vein.

LYMPHATIC DRAINAGE OF LARGE INTESTINES
The lymph channel follows the arteries.

Lymph nodes associated with inferior mesenteric artery
These nodes drain the large intestine from the splenic flexure to sigmoid colon. Some lymphatics in the splenic flexure region also run to the splenic lymph nodes.

Lymph nodes associated with ileocolic artery
These nodes drain the region of the appendix and caecum.

Lymph nodes associated with middle colic artery
These nodes drain the upper part of the ascending colon, hepatic flexure, and transverse colon.

In the case of the transverse colon the primary lymph nodes are close to the transverse colon. In other parts of the large intestine the lymph channel may run directly to more central nodes making attempts to stop the spread of secondaries from a neoplasm.

Lymph drainage of rectum
Lymph nodes are scattered over the walls of the rectum. The upper nodes drain into the superior rectal nodes and follow the inferior mesenteric artery to the pre-aortic nodes. The middle nodes drain to nodes associated with the internal iliac artery. The lower nodes drain to the inguinal nodes in the groin.

ABDOMINAL REGIONS
The abdomen is divided into nine regions by two horizontal and two vertical lines making a pattern like that used for playing noughts and crosses (tic, tac, toe) [FIG. 11.28]. The upper horizontal line is at the level of the lowest part of the costal margin (subcostal line).

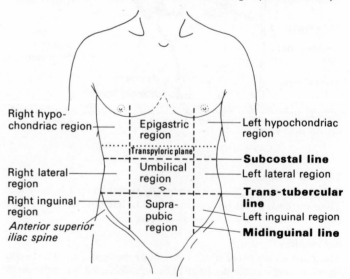

FIG. 11.28. The regions of the abdomen.

The lower horizontal line is at the level of the tubercles of the iliac crests (transtubercular line). The vertical lines pass through the midinguinal point which is half-way between the anterior superior iliac spine and the pubic symphysis.

The midline regions from above downwards are:
1. Epigastric region.
2. Umbilical region.
3. Pubic (hypogastric) region.
The lateral regions from above downwards are the left and right:
1. Hypochondriac region.
2. Lateral (lumbar) region.
3. Inguinal (iliac) region.

The transpyloric plane is a horizontal plane half-way down the body from the jugular notch to the crest of the pubis. It is at the level of the tip of the ninth costal cartilage, where the lateral border of the rectus cuts the costal margin. The plane is also half-way between the lower border of the sternum and the umbilicus, a hand's breadth below the sternum (or above the umbilicus).

The regions of the abdominal wall are most commonly made use of when describing the site of **abdominal pain**. Foregut pain (stomach) is usually referred to the epigastrium, midgut to the umbilicus, and hindgut to hypogastric region. This follows the simple arrangement in the embryo in which the gut runs straight through the abdominal cavity.

Pain originating from the parietal peritoneum, which is a constituent part of the body wall, is experienced 'locally'.

12 The pelvis

The anatomy of the pelvis has to meet the requirements of weight transmission from the vertebral column to the lower limbs. In addition because of the large size of the human foetal head, the female pelvis shows certain special modifications which have also to satisfy the requirements of childbirth.

The word 'pelvis' [FIG. 12.01] means a 'basin'. The upper part is the greater (or false) pelvis. The part below the brim or inlet of the

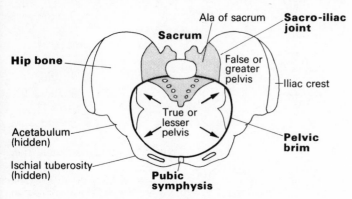

FIG. 12.01. 60° anterosuperior view of the pelvic cavity.
The brim of the pelvis is approximately oval and separates the true pelvis from the false.
The shape of the pelvic inlet and pelvic cavity is markedly encroached on in the male by the sacrum behind and the hip joints at the sides. The oval shaped inlet shown here is typical of the female pelvis.

pelvis, is the 'true' pelvis. It is natural [FIG. 12.02] to hold an articulated pelvis so that the inlet is in the horizontal plane. Look at your own body and you will realize that this must be wrong, because the anterior superior iliac spines and the pubic symphysis are all in the same frontal plane. The anterior abdominal wall is flat. The anterior superior iliac spines and the pubic symphysis lie approximately in the same frontal plane.

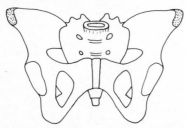

FIG. 12.02. Pelvis with pelvic brim horizontal (in incorrect anatomical position).
If an articulated pelvis is placed on a flat surface, resting on the ischial tuberosities and coccyx, the pelvic brim appears to be horizontal. This is not the orientation when standing (see next Figure).

If you now hold the pelvis correctly in the anatomical position the angle of the inlet of the true pelvis is at an angle of about 60° to the horizon. The lowest part of the sacrum lies above the level of the pubic symphysis [FIG. 12.03].

The pelvis is made up of three major components, the two hip or innominate (= no name!) bones at the sides and the sacrum (really **os sacrum** = sacred bone) behind. The two hip bones are joined to the sacrum by the **sacro-iliac joints**. In front they are joined to each other in the midline at the **pubic symphysis** [FIG. 12.01].

The hip bone is formed by the fusion of three separate components, the ilium, the ischium ('ch' is pronounced hard, as in chemist and not as in the word chinese) and the pubis. These terms are still used to designate the separate regions of the hip bone, even after fusion has taken place [FIG. 12.04].

The sacrum really consists of five separate sacral vertebrae which fuse together. The coccyx, all that is left of our tail, consists of four coccygeal vertebrae which also fuse to each other.

The **crest of the ilium** (iliac crest) can be felt in yourself about 10 cm above the great trochanter of the femur. It supports your trousers or skirt. You sit on the **tuberosity of the ischium** also known as the **ischial tuberosity**. The body of the left and right **pubic bones** can be identified in yourself separating the anterior abdominal wall from the genitalia. They are joined to each other across the midline by the **pubic symphysis**.

The ilium, the ischium and the pubis all participate in the formation of the socket of the hip joint, the **acetabulum**. Up to the age of about sixteen the three components of the acetabulum are separated from each other by the **triradiate cartilage** [FIG. 12.04]. When growth ceases this disappears like an epiphyseal cartilage and the bones finally unite.

Weight is transmitted from the sacro-iliac joint to the head of the femur only by that part of the ilium which runs along the anterior border of the greater sciatic notch [FIG. 12.04]. This bone is very thick. Note that weight transmission does not normally involve the triradiate cartilage. The weight bearing area of the acetabulum consists entirely of ilium. It has an area of only 2–3 cm².

LESSER PELVIS

The boundary between the greater pelvis and the lesser pelvis is termed 'the brim of the lesser pelvis' or more simply 'the brim of the pelvis', or 'the inlet of the pelvis'. This inlet is entirely bony. Below is the cavity of the lesser pelvis which leads to the pelvic outlet.

The pelvic outlet consists of the sacrum and coccyx behind, the sacrotuberous and sacrospinous ligaments, the ischial tuberosity, the ischiopubic rami and the body of the pubis, and pubic symphysis. As part of the pelvic outlet is not bony it follows that it is potentially more flexible than the inlet. The cavity of the pelvis

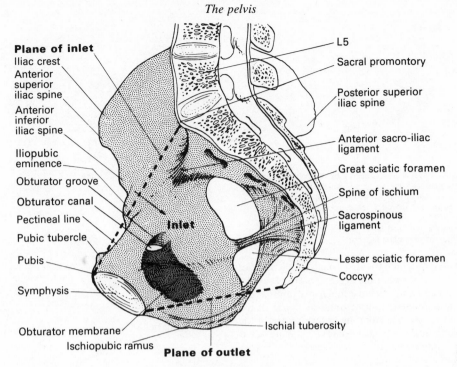

FIG. 12.03. Pelvic cavity.
Note the plane of the pelvic inlet and outlet.

between the inlet and outlet forms the bony wall of the birth canal in the female [FIG. 12.03].

It is normal in discussing obstetric problems in the lesser pelvis to consider:

(i) the inlet;
(ii) the canal itself;
(iii) the outlet.

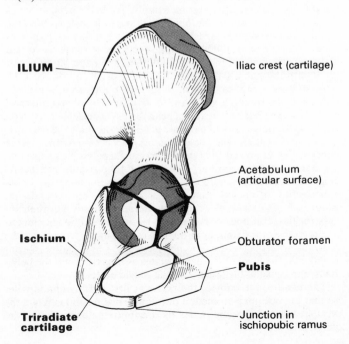

Male and female pelves

The typical male and female pelves are markedly different from each other. The reason is that the female pelvis is not only concerned with locomotion but also with the problem of allowing the baby to pass through during parturition. The most difficult part of the baby to pass through the birth canal is the baby's head.

If available, the instructive thing to do is to take a foetal skull and try to put it through a **male** pelvis. When you try this, you will immediately become aware of the differences which must exist between male and female pelves [FIG. 12.05].

Inlet of pelvis

Looking at the **inlet** of the lesser pelvis of the male you will see that the inlet is somewhat 'heart-shaped'. A full-term foetal skull cannot enter such a shaped pelvis. The inlet is encroached on at the sides by the acetabula of the hip joints and at the back by the position of the promontory of the sacrum.

In the female pelvis the hip joints are smaller than in the male,

FIG. 12.04. Right hip bone—infant.
The hip (innominate) bone develops from three separate components, the ilium, ischium, and pubis. The triradiate cartilage in the acetabulum represents the meeting place of all three elements. The ischiopubic ramus is also interrupted by an epiphyseal plate-like cartilage. The innominate bone unites into one during the early teens. There are also a number of secondary centres, for example, for the ischial tuberosity and iliac crest. The triradiate cartilage itself may contain secondary ossification centres.

At birth the acetabulum has more cartilage than bone because the triradiate cartilage is very extensive. Congenital dislocations of the hip have to be diagnosed at this stage. If the femoral head can be replaced in the acetabulum soon after birth and kept there, the acetabulum is still sufficiently plastic to adapt to the shape of the femoral head. The stage of development shown in the figure would probably not be able to do this adequately.

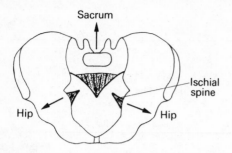

FIG. 12.05. The male pelvic canal is encroached on by the ischial spines and large hip joints at the sides and the sacrum posteriorly. In the female the hip joints are relatively smaller than in the male, and the sacrum is 'displaced' backwards. In the pelvic outlet, the subpubic angle is wide enough in the female easily to accommodate the foetal skull at term. The other differences between the male and female pelves are secondary to these fundamental factors.

and this reduces the degree to which they encroach on the inlet of the pelvis from the sides. Furthermore the acetabula are relatively speaking further apart. When handed a hip bone in an examination, the first thing to do is to compare the horizontal diameter of the acetabulum (x cm) with the distance from the anterior edge of the acetabulum to the pubic symphysis (y cm). In a typical male pelvis these distances are very similar, but in the female pelvis y is very much greater than x [FIG. 12.06].

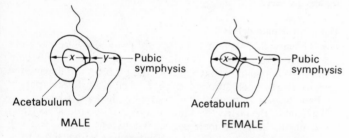

FIG. 12.06. Male and female pelves. Measure the horizontal diameter of the acetabulum (x) with your fingers and compare it with the distance of the acetabulum from the pubic symphysis (y). In the male x is approximately equal to y. In the female y is greater than x.

In the female pelvis, not only is the hip joint reduced in size, but in addition the sacrum is displaced backwards 'en bloc'. When you look at the male pelvis from the side, you will understand that there must be a change in the shape of the greater sciatic notch if the sacrum is to be 'displaced' in this way [FIG. 12.07]. In the male pelvis the notch, often likened to the inverted letter J, i.e. ⌐, while the female notch is like the inverted letter L, i.e. ⌐.

Birth canal

The other major change which follows from the backward displacement of the sacrum in the female is the direction of the spinous process of the ischium (ischial spine). In the male these processes point inwards and encroach on the pelvic canal, while in the female they point in a more backward direction and, as a result, do not encroach on the birth canal [FIG. 12.10]. In both cases the ischial spines point with the sacrospinous ligaments straight at the sacrum. In the female the backward displacement of the sacrum affects the direction of the sacrospinous ligaments and hence the ischial spines.

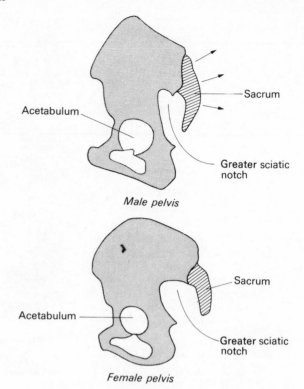

FIG. 12.07. The greater sciatic notch and the position of the sacrum in male and female pelves.

Outlet of pelvis

At the outlet of the pelvis the important distinction between male and female concerns the angle between the ischiopubic rami (the subpubic angle beneath the pubic symphysis). In the male this angle is about 60–70°, but in the female the angle is at least 90° [FIG. 12.08]. In the second stage of labour with the usual occipital presentation of the baby, the baby has its neck flexed. Once the back of the head has reached the pelvic outlet and the nape of the neck is level with the pubic symphysis, the head can extend on the neck. This allows the face to be born so that the baby can now breathe. The wider the subpubic angle, the sooner the baby's head will reach this point. The back of the foetal head fits in very easily beneath the two inferior pubic rami of a female pelvis. In the male the narrower angle means that the head would lie between the ischial tuberosities. In an android (male-like) female pelvis, this angle may be narrower than normal. The ischial tuberosities are nearer together than they should be. The distance separating the ischial tuberosities is always examined when a patient attends the antenatal clinic. The clenched fist should fit easily between the tuberosities. If it does not a difficult labour should be anticipated.

The transverse diameter of the pelvic inlet in the female is wider than the anteroposterior. The reverse is true of the outlet. During childbirth, the head of the baby rotates as it passes from the pelvic inlet, through the birth canal to the pelvic outlet.

The head of the baby in the pelvis at first most commonly lies so that the occiput is in the left or left anterior position. When the head reaches the pelvic outlet the occiput rotates so that it lies anteriorly.

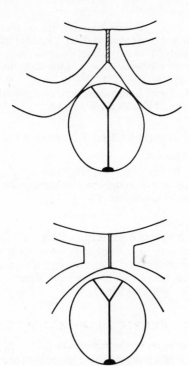

FIG. 12.08. The width of the subpubic angle in a normal female pelvis easily straddles the mature foetal head. In the male pelvis the angle is narrow and the ischial tuberosities lie much closer together.

As soon as the head has actually been born it returns to its original position with the occiput to the left. As the shoulders come down the right shoulder twists to the front and the occiput rotates further to the left [FIG. 12.09].

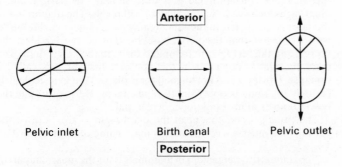

FIG. 12.09. Changes in shape of cross-sectional area of pelvis. The baby's head rotates as it passes through from the pelvic inlet to the birth canal and thence to pelvic outlet. The large arrow shows the sagittal plane of the foetal skull [see FIG. 12.08].

MALE AND FEMALE PELVES—SUMMARY

The female pelvis is wider and shallower than the male [FIG. 12.10].

In the male the anterior superior iliac spine is vertically above the symphysis pubis in the standing position. In the female since the Y shaped ligaments are shorter, the anterior spine often lies slightly anterior to pubic symphysis.

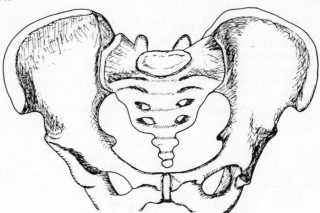

(a) Female pelvis

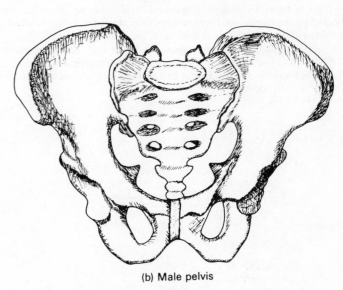

(b) Male pelvis

FIG. 12.10 Pelves in the anatomical position.
(a) Female pelvis.   (b) Male pelvis.

The acetabula are further apart in the female hence the femurs are more oblique.

The differences may be summarized as follows:
1. Female pelvis wider and shallower than male.
2. Female sacrum shorter, wider, and more concave than male.
3. Acetabula smaller and further apart in female.
4. Pubic arch 90° in female, 70° in male.
5. Pubic bone longer, but less height in female.

However, some female pelves approach the male configuration.

PELVIC LIGAMENTS AND MEMBRANES

It is self-evident when looking at the articulated pelvis that additional structures are needed not only to strengthen the bony arrangement but also to fill in the gaps so as to keep the pelvic viscera in place. These structures are the ligaments, membranes, and muscles of the pelvis.

OBTURATOR AND PERINEAL MEMBRANES

When looking at an articulated pelvis it is easy to understand the **obturator membrane**. This simply 'fills in' most of the large gap at the side of the pelvis, the obturator foramen [FIG. 12.11]. Anything deep to this membrane is inside the pelvis; anything superficial is outside the pelvis.

The **perineal membrane** serves the same function inferiorly. It

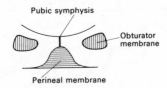

FIG. 12.11. The obturator and perineal membranes.

joins the two ischiopubic rami to each other [FIG. 12.12]. It is in continuity with the bony pelvis and is in the same plane like the obturator membrane. Anything above this membrane is in the pelvis. Anything below it is outside the pelvis.

The perineal membrane apparently has a 'free' edge posteriorly [FIG. 12.11], since it only extends backwards to a line between the two ischial tuberosities. Just as nature abhors a vacuum, so the body

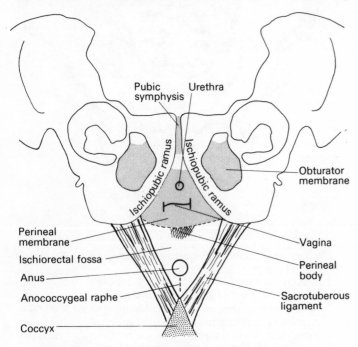

FIG. 12.12. Note the triangular shape of the perineal membrane. Its apex (slightly truncated) is below or behind the pubic symphysis. Its base runs transversely just in front of the ischial tuberosities. It has a free posterior border with the perineal body being attached to it in the midline. It is co-extensive with the urogenital triangle.

Like the obturator membrane, the perineal membrane is in the same plane as the bony pelvis.

In the female, the urethra and vagina pass through the membrane from inside to outside the pelvis. In the male apart from vessels and nerves, the urethra is the only important structure which passes through the perineal membrane.

The triangular area behind the posterior border of the perineal membrane is called the anal triangle.

abhors a 'free edge'. As will be seen below, there is a fibrous membrane attached to it [see FIG. 12.17, p. 231] which folds over and runs anteriorly. This is the continuation downwards into the perineum of the membranous layers of the superficial fascia of the anterior abdominal wall [FIG. 12.35, p. 248].

But the perineal membrane alone would be insufficient to retain the pelvic contents. Muscles are required.

SACROPELVIC LIGAMENTS

Sacrotuberous ligament

The sacrotuberous ligament runs from the dorsal surface of the sacrum and coccyx and from the superior and inferior posterior iliac spines to the ischial tuberosity [FIG. 12.13].

Sacrospinous ligament

The sacrospinous ligament which runs from the lower part of the sacrum and coccyx to the ischial spine crosses the sacrotuberous ligament. The sacrospinous ligament separates the **greater sciatic foramen** (leading to the pelvis) from the **lesser sciatic foramen** (leading to the ischiorectal fossa).

PELVIC DIAPHRAGM

Consider the problems in the pelvis which man has overcome in adapting to an upright posture. There is a need to prevent the pelvic viscera from escaping through the floor of the pelvis. (The perineal membrane, even if strong enough, could not do this, since it only extends as far back as the ischial tuberosities.) The escape is avoided by two factors. Firstly, the pelvic inlet, as has been seen, lies at an angle of about 60° to the horizontal. Secondly, the pelvic outlet is guarded by a muscular pelvic diaphragm which is convex downwards.

The diaphragm of the floor of the pelvis is as important as 'the diaphragm'. The latter is a muscular barrier which (usually) prevents the abdominal viscera from being sucked into the thorax by the 'negative' intrathoracic pressure. Should this barrier fail, a diaphragmatic hernia (hiatus hernia) will result. The diaphragm of the floor of the pelvis in the female may be damaged in childbirth, with the result that the pelvic viscera, that is, the bladder, the uterus, and the rectum, are herniated through the pelvic floor. The word 'hernia' is not used in this context. Instead it is called a **prolapse**. Obstetricians are blamed for prolapses but gynaecologists (usually the same people) aim to repair them. Hernia through the pelvic floor is almost unknown in the male.

The principal component of the diaphragm of the floor of the pelvis is the muscle with the unfortunate name of **levator ani** [FIG. 12.14].

The standard arrangement in mammals is for the muscular part of the diaphragm of the floor of the pelvis to be attached by way of origin along the inlet of the lesser pelvis and for the fibres to sweep medially and backwards towards the midline. Many animals use this muscle to wag their tails (the coccyx), and it is described as consisting of three parts which are continuous with each other, the most anterior is the **pubococcygeus**, the intermediate is the **iliococcygeus**, and the most posterior is the **ischiococcygeus**.

In man the origin of the diaphragm has migrated downwards and instead of running along the brim of the pelvis, it is attached directly to bone only in two places, to the pelvic aspect of the body of the pubis and to the ischium just above the ischial spine. Between these bony points the muscle is attached to a thickening of the already

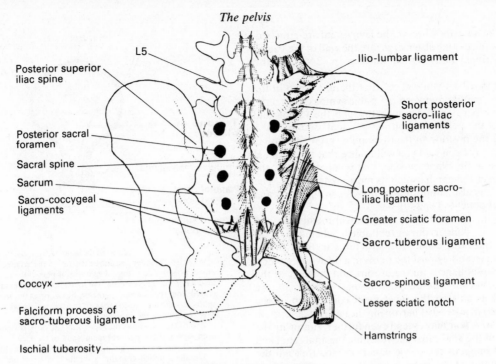

FIG. 12.13. Pelvis from behind.
Note the sacro-iliac, sacrotuberous, and sacrospinous ligaments.

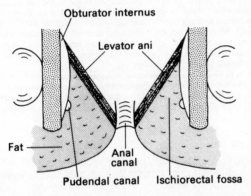

FIG. 12.14. The levator ani muscle, ischiorectal fossa, and pudental canal.

thick obturator internus fascia which covers the pelvic surface of the obturator internus muscle. This is called the white line. There is a well marked cleavage plane between this fascia and the obturator internus itself so the origin of levator ani is rather like the handle of a suitcase, fixed only at each end.

The ischiococcygeus muscle of 'comparative anatomy' is specialized in man so that its dorsal aspect becomes the sacrospinous ligament. The function of this ligament as an accessory ligament of the sacro-iliac joint has already been discussed [p. 61]. It is a specialization associated with the upright posture. The ventral aspect is the coccygeus muscle itself, which merges into the sacrospinous ligament dorsally. It is worth remembering that it is the other two parts of this muscle, that is the pubococcygeus and iliococcygeus which in human anatomy are called the levator ani. This makes it obvious that the last fibres of levator ani will be next to the

first fibres of coccygeus. Levator ani and coccygeus are really the same muscle. Coccygeus is the posterior part of the diaphragm of the floor of the pelvis [FIG. 12.15].

The fibres of levator ani sweep backwards and medially towards the midline. Some of them are attached to the coccyx (recall the tail wagging background of this muscle) and further forward they reach the anococcygeal raphé. Some fibres blend in with the external anal sphincter and others help to form a sling which loops round the anorectal junction from behind. This part of the levator ani is called the **puborectalis**. It is the main factor responsible for the sharp angle between the long axis of the rectum and anal canal when seen from the side [p. 234].

In front of the anal canal and rectum is the 'central' point of the perineum, the **perineal body**. This is a tough nidus of fibrous tissue

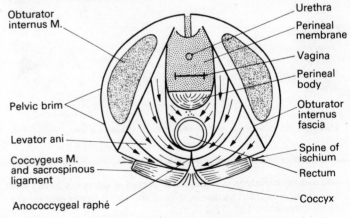

FIG. 12.15. The pelvic diaphragm.

to which most of the anterior fibres of the levator ani are attached. In the female a few fibres may also merge into the wall of the vagina in front of the perineal body. Other muscles [p. 246] are also attached here.

Looking at Figure 12.15 you will see that there is a gap in the muscle fibres in the midline between the pubic symphysis and the central point of the perineum. Through this gap the vagina and urethra pass out of the pelvis through the perineal membrane.

In Figure 12.12 the position of the ischiopubic rami are shown. From what has already been said you will realize that the perineal membrane must lie in the plane outside, that is superficial to, the levator ani, and cover the gap between its most medial fibres. The perineal membrane is also supported by muscles in the deep and superficial perineal pouches [p. 248].

It is easy to see that in the female the perineal membrane and perineal body are in danger during childbirth when the head descends through the birth canal. The structures in front of the vagina may be compressed against the pubis in a difficult delivery, with very serious consequences for the urethra and the neck of the bladder due to ischaemia. Behind the vagina, rupture of the perineal body will mean that the subsequent contraction of the muscles inserted into it, instead of narrowing the gap just described, will actually widen it. A tear may extend even further, rupturing the external sphincter of the anal canal and reaching the anal canal and rectum. When this happens the pelvic viscera escape through the diaphragm of the floor of the pelvis when the patient coughs or stands up. It is a heavy price to pay for the birth of a child.

PELVIC DIAPHRAGM MUSCLES

Levator ani

Origin: 1. Pelvis—posterior surface of pubis
2. Spine of ischium
Insertion: Coccyx and central point of perineum
Nerve supply: Pudendal nerve
Action: Forms floor of pelvic cavity

Its origin is like the handle of a suitcase being attached to bone at its two ends. Between these bony points it fuses with the fascia covering the inner surface of obturator internus. This muscle supports the pelvic organs. Failure to do so results in prolapse of uterus, bladder or rectum.

Coccygeus

Origin: Pelvis—spine of ischium
Insertion: Coccyx and sacrum
Nerve supply: Pudendal nerve
Action: Forms floor of pelvic cavity at back

This muscle is a posterior continuation of levator ani.

THE BLADDER

The urinary bladder lies in the anterior part of the pelvic cavity and is protected in front by the bodies of the pubic bones and by the pubic symphysis [Fig. 12.17].

The bladder receives urine from the kidneys via the **ureters**,

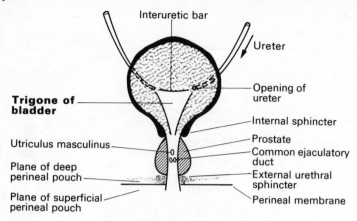

Fig. 12.16. Male bladder from the front.

When the bladder expands the size of the trigone does not increase proportionally. The internal sphincter is part of the circular muscle of the bladder neck. The external sphincter, under voluntary control, is separated from the internal sphincter, by the prostate. In the female the two sphincters are also quite separate structures. In both sexes the sphincters have separate nerve supplies.

This figure also shows the superficial and deep perineal pouches lying on each aspect of the perineal membrane in the male.

The superficial pouch contains the 'roots' of the external genitalia.

The deep pouch contains a number of structures which are passing through it. Note the position of the bladder and prostate to the perineal membrane. Immediately beyond the prostate, the urethra is surrounded by the external sphincter (voluntary). This muscle therefore lies on the upper surface of the perineal membrane and its overall size is co-extensive with the deep pouch. This pouch is enclosed above by the fascia covering the deep surface of the sphincter, which is part of the parietal pelvic fascia.

It is natural to think that the external sphincter urethrae is simply a ring of muscle fibres encircling the urethra, but many of the fibres run from one ischiopubic ramus to the other. They have an important role to play in guarding the midline gap between the fibres of the levator ani, and strengthening the perineal membrane.

The superficial pouch is enclosed on its superficial aspect by a fascial layer which is continuous with the membranous layer of superficial fascia on the anterior abdominal wall. This fascial layer runs over the external genitalia in both sexes and fuses with the posterior border of the perineal membrane.

usually only one from each kidney. The bladder wall consists of smooth muscle fibres, called the **detrusor** muscle whose contraction empties the bladder. The muscle is thickened at the opening of the **urethra**, the tube by which the urine passes from the bladder to the exterior, forming the internal sphincter [Fig. 12.16].

The lining of the bladder is called pseudostratified epithelium. The epithelium gives the impression of being several cells thick (stratified) when the bladder is empty but only one cell thick when it is distended. This type of epithelium is only found in the urinary tract.

The motor nerves which empty the bladder come from the sacral parasympathetic outflow (S2, 3, and 4).

The inner surface of the bladder is covered with small ridges and folds when it is empty but when it is distended the wall appears quite smooth. The bladder has great distensibility and chronic stretching of the wall causes hypertrophy of the muscle wall and an increase in cell multiplication in the mucosa.

The triangular area bounded by the openings of the ureters and the urethra is called the **trigone**. Here the mucous membrane is more firmly fixed to the underlying muscle, and it does not wrinkle when the bladder is empty. Also distension of the bladder wall only stretches the trigone to a limited degree. The trigone is the origin of sensation from the bladder (whether it is full or empty) and also of pain. Thus a bladder stone coming in contact with the trigone is excruciatingly painful. The same stone may not give rise to any pain at all, if it does not come in contact with the trigone. The lithotomy position (cutting for stone position) still used in the operating theatre, commemorates the old practice of removing bladder stones with a knife through the perineum. No patient would have submitted to such an operation, which was carried out without an anaesthetic, had not the pain from the trigone been so unbearable (see Pepys's Diary).

When the bladder is distended, the fundus rises out of the pelvis. It cannot be palpated as such, but its presence can be revealed by percussion. Try this on yourself, percussing just above the pubic symphysis, with the bladder full, and then with the bladder empty. When the bladder is full, the note is dull; when empty the note is resonant because of the gas which is present in the underlying intestine.

In cases of chronic retention of urine, the bladder fundus may reach the umbilicus. The peritoneum is raised up and peeled away from the anterior abdominal wall in such cases. Ultimately the increased pressure in the bladder forces open the bladder sphincters giving rise to 'retention with overflow'.

Note that in babies and small children, the pelvis is relatively small and the bladder lies mostly in the abdomen [see FIG. 11.08, p. 207]. A dull percussion note will be found even when the bladder is relatively empty. The site of the bladder in the young has to be taken into account when planning surgical incisions in the lower abdomen.

The ureters pass through the bladder wall in an oblique direction downwards and medially. The obliquity of the canal (like that of the inguinal canal) helps to prevent reflux of urine up the ureter during micturition. The openings of the ureters are connected by a marked thickening of the submucous tissue, which is the interureteric bar.

This is easily recognized when the bladder is examined with a cystoscope. By travelling along it to left or right the openings of the ureters can be found. Urine passes into the bladder in small squirts at intervals as a result of the peristaltic contractions of the ureter.

Male bladder

In the male the bladder neck lies on the upper surface of the prostate, and the urethra passes downwards from the bladder through the prostate. The bladder neck and prostate are fixed to the pubis by the puboprostatic ligaments [FIG. 12.25].

Above the base of the prostate and behind the posterior surface of the bladder are the ampullae of the two vasa deferentia which lie in juxtaposition on each side of the midline. Lateral to them are the semial vesicles. [FIG. 12.26].

Note that the posterior surface of the prostate can be palpated 'per rectum' ('p.r.' examination). The finger reaches as far as the peritoneum of the rectovesicle pouch and therefore also as far as the ampullae of the vasa and seminal vesicles. Under normal conditions neither of these structures can be identified per rectum because their consistency is similar to that of the surrounding soft parts. If they can be identified individually then this is evidence of disease.

Female bladder

The posterior surface of the bladder in the female [p. 245] is related to the upper part of the vagina, the shallow anterior fornix of the vagina, and the cervix of the uterus. None of the structures referred to in the above section are present in the female.

In the female the neck of the bladder is fixed to the pubis by the pubovesicle ligaments. When the bladder fills, these ligaments ensure that whatever happens to the fundus of the bladder, the neck remains stationary.

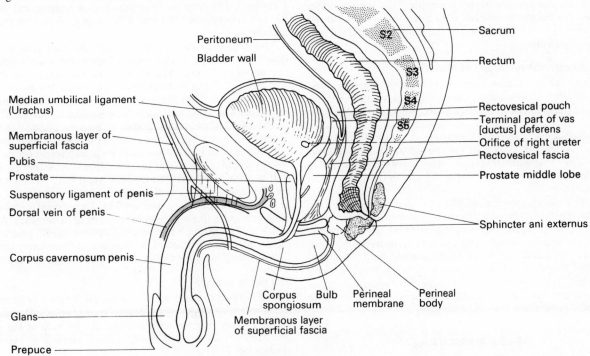

FIG. 12.17. The male pelvis.
The median umbilical ligament, which is attached to the apex of the bladder, represents the remnants of the intra-abdominal part of the allantois (the urachus).

THE BLADDER—SUMMARY

The empty bladder has the shape of a triangular pyramid which becomes more spherical as the bladder fills. In the adult the empty bladder is a pelvic organ covered on its superior surface with peritoneum. This surface is in contact with the coils of small intestine and with the pelvic colon.

In the male this peritoneum is reflected at the back on to the rectum forming the **rectovesical pouch**. In the female it is reflected on to the uterus forming the **uterovesical pouch**.

As the bladder fills it lifts the peritoneum off its superior surface. The bladder then comes in direct contact with the abdominal wall without any intervening peritoneum. Thus a suprapubic drain can be put through the abdominal wall without entering the peritoneal cavity.

In the male the base of the bladder lies in front of the rectum with the seminal vesicles and the vasa deferentia intervening. The neck of the bladder is connected to the prostate which also lies in front of the rectum.

In the female the base of the bladder is attached to the anterior wall of the vagina. The anterior surface of the uterus lies on the bladder. The neck of the bladder lies on the upper layer of the urogenital diaphragm [Fig. 12.29].

The trigone

The trigone of the bladder is an equilateral triangular area with sides of about 2.5 cm. The ureters enter via the ureteric orifices at its two posterior corners (superolateral angles of bladder) whilst the urethra leaves at its anterior corner (inferior angle of bladder). The trigone lies 2 cm behind the pubic symphysis, and is relatively fixed in position.

When viewed through a cystoscope the trigone has a reddish colour and the mucous membrane is smooth. Over the rest of the bladder the mucous membrane is thrown into folds and is yellowish in colour. The mucous membrane of the bladder is covered with transitional epithelium.

Internal and external sphincters of bladder

The bladder has two sphincters, the smooth muscle internal sphincter controlled by the autonomic nervous system, and the striated muscle external sphincter controlled by the voluntary nervous system.

The smooth muscle **internal sphincter** or sphincter vesicae is formed by the smooth muscle of the trigone around the internal urethral orifice. This sphincter is opened by activity in the parasympathetic nervous system and closed by activity in the sympathetic nervous system.

The striated muscle **external sphincter** or sphincter urethrae lies distal to the prostate gland in the male. It surrounds the membranous part of the urethra. With sexual excitement the internal sphincter closes but the external sphincter opens. This prevents the bladder being emptied and thus prevents the seminal fluid being contaminated with urine as it passes along the urethra. It also ensures that the seminal fluid, which largely consists of the secretions of the prostate and seminal vesicles as well as the spermatozoa from the testis, cannot regurgitate into the bladder.

MALE URETHRA

With the penis in the flaccid state the urethra is S-shaped. It is approximately 20 cm in length. The urethra runs from the internal vesical urethral orifice (internal meatus) of the bladder to the external urethral orifice (external meatus) at the tip of the penis [Fig. 12.18]. It is divided into three parts:

1. Prostatic urethra.
2. Membranous urethra.
3. Penile (spongy) urethra.

The prostatic urethra is lined by transitional epithelium (like the bladder). The membranous and penile urethra is lined by stratified columnar epithelium but with numerous mucous glands. The navicular fossa has stratified squamous epithelium.

The prostatic urethra

This is the first part of the urethra. It is the widest and most distensible part. It is about 3 cm long.

The upper part of the floor (posterior wall) of the prostatic urethra has a longitudinal ridge termed the **urethral crest** or **verumontanum** (*verumontanum* (Latin) = mountain ridge). At the summit of this elevation is a small pit about 6 mm deep termed the prostatic utricle (or *utriculus masculinus*). This is the remains of structures which develop into the vagina and uterus in the female.

The ejaculatory ducts open one on each side of the prostatic utricle. The prostate itself sends its secretions into the urethra via 20–30 small ducts which enter along a groove which runs alongside the urethra crest.

The longitudinal elevation produced by the urethral crest gives the urethra in this prostatic region a horseshoe-shaped cross-sectional area.

The membranous urethra

This is the second part of the urethra where it is passing through the deep perineal pouch. It is the least distensible part of the urethra with the possible exception of the external meatus. It is just over 1 cm long. The term *membranous* does not mean that the urethra itself is membranous but that it is about to perforate the perineal membrane.

This region is surrounded by the external sphincter muscle (sphincter urethrae). This external sphincter muscle lies between the upper and lower fascial layers of the urogenital diaphragm which fuse together at the front and back to form a water-tight deep perineal pouch. Thus the extravasation of urine following damage to the membranous urethra will be limited to this small pouch.

The two bulbo-urethra glands (Cowper's glands) are situated behind the lowest part of the sphincter urethrae. Their ducts pierce the lower fascial layer of the urogenital diaphragm (perineal membrane) to enter the penile urethra [Figs. 12.18 and 12.25].

It is always necessary to exercise great care when entering the membranous urethra with an instrument such as a cystoscope because in this place it is anchored by the perineal membrane, and trauma is more likely than in other parts of the urethra. A catheter which will go through the external meatus should pass easily into the bladder (constriction of the external sphincter may temporarily prevent this). An abnormal narrowing of the urethra is a stricture.

The penile urethra

This third part of the urethra is about 16 cm long. It is surrounded along its entire length by the corpus spongiosum and is also known as the spongy or cavernous urethra. In the **bulb of the penis** the direction of the urethra changes from vertical to more or less horizontal.

The penile urethra has two dilatations. At its commencement in the bulb the dilatation is termed the **intrabulbar fossa**. The second

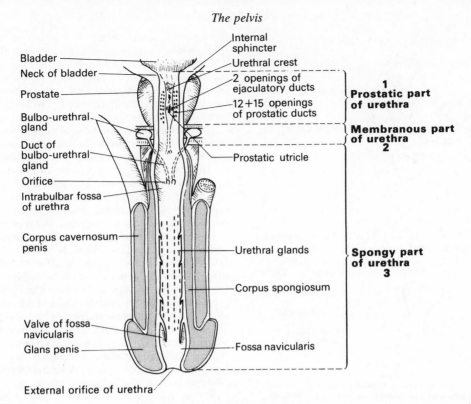

Internal sphincter

Urethral crest

Bladder

Neck of bladder

Prostate

2 openings of ejaculatory ducts

1 Prostatic part of urethra

12+15 openings of prostatic ducts

Bulbo-urethral gland

Membranous part of urethra 2

Duct of bulbo-urethral gland

Prostatic utricle

Orifice

Intrabulbar fossa of urethra

Corpus cavernosum penis

Urethral glands

Spongy part of urethra 3

Corpus spongiosum

Valve of fossa navicularis

Glans penis

Fossa navicularis

External orifice of urethra

FIG. 12.18. The penis.

dilatation is the **navicular fossa** which is close to the external urethral orifice.

The floor of the penile urethra has numerous openings of urethral glands which may catch a bougie or catheter as it is being inserted. These are absent in the roof of the penile urethra and hence the catheter should be kept to this part of the urethra during insertion.

In the bulb of the penis the direction of the urethra changes from vertical to more or less horizontal. The mucous membrane contains a large number of mucous glands which discharge at short intervals. The lumen is also interrupted by pocket like 'lacunae' which may cause problems when instruments are introduced.

The numerous mucous secreting glands associated with the urethra including the bulbo-urethral glands, provide an ideal culture medium for various sexually transmitted disease organisms.

The **external urethral orifice** or external meatus is a vertical slit. It is the narrowest and least distensible part of the urethra.

The urethra is considered further on page 240.

Blood supply
Inferior vesical artery
Internal pudendal artery (branches of)

The venous drainage is into the prostatic venous plexus and into the internal pudendal vein.

Nerve supply
Pudendal nerve

Lymphatic drainage
The prostatic urethra and membranous urethra drain into the internal iliac nodes.

The penile part drains into the superficial inguinal nodes.

FEMALE URETHRA

The female urethra is about 4 cm long. It runs from the internal urethral orifice of the bladder along the anterior vaginal wall to the external urethral orifice in the vestibule between the clitoris and vaginal opening [FIG. 12.29].

Since there is no prostate in the female, there is no prostatic part to the urethra. From the neck of the bladder the urethra passes directly through the deep perineal pouch where it is surrounded by the voluntary striated muscle external sphincter. There is also a smooth muscle internal sphincter surrounding the urethral orifice of the bladder as in the male.

THE RECTUM

The rectum is about 12 cm in length. It commences at the termination of the sigmoid colon opposite S3. The mesentery of the sigmoid colon disappears at this point but the rectum is still covered by peritoneum in front and at the sides. The middle part of the rectum only has peritoneum on its anterior surface. The lowest part has no peritoneal covering. The taeniae coli are not found in the rectum nor are there any appendices epiploicae. Most of the longitudinal muscle runs on the anterior and posterior surface and is not so well represented at the sides.

The rectum follows the concavity of the sacrum and then joins the anal canal at an angle because the anorectal junction is pulled forward by the puborectalis muscle sling [FIG. 12.19].

When viewed from the front the lumen of the rectum does not run, as its name would suggest in a vertical **straight** line. Instead its course is extremely tortuous. Its course is interrupted by three transverse folds (valves of Houston). The middle one is the largest

a)

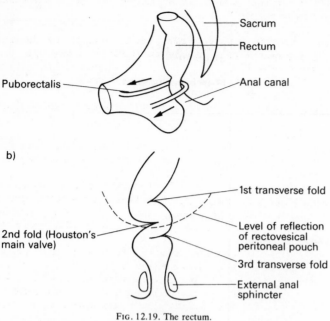

b)

FIG. 12.19. The rectum.
(a) From the side.
The puborectalis is a sling formed by the most anterior fibres of levator
ani which loop around the anorectal junction and pull the gut forward.
(b) From the front.

and is at the level of the reflection of peritoneum forwards at the rectovesical or rectouterine pouch.

Note that the first fold is where you would expect to find it, that is on the left because the rectum is approached from the left by the sigmoid colon. These folds typically contain all the layers of wall of the rectum including the longitudinal coat.

It is essential to understand the complex curves of the rectum and anal canal before an instrument, such as a proctoscope, is used.

Anal canal

The anal canal is about 3 cm long. The upper one-third has mucous membrane which is similar to that of the rectum. The lower part has stratified squamous epithelium rather like epidermis. The upper part is insensitive except to stretch. The lower part is sensitive to the same sensations as the skin [FIG. 12.20].

The internal surface of the anal canal is marked by longitudinally running anal columns in which the terminal branches of the superior rectal artery run with their companion veins. The artery is the direct continuation of the inferior mesenteric artery. The veins drain into the portal vein. The superior rectal artery anastomoses with the inferior rectal artery in the anal canal. The veins anastomose with the inferior rectal veins. This is an important region where portal–systemic venous anastomosis takes place [p. 220].

Distention of the superior rectal veins gives rise to haemorrhoids (piles). Since the veins follow the arteries their position is determined by the way the superior rectal artery (the direct continuation of the inferior mesenteric) is distributed into three branches, one to the left of the rectum and two on the right in the anterior and posterior positions. Clinicians think of the anal canal–rectum like a clock face with the main arteries and veins in positions corresponding to 3 o' clock, 7 o' clock, and 11 o' clock [FIG. 12.20(b)]. This is where you expect the piles to be. In clinical practice, in examining the anal canal and rectum, the patient lies on one side, or in the knee–elbow position. The clock face is therefore rotated and some students become confused as a result.

Morphologically the lower two-thirds of the anal canal wall is an 'in pushing' of the 'surface skin' and has the same kind of nerve supply as the skin itself. The upper part of the anal canal has the same kind of nerve supply as the rest of the intestine.

The anal canal is a complex part of the gut with its dual origin. If it happens that the two sections of the canal fail to unite, then the child is born with an 'imperforate anus'. There is thus an intestinal obstruction which will be fatal unless corrected surgically.

In the normal adult the junction 'line' between rectum (the true intestine) and the proctodeum (surface in-pushing) is marked by

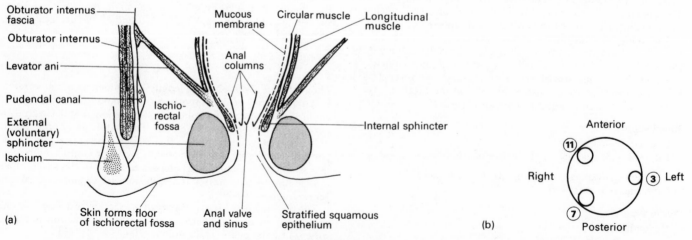

FIG. 12.20. The anal canal.
The anal canal is about 3–4 cm in length. The anal columns contains a terminal branch of the superior rectal artery and accompanying vein. The inferior rectal artery supplies the lower part of the anal canal and anastomoses with the superior rectal artery. The three largest vessels are distributed one to the left (3 o'clock) and two on the right (7 o'clock and 11 o'clock) (b). Since the veins follow the arteries these are the common sites for piles.
The nerve supply of the skin and mucous membrane switches from autonomic to somatic nerves, similar to the switch in nerve supply in passing from the internal to the external sphincters.

the anal valves. Above the anal valves the veins can be injected painlessly in patients with 'piles'. Injection into the anal canal below this level is distressingly painful.

An abscess in the ischiorectal fossa is a common condition. In such cases the fat (which has a poor blood supply and therefore is not able to resist infection) is replaced by a great volume of pus. The infection may come from the skin, but more frequently it comes from the mucous glands of the anal canal whose secretory cells are embedded in the fat of the ischiorectal fossa, just as the secretory parts of the sweat glands lie in the fat of the superficial fascia in other parts of the body.

When an abscess spreads into the anterior recess of the ischiorectal fossa, it may produce retention of urine because of its close proximity to the external sphincter of the bladder.

ISCHIORECTAL FOSSA

The ischiorectal fossa is a space on each side between the rectum and the ischium [FIG. 12.14, p. 229]. It lies between the levator ani muscle (which is inserted below in the anal canal between the internal and external sphincters) and the vertical ischial wall which is covered by obturator internus. The fossa is wedge-shaped and has the skin to the side of the anus as its base. The thin end of the wedge is the junction of the levator ani to the obturator internus fascia [FIG. 12.20].

Medial wall:
Fascia over levator ani and external sphincter of anal canal.
Lateral wall:
Fascia over obturator internus and ischial tuberosity. The obturator muscle is so called because it closes the obturator foramen of the pelvis (*obturare* (Latin) = to stop up).
Posterior wall:
Gluteus maximus and sacrotuberous ligament.
Anterior wall:
Deep layer of superficial perineal fascia winding round superficial transverse perinei muscle and deep layer of urogenital diaphragm [FIG. 12.21].

The fossa is crossed by the inferior rectal artery, vein and nerves which run from the lateral wall to the medial wall. The fossa itself contains fat and bands of fascia.

The **pudendal canal** is formed in the lateral wall of the fossa about 4 cm above the ischial tuberosity by the separation of the layers of the fascia over the obturator internus. It contains the pudendal nerves and internal pudendal blood vessels. Its older name was Alcock's canal.

Note that the **ischiorectal fossa** does not lie as its name suggests between ischium and rectum, but between the muscles which cover them. These are the obturator internus and the levator ani. Furthermore, this fossa lies between the ischium and the **anal canal**, not the rectum. The term *ischiorectal* was introduced when the term 'rectum' included the anal canal which was then considered to be the third part of the rectum.

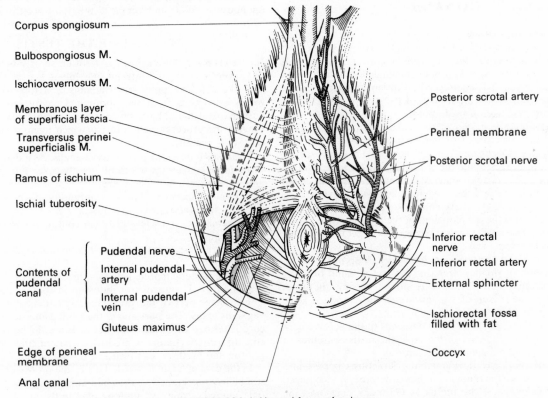

FIG. 12.21. Male ischiorectal fossa and perineum.
On the right side the skin only has been removed. On the left the fat of the ischiorectal fossa has also been removed and the levator ani is seen. In the perineum the membranous layer of superficial fascia (transparent) is suggested and the muscles covering the crus and bulb can be seen through it. [Compare female. FIG. 12.31.]

The fossa is wedge-shaped, with the 'white line' along which the levator ani joins the obturator internus fascia, forming the thin edge of the wedge, while the base of the wedge is simply the skin between the anus and the ischial tuberosity.

In front it is bounded by the posterior edge of the perineal membrane, to which the membranous layer of the superficial fascia of the anterior abdominal wall is attached [FIG. 12.21]. The attachment of the levator ani in front to the body of the pubis carries forwards above the deep perineal pouch, producing the anterior recess of the ischiorectal fossa. Behind, the fossa is delineated by gluteus maximus and also by the sacrotuberous ligament. The two ischiorectal fossae communicate with each other behind the anal canal.

The shape of the ischiorectal fossae, because they are filled with fat which is liquid at body temperature, can be changed from moment to moment when the size and shape of the rectum and anal canal change in defaecation, and when the size of the birth canal changes in parturition.

The vessels and nerves which enter the lateral wall of the ischiorectal fossa do so via the **pudendal canal**. These are the **nerve to obturator internus**, the **internal pudendal artery**, and the **pudendal nerve**. They run in a tunnel in the obturator internus fascia. The branches to the sphincters of the anal canal come off these arteries and veins almost immediately and travel through the fat of the ischiorectal fossa to the external anal sphincter (under voluntary control), to the skin around the anus, and to the lining of the lower two-thirds of the anal canal.

GONADS

Embryological considerations

Every cell in the body is derived from a single fertilized ovum. The genetical information in the daughter cells is replicated at each cell division and is present thereafter in every single cell in the body. This means that as cell division proceeds, differentiation takes place using information from different parts of the genetic code so that different organs and tissues are formed.

In early foetal life, there is no anatomical difference between the sexes and the same structures are common to both. As development continues the 44 + XY chromosome pattern of the male causes the gonad to develop as a testis and the urogenital system to develop as a male. The 44 + XX chromosomes of the female cause the gonad to develop as an ovary and the urogenital system to develop as a female. It is the presence of the Y chromosome that appears to be the decisive factor for the development of the male characteristics, because without it (as in Turner's syndrome, 44 + XO) the female characteristics develop.

Both the ovary and the testis are formed high up in the abdomen in the region of the suprarenal gland and descend bringing their blood supply with them. It is not surprising therefore that tumours in the region of the suprarenal may produce large amounts of sex hormones.

In the case of the female the ovary descends to the pelvis where the descent stops. In the case of the male the testis continues to descend via the inguinal canal to the scrotum.

Remnants of the early developmental structures persist into adult life. For instance the remnant of the vas deferens is recognized in the broad ligament of the uterus as Gartner's duct, and the remnant of the uterus is present in the male as the utriculus masculinus.

The gubernaculum [p. 198] is the structure which guides the gonad in its descent. In the male it is a fibrous cord which extends from the foetal testis down the inguinal canal to the scrotum. In the female it is represented by the ligament of the ovary and the round ligament of the uterus which runs through the inguinal canal to the labia majora which are the female homologue of the scrotum.

Wolffian ducts or mesonephric ducts

In early foetal life the primitive kidney consists of the **mesonephros**. Remnants remain but the definitive kidney develops from a different structure, the **metanephros**. The mesonephros drains into the future bladder by the mesonephric or Wolffian duct.

Müllerian ducts or paramesonephric ducts

These develop as a pair of embryonic ducts on the external surface of the Wolffian ducts.

In the female they become the uterine tubes, uterus and upper two-thirds of the vagina. The lower extremity becomes the hymen.

In the male all that remains is the prostatic utricle (which is the homologue of the female vagina and uterus) and the appendices testis.

Inguinal canal

The inguinal canal [p. 197] is about 4 cm long. It extends from the **deep** (internal) **ring** which is a circular opening in the transversalis fascia about 1 cm above the midinguinal point to the **superficial** (external) **ring** which is an opening in the external oblique aponeurosis above and lateral to the pubic crest.

DESCENT OF THE TESTIS

The testis normally reaches the scrotum shortly before birth. At the seventh month of intrauterine life it lies at the deep ring [FIG. 12.22]. The processus vaginalis is a tube of peritoneum which evaginates from the abdominal cavity into the scrotum during foetal life. The term *processus vaginalis* is Latin for the evaginating process or the process which forms a sheath (*vagina* (Latin) = sheath).

The testis follows the processus vaginalis. As it does so it picks up three fascial coverings which ultimately cover the spermatic cord and line the scrotum. These are:

1. Internal spermatic fascia (or infundibuliform fascia) continuous with the transversalis fascia.
2. Cremasteric muscle and fascia continuous with the internal oblique.
3. External spermatic fascia continuous with the external oblique.

The cremasteric muscle acts as a thermostat for the testis, withdrawing it back towards the abdomen in cold weather.

The lower part of the processus vaginalis becomes the tunica vaginalis testis. The part above the testis alongside the spermatic cord is termed the funicular process. It usually becomes a fibrous cord (*funiculus* (Latin) = a cord). A patent funicular process of peritoneum predisposes towards a scrotal hernia.

In the case of an undescended testis, the processus vaginalis may still be in connection with the abdominal cavity leading to a potential inguinal hernia. An **undescended testis** may be found at any point along its route. An **ectopic testis**, on the other hand, is one that has migrated to the wrong place such as the perineum after it has passed through the inguinal canal.

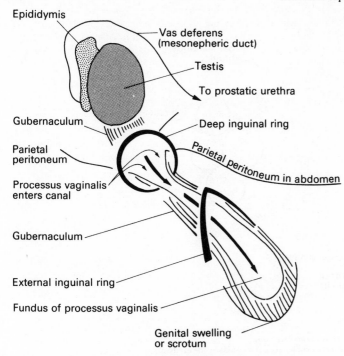

FIG. 12.22. Testis at the inguinal canal.
During month 7 of intra-uterine life, the testis lies near the deep inguinal ring. The processus vaginalis and gubernaculum have already reached the floor of the scrotum. The testis passes through the inguinal canal behind (not through) the processus vaginalis. The inguinal canal has already formed around the processus and gubernaculum which are the harbingers of the testis. In the female the processus, the gubernaculum, and the inguinal canal undergo the same early developmental stages. The vestige of the processus is unimportant, but the gubernaculum gives rise to the round ligament of the uterus.

THE MALE REPRODUCTIVE SYSTEM

The male gonad or **testis** lies in the scrotum [FIG. 12.23]. It is oval in shape with an upper and lower pole. It is about 3 cm long. The epididymis lies along its posterior border. The testis [FIG. 12.24] has three coats:

1. **Tunica vaginalis**. This is a serous membrane with a visceral layer which covers the testis and epididymis and a parietal layer lining the scrotum. It is derived from the processus vaginalis.

2. **Tunica albuginea**. This consists of non-elastic white fibrous tissue surrounding the testis deep to the tunica vaginalis. It sends a septum into the testis from which partitions run which divide the testis up into many compartments. The convoluted tubules in these compartments make the spermatozoa.

Because the tunica albuginea is non-distensible, haemorrhages into the testis may cause pressure symptoms leading to atrophy.

3. **Tunica vasculosa**. This is the blood vessel layer which is inside the tunica albuginea and lines its partitions.

The skin of the **scrotum** contains the dartos muscle and it also fuses on its deep surface with the three coverings of the 'cord' and with the parietal layer of the tunica vaginalis testis. A midline septum separates the scrotum into two separate compartments.

Over its whole surface the testis is covered by the thick layer of

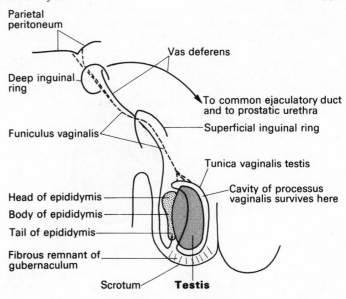

FIG. 12.23. Testis in definitive position.
The testis and epididymis having traversed the inguinal canal shortly before birth now, lie in their definitive position. The processus vaginalis is obliterated except where it surrounds the testis forming the tunica vaginalis.

white fibrous tissue the tunica albuginea. The visceral layer of the tunica vaginalis fuses with the tunica albuginea and covers the testis at the front and sides but not posteriorly.

Behind the testis lies the epididymis [FIG. 12.25]. This has a head, a body, and a tail. It is partly separated from the testis on the lateral side by an inpushing of the tunica to form the sinus of the epididymis. The vas deferens can be seen as a tortuous rather narrow duct on the medial side of the tail of the epididymis. Once the vas leaves the epididymis it becomes straight and increases considerably in thickness.

Sperm formation takes place in the seminiferous tubules of the testis. There are about 500–750 tubules in each testis, packaged two or three at a time by fibrous partitions. The tubules meet near the posterior part of the testis and then run into 6–8 narrow lengthy and convoluted tubules which form the head of the epididymis. The body and tail of the epididymis consists of the convolutions of the vas deferens. We have no clear understanding of the functional changes which occur while the sperm are propelled passively along the complex lumen of the epididymis.

The three blood vessels which lie in the spermatic cord [p. 239], that is the testicular artery (from the aorta), the cremasteric artery from the inferior epigastric, and the artery to the vas from the inferior vesicle, all anastomose with each other when they reach the scrotum.

The testis contains **interstitial cells** which lie in the recesses between the seminiferous tubules. These secrete the male sex hormone (testosterone) into the bloodstream.

The testis, epididymis and vas are all easily identifiable in the scrotum by palpation.

Note that the testis has played an important role before birth in initiating certain changes which take place in the development of the normal male. At birth the testis has already been put to the test as far as hormone secretion is concerned. The testis and scrotum

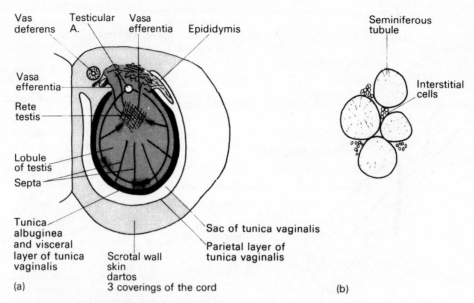

Fig. 12.24. Horizontal section through testis and scrotum.

The testis can move fairly freely within the scrotum because it is surrounded on three sides by the cavity of the processus vaginalis.

The smooth muscle in the wall of the scrotum the dartos, is supplied by the sympathetic. Its reflex contractions in response to cold are adjuvant to the simultaneous contraction of the cremaster muscle. L1 supplies both muscles.

The seminiferous tubules are responsible for the formation of sperm which finally mature outside the testis. The interstitial cells lie in the chinks between the tubules. They secrete male hormones into the bloodstream.

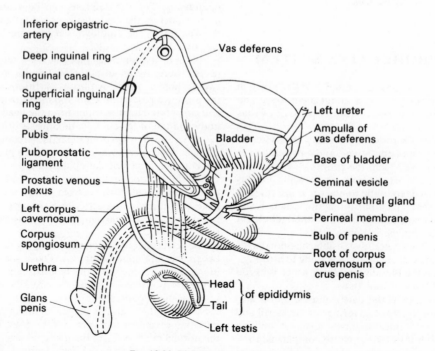

Fig. 12.25. Diagram of male genital organs.

hypertrophy at puberty along with the development of other secondary sexual characteristics.

Hydrocele
A hydrocele is the accumulation of fluid in the tunica vaginalis sac of the testis. The fluid is removed by tapping.

The epididymis
The epididymis, which is a long coiled tube, rather confusingly starts at the top of the testis where it receives the efferent ducts of the testis and finishes at the bottom where it leads into the vas (ductus) deferens. The start of the epididymis is termed the head or globus major. The middle part is termed the body, and the lower part is the tail or globus minor. The head of the epididymis consists of the coiled vasa efferentia of the testis, while the body and tail are formed by a coiled tube which eventually becomes the vas deferens.

Developmental remnants
A small appendage at the upper pole of the testis represents the remains of the Müllerian duct. It is sometimes referred to as the hydatid of Morgagni.

Part of the Wolffian duct system persists as convoluted tubules known as the paradidymis or organ of Giraldès. These are the remains of the paragenital tubules of the mesonephros.

Blood supply to testis
The **testicular artery** [p. 219] arises from the abdominal aorta. It reaches the testis via the spermatic cord and enters the testis at its upper pole medial to the head of the epididymis.

The venous drainage is via the pampiniform plexus of veins which join to form three venous trunks surrounding the vas in the inguinal canal. At the inner ring these join to form the testicular (spermatic) vein.

Note that although both testicular arteries arise from the aorta and the right testicular vein runs usually to the inferior vena cava, the left testicular vein runs to the left renal vein which in turn runs to the inferior vena cava. This is not surprising since the vena cava lies to the right of the midline. There is no vena cava on the left side.

The **lymphatics** draining the testis run with the testicular artery to para-aortic lymph nodes. Tumours from the testis metastasize directly to these lymph nodes high up in the posterior abdominal

Ductus deferns (vas deferens)
The ductus (vas) deferens, or **vas** as it is commonly called, starts at the lower end of the epididymis. It runs upwards to enter the spermatic cord with the testicular blood vessels, nerves, and lymphatics [Fig. 12.25]. It can be felt as a hard round cord at the top of the testis in the scrotum.

It enters the inguinal canal via the superficial ring and enters the abdomen via the deep ring. Here it leaves the testicular artery and runs down the side wall of the pelvis and across the pelvic floor to reach the prostate where it is joined by the duct from the seminal vesicle to form the **common ejaculatory** duct in the substance of the prostate.

It will be seen from Figure 12.25 that the vas passes in between the ureter and the bladder.

SPERMATIC CORD
The spermatic cord [p. 197] which runs from the deep ring to the testis has three coverings:

1. External spermatic fascia. This is derived from the external oblique.
2. Cremasteric fascia and muscle. This is derived from the internal oblique.
3. Internal spermatic fascia, also known as infundibuliform fascia. This is derived from the transversalis fascia.

Contents of the spermatic cord
1. Ductus deferens (vas deferens).
2. Testicular artery (branch of abdominal aorta).
3. Cremasteric artery (branch of inferior epigastric artery).
4. Artery to the vas (branch of inferior vesical artery).
5. Pampiniform plexus of veins.
6. Genital branch of ilio-inguinal nerve.
7. Genital branch of genitofemoral nerve.
8. Autonomic nerves.
9. Lymphatics.
10. Funiculus vaginalis.

The spermatic cord is about half the diameter of the little finger. Its principal constituent is the ductus (**vas**) deferens. This commences on the medial side of the epididymis, which lies immediately behind the testis. The spermatic cord can easily be palpated with finger and thumb just below the superficial inguinal ring, through the skin of the scrotum. The vas deferens lies in a posteromedial position and is easily distinguishable because of its very hard 'whip cord' like consistency. It can easily be ligatured in this region.

Sperm which develop in the testis find their way via the vas deferens to the prostatic urethra when, during ejaculation, they mix with the secretions of the prostate and seminal vesicles to form the seminal fluid.

The vas is supplied by its own artery which is derived from the posterior inferior vesical artery. The cremaster muscle is also supplied by its own branch from the inferior epigastric artery and the testis is supplied as has been seen directly from the aorta at the level of L2. So altogether there are three arteries in the cord.

The veins draining the contents of the scrotum (not just the testis) form a very complex meshwork called the **pampiniform plexus** which is an important constituent of the cord. The plexus is thought to provide a heat-exchange mechanism which ensures that the arterial blood temperature falls appropriately as it passes down the cord before it reaches the testis. The plexus runs into a simple straight vessel once the abdominal cavity has been reached. It is well known that in an undescended testis, sperm formation hardly takes place. It seems that the higher temperature of the abdominal cavity compared with that of the scrotum, is not conducive to spermatogenesis. Even in the undescended testis, interstitial cells, which lie between the seminiferous tubules, function normally and secrete testosterone.

Sometimes the testis, having passed through the inguinal canal, is deflected into an abnormal or ectopic position either alongside the shaft of the penis or in the thigh, abdominal wall or perineum. This is an **ectopic testis**. When a lump is being diagnosed in these regions always look to see whether there are two testes in the scrotum.

Seminal vesicles
The seminal vesicles [Fig. 12.25] are coiled tubes behind the base of the bladder and above the prostate. The apex of each seminal vesicle is covered by peritoneum [Fig. 12.26].

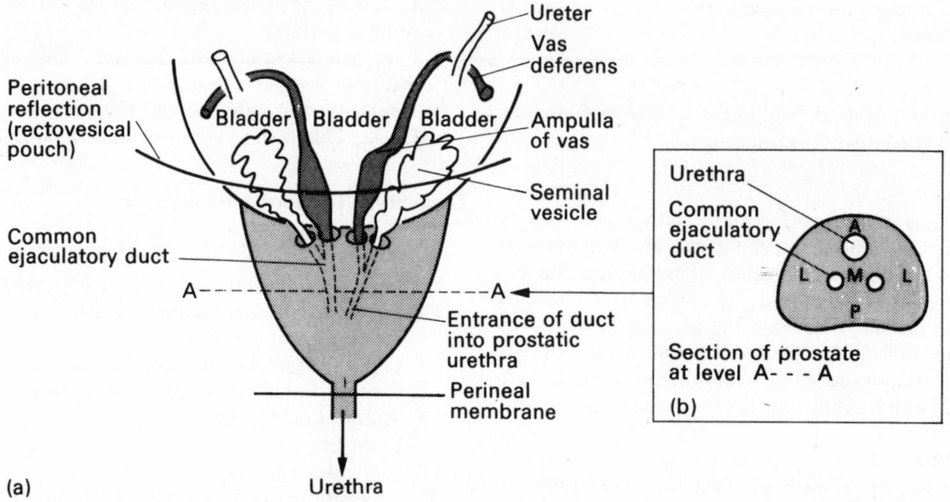

Fig. 12.26. Base of bladder and prostate from behind (a).
The structures shown in the diagram can be palpated by rectal examination. The reflection of peritoneum at the rectovesical pouch is just within reach. The common ejaculatory duct is buried throughout its length in the substance of the prostate, as are its two tributaries.
The middle lobe of the prostate (b) lies between the urethra and the common ejaculatory ducts. It is easy to see that enlargement here, which may not affect the overall size of the prostate, can easily obstruct the urethra. A=anterior, L=lateral, P=posterior, M=middle lobes of prostate.

Common ejaculatory ducts

The common ejaculatory duct is formed on each side by the duct of the seminal vesicle joining the vas deferens. The combined duct runs for a distance of 2 cm through the prostate to open into the prostatic urethra.

Prostate

The prostate is considered anatomically to have the shape of an inverted cone with its *base* in contact with the bladder and its *apex* resting on the urogenital diaphragm. (Actually its shape shows a closer resemblance to that of a large chestnut.) The front of the prostate is joined to the pubic symphysis by the **puboprostatic ligaments**.

The urethra runs for a distance of 3 cm through the prostate from the base to a point just above the apex. The common ejaculatory ducts which have entered the prostatic tissue at its base opens into the urethra at the side of the prostatic utricle (uterus masculinus) [Figs. 12.16 and 12.18].

The prostate is said to have five 'lobes' but they are not separated in any way from each other. The middle lobe is bounded by the urethra and common ducts. The other 'lobes' are situated as shown in Figure 12.26(b).

Enlargement, especially of the middle lobe of the prostate, quickly obstructs the urethra, and gives rise to retention of urine. It is a very common condition in elderly men.

During ejaculation the circular smooth muscle around the neck of the bladder (internal sphincter) closes preventing seminal fluid from being discharged into the bladder by the vigorous contraction of the bulbospongiosus muscle.

The penis

The penis consists of three cylindrical bodies of erectile tissue, the two **corpora cavernosa** and the **corpus spongiosum**† which transmits the urethra. The erect state is the 'anatomical position' hence the corpora cavernosa lie dorsal to the corpus spongiosum and the prominent single median vein is termed the dorsal vein and the paired accompanying arteries and nerves (on each side) the dorsal arteries and veins of the penis. Erection of the penis is brought about by the sacral parasympathetic outflow (S2, 3, 4). For this reason these nerves were previously known as the nervi erigentes (the nerves that cause an erection). The term 'corpus spongiosum' is only appreciated when the erectile tissue around the urethra and glans is compared during erection with the very 'solid' feel of the corpus cavernosum.

Corpora cavernosa

The corpora cavernosa consists of a network of fibrous tissue which becomes filled with blood during erection of the penis. The two corpora cavernosa lie next to each other with an incomplete septum between. Each has a central artery: the artery of the corpus cavernosum, which is the continuation of the deep branch of the internal pudendal artery.

The corpora cavernosa separate posteriorly to form the two crura which are attached to the ischiopubic rami of the pelvis. The crura are covered by the ischiocavernosus muscle. They provide the penis with a secure attachment to the pelvis!

Corpus spongiosum

The corpus spongiosum is expanded to form the **glans penis** (glans (Latin) = acorn) anteriorly and the bulb of the penis posteriorly.

† Remember that *corpus* (Latin) is neuter hence spongios*um*. Its pleural is *corpora* hence the two corpor*a* cavernos*a*. The muscle (masculine) is, however, bulbospongios*us* or ischiocavernos*us*.

The bulb is covered by the bulbospongiosus muscle. The skin of the penis forms a fold, the **prepuce**, which covers the glans for a variable distance. The operation of circumcision removes the greater part of this fold.

The bulbospongiosus muscle runs from the bulb and has many fibres closely associated with the corpus spongiosum which surrounds the urethra. This muscle has an important role in propelling semen along the urethra during ejaculation. The 'bulbospongiosus reflex' is elicited by stimulation of the glans penis and this results in reflex contraction of the bulbospongiosus muscle. The reflex requires the integrity of the sensory and motor branches of the pudendal nerves and of the segments S2, 3, and 4 in the spinal cord.

Pudendal nerves and blood vessels

The principal nerves and blood vessels in this region are the pudendal nerve and internal pudendal arteries (Latin *pudeo* = I am ashamed; in polite society the genitalia were once called 'the pudenda' whence the adjective 'pudendal').

The pudendal nerve (S2, 3, 4) leaves the pelvic cavity with the internal pudendal artery and the nerve to obturator internus by passing through the great sciatic notch just above the ischial spine [FIG. 12.27]. They all return immediately back into the pelvis below the ischial spine by passing into the fascia of the obturator internus in the lateral wall of the ischiorectal fossa. The nerve to obturator internus supplies that muscle. The pudendal nerve gives off branches which run through the ischiorectal fossa to supply the external anal sphincter. Faecal continence depends on the integrity of this nerve. Cutaneous branches also pass into the ischiorectal

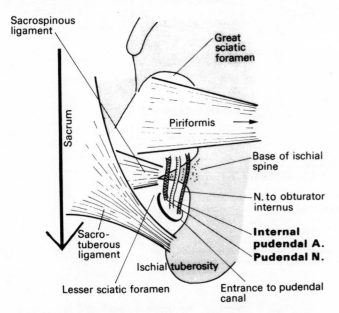

FIG. 12.27. The internal pudendal artery and pudendal nerve leaving and re-entering the pelvis.

This diagram shows the internal pudendal artery and its companion pudendal nerve leaving the pelvis via the great sciatic foramen below piriformis, passing behind the ischial spine and sacrospinous ligament and re-entering the pelvis via the lesser sciatic foramen. They immediately enter the pudendal canal and lie in the lateral wall of the ischiorectal fossa. The nerve to obturator internus accompanies them in the obvious position that is the most lateral position.

Many other structures, including the sciatic nerve, leave the pelvis above or below piriformis [p. 124]. They are all omitted for simplicity here.

fossa on their way to supply the skin of the scrotum or the labia in the female.

The main destinations of the pudendal artery, as its name indicates, are the bulb of the penis and the two crura. The most direct route from the spine of the ischium to the bulb or the crura is through the deep perineal pouch and through the perineal membrane [FIG. 12.21, p. 235]. On each side the pudendal artery gives off the artery to the bulb and then divides into the deep and dorsal arteries. The deep artery pierces the perineal membrane while the latter travels close to the midline and escapes from the pelvis into the dorsum of the penis just below the pubic symphysis. This artery lies superficially in relationship to the corpus cavernosum. The final branches of the internal pudendal artery are the deep and dorsal arteries, but the antithesis is clearer when it is realized that they come to lie deep (in the corpus cavernosum) and superficially but dorsal in position. These vessels are accompanied by the autonomic nerve fibres supplying the erectile tissue. The dorsal arteries are accompanied by the large dorsal nerve of the penis. The dorsal vein of the penis lies in the midline, running just below the pubic symphysis to the prostatic venous plexus.

The pudendal nerves are joined by parasympathetic branches from the sacral outflow (S2, 3, 4) which, as has been seen, cause vasodilatation in the erectile tissue when they are stimulated. Their course lies within the pelvic cavity and they only join the pudendal nerves close to the target organs.

THE FEMALE REPRODUCTIVE SYSTEM

The **broad ligament** [FIG. 12.28], which is really just the mesentery of the uterus and uterine tubes, runs from the side walls of the pelvis to the side of the uterus. It divides the female pelvis into an anterior and posterior compartment. The broad ligament is attached to the pelvis immediately in front of the ureters.

Because of the anteverted position of the uterus the broad ligament, like the uterus itself, lies more or less in the horizontal plane. It consists of two layers of peritoneum which are continuous with each other over the free edge which contains the uterine tube. The upper (often said to be posterior) layer is continuous with the peritoneum which lines the pelvic cavity on each side of the midline and the lower (often said to be anterior) is continuous with the pelvic lining on each side of the bladder.

The **ovary** lies behind the broad ligament. The ovary is often said to lie 'in' the broad ligaments. This is not quite true because the ovary projects from its upper surface and is not actually covered by peritoneum. If it were ovulation would not be possible.

Different parts of the broad ligament are given separate and at first rather confusing names.

The **uterine tube** (Fallopian tube) lies at the upper border of the broad ligament. An alternative name for the uterine tube is the **salpinx**, hence the term **salpingitis** for inflammation of the uterine tube.

The outer quarter of the broad ligament, carrying the ovarian artery and vein, is termed the **suspensory ligament of the ovary**.

The short mesentery which runs from the back of the broad ligament to the ovary is termed the **mesovarium**. The part of the broad ligament below the mesovarium is termed the **mesometrium**.

The part of the broad ligament between the uterine tube and the ovary is termed the **mesosalpinx**.

The ligament connecting the medial pole of the ovary to the

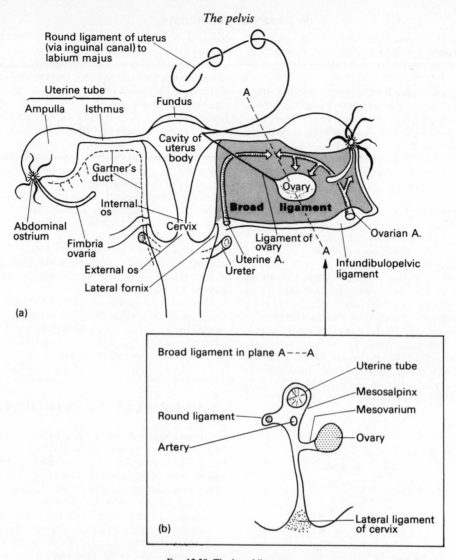

Fig. 12.28. The broad ligament.

(a) The fold of peritoneal mesentery which encloses the uterus and uterine tube is called the broad
      ligament of the uterus. It also includes the ovary, as well as the blood supply of all the contained
      organs and other structures.
          The ovary itself is not covered by peritoneum as it is a special case. This arrangement allows
      the ovum at ovulation to escape from the ovary and reach the abdominal ostium of the uterine
      tube.
      Note the close relationship of the ureter to the lateral fornix of the vagina, and to the uterine artery.
      Infundibulopelvic ligament is alternative name of suspensory ligament of the ovary.

(b) In section the base of the broad ligament encloses the lateral ligament of the cervix, which helps
      to join the cervix to the side wall of the pelvis and to keep the cervix in position.

fundus of the uterus is termed the **ligament of the ovary**. It is about 3 cm long and shows as a ridge on the posterior layer of the broad ligament. It represents the remains of the upper part of the gubernaculum.

The lower part of the gubernaculum is represented by the **round ligament of the uterus**. It is continuous with the ligament of the ovary in the fundus and runs from the uterus to the side wall of the pelvis forming a ridge on the anterior surface of the broad ligament. It crosses the obturator artery, vein, and nerves and the external iliac artery and vein and enters the inguinal canal at the deep inguinal ring. It traverses the inguinal canal and is eventually attached to the labium majus.

The broad ligament contains the blood supply of the uterus, uterine tubes, and the ovaries.

The uterine artery is a branch of the internal iliac artery. It comes off just in front of the ureter. The ureter and the uterine artery run closely together for 2–3 cm and then the artery twists in front of the ureter before entering the base of the broad ligament. It reaches the lower part of the cervix of the uterus by travelling medially in the broad ligament. Then it travels along the side wall of the uterus towards the terminal part of the uterine tube. It gives off branches anteriorly and posteriorly supplying the body of the uterus. It also gives off a branch which runs downwards to the vagina.

The ovarian artery comes from the abdominal aorta at the level of the renal arteries. It travels down the posterior abdominal and pelvic walls before sweeping forwards at the edge of the broad ligament in the suspensory ligament of the ovary (infundibulo pelvic ligament). The artery itself is the only structure which gives this ligament any strength since the connective tissue in it is very frail.

The uterine and ovarian arteries anastomose very freely with each other in the broad ligament. In operations here, it is essential to ligature both arteries.

THE OVARY

The ovary has the shape of an almond. It lies with its long axis vertical in a peritoneal depression at the side wall of the pelvis termed the **fossa ovarica**. This depression has the ureter and hypogastric vessels behind and the external iliac vessels and the obliterated hypogastric artery above. The obturator artery, vein and nerve form the floor.

It should be noted that the outer aspect of the pelvic wall at this point is the **acetabulum** (socket for head of femur). Hence a fracture here could break into the fossa ovarica.

The **mesovarum**, which as has been seen connects the ovary to the back of the broad ligament, brings the blood vessels and nerves to the ovary.

The lower pole of the ovary is connected to the uterus by the ligament of the ovary.

The uterine tube runs up the anterior border of the ovary, over the upper pole, and down the posterior border and medial surface of the ovary. The funnel-shaped dilatation of this part of the uterine tube is termed the **infundibulum** (Latin = funnel). It breaks up at the end into a series of fringe-like processes termed **fimbriae** (Latin = fringe). One is larger than the rest and is attached to the ovary. It is termed the **fimbria ovarica**.

The infundibulum is connected to the side-wall of the pelvis by the **infundibulopelvic ligament**. The ovarian artery, vein and nerves run in this ligament.

Blood supply to ovary
1. **Ovarian artery and vein**
This artery arises from the abdominal aorta just below the renal artery. Remember that although the right ovarian vein drains in to the inferior vena cava, the left drains into the left renal artery.
2. **Uterine artery**
The uterine artery is a branch of the internal iliac artery.

Nerve supply to ovary
T10 hypogastric plexus.

Lymph drainage
With arteries to para-aortic and lumbar lymph nodes.

THE UTERINE TUBE

The uterine or Fallopian tube conveys the ovum from the ovary to the cavity of the uterus. It is about 7 cm long. It runs mainly in the upper border of the broad ligament.

The uterine tube is divided into three parts:
1. Intramural part.
2. Isthmus.
3. Ampullary part.

The intramural part, as its name implies, runs in the wall of the uterus to open in the upper corner of the uterine cavity. The isthmus is the straight inner third of the uterine tube. It has a small diameter.

The ampullary part (*ampulla* (Latin) = bottle; hence ampule as a container for drugs) is the dilated extremity of the uterine tube. It terminates in the infundibulum and fimbriae as has been seen above at the abdominal ostium.

The two uterine (Fallopian) tubes, one on each side of the midline run into the upper lateral corners of the uterine cavity. They separate the fundus of the uterus (above) from its body (below).

Since fertilization takes place as the ovum is passing along the tube and not in the uterus itself, the male sperms must find their way to the tube.

The parts of the tube are shown in FIGURE 12.28. Note the fimbriae, one of which like an extra long finger lies very close to the ovary. When ovulation occurs the egg escapes into the peritoneal cavity and is then guided by the fimbriae towards the abdominal ostium of the tube. The lateral two-thirds of the tube is widely dilated and is called the ampulla. This is where fertilization usually occurs. The fertilized egg is then conveyed through the narrower isthmus of the tube to the intramural (the narrowest) part and then into the uterus itself.

The egg is moved by the peristalsis of the smooth muscle in the wall of the tube and also by the ciliated cells which line parts of its lumen. During this time a fertilized egg starts to divide and will have reached the 8–16-cell stage by the time the uterine cavity is reached. The journey may take as long as seven days.

A fertilized egg which dallies in the uterine tube may implant there and give rise to an **ectopic pregnancy**. Ectopics rupture after 8–10 weeks and give rise to an acute emergency. Very rarely implantation can also occur in the peritoneal cavity. In such a case the foetus must be delivered by a laparotomy. No attempt is made to separate the placenta from the underlying intestines or liver. It is left and is ultimately absorbed.

THE UTERUS

The uterus lies in the pelvis between the rectum behind and the bladder in front [Figs. 12.28 and 12.29]. It is a hollow muscular organ having the shape of an inverted pear flattened from front to back. It is about 7 cm long, 5 cm wide, and 2.5 cm thick. The cavity of the uterus is slit-like seen from the side; from the front it is triangular with base uppermost.

The uterus has a **body** and a **cervix** with a slight constrictor termed the isthmus at the junction between the two parts. The uterine (Fallopian) tubes enter the body of the uterus. The part of the body above these tubes is termed the **fundus** (*fundus* (Latin) = bottom of a bag or part farthest removed from the opening).

The cavity of the uterus leads to the outside via the cervix and vagina. The cervix joins the upper anterior vaginal wall at right angles [Fig. 12.29]. When the bladder is empty, the uterus usually flops still further forwards over the bladder with a bend between the body and cervix. This position is termed anteversion and anteflexion. The **anteversion** refers to the angle between the cervix and the vagina and the **anteflexion** refers to the angle between the body and the cervix.

Sometimes the uterus is bent right back with respect to the vagina so that the fundus is directed backwards and downwards. The

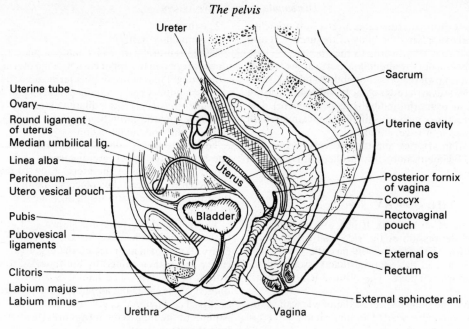

Ureter

Uterine tube
Ovary
Round ligament
of uterus
Median umbilical lig.
Linea alba
Peritoneum
Utero vesical pouch
Pubis
Pubovesical
ligaments
Clitoris
Labium majus
Labium minus
Urethra

Sacrum
Uterine cavity
Posterior fornix
of vagina
Coccyx
Rectovaginal
pouch
External os
Rectum
External sphincter ani
Vagina

Fig. 12.29. Female pelvis.

inexperienced, at a rectal examination, may mistake the fundus of such a retroverted uterus for a tumour.

The **cervix** has a narrow circular lumen running more or less longitudinally. The lumen joins the uterine cavity at the internal os. The cervix projects into the upper part of the vaginal cavity which it joins at the external os.

The cervix consists of a **supravaginal part** which is surrounded by the fibrous **parametrium** and a conical **vaginal part**. The vault of the vagina which surrounds the cervix of the uterus is termed the vaginal **fornix** (*fornix* (Latin) = arch). The dilated upper end of the vagina where the cervix joins is termed the **ampulla** of the vagina.

The opening of the cervix into the vagina usually has an anterior and posterior lip. It is usually in contact with the posterior vaginal wall.

During menstruation the mucosa of the body of the uterus (the endometrium) is shed, but not that of the cervix.

Most of the uterus is covered with **peritoneum**. The peritoneal membrane in front of the uterus is reflected from the posterior surface of the bladder onto the junction of the cervix and body forming the uterovesical pouch [Fig. 12.29]. Below this pouch the cervix is separated from the bladder by fatty fibrous tissue. Passing over the front of the body and fundus the peritoneum then covers the whole of the posterior surface of the uterus, and reaches a point on the posterior wall of the vagina about 1–2 cm below the posterior fornix. It is then reflected on the front of the rectum. The **recto-uterine** (a better name would be rectovaginal) pouch is often referred to as the pouch of Douglas. It is the most dependent part of the peritoneal cavity in the upright female patient. It is a common place for secondary complications of peritonitis to occur. An instrument or other device in the posterior fornix of the vagina can easily penetrate the vaginal wall and enter the peritoneal cavity, setting up peritonitis.

Laterally the two layers of peritoneum run to the side walls of the pelvic cavity forming the broad ligament of the uterus. The broad ligament thus extends over the whole of the lateral surface of the uterus including the supravaginal cervix and encloses the uterine tube, the broad ligament, and parametrium as well as nerves and blood vessels.

Implantation of the fertilized egg takes place in the uterus. During pregnancy the uterus enlarges greatly in size. The fundus is palpable over the pubic symphysis after three months; is level with the umbilicus at six months, and reaches the upper epigastrium towards the end of the pregnancy (nine months).

The parametrium

The parametrium (*para* (Greek) = near; *metra* (Greek) = uterus) is the name given to the connective tissue surrounding the uterus. It lies in the lower part of the broad ligament and consists of three paired ligaments surrounding the supravaginal cervix.

1. Uterosacral ligaments.
2. Lateral cervical ligaments.
3. Pubocervical ligaments.

As their names imply these ligaments run to the sacrum, lateral wall of the pelvis, and the body of the pubis respectively.

The uterine artery passes in front of the ureter close to the cervix.

The round ligament

The round ligament is a fibromuscular band which runs from the upper lateral angle of the uterus (below and in front of the uterine tubes) to the deep inguinal ring. It runs though the inguinal canal to end in the labium majus.

THE VAGINA

The vagina is a highly distensible muscular canal about 8 cm long. The outer muscular coat consists of longitudinal smooth muscle fibres. The inner muscular coat consists of circular smooth muscle fibres. It is lined by a special stratified squamous epithelium through which the transudation of fluid occurs in sexual excitement which lubricates the inside of the vagina and facilitates the entry of the male penis. Normally the anterior and posterior walls are in contact.

The vagina lies between the urethra and bladder in front and the recto-uterine pouch (pouch of Douglas) and the rectum behind. About one-third up the vagina, the levator ani muscle forms a sphincter around the vagina. Unless this sphincter is relaxed, the insertion of the penis into the vagina in sexual intercourse is difficult.

Blood supply

1. Vaginal branch of internal iliac artery.
2. Vaginal branch of uterine artery.

There may also be branches from the pudendal artery and the middle rectal artery.

Lymph drainage

Upper two-thirds drains into internal iliac lymph nodes.

Lower one-third drains to the upper superficial inguinal nodes in the groin.

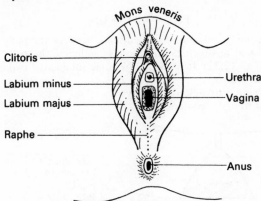

The whole region enclosed between the labia minora is called the vestibule of the vulva

FIG. 12.30. External genitalia in the female.

THE VULVA

The female external genitalia are termed the vulva. An alternative name is the pudenda. It consists of the two labia majora, the two labia minora, and the vestibule [FIG. 12.30].

Labia majora

The labia majora, or major lips, join in front over the lower part of the pubic symphysis and behind in an ill-defined posterior commissure in front of the anus [FIG. 12.31].

These labia are the homologue of the scrotum in the male. They contain fat and smooth muscle (the homologue of the dartos muscle) but no gonad.

Labia minora

The labia minora in front surround the clitoris. Like the penis, the clitoris contains erectile tissue, but no urethra.

Vestibule

The vestibule is that part of the vulva bounded by the labia minora. It contains the openings of the vagina, urethra and the vestibular glands (Bartholin's glands). These glands correspond to the bulbo-urethral glands (Cowper's glands) in the male.

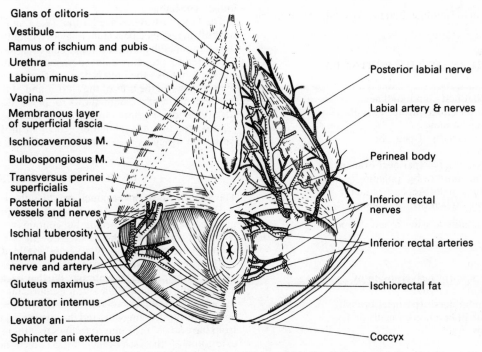

FIG. 12.31. Female perineum. [Compare Male, FIG. 12.21.]

THE PERINEUM

The perineum is a diamond-shaped space between the pubic symphysis, ischial tuberosities and the coccyx below the pelvic diaphragm [FIG. 12.32]. It is divided by a line joining the two ischial tuberosities into:

1. Urogenital triangle (anteriorly).
2. Anal triangle (posteriorly).

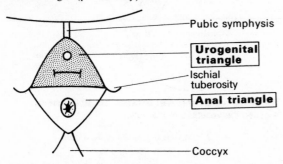

FIG. 12.32. The perineal membrane has a free edge posteriorly which extends between the two ischial tuberosities. It divides the perineum into the urogenital triangle anteriorly and the anal triangle posteriorly.

Urogenital triangle

The urogenital triangle is divided by a triangular layer of fascia which stretches across the pubic arch into:

1. Superficial perineal pouch.
2. Deep perineal pouch.

Anal triangle

The anal triangle lies behind a line joining the ischial tuberosities. It is bounded laterally by sacrotuberous ligaments and has the coccyx behind.

The anal canal lies in the midline. Lateral to the anal canal are the ischiorectal fossae.

MUSCLES OF THE PERINEUM

Bulbospongiosus

Origin:	Perineal body, and midline raphé on under surface of bulb of penis in male
Insertion:	Dorsal surface of penis or clitoris
Nerve supply:	Pudendal nerve (perineal branch)
Action:	Empties urethra by 'pumping' action in male. Acts as sphincter around vaginal orifice in female

The two halves of the muscle wind round the penis.

Ischiocavernosus

Origin:	Ischium
Insertion:	Crus of corpus cavernosum of penis or of clitoris
Nerve supply:	Pudendal nerve (perineal branch)
Action:	Assists in erection of penis or clitoris

These muscles cover the crura of the penis.

Superficial transverse perineal muscle [FIGS. 12.21 and 12.31].

Origin:	In front of ischial tuberosity
Insertion:	Perineal body
Nerve supply:	Pudendal nerve (perineal branch)
Action:	Supports perineum

This is a transverse bundle running along the posterior edge of the perineal membrane.

Deep transverse perineal muscle

Origin:	Ramus of ischium
Insertion:	Perineal body. Blends with sphincter urethrae
Nerve supply:	Pudendal nerve
Action:	Supports perineum

PERINEAL BODY

This is a fibrous mass at the centre of the perineum which forms an attachment for levator ani, superficial and deep transverse perinei, external anal sphincter, and bulbospongiosus.

Urogenital diaphragm

The striated muscle running between the two sides of the pubic arch constitutes the urogenital diaphragm. This muscle (deep transverse perinei and sphincter urethrae) is surrounded by fascia which encloses totally the space termed the deep perineal pouch.

The inferior surface of the fascia is termed the **perineal membrane**. Behind it is attached, in front of the perineal body, to the superficial fascia. The space between the superficial fascia and the perineal membrane forms the superficial perineal pouch.

It is helpful to understand the similarities between the male and female perineum. Note that early in development the two sexes are indistinguishable.

The key illustrations are FIGURE 12.33 and 12.34. Imagine that the bulb in the male perineum is divided longitudinally so as to form the bilateral bulb of the vestibule. This consists of erectile tissue and is covered by the same bulbospongiosus muscle as we described in the male but like the bulb is now split into two halves. The resemblance between the crus of the clitoris and of the penis is obvious.

The division of the bulb allows room for the vagina to reach the surface in the midline. In the male the vagina and uterus are represented only by a small diverticulum in the prostatic urethra called the utriculus masculinus.

PERINEAL POUCHES

The superficial perineal pouch contains the roots of the external genitalia in both sexes. The deep perineal pouch is not really a proper 'space' in its own right, but contains a number of structures which are mostly passing to the superficial pouch. The external sphincter urethrae is the only structure 'confined' to the deep pouch.

The superficial and deep perineal pouches are related immediately to the perineal membrane. Further back in the anal triangle, on each side of the midline, is an ischiorectal fossa [p. 234]. Since the apex of the ischiorectal fossa is the junction of the levator ani and the obturator internus, and the levator ani takes origin in front, from the body of the pubis, it follows that there must be an anterior extension of the ischiorectal fossa which lies above the deep perineal pouch. Infection spreading into the 'anterior recess' of the

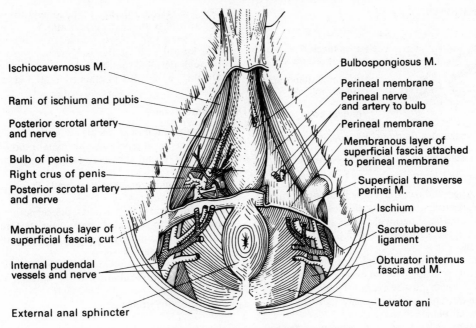

Ischiocavernosus M.

Rami of ischium and pubis

Posterior scrotal artery
and nerve

Bulb of penis

Right crus of penis

Posterior scrotal artery
and nerve

Membranous layer of
superficial fascia, cut

Internal pudendal
vessels and nerve

External anal sphincter

Bulbospongiosus M.

Perineal membrane

Perineal nerve
and artery to bulb

Perineal membrane

Membranous layer of
superficial fascia attached
to perineal membrane

Superficial transverse
perinei M.

Ischium

Sacrotuberous
ligament

Obturator internus
fascia and M.

Levator ani

Fig. 12.33. Male perineum. [Compare Fig. 12.21].

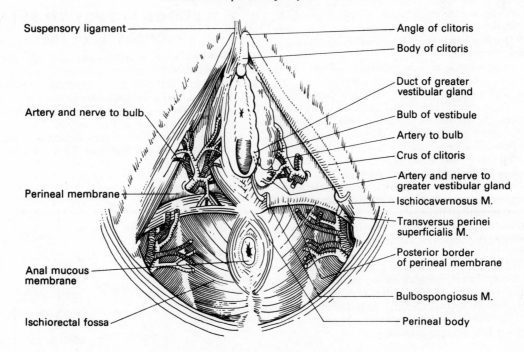

Suspensory ligament

Artery and nerve to bulb

Perineal membrane

Anal mucous
membrane

Ischiorectal fossa

Angle of clitoris

Body of clitoris

Duct of greater
vestibular gland

Bulb of vestibule

Artery to bulb

Crus of clitoris

Artery and nerve to
greater vestibular gland

Ischiocavernosus M.

Transversus perinei
superficialis M.

Posterior border
of perineal membrane

Bulbospongiosus M.

Perineal body

Fig. 12.34. Female perineum. [Compare Fig. 12.31].

ischiorectal fossa (above the external urethral sphincter) may produce spasm of the urethral sphincters and retention of urine.

PERINEAL FASCIAL PLANES

The arrangement of the fascia planes associated with the perineal region is complex.

If a male patient has a ruptured urethra in the region of the bulb, the urine leaks out into the superficial perineal pouch [Fig. 12.35]. The extraversated urine spreads into the scrotum and the root of the penis. It then passes up under the superficial fascia of the anterior abdominal wall. It does not, however, enter the thighs, nor does it pass backwards to the anal triangle. It follows therefore that the superficial fascia must be adherent to the perineal fascia or deep

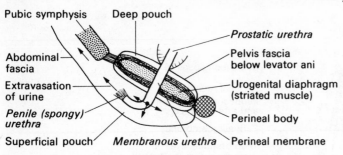

FIG. 12.35. Extravasation of urine following damage to the penile urethra.

fascia at the back of the urogenital triangle, at the ischiopubic rami, and at the inguinal ligament otherwise urine would reach the fascial planes of the thigh. The superficial perineal pouch, however, is in continuity with the fascia in the scrotum and with the space under the superficial fascia of the penis.

If a patient has a ruptured urethra in the region of the membranous urethra, there is no extravasation of urine except into the small deep pouch. Thus the deep pouch must be a totally enclosed space.

So starting with the answer and working backwards we can now consider the perineal fascial planes in more detail.

The **urogenital diaphragm**, as has been seen, consists of striated muscle with its two layers of investing fascia which stretch across the pubic arch. The muscles are the deep transverse perineal muscle and the external sphincter muscle (sphincter urethrae). The upper layer of fascia is the **parietal pelvic fascia**. The lower layer of fascia is termed the **perineal membrane**.

This perineal membrane separates the superficial compartment of the urogenital perineum from the deep compartment.

In front the two layers fuse to form the transverse ligament. Behind they fuse with the superficial perineal fascia. Between these two layers are found the membranous urethra, the two bulbo-urethral glands (Cowper's glands) and the dorsal arteries, veins, and nerves of the penis. The upper layer of fascia (perineal membrane) is pierced by these vessels and nerves as well as by the urethra and the ducts of the bulbo-urethral glands.

The deep membranous layer of the superficial fascia of the lower abdomen (Scarpa's fascia) passes over the pubic bone and becomes the deep layer of the superficial perineal fascia (Colles's fascia). It surrounds the penis and lines the scrotum. Posteriorly it fuses with the perineal membrane. At the sides it is attached to the ischiopubic ramus of the pelvis.

Thus although the extravasated urine can pass into the scrotum, around the penis, and up the anterior abdominal wall, it cannot pass directly into the thigh because of the attachment of the fascia to the ischiopubic ramus. Nor can it enter the thigh from the abdominal wall because the superficial fascia of the abdomen is attached to the deep fascia just below the inguinal ligament.

SUMMARY

Superficial perineal pouch

The superficial perineal pouch is the space between the membranous layer of the superficial fascia and the perineal membrane. The contents are:
1. (a) Bulb of penis or clitoris.
   (b) Crura of penis or clitoris.

2. Their covering muscles: bulbospongiosus, ischiocavernosus.
3. Superficial transverse perinei muscles.
4. Blood supply and nerve supplies from internal pudendal artery (via lateral wall of the ischiorectal fossa and deep pouch, through perineal membrane).
5. Urethra, concealed in male by bulb and corpus spongiosus.

Deep perineal pouch

The deep perineal pouch lies between the perineal membrane and the pelvic fascia. It encloses the membranous urethra and the surrounding external sphincter muscle (sphincter urethrae). The contents are:
1. External sphincter urethrae and deep transverse perinei muscles.
2. Bulbo-urethral glands—ducts pierce perineal membrane and discharge into the bulbar part of the urethra in the male.
3. Internal pudendal artery giving off artery to the bulb and the deep and dorsal arteries.
4. Pudendal nerve supplying sphincter urethrae and via perineal membrane the structures in the superficial pouch and also dorsal nerve.
5. Membranous urethra.

BLOOD SUPPLY TO PELVIS
COMMON ILIAC ARTERIES

The abdominal aorta bifurcates into the two common iliac arteries just to the left of the midline at the level of L4 at a point roughly 1 cm below and to the left of the umbilicus. A line joining this point to the midinguinal point (half-way between the anterior superior iliac spine and the pubic symphysis) gives the surface marking of the common iliac and external iliac arteries. The upper third corresponds to the common iliac artery whilst the lower two-thirds correspond to the external iliac artery [FIG. 12.36].

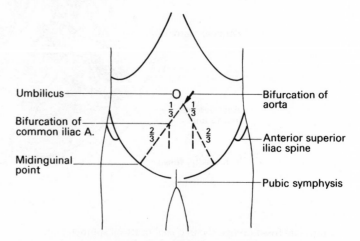

FIG. 12.36. Surface marking of iliac arteries.
The aorta bifurcates 1 cm below and to the left of the umbilicus. The midinguinal point identifies the pulsation of the external iliac artery. One-third of the way between the two points, the common iliac artery divided into internal and external branches. The ureter usually lies just in front of the common iliac artery bifurcation point, and the same procedure identifies its position here.

The relative positions of the abdominal arteries and veins is, at first sight, confusing since at the level of the diaphragm the inferior vena cava is anterior to the aorta whilst the bifurcation of the abdominal aorta lies in front of the interior vena cava. Thus although, higher up, the right renal artery passes **behind** the inferior vena cava, the common iliac artery passes **in front of** the common iliac vein. Furthermore, although the abdominal viscera are supplied by the abdominal aorta, and the kidneys, suprarenals, gonads, and pelvic organs drain into the inferior vena cava (or its tributaries), the digestive tract, and spleen drain into the portal vein which takes the blood to the liver [see p. 221].

Over the sacro-iliac joint the common iliac artery gives off the internal iliac artery and becomes the external iliac artery.

INTERNAL ILIAC ARTERY

The internal iliac artery runs into the pelvis towards the upper border of the greater sciatic notch where it divides into the anterior and posterior divisions [Figs. 12.37 and 12.38].

ANTERIOR DIVISION OF INTERNAL ILIAC ARTERY

Branches:
1. Superior vesical artery. ③
2. Middle vesical artery. ③
3. Inferior vesical artery in male—vaginal artery and uterine artery in female. ④
4. Middle rectal artery.
5. Obturator artery. ①
6. Internal pudendal artery. ⑥
7. Inferior gluteal artery. ⑦

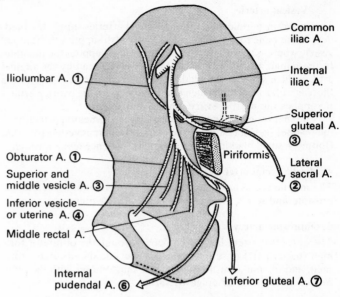

FIG. 12.37. Internal iliac artery and its branches.

The internal iliac artery runs for a short distance before separating into anterior and posterior divisions.

The posterior becomes the superior gluteal artery. It gives off two rather unimportant branches before it reaches piriformis.

The anterior division becomes the inferior gluteal artery. Before doing so it gives off the obturator artery, and branches to the pelvic viscera, bladder, etc., uterus etc., and rectum as well as the internal pudendal artery. This division supplies all the pelvic viscera.

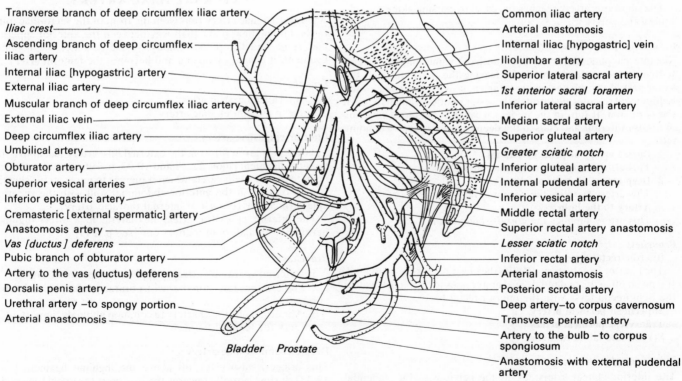

FIG. 12.38. Arteries of pelvic cavity (male). The internal iliac branches are subject to considerable individual variation.

1–3. Vesical arteries

The superior, middle, and inferior vesical arteries supply the bladder. In the male the middle vesical artery also supplies the seminal vesicle whereas the inferior vesical artery also supplies the prostate. In the female the inferior vesical artery is replaced by the vaginal artery which supplies the bladder, vagina, and rectum. In addition there is a uterine artery which anastomoses with the ovarian artery. It supplies the uterus, ovary, ureter, and vagina.

Note that **vesical**, as in *superior vesical artery*, means pertaining to the bladder (*vesica* (Latin) = the bladder) whereas **vesicle** (*vesicula* (Latin) = small bladder) means a small sac, hence *seminal vesicle*.

4. Middle rectal artery

The middle rectal artery supplies the lower rectum as well as the prostate and seminal vesicles. It is not always present.

5. Obturator artery

The obturator artery leaves the pelvis through the obturator foramen (upper part) to supply structures outside the obturator foramen, and the hip joint. In the pelvis it anastomoses with the pubic branch of the inferior epigastric artery.

The **obturator foramen** seen in the bony skeleton as a large aperture in the pelvic bone is not a foramen in life since, as has been seen, it is almost completely covered by the obturator internus muscle and its investing fascia. However, a small canal (about the size of the small finger) persists in this fascia to allow the obturator nerve and obturator blood vessels to leave the pelvis and pass to the inner side of the thighs to supply the adductor muscles [p. 134]. It is situated below the acetabulum close to the anterior pubic ramus.

The obturator artery reaches the canal by passing along the side wall of the pelvis.

6. Internal pudendal artery

The internal pudendal artery is a branch of the internal iliac artery. It follows the pudendal nerve by leaving the pelvis through the lower part of the greater sciatic foramen, crossing the spine of the ischium and re-entering through the lesser sciatic foramen to reach the perineum. It runs in the pudendal canal in the fascia covering obturator internus and ends in the deep perineal space by dividing into:

1. **Dorsal artery of penis or clitoris**
   This artery supplies the prepuce and glans.
2. **Deep artery of penis or clitoris**
   This artery supplies the corpus cavernosum.
3. **Artery to the bulb**
   This artery supplies the corpus spongiosum.

Branches:
   Inferior rectal artery
   The inferior rectal artery supplies the lower part of the anal canal. It is given off in the pudendal canal and crosses to the medial wall of the ischiorectal fossa with the inferior rectal nerve.
   Superficial perineal artery
   Transverse perineal artery
   Scrotal and labial branches

7. Inferior gluteal artery

The inferior gluteal artery leaves the pelvis with the pudendal artery through the greater sciatic notch below piriformis. It supplies the gluteal muscles in the lower part of the buttocks, and forms anastomoses around the hip joint.

POSTERIOR DIVISION OF INTERNAL ILIAC ARTERY

Branches [FIG. 12.37]:
   1. Iliolumbar artery. ①
   2. Lateral sacral artery. ②
   3. Superior gluteal artery. ③

Iliolumbar artery

The iliolumbar artery runs upwards on the sacrum behind the common iliac artery. It supplies the iliac fossa, and sends a spinal branch in between L5 and the sacrum. It supplies iliacus and posterior abdominal wall muscles. It supplies the territory of the 'missing' fifth lumbar artery.

Lateral sacral artery

The lateral sacral artery anastomoses with the median sacral artery (branch of aorta) at the coccyx and supplies the sacral canal.

Superior gluteal artery

The superior gluteal artery leaves the pelvis through the greater sciatic foramen above piriformis at a point one-third of the way from the posterior superior iliac spine and greater trochanter. It supplies the muscles in the gluteal region. Like the inferior gluteal artery it joins the anastomosis around the hip joint. It accompanies the superior gluteal nerve.

EXTERNAL ILIAC ARTERY

The external iliac artery runs from the front of the sacro-iliac joint to the midinguinal point (half-way between the anterior superior iliac spine and the pubic symphysis). Here, under the inguinal ligament, it changes its name and becomes the **femoral artery**.

Branches:
   1. Inferior epigastric artery.
   2. Deep circumflex iliac artery.

Inferior epigastric artery

This artery is given off from the external iliac artery just above the inguinal ligament [FIGS. 12.38 and 10.07, p. 196]. It runs upwards deep to transversalis fascia which it pierces to enter rectus sheath. It anastomoses in the anterior abdominal wall with the superior epigastric artery (branch of internal thoracic artery).

The inferior epigastric artery lies lateral to a direct inguinal hernia but medial to an indirect one [see p. 199].

Branches:
   1. External spermatic artery (supplies the spermatic cord).
   2. Pubic branch (anastomoses with obturator artery behind pubic bone).
   3. Abnormal obturator artery (a common anomaly, important in surgery for femoral hernia).

Deep circumflex iliac artery

This artery is also given off above the inguinal ligament [FIG. 12.38]. It runs laterally towards the iliac crest to supply the neighbouring muscles.

COMMON ILIAC VEINS

The common iliac veins unite on the right side of L5 vertebra to form the inferior vena cava. Both veins lie to the right of and behind the corresponding common iliac arteries. As a result of the veins lying to the right of the arteries the left common iliac vein is longer than the right. The fact that oedema of the left leg is commoner than that of the right, may be associated with the relationship of this vein to the arteries.

Tributaries:

Internal iliac vein.
External iliac vein.
These unite in front of the sacro-iliac joint to form the common iliac vein.

Internal iliac vein

The blood from the pelvic viscera drains first into the venous plexuses surrounding the viscera. These are:

Vesical plexus.
Prostatic plexus (in male).
Uterine plexus (in female).
Vaginal plexus (in female).
Rectal plexus.

These venous plexuses communicate with one another and also with the vertebral venous plexuses. This anastomosis provides a possible route for the spread of secondaries from neoplasms of the pelvis to the vertebrae of the spinal column.

The venous plexuses drain via veins corresponding to the branches of the internal iliac artery to the internal iliac vein.

External iliac vein

This is a continuation of the femoral vein which becomes the external iliac vein as it passes under the inguinal ligament. It lies 1 cm proximal to the femoral/external iliac artery at this point.

Tributaries:

Inferior epigastric vein
Deep circumflex iliac vein
These veins follow the corresponding arteries.

Gonadal veins

In the male the venous drainage of the testis forms the pampiniform plexus (*pampinus* (Latin) = tendril) of the spermatic cord (which has the form of a tendril—a leafless plant organ which attaches itself to another body for support). The gonadal vein formed from this pampiniform plexus drains usually into the inferior vena cava at the level of the renal vein on the right side, and into the renal vein on the left side. The same applies to the ovarian veins.

LYMPHATIC DRAINAGE OF PELVIS

The lymphatic channels follow the external iliac, internal iliac, and gonadal blood vessels.

External iliac lymph nodes
Drainage
1. Lower limb.
2. Lower abdominal wall.
3. Bladder.
4. Prostate.
5. Cervix.

Internal iliac lymph nodes
Drainage
1. Pelvic viscera.
2. Gluteal region.
3. Perineum.

Para-aortic lymph nodes

The gonadal lymph vessels run to the para-aortic lymph nodes which lie alongside the abdominal aorta.

Drainage
Testis.
Ovary.

These para-aortic lymph nodes also drain the posterior abdominal wall, the kidneys and the suprarenals.

Malignant disease in the pelvis is common. The uterus, either body or cervix, is second to the breast as the commonest site of carcinoma in the female.

The lymphatic drainage of individual organs follows their arteries and veins but, in addition, attention must be drawn to the possibility of malignant disease of the body of the uterus giving rise to a 'lump' in the superficial inguinal lymph nodes. The lymphatic vessels run with the round ligament of the uterus through the inguinal canal to reach the superficial inguinal lymph nodes. These lymph nodes also drain the whole of the body surface below the level of the umbilicus including the perineum, the lower part of the anal canal and the vestibule of the vulva.

Malignant disease of the ovary spreads upwards around the ovarian artery to the para-aortic lymph nodes at the back of the abdomen at the level of the transpyloric plane in the same way as testicular disease [p. 239].

NERVE SUPPLY TO THE PELVIS

S2, 3, 4, 5 and the coccygeal nerve form the lower part of the sacral plexus termed the pudendal plexus [p. 124]. Its branches are:

Perforating cutaneous nerve (S2, 3).
Pudendal nerve (S2, 3, 4).
Sacral parasympathetic outflow (nervi erigentes) (S2, 3, 4).
Muscular branches.
Coccygeal plexus (anococcygeal nerve) (S4, 5) Co 1.

Perforating cutaneous nerve

The perforating cutaneous nerve supplies the skin over the lower and medial part of the buttocks.

PUDENDAL NERVE

The pudendal nerve (S2, 3, 4) is derived from the sacral plexus. It leaves the pelvis through the greater sciatic foramen below piriformis, crosses the ischial spine and re-enters via the lesser sciatic foramen to reach the perineum. It runs in the pudendal canal over obturator internus [p. 235] gives off the inferior rectal nerve and ends by dividing into:

1. Perineal nerve.
2. Dorsal nerve of penis or clitoris.

Branches:

1. Inferior rectal nerve

This nerve is given off in the pudendal canal. It crosses to the medial wall of the ischiorectal fossa and supplies:

    (a) levator ani;
    (b) external anal sphincter;
    (c) perianal skin.

2. Perineal nerve

This terminal branch runs into the superficial pouch and supplies:

    (a) perineal muscles;
    (b) scrotal skin;
    (c) perineal skin.

The perineal skin is also supplied by the perineal branch of the posterior cutaneous nerve of thigh (S1, 2, 3).

3. Dorsal nerve of penis or clitoris

This terminal branch runs forwards on the medial side of ischiopubic ramus to the deep perineal pouch. It pierces the perineal membrane to reach the dorsum of the penis. It supplies the skin and glans of the penis.

Coccygeal plexus

The coccygeal plexus supplies coccygeus and part of levator ani. It also supplies the skin over the dorsum of the coccyx and between the coccyx and the anus.

UMBILICAL BLOOD VESSELS

In FIGURE 12.38 the right umbilical artery is shown. This vessel is the largest branch of the internal iliac artery on each side in the foetus. The vessels each give off branches to the bladder fundus and reach the deep aspect of the anterior abdominal wall on each side of the midline. They converge on the umbilicus and then run in the umbilical cord to the placenta. The oxygenated blood returns from the placenta via a single umbilical vein which passes back through the umbilicus and runs in the free border of the falciform ligament of the liver to the left branch of the portal vein.

The umbilical arteries fibrose after birth and become the two lateral umbilical ligaments. The vein becomes the ligamentum teres of the liver.

Other vestiges (*vestigium* L. = footprint) of the foetal circulation are:

    (a) Ligamentum venosum which is the fibrosed ductus venosus which conveyed blood from the portal vein to the inferior vena cava (short-circuiting the liver).
    (b) Fossa ovalis which marks the site of the foramen ovale through which blood flowed through the interatrial septum from right to left.
    (c) Ligamentum arteriosum which is the fibrosed remnant of the ductus arteriousus which conveys blood from the pulmonary artery to the aorta (short-circuiting the lungs).

13 The blood supply to the head and neck

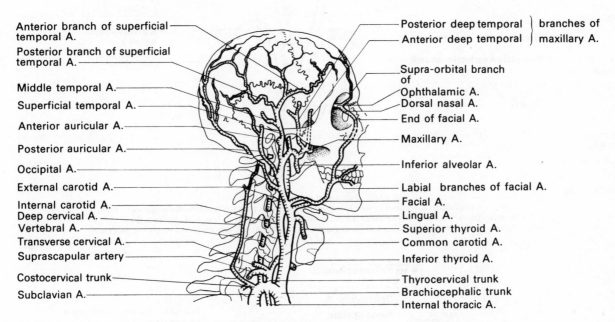

Anterior branch of superficial temporal A.
Posterior branch of superficial temporal A.
Middle temporal A.
Superficial temporal A.
Anterior auricular A.
Posterior auricular A.
Occipital A.
External carotid A.
Internal carotid A.
Deep cervical A.
Vertebral A.
Transverse cervical A.
Suprascapular artery
Costocervical trunk
Subclavian A.

Posterior deep temporal ⎫ branches of
Anterior deep temporal ⎭ maxillary A.
Supra-orbital branch of Ophthalamic A.
Dorsal nasal A.
End of facial A.
Maxillary A.
Inferior alveolar A.
Labial branches of facial A.
Facial A.
Lingual A.
Superior thyroid A.
Common carotid A.
Inferior thyroid A.
Thyrocervical trunk
Brachiocephalic trunk
Internal thoracic A.

Fig. 13.01. Arteries of head and neck.

COMMON CAROTID ARTERY

On the **right side** the common carotid artery commences behind the junction of the clavicle with the sternum as a terminal branch of the **brachiocephalic trunk** [Fig. 13.02].

On the **left side** it arises from the **arch of the aorta** and passes through the superior part of the mediastinum with the lung and pleura to the left and in front. Behind it lies the trachea, oesophagus, recurrent laryngeal nerve, and thoracic duct. It passes behind the junction of the clavicle with the sternum.

The omohyoid crosses the common carotid artery at the level of the cricoid cartilage (C6 vertebra). The upper part of the artery is overlapped by the anterior edge of sternomastoid. The lower part is covered by sternomastoid, sternohyoid, and sternothyroid.

The common carotid artery lies in the carotid sheath in front of the anterior tubercles of cervical vertebrae [p. 316].

Contents of carotid sheath

(i) Common carotid artery, internal carotid artery.
(ii) Internal jugular vein.
(iii) Vagus nerve.

The sympathetic trunk lies behind and medial to the carotid sheath whilst the ansa cervicalis lies in front.

The common carotid artery runs in front of the large anterior tubercle of C6. This fact is used in garotting.

Below C6 it lies in front of the vertebral artery [p. 260] and on the left side in front of the thoracic duct.

The origins of scalenus anterior and other muscles from anterior tubercles of third, fourth, fifth, and sixth cervical vertebrae lie behind the common and internal carotid arteries.

INTERNAL CAROTID ARTERY

The internal carotid artery [Fig. 13.03] commences at the upper border of the thyroid cartilage (level C3–4). It runs upwards in front of the transverse processes of C3, C2, and C1 to the base of the skull. In the skull it takes a very tortuous path. It enters the carotid canal, turns medially and forwards to the apex of the petrous temporal bone, then runs upwards in the upper half of the foramen lacerum to enter the back of the cavernous sinus. It bends forwards and having reached the back of the anterior clinoid it runs upwards, pierces the dura and runs backwards dividing into:

Anterior cerebral artery
Middle cerebral artery

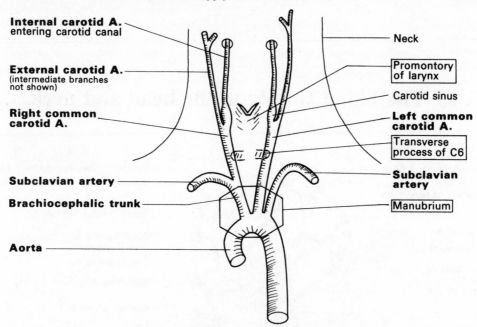

FIG. 13.02. Diagram of common carotid arteries.
The common carotid arises from the brachiocephalic trunk on the right side and from the aortic arch on the left side. It does not give off any branches before dividing into the internal and external arteries at the level of the laryngeal promontory (level of C4). Note the position of the manubrium sterni and transverse process of C6.

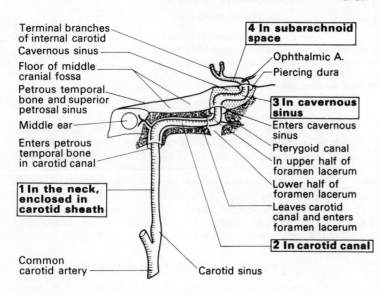

FIG. 13.03. Internal carotid artery.
Diagrammatic representation of the course of the internal carotid artery. The first part (1) runs entirely in the neck from the bifurcation of the common carotid until it enters the carotid canal. It is enclosed in the carotid sheath. In the carotid canal (2) it runs horizontally and reaches the posterior wall of the foramen lacerum. This so called foramen is really a canal about 1 cm in length. The artery enters it about half way up. The pterygoid canal leaves the foramen about half way up the anterior wall and reaches the pterygopalatine fossa. The artery turns upwards to reach the cavernous sinus (3). It leaves the sinus on the medial side of the anterior clinoid process (not shown in diagram) and pierces the dura and then lies in the subarachnoid space (4). It terminates by becoming the anterior and middle cerebral arteries.

Note the middle ear cavity lies behind the commencement of the carotid canal. The artery is surrounded by a dense plexus of postganglionic fibres many of which contribute to the nerve of the pterygoid canal.

Cervical part

In the neck the external carotid artery lies superficial to it. Behind is rectus capitis, superior laryngeal nerve, and the superior cervical ganglion of the sympathetic nervous system.

The superior constrictor separates it medially from the tonsil. At the base of the skull the carotid canal lies in front of the jugular foramen. Thus the inferior petrosal sinus, IXth, Xth, and XIth cranial nerves separate internal carotid artery from jugular vein.

Petrous part

In this part of its course it lies in front of the cochlea and the middle ear. It then runs close to the trigeminal ganglion being separated from it only by thin cartilage.

Intracranial part

The internal carotid artery passes forward through the cavernous sinus before curling upwards on the lateral side of the optic chiasma. It lies in a shallow groove on the body of the sphenoid with the sixth cranial nerve on its outer side. The pituitary gland lies in the pituitary fossa on the medial side of the cavernous sinus. On the lateral wall of the sinus from above down are seen:

1. III (third cranial nerve).
2. IV (fourth cranial nerve).
3. V_1 (ophthalmic division of fifth cranial nerve).
4. V_2 (maxillary division of fifth cranial nerve).

The carotid syphon is the name given to the loop of the internal artery.

The **cavernous sinus** [FIG. 13.04] is an important venous sinus. It lies on either side of the pituitary gland. The internal carotid artery passes through the cavernous sinus. As a result any damage to the artery in this region will produce an arteriovenous anastomosis. All the cranial nerves from the second to sixth, with the exception of V_3 (the mandibular division of the trigeminal nerve), pass through or are close to the cavernous sinus.

BRANCHES OF INTERNAL CAROTID ARTERY

The internal carotid artery does not give off any branches in the cervical region. It enters the skull through the carotid canal and gives off the caroticotympanic artery to the middle ear and the pterygoid artery which accompanies the great petrosal nerve. It runs to the cavernous sinus and gives off branches to the pituitary gland. It gives off the ophthalmic artery and terminates by dividing into the middle cerebral and anterior cerebral arteries.

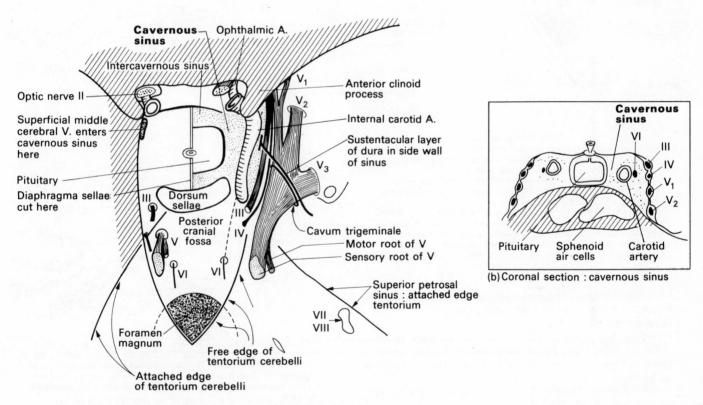

(a) Dorsal view : Cavernous sinus

(b) Coronal section : cavernous sinus

FIG. 13.04. The cavernous sinus.
(a) View looking down on the dura mater which is related to the cavernous sinus. On the left the dura is undisturbed: on the right it has largely been removed to show the right cavernous sinus (stippled) and the structures which lie in it, or in its wall.
(b) Coronal section through the anterior half of the two cavernous sinuses.
   Note that nerves VI and V, which are derived from the hindbrain, emerge below the tentorium in the posterior cranial fossa. VI perforates the inner layer of the dura not far from the foramen magnum, and passes upwards in the same plane as the sinus itself. It traverses the sinus mostly in contact with the lateral wall of the carotid artery.
   The trigeminal motor and sensory roots invaginate in front of them for a considerable distance a pocket consisting of the inner layer of dura belonging to the posterior fossa. This extends around the trigeminal ganglion and is adherent to the deep surface of the dura of the middle fossa which forms the side wall of the sinus. The presence of the cavum trigeminale means that the side wall of the sinus has not one layer of dura but three. A right hand in the trouser pocket makes a model of the cavum. The space between your leg and the trouser material is the cavernous sinus. At the level of your hand however there are three layers of material, two superficial and one deep.
   IV and III emerge from the midbrain which lies in the opening of the tentorium. IV enters the dura just at the point where the free and attached borders of the tentorium cross each other. III enters the dura after the cross over has taken place.
   The cavernous sinus drains in particular the ophthalmic veins and superficial middle cerebral veins. Blood escapes via the superior and inferior petrosal sinuses to the internal jugular veins. The sinuses communicate with each other via the intercavernous sinuses. The hormones of the pituitary feed into these sinuses before reaching the general circulation.

1. Ophthalmic artery

The ophthalmic artery [FIG. 13.05] arises from the internal carotid artery as it loops around forming the carotid syphon.

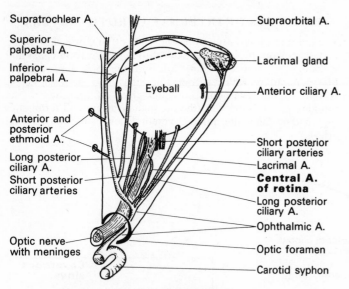

FIG. 13.05. Ophthalmic artery.

The ophthalmic artery enters the orbit on the inferior aspect of the theca of the optic nerve via the optic foramen. It terminates in the anteromedial part of the orbit by becoming the supratrochlear artery.

Note that the short and long posterior ciliary arteries come from the muscular branches of the ophthalmic artery. There are six such branches just as there are six extrinsic eye muscles.

The central artery of the retina is the most important branch of the ophthalmic artery. It passes through the theca about two-thirds of the way along the optic nerve, crosses the subarachnoid space and comes to lie embedded with its two venae comitantes, right in the middle of the optic nerve. The central artery approaches the optic nerve on its ventral aspect and is shown by a dotted line.

The ophthalmic artery passes through the optic foramen with the optic nerve and breaks up into the following branches:

1. Central artery of retina.
2. Anterior and posterior ciliary arteries: the short posterior ciliary arteries enter the choroid coat of the eye. The long posterior ciliary arteries supply the ciliary body and iris diaphragm.
3. Lacrimal artery: this artery supplies the lacrimal gland, upper eyelid, and conjunctiva.
4. Supra-orbital artery: this artery runs with the supraorbital nerve to the scalp.
5. Supratrochlear (frontal) artery: this artery runs with the supratrochlear nerve to the scalp.
6. Anterior ethmoidal artery: this artery follows the nasal nerve.
7. Posterior ethmoidal artery: this artery supplies the upper part of the nose and the posterior ethmoidal air-cells.
8. Dorsal nasal artery.
9. Palpebral branches: these supply the upper and lower eyelids.
10. Muscular branches: these supply the extrinsic muscles of the eyes.

2. Middle cerebral artery

The middle cerebral artery is the direct continuation of the internal carotid artery after it has given off the anterior cerebral artery. It supplies the lateral aspect of the cerebral hemisphere except for a strip of cortex near the anterior border which is supplied by the anterior cerebral artery, and a strip of cortex near the posterior border which is supplied by the posterior cerebral artery [FIG. 13.06].

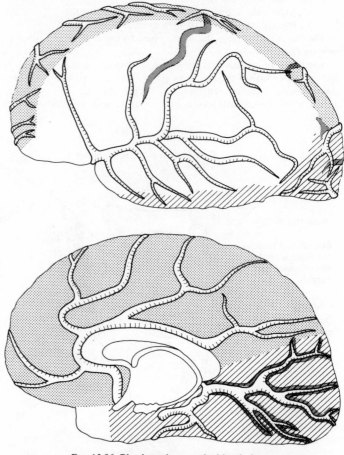

FIG. 13.06. Blood supply to cerebral hemispheres.

The cerebral hemispheres are each supplied by three arteries, all of which begin in the circle of Willis. These are the anterior, middle, and posterior cerebral arteries. They do not seem to be distributed on a functional basis. The deep lying nuclei are supplied by numerous long thin vessels which enter the brain through the anterior and posterior perforated substances.

What different functional disability would you expect following obstruction of one of the three main arteries?

The middle cerebral artery supplies the choroid plexus in the inferior horn of the lateral ventricle.

The **posterior communicating artery** runs backwards from the middle cerebral artery and joins it to the posterior cerebral artery.

Lenticulostriate arteries

The medial striate branch of the middle cerebral artery passes through the anterior perforated substance to supply the lentiform and caudate nuclei of the basal ganglia. The lateral striate branch supplies the caudate nucleus. These lenticulostriate arteries are thin-walled vessels which are very vulnerable to any sudden increases in blood pressure. Their rupture leads to a cerebral haemorrhage in the region of the internal capsule with paralysis of

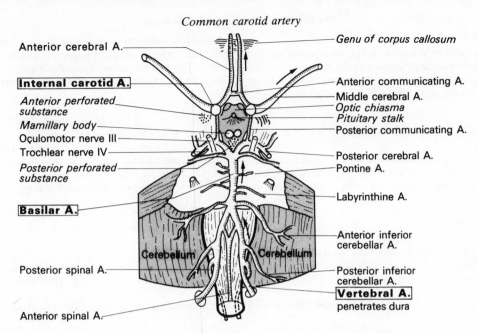

Anterior cerebral A.
Genu of corpus callosum
Internal carotid A.
Anterior perforated substance
Mamillary body
Oculomotor nerve III
Trochlear nerve IV
Posterior perforated substance
Basilar A.
Cerebellum
Posterior spinal A.
Anterior spinal A.
Anterior communicating A.
Middle cerebral A.
Optic chiasma
Pituitary stalk
Posterior communicating A.
Posterior cerebral A.
Pontine A.
Labyrinthine A.
Anterior inferior cerebellar A.
Posterior inferior cerebellar A.
Vertebral A.
penetrates dura

Fig. 13.07. Ventral view of brain to show internal carotid and vertebral arteries.

the other side of the body called hemiplegia. Such symptoms do not always imply a haemorrhage since similar symptoms may result from thromboses or even spasm of these arteries. In the case of spasm the symptoms may be transient.

3. Anterior cerebral artery

The anterior cerebral artery is the other and smaller terminal branch of the internal carotid artery. It runs forward, and communicates with the anterior cerebral artery of the other side by the **anterior communicating artery** which is usually a very short vessel (4 mm) [Fig. 13.07].

The anterior cerebral artery curves round the front of the corpus callosum and supplies the medial aspect of the cerebral hemisphere as far back as the fissure between the parietal and occipital lobes.

As has been seen it also supplies a small strip of cerebral cortex on the lateral surface at the front, and sends branches through the anterior perforating substance to the caudate nucleus.

EXTERNAL CAROTID ARTERY

The external carotid artery [Fig. 13.08] is so named because it supplies the external aspect of the skull. It is the smaller of the two terminal branches of the common carotid artery [Fig. 13.01].

The external carotid artery starts at the upper border of the thyroid cartilage and runs upwards and backwards. Behind the neck of the mandible it divides into the superficial temporal artery and the maxillary artery.

In its lower part it is covered by skin, superficial and deep fascia and sternomastoid. It then passes deep to digastric, stylohyoid, and parotid gland.

The external carotid artery is crossed by venous plexus formed by the facial, lingual, and superior thyroid veins.

The twelfth cranial nerve (XII) crosses superficially to the external and internal carotid arteries just below the posterior belly of the digastric.

The external carotid artery runs lateral to the styloid process and

its muscles whilst the internal carotid runs medial to the styloid process. Thus higher up the external and internal carotid arteries are separated by the styloglossus, stylopharyngeus as well as by the ninth cranial nerve (IX) and the pharyngeal branch of the vagus (X).

Branches of external carotid artery:

1. Ascending pharyngeal artery.
2. Superior thyroid artery.
3. Lingual artery.
4. Facial artery.
5. Occipital artery.
6. Posterior auricular artery.
7. Superficial temporal artery.
8. Maxillary artery.

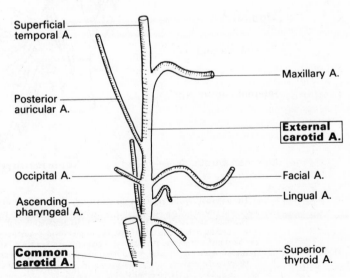

Superficial temporal A.
Posterior auricular A.
Occipital A.
Ascending pharyngeal A.
Common carotid A.
Maxillary A.
External carotid A.
Facial A.
Lingual A.
Superior thyroid A.

Fig. 13.08. Diagram of branches of the external carotid artery.

1. Ascending pharyngeal artery

The ascending pharyngeal artery arises from the external carotid artery about 1 cm from its commencement at the bifurcation of the common carotid artery. It passes deep to styloglossus and runs up on the superior constrictor to the base of the skull [Fig. 13.09].

Branches:

1. Pharyngeal branch—this branch supplies the tonsil, pharynx, and auditory tube.
2. Palatine branch—this branch runs to the soft palate.
3. Tympanic branch—this branch runs to the middle ear.
4. Meningeal branch—this branch enters the skull and supplies the meninges.

2. Superior thyroid artery

The superior thyroid artery arises from the external carotid artery just below the greater horn of the hyoid. It runs to the upper pole of the thyroid gland. To do so it passes over the surface of the middle constrictor and passes deep to the omohyoid, sternohyoid, and sternothyroid.

In the thyroid gland it divides into an anterior branch with anastomoses with the corresponding artery of the other side, and a posterior branch which anastomoses with the inferior thyroid artery. These anastomoses make the thyroid a very vascular endocrine gland especially when over-active (thyrotoxicosis).

Branches:

1. Hyoid branch (subhyoid artery)—this artery runs deep to thyrohyoid just below the hyoid.
2. Superior laryngeal artery—this artery runs deep to thyrohyoid and runs with the internal laryngeal nerve to the upper larynx.
3. Cricothyroid branch—this artery anastomoses with that of the other side and sends a branch through the median hole in the cricothyroid membrane.
4. Sternomastoid branch—this artery crosses the carotid sheath and runs to the sternomastoid.

3. Lingual artery

The lingual artery arises from the external carotid artery at the level of the greater horn of the hyoid. As its name implies it supplies the tongue.

It starts with an upward loop and then passes to the tongue

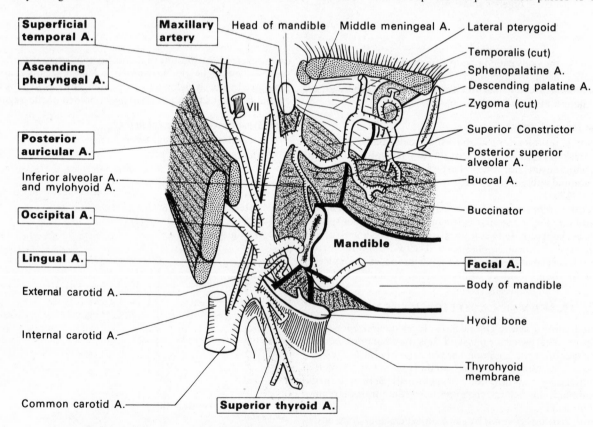

FIG. 13.09. External carotid artery: maxillary artery.

The external carotid begins at the level of the upper border of the thyroid cartilage and terminates by dividing into the superficial temporal and maxillary arteries deeply embedded in the parotid gland.

The maxillary artery runs from its origin to its destination in the pterygopalatine fossa, where it divides into the descending palatine artery, and the sphenopalatine artery. The former divides into the greater and lesser palatine arteries which are the major blood supply of the hard and soft palate; the latter also gives off two branches as it passes through into the nasal cavity one supplying the lateral wall (posterior lateral nasal artery) and the other the septum (posterior septal nasal artery).

Many of the branches of the maxillary artery are reduplicated and there are also many muscular branches which have been omitted here. In addition in the pterygopalatine fossa it gives off branches which accompany all the nerves which either enter or leave the fossa. These unimportant blood vessels have also been omitted here.

between the quadrilateral-shaped hyoglossus muscle and the middle constrictor just above the greater horn of the hyoid. It enters the under surface of the tongue and runs to the tip.

Branches:

1. Hyoid branch (suprahyoid artery)—this artery passes along the upper border of the hyoid bone.
2. Sublingual artery—this artery supplies the sublingual gland. It reaches it by running between the genioglossus and the gland itself.
3. Dorsalis linguae artery—several branches of this artery run to supply the dorsum of the tongue. A branch runs to the anterior pillar of the fauces and the tonsil.

Facial artery

The facial artery (external maxillary artery) arises from the external carotid artery just above the lingual artery. As its name implies it supplies the face. It supplies particularly the chin, lips, and nose. It also sends deeper branches to the submandibular salivary gland, the tonsil, the palate, and the pharynx.

The facial artery runs a tortuous course to reach the face. It first passes upwards deep to stylohyoid and digastric. It forms a loop upwards in the superior constrictor deep to the angle of the jaw and then runs downwards on the submandibular salivary glands before curling round the lower border of the mandible and running upwards along the side of the nose to reach the inner side of the eye (medial angle) where it anastomoses with the ophthalmic artery. This anastomosis enables blood to flow from the external carotid artery to the internal carotid artery in the skull.

The point where the facial artery crosses the lower border of the mandible is used for palpating the **facial pulse**. It lies just in front of the insertion of masseter. The anterior border of this muscle can be readily felt by clenching the jaws and this muscle forms a good landmark for locating the artery.

Branches:

1. **Ascending palatine artery**—this artery runs first between styloglossus and stylopharyngeus and then between the medial pterygoid and superior constrictor to reach the soft palate. It sends a branch to the tonsil.
2. **Branch to tonsil**—this artery arises from the loop of facial artery. It pierces the superior constrictor to reach the tonsil.
3. **Branch to submandibular gland**
4. **Submental artery**—this artery runs on mylo-hyoid and then passes upwards to the chin.
5. **Superior and inferior labial arteries**—these arteries supply the lips. The superior labial artery supplies the upper lips. The inferior labial artery supplies the lower lips. Since these arteries lie deep to the orbicularis oris their pulse is felt more readily through the mucous membrane on the inside of the mouth than from outside, when the lip is held between finger and thumb [FIG. 13.01].
6. **Lateral nasal artery**—this artery supplies the side of the nose.
7. **Angular artery**—this artery runs to the medial angle of the eye and anatomoses with the ophthalmic artery.

5. Occipital artery

The occipital artery arises from the external carotid artery at the same level as the facial artery but from its posterior side. It passes backwards and upwards to reach and supply the back of the head [see p. 294]. In this region it runs with the greater occipital nerve and can usually be palpated as it crosses the superior nuchal line.

Branches:

1. Muscular to muscles of back of neck and skull.
2. Mastoid branch—this artery enters the mastoid foramen (not to be confused with the stylomastoid foramen) and supplies the bone and meninges.

6. Posterior auricular artery

The posterior auricular artery is a small branch which arises from the external carotid artery high up at the level of the superior border of the posterior belly of digastric. It supplies the parotid gland, the ear, and the scalp behind the ear.

Branch:

Stylomastoid artery—this artery enters the stylomastoid foramen to supply the structures in the temporal bone and the facial nerve (VII).

7. Superficial temporal artery

This terminal branch of the external carotid artery is a continuation of the artery. It starts in the parotid gland behind the neck of the mandible and runs upwards over the posterior part of the zygomatic arch where its pulse can be felt. About 2 cm above the zygomatic arch it divides into an anterior and a posterior branch which anastomoses with the other arteries supplying the scalp [see p. 294].

It is the anterior branch of the superficial temporal artery that can be seen running a tortuous course under the skin in the temporal region of the head in elderly people.

Branches:

1. Transverse facial artery—this artery runs above the parotid duct below the zygomatic arch.
2. Middle temporal artery—this artery crosses the zygomatic arch and pierces the temporalis and grooves the temporal bone.
3. Zygomatico-orbital artery—this artery runs anteriorly above the zygomatic arch. It anastomoses with the ophthalmic artery.

8. Maxillary artery

The maxillary artery (internal maxillary artery) arises behind the neck of the mandible in the parotid gland as the larger of the two terminal branches of the external carotid artery. It runs inwards and upwards between the mandible and the sphenomandibular ligament to reach the pterygomaxillary fissure where it divides to supply the internal aspects of the maxilla [FIG. 13.10].

Branches:

1. Branches to foramina and fissures:
     (i) to external auditory meatus;
     (ii) to foramen ovale;
     (iii) to middle ear via petrotympanic fissure.
2. Middle meningeal artery.
3. Inferior alveolar (dental) artery.
4. Muscular branches to buccinator, temporalis, pterygoids, and masseter.
5. Maxillary branches to medial and lateral walls of nose.

Haemorrhage from the anterior septal artery gives rise to epistaxis (nose bleeding).

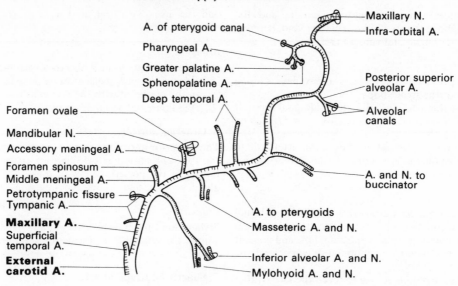

A. of pterygoid canal
Pharyngeal A.
Greater palatine A.
Sphenopalatine A.
Deep temporal A.
Foramen ovale
Mandibular N.
Accessory meningeal A.
Foramen spinosum
Middle meningeal A.
Petrotympanic fissure
Tympanic A.
Maxillary A.
Superficial temporal A.
External carotid A.
Maxillary N.
Infra-orbital A.
Posterior superior alveolar A.
Alveolar canals
A. and N. to buccinator
A. to pterygoids
Masseteric A. and N.
Inferior alveolar A. and N.
Mylohyoid A. and N.

FIG. 13.10. Branches of the maxillary artery.

Middle meningeal artery

The middle meningeal artery passes through the foramen spinosum to the middle cranial fossa. It divides into the anterior and posterior branches at the antero-inferior angle of the parietal bone. This branch of the maxillary artery is clinically the most important because of its involvements in fractures of the skull.

Inferior alveolar artery

This artery enters the inferior canal with the inferior alveolar nerve and supplies the teeth of the lower jaw.

Maxillary branches

The maxillary division of the fifth cranial nerve (V_2) is accompanied by branches from the maxillary artery. These have the same names as the nerves:
  Posterior superior alveolar (dental) artery.
  Infra-orbital artery.
  Descending palatine artery which gives rise to the greater and lesser palatine arteries.
  Sphenopalatine artery (nasopalatine artery).
  Artery of pterygoid canal (vidian artery).

VERTEBRAL ARTERY

The vertebral arteries [FIG. 13.11] on each side may conveniently be divided into four parts:

First part: Arises as branch of first part of subclavian artery and runs to the transverse process of C6. Note that the artery does not pass through the hole (foramen transversarium) in the transverse process of C7. The foramen transversarium of C7 is, however, traversed by some of the veins and sympathetic nerves which accompany the artery higher up.

Second part: Passes through the foramina transversaria in trans-

verse processes of all cervical vertebrae from C6 up to and including C1 which is more lateral than the others. The vertebral artery is accompanied by the vertebral veins and by sympathetic nerve fibres.

Third part: Passes through suboccipital triangle behind the lateral mass of the atlas.

Fourth part: Pierces the dura and arachnoid of the atlas and enters the skull through the foramen magnum in front of the

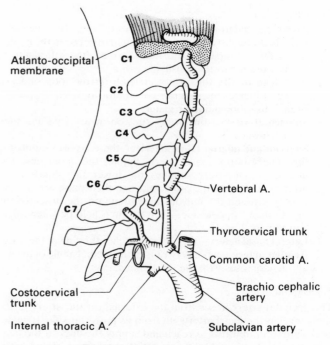

Atlanto-occipital membrane
C1
C2
C3
C4
C5
C6
C7
Vertebral A.
Thyrocervical trunk
Common carotid A.
Brachio cephalic artery
Costocervical trunk
Internal thoracic A.
Subclavian artery

FIG. 13.11. Vertebral artery in the neck.

spinal cord. It runs up over the anterior surface of the medulla and joins its opposite artery to form the basilar artery at the inferior border of the pons.

Branches:

1. Posterior meningeal artery.
2. Anterior and posterior spinal arteries.
3. Posterior inferior cerebellar artery.

BASILAR ARTERY

The basilar artery [FIG. 13.12] is formed at the lower border of the pons by the union of the two vertebral arteries. It passes upwards in a median groove on the **anterior** surface of the pons giving off branches on either side (see below). The basilar artery terminates at the upper border of the pons by dividing into the two **posterior cerebral arteries**.

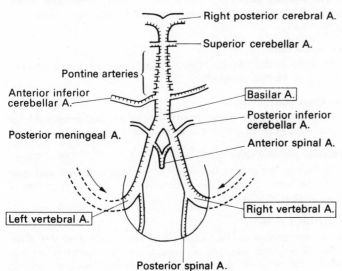

FIG. 13.12. Branches of vertebral and basilar arteries viewed from above. They lie in front of the spinal cord and brainstem.

Branches:

1. Pontine arteries—these arteries supply the pons.
2. Labyrinthine artery—this artery enters the internal acoustic meatus with the seventh and eighth cranial nerves and supplies the inner ear.
3. Anterior inferior cerebellar artery—this artery supplies the anterior part of the ventral surface of the cerebellum.
4. Superior cerebellar artery—this artery runs parallel but slightly inferior to the posterior cerebral artery. Since it arises on the opposite aspect of the brain stem to the cerebellum which it supplies, it winds round the cerebral peduncle. The third and fourth cranial nerves lie between the superior cerebellar artery and the posterior cerebral artery, as you would expect since both nerves emerge from the midbrain.

Posterior cerebral artery

The posterior cerebral artery runs round the cerebral peduncle to supply the medial surface of the occipital lobe. As has been seen above, it also supplies the posterior part of the lateral surface of the occipital lobe. This part of the cortex is associated with vision (visual cortex).

The **posterior communicating branch** joins the posterior cerebral artery to the middle cerebral artery. There is, of course, no need for a communicating branch between the two posterior cerebral arteries because they both arise from the basilar artery and are therefore already joined.

The posterior cerebral artery supplies, in addition to the posterior half of the visual pathway and the visual cortex, the surface of the cerebral hemisphere in contact with the tentorium and deeper structures such as the red nucleus, geniculate bodies, and the posterior part of the thalamus. It supplies the choroid plexus of the third ventricle.

CEREBRAL BLOOD SUPPLY AND HEAD MOVEMENT

The **circle of Willis** [FIG. 13.13] is the name given to the arteries which anastomose between two internal carotid arteries and the basilar artery. It lies in the subarachnoid space at the base of the brain and surrounds the pituitary stalk and optic chiasma [FIG. 13.07].

The presence of the circle of Willis makes it possible in certain patients to ligature one of these arteries.

It is easy to make out a case that the vertebral arteries are more important or more vital than the internal carotids because they supply the brainstem where the vital centres are to be found. The carotid artery bends through six right angles in passing from the carotid canal to reach the circle of Willis while the vertebral artery also bends through six when it passes through the foramen transversarium of the atlas, it bends backwards in the horizontal plane behind the lateral mass and then turns upwards to pierce the spinal theca and enter the skull through the foramen magnum in front of the spinal cord and brainstem. Think, however, about the artery passing from the transverse foramen of the axis (C2) to that of the atlas (C1). In the anatomical position the two foramina are more or less one above the other. But when the head is turned to the side the situation changes completely.

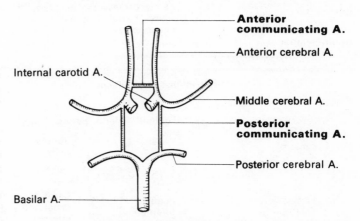

FIG. 13.13. The circle of Willis.

The circle of Willis is really a hexagon. It lies at the junction of the midbrain with the forebrain. Enclosed in the 'circle' are the posterior perforated substance, the mamillary body, the pituitary stalk, and the optic chiasma. Many very fine arteries branch from the posterior and middle cerebrals to enter the perforated substance. The branches which go through the anterior perforated substance supply the lentiform nucleus and corpus striatum and for this reason are called lenticulostriate arteries. They also supply the internal capsule and they are the commonest vessels which are involved in the development of a stroke (see also FIG. 13.07).

If you have these two bones in your hand, you will see that the distance is considerably increased when the atlas rotates on the axis. Think how these arteries twist to and fro as the head turns from side to side. If they have the suppleness characteristic of the blood vessels of youth no problem arises, but in the aged the blood supply of the most important parts of the brain may be in jeopardy, and fainting attacks may occur.

VENOUS SINUSES

The venous sinuses [FIG. 13.14] are venous channels formed by the dura mater. The principal venous sinuses are as follows:

Superior longitudinal sinus

This starts as nose veins which pass through the foramen caecum and runs over the under surface of the frontal and parietal bones. It turns right at the upper part of the occipital bone and becomes the right transverse sinus.

The superior cerebral veins run unprotected from the brain itself to join the superior longitudinal sinus.

The superior longitudinal sinus passes close to the motor and sensory cortex (the area associated with the toes and feet).

Inferior longitudinal sinus

This runs along the lower border of the falx cerebri [FIG. 13.15] to join the straight sinus.

Straight sinus

The straight sinus is formed in the midline at the junction of the tentorium cerebelli and falx cerebri which receives the vein of Galen from the cerebral nuclei.

Lateral sinuses

These start at the internal occipital protuberance. They run outwards and then forwards in the attached margin of tentorium cerebelli in the groove on the occipital bone. They now leave the tentorum and run in a groove on the mastoid, temporal, and jugular part of the occipital bone and drain into the sigmoid sinus.

The **sigmoid sinus** runs behind the mastoid antrum and leaves the skull via the jugular foramen to join the internal jugular vein [FIG. 13.14].

Occipital sinus

This starts around the foramen magnum and runs up the attached margin of the falx cerebelli.

Superior petrosal sinus

This runs along the upper border of the petrous temporal bone attached to the margin of the tentorium cerebelli.

Inferior petrosal sinus

This lies in the groove between the petrous temporal and basi-occipital bones. It unites the cavernous sinus with the beginning of the internal jugular vein.

Cavernous sinus

The cavernous sinus [FIG. 13.15] is formed by the junction of the ophthalmic veins, the sphenoparietal sinus, the superior petrosal sinus, the inferior sinus and the superficial middle cerebral (Sylvian) vein.

VENOUS DRAINAGE OF THE BRAIN

Certain cerebral veins run with the cerebral arteries while others do not. Most of the venous return from the brain eventually finds its way into the internal jugular vein.

On each side the anterior and middle cerebral arteries are accompanied by the **anterior** and **deep middle cerebral veins**. The former runs forwards above the corpus callosum while the latter lies deep in the lateral sulcus on the surface of the insula. They meet near the termination of the internal carotid artery and the circle of Willis, and form the **basal vein**. The two basal veins then run backwards skirting around the midbrain and meet dorsal to the brain stem in the midline. to form the **great cerebral vein** (of Galen) [FIG. 13.16]. This empties straight into the straight sinus, and the blood then passes into the transverse sinuses (left more than right usually) and the sigmoid sinuses to reach the jugular veins.

Recall that the posterior cerebral and superior cerebellar arteries to begin with, run from their origins from the basilar artery around the midbrain, also from ventral to dorsal. Their companion veins can readily drain into the basal veins.

The great cerebral vein is also supplied by the **inferior sagittal**

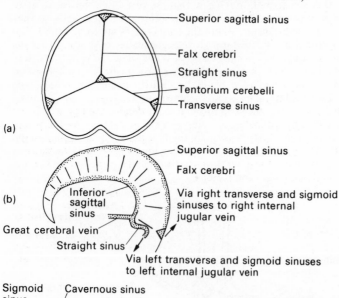

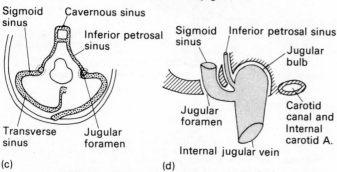

FIG. 13.14. Venous sinus and the jugular bulb.
(a) Sagittal section near back of skull.
(b) Falx cerebri from the left.
(c) Sinuses in posterior fossa.
(d) Lateral view of right jugular bulb.

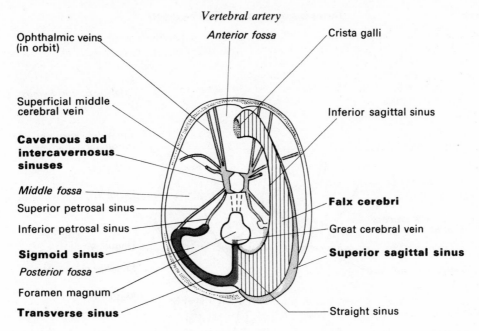

FIG. 13.15. Venous sinuses. View from above.

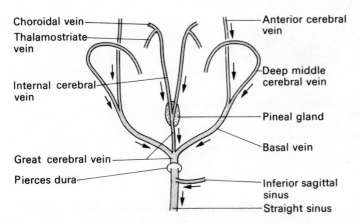

FIG. 13.16. Great cerebral vein (of Galen) and its tributaries.

joining them to the superior cerebral veins. They drain into the transverse and superior petrosal sinus.

The **superficial middle cerebral vein** needs special mention. It lies superficially in the lateral sulcus. Look at a skull and it is easy to understand that it runs close to the lesser wing of the sphenoid and that it will drain into the cavernous sinus [FIG. 13.15].

The cavernous sinus drains into the extremities of the sigmoid sinus via the superior and inferior petrosal sinuses, and then into the internal jugular vein. It also drains via the ophthalmic veins into the facial veins.

The most important clinical consideration concerns the superior cerebral veins as they pass through the arachnoid membrane and subdural space. In the commotion which occurs momentarily in head injury these veins, which have very thin walls, move with the cerebral cortex and arachnoid membrane while their openings into the superior sagittal sinus remain fixed. The resultant stretching may rupture the veins and gives rise to extensive subdural bleeding.

Emissary veins

Emissary veins connect the **veins outside** the skull to the **venous sinuses** inside the skull. Their importance stems from the fact that they provide a direct route whereby infection can spread from the face, and scalp to the meninges setting up meningitis. They do not have valves and the blood can flow in either direction.

The **ophthalmic veins** are the best example. They drain the contents of the orbit. They anastomose with the facial veins and therefore may receive blood from the mask area of the face. They drain into the cavernous sinus. An infection spreading say from a boil on the face to the cavernous sinus causes the blood to clot and causes **cavernous sinus thrombosis**. If you examine a skull, foramina are often found near the midline of the calvarium connecting scalp veins with the superior longitudinal sinus. A foramen is usual in the mastoid region draining into the sigmoid sinus. A large posterior condylar foramen is common also draining into the sigmoid sinus. Emissary veins are very variable in size and position.

sinus and by the paired **internal cerebral veins**. Each of these veins commences deep inside the brain close to its own interventricular foramen, **below** the corpus callosum in the tela choidea of the third ventricle. One is the **choroidal vein** which runs with the choroid plexus and drains it, while the other is the **thalamostriate vein**. This vein runs with the stria terminalis in the groove between the thalamus and caudate nucleus in the floor of the body of the lateral ventricle.

The superolateral surface of the cerebral cortex is drained by the **superior** and **inferior cerebral veins** which lie in the cerebral sulci.

There are about twelve **superior cerebral veins** which run upwards and backwards. They penetrate the arachnoid and cross the subdural space turning anteriorly as they do so. Their blood is discharged against the stream into the superior sagittal sinus. They drain the upper part of the cortex.

The **inferior cerebral veins** drain the temporal and occipital and lower parts of the frontal lobes. They have anastomotic channels

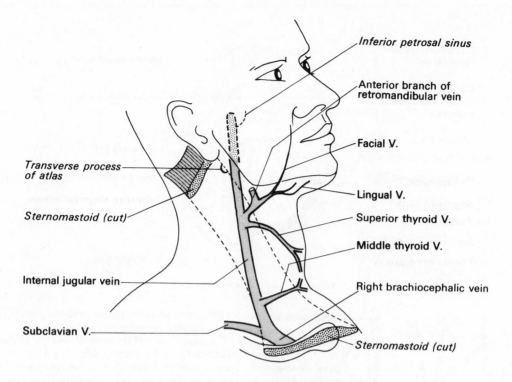

Fig. 13.17. Internal jugular vein.
The internal jugular vein lies in the carotid sheath. It commences at the base of the skull, level with the external auditory meatus. It is joined immediately by the inferior petrosal sinus. It passes just in front of the transverse process of the atlas, and from there downwards is covered by the sternomastoid.
A large tributory often conveys blood from the anterior branch of the retromandibular from the facial, lingual and inferior thyroid veins. Blood from the pterygoid and pharyngeal plexuses also flows into the internal jugular vein.

INTERNAL JUGULAR VEIN

The sigmoid (transverse) venous sinus empties into the side of the internal jugular vein at the jugular foramen [Fig. 13.14(d)]. The internal jugular vein [Fig. 13.17] runs down the neck to join the subclavian vein behind the medial end of the clavicle and become the brachiocephalic vein.

The internal jugular passes vertically down the neck on rectus capitis lateralis and the muscles of the posterior triangle. It passes behind the styloid process and the muscles that arise from it. It lies deep to sternomastoid in the carotid sheath.

Tributaries:
1. Inferior petrosal sinus.
2. Facial vein.
3. Lingual vein.
4. Superior thyroid vein.
5. Middle thyroid vein.

EXTERNAL JUGULAR VEIN

The external jugular vein [Fig. 13.18] starts by the union of the posterior auricular vein with the posterior branch of the retromandibular vein.

It can be seen under the skin at the side of the neck running superficial to sternomastoid from the midpoint between the mastoid process and the angle of the jaw to just above the midpoint of the clavicle. It pierces deep fascia 1 cm above the clavicle, crosses in front of the subclavian artery and joins the subclavian vein.

ANTERIOR JUGULAR VEIN

The anterior jugular vein starts by the union of veins in the submental region. It runs down the front of the neck to the side of the midline. It passes deep to sternomastoid and enters the external jugular vein.

POSTERIOR JUGULAR VEIN

The posterior jugular vein runs downwards from the upper angle of the posterior triangle along the edge of the trapezius. It then runs forwards and enters the external jugular vein just above the clavicle.

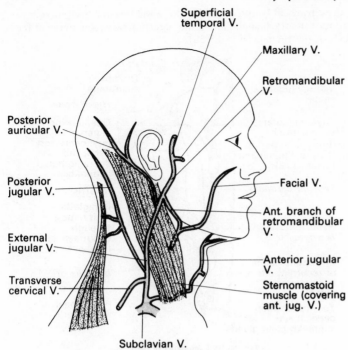

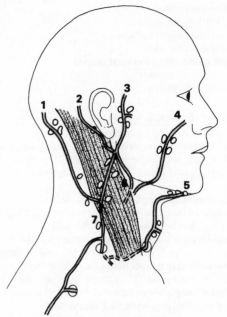

(a) Superficial cervical lymph nodes

FIG. 13.18. The external jugular vein.

The posterior jugular vein lies in the posterior triangle. the anterior jugular in the anterior triangle. The external jugular lies throughout most of its course superficial to the sternomastoid muscle, that is between the triangles.

The retromandibular vein is the companion vein of the terminal part of the external carotid artery. It therefore commences at the junction of the venae comitantes of the superficial, temporal, and maxillary arteries. The retromandibular vein has a short course and divides into anterior and posterior branches before it leaves the parotid gland. The posterior branch is the principal tributary of the external jugular vein. The anterior branch joins the internal jugular.

Both posterior and anterior jugular veins drain into the external jugular. which joins the terminal part of the subclavian vein. Note the anterior jugular passes deep to the sternomastoid.

The external jugular vein is guarded by a valve as it is about to pass through the investing fascia just above the clavicle. Severance of the vein in this region may be fatal as the investing fascia may hold the vein's lumen open. and the negative pressure in the thorax may then suck air into the right heart. An air 'embolus' may result. A finger on the wound will immediately prevent further air from entering and will arrest any bleeding.

LYMPHATICS OF THE HEAD AND NECK

The lymphatics of the head and neck form one continuous group of lymph nodes and lymph channels with numerous interconnections [Fig. 13.19].

The chains of lymph nodes may be conveniently divided as follows:

    1. **Collar chain**
        (i) submental nodes;
        (ii) submandibular nodes;
        (iii) parotid nodes;
        (iv) Posterior auricular nodes;
        (v) occipital nodes.

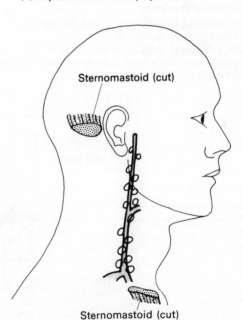

(b) Deep cervical lymph nodes

FIG. 13.19. Lymph drainage of the head and neck.
(a) The 'collar' chain of five groups of lymph nodes is shown. All these lymph nodes are readily palpable when they are enlarged. Superficial nodes also accompany the external jugular vein.
(b) The deep cervical nodes are close to the internal jugular vein. The lymph nodes below and behind the angle of the jaw, at the point where usually the facial, lingual, and superior thyroid veins converge on the internal jugular are the 'tonsilar nodes' which enlarge in tonsilitis. Note that the sternomastoid has been removed. To palpate deep cervical lymph nodes push the upper part of the muscle backwards or the lower part forwards.

2. Anterior cervical nodes
    (i) infrahyoid nodes;
    (ii) prelaryngeal nodes;
    (iii) tracheal nodes.

3. Jugular chain (deep cervical nodes)
    (i) superior group;
    (ii) inferior group.

Collar chain of the lymph nodes

There is a 'collar' of lymph glands between the head and neck [Fig. 13.19]. The **submental lymph nodes** at the front are situated on the mylohyoid muscle above the hyoid bone. They drain the anterior part of the tongue, the lower lip, and the anterior part of the floor of the mouth.

The **submandibular lymph** nodes lie in and on the submandibular salivary gland. These glands drain the side of the tongue, the floor of the mouth, and the lower part of the face.

The **parotid lymph nodes** lie in, on and deep to the parotid gland. They drain the temporal region, the nose, and the nasopharynx.

The **retro-auricular lymph nodes** lie behind the ear at the insertion of the sternomastoid. They drain the side of the scalp.

The **occipital lymph nodes** lie in the region of the superior nuchal line. They drain the posterior part of the scalp.

Anterior cervical lymph nodes

The **anterior cervical lymph** nodes drain the larynx, the thyroid gland and upper part of the trachea.

The **infrahyoid lymph** nodes are found along the anterior jugular vein. The prelaryngeal lymph nodes are found in the centre part of the cricothyroid ligament.

The **tracheal lymph nodes** lie along the inferior thyroid vein in front of the trachea and along the recurrent laryngeal nerve at the side of the trachea [Fig. 13.20].

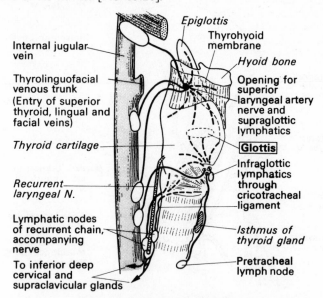

FIG. 13.20. Lymphatics of the larynx.
The lymphatic drainage of the upper half of the larynx above the vocal folds passes through the thyrohyoid membrane with the superior laryngeal vessels. From the lower half the lymph passes through the cricothyroid membrane, or below the cricoid with the inferior (recurrent) laryngeal vessels. The vocal folds form the boundary between the two lymphatic pathways.

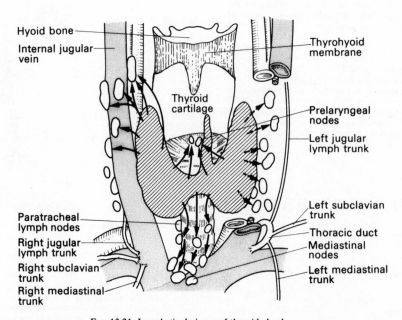

FIG. 13.21. Lymphatic drainage of thyroid gland.
The lymph from the thyroid gland is mostly conveyed to the deep cervical lymph nodes both in front and behind the internal jugular veins. Lymph is also conveyed downwards to the pretracheal and mediastinal nodes, the lymph vessels running with the inferior thyroid veins.

Jugular chain of lymph nodes

All the lymphatics of the head drain into the **jugular chain of lymph nodes** also known as the **deep cervical lymph nodes**. They are subdivided into the superior group and the inferior group.

The **superior group** of lymph nodes lies alongside the upper part of the internal jugular vein. They extend up to the transverse process of the first cervical vertebra (atlas).

The **inferior group** lie alongside the lower part of the carotid sheath mostly under the cover of the sternomastoid. The **supraclavicular glands** along the posterior belly of the omohyoid and which drain the upper part of the chest and arms also form part of this group.

All the lymph channels drain into the **thoracic duct** on the left side, and the **right lymphatic duct** on the right. These lymphatic ducts enter the brachiocephalic veins via one or more each guarded by a valve to prevent the reflux of venous blood [FIG. 13.21].

The right lymphatic ducts drain the right side of the head and neck as well as the right side of the thoracic cavity, the upper surface of the liver and the right upper limb. The thoracic duct drains the rest of the body.

SUMMARY

LYMPHATIC DRAINAGE OF THE NECK, FACE, AND SCALP

As in other parts of the body the lymphatics accompany the blood vessels. In the superficial fascia the lymphatics run with the veins which are the only vessels which run any distance in this plane. These veins are the anterior, external, and posterior jugular veins. The strategy is exactly the same as that of the superficial lymphatics in the two limbs.

The deep lymphatics run with the arteries (and companion veins).

In FIGURE 13.19 notice the nodes associated with:
1. The occipital artery.
2. The posterior auricular artery.
3. Superficial temporal artery.
4. The facial artery.

Group 3 nodes are called parotid nodes as they lie embedded in the parotid gland with the commencement of the artery. All the other nodes can easily be palpated when they are enlarged.

Group 5 are the submental nodes (*mens* (Latin) = the chin) which are associated with the anterior jugular vein. Group 6, the submandibular and Group 7 the superficial cervical nodes, are associared with the external jugular vein.

The submandibular group of lymphatics which drain the tonsil lie in the course of the facial vessels and accompany the veins to the carotid sheath where they communicate with the jugulodigastric lymph nodes. All outlying lymphatics drain ultimately into the deep cervical lymph nodes and then into the jugular lymph trunk, which opens into the thoracic duct on the left side and independently into the venous system on the right.

Normally the lymph nodes are not palpable and pathological significance is attached to them when they are. But no special significance is necessarily attached to enlargement of submental lymph nodes. Like palpable superficial inguinal lymph glands, this is usually just evidence of a low grade infection in the catchment area of the nodes, in this case from the lower teeth and gums.

14 The nerve supply to head and neck

The head and neck are supplied by cranial nerves which arise from the brain and cervical nerves which arise from the upper part of the cord [FIG. 14.01]. The twelve cranial nerves have both names and numbers. To prevent confusion with the cervical nerves, which also have numbers the roman numerals I–XII will be used as abbreviations for the cranial nerves whilst 'C' with the appropriate arabic numeral 1–8 will be used for the cervical nerves.

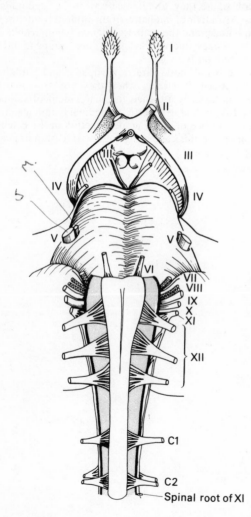

Spinal root of XI

I OLFACTORY NERVE

The first cranial nerve or olfactory nerve (I) is concerned with the sense of smell. The **olfactory receptors** are found in the mucous membrane of the nose above the superior concha. About twenty nerve filaments pass upwards through the cribriform plate of the ethmoid on each side to reach the **olfactory bulb**. The olfactory bulb can be considered to be an extension of the brain. It lies at the side of the crista galli.

The olfactory bulb acts as a relay station linking the sensory olfactory receptors in the nose with the olfactory cortex of the brain. It appears to be more than the site of simple nerve synapses.

FIG. 14.01. The cranial nerves and brainstem.

I Olfactory nerves pass through the cribriform plate and relay in the olfactory bulb. Impulses then travel in the olfactory tract and pass medial to the 'septal' region near the midline, or laterally towards the hippocampus. Function: smell.

II Optic nerves have their cells of origin in the retina. They are continuous with the optic chiasma and then with the optic tract. Note the chiasma just in front of the pituitary, and the tracts which embrace the midbrain on their way to the lateral geniculate bodies. Function: sight.

III Oculomotor nerves emerge close to each other on each side of the midline in the interpeduncular fossa, which is the deep recess between the two cerebral peduncles. Unlike I and II it is a motor nerve. Function: moves eyeball, raises eyelid, constricts pupil.

IV Trochlear nerve is unique in coming out of the dorsum of the brainstem. However, it soon joins III. Function: moves eyeball (superior oblique muscle only).

V The trigeminal nerve is a mixed nerve with large sensory and small motor components which are exactly like the sensory and motor components of an ordinary spinal nerve. Function: sensory to most skin of the face, eye and eye muscles, parts of mouth and tongue: motor to muscles of mastication.

VI Abducent nerve like III and IV a motor nerve. Function: moving eyeball. It supplies the lateral rectus.

VII The facial nerve has two components, the main trunk and the fine thread-like nervus intermedius. The former is motor, while the latter is sensory (taste) to the anterior two-thirds of the tongue. Surprisingly it also contains motor fibres, which are parasympathetic secretomotor to the lacrimal gland as well as the sublingual and submandibular glands. Main function: supplies muscles of facial expression.

VIII The vestibulocochlear nerve is in two separate segments; one is the vestibular, the other the acoustic nerve. Function: afferents from vestibular apparatus; hearing.

IX, The glossopharyngeal, vagus and accessory nerves all emerge close
X, together from the upper medulla. The vagus usually has several
and small radicles. The accessory nerve is remarkable for its spinal root,
XI which begins in the spinal cord at C5, runs upwards between the anterior and posterior nerve roots and joins the cranial root. Function: IX motor to pharyngeal and palatal muscle: sensory pharynx tonsil, carotid body and sinus. X motor to pharynx, and voluntary muscle of larynx: also heart, lung, and gut. XI accessory to vagus: motor to pharynx and palate: sternomastoid and trapezius: afferent component from neck muscles in spinal root.

XII The hypoglossal nerve.

[handwritten top margin mnemonic: "Oh oh Oh To Try And Feel Very Girls Vagina And Hymen" with nerve numbers I–XII annotated below]

THE CRANIAL NERVES

Nerve	No.	Main action	Disability when paralysed	Additional symptoms and signs
Olfactory	I	Smell	Agnosia (loss of smell)	
Optic	II	Vision	Blind in part or whole visual field	Visual field defect partial or complete
			Optic chiasma Tunnel vision with loss of both temporal visual fields	Bitemporal hemianopia
			Optic tract Loss of temporal field on contralateral side and nasal field on ipsilateral side	Homonymous hemianopia
			Optic radiation visual cortex Like optic tract, or partially like optic tract	Quadrantic or complete homonymous hemianopia
Oculomotor	III	Supplies levator palpebrae superiosis, four external oculomotor muscles. Constrictor pupillae. Ciliary muscle	Ptosis—upper lid. Eye looks down and out. Pupil dilated. Accommodation paralysed	Diplopia (double vision). Squint
Trochlear	IV	Supplies superior oblique	Cannot look down and out	Diplopia when looking down and out on affected side. Squint on affected side
Trigeminal nerve	V	Supplies muscles of mastication. Sensation to skin of face	Jaw paralysed. Anaesthesia above orbit (V_1) below orbit (V_2) below mouth (V_3)	Chin deviates towards affected side when mouth opened. No blink when cornea stimulated
Abducent	VI	Supplies lateral rectus	Cannot look laterally on affected side	Diplopia when looking laterally. Squint when looking laterally
Facial	VII	Supplies muscles of facial expression: taste anterior two-thirds of tongue	One side of face immobile. Cannot shut eye. Loss of taste in lesion above stylomastoid foramen	Upper and lower face equally affected: no 'emotional' movements. Both distinguish from supranuclear lesion
Vestibulocochlear	VIII	Afferents from vestibular apparatus. Afferents from cochlea	Nystagmus. Giddiness (vertigo). Complete deafness	Distinguish eighth nerve deafness from conducting deafness
Glossopharyngeal and Vagus	IX } X }	Supply soft palate, pharynx, larynx: heart, lungs, gut	No obvious effect on viscera. Paralysis of palate and pharynx: loss of sensation in pharyngeal wall: paralysis and anaesthesia of larynx on one side	Dysphagia with nasal regurgitation. Inability to puff cheeks without holding nose. Nasal voice. Hoarse voice. Palate deviates towards normal side when palatal and pharyngeal membranes stimulated (gagging)
Accessory	XI	Pharynx and soft palate. Sternomastoid, trapezius	Palate, sternomastoid, and trapezius paralysed	When soft palate stimulated gagging does not occur on one side
Hypoglossal	XII	Supplies intrinsic and extrinsic muscles of tongue except palatoglossus	Paralysis of tongue with wasting on affected side	Tip of tongue deviates towards side of lesion when stuck out. Speech affected. Swallowing unaffected unless bilateral lesion

[handwritten left-margin notes:]
- S — Olfactory Foramen
- S — Optic Foramen
- M — Supra orbital foramen?
- M — Only one to come from dorsum of brain stem
- S+M — 1 Ophthal – Ant Ethmoidal, 2 Max – Rotundum, 3 Mand – Ovale
- M
- Facial = M + nervus intermedius = S → Auditory Foramen + parasymp secretomotor to lacrymal, sublingual & submandibular glands
- S
- Jugular Foramen
- Magnum
- Hypoglossal canal foramen

Note: a clinical *sign* is an objective finding: example squint. A *symptom* is subjective and can only be experienced by the patient: example diplopia.

The sensory information is processed before being sent to the brain by interneurones which form microcircuits between the olfactory receptors.

The nerve impulses pass from the olfactory bulb along the olfactory tract. This is a flattened band lying under the frontal lobe of the brain to the olfactory trigone in front of the anterior perforating substance (*trigmos* (Greek) = triangular). From here the fibres run either over the corpus callosum (as striae longitudinales) or across the hippocampal fissure to reach the hippocampal area.

The **hippocampus** is the curved elevation in the floor of the inferior horn of the lateral ventricle. It consists mainly of grey matter [FIG. 14.02].

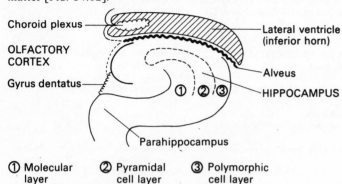

Choroid plexus — Lateral ventricle (inferior horn)

OLFACTORY CORTEX

Gyrus dentatus — Alveus — HIPPOCAMPUS

① ② ③

Parahippocampus

① Molecular layer   ② Pyramidal cell layer   ③ Polymorphic cell layer

FIG. 14.02. The olfactory cortex in the hippocampus.
1. Molecular layer.
2. Pyramidal cell layer.
3. Polymorphic cell layer.
    The hippocampus or sea-horse in ancient Greece was a monster with a horse's body and a fish's tail. It is the name given to the curved elevation in the floor of the inferior horn of the lateral ventricle [p. 369] consisting mainly of grey matter. It is the site of the conscious appreciation of smell.
    The alveus hippocampi is the name given to the bundle of nerve fibres investing the convexity of the hippocampus. The fascia dentata (dentate gyrus) has a serrated edge. The fimbria of the hippocampus is a flattened band of white fibres which runs along the medial margin of the hippocampus. It is continuous with the crus of the fornix.

The **fascia dentata** is a serrated band of grey matter on the medial side of the hippocampus. Dentate means having a toothed or serrated edge. The fascia dentata is the receptive area for smell (olfactory cortex). There are association smell areas in the hippocampus itself and in the hippocampal convolution.

II OPTIC NERVE

The second cranial nerve or optic nerve (II) is concerned with vision. The optic nerve leaving the eye is the third neurone in the sensory pathway for sight. The first neurone is the cone or rod of the retina. The second neurone is the bipolar cell of the retina which acts as an internuncial neurone between the cone or rod and the optic nerve. The optic nerve is thus the third neurone in the pathway. It runs to the **lateral geniculate body** where it synapses with the fourth neurone which runs to the visual cortex surrounding the calcarine sulcus of the occipital lobe of the brain. Optic nerve fibres also run to the superior colliculus and the midbrain for visual reflexes.

The nerve impulses from the lateral part of the retina run to the visual cortex of the same side. The nerve impulse from the nasal side of the retina cross and run to the visual cortex of the other side. This decussation takes place at the **optic chiasma** which is a characteristic feature of the optic nerve. It lies in front of the pituitary gland. A pituitary tumour which presses on the crossing fibres will cause blindness of the nasal part of each retina which receives light from the temporal fields of vision. This vision loss is termed bitemporal hemianopia.

The optic nerve can be considered to be an outgrowth of the brain since it is surrounded by the three membranes—pia, arachnoid, and dura as far as the eyeball. The central artery of the retina, which is given off by the ophthalmic artery near the optic foramen, enters the dura and arachnoid and runs in the subarachnoid space surrounding the optic nerve with its ophthalmic vein before entering the optic nerve itself 1 cm behind the eye.

The optic nerve leaves the back of the eyeball medial and just below the posterior pole. The nerve has some slack in it to allow for the movements of the eye. It leaves the orbit through the optic foramen (optic canal) which it shares with the ophthalmic artery. Immediately below are the sphenoidal air sinuses and if a sinus infection spreads through the very thin plates of bone retrobulbar neuritis may result.

As the optic nerve runs backwards and medially to join the optic chiasma, it has the internal carotid artery below and the anterior cerebral artery above. The internal carotid artery has a very tortuous pathway in this region. It runs forwards in the cavernous sinus at the side of the pituitary gland and then loops upwards and backwards before dividing into the anterior and middle cerebral arteries. It is this upwards and backwards loop that comes close to the optic nerve. An aneurysm may compress the optic nerve.

The optic chiasma is a flattened band lying in the midline over the anterior part of the pituitary fossa and the optic groove of the sphenoid bone. The optic chiasma, like the pituitary gland, is lying on the inferior surface of the brain and immediately above is the third ventricle. The optic chiasma in fact forms the floor of the third ventricle. The two internal carotid arteries are now lying on either side of the optic chiasma and the infundibulum or pituitary stalk lies immediately behind.

From the optic chiasma, the **optic tracts** pass outwards and backwards around the cerebral peduncles as flattened bands. The lateral part of the optic tract runs to the **lateral geniculate body** where the fibres relay. The remaining fibres continue on without relaying in a tract termed the **superior brachium** or brachium of the superior colliculus to the **superior colliculus** and the **pretectal area**. The alternative name for the superior colliculus is the superior corpus quadrigeminum of the midbrain.

From the lateral geniculate body the optic radiation extends to the **visual cortex** in the occipital lobe. These geniculocalcarine tracts relay the visual impulses from the lateral geniculate body to the calcarine cortex. The fibres sweep backward round the lateral ventricle and its roof which is formed by fibres of corpus callosum termed the **tapetum** (Latin = carpet). The fibres run between the lentiform nucleus and the tail of the caudate nucleus to reach the superior and inferior wall of the calcarine sulcus.

The fibres from the central area of the retina (the macula) reach the most posterior part of the calcarine sulcus. The lateral retinal fibres only reach the more anterior part of this sulcus.

Visual field defects are considered on page 359.

III OCULOMOTOR NERVE

The oculomotor nerve supplies the extrinsic and intrinsic muscles of the eye with the exception of the superior oblique (IV), the lateral rectus (VI) and the levator palpebrae superioris.

It emerges from the front of the midbrain from its nucleus close to the cerebrospinal fluid aqueduct between the third and fourth ventricles (see brainstem section) at level of the red nucleus and superior colliculus [FIG. 14.03]. The parasympathetic nucleus (Edinger–Westphal nucleus) or **pupilloconstrictor centre** lies close by (dorsomedially) and the parasympathetic fibres to the constrictor pupillae and the ciliary muscle from this nucleus join the third cranial nerve.

The oculomotor nerve passes between the posterior cerebral and superior cerebellar arteries and pierces the dura in the roof of the cavernous sinus and runs in the lateral wall of the cavernous sinus [p. 255] above all the other nerves.

The oculomotor nerve divides into a superior and an inferior branch. These enter the orbit through the **superior orbital fissure** between the two heads of the lateral rectus. The nasociliary nerves lie between two branches [FIG. 14.07, p. 275].

The **superior branch** of the oculomotor nerve supplies:
1. Superior rectus.
2. Levator palpebrae superioris.

The nerve runs to superior rectus, and passes through this muscle to reach levator palpebrae superioris.

The **inferior branch** of the oculomotor nerve supplies:
1. Medial rectus.
2. Inferior rectus.
3. Inferior oblique.

The nerve to inferior oblique is a long nerve which runs along the floor of the orbit. A branch from this nerve to the ciliary ganglion carries the preganglionic parasympathetic fibres to this ciliary ganglion where they synapse with postganglionic parasympathetic fibres which run to the iris and ciliary muscle of the eye. The postganglionic fibres reach the eye via the short ciliary nerves.

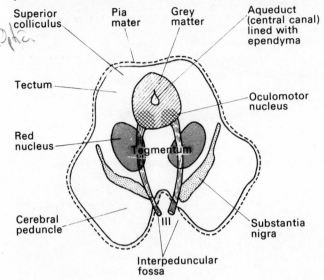

FIG. 14.03. Cross-section of brainstem at level of midbrain (superior colliculus).

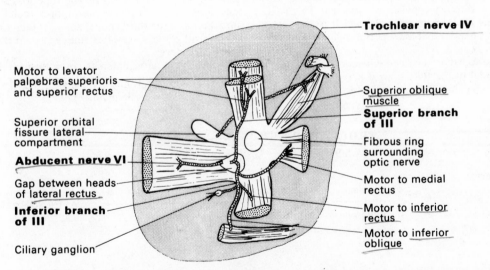

FIG. 14.04. The nerve supply to the external ocular muscles.

Diagram of right orbit from the front. The eyeball has been removed to show the oculomotor III and abducent nerve VI which, having reached the orbit through the superior orbital fissure, pass through the hiatus between the two 'heads' of the lateral rectus muscle. These supply their target muscles on their deep surfaces. The trochlear nerve avoids the lateral rectus, and passes superior to levator palpebrae to reach the superior oblique muscle.

Oculomotor nerve: the oculomotor nerve divides into superior (to superior rectus and levator palpebrae superioris) and inferior branches. The latter supplies the medial and inferior rectus muscles and inferior oblique. The branch to the last muscle supplies the preganglionic parasympathetic fibres to the ciliary ganglion. Having relayed in the ganglion the postganglionic fibres pass in the short ciliary nerves to the eye ball. The ganglion also transmits (without relaying) sympathetic and sensory fibres which are not shown.

OCULOMOTOR NERVE III—SUMMARY

Motor nerves

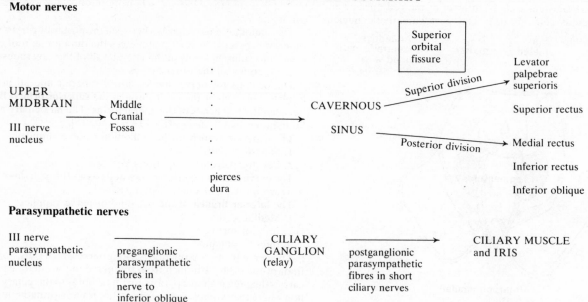

Parasympathetic nerves

III nerve parasympathetic nucleus — preganglionic parasympathetic fibres in nerve to inferior oblique — CILIARY GANGLION (relay) — postganglionic parasympathetic fibres in short ciliary nerves → CILIARY MUSCLE and IRIS

IV TROCHLEAR NERVE

The trochlear nerve (IV) supplies that extrinsic muscle of the eye whose tendon passes through a pulley or trochlea. This is the superior oblique muscle. Remember (SO_4) which is *sulphate* in chemistry and also a reminder that the superior oblique is supplied by the fourth cranial nerve.

The nucleus of the fourth cranial nerve lies in the floor of the fourth ventricle, caudal to the third cranial nerve nucleus at the level of the inferior colliculus. This nerve emerges from the *wrong* aspect of the brain and as a result it has to curl round the sides of the brainstem, crus cerebri, and the upper border of the pons. It runs between the posterior cerebral and superior cerebellar arteries. It runs along the free border of the tentorium cerebelli and pierces the dura lateral to the third cranial nerve. It then runs forwards in the lateral wall of the cavernous sinus between the oculomotor and ophthalmic nerve.

The trochlear nerve enters the orbit through the superior orbital fissure and crosses the origin of the superior rectus and levator palpebrae superioris to reach the superior oblique.

TROCHLEAR NERVE IV—SUMMARY

Supplies superior oblique only (—SO_4)

Motor nerve

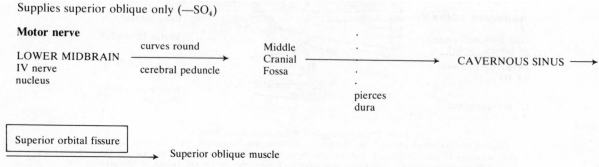

V TRIGEMINAL NERVE

The trigeminal or fifth cranial nerve [Fig. 14.05] provides the main sensory innervation (touch, pain, heat and cold, proprioception) of the head. It is so called because it divides at the trigeminal ganglion into three parts (*trigeminus* (Latin) = threefold). These divisions of the trigeminal nerve are often referred to as nerves in their own right:

1. Ophthalmic nerve (ophthalmic division of trigeminal nerve).
2. Maxillary nerve (maxillary division of trigeminal nerve).
3. Mandibular nerve (mandibular division of trigeminal nerve).

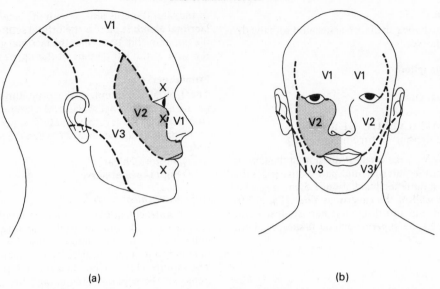

(a)   (b)

Fig. 14.05. The trigeminal nerve (V).
The three areas of the face and scalp supplied by three branches of the trigeminal nerve are shown.
Note that the opening between the eyelids (palpebral fissure) separates V_1 from V_2. The opening of the
lips separates V_2 from V_3.
The skin over the angle of the jaw is supplied not by V_3, but by the great auricular nerve, which also
supplies the lower part of the pinna of the ear.

These divisions may be referred to by the abbreviations V_1, V_2, and V_3 respectively.

The upper two divisions are made up entirely of sensory nerve fibres. The mandibular division contains both sensory and motor nerve fibres. In addition all three divisions provide pathways for sympathetic and parasympathetic nerve fibres which are running to the lacrimal and salivary glands.

The trigeminal or fifth cranial nerve has a small motor root (motor nucleus) and a large sensory root which extends from the midbrain down to the second cervical nerve segment.

The motor nucleus is found in the upper part of the floor of the fourth ventricle under cover of the superior cerebellar peduncle. It supplies the muscles which have been derived from the 1st branchial arch, namely the muscles of mastication—masseter, temporalis, medial pterygoid, lateral pterygoid, plus the tensor tympani and tensor palati.

The trigeminal nerve leaves the anterolateral surface of the pons well clear of any other nerve [Fig. 14.01, p. 268]. The two roots (motor and sensory) are at first separate. The nerves passes forward across the posterior cranial fossa and pierces the tentorium cerebelli [p. 331] to reach the trigeminal ganglion.

Trigeminal ganglion

The trigeminal ganglion (semilunar or Gasserian ganglion) lies in the dural cavum trigeminale in the lateral wall of the cavernous sinus. Immediately lateral to the trigeminal ganglion are the greater and lesser petrosal nerves. The middle meningeal artery, having entered the skull through the foramen spinosum, also lies lateral to the ganglion. The cavernous sinus [p. 255] with the third, fourth, and sixth cranial nerves lies medial to the ganglion. More medial again is the pituitary gland. The uncus of the brain lies above.

The trigeminal ganglion is associated with the sensory nerve fibres only. It corresponds to the spinal ganglia of the sensory spinal nerves. It is not a synapse but a T-junction connecting the cell of origin to the sensory nerve fibre.

The trigeminal ganglion, like the spinal ganglia, may be infected with the virus of chicken-pox, which having lain dormant for many years, flares up and produces the symptoms of herpes zoster (shingles). The triple response of the skin supplied by the cutaneous nerve is often followed by a severe neuralgic pain. It is usually unilateral. Herpes of the ophthalmic division is particularly serious since it may lead to blindness due to corneal ulceration. Trigeminal neuralgia (tic douloureux) which may be unassociated with herpes, is paroxysmal neuralgia of the trigeminal nerve.

Although the sensory fibres conveying touch, pain and temperature have their cells of origin in the trigeminal ganglion, the proprioceptive fibres from the muscles of mastication have their cells of origin in the brain stem itself. Thus:

Cells of origin

Stretch receptors in muscles of mastication	Mesencephalic nucleus →	Mesencephalic nucleus
Touch	Trigeminal ganglion →	Main sensory nucleus
Pain and temperature	Trigeminal ganglion →	Spinal nuclei

The sensory fibres from these nuclei cross to the other side via the trigeminal lemniscus and run to the superior colliculus and the thalamus of the other side.

Parasympathetic ganglia

The following parasympathetic ganglia are associated with the divisions of the trigeminal nerve:

V_1—ciliary ganglion;
V_2—sphenopalatine ganglion;
V_3—otic ganglion;
V_3—submandibular ganglion.

V₁ OPHTHALMIC NERVE
(ophthalmic division of trigeminal nerve)

The ophthalmic nerve or V_1 is the smallest division of the trigeminal nerve. It arises from the anteromedial part of the trigeminal ganglion [FIG. 14.06] as a flattened band about 2.5 cm long which passes along the lateral wall of the cavernous sinus [FIG. 13.04; p. 255] below the third and fourth cranial nerves. Just before entering the orbit through the **superior orbital fissure**, it divides into its three branches:

1. Lacrimal nerve.
2. Frontal nerve.
3. Nasociliary nerve.

The ophthalmic nerve is joined by sympathetic fibres from the adventitia of the internal carotid artery.

Lacrimal nerve

The lacrimal nerve enters the orbit through the lateral part of the superior orbital fissure [FIG. 14.07]. It runs along the upper border of the lateral rectus muscle with the lacrimal artery and enters the **lacrimal gland**. It pierces the orbital septum and ends in the skin of the upper eyelid. It supplies the lacrimal gland, the conjunctiva and the skin of the lateral part of the upper eyelid.

Frontal nerve

The frontal nerve enters the orbit through the superior orbital fissure above the upper head of the lateral rectus muscle. The nerve runs between the levator palpebrae superioris muscle and the roof of the orbit. About midway between the apex and base of the orbit it divides in:

(i) Supratrochlear nerve;
(ii) Supra-orbital nerve.

(i) *Supratrochlear nerve*

The supratrochlear nerve is the smaller of the two branches of the frontal nerve. It runs medially and forwards, passing above the pulley of the superior oblique muscle and gives off a descending filament to join the infratrochlear branch of the nasociliary nerve. The supratrochlear nerve then emerges from the orbit between the pulley of the superior oblique and the supra-orbital foramen. It curves upwards on the forehead close to bone with the supratrochlear branch of the ophthalmic artery. It sends filaments to the conjunctiva and the skin of the upper eyelid. It ascends under the cover of corrugator and frontalis and divides into branches which pierce these muscles and supply the skin of the lower part of the forehead.

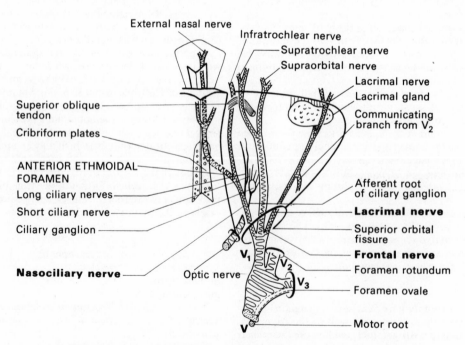

FIG. 14.06. The ophthalmic nerve (V_1).
The right ophthalmic nerve is seen diagrammatically from above. It divides before reaching the orbit into its three branches nasociliary, frontal, and lacrimal. The nasociliary nerve gives off the afferent nerves to the eyeball (hence ophthalmic nerve).
The fibres run in the long and short ciliary nerves. The frontal nerves are cutaneous in distribution. The lacrimal nerve picks up parasympathetic secretomotor fibres destined for the lacrimal gland from a communicating branch from the zygomatic nerve (V_2).

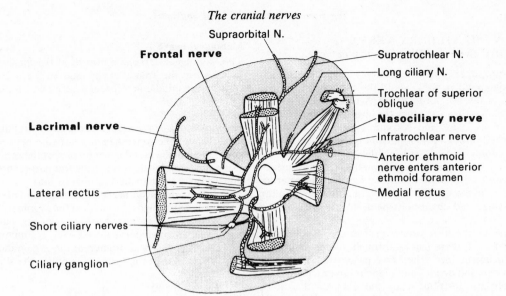

FIG. 14.07. The superior orbital fissure.
This diagram shows the three branches of V₁ as they pass through the superior orbital fissure. Both the lacrimal and frontal nerves lie next to the periosteum of the orbit as soon as they pass into the orbit and they remain in this place. The nasociliary nerve however passes between the two heads of the lateral rectus, in company with III and IV. It swings medially and lies between the superior and inferior branches of III. This branch is the most important because its ciliary component goes to the eye. The nasal component is one of the two terminal branches called the anterior ethmoid nerve. It supplies part of the nasal cavity and finally the skin on the external aspect of the nose.

(ii) *Supra-orbital nerve*

The supra-orbital nerve runs forwards between the levator palpebrae superioris and the roof of the orbit. It passes through the **supra-orbital notch** and gives off palpebral filaments to the upper eyelid and conjunctiva. It ascends the forehead with the supraorbital artery and divides into a smaller medial and a larger lateral branch which supplies the skin of the scalp as far back as the lambdoid suture (junction between the parietal and occipital bones).

Nasociliary nerve

The nasociliary nerve enters the orbit through the medial part of the superior orbital fissure between the two heads of lateral rectus [FIG. 14.07]. It crosses the optic nerve and runs obliquely below the superior rectus and superior oblique to the medial wall of the orbit. It leaves the orbit through the anterior ethmoidal foramen to appear in the anterior cranial fossa for a short distance lateral to crista galli [FIG. 14.06]. It reaches the nose through the front of the cribriform plate and supplies the nasal mucous membrane. Finally it passes between the nasal bone and lateral nasal cartilage to reach the face.

Branches:

1. **Long ciliary nerves**
   The two long ciliary nerves are given off by the nasociliary nerve as it passes over the optic nerve. They pierce the sclera and running on the choroid supply the iris, ciliary muscle, and cornea with sensory fibres.

2. **Infratrochlear nerve**
   The infratrochlear nerve is given off by the nasociliary nerve as the latter leaves the orbit [FIG. 14.06]. The infratrochlear

nerve (as its name implies) passes below the pulley (trochlea) of the superior oblique. It supplies the skin and conjunctiva around the inner angle of the eye, the caruncle, and the lacrimal sac which is draining tears from the eye and passing them down to the nose.

3. **Sensory root of the ciliary ganglion**
   The long root of the ciliary ganglion is given off by the nasociliary nerve as it passes between the two heads of the lateral rectus. This root contains only sensory fibres.

CILIARY GANGLION

The ciliary ganglion is a small reddish body about 2–3 mm in size. In addition to its connection with the nasociliary nerve, the ganglion receives **parasympathetic fibres** from the nerve to the inferior oblique (inferior division of the third cranial nerve) and **sympathetic fibres** from the plexus surrounding the internal carotid artery. These sympathetic fibres come from the first thoracic preganglionic outflow. These postganglionic fibres have originated in the superior cervical ganglion in the neck and have reached the eye by passing up in the outer coat (adventitia) of the internal carotid and ophthalmic arteries.

The short ciliary nerves run from the ganglion to the back of the eyeball where they pierce sclera and supply the iris, ciliary body, and cornea.

The ciliary ganglion is the site of the parasympathetic synapses for the third cranial nerve parasympathetic fibres supplying the eye. From the ciliary ganglion the postganglionic parasympathetic fibres run to the ciliary muscle for focusing, and to the iris (parasympathetic activity constricts the pupil). The sympathetic fibres dilate the pupil; they play no part in focusing.

V₂ MAXILLARY NERVE
(maxillary division of the trigeminal nerve)

The maxillary nerve (superior maxillary nerve) or V₂ is the second division of the trigeminal nerve [FIG. 14.08]. It arises from the middle of the convex anterior border of the trigeminal ganglion as a flattened band. It passes forwards along the lower part of the lateral wall of the cavernous sinus, and becoming more cylindrical in shape, leaves the skull through the **foramen rotundum** in the greater wing of the sphenoid bone. It crosses the upper part of the pterygopalatine fossa and inclines laterally on the posterior surface of the orbital process of the palatine bone, and on the upper part of the posterior surface of the maxilla. It enters the orbit through the **inferior orbital fissure** and changes its name to the **infra-orbital nerve**.

The nerve traverses the infra-orbital groove and canal in the floor of the orbit and then passes through the **infra-orbital foramen** to appear on the face. The nerve passes under cover of levator labii superioris and divides into branches (nasal, palpebral, and labial) which are distributed to the side of the nose, lower eyelid, skin, mucous membrane of cheek and upper lip [FIG. 14.05, p. 273].

Branches of the Maxillary nerve (V₂)
1. Meningeal nerve.
2. Zygomatic nerve.
3. Ptyergopalatine nerves.
4. Posterior superior alveolar nerve.
5. Middle superior alveolar nerve.
6. Anterior superior alveolar nerve.
7. Infra-orbital nerve (terminal branch).

Meningeal nerve

The meningeal nerve is a branch of the maxillary nerve which is given off in the cranial cavity near to the trigeminal ganglion. It follows the middle meningeal artery and supplies the dura.

Zygomatic nerve

The zygomatic nerve (temporomalar or orbital nerve) enters the orbit through the sphenomaxillary fissure between the lateral margin of the maxilla and the orbital part of the sphenoid bone. It runs along the lateral wall of the orbit and passes through a foramen in the zygomatic bone where it divides into:

1. **Zygomaticofacial nerve**—the zygomaticofacial nerve reaches the face through a foramen in the zygomatic bone [FIG. 14.08] in front and supplies the skin over this part of the face.
2. **Zygomaticotemporal nerve**—the zygomaticotemporal nerve passes through a foramen in the zygomatic bone in the temporal fossa at the side and supplies the skin in this region.

Pterygopalatine nerve

Two pterygopalatine nerves run from the maxillary nerve downwards to the pterygopalatine ganglion which is therefore suspended below the maxillary nerve (see below).

Posterior superior alveolar nerve

The posterior superior alveolar (dental) nerve arises from the maxillary nerve before it enters the infra-orbital groove. It runs down the posterior surface of the maxilla and enters the posterior alveolar canals in the maxilla to reach the alveoli or tooth sockets of the upper molar teeth [FIG. 14.08].

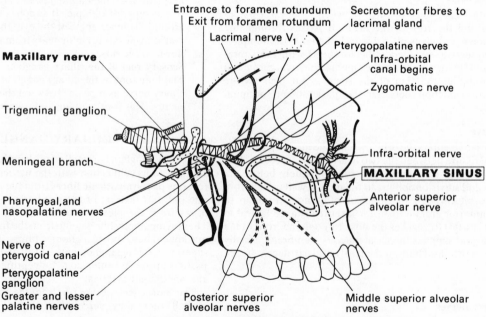

FIG. 14.08. The maxillary nerve (V₂).

The maxillary nerve commences at the trigeminal ganglion. It runs through the foramen rotundum (really a short canal) into the pterygopalatine fossa, where it gives off most of its branches. When it lies in the infra-orbital canal it changes its name to infra-orbital nerve. This nerve supplies the lower eyelid, upper lip and the cheek between them.

The superior alveolar and palatine nerves are all closely related to the maxillary antrum to which, in passing, they supply sensory fibres. The teeth are supplied indirectly but by means of a nerve plexus (not shown in the diagram) formed by the terminal branches of the alveolar nerves.

The term alveolus is the diminutive of alveus which in Latin meant a trough or excavated cavity. It is used for the bony socket of a tooth, lung air cell [p. 178], the terminal acini of a racemous gland (milk alveoli, p. 167) and cells of gastric mucosa.

Middle superior alveolar nerve

The middle superior alveolar (dental) nerve arises from the maxillary nerve in the infra-orbital canal. It runs down the lateral wall of the maxillary antrum and supplies the two premolar teeth. It also supplies the mucous membrane of the maxillary antrum and the mouth.

Anterior superior alveolar nerve

The anterior superior alveolar (dental) nerve arises from the maxillary nerve just before it passes through the infra-orbital foramen. It runs down the anterior wall of the maxillary antrum and supplies the canine and upper incisor teeth. It also supplies the mucous membrane of the maxillary antrum and the mouth.

Infra-orbital nerve

Having passed through the infra-orbital foramen the maxillary nerve becomes the infra-orbital nerve which divides into the terminal branches:

1. Palpebral nerve.
2. Nasal nerve.
3. Labial nerve.

These supply the skin of the face [FIG. 14.05, p. 273].

PTERYGOPALATINE GANGLION

The pterygopalatine ganglion is also known as Meckel's ganglion after the German eighteenth-century anatomist Meckel (1724–74) who described it. It was previously known as the sphenopalatine ganglion†

The pterygopalatine ganglion [FIG. 14.08] lies in the upper part of the pterygopalatine fossa lateral to the sphenopalatine foramen. As has been seen it is suspended from the maxillary nerve by the two pterygopalatine nerves. Since the ganglion is surrounded by bone, the nerves reach it by passing through bony canals and foramina.

The **pterygopalatine canal** runs from the pterygopalatine fossa, where the pterygopalatine ganglion is situated, downwards towards the mouth. The greater palatine canal is a continuation of this pterygopalatine canal. Its opening is termed the greater palatine foramen. The lesser (smaller) palatine canals branch from the greater palatine canal and open separately.

Although the pterygopalatine ganglion is an anatomical junction for a number of nerves, most of these nerves are simply passing through the ganglion. The only nerves which relay in the ganglion are the parasympathetic fibres to the lacrimal gland and mucous membrane of the nose and mouth which have arrived via the greater (superficial) petrosal nerve—this is a branch of the facial nerve (VII). The **greater petrosal nerve** is joined in the foramen lacerum by the **deep petrosal nerve** which is bringing sympathetic fibres from the sympathetic plexus around the internal carotid artery. These sympathetic fibres have relayed in the superior cervical ganglion. The two nerves form the **nerve of the pterygoid canal** so named because it passes through the pterygoid canal in the pterygoid bone. The pterygoid canal is also termed the Vidian canal

† In current terminology there is still a sphenopalatine foramen, but all the other terms have been changed to pterygopalatine.

after the Italian sixteenth-century anatomist Guido (Vidius) who described it. [*Gu* (French) = W = V (Latin) since there is no W in Latin].

Nerves which run to (or pass through) the pterygopalatine ganglion:

1. Pterygopalatine nerves (from maxillary nerve to pterygopalatine ganglion).
2. Nerve of pterygoid canal (sympathetic and parasympathetic).
3. Greater palatine nerve.
4. Lesser palatine nerve.
5. Nasopalatine nerve.
6. Nasal branches.
7. Pharyngeal branch.

Greater palatine nerve

The greater palatine nerve (anterior palatine nerve [FIG. 14.08] runs down the greater palatine canal with the greater palatine arteries and veins in the lateral wall of the nasal cavity. It gives off branches to the back of the nasal cavity. It then passes through the greater palatine foramen to reach the hard palate. It supplies the mucosa of the hard and soft palates (see also p. 307).

Lesser palatine nerve

The lesser palatine nerves (middle and posterior palatine nerves) run down behind the greater palatine nerve as a series of filaments in the lesser palatine canals. They pass through the lesser palatine foramina and supply the mucosa of the soft palate, the uvula, and the tonsil.

Nasopalatine nerve

The nasopalatine nerve (or long sphenopalatine nerve) takes an entirely different route. It enters the nasal cavity via the sphenopalatine foramen and crosses the roof to reach the midline septum. It passes downwards and forwards in a groove in the septum and supplies the mucosa of the nose. It passes through the **incisive foramina** (incisive = pertaining to the front incisor teeth) near the midline at the front of the hard palate to supply the mucous membrane of the hard palate and gums [FIG. 14.09]. The incisive canal which runs from the floor of the nasal cavity to the bony pit behind the upper incisors (incisive fossa) bifurcates and opens into the incisive fossa by two foramina. The medial foramen transmits the nasopalatine nerve. The lateral foramen transmits a branch of the greater palatine artery [p. 330].

Nasal branches

The nasal branches supply the mucous membrane over the superior and middle conchae, the posterior ethmoidal air cells and the upper part of the nasal septum. They enter the nasal cavity via the sphenopalatine foramen between the sphenoid and the orbital process of the palate bone.

Pharyngeal branch

The pharyngeal branch passes posteriorly in the **pharyngeal canal** to supply the pharynx behind the opening of the auditory tube.

V₃ MANDIBULAR NERVE

The mandibular nerve‡ or V_3 is the lowest division of the trigeminal

‡ This is mandibular nerve of trigeminal. The mandibular nerve of the facial will be found on p. 282.

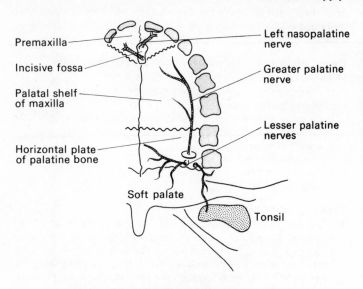

Premaxilla

Incisive fossa

Palatal shelf of maxilla

Horizontal plate of palatine bone

Soft palate

Left nasopalatine nerve

Greater palatine nerve

Lesser palatine nerves

Tonsil

FIG. 14.09. The palatine nerves.

The nasopalatine nerve reaches the palate by a circuitous route. Its fibres pass through the pterygopalatine ganglion to reach the sphenopalatine foramen which is in the lateral wall of the nose. They reach the septum of the nasal cavity by passing across the roof and running down in the septum to reach the incisive canal. This widens out into the incisive fossa. Curiously the nerves do not lie side by side, but one in front of the other.

Note that IX also supplies the soft palate and tonsil and overlaps the area shown which is supplied by V₂.

The obliquely running suture which joins the 'premaxilla' to the rest of the hard palate runs from the incisive fossa to the gap between the lateral incisor and canine teeth. In cleft palate, the gap commences between these two teeth, runs to the incisive fossa and then in the mid-line backwards to split the uvula (p. 295).

nerve. It is motor to the muscles of mastication as well as supplying the lower part of the face with sensory nerves. The larger sensory component has its cells of origin in the trigeminal ganglion whilst the smaller motor component from the motor nucleus of V bypasses the ganglion. The two components fuse into a single nerve trunk shortly after passing through the **foramen ovale**.† The mandibular nerve runs downwards between the lateral pterygoid muscle on its outer side, and the tensor palati muscle and auditory tube on its inner side [FIGS. 14.10 and 14.11]. It gives off:

Motor nerves:

1. **Nerve to medial pterygoid** (and tensor tympani and tensor palati)—this nerve supplies the medial pterygoid muscle. It gives off branches to tensor tympani and tensor palati which pass through the otic ganglion but do not synapse there.
2. **Nerve to lateral pterygoid.**
3. **Nerve to masseter.**
4. **Nerve to temporalis.**

Sensory nerves:

5. **Spinosus nerve** (nervus spinosus)—this is a recurrent nerve which re-enters the skull through the foramen spinosum. It follows the middle meningeal artery and supplies and dura of the middle cranial fossa. It also supplies the mastoid air cells.
6. **Buccal nerve**—the (long) buccal nerve runs between the two heads of the lateral pterygoid medial to the mandible. It runs across the lower head of lateral pterygoid, pierces the buccinator and supplies the skin and mucous membrane behind the angle of the mouth [FIG. 14.10].

† This is the foramen ovale of the sphenoid bone. The other foramen ovale is the foramen ovale of the heart which is the foetal opening between the two atria allowing blood to be shunted from the right to the left side of the heart [p. 186].

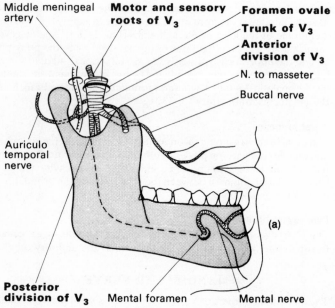

Middle meningeal artery

Motor and sensory roots of V₃

Foramen ovale

Trunk of V₃

Anterior division of V₃

N. to masseter

Buccal nerve

Auriculo temporal nerve

(a)

Posterior division of V₃

Mental foramen

Mental nerve

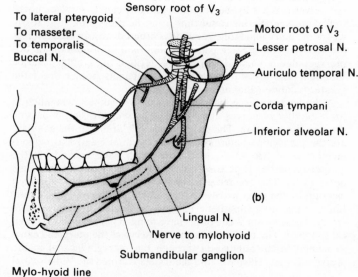

To lateral pterygoid
To masseter
To temporalis
Buccal N.

Sensory root of V₃

Motor root of V₃

Lesser petrosal N.

Auriculo temporal N.

Corda tympani

Inferior alveolar N.

(b)

Lingual N.

Nerve to mylohyoid

Submandibular ganglion

Mylo-hyoid line

FIG. 14.10. The nerve supply to the mandible.
(a) Lateral view.   (b) Medial view.

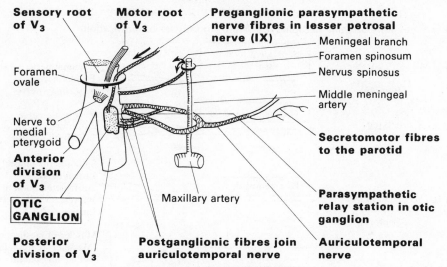

FIG. 14.11. Mandibular nerve (V₃) (view as in FIG. 14.10(b)).

The motor and sensory components unite below the foramen ovale (compare spinal nerve roots).

The parasympathetic supply (from IX) to the parotid gland relay in the otic ganglion. The postganglionic fibres run to the parotid gland in the auriculotemporal nerve.

The anterior division is mainly motor supplying three important muscles but it has the sensory buuccal nerve as its terminal branch. The posterior division is mainly sensory having three important sensory branches, but it also has a motor branch supplying the mylohyoid and anterior belly of digastric.

It is **not** motor to the buccinator muscle. Buccinator is supplied by the facial nerve (VII).

7. **Auriculotemporal nerve**—the two roots of the auriculotemporal nerve usually surround the middle meningeal artery. This nerve runs backwards deep to the lateral pterygoid and then turns upwards in front of the ear to run with the superior temporal artery to supply the temporal region of the scalp. It supplies the anterior part of the external auditory meatus and the auricle [p. 295]. It conveys the parasympathetic fibres from the otic ganglion and the sympathetic fibres from the sympathetic plexus around the middle meningeal artery to the **parotid salivary gland**.

8. **Inferior alveolar nerve**—the inferior alveolar nerve (inferior dental nerve) runs downwards below the inferior head of the lateral pterygoid muscle and gives off:

Nerve to mylohyoid

This motor nerve descends in a groove on the deep surface of the mandible and supplies the mylohyoid and anterior belly of digastric.

Dental and Incisive nerves

The inferior alveolar nerve then runs between the mandible and the sphenomandibular ligament and enters the **mandibular canal**. This canal lies approximately at the middle of the deep surface of the ramus of the mandible [FIG. 14.11].

The nerve runs along the mandibular canal (inferior dental canal) and sends **dental branches** to the roots and gums of the molar and premolar teeth. It may be damaged when wisdom teeth are removed. The **incisive nerve branch** supplies the lower canine and incisor teeth.

Mental nerve

The mental nerve is the terminal branch of the inferior alveolar nerve. It passes out through the mental foramen [FIG. 14.10(a)] and supplies the skin of the chin and lower lip.

9. **Lingual nerve**—the lingual nerve is joined by the **chorda tympani** as it passes deep to the lateral pterygoid. It runs forwards between the medial pterygoid and the mandible under the mucous membrane of the mouth. It then runs between mylohyoid and hyoglossus superior to the deep part of the submandibular gland [p. 301]. It twists round the outside of the submandibular duct, passes inwards under the duct to lie under the mucous membrane of the tongue, between the sublingual gland and genioglossus. It supplies the anterior two-thirds of the tongue (general sensation V₃, taste chorda tympani VII) [p. 299].

SUBMANDIBULAR GANGLION

The submandibular ganglion is suspended from the lingual nerve above the deep part of the submandibular gland. It is the parasympathetic ganglion for the secretomotor parasympathetic fibres supplying the **sublingual gland**. The majority of parasympathetic fibres to the **submandibular gland** relay in the gland itself [FIG. 14.00].

OTIC GANGLION

The otic ganglion is situated on the deep surface of the mandibular nerve just below the foramen ovale [FIG. 14.10]. It lies lateral to the tensor palati and the auditory tube. The middle meningeal artery lies behind.

The otic ganglion is the parasympathetic ganglion for the secretomotor parasympathetic fibres supplying the **parotid gland** [p. 306]. The preganglionic parasympathetic fibres come from the tympanic branch of the glossopharyngeal nerve IX which joins a branch of the facial nerve to form the lesser (superficial) petrosal nerve [p. 287]. The postganglionic fibres run in the auriculotemporal nerve to the parotid gland.

TRIGEMINAL NERVE V—SUMMARY

Sensory to face, mouth, nose, and nasopharynx. Motor to muscles of mastication.

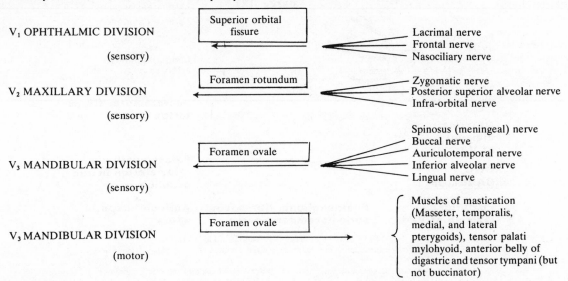

V₁ OPHTHALMIC DIVISION

(sensory)

Superior orbital fissure

Lacrimal nerve
Frontal nerve
Nasociliary nerve

V₂ MAXILLARY DIVISION

(sensory)

Foramen rotundum

Zygomatic nerve
Posterior superior alveolar nerve
Infra-orbital nerve

V₃ MANDIBULAR DIVISION

(sensory)

Foramen ovale

Spinosus (meningeal) nerve
Buccal nerve
Auriculotemporal nerve
Inferior alveolar nerve
Lingual nerve

V₃ MANDIBULAR DIVISION

(motor)

Foramen ovale

Muscles of mastication (Masseter, temporalis, medial, and lateral pterygoids), tensor palati mylohyoid, anterior belly of digastric and tensor tympani (but not buccinator)

The sympathetic fibres passing through the otic ganglion come from the sympathetic plexus around the middle meningeal artery. These are postganglionic sympathetic fibres that have come from the superior cervical ganglion and have reached the plexus in the outer coats of the external carotid artery, the maxillary artery and the middle meningeal artery.

The motor and sensory nerves to the tensor tympani and tensor palati which come from the nerve to medial pterygoid pass through the otic ganglion.

VI ABDUCENT NERVE

The abducent (abducens) nerve or sixth cranial nerve is a slender nerve which is the motor nerve supply to the lateral rectus muscle of the eye. The lateral rectus muscle of one eye works with the medial rectus of the other eye (supplied by oculomotor nerve III) to move the eyes to one side hence the name abducens (Latin = to lead away).

The nucleus of the sixth cranial nerve is a small spherical mass of grey matter in the floor of the fourth ventricle close to the midline. The nerve emerges from the lower border of the pons and crosses the antero-inferior cerebellar artery. The sixth nerve has to run upwards along the whole length of the pons between it and the occipital bone to reach the eye and, as a result, is very vulnerable to injury. The nerve pierces the dura on the basilar part of the sphenoid bone 1 cm below the root of the dorsum sellae. It runs up the posterior aspect of the petrous temporal bone, makes a right

angle bend as it turns horizontally to enter the lateral wall of the cavernous sinus with the internal carotide artery [FIG. 13.04, p. 255]. The sixth nerve enters the orbit through the superior orbital fissure between the two heads of lateral rectus to reach the body of the lateral rectus muscle which it supplies.

In the cavernous sinus it receives sympathetic fibres from the carotid plexus.

VII FACIAL NERVE

The facial nerve or seventh cranial nerve supplies the structures which have developed from the second branchial arch. Although the seventh cranial nerve is usually thought of as the nerve which innervates the muscles of the face (hence its name—facial nerve) it spends the part of its course from the brainstem to the periphery in close proximity to the eighth nerve and hence in close proximity to the inner ear which lies deep in the petrous temporal bone.

The facial nerve arises by two roots from the lateral part of the sulcus between the pons and the medulla medial to the eighth nerve [FIG. 14.01]. The larger of the two roots is motor. The smaller (nervus intermedius) is sensory and also contains the parasympathetic fibres. It is termed the nervus intermedius because it lies between the main seventh nerve root and the roots of the eighth nerve.

The seventh nerve runs laterally and forwards with the eighth nerve across the posterior cranial fossa and the two enter the internal acoustic (auditory) meatus with the labyrinthine artery which is a branch of the basilar artery. The seventh nerve pierces

ABDUCENT NERVE VI—SUMMARY

Supplies lateral rectus only (LR₆)

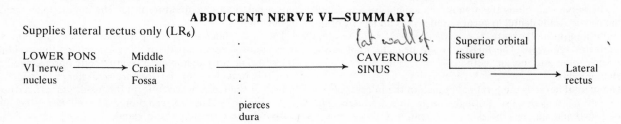

LOWER PONS
VI nerve
nucleus

Middle
Cranial
Fossa

pierces
dura

lat wall of
CAVERNOUS
SINUS

Superior orbital
fissure

Lateral
rectus

the dura and passes through the perforated plates of bone (lamina cribrosa) at the lateral end of the internal acoustic meatus and continues to run laterally for a short distance in the bony facial canal which runs between the cochlear and the vestibule of the inner ear [FIG. 14.12].

This canal becomes enlarged by the sensory **geniculate ganglion** (*geniculum* Latin = small knee) of the facial nerve which contains the cells of origin of the sensory taste fibres and gives off:

1. Greater (superficial) petrosal nerve.
2. Branch to tympanic plexus.

3. Branch to sympathetic plexus on middle meningeal artery.

The geniculate ganglion is so named because the nerve makes a sharp bend here. It turns sharply backwards (**geniculum** of facial nerve) in the bone of the upper part of the medial wall of the middle ear. The facial nerve makes a second bend, this time downwards on reaching the posterior wall of the middle ear and gives off:

1. Nerve to stapedius.
2. Chorda tympani.
3. Communicating nerve to auricular branch of the vagus.

The facial nerve emerges from the skull through the stylomastoid

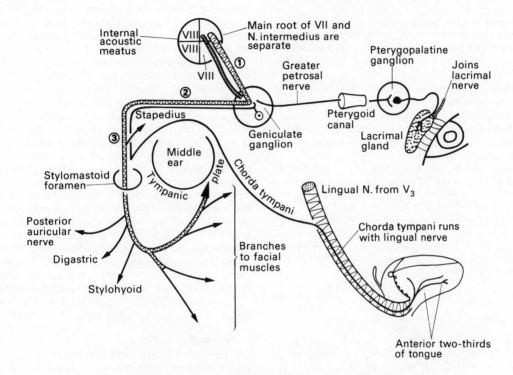

FIG. 14.12. Diagram of facial nerve (VII) and its branches.
The facial nerve runs laterally and enters the anterosuperior quadrant of the internal acoustic meatus. It *continues* ① in the same lateral direction to the genu. At the genu it runs horizontally backwards ② in the medial wall of the middle ear. It passes just below the lateral semicircular canal of the inner ear. Having reached the posterior extremity of the middle ear it turns downwards ③ in the vertical direction and escapes from the skull through the stylomastoid foramen. The canal for the facial nerve, which commenced at the internal acoustic meatus terminates here. The stapedius muscle belly runs parallel and close to section ③ of the nerve, and receives its nerve supply from it.

The chorda tympani is a branch of part ③. In facial paralysis due to a lesion at the stylomastoid foramen or more distally, taste in the anterior two-thirds of the tongue is unaffected. A lesion proximal to a point a few millimetres above the foramen will produce loss of taste owing to chorda tympani paralysis.

Note that the geniculate ganglion is like a dorsal spinal ganglion in that it contains the cells of origin of the sensory fibres of VII. Passing through it without relaying is the parasympathetic secretomotor pathway to the lacrimal gland. These fibres relay in the pterygopalatine ganglion. The fibres pass along the floor of the middle cranial fossa into the foramen lacerum, pterygoid canal, and pterygopalatine fossa. Having relayed, the postganglionic pathway passes with the zygomatic branch of the maxillary nerve (V₂) and then changes horses and joins the lacrimal nerve a branch of the ophthalmic nerve (V₁). Activity in this pathway produces copious tears.

The nerve gives off the posterior auricular nerve and branches to the posterior belly of digastric (origin from the deep surface of the mastoid process) and the stylohyoid (from the styloid process) which are adjacent to the nerve as it emerges from the stylomastoid foramen. When it enters the parotid gland it breaks up into its several branches. In the parotid the nerve is traditionally referred to as the pes anserinus (=a goose's foot—now unofficial and fanciful jargon).

In the face the named branches break up into an elaborate plexus of fine nerves which communicate freely with the local branches of the trigeminal nerve V. Nerves as they enter an individual muscle of facial expression therefore consist of a mixture of motor and sensory fibres, just like any other 'motor' nerve in other parts of the body.

foramen under the cover of the mastoid process. It then runs anterolaterally between the styloid process and the posterior belly of **digastric** and gives off:

1. Nerves to posterior belly of digastric and stylohyoid.
2. Posterior auricular nerve.

It enters the posteromedial aspect of the parotid gland (the gland is actually wrapped around the nerve) and after passing superficially to the external carotid artery breaks up in the parotid gland [Fig. 14.13] into its terminal branches:

1. **Temporal branch**—this branch passes over the zygomatic arch to supply the facial muscles above the level of the palpebral fissure.
2. **Zygomatic branches**—the smaller upper branches cross the zygomatic arch and run towards the orbicularis oculi muscle. The larger lower branches run with the parotid duct over the masseter to form the infra-orbital plexus with the infra-orbital nerve(V_2).
3. **Buccal branches**—these branches run towards the angle of the mouth and form a plexus with the buccal nerve. This branch supplies the buccinator.
4. **Mandibular branch**—the (marginal) mandibular branch runs along the inferior border of the mandible and supplies the muscles of the chin and lower lip.
5. **Cervical branch**—this branch leaves the lower border of the parotid gland and runs forward under platysma, which it supplies.

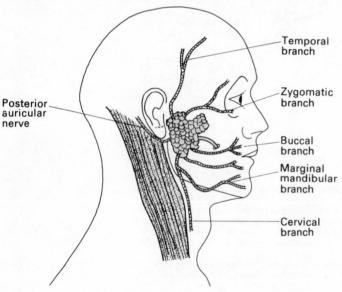

FIG. 14.13. The branches of the facial nerve (VII).

Greater petrosal nerve

The greater (superficial) petrosal nerve is given off by the facial nerve at the geniculate ganglion. It passes through the hiatus of the facial canal (hiatus Fallopii) and runs in a groove on the surface of the petrous temporal bone passing under the trigeminal ganglion to reach the foramen lacerum. Here it meets the **deep petrosal nerve** bringing sympathetic fibres from the internal carotid artery. Together they become the **nerve of the pterygoid canal**. This nerve, as its name implies, runs through the pterygoid canal to reach the pterygopalatine ganglion.

The greater petrosal nerve contains parasympathetic fibres to the lacrimal glands, sensory fibres supplying the mucous membrane of the soft palate and secretory fibres to the mucous glands.

Chorda tympani

The chorda tympani fibres run in the nervus intermedius. It is given off by the facial nerve just above the stylomastoid foramen. It runs in a canal in the temporal bone to the middle ear, crosses the tympanic membrane and neck of the malleus, and enters another canal in the temporal bone (Huguier's canal) which leads into the petrotympanic fissure or Glaserian fissure named after the Swiss seventeenth-century anatomist Glaser. Huguier was a nineteenth-century French surgeon. The nerve forms a groove in the spine of the sphenoid bone and joins the lingual nerve, which is a branch of the mandibular division of the fifth cranial nerve [Fig. 14.10(b)].

The chorda tympani contains the taste fibres from the anterior two-thirds of the tongue. The other sensations from this region of the tongue reach the brain via the trigeminal nerve.

The chorda tympani also supplies the submandibular and sublingual salivary glands with parasympathetic fibres. These fibres are secretomotor to the glands. They produce salivation and also increase the blood flow. The parasympathetic fibres in the chorda tympani are preganglionic. Rather confusingly, the synapses for the parasympathetic fibres to the sublingual salivary glands are found in the submandibular ganglion. The synapses for the submandibular gland are found mainly in the gland itself. From these synapses the postganglionic fibres run to the secreting cells. These postganglionic fibres are cholinergic (they release acetylcholine) and this action is blocked by atropine and hyoscine [p. 11].

Although the motor nucleus of the facial nerve lies close to its point of exit (as an upward continuation of the anterior horn cells of the spinal cord), the nerve fibres first run backwards to the floor of the fourth ventricle, loop around the nucleus of the abducent nerve and pass forwards again to run between their own nucleus and the spinal root of the trigeminal nerve [Fig. 19.22, p. 380].

Facial canal

The path taken by the **facial nerve** in the temporal bone, although appearing at first sight to be extremely complex, is easy to deduce. It runs laterally, backwards, then downwards [Fig. 14.15(b)].

It first runs laterally until it comes to the medial wall of the middle ear (not the lateral wall because this would mean dangling in mid-air as it passed across the middle ear itself). It turns **backwards** and runs in its bony canal on the medial wall of the middle ear (not forwards because it has to reach its exit from the skull, the stylomastoid foramen, which lies **behind**, not in front of, the external acoustic meatus). The geniculate ganglion is found at this first bend. When it reaches the posterior wall of the middle ear (not anterior wall because it is running backwards), it turns **downwards** (obviously not upwards) and leaves the skull via the stylomastoid foramen which as its name implies, is close to the mastoid process [Fig. 14.14].

The **chorda tympani** does cross the **lateral wall** of the middle ear. Since it runs horizontally across the inside of the ear drum, it is running parallel to the facial nerve on the opposite wall. Since it is a branch of the facial nerve, it must have originated as the facial nerve was running down the posterior wall—as this is the only point where the nerves could meet.

The **greater petrosal nerve**, on the other hand, is running to the pterygopalatine ganglion which lies anteriorly and more medially at the front of the skull. The nearest point on the route of the facial

FACIAL NERVE VII—SUMMARY

Motor to muscles of face

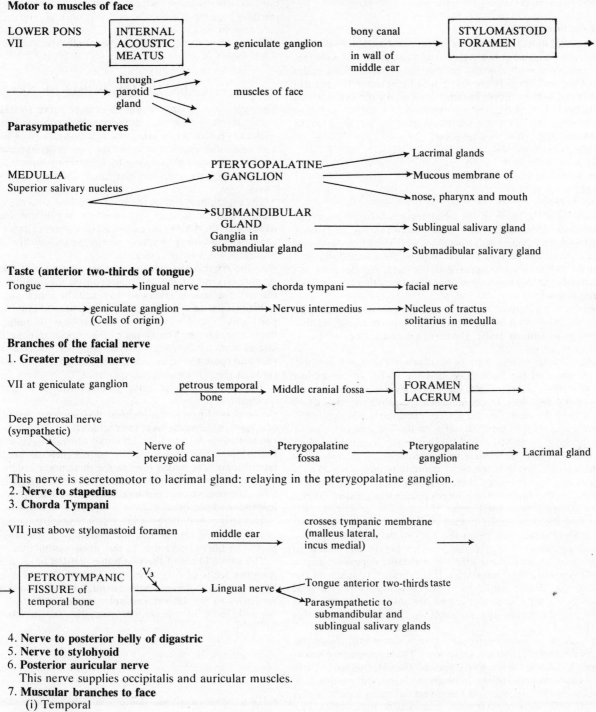

LOWER PONS VII → INTERNAL ACOUSTIC MEATUS → geniculate ganglion

bony canal in wall of middle ear → STYLOMASTOID FORAMEN →

→ through parotid gland → muscles of face

Parasympathetic nerves

MEDULLA
Superior salivary nucleus → PTERYGOPALATINE GANGLION → Lacrimal glands
→ Mucous membrane of
→ nose, pharynx and mouth

→ SUBMANDIBULAR GLAND
Ganglia in submandiular gland → Sublingual salivary gland
→ Submadibular salivary gland

Taste (anterior two-thirds of tongue)

Tongue → lingual nerve → chorda tympani → facial nerve

→ geniculate ganglion (Cells of origin) → Nervus intermedius → Nucleus of tractus solitarius in medulla

Branches of the facial nerve

1. Greater petrosal nerve

VII at geniculate ganglion —petrous temporal bone→ Middle cranial fossa → FORAMEN LACERUM →

Deep petrosal nerve (sympathetic)
→ Nerve of pterygoid canal → Pterygopalatine fossa → Pterygopalatine ganglion → Lacrimal gland

This nerve is secretomotor to lacrimal gland: relaying in the pterygopalatine ganglion.

2. Nerve to stapedius

3. Chorda Tympani

VII just above stylomastoid foramen —middle ear→ crosses tympanic membrane (malleus lateral, incus medial) →

→ PETROTYMPANIC FISSURE of temporal bone —V₃→ Lingual nerve ← Tongue anterior two-thirds taste
← Parasympathetic to submandibular and sublingual salivary glands

4. Nerve to posterior belly of digastric

5. Nerve to stylohyoid

6. Posterior auricular nerve
This nerve supplies occipitalis and auricular muscles.

7. Muscular branches to face
  (i) Temporal
  (ii) Zygomatic
  (iii) Buccal
  (iv) Mandibular
  (v) Cervical
These nerves supply buccinator, platsyma, and the muscles of facial expression [p. 297].

nerve will be first bend at the geniculum before the facial nerve starts running backwards. It is not surprising therefore that the greater petrosal nerve arises from the geniculate ganglion.

Testing the facial nerve

All the muscles of facial expression are supplied by the facial nerve which leaves the skull at the stylomastoid foramen, and runs forwards through the substance of the parotid gland and across the ramus of the mandible. It breaks up into branches in the parotid which fan out as they supply muscles from above the eyebrows to below the chin [Fig. 14.13].

When testing this nerve it is usual to start at the forehead and to work down. Make your own schedule:
1. Raise your eyebrows.
2. Screw up your eyes.
3. Show your teeth.
4. Whistle (or try to).
5. Etc.

A paralysis of the trunk of the facial nerve sometimes occurs at the stylomastoid foramen for no obvious reason; we say the cause is 'idiopathic'. This is called Bell's paralysis which was first described by Charles Bell. The patient's face droops on the affected side and the skin wrinkles disappear because all the facial muscles are completely paralysed. No movement, emotional or otherwise, is possible [see below].

In facial nerve paralysis from any cause, the eyesight is in danger, because the eye cannot be closed in sleep, nor is blinking possible (paralysis of **orbicularis oculi**). The cornea becomes dry and severe infection can set in.

In addition, saliva collects in the vestibule of the mouth because of the paralysis of the **buccinator**, and the corner of the mouth droops (paralysis of **levator labii superioris**, **levator anguli oris**, and the **zygomatic muscles**). In consequence the saliva overflows and dribbles down the chin. The paralysis of **orbicularis oris** means it is impossible to make an airtight seal with the lips and consonants such as 'p' and 'f' cannot be enunciated with normal emphasis. Most cases of Bell's paralysis recover spontaneously.

Besides Bell's paralysis the facial nerve may be paralysed as a complication of other lesions along its course either in the medulla oblongata of the brain, at the internal acoustic meatus, in the middle ear, or in the parotid gland.

Note that the mandibular branch of the seventh cranial nerve strays down into the neck below the body of the mandible before sweeping upwards to supply the muscles below the lateral to the lips (part of **orbicularis oris**, the **mentalis**, **depressor anguli oris**, and **depressor labii inferioris**). If this branch is damaged the corner of the mouth turns up in an unsightly sneer. Surgical incisions in the submandibular region are always made at least a thumb's breadth below the mandible to avoid damage to this nerve [Fig. 14.13].

Finally, look in your mirror and give a sharp sniff. Note the flaring of the nostrils (levatores alae nasi). These muscles are accessory muscles of respiration which prevent the external nares from being sucked inwards during inspiration. In normal people it is inactive except in exercise and emotional stress. In a patient it is usually a sure sign of dyspnoea and occurs in diseases like pneumonia.

Paralysis of one half of the face may occur in patients with hemiplegia, when the brain rather than the peripheral nervous system has been damaged. The disability often differs from that which follows damage to the seventh cranial nerve itself in a most interesting way. Such patients are unable to smile politely when you say something which does not really amuse them, but if you can produce a 'belly laugh', which is an emotional response, the movements of the face may then be perfectly normal. This means that the pathways in the brain for deliberate facial movements (acting) are different from those which are essentially spontaneous and emotional in origin.

VIII VESTIBULOCOCHLEAR NERVE

The vestibulocochlear or eighth cranial nerve (VIII), also commonly referred to as the auditory nerve, consists of two parts, a cochlear division for hearing and vestibular division for balancing. The vestibular division supplies the vestibular apparatus [p. 363] which consists of the saccule, utricle, anterior (superior) semicircular canal, posterior semicircular canal, and the lateral semicircular canal.

The eighth nerve joins the brainstem lateral to the nervus intermedius and the seventh cranial nerve. It runs outwards and forwards with these two nerves and all three nerves enter the internal acoustic (auditory) meatus, accompanied by the labyrinthine branch of the basilar artery.

In the meatus the eighth nerve is separate in its two divisions. At the far end of the internal acoustic meatus the nerves pass through the **lamina cribrosa**† to reach the inner ear. The lamina cribrosa is partially divided into four quadrants [Fig. 14.15] by bony partitions. The fibres to the cochlea pass through the anteroinferior quadrant. The fibres to the utricle and the anterior and lateral semicircular canals pass through the posterosuperior quadrant and those to the saccule and posterior semicircular canal pass through the postero-inferior quadrant.

The seventh cranial nerve passes through the anterosuperior quadrant.

The nerve fibres of the **cochlear division** have their ganglia (cells of origin) in the **modiolus** of the cochlea (spiral ganglion). The fibres run to two nuclei, the dorsal and the ventral cochlear nuclei in the brainstem. The fibres from the **ventral cochlear nucleus** join the lateral lemniscus which runs to the medial geniculate bodies, the inferior colliculi and the anterior transverse gyri of the temporal lobe. The fibres from the **dorsal cochlear nucleus** cross to the lateral lemniscus of the other side.

The striae medullares is used as a landmark which divides the fourth ventricle into an inferior medullary and a superior pontine part. The fibres lie ventral to the striae medullares.

The nerve fibres of the **vestibular division** have their **vestibular ganglion** (cells of origin) in the internal acoustic meatus. These fibres run to the **dorsal vestibular nucleus** in the floor of the fourth ventricle, to the **lateral vestibular nucleus** (Deiter's nucleus) in the side wall of the fourth ventricle and to the **cerebellum** [p. 377].

---

† The term lamina cribrosa (Latin = thin plate perforated like a sieve) is used for:
1. In the thigh for the fascia covering the saphenous opening.
2. In the eye for the part of the sclera through which the optic nerve passes.
3. In the brain for the anterior and posterior perforated spaces.
4. In the ear for the perforated plates of bone through which the seventh and eighth cranial nerves pass.

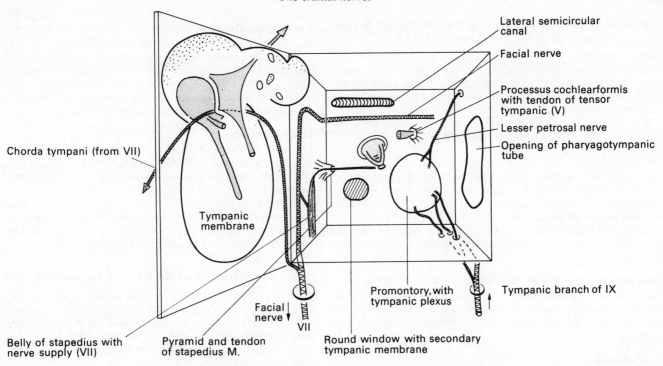

Chorda tympani (from VII)

Tympanic membrane

Lateral semicircular canal

Facial nerve

Processus cochlearformis with tendon of tensor tympanic (V)

Lesser petrosal nerve

Opening of pharyagotympanic tube

Facial nerve VII

Promontory, with tympanic plexus

Tympanic branch of IX

Belly of stapedius with nerve supply (VII)

Pyramid and tendon of stapedius M.

Round window with secondary tympanic membrane

FIG. 14.14. Diagrammatic representation of middle ear as a small box. The lateral wall is the lid. It has been opened to reveal the contents. Note the facial nerve, chorda tympani and tympanic plexus.

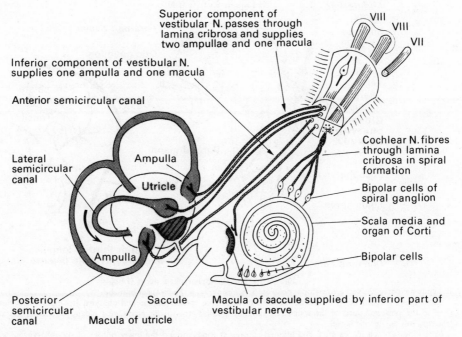

Superior component of vestibular N. passes through lamina cribrosa and supplies two ampullae and one macula

Inferior component of vestibular N. supplies one ampulla and one macula

Anterior semicircular canal

Lateral semicircular canal

Ampulla

Utricle

Posterior semicircular canal

Ampulla

Macula of utricle

Saccule

Macula of saccule supplied by inferior part of vestibular nerve

VIII VIII VII

Cochlear N. fibres through lamina cribrosa in spiral formation

Bipolar cells of spiral ganglion

Scala media and organ of Corti

Bipolar cells

FIG. 14.15. The vestibulocochlear nerve (VIII).

GREY MATTER AND TRACTS ASSOCIATED WITH VESTIBULOCOCHLEAR NERVE

Lateral lemniscus

The lateral lemniscus (or fillet) is a bundle of nerve fibres which arises from the cochlear nuclei of both sides, crosses to the other side, and terminates in the inferior colliculus and medial geniculate body of the other side.

The decussation takes place ventrally in the pons and the transverse decussating fibres form the **trapezoid body**.

Medial lemniscus

The lateral lemniscus (above) should not be confused with the medial lemniscus which is a band of nerve fibres which arise from the nucleus cuneatus and nucleus gracilis in the medulla, cross as the **internal arcuate fibres** and run upwards through the pons to the ventrolateral nucleus of the thalamus. This is a continuation of the posterior columns sensory pathway of the spinal cord for touch and proprioception.

Medial longitudinal bundle (fasciculus)

This is an intersegmented bundle of nerve fibres running from the medulla to the upper part of the midbrain adjacent to the midline. It receives contributions from both the vestibular and cochlear divisions.

Inferior colliculus

The inferior colliculus, alternatively known as inferior quadrigeminal body, is a centre for auditory reflexes.

The two inferior colliculi and the two superior colliculi (centres for visual reflexes) form rounded eminences on the back of the midbrain. Together they are termed the corpora quadrigemina.

Medial geniculate body

The geniculate bodies are four flattened oval bodies on the posterior inferior aspect of the thalamus. The lateral geniculate body is a relay station for visual impulses from the retina to the occipital (visual) cortex. The medial geniculate body receives auditory impulses from the inferior colliculus and relays them via the auditory radiations to the temporal (auditory) cortex.

Temporal cortex

The auditory area (area 41) lies in a transverse gyrus which runs across the middle of the superior surface of the superior temporal gyrus of the temporal lobe of the brain. This area is mostly buried in the floor of the lateral fissure.

The ears are bilaterally represented in the auditory cortices so that removal of one temporal lobe produces very little hearing loss.

Stimulation of the auditory cortex in a conscious patient produces the sensations of buzzing and roaring sounds.

IX GLOSSOPHARYNGEAL NERVE

The glossopharyngeal nerve or ninth cranial nerve [Fig. 14.15] is the nerve supply to the third branchial arch. It arises from a series of rootlets from the posterolateral sulcus of the medulla between the olive and the inferior cerebellar peduncle [Fig. 14.01, p. 268]. These rootlets are continuous with those of the vagus (tenth cranial) nerve (see below).

The glossopharyngeal nerve passes superiorly and laterally below the flocculolobular lobe of the cerebellum to reach the **jugular foramen** through which it leaves the skull. This foramen is shared with the **vagus** and the **accessory nerve**, the **inferior petrosal sinus** and the **sigmoid sinus**.

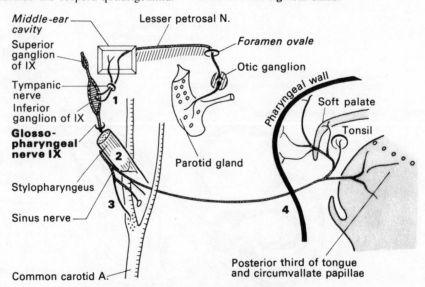

FIG. 14.16. The glossopharyngeal nerve (IX).
1. Tympanic nerve. 2. Nerve to stylopharyngeus. 3. Sinus nerve. 4. Pharyngeal branches.

As its name suggests the principal destinations of this nerve are the tongue and pharynx. It supplies sensation including taste to the posterior third of the tongue. It supplies motor fibres to the pharyngeal plexus as well as sensation to the pharynx, soft palate, and tonsil.

In passing between the external and internal carotids it winds around the lower border of stylopharyngeus supplying it. The nerve can be found here quickly and easily in a dissection. The sinus nerve comes off in this region.

The tympanic nerve contains sensory fibres to the middle-ear cavity and pharyngotympanic tube as well as the preganglionic parasympathetic secretomotor fibres to the parotid gland. These, having traversed the middle ear, pass in the lesser petrosal nerve to the foramen oval through which they pass to relay in the otic ganglion. The postganglionic fibres accompany the auriculotemporal nerve to the parotid.

The glossopharyngeal nerve has two ganglia. The **superior ganglion** lies in the jugular foramen. The **inferior ganglion** lies below the foramen.

The glossopharyngeal nerve passes downwards in the neck between the internal and external carotid arteries. It curls round the stylopharyngeus which it supplies; these motor fibres probably come from the **nucleus ambiguus**. The glossopharyngeal nerve then passes between the superior and middle constrictors [FIG. 16.11, p. 319] to reach the tongue and pharynx.

It supplies the posterior third of the tongue with taste and general sensation. The taste fibres probably run to the **tractus solitarius** in the brainstem. The glossopharyngeal nerve supplies the soft palate. tonsil. and oropharynx with general sensation. It supplies the parotid gland with secretomotor parasympathetic fibres. It intervates the carotid sinus and carotid body (sinus nerve).

Branches:

1. **Tympanic nerve**—the tympanic branch of the glossopharyngeal nerve comes from the inferior ganglion. It joins the sympathetic nerve from the superior cervical ganglion and enters the foramen in the petrous temporal bone to reach the middle ear where it forms the **lesser (superficial) petrosal nerve**. This nerve leaves the skull through the foramen ovale to reach the **otic ganglion**. As has been seen [p. 278] the parasympathetic fibres reach the parotid gland from the otic ganglion via the auriculotemporal branch of the submandibular division of the trigeminal nerve.

2. **Nerve to stylopharyngeus**

3. **Sinus nerve**—the baroreceptors in the region of the carotid sinus and the chemoreceptors in the carotid body share the same nerve, the sinus nerve, which joins the glossopharyngeal nerve before it enters the skull.

4. **Pharyngeal branches**—the glossopharyngeal nerve and the vagus nerve form the pharyngeal plexus which supplies the constrictors muscles and mucous membrane.

5. **Glossal branches**—supply mucous membrane of posterior third of tongue with general sensation and taste.

GLOSSOPHARYNGEAL NERVE IX—SUMMARY

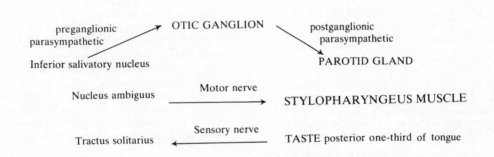

Nerve supply to third branchial arch.

Series of fine rootlets in posterolateral sulcus of medula

JUGULAR FORAMEN — Tympanic nerve

— Sinus nerve

— Sensory to mucous membrane of
oropharynx
tonsil
soft palate
posterior one-third of tongue

preganglionic parasympathetic → OTIC GANGLION → postganglionic parasympathetic

Inferior salivatory nucleus → PAROTID GLAND

Nucleus ambiguus —Motor nerve→ STYLOPHARYNGEUS MUSCLE

Tractus solitarius ←Sensory nerve— TASTE posterior one-third of tongue

X VAGUS NERVE

The vagus or tenth cranial nerve [FIG. 14.17] arises by a series of rootlets from the posterolateral sulcus of the medulla. These rootlets are continuous with those of the ninth cranial nerve. Like the glossopharyngeal nerve, the vagus passes outwards beneath the flocculonodular lobe of the cerebellum and leaves the skull through the **jugular foramen**. Again like the glossopharyngeal it has two ganglia, the **superior vagal ganglion** (jugular ganglion) and the inferior vagal ganglion (nodosum). These ganglia contain the cells of origin of the sensory nerve component of the vagus nerve. The motor nerves have their origin in the **dorsal nucleus of the vagus** in the floor of the fourth ventricle and in the **nucleus ambiguus** which it shares with the glossopharyngeal nerve.

As its name implies (*vagus* (Latin) = the wanderer) the vagus supplies the neck, thorax, and most of the abdomen with parasympathetic and sensory fibres. It also supplies the muscles of the pharynx and soft palate (except tensor palati) via the pharyngeal plexus which it forms with the glossopharyngeal nerve. Via the laryngeal nerves it supplies the muscles of the larynx which make speech possible.

The vagus passes down the neck in the carotid sheath with first the internal carotid artery and then the common carotid artery and the internal vein.

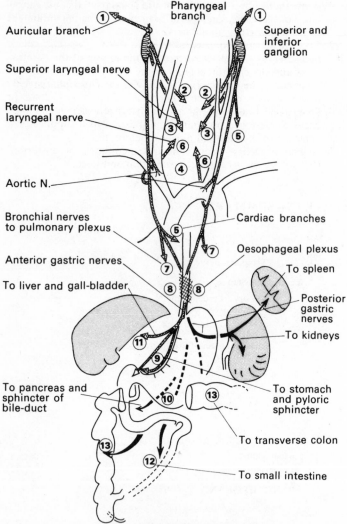

Pharyngeal branch

Auricular branch

Superior and inferior ganglion

Superior laryngeal nerve

Recurrent laryngeal nerve

Aortic N.

Bronchial nerves to pulmonary plexus

Cardiac branches

Oesophageal plexus

To spleen

Anterior gastric nerves

Posterior gastric nerves

To liver and gall-bladder

To kidneys

To pancreas and sphincter of bile-duct

To stomach and pyloric sphincter

To transverse colon

To small intestine

FIG. 14.17. The vagus nerve (X).

The two vagi are not anatomically symmetrical. The recurrent laryngeal nerve hooks round the aortic arch and ligamentum arteriosum on the left side, but on the right, as there is no aorta, it hooks round the subclavian artery instead.

There are several cardiac branches of both vagi which come off at different levels. The right vagus goes mainly to the deep cardiac plexus while the left goes mainly to the superficial. This correlates with the 'rotation' of the heart. The right vagus mainly contributes to the posterior oesophageal plexus, the left to the anterior. At the cardia the plexus forms a few fairly large branches, the anterior and posterior gastric nerves.

The anterior gastric nerve supplies mainly the stomach, liver, and gallbladder. The main branch supplying the pyloric sphincter passes on its own through the lesser omentum. When the branches running along the lesser curvature are out, in patients with gastric or duodenal ulceration, this nerve is spared.

The twisting of the vagi as they approach the oesophagus, so that the right is behind and the left in front, is correlated with the rotation of the stomach. The numbers correspond to those in the text.

On the right side the vagus passes in front of the first part of the subclavian artery [p. 181] and behind the right brachiocephalic vein [p. 181]. It gives off the recurrent laryngeal nerve which runs round the subclavian artery and returns to the neck to supply all the muscles of the larynx with the exception of the cricothyroid [p. 320]. The vagus continues down through the mediastinum on the right side of the trachea to the oesophageal plexus. From this plexus the right vagus continues down behind the oesophagus to reach the abdomen.

On the left side the vagus passes down between the left common carotid artery and the left subclavian artery [p. 184] and crosses the arch of the aorta. It runs behind the left brachiocephalic vein [p. 181]. The left recurrent laryngeal runs round the arch of the aorta (and the ductus arteriosus which in foetal life joins the pulmonary artery to the aorta). The vagus continues down through the mediastinum to join the oesophageal plexus. From this plexus the left vagus continues down in front of the oesophagus to reach the abdomen.

Branches of the vagus:

1. **Auricular nerve**—the auricular branch of the vagus nerve arises in the jugular foramen. It reaches the ear through a canal in the petrous temporal bone and emerges through the tympanomastoid fissure to supply the skin behind the ear. It supplies the posterior wall of the external auditory meatus. Stimulation of this region may bring about a variety of vagal reflexes.

2. **Pharyngeal nerves**—the pharyngeal branches of the vagus nerve arise just below the jugular foramen. They join the pharyngeal branches of the glossopharyngeal nerve to form the pharyngeal plexus. The pharyngeal plexus supplies the muscles of the pharynx and soft palate with the exception of the tensor palati which is supplied by the mandibular division of the trigeminal nerve (V_3).

3. **Superior laryngeal nerve**—the superior laryngeal nerve also arises just below the jugular foramen. It runs behind the internal carotid artery and divides into:
   (i) **External laryngeal nerve**—this nerve supplies the cricothyroid muscle.
   (ii) **Internal laryngeal nerve**—this is a sensory nerve supplying the mucous membrane of the larynx above the vocal folds, the epiglottis, and the back of the tongue.

4. **Aortic nerve**—the aortic nerve supplies the baroreceptors of the aortic arch and brachiocephalic trunk and the chemoreceptors of the aortic bodies.

5. **Cardiac nerves**—the cardiac nerves of the two sides run to the cardiac plexus. From this plexus inhibitory parasympathetic activity passes to the sinu-atrial and atrioventricular nodes of the heart. These fibres originate in the nucleus ambiguus in the medulla.

6. **Recurrent laryngeal nerves**—the right recurrent laryngeal nerve passes round the right subclavian artery and enters the larynx by passing beneath the lower border of the inferior constrictor muscle. The left recurrent laryngeal nerve passes round the arch of the aorta and runs upwards behind the left common carotid artery to the groove between the left side of the trachea and the oesophagus and like the right recurrent laryngeal nerve enters the inferior aspect of the larynx by passing under the lower border of the inferior constrictor muscle.

The recurrent laryngeal nerves supply all the intrinsic muscles of the larynx except cricothyroid [see p. 323]. These are striated muscles and these vagal fibres are part of the voluntary nervous system (special visceral efferent fibres).

7. **Bronchial nerves**—the vagal fibres to the lungs contain efferent parasympathetic bronchoconstriction fibres to the smooth muscle of the bronchi. The Hering–Breuer afferent fibres from the stretch receptors of the lungs and bronchi run in these branches of the vagus.

8. **Oesophageal nerves**—the peristaltic waves in the oesophagus, which will even allow water to be swallowed and passed to the stomach when standing on one's head, depend on the vagal branches to the smooth muscle of the oesophagus.

9. **Gastric branches**—the vagus plays an important part in stimulating both the motility of the stomach (smooth muscle) and the secretion of gastric juice with its pepsin and hydrochloric acid for the digestion of protein. This parasympathetic secretory activity is augmented by the hormone gastrin [p. 215]. In the selective vagotomy operation for peptic

ulcer only the vagal fibres to the gastric secreting glands are cut [p. 215].

10. **Pancreatic branches**—the vagus stimulates pancreatic secretion. This parasympathetic activity is augmented by the hormones secretin and cholecystokinin (pancreozymin).

11. **Branches to gall-bladder**—the vagal parasympathetic activity causes contraction of the gall-bladder. The principal mechansim for gall-bladder contraction is, however, the release of cholecystokinin (pancreozymin).

12. **Branches to small intestine**—the vagal parasympathetic activity increases the motility of the small intestine and relaxes the sphincters.

13. **Branches to large intestine**—the vagal parasympathetic activity extends as far as the transverse colon of the large intestine. The remainder of the digestive tract is supplied by the sacral parasympathetic outflow from the sacral 2, 3, and 4 segments.

VAGUS NERVE X—SUMMARY

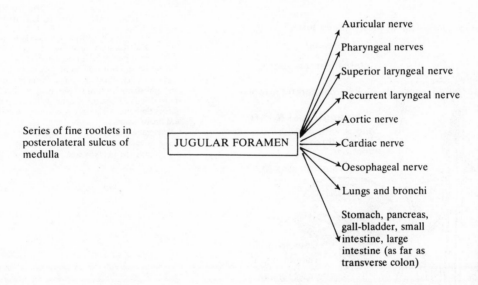

Series of fine rootlets in posterolateral sulcus of medulla

JUGULAR FORAMEN

Auricular nerve

Pharyngeal nerves

Superior laryngeal nerve

Recurrent laryngeal nerve

Aortic nerve

Cardiac nerve

Oesophageal nerve

Lungs and bronchi

Stomach, pancreas, gall-bladder, small intestine, large intestine (as far as transverse colon)

XI ACCESSORY NERVE

The spinal accessory or eleventh cranial nerve [FIG. 14.18] supplies the trapezius and sternomastoid.

It arises mainly from the **nucleus ambiguus** in the medulla (accessory or medullary part) and from the **anterior horn cells** of C1, C2, C3, C4, and C5 (spinal part). The spinal nerves run upwards behind the ligamentum denticulum of the spinal cord through the foramen magnum behind the vertebral artery into the posterior cranial fossa of the skull. Here the spinal fibres join the medullary part and they leave the skull again as the eleventh cranial nerve through the **jugular foramen** in the same dural compartment as the vagus. Below the jugular foramen most of the eleventh cranial nerve joins the vagus but the remainder passes downwards in front of the transverse process of the atlas into the sternomastoid muscle which it supplies. It reappears at the middle of the posterior border of this muscle and crosses the posterior triangle [p. 312] to reach trapezius which it also supplies. The lower part of trapezius is also supplied by fibres running directly from C3 and C4.

XII HYPOGLOSSAL NERVE

The hypoglossal nerve or twelfth cranial nerve [FIG. 14.19] supplies the extrinsic and intrinsic muscles of the **tongue** with the exception of the palatoglossus which is supplied by the pharyngeal plexus.

The nucleus of the hypoglossal nerve lies in the floor of the fourth ventricle. The nerve emerges from the anterior surface of the medulla as sets of rootlets between the pyramid and the olive. These unite and the hypoglossal nerve leaves the skull through the **hypoglossal canal** (anterior condylar foramen) medial to the internal carotid artery and the vagus. Here it is joined by a branch from C1.

The hypoglossal nerve runs deep to the styloid process and the posterior belly of digastric and then loops around in front of the internal and external carotid arteries. It crosses the loop of the lingual artery and passes deep to the posterior belly of digastric [FIG. 14.00]. It passes medially to reach the tongue between hyoglossus and mylohyoid.

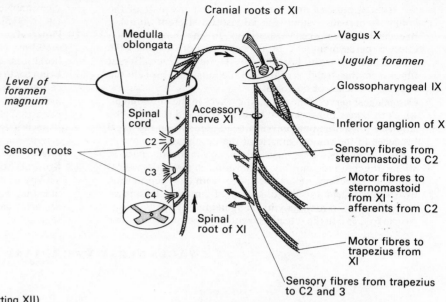

FIG. 14.18. The accessory nerve (XI).

The accessory nerve on the right side is seen from behind.

The spinal part of XI is motor in function and comes from anterior horn cells in the anterior column of grey matter from C1 to C5.

The sensory fibres from the sternomastoid and trapezius enter the posterial spinal roots at C2–4 having their cells of origin in the posterior spinal ganglia. The sensory pathway is quite orthodox, but the motor fibres pursue their curiously circuitous route, up through the foramen magnum and down again.

The cranial part of XI, from the cells of the nucleus ambiguus, joins the vagus by means of a branch to the inferior ganglion of X. These fibres supply the pharynx and soft palate via the pharyngeal plexus.

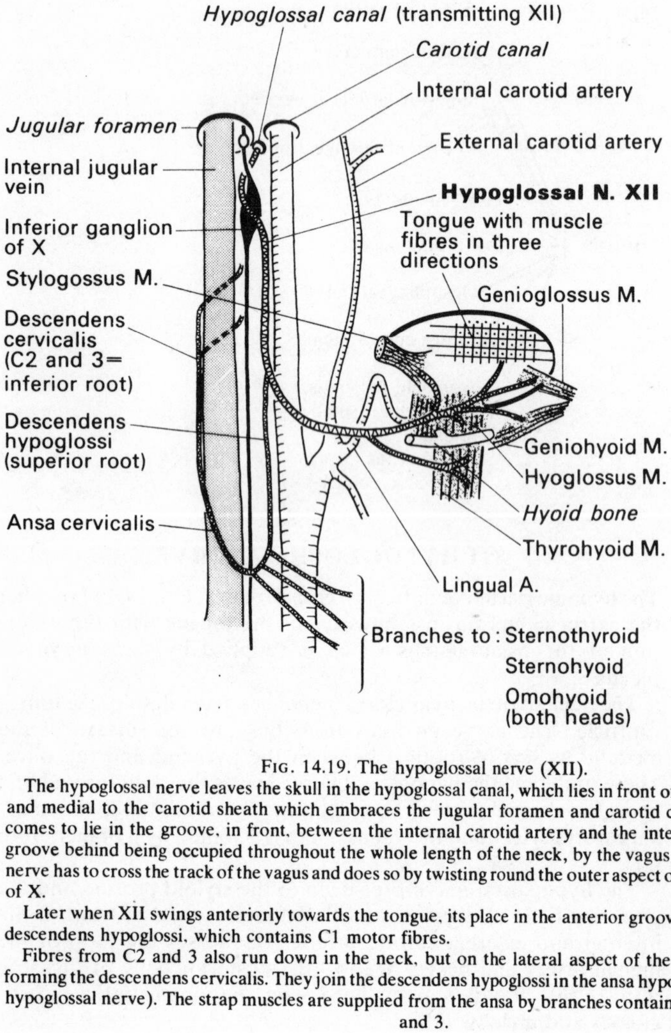

FIG. 14.19. The hypoglossal nerve (XII).

The hypoglossal nerve leaves the skull in the hypoglossal canal, which lies in front of the occipital condyle and medial to the carotid sheath which embraces the jugular foramen and carotid canal. The nerve soon comes to lie in the groove, in front, between the internal carotid artery and the internal jugular vein, the groove behind being occupied throughout the whole length of the neck, by the vagus (X). The hypoglossal nerve has to cross the track of the vagus and does so by twisting round the outer aspect of the inferior ganglion of X.

Later when XII swings anteriorly towards the tongue, its place in the anterior groove is then taken by the descendens hypoglossi, which contains C1 motor fibres.

Fibres from C2 and 3 also run down in the neck, but on the lateral aspect of the internal jugular vein forming the descendens cervicalis. They join the descendens hypoglossi in the ansa hypoglossi (= loops of the hypoglossal nerve). The strap muscles are supplied from the ansa by branches containing fibres from C1, 2, and 3.

Branches of the hypoglossal nerve:
1. Meningeal branch.
2. Nerves to thyrohyoid and geniohyoid (with fibres from C1).
3. Nerve to styloglossus.
4. Nerve to hyoglossus.
5. Nerve to genioglossus.
6. Nerve to geniohyoid (C1 fibres travelling in hypoglossal nerve).
7. Nerves to intrinsic muscles of the tongue.
8. Descendens hypoglossi nerve which forms the superior root of the ansa.

Ansa cervicalis

The ansa cervicalis is formed by the superior root (which includes fibres from C1) joining the inferior root from C2 and C3. The ansa supplies the infrahyoid muscles with:
1. Nerve to sternohyoid.
2. Nerve to sternothyroid.
3. Nerve to omohyoid.

THE CERVICAL PLEXUS

The cervical plexus [FIG. 14.20] consists of the anterior rami of the first four cervical nerve roots and their branches. The accessory (XI) and especially the hypoglossal (XII) cranial nerves are also involved in the plexus which lies above the brachial plexus.

The ventral muscles of the vertebral column in this region are supplied by these nerves.

The infrahyoid muscles, or strap muscles, are supplied by C1, and by C2 and 3. The C1 fibres join the hypoglossal nerve for a short distance and some of them supply the thyrohyoid muscle. Their cells of origin are not in the motor nucleus of XII but in the first segment of the spinal cord and they are therefore not cranial but spinal nerve fibres, which are travelling for part of their way in cranial nerve XII.

Other C1 fibres also leave the hypoglossal nerve high up in the neck and pass downwards in the carotid sheath, continuing the path taken initially by the hypoglossal nerve itself in the anterior groove between the internal carotid artery and internal jugular vein. This branch is the **superior root** of the **ansa cervicalis**. It is joined by another branch from C2 and 3 which descends just lateral to the carotid sheath, the **inferior root**. The loop formed low down in the neck by the junction of the two roots is called the **ansa cervicalis** [FIG. 14.20]. It gives rise to several branches which supply the omohyoid (both heads) the sternothyroid and sternohyoid. These muscles are involved in movements of the larynx and hyoid bone in swallowing, speaking, etc. They are relevant to the surgeon operating in the front of the neck.

The cutaneous branches of the cervical plexus fan out as shown in FIGURE 14.21, the lesser occipital nerve (C2), the great auricular nerve (C2, 3), the anterior cutaneous nerve of the neck (C2, 3) and the supraclavicular nerves (C3, 4). There are three supraclavicular nerves. The middle one can easily be palpated against the middle-

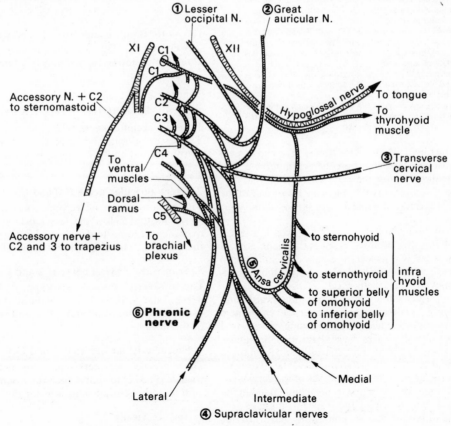

FIG. 14.20. The ansa cervicalis and the cervical plexus.

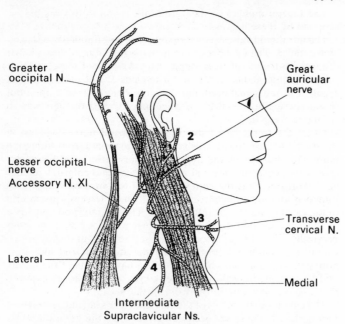

Fig. 14.21. The nerves of the cervical plexus.

    The cervical nerves give off dorsal rami which contribute to the segmentally organised supply of the erectores spinae in the neck. The cervical plexus consists of the ventral rami of C1–4 which immediately give off branches to the ventral intervertebral muscles that is to the anterior and lateral rectus capitis muscles (from atlas to base of skull) to the longus capitis and cervicis as well as the scalene muscles and levator scapulae. Sensory branches from the sternomastoid (C2 and 3) and from the trapezius (C3 and 4) also join the ventral rami. The phrenic nerve is the most important branch of this plexus. If an accessory phrenic is present it is a branch of the brachial plexus.

    The cutaneous branches twist round the posterior border of sternomastoid, all within about 2 cm of each other. The lesser occipital also twists round the accessory nerve.

    The superior cervical ganglion gives off branches which supply the cervical plexus nerve roots (not shown in the diagram). These are distributed mainly to the skin.

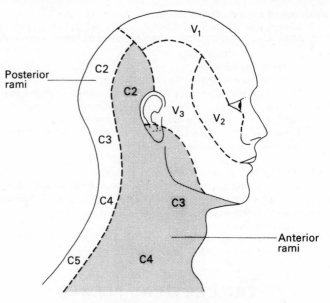

Fig. 14.22. Cervical dermatomes.

    The regular segmentation of the posterior rami of the cervical nerves is easy to understand. Note that the posterior rami of C6, 7, and 8 may not reach the skin surface.

    The anterior rami of C2, 3, and 4 supply from above the ear to the point of the shoulder. C4 extends below the clavicle and supplies as far down as the T2 dermatome. The intervening segments C5–T1 supply the upper limb.

third of the clavicle just beneath the skin. It may give rise to pain following a fracture of the clavicle. These nerves innervate skin down to the T2 dermatome below the clavicle.

    Note that the great auricular nerve supplies the lower half of the pinna of the ear, and also a considerable area of skin over the angle of the mandible [Fig. 14.22].

    The most important branch of the cervical plexus is the C4 contribution by which the phrenic nerve supplies the diaphragm. If the neck is broken just below this level the patient is still able to breath and may survive with a quadriplegia (all four limbs paralysed) but a lesion above C4 is usually fatal.

    The accessory nerve (XI), although it passes downwards and laterally parallel to the nerves of the brachial plexus, never encounters the vertebral compartment of the neck or its enclosing prevertebral fascia [p. 316]. This is because it leaves the skull through the jugular foramen and passes straight into the carotid sheath, and from there it lies in the substance of the sternomastoid muscle. It lies therefore on a much more superficial plane than the nerves of the cervical or brachial plexus, and crosses the posterior triangle embedded in the fascia of the roof on its way to supply the trapezius muscle.

    The sensory nerve supply of the sternomastoid and trapezius muscles reaches them from C2 and 3 and from C3 and 4 respectively.

The motor supply comes from the cervical component of the accessory nerve (XI). This is another interesting case of the segregation of the motor and sensory supply of a muscle (see also face, tongue, and diaphragm).

BRANCHES OF CERVICAL PLEXUS

1. Lesser occipital nerve (C2)
The lesser occipital nerve appears at the posterior border of the sternomastoid and runs upwards to supply the skin at the back of the ear.

2. Great auricular nerve (C2 and C3)
The great auricular nerve appears at the posterior border of the sternomastoid below the lesser occipital nerve. It runs upwards over the sternomastoid and supplies the skin over the lower part of the ear, the parotid gland and angle of jaw.

3. Transverse cervical nerve (C2 and C3)
The transverse cervical nerve appears below the lesser occipital nerve at the posterior border of the sternomastoid. It runs forwards over the sternomastoid and supplies the skin over the side and front of the neck.

4. Supraclavicular nerves (C3 and C4)
The supraclavicular nerves pass downwards through the posterior triangle [p. 312] to supply the skin around the clavicle and over the deltoid and pectoralis major as far down as the third rib.

5. Ansa cervicalis
The hypoglossal nerve receives a branch from C1 and gives rise to

the superior root. This joins the inferior root from C2 and C3 forming the ansa cervicalis which supplies the sternothyroid, sternohyoid and omohyoid [see p. 312].

6. Muscular branches

The cervical plexus supplies the trapezius (C3, C4, and XI), levator scapulae (C3, and C4), sternomastoid (C2 and XI), the prevertebral muscles and the diaphragm (phrenic nerve C3, 4, and 5).

Phrenic nerve

The phrenic nerve supplies the diaphragm. It arises from the anterior primary rami of C3, 4, and 5, the largest component coming from C4. The phrenic nerve in the neck runs down in front of scalenus anterior and deep to sternomastoid. It passes behind the transverse cervical artery, the suprascapular artery and omohyoid muscle.

On the left side the phrenic nerve passes behind the thoracic duct and crosses in front of the *first* part of the subclavian artery to enter the thorax in front of the cervical pleura. It passes behind the left brachiocephalic vein and runs down through the mediastinum in front of the arch of the aorta, in front of the hilus of the left lung and along the lateral border of the pericardial sac to reach the diaphragm [p. 181].

On the right side the phrenic nerve crosses in front of the insertion of scalenus anterior to the first rib and hence in front of the *second* part of the subclavian artery. Like the left phrenic nerve, it enters the thorax in front of the cervical pleura. It runs down through the mediastinum lateral to the venous channels, that is, the right brachiocephalic vein, superior vena cava, right atrium and inferior vena cava to reach the diaphragm. Like the left phrenic nerve it passes in front of the hilus of the lung. An accessory phrenic nerve may be present and is usually derived from the nerve to subclavius.

15 The scalp and face

THE SCALP

Examine the skin of the scalp. It is possible to move the scalp backwards and forwards by means of the occipitofrontalis muscle. This consists of a thin sheet of muscle covering the back of the head (occipital part) and a second sheet covering most of the forehead (frontal part). They are joined to each other by a thick aponeurosis. When one or other part of the muscle contracts the aponeurosis moves anteriorly or posteriorly and takes the skin of the scalp with it as they are firmly adherent to each other.

The skull itself is covered with periosteum (the pericranium) [FIG. 15.01] and an intervening layer of loose connective tissue allows the aponeurosis to slither to-and-fro over the periosteum. The word 'scalp' has five letters, and the scalp itself has five layers:

1. **S** = skin.
2. **C** = connective tissue. ⎱ *Firmly attached to each other*
3. **A** = aponeurosis. ⎰
4. **L** = loose fascia (plane of cleavage).
5. **P** = periosteum.   *Firmly attached to skull only at sutures*

There is also a plane of cleavage between layer 5 and the bone, where blood may collect [p. 329] after a head injury (cephalhaematoma).

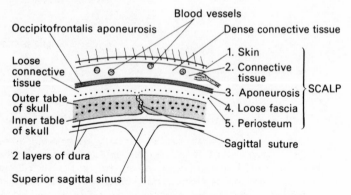

FIG. 15.01. Layers of the scalp.
The diagram shows a coronal section through the sagittal suture of the skull. Notice the blood vessels which bleed copiously when the scalp is damaged.

The blood supply is very rich [p. 259] with copious anastomoses between neighbouring blood vessels and across the midline. A scalp wound bleeds from both cut edges, and gapes open because of the pull of the occipitofrontalis.

When the scalp is torn off the head (this is usally caused by allowing long hair to become enmeshed in machinery) it is layer 4 which separates, so the aponeurosis is pulled away with the skin.

The pericranium becomes dry and the outer table of the skull loses its source of nutrition, and this bone dies and separates maybe 12–18 months later. During the whole of this time the danger that infection may spread via emissary veins and give rise to meningitis threatens the patient's life.

Arteries of the scalp

The scalp is supplied by six arteries on each side. These arteries anastomose freely with each other [FIG. 15.02].

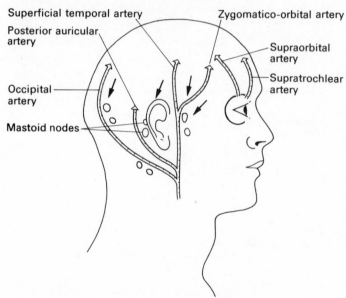

FIG. 15.02. Arterial supply and lymph drainage of scalp.

These are:
1. Supratrochlear artery (from ophthalmic artery from internal carotid artery).
2. Supraorbital artery (from ophthalmic artery from internal carotid artery).
3. Zygomatico-orbital artery (from superficial temporal artery from external carotid artery).
4. Superficial temporal artery (from external carotid artery).
5. Posterior auricular artery (from external carotid artery).
6. Occipital artery (from external carotid artery).

Veins of the scalp

The veins of the scalp are very variable. The principal veins are:
1. Supratrochlear veins.
2. Supra-orbital veins—the supratrochlear and the supraorbital

294

veins drain via the ophthalmic veins to the cavernous sinus in the skull. They also drain into the facial vein.

3. Superficial temporal veins—the superficial temporal veins drain to the retromandibular vein. The anterior division of this vein runs to the internal jugular vein. The posterior division runs to the external jugular vein.

4. Posterior auricular vein—the posterior auricular vein drains to the external jugular vein. It also drains via the mastoid emissary vein to the lateral sinus.

5. Occipital vein—the occipital vein does not follow the artery. Instead it runs to the suboccipital venous plexus.

6. Parietal emissary vein—this vein runs to the superior longitudinal sinus.

Lymphatic drainage of the scalp

The anterior and side regions of the scalp drain to the pre-auricular and mastoid lymph nodes behind the ear. The posterior region drains backwards to the occipital nodes at the origin of trapezium [FIGS. 15.02 and 13.19, p. 265].

Nerves of the scalp

There are ten nerves on each side, one motor nerve and four sensory nerves in front of the ear, and one motor nerve and four sensory nerves behind the ear [FIG. 15.03].

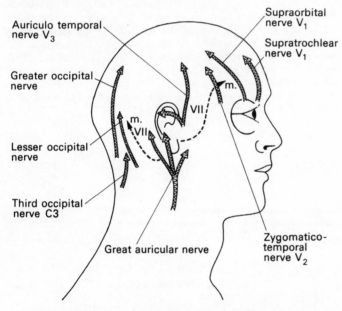

FIG. 15.03. Sensory nerves supplying the scalp. The muscles are supplied by the facial nerve VII shown dotted.

In front of the ear

The motor nerve is the temporal branch of the facial nerve. It is the motor supply to frontalis and corrugator supercilli. This nerve crosses the zygomatic arch and runs upwards and forwards. If it is accidentally cut, drooping of the eyebrows and smoothness of the forehead will result on the affected side.

The sensory nerves are:
1. Supratrochlear nerve V_1.
2. Supra-orbital nerve V_1.
3. Zygomaticotemporal nerve V_2.

4. Auriculotemporal nerve V_3.

Behind the ear

The motor nerve is the posterior auricular branch of the facial nerve. The sensory nerves are:
1. Great auricular nerve — ventral rami C2, 3.
2. Lesser occipital nerve — ventral ramus of C2.
3. Greater occipital nerve — dorsal ramus of C2.
4. Third (small) occipital nerve — dorsal ramus of C3.

THE FACE

Run your fingers over your forehead. The prominence between the eyebrows is the **glabella**, and the depression below it is the **nasion**. With your thumb find the supraorbital notch, 2 cm lateral to the glabella [FIG. 15.04]. Press firmly on the supraorbiatal nerve, which runs in this notch. It is painful. Pressure on this nerve is used clinically to assess the depth of coma in unconscious patients.

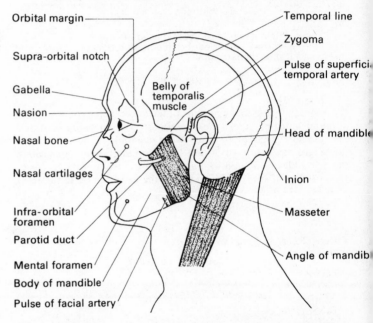

FIG. 15.04. Important landmarks of the side of the face and skull.

In a good light examine the relationship of the upper eyelid to the eyeball. The upper part of the cornea is normally hidden by the eyelid [FIG. 15.05]. In excitement and fear the lid is retracted and the sclera is exposed above the cornea. In certain diseases, such as exophthalmic goitre, the sclera may be permanently exposed in this way.

Feel the nose and note the junction of the bony and cartilaginous parts. The nasal bones are the most exposed bones and they are flattened when the nose is 'broken' [FIG. 15.06].

In the upper lip the central region, the **philtrum**, is separated by two vertical ridges from the rest of the lip. The ridge is present because the philtrum of the lip develops from a different source to the rest of the lip. The ridge represents the region where failure of fusion produces a hare lip (please, not 'hair'). A cleft palate is a similar defect concerning the palate [FIG. 15.07]. Hare lip and cleft palate are often associated, but each may occur independently.

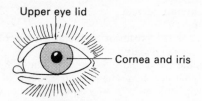

Upper eye lid

Cornea and iris

Left eye: normal relations of
upper eyelid and cornea

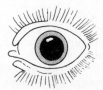

Left eye: sympathetic overactivity
(fear etc):

Sclera (white of eye) exposed between
cornea and upper lid due to lid retraction,
plus dilated pupil.

FIG. 15.05. The pupil of the eye.

Failure of fusion in the midline of the lower lip is extremely rare, although hare lip is one of the commonest congenital abnormalities.

Feel your way along the **body of the mandible** to the angle of the jaw. Here the bone turns upwards to form the **ramus**. The **head of the mandible** is palpated with a finger just in front of the tragus of the ear. Open your mouth wide and the head can be felt moving forwards. Below the head is the **neck** of the mandible which is easy to feel with the mouth wide open.

Notice that when the mouth opens only the mandible moves. Also the angle of the mandible moves backwards while its head moves forwards. The axis of this movement, where neither backward nor forward movement occurs, must lie about half way between the angle and the head [p. 303].

When the jaw is clenched the masseter muscle running downwards and backwards towards the angle of the jaw can be seen and felt. Just in front of the lowest part of this muscle and superficial to the mandible lies the **facial artery** and between the tragus of the ear and the head of the mandible is the **superficial temporal** pulse. Both these pulses are commonly employed by anaesthetists. Find them in yourself.

Palpate the **parotid duct** as it passes across the anterior edge of the masseter. Its termination is best felt in the cheek by rolling it between finger and thumb (thumb inside the mouth).

Skin sensation of the forehead, upper lid, and nose is supplied by the ophthalmic division of the trigeminal nerve (V_1). We have already identified its main branch the supraorbital branch in the supraorbital notch [FIG. 14.05, p. 273].

Below the inferior orbital margin is the infra-orbital foramen which transmits the infra-orbital nerve which is the main terminal branch of the maxillary division (V_2). This nerve supplies the skin of the whole of the lower eyelid, the whole of the upper lip and the intervening cheek. Below the corner of the mouth the main terminal branch of V_3, the mandibular division, is the **mental nerve** (*mentum* (Latin) = the chin). All three foramina lie along approximately the same vertical line about 3 cm from the midline. These are the skin regions examined when the three component parts of this cranial nerve are being studied. Note that the skin over a large area adjacent to the angle of the jaw is not supplied by any cranial nerve (but by the great auricular nerve).

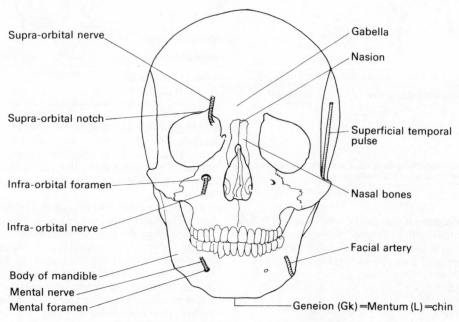

FIG. 15.06. Important landmarks of the front of the face.

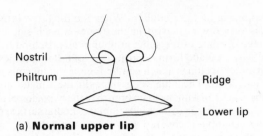

(a) **Normal upper lip**

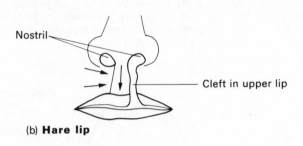

(b) **Hare lip**

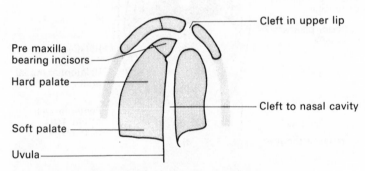

(c) **Hare lip with cleft palate (from below)**

FIG. 15.07. Hare lip and cleft palate.
The cleft in hare lip is to the side. The cleft in the palate starts to the side and reaches the midline at the incisive foramen. It then runs backwards in the midline. The defects are often much more severe than shown here.

The facial muscles

The face differs from other parts of the surface of the body in that voluntary muscles are attached to the skin. These are the **muscles of facial expressions** which are one of the special distinguishing anatomical features of man. These muscles are mostly given names which tell what they do. For example, **levator anguli oris** elevates the corners (angles) of the mouth. Its antagonist is **depressor anguli oris**. These muscles are shown in FIGURE 15.08.

Look in the mirror. Screw up your eye. Purse your lips. There must be a sphincter-like muscle around both the palpebral fissure (the eye) and the mouth for these movements to be possible. These muscles are called **orbicularis oculi** and **orbicularis oris**. Raise your eyebrows (**occipitofrontalis** muscle). Breathe in vigorously and the nostrils will flare open (**levator anguli oris alaeque nasi muscle**). Smile (**risorius muscle**).

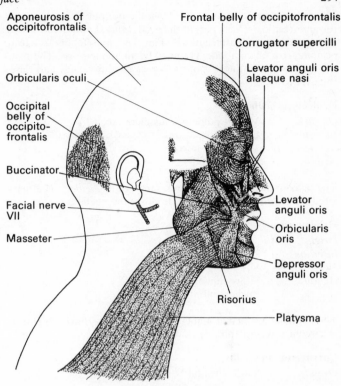

FIG. 15.08. The muscles of facial expression.

The muscle in the cheek [FIG. 15.08] the buccinator, comes from the anterior border of the pterygomandibular ligament and joins the orbicularis oris [FIG. 15.08].

All the **muscles of facial expression** are supplied by the **facial nerve** [p. 282].

MUSCLES OF FACIAL EXPRESSION

Buccinator

Origin:	Skull—pterygomandibular raphé, maxilla, and mandible
Insertion:	Sides of mouth. Blends with orbicularis oris
Nerve supply:	VII cranial nerve
Action:	Compresses cheek and retracts angle of mouth. Used when smiling and blowing a trumpet. Empties the vestibule of the mouth

The name indicates that it is the cheek muscle used for blowing a trumpet (*buccinator* (Latin) = trumpeter; *bucca* (Latin = cheek). It is used to keep the food between the teeth when chewing.

Occipitofrontalis (Epicranius)

Origin:	Occipital bone
Insertion:	Eyebrow tissue
Nerve supply:	VII cranial nerve
Action:	Wrinkles forehead Raises eyebrows

The name indicates that it runs from the occiput to the forehead (*occiput* (Latin) = back part of the head; *frons* (Latin) a forehead). Its aponeurosis is attached to skin. The muscle has separate occipital-frontal components joined by an aponeurosis. The frontalis part wrinkles the skin of the forehead transversely and raises the eyebrows.

Orbicularis oculi

Origin: Medial pulpebral ligament etc.
Insertion: Encircles eye
Nerve supply: VII cranial nerve
Action: Closes eyelid

The name indicates that it encircles the eye (*orbicularis* (Latin) = circular; *oculus* (Latin) = eye). It closes the eyelids.

Orbicularis oris

Origin: } Encircles mouth
Insertion: }
Nerve supply: VII cranial nerve
Action: Draws lips together
Puckers mouth

The name indicates that it encircles the mouth (*orbicularis* (Latin) = circular; *os*, *ora* (plural) (Latin) = mouth).

Corrugator supercilli

Origin: Skull—frontal bone
supercillary ridge
Insertion: Skin of eyebrow
Nerve supply: VII cranial nerve
Action: Wrinkles forehead vertically

The name indicates that it wrinkles the eyebrows (*corrugare* (Latin) = to wrinkle; *supercilium* (Latin) = eyebrow) as when frowning.

Platysma (Platysma myoides)

Origin: Fascia over upper part of
deltoid and pectoralis major
Insertion: Mandible—lower border
Skin—around corner of mouth
Nerve supply: VII cranial nerve
Action: Pouting. Draws down
corner of mouth

The name indicates that it is flat (*platysma* (Greek) = flat).

Platysma is a broad sheet of muscle in the superficial fascia which runs upwards from the chest along the side of the neck to be inserted into the mandible. It draws the corners of the mouth downwards and is used in pouting. When it is contracting it produces vertically running 'strings' of the neck.

THE MOUTH

The space between the inner surface of the lips and cheek and the teeth is the **vestibule** of the mouth.

The mouth itself is roofed by the **hard and soft palate**. Part of the floor of the mouth lies below the anterior one third of the tongue. It is covered with mucous membrane and joined to the inferior sur-

face of the tongue by the frenulum. When the tongue is lifted the frenulum is very obvious. Also in the floor on each side of the midline is the opening of the submandibular duct. Behind this the sublingual salivary gland forms a projecting fold. It discharges by a number of small openings which can be seen if the mucous membrane is first dried. The blue veins beneath the tongue are the ranula veins. They run with the main branch of the hypoglossal nerve.

The teeth and gums lie to each side, and the medial surface of the body of the mandible is covered with mucous membrane.

The mouth communicates behind with the pharynx. Examine the back of your own throat in the mirror. If your tongue is in the way, say 'aaah' or pant through your mouth. This will raise the soft palate as well as depressing the tongue. Note the two palatine arches (pillars of the fauces, p. 305) between which is the boundary between mouth and pharynx. The tonsil is therefore in the lateral wall of the oropharynx [FIG. 15.09].

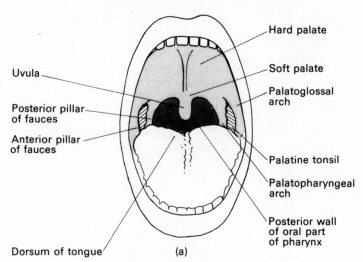

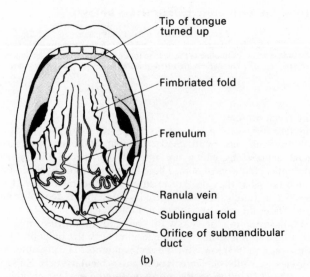

FIG. 15.09. The mouth and tongue.
(a) The open mouth showing the palate and fauces.
(b) The inferior surface of the tongue.

THE TONGUE

The tongue is the only muscular organ in the body that can be moved in any direction: up, down, side-to-side, in-and-out. The **intrinsic muscle fibres** do not run randomly, but can be shown with the microscope to be arranged in three planes at right angles running vertically, longitudinally, and transversely.

The motor nerve is the **hypoglossal** or twelfth cranial nerve (XII).

The tongue is covered with mucous membrane on its upper and lower surfaces. The mucous membrane of the anterior two-thirds of the upper surface of the tongue is supplied by the **lingual nerve**, a branch of the fifth cranial nerve (V) for general sensation and by the seventh cranial nerve (VII) for taste. The posterior third is supplied by the ninth cranial nerve (IX) for both general sensation and taste, and also has a small contribution from the superior laryngeal nerve (X).

The extrinsic muscles connect the tongue to the lower jaw (mandible) as well as the hyoid bone, palate, and styloid process. The fact that the tongue is attached to the lower jaw is used to maintain a clear air way in a supine unconscious person. Gravity tends to make the tongue fall backwards blocking the airway. To prevent this the lower jaw has to be protracted. This pulls the tongue forwards away from the pharyngeal wall.

There are four **extrinsic muscles**:
1. The hyoglossus which runs upwards.
2. The genioglossus which runs backwards.
3. The palatoglossus which runs downwards.
4. The styloglossus which runs forwards (and downwards).

Between them these four muscles can move the tongue as a whole in any direction. The intrinsic muscles can alter the shape of the tongue.

EXTRINSIC MUSCLES OF TONGUE

The extrinsic muscles of the tongue [FIG. 15.10], with the exception of the palatoglossus, are supplied by the motor nerve to the tongue, the hypoglossal or twelfth cranial nerve.

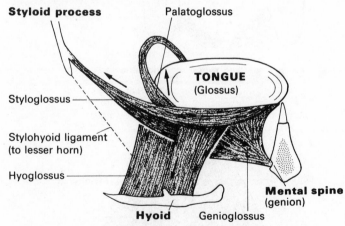

FIG. 15.10. The extrinsic muscles of the tongue.

Hyoglossus

Origin:	Body and greater cornu of hyoid
Insertion:	Side of tongue
Nerve supply:	Hypoglossal—XII
Action:	Makes dorsum convex
	Depresses tongue

The name indicates that it runs from the hyoid bone to the tongue (*glossa* (Greek) = tongue).

Genioglossus (Geniohyoglossus)

Origin:	Mental spines
Insertion:	Tongue and hyoid
Nerve supply:	Hypoglossal—XII
Action:	Makes dorsum of tongue concave in sucking. Posterior fibres protrude. Anterior fibres retract

The name indicates that it runs from the chin (*geneion* (Greek) = chin) to the tongue. Mental from the Latin *mentum* = chin.

Palatoglossus (Glossopalatine)

Origin:	Soft palate—anterior
Insertion:	Side of tongue
Nerve supply:	Accessory—XI (pharyngeal plexus)
Action:	Elevates tongue
	Constricts anterior fauces

The name indicates that it runs from the palate to the tongue.

Styloglossus

Origin:	Styloid process
Insertion:	Side of tongue
Nerve supply:	Hypoglossal—XII
Action:	Elevates and retracts tongue

The name indicates that it runs from the styloid process to the tongue.

Attachments of the tongue

The tongue is attached to a number of surrounding structures. If the tongue were removed the following attachments would have to be cut.

The tongue is attached to:
1. Mandible — Mucous membrane and geniohyoglossus
2. Hyoid bone — Hyoglossus and geniohyoglossus
3. Epiglottis — Glosso-epiglottic folds
4. Pharynx — Mucous membrane and superior constrictor
5. Soft palate — Anterior pillars of fauces and palatoglossus
6. Styloid process — Styloglossus

Functions of the tongue

The tongue is used in the following actions:
1. Tasting.
2. Speech.
3. Mastication.
4. Deglutition (the act of swallowing).

The tongue has an upper surface (dorsum), and under surface, two sides, an apex and a 'root'.

Dorsum of tongue

The dorsum or upper surface of the tongue is divided into an anterior two-thirds and a posterior third by the terminal sulcus (sulcus terminalis) which is a V-shaped groove with its apex directed backwards [FIG. 15.11].

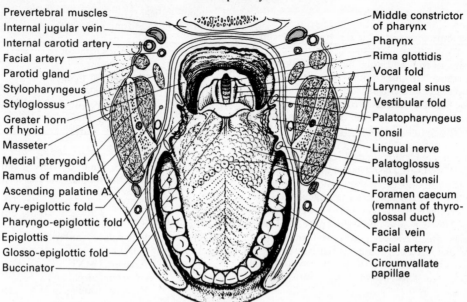

FIG. 15.11. The tongue and surrounding structures.

The tongue also has a shallow central groove running anteroposterior in the midline which terminates at the apex of the terminal sulcus. The foramen caecum (*caecum* (Latin) = blind) is a blind pit situated at the posterior termination of the median raphe of the tongue at the apex of the terminal sulcus. It represents the remains of the thyroglossal duct [p. 325]. The whole or part of the thyroid gland may be situated here forming a lingual thyroid.

The anterior two-thirds constitutes the oral part of the tongue.

The posterior one-third is the pharyngeal part of the tongue.

The surface of the anterior two-thirds is rough. It is covered with four types of papillae which are associated with the sensation of taste.

The circumvallate papillae which are easily visible to the naked eye and lie just in front of the sulcus limitans. They can be seen to project above the surrounding tongue surface and are surrounded by a groove 2–3 mm deep. Their name is derived from their fancied resemblance to a miniature fort surrounded by moat and palisade (*circum* = around; *vallum* = palisade).

The blood supply comes mainly from the lingual artery [p. 258]. The arteries to the face and pharynx make small contributions.

Lymphatic drainage of tongue

Apex of tongue

Drains to submental nodes.

Side of tongue

Drains to lymphatics in and around submandibular salivary gland. Hence this gland is removed in surgery for tongue carcinoma.

Back of tongue

Drains to superior deep cervical nodes.

Middle of tongue

Drains to submandibular and deep cervical nodes.

All regions drain to a large lymph node situated near internal jugular vein below the posterior belly of digastric.

SUBMANDIBULAR REGION

The submandibular region includes the digastric and submental triangles [p. 310].

The key muscle is the **mylohyoid** [FIG. 15.12]. It is attached to the

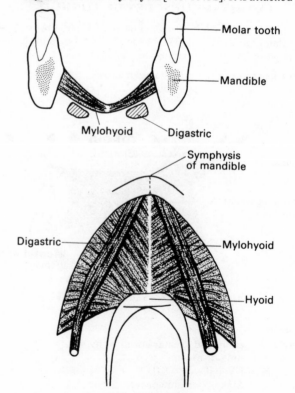

FIG. 15.12. The mylohyoid muscle.

whole length of the mylohyoid line of the **mandible** and to the body of the **hyoid bone**. The two muscles meet in the midline raphe running from the mandible to the hyoid and in doing so form the **diaphragm of the floor of the mouth**.

Deep to this muscle is the **hyoglossus** [FIG. 15.13]. The mylohyoid [FIG. 15.12] should be considered to be in the same plane as the mandible, while the hyoglossus is in the plane of the root of the tongue. It follows that anything superficial to mylohoid is in the neck, anything between the two muscles is in the plane of the floor of the mouth, and anything deep to hyoglossus is in the tongue.

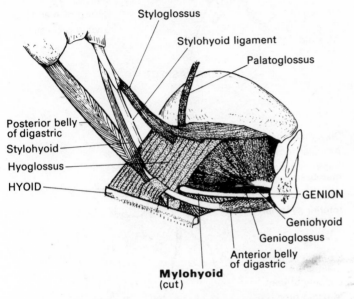

FIG. 15.13. The hyoglossus and digastric muscles.

The **submandibular gland** can be palpated in the neck in the digastric triangle. It is superficial to mylohyoid in this region. Its deep part runs behind the free posterior border of mylohyoid and gets into the plane of the floor of the mouth, where its duct discharges at the sublingual papilla [FIG. 15.14]. The **sublingual gland** lies in this plane also. The **lingual nerve** loops round the submandibular duct (Wharton's duct) from lateral to medial. This nerve is a branch of the mandibular division of V, descends from the infratemporal region and enters the floor of the mouth by slipping

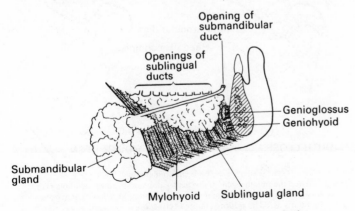

FIG. 15.14. The submandibular and sublingual salivary glands.

between the lowest fibres of the superior constrictor where they are attached to the mandible just behind the last molar tooth and the mylohyoid line. The nerve touches the bone here.

The **hypoglossal nerve** [p. 290] lies superficial to the external carotid in the carotid triangle and sweeps forward from here and enters the same plane of cleavage as the lingual nerve, with which it exchanges numerous branches, superficial to hyoglossus.

The **lingual artery**, as it is a branch of the external carotid, lies deep to the hypoglossal nerve, and the **glossopharyngeal nerve** is in an even deeper plane, deep to the external carotid. When they reach the submandibular region they both lie deep to the hyoglossus.

The lingual artery has a characteristic upward loop on it, which presumably provides some slack which can be taken up when the tongue is protruded or elevated [see FIG. 13.09, p. 258].

THE HYOID

The hyoid bone is not normally part of an articulated skeleton and its existence therefore tends to be forgotten.

The hyoid lies between the root of the tongue and the larynx. It supports the tongue and gives attachment to some of its muscles. It can also be considered to be part of the larynx (see infrahyoid muscles, p. 312).

The hyoid is a U-shaped bone consisting of a horizontal anterior body to which is attached two greater horns (greater cornua) which project backward. In addition there are two lesser horns (lesser cornua) which project upwards and slightly backwards [FIG. 15.15].

The hyoid is slung from the styloid processes by the stylohyoid ligaments. These are attached to the lesser horns. The hyoid is also attached below to the larynx by the thyrohyoid membrane.

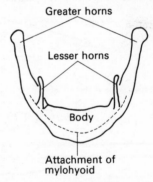

FIG. 15.15. The hyoid bone.

SUPRAHYOID MUSCLES

Digastric

Origin:	Posterior belly: mastoid process
Insertion:	Intermediate tendon—hyoid (cornu) sling
	Anterior belly: symphysis menti
Nerve supply:	Posterior belly: VII
	Anterior belly: V_3 (nerve to mylohyoid)
Action:	Elevates hyoid bone
	Fixes hyoid bone
	Opens mouth

The name indicates that it has a double belly with an intermediate tendon (*dis* (Greek) = double; *gaster* (Greek) = belly).

Mylohyoid

Origin: Mandible (mylohyoid line)
Insertion: Hyoid (body)
Nerve supply: V_3
Action: Elevates hyoid bone
                  Supports floor of mouth

The name indicates that it runs from the region of the molar teeth (*myle* (Greek) = mill) to the hyoid bone [FIG. 15.12].

Stylohyoid

Origin: Styloid process
Insertion: Hyoid (cornu)
Nerve supply: VII
Action: Elevates hyoid

The name indicates that it runs from the styloid process to the hyoid bone (*hyeides* (Greek) = U-shaped) [FIG. 15.12].

Geniohyoid

Origin: Mental spines
Insertion: Body of hyoid
Nerve supply: XII (C1)
Action: Elevates hyoid
                  Draws hyoid forwards

The name indicates that it runs from the chin (*geneion* (Greek) = chin) to the hyoid bone.

THE MANDIBLE

The mandible [FIG. 15.16 and FIG. 15.17] has a horizontal **body** and

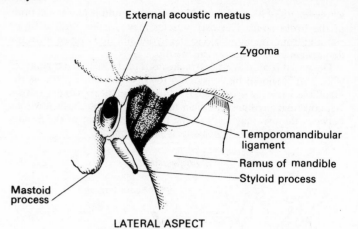

LATERAL ASPECT

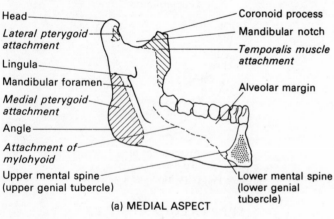

(a) MEDIAL ASPECT

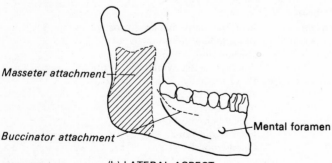

(b) LATERAL ASPECT

FIG. 15.16. The mandible and its muscle attachments.

SAGITTAL SECTION

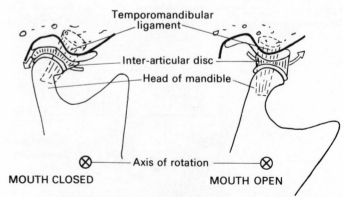

FIG. 15.17. The temporomandibular joint.
When the mouth opens the articular disc and head of mandible slide forwards on the temporal bone. This is termed protraction when it occurs on both sides. The return is termed retraction.
Alternate movements on the two sides produce a grinding action.

a **ramus** which meet at the angle of the mandible. The ramus, when followed upwards, forms the neck and then the **head** of the mandible and the **coronoid process** anteriorly. The **mandibular notch** separates them.

The **alveolar margin** bears the teeth. It is absorbed if the teeth are removed. The mandible is traversed from the **mandibular foramen** to the **mental foramen** by the main branch of the mandibular division of the fifth cranial nerve (V_3). It emerges as the **mental nerve** and then supplies the skin below the mouth.

The coronoid process is the insertion of the **temporalis muscle**. The insertion of the **masseter** is well marked on the lateral aspect of the angle of the mandible. On the deep aspect of the angle there is another well marked area for the insertion of the **medial pterygoid muscle**. This muscle runs downwards and backwards like the masseter. It can be palpated.

Just below the head of the mandible, is the neck. Anteriorly, there is a well marked depression, the pterygoid fovea which marks part of the insertion of the **lateral pterygoid muscle**.

Temporomandibular joint

The temporomandibular joint is a complex synovial hinge joint. Its main feature is that it is in two parts, divided by an interarticular disc. The head of the mandible is convex while the articular surface of the temporal bone is concave behind and convex in front [FIG. 15.17]. The articulation between the head of the mandible and the disc is a pure **hinge joint**, while that between the disc and the concavoconvex surface of the temporal bone is a **gliding joint** in which the flexible disc adapts its shape readily to the variations in shape in the temporal bone.

The movements of the mandible are seen when the mouth is opened or closed, when the chin swings from side to side, and when it is protracted and retracted.

All these movements can be produced by the two lateral and two medial pterygoid muscles [FIG. 15.18]. The medial pterygoid muscles working together **close** the mouth, the lateral pterygoids open it, the ipsilateral muscles together swing the chin to the opposite side, all four muscles **protrude** the chin. The closing movement is powerfully assisted by the masseter and temporalis, and is far the strongest movement. **Mouth opening** is assisted by the digastric and supra and infrahyoid muscles. **Retraction** is carried out by the posterior horizontal fibres of the temporalis muscle, assisted by the digastric whose line of pull is just right for retraction.

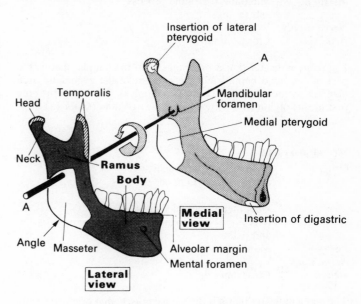

FIG. 15.19. The axis of rotation of the mandible

To understand the action of these muscles it is important to grasp the fact that the axis around which the temporomandibular joint movements take place is as shown in FIGURE 15.19.

The joint is strengthened on its lateral side by the **lateral ligament**. In addition we mention the **sphenomandibular ligament** (spine of sphenoid to the lingula of the mandible) and the **stylomandibular ligament** (lateral aspect of styloid process to the back of the ramus and angle of the mandible). The former, as it is attached at the axis of movement can have little effect on joint movement. The latter is merely thickening in the capsule of the parotid gland. It is no more a separate structure in its own right than are the extensor retinacula of the wrist or ancle joints.

The joint is supplied by the auriculotemporal nerve. The muscles of mastication as well as the anterior belly of digastric and the mylohyoid are all supplied by the mandibular division of the fifth cranial verve (V_3). Further details regarding mastication will be found on page 344.

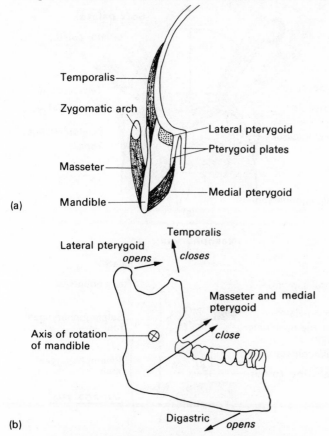

(a)

(b)

FIG. 15.18. Opening and closing the mouth.
(a) The pterygoids, massetery and temporalis muscles of mastication.
(b) The muscles which open and close by rotating the mandible around its axis.

SUMMARY

MUSCLES OF MASTICATION

The mastication muscles are:
  Masseter
  Temporalis
  Medial pterygoid
  Lateral pterygoid
  Mylohyoid
  Anterior belly of digastric
They are all supplied by the mandibular division of the trigeminal nerve, that is the fifth cranial nerve (third division).

Temporalis

Origin: Temporal bone
Insertion: Mandible (coronoid process
                and ramus)
Nerve supply: V$_3$
Action: Closes mouth

The name indicates that it originates from the temple, that is, the part of the head behind the eye and above the zygomatic arch (*tempus, tempora* (*plural*) (Latin) = temples). The hair turns grey first in this region so it shows the passing of time (*tempus* (Latin) = time).

Medial pterygoid

Origin: Superficial head: maxilla
                and palate (pyramidal
                process)
                Deep head: medial side of
                lateral pterygoid plate
                and palate
Insertion: Mandible
Nerve supply: V$_3$
Action: Closes mouth

The name indicates that it is the medial muscle that originates from the wing-shaped pterygoid process of the sphenoid bone (*pterryx* (Greek) = wing).

The two heads of medial pterygoid embrace lower head of lateral pterygoid. It acts like masseter but lies on the inner side of the jaw.

Lateral pterygoid

Origin: Upper head: sphenoid
                  below temporal crest
                Lower head: lateral side of
                  lateral pterygoid plate
Insertion: Mandible (front of neck)
                and temperomandibular
                joint
Nerve supply: V$_3$
Action: Opens mouth, protrudes
                mandible, moves jaw to
                opposite side

The name indicates that it is lateral muscle originating from the pterygoid process of sphenoid.

Masseter

Origin: Zygomatic arch
Insertion: Mandible (ramus)
Nerve supply: V$_3$
Action: Closes mouth

The name indicates that it is a chewing muscle (*maseter* (Greek) = chewer).

PALATE

The palate (adjective palatine) is the roof of the mouth. It is subdivided into the hard palate and the soft palate [FIG. 15.20]. The **hard palate** is formed by the palatal process of the maxillary bones and the palatine bones. These are covered with mucous membrane. The **soft palate** or velum palati consists of an aggregation of muscle fibres covered with mucous membrane. The muscles constituting the soft palate are:
  Tensor palati
  Levator palati
  Palatoglossus
  Palatopharyngeus
  Uvula (azygos uvula)

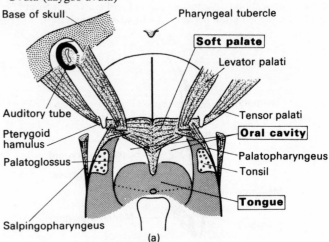

(a)

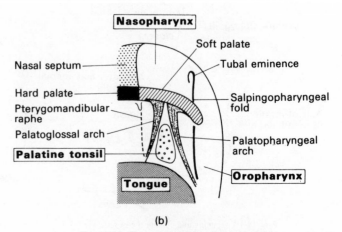

(b)

FIG. 15.20. The palate and associated structures.
(a) viewed from the back.
(b) viewed from the side.

The uvula is a single (non-paired) muscle running from the posterior nasal spine to the aponeurosis of the soft palate.

Fauces

The term fauces is Latin for the upper part of the throat. It is used to indicate the space bounded by the back of the tongue, the palatine arches, and the soft palate.

Tensor palati

The base of the medial plate of the pterygoid process of the sphenoid bone has a depression termed the **scaphoid fossa**. The tensor palati muscle arises from here and from the lateral aspect of the auditory tube. The tendon of the muscle changes direction by curling round the hamulus, which is the hook-like termination of the medial pterygoid plate, to be inserted into the aponeurosis of the soft palate or velum palati. The tendon turns into the horizontal plane and fans out forming the aponeurosis of the soft palate, by joining in the midline the aponeurosis of the muscle from the opposite side. The aponeurosis is attached to the posterior edge of hard palate. When both muscles contract they pull the aponeurosis sideways and tense the palate. Tensor palati is supplied by the mandibular division of the fifth cranial (trigeminal) nerve.

The tensor palati opens the auditory tube when it contracts hence its old name *dilator tubae*. Its main action is to tense the soft palate.

Levator palati

The levator palati muscle arises from the apex of the petrous part of the temporal bone and from the medial surface of the cartilaginous part of the auditory tube. It is inserted into the aponeurosis of the soft palate. Its action is to elevate the soft palate. It is supplied by the eleventh cranial nerve (accessory nerve) via the pharyngeal plexus.

SUMMARY

Tensor palati (tensor veli palatini)

Origin:	Scaphoid fossa and wall of auditory tube
Insertion:	Soft palate (velum palati)
Nerve supply:	V_3
Action:	Tenses soft palate and opens auditory tube

The name indicates that this muscle tenses the soft palate.

This muscle curls round the hamulus which is the hook-like termination of the medial plate of the pterygoid process of the sphenoid bone.

Levator palati (levator veli palatini)

Origin:	Apex of petrous temporal bone and cartilaginous part of auditory tube
Insertion:	Soft palate (velum palati): upper surface of aponeurosis of the soft palate
Nerve supply:	XI via pharyngeal plexus
Action:	Elevates soft palate

The name indicates that this muscle elevates the soft palate.

Palatine arches

The **palatoglossal arch** is formed by the projection of the palatoglossus muscle covered by mucous membrane. Its old name was anterior pillar of the fauces [Fig. 15.09].

The **palatoglossus** runs from the inferior surface of the soft palate to the side of the tongue. It elevates the tongue and constricts the anterior fauces. This muscle is supplied by the eleventh (accessory) cranial nerve.

The **palatopharyngeal arch** is formed by the projection of the pharyngopalatinus muscle covered by mucous membrane. It is also termed the pharyngopalatine arch. Its old name was posterior pillar of the fauces.

The **palatopharyngeus** runs from the soft palate and auditory tube to the aponeurosis of the pharynx. It aids swallowing. This muscle is supplied by the eleventh (accessory) cranial nerve via the pharyngeal plexus.

TONSIL

The palatine tonsil lies between the palatoglossal and palatopharyngeal palatine arches (anterior and posterior pillar of the fauces). Palatine means simply pertaining the palate. Above the tonsil is a triangular area covered by mucous membrane called the **plica triangularis**.

On the surface the position of the tonsil corresponds to a point 1 cm above and 1 cm anterior to the angle of the jaw.

The tonsil has the structure of a lymph node but with a free medial surface covered by stratified squamous epithelium with many deep epithelium-lined invaginations surrounded by lymphoid tissue termed tonsillar crypts.

The tonsil has a very rich blood supply with branches from five arteries [Fig. 15.21]. These present a problem of both primary and secondary haemorrhage in a tonsillectomy operation:

1. Facial artery.
2. Ascending pharyngeal artery.
3. Dorsal lingual artery.
4. Ascending palatine artery.
5. Descending palatine artery.

Since the tonsil is often dissected out in an adult tonsillectomy operation, the relations of its lateral or attached surface are of importance.

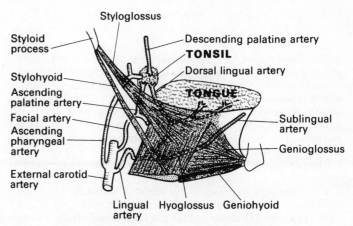

Fig. 15.21. The blood supply to the tongue and tonsil.

The tonsil is enclosed in fascia derived from pharyngeal aponeurosis which forms a capsule for the tonsil. This capsule is in direct contact with the superior constrictor muscle of the pharynx. The pharyngeal and ascending palatine arteries are outside this muscle. Occasionally the facial nerve VII runs close to the lower part of the tonsil.

Lymphatic drainage
The tonsil itself receives lymphatics from the mouth, molar teeth, and nose. It drains into the tonsillar lymph node close to the internal jugular and facial veins and then into the deep cervical lymph chain.

The palatine tonsils on either side, the lingual tonsil [FIG. 15.11; p. 300] and the adenoids (pharyngeal tonsil) form a ring of lymphatic tissue around the entry to the respiratory and alimentary tracts.

PAROTID GLAND

The term parotid is derived from the Greek (*para* = beside; *otis* = the ear). It is a serous salivary gland. The body of the gland is wedge-shaped. The flat end of the wedge [FIG. 15.22] lies just deep to the skin in front of and below the ear. The anteromedial surface is deeply grooved by the ramus of the mandible; the posteromedial surface is grooved first by the mastoid process and sternomastoid muscle and then by the styloid process which lies deep. The thin edge of the wedge passes deeper still and usually comes in contact with the pharyngeal recess (fossa of Rosenmüller) by crossing in front of the upper part of the carotid sheath.

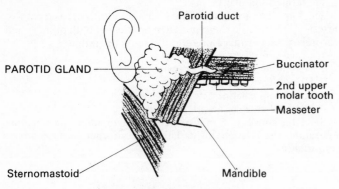

FIG. 15.22. The parotid salivary gland.

The gland passes forwards between the skin and the masseter, and its duct (Stensen's duct) opens as a papilla in the cheek opposite the neck of the second upper molar tooth. With torch and mirror find the papilla in yourself. The terminal part of the duct can easily be felt in the cheek between finger and thumb as it enters the mouth. It can also be rolled against the anterior border of the tensed masseter. This part of the duct lies along the middle third of the line joining the lobule of the ear to the nostril.

The parotid encloses the following structures from superficial to deep:
1. Facial nerve root and its five terminal branches.
2. The entire length of the retromandibular vein.
3. The terminal branches of the external carotid artery, and also the posterior auricular and occipital branches.

There are numerous parotid lymph nodes embedded in the gland.

Nerve supply
The parasympathetic secretomotor supply of the parotid comes from the inferior salivatory nucleus in the medulla via the glossopharyngeal nerve. The preganglionic fibres relay in the otic ganglion, and the postganglionic fibres then join the auriculotemporal nerve to be distributed in the gland [p. 287].

Sympathetic nerves can also be traced into the gland. The preganglionic fibres come from T1 and 2 (like the rest of the sympathetic supply to the head) pass upwards in the sympathetic chain and relay in the superior cervical ganglion. The postganglionic fibres form a plexus around the external carotid artery, which as it is embedded in the gland, leads them to their destination.

It is interesting that following surgery on the parotid, secreto–motor fibres normally destined for the parotid glands or for sweat glands in the skin may regenerate in the wrong pathways. This gives rise not only to gustatory sweating (profuse sweating of parotid skin at meal times) but also to thermal salivation.

The gland is enclosed by a loosely defined capsule, part of which forms the stylomandibular ligament. The gland is enlarged and painful in mumps, and may be infected by bacteria from the mouth in patients who are seriously ill for any reason. Stones sometimes form in the duct and can easily be palpated. The parotid can be demonstrated on X-ray, by slipping a cannula into the papilla and injecting radiopaque fluid into the duct system. Tumours of the parotid are common.

THE NASAL CAVITIES

The nasal cavities lie between the base of the skull and the roof of the mouth. Their anterior apertures, the anterior nares, are pear-shaped with the narrow end upwards. The dilatation just inside the external nose is termed the vestibule. Hairs grow from the lower part of the vestibule. The expanded lateral wall of the vestibule is the **ala** of the nose. The small muscles in this region can flare these **alae nasi**.

The **posterior nares** or **choanae** communicates with the pharynx.

The nasal cavity has a medial wall, a roof, a floor, and a lateral wall.

The nasal cavities on the two sides are separated by the septum which forms the medial wall of each cavity. The **septum** is formed by the perpendicular plate of the ethmoid bone (one-third), the cartilage of the septum (one-third), and the vomer bone (one-third). A deviated septum is the expression used when the septum does not lie exactly in the midline. It may obstruct breathing on one or both sides.

The **roof** of the nasal cavity includes the cribriform plate of the ethmoid through which the olfactory nerves pass.

The **floor** of the nasal cavity is formed by the hard palate.

The **lateral wall** of the nose has three **conchae** (turbinate bones), each shaped like a scroll, which divide the lateral wall into four meatuses [FIG. 15.23].

The **spheno-ethmoidal recess** is the highest meatus. It lies between the roof of the nose and the superior concha. The sphenoidal sinus opens into it [FIG. 15.23].

The **superior meatus** lies between the superior and middle chonchae. The posterior ethmoidal air-cells open into it. The nasal and

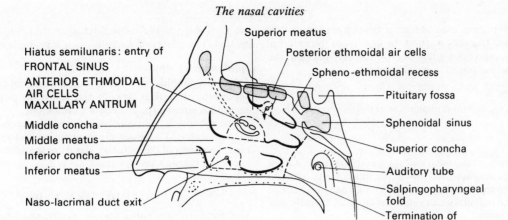

FIG. 15.23. The lateral wall of the nose with parts of superior, middle, and inferior concha removed.

nasopalatine branches of the maxillary division of the trigeminal nerve (V_2) enter the superior meatus through the sphenopalatine foramen.

The **middle meatus** lies between the middle and inferior conchae. The frontal sinus, the maxillary sinus and the anterior ethmoidal air-cells enter into the middle meatus.

The **inferior meatus** lies between the inferior concha and floor of the nose. The nasolacrimal duct opens into the inferior meatus.

The **epithelium** lining the nostrils is stratified squamous like that of the skin. It changes to respiratory epithelium (columnar ciliated cells with mucous and serous glands) within about 1 cm. The mucous membrane is highly vascular and warms and humidifies the air. In many animals the nose plays an important role in salvaging the water in the expired air. This is done by having a cold nose (e.g. elephant rats) which cools the warm air and causes water vapour to precipitate out. A similar phenomenon occurs in man on a frosty day. The nose runs due to the condensation of the water vapour.

The mucous membrane is tightly bound to the periosteum which covers the bones to form the mucoperiosteum. The periosteum does not therefore exist here as a separate membrane. It strips very easily from the bone and surgeons often take advantage of the existence of this plane of cleavage.

Nerve supply of nose [Fig. 15.24]
Olfactory nerve (I)—upper one-third only
    Nasociliary nerve V_1
    Greater palatine nerve V_2
    Nasopalatine nerve V_2
    Nasal nerves V_2

Blood supply of nose
    Sphenopalatine artery
    Great palatine artery
    Ethmoidal artery
    Others

An area about 1.5 cm from the external nares on each side of the nasal septum is a frequent cause of nose bleeding. There is a venous plexus here called Little's area. It is easy to stop. But bleeding

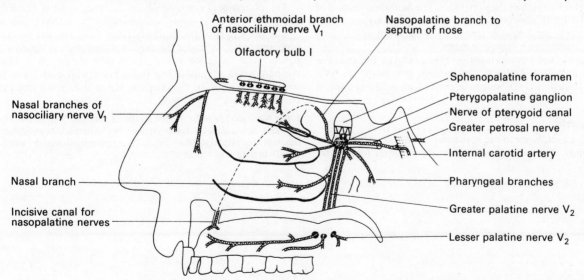

FIG. 15.24. The nerve supply of the lateral wall of the nose with the pterygopalatine (sphenopalatine) ganglion exposed.

sometimes occurs in hypertensive elderly patients from some other part of the cavity and this may warrant the ligation of the external carotid artery. Success is not guaranteed, however, because the internal carotid also supplies the nasal cavity.

Lymphatic drainage

The nose drains to the submandibular and retropharyngeal lymph nodes.

Paranasal air sinuses

The frontal, maxillary, ethmoidal, and sphenoid bones are hollow and contain air sacs or sinuses all of which communicate with the nose through apertures in the lateral wall. The air sacs are termed the frontal sinus, the maxillary antrum, the ethmoidal air-cells, and the sphenoidal air sinus.

To see the outline of these sinuses, stand in front of a mirror in a darkened room with the end of an electric torch in the mouth. The air sacs in the frontal and maxillary bones show up with a red glow.

Frontal air sinus

The frontal sinus is an air-sac in the frontal bone of the forehead. It is variable in extent and is usually asymmetrical on the two sides. A thin plate of bone separates it from the ethmoidal air sacs below and behind. The frontal sinus develops, not as a direct outgrowth from the nasal cavity like the maxillary antrum, but as an outgrowth from an anterior ethmoid air-cell.

The frontal sinus is connected by a narrow canal to the middle meatus where it opens with an anterior ethmoid air-cell near to the opening of the maxillary antrum into a semicircular groove, the **hiatus semilunaris**. This lies below a bulge in the middle meatus termed the bulla ethmoidalis. You cannot see the hiatus semilunaris without removing the middle concha.

Maxillary antrum

The maxillary air sinus or maxillary antrum is situated in the maxillary bone lateral to the nasal cavity and superior to the upper jaw. The roof of the antrum is the floor of the orbit. The floor of the antrum may have elevations produced by the roots of the canine, premolar and molar teeth of the upper jaw. The dental extraction of an upper molar tooth may break into the maxillary antrum. Infections of the tooth sockets may spread to the antrum as in many skulls no bone intervenes between them.

The maxillary antrum drains into the middle meatus. Because the exit from the antrum is high up, the drainage may be poor. As a result an antral washout may have to be carried out in cases of antral infections.

Ethmoidal air-cells

The ethmoidal air-cells are found between the frontal sinus and the sphenoidal sinus. They lie below the anterior cranial fossa. The anterior and middle ethmoidal air-cells open into the middle meatus. The posterior air-cells open into the superior meatus.

Sphenoidal air sinus

The two sphenoidal air sinuses lie in the body of the sphenoid bone behind the ethmoidal air-cells and below the pituitary gland. The cavernous sinuses [p. 255] lie on either side.

The sphenoidal air sinus opens into the spheno-ethmoidal recess (highest meatus) between the superior concha and the roof of the nose. Tumours of the pituitary gland are sometimes removed by an approach through the sphenoid air cells.

AUDITORY TUBE

The auditory tube, also known as the pharyngotympanic tube, Eustachian tube, or pharyngeal salpinx†, enables the air in the middle ear to remain at atmospheric pressure. This is important for hearing since the tympanic membrane (ear-drum) will only be in its most sensitive position if the mean air pressure on its two sides are the same.

The auditory tube runs from the pharynx to the middle ear. It is about 4 cm long and consists of an anterior cartilaginous part 2.5 cm long and a posterior bony part 1.5 cm long. The tube is lined with ciliated mucous membrane which moves mucus continuously towards the pharynx.

The **cartilaginous** part consists of a triangular plate which forms the superior and medial parts of the tube, the tube being closed by fibrous tissue.

The **pharyngeal opening** lies at the back of the lateral wall of the nasal cavity indenting the upper half of the medial pterygoid plate. The opening is surrounded above and behind by the salpingo-pharyngeal fold (Eustachian cushion). This pharyngeal opening lies above the superior constrictor in the space between the upper border of the levator veli palatini and the base of the skull. The cartilaginous part of the tube has the **tensor palati** lying laterally. The **levator palati**, which also takes origin from the tube, lies medially. The mandibular division of the fifth cranial nerve lies lateral to the auditory tube.

The pharyngeal opening of the auditory tube is normally kept closed. It is opened when swallowing by the tensor palati muscle.

The **bony part** of the auditory tube lies in the petrous temporal bone immediately under the canal for the tensor tympani muscle. It opens on the anterior wall of the middle ear [p. 361]. The tensor tympani tendon runs along the medial wall and turns laterally towards the neck of the malleus when it reaches the processus cochlearformis.

† The word *salpigx* is Greek for a tube or trumpet. The terms *salpinx* and *salpingo-* are used in connection with both the auditory and the uterine tubes. Hence salpingitis may mean either inflammation of the auditory tube or inflammation of the uterine tube.

16 The neck, pharynx, and larynx

THE NECK

Surface anatomy

On the back of your neck, find the groove which runs vertically in the midline. It is produced by the **semispinalis capitis** muscles on each side [p. 64]. Lean forward and extend your neck and the groove will become more obvious. When the finger runs downwards the first vertebral spine which is easily defined is that of C7 the **vertebra prominens**. In many people the spine of T1 projects beyond the vertebra prominens. At the upper end of the groove find the occipital bone and the projecting knob of bone in the midline, the external occipital protuberance or **inion** [FIG. 15.04; p. 295].

The semispinalis capitis muscles are the most important preventing the head from falling forwards. They are attached to the back of the skull between the middle and superior nuchal lines, so that they get the best possible leverage. They are covered superficially by the splenius capitis and trapezius muscles, neither of which can be identified with the fingers in this part of the neck.

At the chin feel the **diaphragm of the floor of the mouth**. It lies more or less in the horizontal plane. Put your finger beneath the tongue on the floor of the mouth, and your thumb outside, and move the intervening tissues up and down. At the angle where the horizontal floor joins the vertically running front of the neck you can find the **hyoid bone** [FIG 16.01 and FIG. 15.15; p. 301]. Flex the neck slightly. Put your left thumb in the angle we have just mentioned 2–3 cm to the left of the midline. Push your thumb medially, and as you do so feel with your right hand just below and in front of the angle of the mandible. You will feel the greater cornu of the hyoid bone being displaced to the right. Your left thumb is pushing against the body of the hyoid.

Below the body of the hyoid find the **thyroid cartilage** which is the shield (*thyroid* = shield) which protects the mechanism of the larynx. Below the thyroid cartilage is the **cricoid cartilage** and below this the trachea is easily made out. The **thyrohyoid** and **cricothyroid membranes** cannot be felt as individual structures, but their positions are obvious.

The vocal folds are attached to the thyroid cartilage on its deep surface just below the promontory of the larynx (the Adam's apple). Notice that when the pitch of the voice is raised the thyroid cartilage moves forward. This increases the tension in the cords.

In respiratory obstruction above the larynx a new opening may be made in the trachea, a tracheostomy (*ostium* = a mouth) or in the larynx through the cricothyroid membrane (laryngostomy).

When you swallow, the larynx moves upward about 3–4 cm (about two cervical vertebrae on an X-ray). The hyoid moves forwards to make room for it. At the same time the trachea stretches. This is possible because the tissue which joins the 'rings' of the trachea to each other consist principally of yellow elastic tissue.

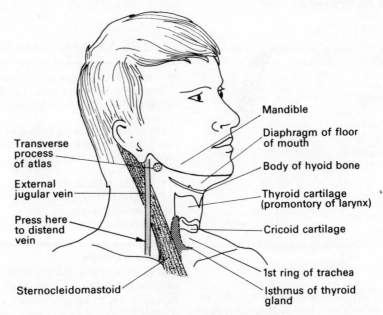

FIG. 16.01. Important landmarks of the neck.

309

Attached to the larynx and trachea on each side of the midline is the **thyroid gland**. You may be able to see this in yourself, when you swallow if you have a long thin neck and your gland is maybe larger rather than smaller than average. It is clinically easier to palpate the larynx, trachea and thyroid gland by standing behind a patient who is sitting in a chair. The Thyroid gland moves in unison with the larynx in swallowing. You should be able to feel your own thyroid.

At the side of the neck the major landmark is the **sternocleidomastoid** (from sternum and clavicle to the mastoid process, p. 328). Trace the muscle up and down. Identify the mastoid process behind the ear (*mastoid* = breast-like!). The muscle is also attached behind the process to the superior nuchal line. We will call the muscle *sternomastoid* for short in accord with general custom.

Both muscles contract when the neck is flexed against resistance, as in raising the head from a pillow. The left sternomastoid contracts when the chin turns to the right. Each muscle also produces side flexion. When one muscle contracts on its own without interference from other muscles in the neck the head adopts the wry-neck or torticollis position. Imitate this by putting your ear as close to the manubrium as you can. Try to do so when lying supine, and one sternomastoid will stand out very clearly.

Put your finger just above and behind the thyroid cartilage in front of the sternomastoid muscle and identify the carotid pulse.

Note the **external jugular vein**. It can be seen distended when a person is lying down. In a normal person sitting upright this vein is collapsed. It runs from just behind the ear to the point above the medial third of the clavicle behind the attachment of the sternomastoid where it passes through the investing fascia. Place your finger over this region and press gently. If you are in the right place the vein will fill up and become distended. When you remove your finger it will empty immediately. Alternatively the vein may be distended by increasing the intrathoracic pressure such as by blowing hard against a closed glottis (Valsalva manoeuvre). In heart failure this vein is used as a manometer by observing the height above the clavicle of the distended part of the vein.

Sternocleidomastoid (sternomastoid)

Origin:	Sternal head—upper anterior surface of manubrium Clavicular head— upper surface of medial third
Insertion:	Skull mastoid process (lateral surface) and superior nuchal line (lateral half)
Nerve supply:	XI motor C2, 3 sensory
Action:	Muscle of one side: head to own side and rotates face to opposite side Muscles of both sides: flexes the neck

If the head is fixed by contracting the postvertebral muscles, the sternomastoids act as important accessory muscle of respiration by raising the sternum and upper ribs.

TRIANGLES OF THE NECK

A knowledge of the boundaries of the triangles of the neck facilitates the clinical description of the position of a lump or of a pain.

The sternomastoid muscle divides the neck into the anterior triangle which lies between the front of the muscle and the midline and the posterior triangle which lies between it and the trapezius [FIG. 16.02].

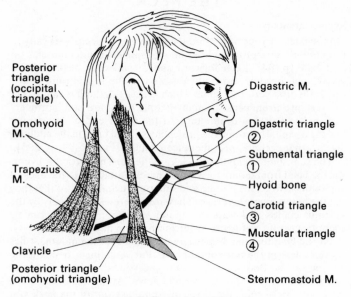

FIG. 16.02. The triangles of the neck.

THE ANTERIOR TRIANGLE OF THE NECK

The anterior triangle is the area at the front of the neck bounded by the sternomastoid muscle, the mandible and the midline.

The sternomastoid muscle can be seen standing out as a well-defined ridge by taking a deep breath or flexing the head forwards against resistance.

There are two double bellied muscles associated with the hyoid bone. These are the digastric and omohyoid [FIG. 16.02]. Using these muscles as boundaries the region is conveniently further sub-divided into:

1. Submental triangle.
2. Digastric triangle.
3. Carotid triangle.
4. Muscular triangle.

Submental triangle

The submental triangle lies above the hyoid bone medial to the anterior belly of the digastric [FIG. 16.03]. It is separated from the mouth by the mylohyoid muscle. At the medial boundary of the triangle in the midline a fibrous raphe joins the symphysis menti to the hyoid bone (*menti* (Latin) = of the chin).

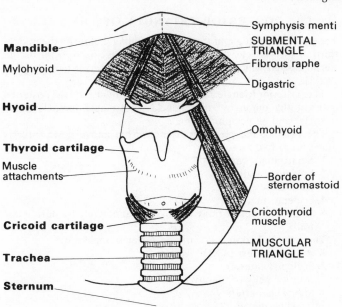

FIG. 16.03. The submental and muscular triangles.

This triangle contains the carotid sheath with the common carotid artery, internal jugular vein, and vagus nerve. The common carotid artery divides in the triangle into the internal and external carotid arteries. The external carotid artery gives off the superior thyroid artery, the ascending pharyngeal artery, the lingual artery, the facial artery, and the occipital artery in the triangle.

Muscular triangle

The muscular triangle is bounded by the superior belly of omohyoid, sternomastoid, and the midline. The anterior jugular veins descends and vertically through this triangle.

The strap-like infrahyoid muscles can raise or depress the larynx and hyoid bone [FIG. 16.04]. They are used in speech and when swallowing. They are extrinsic laryngeal muscles.

The strap-like infrahyoid muscles are:

    Sternohyoid
    Sternothyroid
    Thyrohoid
    Omohyoid

The **sternohyoid** runs from the back of the manubrium and the medial end of the clavicle to the lower border of the hyoid bone.

The **sternothyroid** is deep to the sternohyoid. It runs from the back of the manubrium (lower down) and first costal cartilage to the oblique line of the thyroid cartilage. It passes in front of the thyroid gland.

The **thyrohyoid** is a continuation of the sternothyroid from the oblique line of the cartilage to the lower border of the greater horn of the hyoid bone.

The **omohyoid** (*omos* (Greek) = shoulder) runs from the upper border of the scapula to the inferior surface of the body and greater horn of the hyoid bone. It pulls the hyoid downwards. The intermediate tendon is held in place by the omohyoid fascia which connects it to the clavicle. In a thin person it can be seen twitching above the clavicle as the person talks. This muscle also helps to prevent bulging of the apex of the lung in forced expiration.

Digastric triangle

The digastric triangle is situated above the hyoid bone between the two bellies of digastric and the mandible.

The digastric triangle contains the submandibular salivary gland and submandibular lymph nodes. The facial artery and vein cross the lower border of the mandible.

Carotid triangle

The carotid triangle is bounded by the superior belly of the omohyoid, the posterior belly of digastric and the anterior border of the sternomastoid.

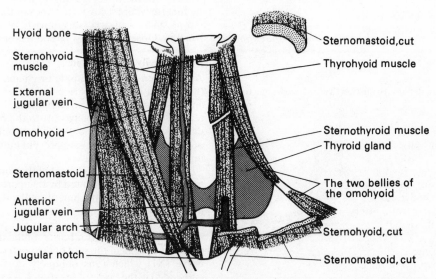

FIG. 16.04. Muscles of the front of the neck.
The sternomastoid and sternohyoid muscles have been removed on the left side to reveal the thyrohyoid, sternothyroid, and omohyoid muscles.

INFRAHYOID MUSCLES

There are four pairs of strap-like muscles lying anteriorly in the neck [FIG. 16.04]. They are enclosed in a fascial sheath.

Sternohyoid

Origin: Clavicle—inner end
               Sternum—(manubrium) back
Insertion: Hyoid—lower border
Nerve supply: Ansa cervicalis—C1, 2, 3
(separate below but fused above)
Action: Pulls down hyoid bone

The name indicates that it runs from the sternum to the hyoid bone.

Sternothyroid

Origin: Sternum—manubrium
Insertion: Thyroid cartilage—
               oblique line
Nerve supply: Ansa cervicalis—C1, 2, 3
Action: Pulls down thyroid cartilage

The name indicates that it runs from the sternum to the thyroid cartilage.

Thyrohyoid

Origin: Thyroid cartilage—
               oblique line
Insertion: Hyoid—lower border of
               cornu
Nerve supply: C1 via hyoglossal nerve
Action: Pulls down hyoid bone

The name indicates that it runs from the thyroid cartilage to the hyoid bone.

Omohyoid

Origin: Posterior belly: Scapula
               (upper border)
               Intermediate tendon
               —over int. jugular vein
               Anterior belly: crosses
                    common carotid artery
                    level of cricoid
Insertion: Hyoid
Nerve supply: Ansa cervicalis
Action: Pulls down hyoid bone

The name indicates that it runs from the shoulder (*omos* (Greek) = shoulder) to the hyoid bone.

THE POSTERIOR TRIANGLE OF THE NECK

The posterior triangle of the neck is bounded below by the middle third of the **clavicle**, behind by the **trapezius** and in front by the **sternomastoid** which separates it from the anterior triangle [FIG. 16.05].

The posterior triangle can be further sub-divided by the posterior belly of the omohyoid in a subclavian triangle (omoclavicular triangle) and an occipital triangle.

The floor of the posterior triangle from above downwards is made up of five muscles:
1. Semispinalis capitis.
2. Splenius capitis.
3. Levator scapulae.
4. Scalenus medius.
5. Scalenus anterior (may be covered completely by sternomastoid)

The principal contents of the posterior triangle are:
1. Accessory nerve (XI).
2. Scalenus anterior.
3. Inferior belly of omohyoid.
4. Subclavian artery.
5. Brachial plexus.
6. External jugular vein.
7. Cervical plexus (branches of).

Accessory nerve (XI)

The eleventh cranial nerve (accessory nerve) leaves the skull with the vagus (X) through the jugular foramen. It crosses the transverse process of the atlas to enter sternomastoid which it supplies. It leaves the cover of the sternomastoid above the middle of its posterior border. It runs across the posterior triangle parallel to the levator scapulae embedded in the fascia of the roof. It disappears deep to the trapezius and supplies it.

Scalenus anterior

This muscle runs from the anterior tubercles of the transverse processes of the third, fourth, fifth, and sixth cervical vertebrae to the first rib where it covers the apex of the lung. It raises the first rib and as such is an accessory muscle of respiration. Reversing its origin and insertion, it is a lateral flexor of the neck [FIG. 16.06].

Inferior belly of omohyoid

As has been seen this muscle originates from the upper border of the scapula and the superior transverse scapular ligament. It runs anterosuperiorly deep to sternomastoid to its own intermediate tendon. The superior belly runs to the hyoid bone. It steadies and depresses the hyoid bone. When both bellies contract it pulls on the omohyoid fascia and help to prevent bulging of the apex of the lung into the root of the neck.

The apex of the triangle is slightly truncated, as it does not come to a point but is represented by that part of the superior nuchal line which separates the attachments of the sternomastoid and trapezius to the back of the skull.

The investing fascia runs from the posterior edge of the sternomastoid to the anterior border of the trapezius. In this region it is called 'the fascia of the roof of the posterior triangle'. The **accessory nerve** (XI) which runs obliquely downwards and backwards is embedded in the fascia of the roof. It is essential before undertaking the most minor surgical procedure in the posterior triangle to consider the position of this nerve and then to make an incision

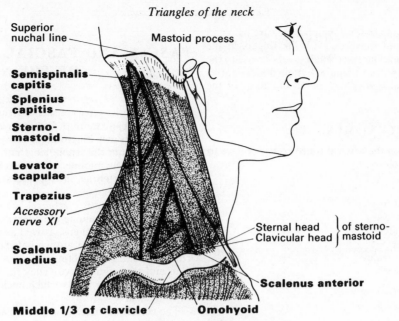

FIG. 16.05. Muscles of the posterior triangle.

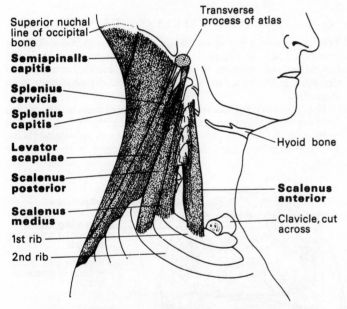

FIG. 16.06. Muscles of the neck beneath sternomastoid and trapezius. The nerve roots of the cervical and brachial plexuses run laterally between scalenus anterior and scalenus medius. The inferior trunk of the brachial plexus (C8 and T1) and the subclavian artery lie directly on the first rib.

somewhere else. Any incision in the vicinity of the nerve must be made parallel to it.

The XI nerve passes over the transverse process of the atlas and reaches the posterior border of the sternomastoid at the point of union of the upper and middle thirds of a line drawn from the mastoid process to the sternoclavicular joint. It runs across the posterior triangle and reaches the trapezius at the junction of its middle and lower thirds.

Floor

The floor of the posterior triangle is made up principally by three muscles. From medial to lateral they are: scalenus medius, levator scapulae, and splenius capitis.

Scalenus medius runs downwards and laterally from the posterior tubercles of the transverse processes of the cervical vertebrae from C2 downwards (sometimes from C1). It is attached to the first rib behind the subclavian artery.

The levator scapulae is attached to the transverse processes of the upper four cervical vertebrae and runs downwards towards the superior angle of the scapula.

The splenius runs upwards and laterally from its origin from cervical and upper thoracic vertebrae to its insertion into the occipital bone just deep to the lateral third of the superior nuchal line. The word splenius means a bandage, and it appears to 'hold down' almost like a retinaculum, the semispinalis capitis muscle.

Note that above the splenius capitis part of the semispinalis is exposed and should be included in a description of the muscles in the floor of this triangle. Similarly on the medial side of the triangle, the scalenus anterior can often be seen.

These muscles lie in the vertebral compartment and are covered by the prevertebral fascia. The prevertebral fascia is 'the fascia of the floor of the posterior triangle'.

The most important structure in the floor of the posterior triangle is the **brachial plexus** [p. 81] which is deep to the prevertebral fascia. Palpate the upper roots about three or four fingers breadths above the medial third of the clavicle. The upper trunk and its division are all exposed in the triangle here. The lower trunk, however, is completely protected by the clavicle.

The plexus runs downwards and laterally with scalenus anterior in front and scalenus medius behind. As the nerves pass between these two muscles, with the subclavian artery, they take with them an extension downwards from the prevertebral fascia which is the axillary sheath. This surrounds the plexus and blood vessels for a few centimetres in the axilla.

Superficial cervical lymph nodes may be seen along the course of the external jugular vein. Deep cervical lymph nodes are palpated by pushing the sternomastoid forwards, with the fingers in the posterior triangle. The thoracic duct on the left side lies close to the lower medial angle of the posterior triangle. Lymph nodes near its termination are a traditional sign of the spread of cancer of the stomach.

THE SCALENE MUSCLES

The scalene muscles run from the cervical transverse processes to the first and second ribs [FIG. 16.06].

Scalenus anterior

Origin: C3, 4, 5, and 6
                 anterior tubercles
Insertion: First ribs (scalene
                 tubercle)
Nerve supply: C3–6 anterior rami
Action: Elevates first rib

Scalenus medius

Origin: C2, 3, 4, 5, 6, and 7
                 posterior tubercles
Insertion: First rib (upper surface)
Nerve supply: C2–7 anterior rami
Action: Elevates first rib

Scalenus posterior

Origin: C2, 3, 4, 5, 6, and 7
                 posterior tubercle
Insertion: Second rib
Nerve supply: C2–7 anterior rami
Action: Elevates second rib

The scalenes, like sternomastoid, act as accessory muscles of respiration by elevating the rib cage.

FASCIA AND FASCIAL COMPARTMENTS OF THE NECK

In FIGURE 16.07 you will see that the neck can be regarded as containing three separate sets of structures which are functionally and anatomically more or less independent of each other. Each is enclosed in its own longitudinal sleeve of fascia which runs in every case throughout the whole length of the neck.

Behind, in the midline, is the **vertebral compartment** containing the cervical vertebrae, the muscles attached to them as well as the cervical nerve roots.

In front of the vertebral compartment is the **visceral compartment**, containing the pharynx, the larynx (leading into the trachea), the oesophagus, as well as the thyroid gland.

On each side in the groove between the vertebral and visceral compartments you will find the carotid artery, the internal jugular vein and the vagus nerve. They lie in what we call the **vascular compartment**. The fascia which encloses this compartment is the carotid sheath.

The vertebral, visceral, and vascular compartments are together enclosed in a common layer of deep fascia, which lies just deep to the skin. This is the **investing fascia of the neck**.

INVESTING FASCIA OF THE NECK

The layer of deep fascia which lies deep to the skin is called the investing fascia of the neck. At the back of the neck it covers the superficial surface of the **trapezius** muscle and reaches the skull at the superior nuchal line of the occipital bone with trapezius.

The superficial surface of the **sternomastoid** muscle, which runs from the manubrium sterni and medial third of the clavicle upwards to be attached to the skull, is also covered by the investing fascia. The gap between sternomastoid and trapezius is the **posterior**

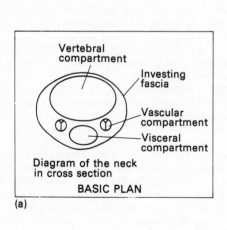

(a) Diagram of the neck in cross section
BASIC PLAN

- Vertebral compartment
- Investing fascia
- Vascular compartment
- Visceral compartment

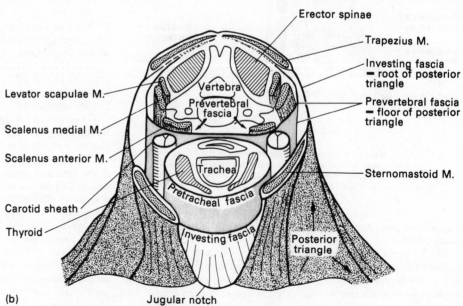

(b)

- Erector spinae
- Trapezius M.
- Investing fascia = root of posterior triangle
- Prevertebral fascia = floor of posterior triangle
- Sternomastoid M.
- Posterior triangle
- Jugular notch
- Levator scapulae M.
- Scalenus medial M.
- Scalenus anterior M.
- Carotid sheath
- Thyroid
- Vertebra
- Prevertebral fascia
- Trachea
- Pretracheal fascia
- Investing fascia

FIG. 16.07. The fascial compartments of the neck.

triangle of the neck [p. 312]. The investing fascia bridges this gap forming the fascia of the roof of the posterior triangle.

The roof fascia is attached, therefore, to the occipital bone, to the middle third of the superior nuchal line, and to the middle third of the clavicle, in both cases bridging the gap between the origins or insertions of these muscles.

The investing fascia also covers the **anterior triangle** of the neck, that is the area bounded by the sternomastoid, the mandible, and the midline. In front of the mastoid process the fascia covers the parotid gland and runs over the **masseter** muscle to the zygomatic arch. In front of the masseter the muscles of facial expression are attached directly to the skin. Hence in this region one of the principal functions of deep fascia which is to allow the skin to slide over the muscles, does not apply. The fascia gets no further than the anterior border of the masseter.

Below the chin and in the submandibular region the fascia blends with the periosteum of the body of the mandible. It also blends with the superficial surface of the hyoid bone and with those parts of the laryngeal skeleton which are exposed. Below the larynx the fascia splits to enclose a space immediately above the manubrium called the **suprasternal space**.

The fascia on the deep surface of the sternomastoid and trapezius muscles must obviously be continuous with the fascia on their superficial surfaces. It is customary to say that the investing fascia 'splits' to enclose them. It also splits to enclose a space (like the suprasternal space) above the middle third of the clavicle. The omohyoid muscle runs across the deep surface of the sternomastoid muscle and is therefore in the plane of the deeper layer giving to it the name of **omohyoid fascia**.

The investing fascia has an important relationship to the accessory nerve, the eleventh cranial nerve. This nerve supplies both sternomastoid and trapezius. It first lies embedded in the substance of the upper third of the sternomastoid and passes downwards and laterally to enter the lower third of the trapezius. Because of its relationship to sternomastoid the nerve appears at its posterior border embedded in the substance of the investing fascia, and it traverses the posterior triangle in this plane. Other motor nerves in the body lie deep to the deep fascia. This special relationship means that the nerve is in danger of being cut whenever the skin of the posterior triangle is incised [p. 313, surface markings].

In most parts of the body only superficial fascia intervenes between the skin and deep fascia. In the front of the neck the platsyma muscles runs down from the chin as far as the clavicles. It is classified as a muscle of facial expression and is supplied by the seventh cranial nerve.

PREVERTEBRAL MUSCLES

The prevertebral muscles [Fig. 16.08] which lie beneath the prevertebral fascia are:
Rectus capitis anterior.
Rectus capitis lateralis.
Longus capitis.
Longus colli.
They flex the neck and flex the head on the neck.

Longus colli

Origin:	Anterior tubercle of atlas
Insertion:	Vertebral bodies C2 to T3 Cervical transverse processes C3 to C6
Nerve supply:	Cervical nerves (ventral rami)
Action:	Flexes the neck Flexes the head

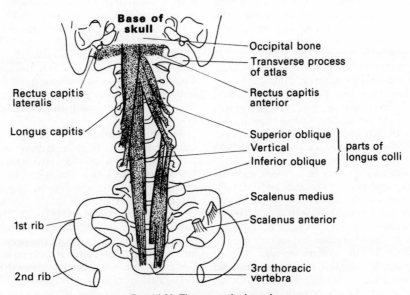

FIG. 16.08. The prevertebral muscles.
These muscles exert very little leverage. Their proprioceptor information may be an important afferent input concerning the position of the head and neck.

Longus capitis

Origin: Cervical transverse processes C3 to C6
Insertion: Base of skull
Nerve supply: Cervical nerves (ventral rami)
Action: Flexes the head on the neck

Rectus capitis anterior

Origin: Atlas—lateral mass
Insertion: Base of skull
Nerve supply: C1
Action: Flexes the head on the neck

Rectus capitis lateralis

Origin: Atlas—transverse process
Insertion: Base of skull
Nerve supply: C1
Action: Flexes the head on the neck

VERTEBRAL COMPARTMENT

The vertebral compartment of the neck contains the seven cervical vertebrae [p. 56] and the muscles attached to them. It is enclosed in its own fascia. [FIG. 16.07].

Anteriorly this fascia covers the bodies of the vertebrae, the intervertebral discs, the longus capitis and longus cervicis muscles. It is called the **prevertebral fascia**. It forms the posterior lining of the retropharyngeal space.

Tracing the prevertebral fascia laterally, it first covers the scalenus anterior, the nerves of the cervical and brachial plexuses, the scalenus medius, the levator scapulae, splenius capitis, and semispinalis capitis. As it covers these muscles, it is a carpet covering the floor of the posterior triangle. Even here, the fascia is still called the prevertebral fascia, although in the posterior triangle it is lateral rather than anterior to the cervical vertebrae.

At the back of the neck deep to the trapezius muscle the prevertebral fascia loses its identity and fuses with the very dense fascia which is found between all the muscles in this region.

Tracing the prevertebral fascia upwards it is eventually attached to the base of the skull just behind the pharyngeal tubercle and in front of rectus capitis anterior. In the midline it passes downwards into the thorax and loses its identity about the level of T6. At the sides it is carried down to the first rib by the scalenus anterior and scalenus medius.

The brachial plexus and subclavian artery lie between scalenus anterior and scalenus medius and pass downwards and laterally over the first rib through the apex of the axilla to supply the upper limb. They are inside the vertebral compartment to start with and make their way out of the enclosing fascial sheath, by pushing out laterally a fascial diverticulum which becomes the axillary sheath, surrounding the brachial plexus and the axillary vessels.

The formation of such a diverticular sleeve is the standard method whereby many structures pass across or through fascia or fibrous 'barriers'. Recall the femoral sheath, the coverings of the spermatic cord in the inguinal canal, and the way the spinal nerves escape from the spinal theca, which are examples of the same mechanism.

VASCULAR COMPARTMENT

The carotid sheath of cervical fascia encloses the **vascular compartment** of the neck. The sheath runs from the **base of the skull**, where it surrounds the jugular foramen and the beginning of the carotid canal, to the **inlet of the thorax**, where it blends with the adventitia of the great vessels. It loses its identity in the superior mediastinum.

The carotid sheath lies in the groove between the vertebral and visceral compartments [FIG. 16.09].

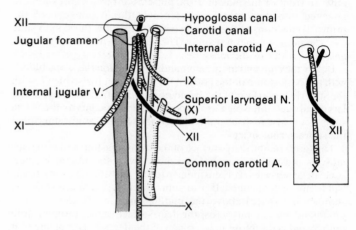

FIG. 16.09. The four nerves in the carotid sheath.

The principal artery of the carotid sheath is the **common carotid artery**. It is a direct branch of the aortic arch on the left and of the brachiocephalic trunk on the right. Its pulsation can be felt through the flesh of the sternomastoid muscle just above the clavicle. It runs vertically upwards from here and branches into the **internal carotid** artery, which is principally destined for the brain and the eyeball, and the **external carotid** artery which is distributed mainly to structures in the neck and face outside the skull. The bifurcation takes place just in front of the sternomastoid muscle level with the upper border of the thyroid cartilage [p. 253].

The principal vein in the carotid sheath is the **internal jugular vein**. This runs downwards from the jugular foramen to its junction with the subclavian vein where, behind the sternoclavicular joint, it terminates by forming the brachiocephalic vein. The internal jugular vein is the companion vein of the internal and common carotid arteries [p. 264].

The internal jugular vein receives the venous blood from the **sigmoid sinus**, one of the large dural sinuses inside the skull, and it is joined just below the jugular foramen outside the skull by the **inferior petrosal sinus**. The vein subsequently receives directly or indirectly the companion veins of most of the branches of the external carotid artery.

The **external carotid artery** [p. 257] leaves the carotid sheath and comes to lie eventually deeply embedded in the parotid gland. Here it terminates, level with the neck of the mandible by dividing into the **superficial temporal** and **maxillary arteries**. It only runs for 7–8 cm before it reaches its terminal bifurcation.

The first branch, the ascending pharyngeal artery, comes off the medial side. It also gives off three branches which run forwards, the superior thyroid, the lingual and facial arteries, and two which run backwards, the **occipital** and **posterior auricular** arteries, making eight branches in all. Branches of the external carotid anastomose freely both with each other as well as across the midline.

The **occipital artery** is palpated 4–5 cm from the midline as it reaches the superior nuchal line at the back of the skull. It is joined on its medial side by the large cutaneous nerve from C2, the greater occipital nerve. Pressure against the back of the head in this region (such as a badly designed dentists' headrest) will give rise to headache.

The superficial temporal, facial, and carotid pulses are often employed by anaesthetists. Be careful if you use the carotid pulse not to compress the **carotid sinus**.

The increase in baroreceptor activity produced by such a carotid sinus compression will reflexly cause a cardiac slowing and an arteriolar vasodilatation leading to a fall in blood pressure and possible fainting. Intentional carotid sinus compression is used clinically. The increase in vagal tone to the heart produced by carotid sinus compression may, in cases of paroxysmal tachycardia, cause the heart to revert back to normal rhythm.

The internal carotid artery does not give off any branches in the neck, nor (apart from its terminal branches) does the common carotid artery.

In addition to the internal jugular vein, the jugular foramen also transmits the **glossopharyngeal nerve** (IX), the **vagus** (X), and the **accessory nerve** (XI). The last two share the same sheath of dura. As all three pass through the jugular foramen, they run straight into the carotid sheath with the vein [FIG. 16.09].

Notice the symmetry of the position of these three nerves in the upper part of the carotid sheath. The vagus runs straight down inside the sheath, in the groove behind, between the artery and the vein. The accessory nerve (XI) passes *backwards* superficial (usually) to the internal jugular vein, across the transverse process of the atlas on its way into the sternomastoid muscle. Find the transverse process of your own atlas, half way between your mastoid process and the angle of the jaw, and turn your head from side to side for confirmation. Pressure here is very unpleasant because the accessory nerve is compressed between your finger and the bone.

The **glossopharyngeal nerve** (IX) runs *forwards* across the internal carotid artery in mirror-image fashion to the accessory nerve running backwards. At first it lies deep to the stylopharyngeus muscle, then twists around its lower border, and accompanies the muscle between the two carotid arteries. The pharyngeal branch of the vagus also runs between the two carotid arteries in the same plane as the glossopharyngeal nerve but lower down.

The **superior laryngeal nerve** (which supplies the upper half of the mucous membrane of the larynx as well as the cricothyroid muscle) comes off the vagus just below its pharyngeal branch, but it lies deep to the internal carotid artery.

Lastly the **hypoglossal nerve** (XII) also lies in the upper part of the carotid sheath. The nerve comes to lie in the anterior groove between the internal carotid artery and the internal jugular vein. The nerve leaves the skull through the hypoglossal canal which is just in front of the occipital condyle. The most direct route (check on a skull) is to go between these vessels. The nerve crosses the vagus (already in the groove behind) on its lateral side. About level with the hyoid bone this nerve swings forwards towards its destination in the tongue and lies superficial to the external carotid artery. A branch (descendens hypoglossi) continues downwards as the superior root of the ansa cervicalis [p. 291].

Note that the last four cranial nerves lie close together in the upper 2–3 cm of the carotid sheath.

Throughout its whole length the carotid sheath contains a large number of **deep cervical lymph nodes** which are especially associated with the vein rather than the arteries. The sheath is crossed by the digastric and omohyoid muscles and in its upper part we speak of jugulodigastric and in its lower jugulo-omohyoid lymph nodes.

In ordinary medical practice the lymph nodes cause more concern than any of the other contents of the vascular compartment which we have mentioned because of their involvement with the spread of infection and malignant disease [p. 265].

VISCERAL COMPARTMENT

Visceral compartment and its fascia

The visceral compartment encloses the nasopharynx, oropharynx, and laryngeal pharynx as well as the larynx and trachea, with the thyroid gland lying anteriorly; the oesophagus takes over from the laryngeal pharynx. It is enclosed in its own fascia like the other compartments in the neck [FIG. 16.07].

First there is the fascia covering the pharynx on its outer surface which allows it to glide easily over the vertebral compartment structures covered by their own prevertebral fascia. This is the retropharyngeal fascia.

The **retropharyngeal fascia** is attached to the base of the skull with the **superior constrictor** muscle and can be traced downwards behind the rest of the pharynx on to the back of the oesophagus where it eventually disappears as a distinct layer in the thorax at the level of about T6. The upper part of the retropharyngeal fascia which clothes the superior constrictor muscle can be followed forwards superficial to the pterygomandibular raphé when it will cover the outer surface of the **buccinator** muscle in the cheek, and is then called the buccopharyngeal fascia.

The fascia covering the back of the posterior part of the **inferior constrictor** and upper part of the oesophagus can be followed round on to the front of the trachea where it is called the **pretracheal fascia**. This fascia also encloses the **thyroid gland**. The pretracheal fascia can be followed downwards into the superior mediastinum where it runs as far as the bifurcation of the trachea where it blends with the posterior aspect of the fibrous pericardium.

Tracing the pretracheal fascia upwards it lies in front of the larynx and the key fact is the attachment of the **sternothyroid muscle**. This muscle is attached to the sternum below (that is, in front of the pretracheal fascia) and to the oblique line on the thyroid cartilage above. It follows that the pretracheal fascia must also be attached to the oblique line immediately deep to the sternothyroid.

The pretracheal fascia is very well developed, suggesting that there must be a considerable amount of sliding between the larynx and trachea and the structures in front of them, not only in the neck but also in the superior mediastinum. Indeed in swallowing, the larynx rises a considerable distance. The bifurcation of the trachea, however, remains in the same position and the trachea is stretched out like a concertina when the larynx is raised. There is a considerable amount of elastic tissue on the wall of the trachea and this stretches when the larynx is raised in swallowing and then causes it to spring back to its initial length immediately afterwards.

The movements of the structures in the visceral compartment of the neck are associated with swallowing and phonation. It is these movements, which take place quite independently, which induce the development of these fascial layers.

This **visceral compartment** contains:

(i) pharynx;   (iv) thyroid and parathyroids;
(ii) larynx;   (v) oesophagus [p. 177].
(iii) trachea and thyroid;

THE PHARYNX

In the pharynx [FIG. 16.10] the upper part of the alimentary tract crosses the pathway of the respiratory tract. Thus the food passes from the mouth through the *pharynx* to the oesophagus whilst the air passes from the nasal cavity through the *pharynx* to the trachea.

The pharynx and swallowing

The principal functional requirement of the pharynx is that in swallowing no foreign substance should enter the respiratory tract.

The swallowing mechanism is amazingly efficient. We swallow some 3–4 million times a year and can thus expect to do so about 200–300 million times in a life-time, yet most people cannot remember the last time they choked.

The ability to swallow develops very early in uterine life. Even the most premature baby can swallow.

The complex co-ordinated movements of the pharynx in swallowing are reflexly controlled by the brainstem (via cranial nerves IX, X, and XI). Swallowing occurs during sleep and also in light coma.

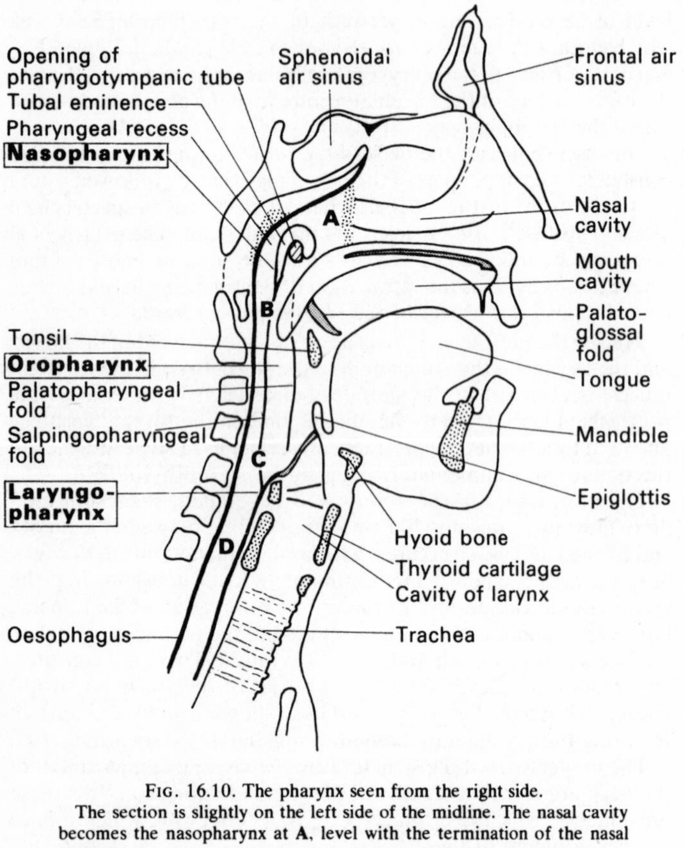

If the swallowing mechanism should fail and something succeeds in entering the larynx or trachea the body is equipped with a further line of defence in the form of the **cough reflex**. Coughing occurs in patients in light coma or light anaesthesia but in deep coma both coughing and swallowing disappear. This is dangerous because regurgitated gastric contents having flooded the pharynx may then be drawn into the lungs causing 'inhalation pneumonia' or even drowning.

In certain skull-base fractures blood may collect in the pharynx and be inhaled into the lungs.

Unconscious patients should always be nursed in the semi-prone coma position so that gravity helps to keep the airways open and also hinders the entry of fluids into the lungs.

Compartments of the pharynx

The pharynx communicates from above downwards with the nose, the mouth, the larynx, and oesophagus [FIG. 16.10]. For this reason it is convenient to sub-divide it into three compartments:

  (i) nasopharynx;
  (ii) oropharynx;
  (iii) laryngeal part of the pharynx whose lower part is often called the hypopharynx.

Nasopharynx

The nasopharynx is box-like. In front it is directly in communication with the nasal cavities via the two **anterior nares** which lie on each side of the **nasal septum**. Its sloping roof lies against the inferior surface of the base of the skull (basi-occiput and basisphenoid), and the posterior wall is against the anterior arch of the atlas. The floor of the box is the **soft palate**. The side wall is largely formed by the medial surface of the **medial pterygoid plate** and the **pharyngotympanic tube**.

The direction of the air stream changes in the nasopharynx from horizontal to vertical. Like most of the respiratory tract the nasopharynx is permanently open.

Just behind the opening of the pharyngotympanic tube is a projection, the **tubal eminence**. This is produced by the cartilaginous part of the tube, which in both the foetal and adult skull approaches the side wall of the nasopharynx (that is the medial pterygoid plate) at an angle of about 45°. Between the tubal eminence and the posterior wall of the nasopharynx is the deep **pharyngeal recess** (the fossa of Rosenmüller). This recess would not exist if it were not for the projection of the tubal eminence.

The following structures can all be seen in the living by examining the nasopharynx with a suitable mirror:

  (i) the nasal septum;
  (ii) inferior turbinate;
  (iii) tubal eminence and opening of the tube;
  (iv) commencement (only) of the pharyngeal recess;
  (v) posterior pharyngeal wall;
  (vi) pharyngeal tonsil (adenoids).

It is remarkable that the distance which separates the tubal eminences on each side is about equal to the depth of each pharyngeal recess so the pharynx is really three times as wide as it looks at this level.

Unfortunately the epithelium of the recess sometimes gives rise to a 'carcinoma of the base of the skull' (really a carcinoma *invading* the base of the skull). Look at the skull and locate the position of the recess. You will realize that as soon as such a tumour begins to

FIG. 16.10. The pharynx seen from the right side.

The section is slightly on the left side of the midline. The nasal cavity becomes the nasopharynx at **A**, level with the termination of the nasal septum.

The nasopharynx leads into the oropharynx at **B**. At **C**, opposite the inlet to the larynx, the oropharynx becomes the laryngeal pharynx. At **D** the pharynx leads into the oesophagus at the level of the lower border of the cricoid cartilage.

The palatoglossal arch divides the mouth cavity from the oropharynx.

Labels on figure:
Opening of pharyngotympanic tube
Tubal eminence
Pharyngeal recess
Nasopharynx
Tonsil
Oropharynx
Palatopharyngeal fold
Salpingopharyngeal fold
Laryngo-pharynx
Oesophagus
Sphenoidal air sinus
Frontal air sinus
Nasal cavity
Mouth cavity
Palato-glossal fold
Tongue
Mandible
Epiglottis
Hyoid bone
Thyroid cartilage
Cavity of larynx
Trachea

spread laterally it will invade the carotid sheath and paralyse the IX, X, XI, and XII cranial nerves; if it tracks forwards it may reach the foramen lacerum and invade the cavernous sinus where nerves III, IV, V, and VI are also in jeopardy.

The pharyngeal tonsil (adenoid) is at the junction of the roof of the nasopharynx with the posterior wall. Enlargement in children obstructs the nasal airway, giving rise to mouth breathing and to characteristic changes in the voice. The pharyngotympanic tube may be blocked with ensuing deafness and otitis media.

Oropharynx

Open your mouth wide and look into a mirror. You will see your mouth and oropharynx. If your tongue is in the way try saying 'aaah' or breathing quickly in and out through the mouth. Your tongue will then descend and the soft palate will arch upwards and you will be able to see the back of your throat. You are now looking into the oropharynx [p. 298].

The oropharynx is partly separated from the nasopharynx by the soft palate which causes the narrowing of the isthmus of the pharynx.

On each side is the **tonsil** (palatine tonsil). It lies between two more or less vertically running folds, the **palatoglossal** and **palatopharyngeal arches** [p. 305]. The former runs from the inferior buccal surface of the soft palate whilst the latter is attached to its upper nasopharyngeal surface. They contain the palatoglossus and the palatopharyngeal muscles.

The palatoglossal arch marks the boundary between the mouth and oropharynx. The tonsil is in the pharynx. A finger sliding along the inside of the mouth will produce nausea in most people when this region is reached.

The **tonsil** [p. 305] has the structure of a lymph node but with a free medial surface covered by stratified squamous epithelium with many deep epithelium-lined invaginations surrounded by lymphoid tissue termed tonsillar crypts.

Laryngeal part of the pharynx

The laryngeal part of the pharynx begins at the level of the opening of the larynx. It runs downwards into the oesophagus. The uppermost part of the larynx is the epiglottis whose tip can just be seen through the mouth in some people when the tongue is depressed and flattened.

The laryngeal part of the pharynx can be seen with a special mirror. The **piriform fossa** lies at the upper lateral part of the laryngeal pharynx. The junction of the oesophagus with the pharynx is guarded by the **cricopharyngeal sphincter**. The sphincter relaxes when swallowing. It is the lowest part of the inferior constrictor muscle and not part of the oesophagus. If swallowed air returns from the fundus of the stomach with a noisy belch it is because it has had to force apart the reluctant fibres of this sphincter.

MUSCULATURE OF THE PHARYNX

The musculature of the pharynx is striated muscle. Although these muscles are not under voluntary control in the ordinary sense of the word the movements of swallowing and of speech take place with great rapidity, and for this reason alone smooth muscle could not be suitable because it contracts so slowly.

The musculature of the pharynx is not arranged as a continuous tube like the muscle in the wall of the oesophagus or small intestine. This is because the pharynx opens *in front* directly into:

(i) the nasal cavity above;
(ii) the mouth cavity;
(iii) the larynx.

Below the opening of the larynx, the larynx itself forms the front of the lowest part of the pharynx.

Thus the muscular wall of the pharynx is to be found only at the sides and at the back of the pharynx.

The musculature of the pharynx is divided into three separate component parts: the first is attached to 'the head' the second to the hyoid, the third to the larynx.

There are gaps between these three muscles in the side walls of the pharynx because of the discontinuity of the sites of origin. At the back of the pharynx the muscles are not only continuous but they actually overlap each other [FIG. 16.11].

The three components of the pharyngeal musculature are termed:

(i) Superior constrictor;
(ii) Middle constrictor;
(iii) Inferior constrictor.

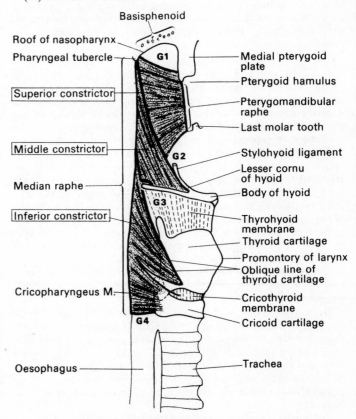

FIG. 16.11. Lateral view of constrictor muscles of pharynx—seen from the right.

G1 Triangular gap above superior constrictor. Transmits pharyngotympanic tube and levator palati. Pharyngobasilar fascia joins muscle to base of skull.

G2 Gap between superior and middle constrictors: traversed by nerves and vessels to tongue and styloglossus and stylopharyngeus.

G3 Between middle and inferior constrictors the internal laryngeal vessels and nerves, approach the larynx from above.

G4 Between inferior constrictor and oesophagus: inferior laryngeal vessels and nerve enter larynx from below.

All four gaps are points of weakness in the pharyngeal wall, compensated by appropriate thickening of the pharyngobasilar fascia.

The **superior constrictor** is principally attached along the whole length of the pterygomandibular ligament. This is the key fact. The ligament runs from the pterygoid hamulus to the mandible, where it is attached just posteromedial to the lower wisdom tooth. The superior constrictor overflows at each end of the ligament onto both the medial pterygoid plate and the mandible. The uppermost fibres sweep steeply upwards and are attached to the pharyngeal tubercle of the basi-occiput just in front of the foramen magnum. The middle fibres are at the level of the soft palate and run horizontally. The lower fibres run downwards [FIG. 16.11]. All the fibres are attached at the back of the pharynx to the median raphe which begins at the pharyngeal tubercle and runs vertically downwards in the posterior wall of the pharynx as far as the oesophagus.

The **middle constrictor** fibres are attached to the hyoid bone [FIG. 16.11] in the angle between the stylohyoid ligament, lesser cornu, and the greater cornu. Again the fibres fan out both upwards and downwards; only a few fibres run horizontally.

The **inferior constrictor** is attached to the larynx on each side. The uppermost fibres sweep upwards but the remainder are crowded together so as to form the cricopharyngeal sphincter whose fibres are mainly horizontal in direction and continuous with the circular (striated muscle) coat in the wall of the oesophagus.

It will be seen in FIGURE 16.11 that there are gaps above and below each of the three constrictor muscles of the pharynx. First there is the triangular gap between the uppermost fibres of the superior constrictor and the base of the skull in front of the pharyngeal tubercle and behind the medial pterygoid plate. The pharyngotympanic tube [FIG. 16.10] passes from the scaphoid fossa into the pharynx through this opening. The **levator palati** muscle, which takes origin from the cartilaginous part of the tube, also traverses this gap as it passes towards the upper surface of the soft palate. The **tensor palati** muscle also guards the opening but does not pass through it. The gap is closed, however, essentially by the **pharyngobasilar fascia**, which lies on the deep surface of the muscle, and gets its name because it joins the pharynx to the base of the skull. It is strong enough to withstand the pressures exerted in the nasopharynx (for example when you blow your nose) but it becomes much thinner in the lower part of the pharynx, and here bulging readily occurs in the gaps above and below the middle constrictor.

Demonstrate this in yourself. First identify the hyoid bone [p. 301] to which the middle constrictor is attached. Place your finger between the hyoid and the mandible [FIG. 16.01], hold your nose with the other hand and try to breathe in and out. Repeat this with your finger between the hyoid bone and the thyroid cartilage. In both cases the pharynx can be felt billowing out and in. In these places diverticulae of the pharynx may occur, for example in glass-blowers.

The gap between the superior and middle constrictors allows the nerves supplying the tongue and the lingual artery to gain access. The gap between the middle and inferior constrictors transmits the superior laryngeal vessels and nerves. Below the inferior constrictor the inferior or recurrent laryngeal nerve gains access to the larynx.

The stylopharyngeus muscle fuses with the pharyngobasilar fascia between the superior and middle constrictor muscles and then is attached to the thyroid cartilage. It raises the larynx in swallowing and is assisted in this by palatopharyngeus and by salpingo-pharyngeus.

THE LARYNX

Feel the larynx in yourself. Notice the **thyroid cartilage** and **cricoid cartilage**. Swallow and notice the upward movement. Make sounds of varying pitch and notice that the thyroid cartilage moves forwards when the voice rises, and backwards when it is lowered. You will also see (unless your voice has been 'trained') that the whole larynx moves upwards as the pitch of the voice rises [FIG. 16.12].

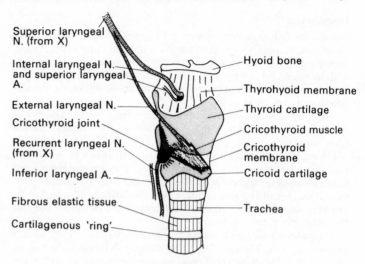

Superior laryngeal N. (from X)

Internal laryngeal N. and superior laryngeal A.

External laryngeal N.

Cricothyroid joint

Recurrent laryngeal N. (from X)

Inferior laryngeal A.

Fibrous elastic tissue

Cartilagenous 'ring'

Hyoid bone

Thyrohyoid membrane

Thyroid cartilage

Cricothyroid muscle

Cricothyroid membrane

Cricoid cartilage

Trachea

FIG. 16.12. The larynx seen from the right side.
All the structures named to the right of the figure can easily be identified in yourself.

The larynx consists of the **cricoid cartilage** (Greek for a ring) which is attached to the upper end of the trachea. The rings of the trachea are joined to each other by elastic tissue and there is a heavy component of elastic tissue in the submucosa of the trachea. This elastic tissue extends upwards beyond the upper border of the cricoid narrowing as it does so, forming the **conus elasticus**. The opening of the conus has an area of less than 10 mm^2 (about half the cross-sectional area of the nostrils: it is the narrowest part of the respiratory tract), compared with the cross-sectional area of the trachea of greater than 100mm^2. The air passes through this opening with a high velocity even in quiet breathing. The free edges of the conus are the vocal folds. The term 'vocal cords' is misleading, as the vibrating elements in the larynx do not hang free in the moving air stream like the strings of an Aeolian harp.

The conus is shielded anteriorly by the **thyroid cartilage** and in fact the anterior extremity of the opening is attached to the thyroid cartilage itself.

The posterior extremity of each vocal fold is attached to one of the **arytenoid cartilages**. The conus elasticus therefore opens into the upper part of the larynx via a triangular shaped slit, narrow in front and wider behind, the **rima glottidis**. (*Rima* = cleft or slit: *glottidis* = of the glottis.)

The pitch of the voice would be changed by altering the tension of the folds, by altering their length, or their density, and by changing the length and width of the gap separating them from each other. The larynx uses all these mechanisms.

FIGURE 16.13 shows diagrammatically the attachments of the vocal folds. In front they are attached to the thyroid cartilage very

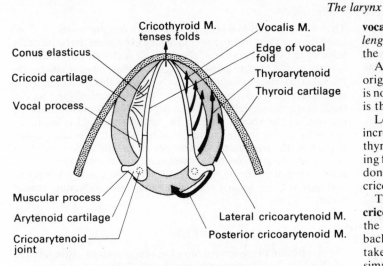

Fig. 16.13. The larynx seen from above.

The inlet of the larynx is shown diagrammatically from above. The vocal folds are attached to the thyroid cartilages anteriorly and behind they are joined to the vocal processes of the arytenoids.

Forward movement of the thyroid cartilage with respect to the cricoid is brought about by the cricothyroid muscle, which is not shown in the Figure (but see FIGURE 16.12).

As the arytenoids are positioned on the cricoid cartilage, this forward movement also takes place with respect to the arytenoids, when this muscle contracts. By this roundabout method the folds are tensed.

Fine adjustment of the vocal fold is brought about mainly by the vocalis muscle. The posterior and lateral cricoarytenoids abduct and adduct the folds respectively. The thyroarytenoid muscle fills the gap between the vocalis and lateral cricoarytenoid muscles.

close to the midline, just below the promontory of the thyroid cartilage. You can feel this promontory in yourself.

Posteriorly, each vocal fold is attached to the **vocal process** of one of the arytenoid cartilages. The arytenoid on each side is in contact with the cricoid cartilage below, through the medium of the **crico-arytenoid joint**. This is a plain synovial joint with surfaces which are slightly concavoconvex. We will assume that this joint has a fixed vertical axis, and that the arytenoid can swivel around this axis so that the vocal process points either inwards or outwards. Inward movement of a vocal process will carry its own vocal fold towards the midline and bring about **adduction** of the folds. Outward movement of the vocal process will **abduct** the folds and widen the rima glottidis.

Another feature of the arytenoid is a second process which projects laterally. This is the **muscular process**. If the muscular process is pulled forwards, the vocal process will adduct, and if it is pulled backwards it will abduct.

Given the skeletal arrangement of the larynx and the attachments of the folds we can not think about what muscles are necessary in order to produce these movements.

Forward movement of the muscular process must be brought about by a muscle mass starting anteriorly [FIG. 16.13]. In fact forward traction is brought about principally by a muscle which is attached to the upper edge of the cricoid (the **lateral crico-arytenoid muscle**) [FIG. 16.13]. But the origin of this muscle mass in this area is not restricted to the cricoid as it spreads onto the deep surface of the adjacent part of the thyroid cartilage to form the **thyro-arytenoid muscle**. The most anterior component of the mass, which is also coming from the thyroid cartilage, does not reach the arytenoid at all, but is inserted into the vocal fold itself. This is the

vocalis muscle. This is an important muscle which can influence the *length* of the fold left free to vibrate and also the *width* and *density* of the vibrating membrane.

A muscle pulling the muscular process backwards can only take origin from the back of the cricoid cartilage near the midline; there is no alternative. This is the **posterior crico-arytenoid muscle**, which is the abductor of the vocal folds.

Let us now think about the tension in the folds. This can be increased either by displacing the arytenoid backwards, or the thyroid cartilage forwards. You have already felt the thyroid moving forwards when the voice is raised so this is obviously the way it is done. In front the thyroid moves forwards with respect to the cricoid which is stabilized on the top of the trachea.

The muscle which is responsible for tensing the vocal folds is the **cricothyroid** [FIG. 16.03]. This takes origin from the outer surface of the cricoid near the midline. The main direction of the fibres is backwards and upwards. The articulation between the cartilages takes place at the two cricothyroid joints. The movement is not simply a forward gliding of the thyroid, because it is also tilted forward [FIG. 16.12] the axis of the tilt being a transverse one passing through the two joints.

The mechanism of increasing vocal fold tension is to increase its length. Now increasing the tension (T) will raise the voice (frequency is proportional to $\sqrt{T}$) while increasing the length (L) will lower it (frequency is proportional to $1/L$). The net effect depends on the elastic modulus of the fold itself. Thus if its elasticity is high then a small increase of length will give rise to a considerable increase in tension, while if it is low a similar increase in length will only give rise to a slight increase in tension. Thus in the second case the pitch of the voice will not rise so much.

The two factors of length and tension work against each other. In a given larynx, there is a limit to the degree to which the voice can be raised or lowered. One of the inevitable consequences of aging is the decay in elasticity which takes place in all the connective tissues of the body (wrinkled face, etc.) including the vocal folds. This is one of the reasons why the quality of the voice deteriorates with age.

Above and parallel to the vocal folds inside the larynx are the **false folds**, separated by the sinus of the larynx [FIG. 16.14]. There is a mucous lined sac-like gland which discharges into the sinus.

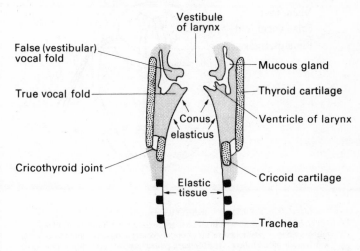

Fig. 16.14. The larynx in coronal section.

The whole of the respiratory tract is covered by **ciliated mucus-secreting epithelium** with the exception of the vocal folds themselves which are covered with **stratified squamous epithelium**. The vocal folds therefore have no built-in lubricating system like the rest of the respiratory mucous membrane, but rely instead on the secretion which collects in the sinus.

The **blood supply** of the larynx (see FIG. 16.00) is derived from branches from the thyroid arteries.

The **nerve supply** is important. There are two branches on each side from the vagus [FIG. 16.12], the superior and recurrent laryngeal nerves.

The **superior laryngeal nerve** divides [FIG. 16.12] into an internal (large branch) and an external branch. The internal branch supplies the vocal fold and the mucous membrane above it. The external branch only supplies the **cricothyroid muscle** which is responsible for tensing the fold.

The **recurrent laryngeal nerve** leaves the vagus at the root of the neck on the **right side** and hooks round the right subclavian artery and reaches the groove between the upper ends of the trachea and the oesophagus. It then turns upwards and enters the larynx just behind the cricothyroid joint. On the **left side** the nerve comes off the vagus in the thorax and hooks round the arch of the aorta (below the ligamentum arteriosum). It runs in the tracheo-oesophageal groove into the neck and enters the larynx in the same way as the right nerve.

The recurrent laryngeal nerve supplies all the other intrinsic muscles of the larynx as well as the vocal fold and the mucous membrane in the lower half of the larynx.

The superior laryngeal nerve is not often paralysed, but the recurrent laryngeal nerve, on the left side especially, because of its position in the mediastinum and neck (it runs very close to or in the thyroid gland) is paralysed as a complication of diseases here: aortic aneurysm, carcinoma of the bronchus, goitre, etc. In such cases the voice is hoarse and coughing is difficult.

The paralysis of this nerve is established by looking at the folds themselves by means of a mirror [FIG. 16.15] held beneath the soft palate. Normally the chords lie at an angle of 20–30° to each other.

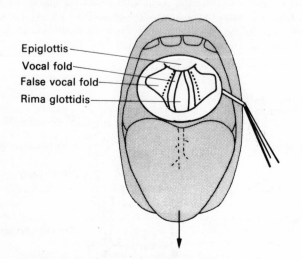

Epiglottis
Vocal fold
False vocal fold
Rima glottidis

FIG. 16.15. Indirect laryngoscopy.
The larynx can be viewed in an unanaesthetized patient by means of a mirror which is positioned in the oropharynx as far upwards and backwards as the palate follows. The tongue is pulled forwards as far as possible in the direction of the arrow.

They move outwards a little in inspiration (cf. the nostril, p. 306) and return to the starting position during expiration. Because of the form of the conus elasticus, the air as it is drawn down into the trachea would tend to push the folds together, were it not for the lateral cricoarytenoid muscles which prevent this by pulling them apart every time a normal person breathes in. Normally the edge of the fold is not a straight line, but is slightly convex outwards due to the tone of the vocalis muscle. In paralysis the fold is immobile during breathing and has a straight edge. When the patient is asked to phonate the paralysed fold does not move, and if the paralysis is of long standing the other fold will come right across the midline. If the external laryngeal nerve is still functional, as if often is, the fold can still be tensed, but adduction and abduction cannot occur.

SPEECH

The acquisition of speech is a most important feature distinguishing mankind. It is an extremely complex skill controlled by correspondingly large areas of brain, and monitored by the sense of hearing.

The larynx is responsible for **vocalization**, that is it can emit a continuous humming sound. The **pitch** of this sound can be varied by changes in the larynx involving the vocal folds, whilst the **loudness** is varied by the response of the larynx to changes in respiratory pressure.

The sound welling out of the larynx passes upwards into the oropharynx and, depending on the position of the soft palate, will then flood either the mouth, or nasal cavities or both.

The sounds from the larynx consist of a basic frequency with a particular series of accompanying harmonics or overtones. The individuality of every person's voice depends on the pattern of these overtones. The pattern can be analysed electronically so as to produce a **voice print** which may be as individual as a fingerprint.

The colour or timbre of the voice has to be changed when different **vowel sounds** are used. Different vowel sounds are produced by changing the shape of the resonating space through which the sound from the larynx must pass. A particular shape will allow certain overtones to resonate and will suppress others. Thus, one particular shape is appropriate for 'AAH' and another one for 'EE'.

Everyone is familiar with the idea that daylight, by passing through a filter, can be converted into any one of the different colours of the rainbow. In a roughly analogous manner the sound which comes out of the larynx by passing through an appropriately shaped resonator can be converted into any of the different vocal colours which we recognize as vowel sounds.

Putting the chambers into the right shape has to be learnt by a child by trial and error. A normal child will get it right in the end by comparing the sound of its own voice with that of other people. The deaf child cannot do this and is therefore dumb.

Adults who become deaf and can no longer hear themselves speak become more and more difficult to understand.

There is a great deal more to speech than the production of vowel sounds. Indeed in English clarity depends to a major degree on the way the **consonants** are enunciated. The sounds of consonants such as /L/, /M/, /N/, and /R/ are vocalized, that is they are accompanied by a humming sound from the larynx. They can be produced continuously (or sung). The sounds of other consonants such as /B/, /D/, and /G/ are also vocalized. But they are characterized by a momentary explosion between the lips, between the tongue and hard palate, or between the back of the tongue and soft palate respectively. They cannot be produced continuously.

Certain unvocalized sounds such as /sh/, /ss/, and /ff/ can also be produced continuously. Sounds such as /P/, /T/, and /K/ are simply unvocalized versions of /B/, /D/, and /G/. Try for yourself /P/ and /B/, /T/ and /D/, and /K/ and /G/.

Many consonants require constant pressure in the mouth—try a long drawn out /sh . . . sh/ or its phonated partner, the /s/ in the words vision or collision. Whenever such sounds are made it is essential to close off the nasopharynx by raising the soft palate and contracting the superior constrictor, otherwise the breath will escape into the nose and no pressure build-up in the mouth will take place. Try saying /sh . . . sh/ again but hold your nose as you do so. It will make no difference whatever to the sound because your soft palate has already closed off your nasal cavity. Try nose holding with /M/, /N/, or /'ng/ and you will notice the difference.

A patient with a paralysed palate may pronounce /M/, /N/, and /'ng/ normally but will be unable to enunciate any of the other sounds. He suffers from a severe speech defect. A child with enlarged adenoids and obstruction in the nasopharynx, by contrast, cannot pronounce /M/, /N/, or /'ng/.

A patient with a facial nerve paralysis in which one half of the orbicularis oris muscle is affected is unable to make a forceful utterance of /B/ or /P/. This defect is, however, of much less importance.

Whispering is talking without using the larynx. Great care is needed when whispering in a language like English which relies heavily on unvocalized consonants, lest the message be overheard at a great distance.† Even a patient who has had his larynx removed surgically can still whisper. In whispering so as not to be overheard it is not enough merely to switch off the larynx because this has no effect on the unvocalized consonants may of which such as /S/, /F/, /Sh/, and /T/ have frequency ranges to which the ear is very sensitive: they have great carrying power. It is thus difficult to whisper secretly in English and eavesdroppers always hear more than you think.

The larynx and swallowing

The vital function of the larynx is to prevent anything other than air from entering the bronchial tree. The inlet of the larynx is guarded by the epiglottis in front and the pair of arytenoid cartilages on either side of the midline behind. The opening is completed by the two ary-epiglottic and the interarytenoid folds. These folds of mucous membrane cover the muscles which really form the sphincter of the inlet of the larynx. These are the aryepiglotticus and interarytenoideus muscles. When the sphincter contracts the arytenoids are drawn forwards, against the projection at the base of the epiglottis, so as to make a water-tight seal [FIG. 16.16].

During swallowing the larynx is drawn upwards by a distance that corresponds to about two cervical vertebrae. It comes to lie closely tucked in behind the overhanging base of the tongue so that the stream of swallowed material is projected over the top of the opening towards the laryngeal pharynx. The upwards movement also pushes against the base of the epiglottis and causes it to tip over

† The consonants underscored are unvocalized, that is, the larynx is not used to produce them even when speaking normally. As a result they will tend to be just as loud when whispered.

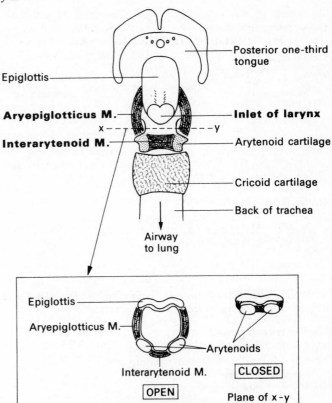

FIG. 16.16. Inlet of larynx from behind.
Contraction of the aryepiglotticus and interarytenoid muscles can cause the arytenoids to completely close the larynx (inset).

and cover the inlet of the larynx. However, cine-radiography has shown that this does not happen until after the liquid has passed the inlet. The epiglottis can be removed surgically but this does not seem to interfere significantly with swallowing.

Coughing

The cough reflex is excited when the lining of the pharynx or respiratory tracts is irritated. A quick grasp of air is drawn into the lungs, the larynx is then closed off by adduction of the vocal folds and a strong expiratory effort is made which builds up the positive pressure in the thoracic cavity. The vocal folds are then suddenly drawn apart, and the air in the lungs is driven out at an average velocity of 100 km/s or more. Foreign material in the respiratory tract is carried away in the blast. The cough reflex, like the swallowing reflex, is controlled by the brainstem.

It is impossible to cough normally if one vocal fold is paralysed, and if both are paralysed, or the larynx has been removed, coughing is impossible. Lifting heavy weights is also impossible.

The expiratory muscles in the abdominal wall provide the expiratory drive, and if they are painful (e.g. after an abdominal operation) coughing is severely inhibited with serious consequences. This is an important factor in the causation of post-operative chest disorders.

SUMMARY

MUSCLES OF THE LA   NX

Cricothyroid

Origin: Cricoid—arch: external aspe
Insertion: Thyroid—lamina: external a ect
Nerve supply: Superior laryngeal nerve
                   (external branch)
Action: Tenses vocal cords

The name implies that it runs from the cricoid cartilage to the thyroid cartilage. It is the only intrinsic muscle which is on the outer aspect of the larynx.

Lateral crico-arytenoid

Origin: Cricoid—lateral surface
Insertion: Arytenoid—muscular process
Nerve supply: Recurrent laryngeal nerve
Action: Approximates vocal cords:
                   closes the rima glottidis

The name implies that it is the muscle running from the lateral part of the cricoid cartilage to the arytenoid cartilage.

Posterior crico-arytenoid

Origin: Cricoid—posterior surface
Insertion: Arytenoid—muscular process
Nerve supply: Recurrent laryngeal nerve
Action: Separates vocal cords:
                   opens the rima glottidis

The name implies that it is the muscle running from the posterior part of the cricoid cartilage to the arytenoid cartilage.

Thyro-arytenoid

Origin: Thyroid—lamina
Insertion: Arytenoid
Nerve supply: Recurrent laryngeal nerve
Action: Closes vestibule of larynx
                   Relaxes vocal cords

The name implies that it unites the two arytenoids. It has transverse and oblique fibres.

Interarytenoid

Origin: Arytenoid—muscular process
Insertion: Opposite arytenoid
Nerve supply: Recurrent laryngeal
Action: Approximates arytenoids

The name implies that it is the transverse muscle uniting the two arytenoids.

Aryepiglotticus

Origin: Arytenoid
Insertion: Edge of epiglottis
Nerve supply: Recurrent laryngeal
Action: With interarytenoid completes
                   Sphincter at inlet to larynx

The name implies that it runs from the arytenoid to the epiglottis.

Vocalis

Origin: Thyroid cartilage
Insertion: Along length of vocal folds
Nerve supply: Recurrent laryngeal
Action: Modifies vibration of
                   vocal fold

The name implies its importance in voice production.

Superior laryngeal nerve

Branch of Vagus (X)
Branches:
  1. External laryngeal nerve.
  2. Internal laryngeal nerve.

The superior laryngeal nerve supplies the inferior constrictor of the pharynx, the cricothyroid, the mucosa of the larynx including the epiglottis and the adjacent part of the base of the tongue.

Recurrent laryngeal nerve

Branch of Vagus (X)
Branches:
  1. Inferior laryngeal nerve—supplies the intrinsic muscle of larynx with the exception of cricothyroid.
  2. Cardiac nerves.
  3. Muscular nerves to oesophagus and hypopharynx.

THE TRACHEA

The trachea [p. 178] runs downwards in direct continuity with the lumen of the lower part of the larynx. It is easily palpable in the neck. It runs downwards approximately in the midline into the superior mediastinum, where it divides into right and left main bronchi at the level of the sternal angle. The pretracheal fascia disappears at this level as it fuses with the fibrous pericardium.

The trachea is lined with ciliated mucous and serous-secreting columnar epithelium. The characteristic feature of the wall (which is always maintained open) is the presence of so-called tracheal rings. These are really horseshoe-shaped 'rings' of hyaline cartilage, in the anterior and side walls of the trachea but not the back. The submucosa contains elastic tissue and the rings are joined together by the same tissue, so when the larynx rises in swallowing (the bifurcation of the trachea remaining in the same position) the trachea can be stretched out and then afterwards recoil to its initial length.

Between the trachea and the pretracheal fascia is the thyroid gland.

The recurrent laryngeal nerves lie in the groove between the trachea and oesophagus. They usually pass behind the thyroid gland, but sometimes they are embedded in it. Paralysis of this nerve is quite common in patients with goitre even before they have been operated on. The external laryngeal nerve runs with the superior thyroid artery but is never embedded in the gland.

OESOPHAGUS

The oesophagus [p. 177] runs from the pharynx at the level of the lower border of the cricoid as far as the cardia of the stomach whose surface marking is just to the left of the lower end of the sternum. In the neck it lies slightly off centre to the left of the midline behind the trachea [p. 177].

Swallowing

When food is taken into the mouth, salivation begins reflexly, and the mouth 'waters'. Saliva is mixed with the food as it is chewed and a rounded slippery semi-liquid bolus is formed between the tongue and hard palate. The bolus is pushed backwards by the tongue. When it comes in contact with the mucous membranes of the oropharynx and the posterior third of the tongue, the swallowing reflex is excited. When drinking from a cup it is usual to take in a quick gasp of air as the cup approaches the lips. The breath is then held until the swallow is complete.

The bolus is propelled rapidly down the oropharynx into the laryngeal pharynx and onwards into the oesophagus by the peristaltic contraction of the striped muscle in their walls.

There are two essential features of swallowing. The first, and of lesser importance, is the closure of the **pharyngeal isthmus**. This stops the bolus from entering the nasal cavity. It is brought about by raising the soft palate with **levator palati** and at the same time by contracting the horizontal fibres of the **superior constrictor**. The posterior pharyngeal wall then comes forward and meets the soft palate halfway. The result in a typical watertight spincter (of Passavant). Defects of the soft palate (cleft palate in a baby) or paralysis of the relevant muscles (bulbar palsy) will allow the bolus to pass upwards from the oropharynx into the nasopharynx.

The second, and more critical, indeed vital, feature of swallowing is that which ensures that the liquid bolus cannot pass through the inlet of the larynx, into the trachea and bronchial tree, causing respiratory obstruction or drowning. To this end the larynx is drawn upwards so as to shorten the oropharynx by the **stylo-, palato-,** and **salpingopharyngeus muscles**. The opening of the larynx is tucked in under cover of the now overhanging posterior third of the tongue. At the same time the base of the epiglottis is pushed upwards and it then tilts over and covers the inlet. But the important mechanism is that which causes the base of the epiglottis and the two arytenoid cartilages to be squeezed tightly together by the **interarytenoid** and **ary-epiglotticus muscles**. This is another typical sphincter mechanism. The sphincter works perfectly even when the epiglottis has been removed surgically.

The raising of the larynx obviously brings about a corresponding upward movement of the lower part of the laryngopharynx. This means that the functional length of the oropharynx is reduced and the bolus can pass almost directly from the mouth into the laryngopharynx or hypopharynx. The very rapid downward passage of the bolus is contributed to not only by the peristalsis of the pharyngeal wall, but also by the subsequent descent of this part of the pharynx itself as it returns to its initial position.

Air is always present in the pharynx and it is swallowed along with the bolus. It reaches the stomach. A considerable volume of air can be swallowed in this way. In babies when feeding from the breast or bottle, a lot of air is swallowed and the stomach becomes painfully distended. Babies will not settle peacefully after a feed until they have 'brought up their wind'.

Swallowing can be observed by means of diagnostic cine-radiography.

THE THYROID GLAND

The thyroid gland consists of two lobes and a connecting isthmus. Each lobe is conical in shape with its base at the level of the sixth tracheal ring and its apex extending up to the middle of the thyroid cartilage. The isthmus lies in front of the second, third, and fourth tracheal rings. The pyramidal lobe (if present) runs upwards towards the body of the hyoid bone. It is usually to one side of the midline [FIG. 16.17].

The thyroid gland develops mainly from a tubular epithelial downgrowth of the tongue which grows down in front of the hyoid bone and the thyroid cartilage. This develops into most of the isthmus and two lateral lobes of the thyroid. The embryological remnants of this outgrowth are the **foramen caecum** on the dorsum of the tongue, the **thyroglossal duct** and the **pyramidal lobe** of the thyroid gland. This may contain muscle tissue (levator glandulae thyroidae) which runs to the hyoid bone above. The thyroglossal duct normally disappears, but it may not do so, in which case it gives rise to a thyroglossal cyst or a fistula which requires surgical treatment.

The thyroid gland has the sternohyoid and sternothyroid muscles in front, the carotid sheath (with the common carotid artery, internal jugular vein, and vagus) posterolaterally, and the trachea with the two recurrent laryngeal nerves on either side behind.

The thyroid gland is surrounded by a closely adherent capsule. It is also surrounded by a sheath derived from the pretracheal fascia [p. 314]. The plane of separation is between these two layers.

Enlargement of the thyroid from any cause is called a **goitre**. The swelling may be evident in the neck and moves upwards in swallowing. The tumour is prevented from extending upwards by the attachment of the sternothyroid muscle to the oblique line of the thyroid cartilage. But it can very easily extend downwards in the plane of cleavage between the trachea and the pretracheal fascia through the inlet of the thorax into the superior mediastinum. This compresses and distorts the trachea. An iceberg-like situation is very common with a small part of the goitre clearly visible in the neck and a large part hidden in the mediastinum. Sometimes the whole tumour may be in the mediastinum.

Arterial supply to thyroid gland [FIG. 16.17]

1. Superior thyroid artery.
2. Inferior thyroid artery.
3. Thyroidea ima (if present) from the brachiocephalic trunk.

Venous drainage of thyroid gland [FIG. 16.17]

1. Superior thyroid vein drains upwards to internal jugular vein.
2. Middle thyroid vein drains horizontally to internal jugular vein.
3. Inferior thyroid veins drain downwards to left brachiocephalic vein.

Nerve supply of thyroid gland

Sympathetic nerves from the superior and middle cervical ganglia.

The normal thyroid secretions, tri-iodo thyronine (T3) and thyroxine (T4), are controlled by the thyroid-stimulating hormone (thyrotrophin) from the anterior pituitary which in turn is controlled by the thyrotrophin-releasing hormone from the hypothalamus. The thyroid gland also produces calcitonin.

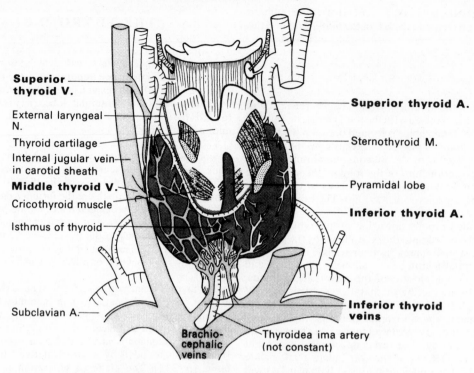

Superior thyroid V.

External laryngeal N.

Thyroid cartilage

Internal jugular vein in carotid sheath

Middle thyroid V.

Cricothyroid muscle

Isthmus of thyroid

Subclavian A.

Brachio-cephalic veins

Thyroidea ima artery (not constant)

Superior thyroid A.

Sternothyroid M.

Pyramidal lobe

Inferior thyroid A.

Inferior thyroid veins

FIG. 16.17. Thyroid gland and its blood supply.
The anterior surface of the thyroid is shown. The strap or infrahyoid muscles have been removed, except for part of sternothyroid.
The carotid sheath and its contents have been removed on the left side, to show the inferior thyroid artery in continuity.

Lymphatic drainage

Lymphatic drainage is to:
1. Deep cervical lymph nodes.
2. Pretracted lymph nodes.
3. The thymus.

THE PARATHYROID GLANDS

The four parathyroid glands are embedded at the back of the thyroid gland. They are about the size of a small pea and are yellowish-brown in colour. They are outside the capsule of the thyroid gland but inside its sheath. The upper parathyroid glands are usually at the level of the cricoid cartilage. The lower parathyroid glands are more variable in position. They are usually found behind the lower part of the lobes at about the level of the fourth tracheal ring. They may lie anywhere between the thyrohyoid membrane and the superior mediastinum when they are often embedded in the thymus.

A possible functional link exists however between the parathyroids and the 'C' (calcitonin) cells of the thyroid and the dihydroxy-vitamin D (1:25 dihydroxy chole calciferol) formed by the kidneys. The parathyroid hormone raises the blood calcium level by withdrawing calcium from the skeleton. Calcitonin has the reverse effect. Dihydroxy-vitamin D increases the absorption of calcium from the gut. Surgery on the thyroid gland may result in removal of or ischaemia in the parathyroids, in which case the patient will develop tetany, with carpopedal and laryngeal spasms.

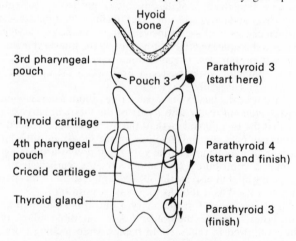

Hyoid bone

3rd pharyngeal pouch

Pouch 3

Thyroid cartilage

4th pharyngeal pouch

Cricoid cartilage

Thyroid gland

Parathyroid 3 (start here)

Parathyroid 4 (start and finish)

Parathyroid 3 (finish)

FIG. 16.18. The parathyroid glands.
This shows the considerable translation of the epithelial cells derived from the 3rd pharyngeal pouch which give rise to parathyroid 3. Parathyroid 4 on the other hand sticks around very close to its origin. Parathyroids 3 sometimes fail to reach their normal destination. They are found away from the thyroid gland in the neck or in the mediastinum usually associated with the thymus, which is also derived from the 3rd pouch.

The upper pair of parathyroids [FIG. 16.18] are often referred to as parathyroids 4, the lower pair as parathyroids 3. There are no parathyroids 1 or 2 since the 3 and 4 refer to their origins in the embryo from the epithelium lining the third and fourth pharyngeal pouches. Parathyroid 4 was originally continuous with the epithelium which comes to lie in the adult between the thyroid and cricoid cartilages. So parathyroid 4 hardly changes its position in the neck. But parathyroid 3 comes from the epithelium between the hyoid and thyroid cartilages (the piriform fossa) and shares this site of origin with the thymus. As the thymus migrates downwards into the thorax, it drags parathyroid 3 along with it so that parathyroid 3 comes to lie below parathyroid 4. The position of parathyroid 3 is very variable because of its morphogenetic background. As the thyroid first develops from the epithelium around the foramen caecum of the tongue, the anatomical association of the thyroid and parathyroid glands looks to be quite fortuitous.

17 Exploring the skull for yourself

Before studying the eye, ear, and brain (Chapters 18 and 19) the principal nomenclature of the skull will be considered. This can be very confusing until it is realized that the names used for many structures have been derived from the Latin and Greek and sometimes Arabic. They were originally chosen because of the apparent similarity of these structures to objects in everyday life. In some cases this relationship has now become rather obscure.

If a foetal skull and an adult skull are available, they may be usefully studied at the same time, otherwise photographs and diagrams may be used [FIGS. 17.01 and 17.02].

BONES OF THE SKULL

FRONTAL	Ossifies in membrane
PARIETAL	Ossifies in membrane
OCCIPITAL	Ossifies partly in cartilage and partly in membrane
TEMPORAL	Ossifies partly in cartilage and partly in membrane
SPHENOID	Ossifies partly in cartilage and partly in membrane

ETHMOID	
SUPERIOR CONCHA	} Ossifies in cartilage
MIDDLE CONCHA	
INFERIOR CONCHA	Ossifies in cartilage
LACRIMAL	Ossifies in membrane
VOMER	Ossifies in membrane
NASAL	Ossifies in membrane
MAXILLA	Ossifies in membrane
PALATINE	Ossifies in membrane
ZYGOMATIC	(*zygoma* (Greek); *malar* (Latin)). Ossifies partly in cartilage and partly in membrane

Zygomatic arch

The zygomatic arch also known as the zygoma or arcus zygomaticus is the arch formed by the union of zygomatic process of the temporal bone with the zygomatic (malar) bone.

INTRODUCTION

The **skull** provides the principal protection for the brain. The extreme necessity for such protection is made clear as soon as you realize that the brain is so soft in consistency that, when it is removed from the inside of the skull, it collapses under its own

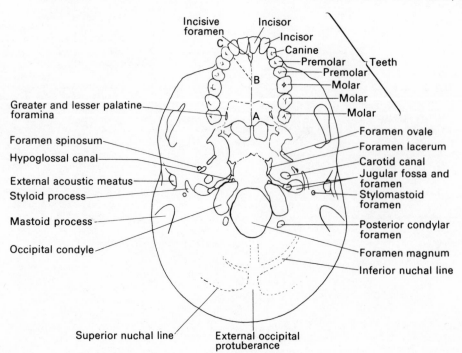

FIG. 17.01. Foramina at the base of skull.

328

FORAMINA OF THE SKULL

Foramen	(Bone)	Gives passage to
Hypoglossal canal foramen Foramen anterolateral to occipital condyles	(Occipital)	Hypoglossal nerve (XII)
Anterior ethmoidal foramen Canal between ethmoid and frontal bones	(Medial wall of orbit)	Nasal branch of ophthalmic nerve Anterior ethmoidal artery and vein
External acoustic (auditory) meatus	(Temporal)	Sound. No anatomical structure goes right through
Foramen lacerum	(Sphenoid)	Greater superficial petrosal nerve
Foramen magnum	(Occipital)	Spinal cord Venous plexuses Spinal accessory nerves Vertebral arteries
Foramen ovale	(Sphenoid)	V_3 (mandibular branch of trigeminal nerve) Small meningeal artery
Foramen rotundum	(Sphenoid)	V_2 (maxillary branch of trigeminal nerve)
Foramen spinosum	(Sphenoid)	Middle meningeal artery
Infra-orbital foramen	(Maxilla)	Infra-orbital artery and nerve
Internal acoustic meatus (auditory foramen)	(Temporal–petrous part)	VII facial nerve VIII vestibulocochlear (auditory) nerve
Jugular foramen	(Junction of temporal and occipital bones)	*Anterior part* IX glossopharyngeal nerve X vagus nerve XI accessory nerve Inferior petrosal sinus *Posterior part* Internal jugular vein
Mandibular foramen Internal aperture of inferior dental canal	(Mandible)	Inferior dental artery, vein, and nerves
Mental foramen External aperture of inferior dental canal	(Mandible)	Mental blood vessels and nerves to skin of face
Olfactory foramina	(Ethmoid)	I Olfactory nerves
Optic foramen	(Sphenoid)	II Optic nerve Ophthalmic artery
Palatine foramina	(Palatine)	Descending palatine blood vessels and nerves
Posterior condyloid foramen Foramen posterior to occipital condyles	(Occipital)	Emissary vein to lateral sinus
Posterior ethmoidal foramen	(Medial wall of orbit)	Posterior ethmoidal artery, vein, and nerve
Sphenopalatine foramen	(Sphenoid–palatine)	Nasal branch of maxillary artery Nerves from sphenopalatine ganglion
Stylomastoid foramen	(Temporal)	VII facial nerve
Supra-orbital foramen Notch in superior orbital margin	(Temporal)	Supra-orbital artery, vein, and nerve

FISSURES OF THE SKULL

Fissure	(Bone)	Gives passage to
Inferior orbital fissure		Inferior orbital arteries and veins Nerves from sphenopalatine ganglion
Petrotympanic fissure Narrow slit in floor of glenoid fossa of temporal bone	(Temporal)	Chorda tympani
Superior orbital fissure Between greater and lesser wings of sphenoid	(Sphenoid)	Ophthalmic vein III IV VI V_1 lacrimal nerve V_1 frontal nerve V_1 nasociliary nerve
Petrosquamous fissure	(Temporal)	Nothing
Pterygomaxillary fissure Cleft between posterior wall of maxillary sinus and pterygoid plate		Maxillary artery Gives entry to pterygopalatine fossa

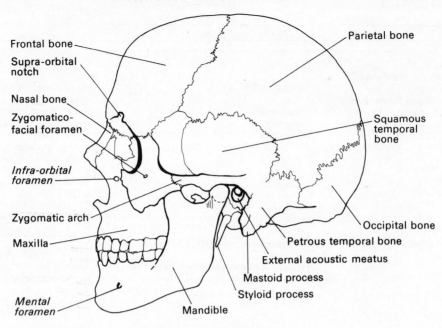

FIG. 17.02. Bones of the skull.

weight. But this problem concerning the softness of the brain has been solved by the body because inside the skull the brain actually floats, or almost does so, in the cerebrospinal fluid like a jellyfish in the sea.

Some people find it hard to believe that the adult brain, weighing say 1250 g, is apparently able to float in 20–30 ml of cerebrospinal fluid and thus to defy the law of Archimedes! A moment's thought, however, shows that the few millilitres of subarachnoid fluid is all that remains *after* the brain has in effect already displaced its own volume of fluid.

The vault of the skull consists of bone which is in three layers—the **diploic venous sinuses** and some **red bone marrow** being sandwiched between two layers of cortical bone called the **inner and outer tables of the skull**. The inner is sometimes also referred to as the vitreous table because, after head injuries, it sometimes cracks and splinters like glass.

Both outer and inner tables, like any bone surface, are covered by periosteum [FIG. 17.03]. The periosteum outside the skull is called the **pericranium**. It is not very firmly attached to the underlying bone except at the suture lines which join adjacent bones together, e.g. at the coronal or sagittal sutures. Bleeding deep to the pericranium (a **cephalhaematoma**) produces a swelling which exactly describes the shape of the underlying bone—usually the parietal, and the diagnosis is easy for this reason. A cephalhaematoma is a common birth injury. It will resolve on its own, in time.

The inner surface of the inner table of the skull is lined by the **dura mater**. This is in two layers, but the *outer layer of the dura* is just another name for the periosteum inside the skull. Like the periosteum covering any other bone, this layer of the dura is always in contact with the skull bones, and goes in and out of every nook and cranny. It also contains the **meningeal arteries** which are really just *specialized periosteal blood vessels* responsible for the blood

supply of the overlying bone. Unlike the pericranium the periosteum inside the skull is not fixed to the bones or to the sutures in any particular places and bleeding between the bone and the dura (**extradural haemorrhage**) can spread very widely, stripping the dura off the bone and compressing the brain.

It is obvious that the inner and outer surfaces of the skull are continuous with each other at all the foramina. The **foramen magnum** is the best example to study because you can get your fingers through it. If the surfaces are continuous, it is obvious that the pericranium and the outer layer of the dura must also be continuous with each other here, and at all the other foramina.

The inner layer of the dura is a different matter. Its function has nothing to do with the periosteum, but is to support the brain and other parts of the central nervous system. It is called the **sustentacular layer of the dura**. For this reason, it follows the general shape of the brain and ignores the skull.

When, at the foramen magnum, the CNS emerges from the skull and descends as the **spinal cord** into the **vertebral canal**, the inner layer of the dura forms a sleeve which extends almost the whole length of the canal, as far as the lower border of the second sacral vertebra.

The spinal cord is not only enclosed by the inner layer of the dura, but also of course by the **arachnoid** and **pia mater**, with the sub-arachnoid fluid (cerebrospinal fluid) in between [p. 381]. The arachnoid is fairly closely applied to the dura and these two mem-branes are often referred to collectively as the **theca of the spinal cord**.

A very similar arrangement is found in relation to the optic nerve (which with the retina, is part of the CNS and not a peripheral nerve). At the optic foramen the outer layer of the dura is continu-ous with the periosteum of the orbit, as we should expect, and the inner layer of the dura (as well as *arachnoid, cerebrospinal fluid,*

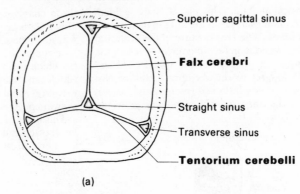

Superior sagittal sinus

Falx cerebri

Straight sinus

Transverse sinus

Tentorium cerebelli

(a)

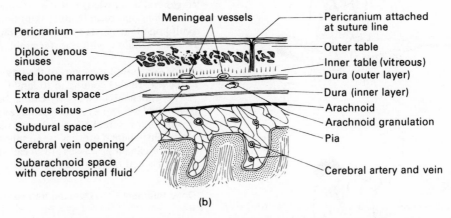

Meningeal vessels

Pericranium

Diploic venous sinuses

Red bone marrows

Extra dural space

Venous sinus

Subdural space

Cerebral vein opening

Subarachnoid space with cerebrospinal fluid

Pericranium attached at suture line

Outer table

Inner table (vitreous)

Dura (outer layer)

Dura (inner layer)

Arachnoid

Arachnoid granulation

Pia

Cerebral artery and vein

(b)

FIG. 17.03. The meninges of the skull. The cerebral arteries and veins lie in the subarachnoid space. The meningeal arteries and veins lie between the skull and the outer layer of dura.

and *pia mater*) extends the whole length of the nerve and fuses with the *sclera* at the back of the eyeball. By analogy with the spinal cord, it is perfectly correct to speak of the theca of the optic nerve.

The outer layer of the dura is not continued into the *vertebral canal*, but it is represented by the periosteum which lines the appropriate parts of the individual vertebrae.

There are many places where the shape of the inside of the skull and that of the brain differ markedly from the other. The situation is similar to that of a square peg in a round hole. A good fit is impossible unless an adaptor is specially devised. The two layers of the dura constitute just such a device: together they constitute the *intracranial adaptor* which ensures that the skull and brain do fit snugly together.

Thus whenever the brain and skull shapes differ, the outer layer of the dura follows the bone, while the inner layer follows the brain. In all such places the two layers part company. The interspaces, however, are not just filled with 'packing tissue' but, instead, with the frugal economy of the body they are made use of to provide room for the venous return from the brain. In this way the dural venous sinuses come to be situated between the outer or periosteal, and inner or sustentacular layers of the dura. Most of the blood which flows in these sinuses finally runs into the internal jugular vein.

The inner layer of the dura forms the *falx cerebri*, a partition in the sagittal plane which separates one cerebral hemisphere from the other. This layer also forms the *tentorium cerebelli*. The occipital lobes of the cerebrum lie on top of the tentorium, more or less in the horizontal plane [FIG. 13.15; p. 263].

The inside of the skull is therefore partly divided into three compartments: one in the posterior cranial fossa beneath the tentorium cerebelli contains the pons, medulla and cerebellum (i.e. the whole of the hindbrain); the rest of the space, enclosed by the anterior and middle fossae and upper surface of the tentorium itself, is divided by the falx cerebri into right and left halves. The need for these partitions is suggested by consideration of what happens when the head is turned quickly from side to side. The brain, because it is floating and because of its inertia, gets 'left behind' and undergoes 'swirling' movements inside the skull. It is obvious that the falx and tentorium help to limit the degree to which 'swirling' can occur. Clearly, in violent changes of acceleration (e.g. a bang on the head), the brain may be damaged as severely by the falx or the edge of the tentorium, as directly by the blow itself. Also, in severe head injuries, the brain may 'bounce' to and fro inside the skull, resulting in various aspects of the 'contre coup' effect.

These swirling movements take place essentially between the brain and the inner layer of the dura. We have seen that the spaces between the two layers of the dura convey the venous return from the brain. It is thus obvious that there must be veins running from the surface of the brain to the dural sinuses [FIG. 17.03]. If the swirling movements of the brain are excessive, these veins will be ruptured, and subdural and subarachnoid bleeding will occur. Such haemorrhage usually takes some time before symptoms appear, and the effect of the expanding haematoma will be added to whatever initial damage the brain may already have suffered.

The spinal cord, as we have seen, lies within the spinal theca and is floating in the cerebrospinal fluid. The pia is 'spot welded' through the arachnoid to the inner surface of the dura in the coronal plane between the spinal nerves. When the posture of the vertebral column and hip joints are changed (e.g. bending down to touch the toes or 'straight leg raising'), the position of the spinal cord and theca change slightly with respect to the vertebral column. Not surprisingly, when the meninges are inflamed (meningitis) flexion of the hip, especially with the knee straight, produces pain and other signs of 'meningism' [FIG. 17.04].

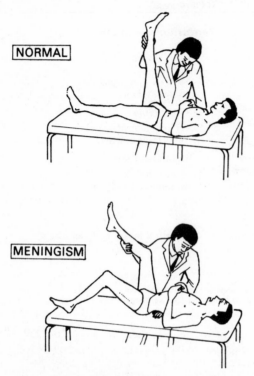

FIG. 17.04. Straight leg raising.
Passive hip flexion with the knee in extension (straight leg raising) in normal people, takes place with the lumbar spines going willingly into flexion. In patients with 'meningism' straight leg raising produces pain with reflex extension of the vertebral column and hip joint.

But the question we now ask is whether there is anything between the periosteum of the inside of the vertebral canal and the sustentacular layer of the dura which could be said to correspond to the venous channels which separate the two layers of the dura inside the skull. There is in fact an extensive internal vertebral venous plexus, which communicates with, and is fed by, the segmentally arranged veins throughout the whole length of the body, and also by the external plexus which lies immediately on the outer surfaces of the vertebrae. The veins in the internal plexus have no valves, and blood can obviously flow instantly from one part of the plexus to another. This means that the spinal cord has further protection; not only is it floating in the cerebrospinal fluid but the theca itself, the cerebrospinal fluid's container, is surrounded by a cushioning plexus of veins. When the position of the theca moves in the vertebral canal with flexion, extension or twisting movements, the blood in the anterior part of the plexus can be shunted instantly to the posterior part, or from side to side. Because the position of the

theca changes, along with the shape of the vertebral canal when movements are carried out, it is necessary to have a 'spinal adaptor' whose shape can easily accommodate to such change. The interstices between the vessels of the plexus are in their turn filled with fat which is liquid at body temperature. The need for an adaptor which can change its shape does not arise inside the skull, but in the vertebral canal the combination of a *valveless venous plexus* and fat answers perfectly the problem of developing an *adaptable adaptor*.

This layer of tissue, consisting of the vertebral venous plexus and fat, is often called the **epidural space**. But we now know that 'space' is a very misleading word in this context.

THE SPHENOID BONE

In general a knowledge of the articulated skull is of more importance than that of individual skull bones. There is, however, one skull bone that needs to be considered in more detail in order to become familiar with the terminology used. This bone is the **sphenoid**.

The sphenoid forms part of the base of the skull. It resembles in shape a barn-owl coming in to land [FIG. 17.05].

The **sphenoid bone** is so called because it is wedge-shaped (*sphenoeides* (Greek) = wedge-shaped). **Pterygoid** (*pteryx* (Greek) = wing; *eidos* (Greek) = form) is the name given to the greater and lesser wing-shaped process of the sphenoid bone. The anterior and posterior **clinoid** processes are so called because they resemble the posts of a bed (*kline* (Greek) = bed).

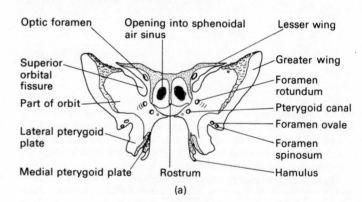

(a)

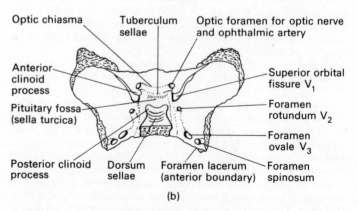

(b)

FIG. 17.05. The sphenoid bone.
(a) Viewed from front.   (b) Upper surface.

The upper surface of the sphenoid bone has a deep fossa for the pituitary gland termed the **sella turcica** because it resembles a Turkish saddle for a horse with a high front and back. The **dorsum sellae** (Latin = back of the saddle) slopes up and overhangs the fossa. It extends laterally as the posterior clinoid processes. Immediately below the pituitary fossa are the sphenoidal air sinuses which open into the spheno-ethmoidal recess of the nose.

The sphenoid bone has a greater and a lesser wing. Between the two is the superior orbital fissure [FIG. 17.05 and FIG. 1802, p. 348]. The posterior margin of the greater wing ends in a sharp angle which projects downwards as the **spine** of the sphenoid. Nearby is the **foramen spinosum** (foramen of the spine) and the **foramen ovale** [FIG. 17.05].

The **foramen rotundum** pierces the wing below the superior orbital fissure and below this is the **pterygoid canal** which runs from the foramen lacerum to the **pterygopalatine fossa**.

The **foramen lacerum** is the irregular aperture between the apex of the petrous part of the temporal bone and the body and greater wing of the sphenoid (*lacerum* (Latin) = torn, hence the word lacerations). The greater petrosal nerve passes through this foramen. The **pterygopalatine fossa** lies in the gap between the pterygoid process of the sphenoid bone and perpendicular plate of the palatine bone which covers the posterior part of the maxillary bone.

The **lateral pterygoid plate** or lamina slopes laterally and backwards. It gives origin to the lateral and medial pterygoid muscles. The lateral pterygoid muscle arises from its lateral surface, the medial pterygoid muscle arises from its medial surface.

The **medial pterygoid plate or lamina** lies anterioposteriorly. It forms the most posterior part of the lateral wall of the nasopharynx. The anterior part of its lateral wall is formed by the palatine bone. The medial pterygoid plate gives origin to the superior constrictor (from lower half of its posterior margin).

The **scaphoid fossa** (*skaphe* (Greek) = boat; *eidos* (Greek) = form, hence boat-shaped) is a depression found at the base of the medial pterygoid plate where it joins the base of the skull. This fossa is medial to the foramen ovale and lateral to the auditory tube. The tensor palati arises from this fossa, and the auditory tube lies in it.

The lower end of the medial plate has the hook-like projection, the **hamulus**, at its posterior border. The tendon of tensor palati curls round the hamulus to reach the aponeurosis of the soft palate.

THE CRANIAL CAVITY

The interior of the cranium [FIG. 17.06] is divided into:
1. Anterior cranial fossa.
2. Middle cranial fossa.
3. Posterior cranial fossa.

These fossae lie at progressively deeper levels. The word fossa is Latin for a ditch or depression.

Anterior cranial fossa

The anterior cranial fossa is the elevated part of the internal base of the skull. It lodges the frontal lobes of the brain. The anterior cranial fossa is formed by the orbital part of the frontal bone, the cribriform plate of the ethmoid bone and the lesser wings of the sphenoid bone. The cribriform plate (*cribrum* (Latin) = sieve) is perforated with holes for the olfactory nerves. In the midline is the ethmoid ridge termed the **crista galli** [FIG. 17.06].

Middle cranial fossa

The middle cranial fossa is more deeply concave and lies at a lower level than the anterior cranial fossa. The lateral recess is for the temporal lobe of the brain. The middle meningeal artery [p. 259] runs across the floor of the lateral part of this fossa. In the midline is the fossa in the sphenoid bone for the pituitary gland. Since this acts as a useful landmark it will be considered further.

Pituitary fossa

The fossa for the pituitary gland in the sphenoid bone is termed, as has just been seen, the **sella turcica**. The posterior wall is termed the **dorsum sellae**. The top of the fossa is closed by dura, the **diaphragm sellae** (Latin = diaphragm of the saddle) which stretches like the canopy over a four-poster bed from the two anterior clinoid processes to the two posterior clinoid processes of the sphenoid bone. Clinoid means resembling a bed, as has been seen, hence clinical medicine = bed-side medicine).

The pituitary gland hangs from the base of the brain by the pituitary stalk. It is thus the undergrowth of the brain or *hypo-*

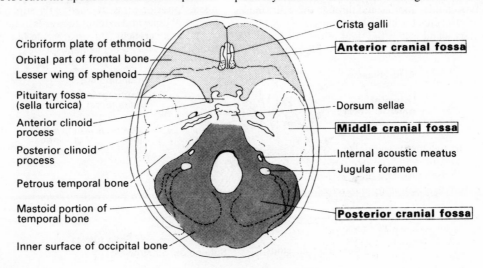

FIG. 17.06. The fossae of the cranial cavity.

physis cerebri. The operation for its removal is termed an **hypophysectomy**. The pituitary stalk, which passes through the centre of the diaphragm sellae, is also known as the **infundibulum** (Latin = funnel shaped) [see FIG. 19.44, p. 402].

To either side of the pituitary gland lie the **cavernous sinuses**. These are venous sinuses between layers of dura. These are joined in front of and behind the pituitary gland by the intercavernous venous sinuses. The cavernous sinus [FIG. 17.07] is very important clinically since it joins the extra cranial veins on the face to the intracranial veins of the skull. It is also one of the most interesting regions in the skull since so many important structures, including the internal carotid artery, pass through it. The cavernous sinus is described in more detail on page 255.

The optic nerve and ophthalmic artery enters the orbit through the optic canal. Most of the other structures entering the orbit pass through the superior orbital fissure. The optic chiasma lies just in front of the pituitary gland hence a pituitary tumour may lead to visual defects and ultimately blindness [p. 359].

Relations of the pituitary gland

The pituitary gland which is the size of a large pea is situated in the pituitary fossa (sella turcica) on the upper part of the body of the sphenoid bone. When it is enlarged it may extend out of its fossa and press on neighbouring structures. Behind is the bony dorsum sellae. In front is the tuberculum sellae (olivary eminence) and the chiasmatic (optic) groove.

On either side of the pituitary gland separated only by dura is the cavernous sinus through which runs the internal carotid artery and the sixth cranial nerve. On the outer wall of the cavernous sinus are from above downwards, the third, fourth, ophthalmic division of the fifth (V_1), and maxillary division of the fifth (V_2) cranial nerves. The third and fourth cranial nerves are more likely to be affected by pituitary enlargements than the sixth which is protected by the internal carotid artery and the bony cavernous groove.

The cavernous groove is a broad groove on the superior surface of the sphenoid bone lodging in the internal carotid artery and which forms part of the medial wall of the cavernous sinus.

Immediately below the pituitary fossa are the sphenoidal air sinuses which have a median septum which is often deviated to one side. In front of the sphenoidal air sinuses are the ethmoidal air sinuses. The pituitary gland is approached through these air sinuses in the nasal and orbit routes for hypophysectomy.

Above the pituitary gland will be the floor of the third ventricle and posteriorly the space between the two cerebral peduncles. The

circle of Willis [p. 261] will be encircling the pituitary stalk.

To the outer side of the cavernous sinus lies the trigeminal ganglion and above this the hook-like extremity of the hippocampal gyrus of the temporal lobe, the uncus. If the pituitary tumour presses on this region of the brain, olfactory hallucinations may result.

The **tuber cinereum** is an area of grey matter which extends from the optic chiasma to the mammillary bodies. The pituitary stalk (infundibulum) is attached to it. It forms part of the floor of the third ventricle.

Posterior cranial fossa

The posterior cranial fossa is the lowest of the three fossae. It lodges the cerebellum, pons, and medulla. It is formed by the inner surface of the occipital bone below the horizontal limb of the confluence of the venous sinuses, the posterior surface of the petrous temporal bone and the inner surface of the mastoid portion of the temporal bone. It lies beneath the tentorium cerebelli.

SUMMARY

FALX AND TENTORIUM

The falx cerebri and tentorium cerebelli are folds of dura mater attached to the bony skull. They divide the cranial cavity into compartments which support different parts of the brain.

The **falx cerebri** is a sickle-shaped process of dura mater separating the two cerebral hemispheres [FIG. 17.03]. Falx is Latin for a sickle hence *falciform* means having the shape of a sickle.

The **falx cerebri** is attached in front to the **crista galli**. This is the superior projection of the ethmoid bone in the midline so called because it resembles a cock's comb (*crista* (Latin) = crest; *galli* (Latin) = of a cock). The falx cerebri is attached superiorly to the vault of the skull. It becomes deeper as it passes backwards and has a free lower edge which lies over the corpus callosum connecting the two cerebral hemispheres. Posteriorly it is continuous with the horizontal folds of dura which form the tentorium cerebelli.

The **tentorium cerebelli** forms a sloping roof over the posterior cranial fossa. Tentorium is Latin for a tent or covering. The tentorium cerebelli is attached to the margin of the posterior cranial fossa. Anteriorly it is attached to the superior border of the petrous temporal bone and extends forwards as far as the cavernous sinus.

The **falx cerebelli** is a small sickle-like process of the dura between the two cerebellar hemispheres.

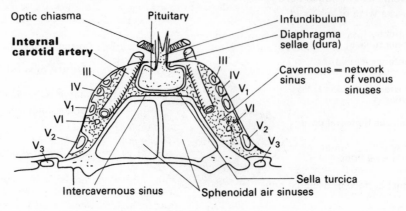

FIG. 17.07. The pituitary fossa.

SUMMARY

VENOUS SINUSES

The venous sinuses are triangular-shaped venous channels lined with endothelium between layers of dura mater forming the falx and tentorium. The largest venous sinuses are found where the dura is attached to the skull but they are also found along the free edges of the falx and tentorium. The shallow grooves formed by the venous sinuses can often be seen on the inner surface of the cranial cavity.

These venous channels drain mainly into the internal jugular vein but they also drain into the vertebral venous plexuses through the foramen magnum. The spread of inflammation and infection from adjacent structures may lead to thromboses in these venous sinuses. Thrombosis may also spread from their venous tributaries.

The **superior longitudinal (sagittal) sinus** runs in the attached margin of the falx cerebri. It starts in front of the crista galli at the foramen caecum where a small vein from the nose enters it. The superior longitudinal sinus curves backward in the vault of the skull in the midline grooving successively the frontal, parietal, and occipital bones. The superior cerebral veins join the superior longitudinal sinus from the surface of the brain.

On either side of the superior longitudinal sinus are irregular diverticulae of the dura filled with venous blood and which communicate with the superior longitudinal sinus through small apertures. The arachnoid granulations, which are associated with these venous lacunae, play an important part in the reabsorption of cerebrospinal fluid [p. 381].

At the back of the skull at the confluence of the sinuses (confluens sinuum), previously known as torcular Herophili after Herophilus (335–280 BC), the superior longitudinal sinus usually turns right to become the right lateral sinus.

The arachnoid granulations, arachnoid villi, or Pacchionian bodies were first described by the Italian anatomist Pacchioni (1665–1726). They are prolongations of the sub-arachnoid space into dural venous diverticulae. The arachnoid granulations allow the surplus cerebrospinal fluid in the sub-arachnoid space to be absorbed into the venous blood.

The **inferior longitudinal (sagittal) sinus** which runs along the inferior border of the falx cerebri joins the **straight sinus** at the junction of the falx cerebri with the tentorium cerebelli. Here they receive the great cerebral vein (of Galen) from the brain. The straight sinus usually turns left and becomes the left lateral sinus.

The **lateral sinuses** which start at the internal occipital protuberance run outwards first as the **transverse sinus** in the attached margin of the tentorium cerebelli and then as the **sigmoid sinus** in a groove on the posterior surface of the temporal bone. The sigmoid sinus runs very close, but posterior to the mastoid (tympanic) antrum which may be the site of an infection.

The **lateral sinuses** pass through the jugular foramen and become the **internal jugular vein**. The **inferior petrosal sinus**, which runs from the cavernous sinus in the groove between the petrous temporal and basi-occipital bones, joins the internal jugular vein.

The **superior petrosal sinus** which runs from the sinus in the margin of the tentorium cerebelli along the upper border of the petrous temporal bone joins the lateral sinus.

The **occipital sinus** which starts on the union of several small veins around the foramen magnum runs up the attached border of the falx cerebelli to join the lateral sinus.

The lateral sinus also receives a mastoid emissary vein. As a result a thrombosis in the lateral sinus may give a swelling behind the ear.

The **cavernous sinus** [p. 255] on either side of the pituitary gland has the internal carotid artery passing through it. Since the cavernous sinus is filled with venous blood, damage to the internal carotid artery will result in an arteriovenous anastomosis. Apart from its importance as an anatomical landmark (the third, fourth, sixth, and upper two divisions of the fifth cranial nerves pass through it, p. 272), the cavernous sinus is of importance since it receives venous blood from the face via the ophthalmic veins. Thus a septic abcess on the face in the *danger area* around the nose may cause a cavernous sinus thrombosis.

The venous channels which join the cavernous sinus are:
1. Ophthalmic veins (superior and inferior).
2. Sphenoparietal sinus.
3. Vein from lateral cerebral fissure (sylvian vein).
4. Emissary vein (of Vesalius) which passes through foramen in front of foramen ovale to join pterygoid veins.
5. Anterior and posterior intercavernous sinuses. These sinuses unite the right and left cavernous sinuses in front of and behind the pituitary gland and thus forms a **circular sinus**.
6. Superior petrosal sinus.
7. Inferior petrosal sinus.

The cavernous sinus drains into the superior and inferior petrosal sinuses [p. 263].

THE SKULL AT BIRTH

The skull at birth is a fascinating example of anatomical adaptation to special functional needs. Before birth much of the skull, including the vault, the face, and mandible are **ossified in membrane**. Only the base of the skull is **ossified in hyaline cartilage**. Individual skull bones which participate in both the base and vault are ossified therefore partly in membrane and partly in cartilage. The fact that those parts of the skull and face which are exposed on the surface are basically membranous, gives them a flexibility and resilience which is advantageous during childbirth.

Look at the vault of the foetal skull [FIG. 17.08(a)] from above and identify the **coronal** and **sagittal sutures**. The former separates the two **parietal bones** from the **frontal bone**. Notice the **metopic** (or frontal) **suture** which divides this part of the frontal bone in the forehead region into right and left halves.

At the region where the two halves of the frontal bone and the parietal bones meet, you will see a diamond-shaped, as yet unossified region, called the **anterior fonticulus** (Latin = *little fountain*) or **fontanelle** (French = *little fountain*). This region of the skull is called the **bregma** (Greek = *forehead*). The **posterior fonticulus** lies at the posterior end of the sagittal suture where the squamous part of the occipital bone in the midline is wedged in between the two parietal bones each side. The three sutures have a shape not unlike the Greek letter *lambda* (λ), and for this reason we speak of the **lambdoid suture**.

Look at the skull from the side and notice that there are two additional fonticuli at the antero- and posterolateral corners of the parietal bone. All the fonticuli, with the exception of the anterior, are ossified soon after birth.

Turn the skull over and look at the **occipital bone** from below [FIG. 17.08(d)]. Instead of being in one piece and fused with the body of the sphenoid bone as it is in the adult skull, the bone is separated from the sphenoid by a suture and is itself in *four separate parts*. The

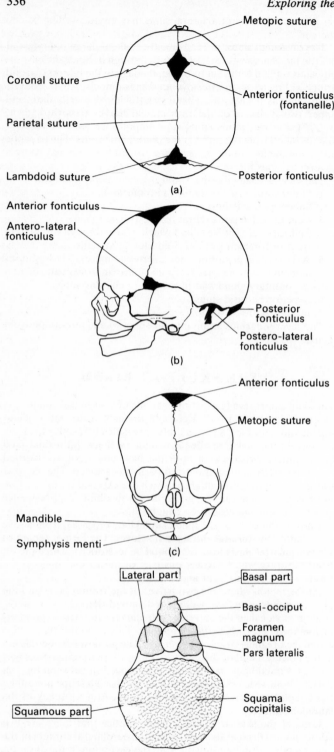

FIG. 17.08. The skull at birth.
(a) View from above.
(b) View from left side.
(c) View from in front.
(d) Components of occipital bone from below.

part lying in front of the **foramen magnum** is the **basal part** of the occipital bone (pars basalis or basi-occiput). On each side of the foramen is the **lateral part** (pars lateralis), which forms about two-thirds of the **occipital condyle**, while behind the foramen magnum is the thin squamous part of the occipital bone (squama occipitalis). You will realize that the **squamous part** of this bone is ossified in membrane, while the rest ossifies in cartilage.

The articulation between the basi-occiput and the body of the sphenoid is not a fibrous but a primary cartilagenous joint. The intervening hyaline cartilage is like an *epiphyseal cartilage* and it is responsible for much of the longitudinal growth of the base of the skull. It has a major role in the forward movement of the facial skeleton which takes place during childhood and adolescence with the growth of the jaws associated with the change from milk to permanent teeth. This important growth cartilage disappears soon after the eruption of the third molars (about 20 years), that is once the teeth and jaws have grown fully.

During childbirth it may be important to recognize the anterior or posterior fontanelles by palpation via the vagina and cervix uteri. This is done to determine whether the head is 'flexed' or not. The point is that in a normal delivery the head is born first. But the circumference of the presenting vertex varies markedly according to the degree of flexion. In **full flexion** of the head (circumference 30 cm), the occiput presents and the **posterior fonticulus** is palpable. If the head is **poorly flexed** (circumference 36 cm) the **anterior fonticulus** can be felt [FIG. 17.09].

In a **face presentation**, the circumference may increase to 40 cm. Clearly the damage done to the foetal skull (and brain) and to the mother's birth canal is influenced markedly by the orientation of the head as it descends during labour.

Up to the time of birth since the bones of the vault of the skull are joined together by flexible fibrous tissue, they can slide over each other, a process called **moulding**. This reduces the overall size of the head. The suture between the squamous part and the rest of the occipital bone is called the *obstetric hinge*, as it allows the squamous part of the occipital bone to slide in beneath the two parietals, at the lambdoid suture.

If the baby is postmature (more than 42 weeks) not only is the head bigger, but because of the rapid ossification of the membranous sutures, moulding is minimal and, for both these reasons, a fracture of the skull is more likely during childbirth.

The anterior fonticulus

The **anterior fonticulus** is of importance in paediatrics. In normal development it will have disappeared by about 18 months. Most normal children are able to get about by walking or tottering by 18 months of age. At this time they subject themselves to a good deal of physical trauma, with frequent falls. At least the skull vault is completed by this time.

There is convincing evidence showing that it is the growth of the brain which regulates the size of the skull. When the brain is small, the skull is also small (*microcephaly*), and the anterior fonticulus closes prematurely. When the brain is enlarged (say in *hydrocephalus*) the fonticulus closes very late or not at all. The size of the head and of the anterior fonticulus provides important evidence as to the growth and development of the brain.

The pulsation of the arteries on the surface of the brain is transmitted to the cerebrospinal fluid, and thence to the anterior fontanelle, so that the heart-rate can easily be measured here.

The fonticulus also acts as an intracranial pressure gauge. An increase in pressure causes the fonticulus to bulge outwards (say in

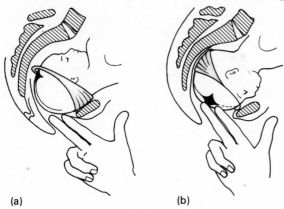

FIG. 17.09. Birth of a baby.
(a) The head if flexed with chin on chest. The posterior fonticulus is palpable whilst the anterior is not. Circumference of head 30 cm.
(b) Head poorly flexed with chin not on chest. The anterior fonticulus is palpable whilst the posterior is not. Circumference of head 36 cm (+ 20 per cent).
Note: when putting on a T-shirt with a very tight collar try starting with it at nape of neck and pull it over from the front.

meningitis), but when there is a fall (say in dehydration) this causes the fonticulus to sink inwards.

It is possible to withdraw c.s.f. from the lateral angle of the fonticulus, or blood by putting a needle into the superior sagittal sinus in the midline.

Facial skeleton

Look now at the new-born skull from the front and examine the facial skeleton [FIG. 17.08(c)]. Notice that the floor of each orbit lies only slightly higher than the floor of the nose. The relative position of the floor of the nose will migrate downwards later in life because of the increasing development of the nasal cavity and of the maxillary sinuses. At birth the facial structures occupy about one-ninth of the volume of the whole head, while in the adult they occupy about one-third.

Finally turn your attention to the **mandible**. At this stage it is in two halves, the sides being connected anteriorly by the **symphysis menti**. The symphysis menti disappears at about twelve months when most babies are eating solids.

The shape of the mandible is quite different from that in the adult, with a wider angle between the body of the mandible and the ramus. The squareness of the angle of the mature jaw makes room for the eruption of the permanent molar teeth. When the teeth fall out, as they may well do in old age, the shape of the mandible reverts to something very close to the foetal shape. There is no part of the skeleton which demonstrates the plasticity of bone more clearly than the mandible [FIGS. 17.10(a) and 17.14].

The ear

Look at the region of the ear [FIG. 17.10(a)]. Notice that at birth (and during infancy) the bony part of the external acoustic meatus (e.a.m.) is absent. It is amazing that in the skull at birth the size of the tympanic membrane (ear-drum) and middle ear is so similar to that in the adult. The external acoustic meatus is no shorter relatively than in the adult, as many books claim, indeed the opposite is true. But as the whole length is cartilaginous, the ear-drum is more liable to injury when an auriscope is clumsily used than it is in adults.

The **pars tensa** of the ear drum is attached to the so-called *tympanic ring*. The ring is incomplete however and only describes about three-quarters of a circle, the upper quadrant being missing. The corresponding part of the drum is the **pars flacida**. During early childhood the deep part of the cartilagenous external auditory meatus begins to ossify as a lateral extension of the tympanic ring. The bony tympanic plate now appears. This is a misleading term as this plate is not flat as you might expect, but gutter-shaped, forming the floor and anterior and posterior walls of this part of the external auditory meatus. The roof is completed by the petrous part of the temporal bone.

If you have a skull in which the ear drum has been removed, you can see into the **middle ear** and its whole anatomical layout is very clearly shown. The handle of the **malleus** points downwards and backwards, and its tip is at the centre of the ear drum. The long process of the **incus** also points downwards and backwards parallel to the handle of the malleus but further back. It is about two-thirds of the length of the malleus. The head of the malleus and the body of the incus articulate with each other by means of a synovial joint but are held firmly together by a strong elastic capsule, and normally these two bones move as one.

You can see the **stapes** attached to the tip of the long process of the incus (also by an elastic capsule) and further in you will see the footpiece (foot plate) of the stapes fitting into the **oval window**. The malleus must vibrate in the same plane as the ear drum because it is embedded in it and the same applies to the long process of the incus because the malleus and incus move as one. The vibration of the incus causes the footpiece of the stapes to vibrate in the oval window. Due to the difference in length of the handle of the malleus and the long process of the incus, the amplitude of movement of the footpiece of the stapes is two-thirds that of the centre of the ear drum, i.e. the movement is geared down, which increases its force.

The middle ear mechanism increases the sharpness of hearing for another reason: sound is collected over a wide area, the ear-drum, and focused onto a small area (the oval window). The ratio is about 15:1, that is, about fifteen times more energy is available to the inner ear, than would be the case if the sound fell directly onto the oval window itself. The same focusing effect is achieved by other vertebrates, such as reptiles, by means of a single bone connecting the large drum directly to the small oval window; but there is no gearing down. A further advantage of a *chain* of ossicles is that when a loud noise is experienced, it is possible for the incus and malleus to subluxate (i.e. dislocate and then return to their original position). In an ordinary articulation, in any other part of the body, it is impossible for subluxation to occur without extensive damage to the capsular ligament; but in the ear, because the capsules are elastic, no significant damage is done.

Note the position of the **round window** which is in the medial wall below and behind the long process of the incus. The window looks almost directly backwards and when seen from the side via the external acoustic meatus, is subject to considerable perspective effect so that it, in fact, looks eliptical. In front of the round window is the promontory produced by the first turn of the cochlea.

On the medial wall, level with the neck of the malleus is the **processus cochleaformis**, which (as its name suggests) looks just like the opening of a snail shell. The tendon of the **tensor tympani** emerges here, and runs laterally across the air space of the middle ear into the **neck of the malleus**. This muscle contracts reflexly when a loud sound is heard and reduces the amplitude of vibration. It is supplied by the mandibular division of the trigeminal nerve (V_3).

The easiest way to detect the contraction of the tensor tympani is

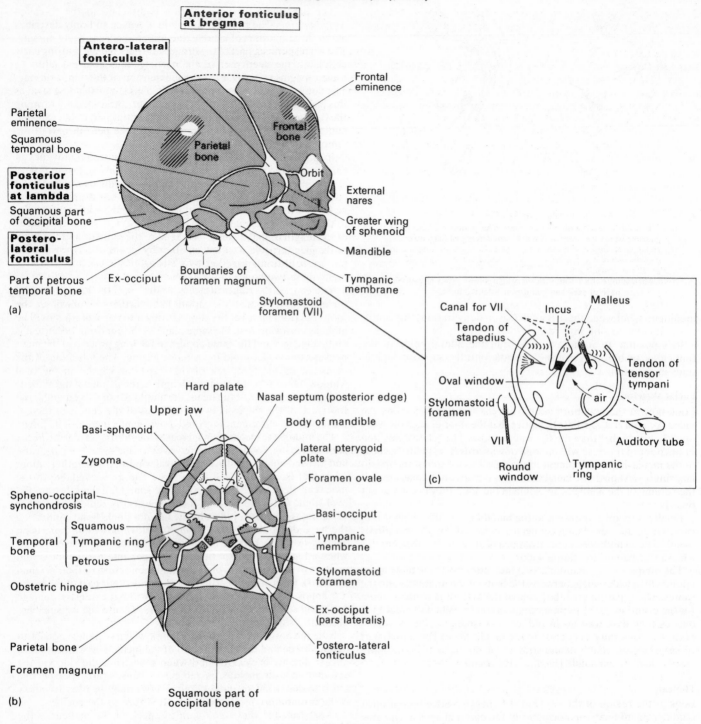

Fig. 17.10. The skull at birth.
(a) View from right side.
(b) View from below.
   The squamous part (squama occipitalis) of the occipital bone ossifies in membrane. The lateral part (pars lateralis) and the basal part (basi-occiput) ossify in cartilage.
(c) View through external acoustic meatus with tympanic membrane removed.

other. The indrawing of the ear drum can easily be seen with an auriscope. This is a very useful fact which makes it possible to assess whether a small baby is deaf or not, or to refute an adult, who claims to be deaf.

On the posterior wall of the middle ear you will see the **apex of the pyramid** at whose tip is a small opening which transmits the tendon of the **stapedius muscle**. This runs forwards and is attached to the neck of the stapes. Its reflex contraction also helps to limit excessive movement. It is supplied by the facial nerve (VII).

Look at the anterior wall, and pass a bent wire through the bony part of the auditory **(pharyngotympanic) tube**. The end comes to lie in a scooped out depression (the scaphoid fossa). In life, this fossa, which reaches as far as the medial pterygoid plate, is occupied by the cartilagenous part of the tube [FIG. 17.10(c)].

At birth the middle ear is filled with fluid, but soon afterwards air enters via the auditory tube. The tube is normally closed but the flexible cartilagenous part in the scaphoid fossa is pulled open at frequent intervals by muscles during swallowing, yawning, etc. As soon as air enters the middle ear it begins to be absorbed by the blood capillaries lining the cavity (whose oxygen tension must be lower than that of atmospheric air) so it is necessary for it to be replaced at frequent intervals. Deafness ensues in a few minutes when the tube is obstructed. The tube provides the route whereby infection spreads from the pharynx giving rise to otitis media.

Stylomastoid foramen

Look at the **stylomastoid foramen** [FIG. 17.10(b)], where the facial nerve (VII) emerges, and push a wire upwards into it.

It is conventional to describe the facial (VII) nerve in the reverse direction. The important point is that the nerve runs:

1. Laterally through the internal acoustic meatus as far as the anterior part of the medial wall of the middle ear [FIG. 14.12; p. 281].
2. Backwards in the medial wall in the same horizontal plane as 1.
3. Downwards (having reached the posterior extremity of the medial wall) so as to emerge at the stylomastoid foramen.

Damage to the nerve when it lies close to the middle ear can paralyse the stapedius, which gives rise to tinnitus, and the chorda tympani, which will result in loss of taste from the anterior two-thirds of the tongue. It is easy to see that the canal in which the wire is lying, is in continuity with a horizontal canal high up on the medial wall of the middle ear. The thin bony partition separating the nerve from the middle ear is often deficient, and paralysis of the facial muscles is an important complication of middle-ear disease.

Notice that, at the time of birth, there is no **mastoid process**. It appears at about six months, when the infant is learning to keep its head upright, which involves the sternomastoid muscle. Usually the mastoid process is not a solid bony structure. It is invaded and made pneumatic, by epithelial lined air cells which are outgrowths from the tympanic antrum. Explore the **epitympanic recess** (which houses the head of the malleus and body of the incus) with your wire and feel your way into the antrum, which lies just behind it. Air cells are not confined to the mastoid process but may invade almost any part of the petrous part of the temporal bone.

The stylomastoid foramen is very exposed on the side of the head at birth instead of being safely protected by the mastoid processes as in the child and adult. Paralysis may occur during childbirth when obstetric forceps are incorrectly placed on the head, in such a way that the tips of the forceps on one or both sides squeeze on the nerve. Fortunately recovery usually occurs quite soon. Also facial paralysis can easily occur if a baby is hit on the side of the head.

THE ADULT SKULL

Look at the upper surface of the vault of the adult skull which is the **calvaria**. Compare it with the vault of the foetal skull. Not only have the fonticuli disappeared but the metopic or frontal suture has also gone and the frontal bone is now in one piece united across the midline. The frontal suture has disappeared by the age of six or seven usually, but in a small number of people it persists. It is important when you X-ray a head injury not to mistake a normal or persistent frontal suture for a fracture.

Sometimes in the parietal bone near the midline you can see an opening in the bone which actually goes right through into the inside of the skull. These openings are fairly irregular in position and they provide a passage way for the emissary veins [p. 263]. These veins provide a route whereby infection can spread from the outer surface of the skull into the inside.

The sutures connecting the bones of the calvaria are strikingly irregular. The bones fit together like a rather complicated jig-saw puzzle. The irregularities of the suture lines often meander to such a degree that little bones get cut off and form what are called sutural or wormian† bones. Again it is important not to mistake the appearance of a very irregular suture line for a fracture.

When you look at the calvaria from below you will notice a shallow groove running in the sagittal plane. This is for the superior sagittal sinus. Slightly to one side of the midline you can often see quite well marked pits which are produced by the arachnoid granulations in the frontal or parietal bones near the midline. This again is a feature which is not to be seen in a young skull. Notice also the tree-like branching of the meningeal vessels which groove the inner surface of the calvaria.

As has been seen the base of the inside of the skull is divided into three separate regions called the **anterior**, **middle**, and **posterior cranial** fossae. The watershed separating the anterior fossa from the middle fossa is formed by the lesser wing of the sphenoid bone. The watershed between the middle fossa and the posterior fossa is formed by the petrous part of the temporal bone [FIG. 17.11].

Put a finger into the orbit and with your thumb in the anterior cranial fossa and note how thin the orbital plane is which separates these two cavities.

The floor of the middle cranial fossa is formed by the temporal bone and in front by the greater wing of the sphenoid. There is a cleft between the greater and lesser wings of the sphenoid bone passing from the inside of the skull to the orbit. This is the superior orbital fissure.

In the posterior fossa put your finger on the internal occipital protuberance. Running down from here towards the foramen magnum is the **occipital crest**. Running transversely is a groove called the **transverse sulcus** in which the transverse sinus of the dura lies. Continue laterally along the transverse sulcus and you will come to a deep groove in the posterior part of the temporal bone called the **sigmoid** (*Greek* = S) **sulcus** in which lies the sigmoid sinus which takes you to the **jugular foramen**. The blood running in these sinuses empties into the internal jugular vein.

The **superior petrosal sinus** runs along the edge of bone separating the middle fossa from the posterior fossa. The posterior fossa is roofed over by **tentorium cerebelli** which is attached so as to enclose at the side the superior petrosal sinus and behind the transverse sinus.

The frontal lobes of the cerebral hemispheres lie on the floor of

† Named after the Danish anatomist Olaus Worm 1588–1654.

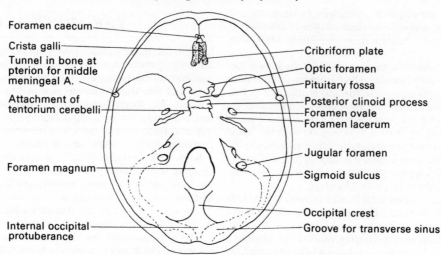

Foramen caecum

Crista galli

Tunnel in bone at
pterion for middle
meningeal A.

Attachment of
tentorium cerebelli

Foramen magnum

Internal occipital
protuberance

Cribriform plate

Optic foramen

Pituitary fossa

Posterior clinoid process
Foramen ovale
Foramen lacerum

Jugular foramen

Sigmoid sulcus

Occipital crest
Groove for transverse sinus

FIG. 17.11. Important landmarks inside base of skull.

the anterior fossa, the temporal lobes lie in the middle fossa, and the cerebellum and the brainstem lie in the posterior fossa. The **occipital lobes** of the cerebrum lie on the upper surface of the tentorum cerebelli which forms a shelf separating the occipital part of the cerebrum from the cerebellum.

Look at the skull in the midline [FIG. 17.11]. In front you will see the frontal crest and if you run your finger down this crest you will notice the position of the foramen caecum (= a *blind foramen*) which marks the point where the superior sagittal sinus begins. Behind the foramen caecum is a projecting wing of bone called the crista galli (cock's comb). On each side is the perforated **cribriform plate** of the ethmoid bone. The cribriform plate transmits the fibres of the **olfactory nerve**. The cells of origin of these nerves are actually in the olfactory mucosa of the nose and the fibres will eventually synapse with new neurones in the olfactory bulb. In the periphery, that is in the mucous membranes of the nose, the nerve cells and their fibres are lying in the tissue fluid. As they pass upwards they come to be surrounded by a diverticular sleeve of arachnoid and are surrounded by subarachnoid fluid. In fact subarachnoid fluid and tissue fluid in the nose are in direct continuity with each other. This provides a route whereby infections can spread from the nasal cavity to the brain and produce, say, meningitis or poliomyelitis.

Behind the cribriform plate you come to the body of the sphenoid and behind that is the **pituitary fossa** or sella turcica in which the **pituitary gland** lies. The dorsum sellae bounds the pituitary fossa posteriorly. Identify the posterior clinoid processes to which the free border of the tentorium cerebelli is attached. Immediately in front of the anterior clinoid process is the **optic foramen** which transmits the **optic nerve** and the ophthalmic artery. The layer of pia which lines the nerve is surrounded by cerebrospinal (subarachnoid) fluid, the subarachnoid membrane and the inner sustentacular layer of the dura. The retina may be regarded as part of the central nervous system and the optic nerve as part of the brain, equivalent to a white fibre tract. As part of the central nervous system it is surrounded by three meninges. This also means that when the nerve is damaged no regeneration can take place. In addition its cells of origin are peripheral rather than central, that is to say, in the retina and not in the brain.

Those structures which enter the orbit, apart from the optic nerve and the ophthalmic artery, do so by passing through the **superior orbital fissure**. This means the **oculomotor nerve** (III), the **trochlear nerve** (IV), and the **abducent nerve** (VI) which between them supply all the extrinsic muscles of the eyeball, as well as the levator palpabrae superioris. The big sensory nerve which supplies the eyeball and the orbital structures is the **ophthalmic division** of V which passes through the superior orbital fissure with the ophthalmic veins. These veins drain backwards into the cavernous sinus. The ophthalmic veins anastomose with the veins on the superficial surface of the face and nose and they provide a direct connection between the surface of the face and the inside of the skull. They are the best example of emissary veins. It is perhaps also worth mentioning that the ophthalmic veins have no valves in them and blood can flow in either direction according to the orientation of the head.

Let us come back to the **pituitary fossa** [FIG. 17.11]. Needless to say the pituitary actually lies in the pituitary fossa and the fossa is separated from the rest of the inside of the skull by a dural diaphragm called the diaphragma sellae. On each side of the pituitary fossa there is the **cavernous sinus**—a large space filled with blood which lies between the inner (or sustentacular) layer of the dura which supports the brain and the outer layer of the dura which is closely associated with the bone.

If you look at the base of the skull just behind and lateral to the pituitary fossa you will see the **foramen lacerum**. This is bounded in front by the sphenoid bone and behind by the temporal bone. You will notice that the upper part of the foramen lacerum, as you are looking at it from above, is quite smooth and circular in shape. In fact the upper part of the foramen has the carotid artery lying in it. If you turn the skull over and look at it from below you will see that the same foramen has a very jagged or, if you like, *lacerated* appearance and it is in fact filled with fibrous tissue and cartilage and nothing of any significance goes right through this foramen. The carotid canal enters the foramen lacerum from behind, and the pterygoid canal commences in or very close to the anterior edge of the foramen. These canals can easily be demonstrated with a bent wire.

Immediately on the lateral side of the foramen lacerum you will see the **foramen ovale** completely surrounded by sphenoid bone. In front of the foramen ovale is the **foramen rotundum**. It is through the foramen ovale that the **mandibular division** of V will escape from the skull; through the foramen rotundum the **maxillary division** will escape and we already know that the **ophthalmic division** escapes through the superior orbital fissure.

Immediately behind and slightly lateral to the foramen ovale you will see an opening 1.5–2 mm across which is the **foramen spinosum**. This foramen gets its name from the spine of the sphenoid which you will see if you look at the skull from below. The foramen transmits the **middle meningeal artery**, a branch of the maxillary artery. It runs laterally on the floor of the middle fossa for about 1.5 cm and then divides into anterior and posterior branches. The anterior branch lies in a tunnel of bone in the region of the pterion.† If the skull is fractured in this region, then the middle meningeal vessels may rupture and bleed between the bone and the outer layer of dura. This produces an extradural haemorrhage. Such a haemorrhage usually produces a bad headache because the dura has a sensory nerve supply, like the periosteum of any other bone.

Look now in the **posterior fossa** and identify the position of the internal acoustic meatus. This opening transmits the seventh and eighth cranial nerves. Both these nerves are really in two parts. The two parts of the eighth nerve are two fairly large nerves joined together by some loose connective tissue. The secondary part of the facial nerve, the seventh nerve, is called the **nervus intermedius** and it is a very fine strand of nervous tissue.

So far as the eighth nerve is concerned the two parts of the nerve serve first the acoustic part of the inner ear and secondly, the vestibular part and we speak of the acousticovestibular nerve in the PNA. The major part of the facial nerve is motor and supplies the muscles of facial expression and a few other muscles. The nervus intermedius is a small but not unimportant part of the nerve. In it lie the parasympathetic nerve fibres which supply the lacrimal gland, the sublingual gland and the submandibular gland as well as the sensory fibres from the anterior two-thirds of the tongue, which subserve the sense of taste.

The next foramen you need to identify is the **jugular foramen**. Come back to the transverse sulcus which we mentioned earlier which houses the transverse sinus leading into the sigmoid sinus with the S-shaped groove running on the back of the temporal bone. This takes you down to the jugular foramen. This foramen transmits not only this very large venous channel but also the **glossopharyngeal** IX, the **vagus** X, and the **accessory** nerve XI. The · glossopharyngeal nerve, runs through its own sleeve of dura whereas the vagus and accessory nerve share a common sleeve. This is how the accessory nerve got its name—it was always regarded as being accessory to the vagus.

Finally, find the **hypoglossal canal**. When you look at the skull from above you see this foramen on the lateral side of the foramen magnum. Look from behind and you can see straight through it.

This foramen transmits the last of the cranial nerves, the hypoglossal nerve, which supplies muscle of the tongue. The canal is often double-barrelled, each part transmitting part of the nerve XII.

† The pterion is the region where the frontal, parietal, temporal, and sphenoid (greater wing) bones join. It lies 3 cm behind the external angular process of the orbit. It gets its name from *pterion* (Greek) = a wing, the region of the greater wing of the sphenoid.

There is often a large opening in the bone just behind the occipital condyle. This is the posterior condylar canal, which transmits another emissary vein.

Look at the skull from the front [FIG. 17.12]. Begin by looking at the frontal bone. This forms the forehead and the roof of the orbit. Look into the orbits and identify the optic canal and the superior orbital fissure which we have already seen from the inside of the skull and also the inferior orbital fissure which we have not mentioned. Identify the greater wing of the sphenoid and notice that it separates the superior and inferior orbital fissures from each other. Now put a finger and thumb into each orbit across the midline. The bone which separates them is the **ethmoid**. There is an extremely thin layer of bone separating the ethmoid, nasal sinuses from the orbit itself, which used to be called the **lamina papyracea**. Infection can spread from the inside of the nose into the orbit through this very thin bone. Just in front of the ethmoid you will see a large opening running vertically downwards. If a seeker were pushed into it, it emerges beneath the inferior nasal concha in the inferior meatus of the nose. This is the canal for the **nasolacrimal duct** which allows the tears to drain into the nasal cavity. In the upper margin of the orbit slightly nearer the midline, notice the supra-orbital groove or supra-orbital foramen which transmits the **supra-orbital nerve** which is the longest terminal branch of the ophthalmic division of V. Almost in line vertically with the supra-orbital foramen but below the orbit is the infra-orbital foramen. This is a much bigger opening than the other one and transmits the maxillary division of V and a fairly large blood vessel. If you push a seeker up the infra-orbital foramen you will see that it takes you to a groove which lies in the floor of the orbit—the infra-orbital groove—which in its turn leads to the inferior orbital fissure.

We last left the **maxillary division** of V as it entered the foramen rotundum in the middle cranial fossa. It runs in that foramen until it gets into the **pterygopalatine fossa**. It has to turn laterally in the pterygopalatine fossa and eventually enters the orbit by running into the infra-orbital groove.

Notice the position of the mental foramen in the mandible. It lies 2 or 3 cm from the midline. The mental nerve emerges through the foramen in an upward and backward direction as you can see by just looking at the bone before it turns forwards towards the midline. This the principal termination of the **mandibular division** of V.

Notice that the principal terminal branches of the three divisions of the trigeminal nerve, the ophthalmic coming out at the supra-orbital foramen or groove, the maxillary coming out at the infra-orbital foramen, and the mandibular coming out through the mental foramen are all more or less on the same vertical line. These are the places to go for when testing for sensation of the trigeminal nerve. Make sure you can identify these three nerves in yourself.

Look at the **base of the skull** from below [FIG. 17.01]. First of all beginning anteriorly notice the arrangement of the teeth. There should normally be eight teeth in each jaw in the adult; two incisors, a canine, two premolars, and three molars.

Each tooth consists of a root, neck and crown. The dental pulp consists of vessels and nerves, surrounded by dentine. The crown is covered by enamel. Special bone-like material, cementum, surrounds the roots. The roots are fixed to the alveolar bone by the periodontal membrane. The upper molars have three roots (trifid), the lower molars have two. All other teeth have a single root.

Notice just behind the palate the position of the two **pterygoid plates**. The lateral pterygoid is really just ossification in the intermuscular septum which separates the medial and lateral pterygoid

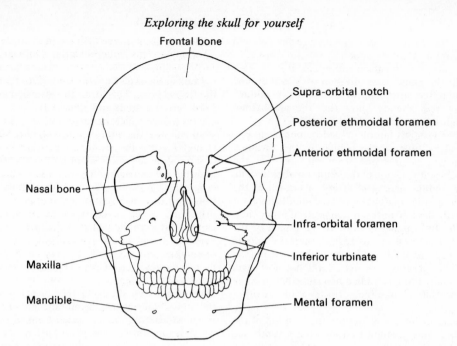

FIG. 17.12. The skull from the front.

muscles from each other. The medial surface of the medial pterygoid plate forms the side wall of the nasopharynx.

Notice the pterygoid hamulus which is attached to the inferior border of the medial pterygoid plate. When you open your mouth wide you can sometimes see, and you can always feel, the hamulus projecting downwards just behind the terminal part of the hard palate.

Notice the position of the **foramen ovale** just opposite the edge of the lateral pterygoid plate, with the **foramen spinosum** through which the middle meningeal artery runs lying just behind it.

Hard palate

The hard palate is made up of the palatal shelves of the maxilla and by the horizontal component of the palatine bone posteriorly. The incisive foramen is in the midline immediately behind the 'premaxillary' part of the maxilla. A suture runs from here towards the gap between the lateral incisor and the canine [see hare lip, p. 297].

The nasopalatine nerves go through this foramen and supply the premaxilla. Two other foramina on each side transmit posteriorly the greater and lesser palatine nerves. They supply most of the rest of the hard palate.

Note that in cleft palate the cleft follows the suture marked ABC in FIGURE 17.02.

Nasal cavity

Identify the **anterior** and **posterior nares**. The former are bounded by the maxilla except for the two nasal bones [FIG. 17.12]. It is possible to see the **vomer** (plough share), which forms about one third of the septum. The remainder is hyaline cartilage and will no longer be present [p. 306].

The **posterior nares** [p. 304] are bounded by the horizontal shelves of the palatine bones below, by the medial pterygoid plates laterally, and by the body of the sphenoid. The vomer forms the septum. In the lateral wall of the nose identify the inferior, middle, and superior **conchae** (tubinate bones) which divide the nasal cavity

into four regions, known as the **inferior meatus** (between inferior concha and floor), the **middle meatus** (between inferior and middle conchae), the **superior meatus** (between middle and inferior conchae), and the **spheno-ethmoid recess** which is roofed over by the ethmoid, bounded behind by the body of the sphenoid and limited below by the superior turbinate.

The **nasolacrimal duct** drains into the inferior meatus [p. 307].

The **frontal and anterior ethmoid** air cells have a common opening and drain into the middle meatus. The **maxillary sinus** also drains here. The superior and spheno-ethmoid recess drain other **ethmoid air cells** and the often large **sphenoid sinuses** also drain into the recess.

The sphenopalatine foramen is easy to see through the posterior nares high up in the posterosuperior aspect of the lateral wall. Pass a bent wire through the foramen into the pterygopalatine fossa. It transmits from the maxillary nerve branches which supply sensation and secretomotor fibres to the medial and lateral walls of the nasal cavity.

Note [p. 308] that drainage of the maxillary sinus is not assisted by gravity. It depends on the presence of ciliated epithelium which propels mucous towards the opening or openings into the middle meatus. If the action of the cilia is damaged say by infection then mucopus will collect in the sinus.

Pterygopalatine fossa

The pterygopalatine fossa [FIG. 17.13] can be seen in the intact skull looked at from the side. It is a narrow cleft between the pterygoid plates behind and the maxilla in front. It is a few millimetres wide superiorly but inferiorly the anterior and posterior walls meet each other and fuse. The gap between the pterygoid plates and the maxilla, by which the fossa can be entered is the **pterygomaxillary fissure.**

The depth of the fossa is a thin plate of bone in the sagittal plane, which is the perpendicular plate of the palatine bone (the horizontal part of the palatine bone forms the posterior part of the hard palate). The superior part of this plate has in it a foramen 2–3 mm in diameter which is roofed over by part of the body of the sphenoid,

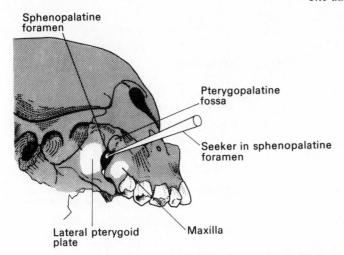

Sphenopalatine
foramen

Pterygopalatine
fossa

Seeker in sphenopalatine
foramen

Lateral pterygoid   Maxilla
plate

FIG. 17.13. The pterygopalatine fossa and the sphenopaltine foramen.

and is the **sphenopalatine foramen.**† Pass a seeker through this foramen, and then examine the nasal cavity through the posterior nares. The position of the foramen is evident, and the perpendicular plate of the palatine bone is obviously the partition separating the nasal cavity from the fossa. Confirm this by shining a light into the fossa.

The medial part of the anterior wall is covered by an extension from the palatine bone, and is called the pterygo*palatine* fossa for this reason.

Look at the skull from in front. At the back of the orbit the superior orbital fissure leads into the middle cranial fossa directly, but the inferior orbital fissure leads into the pterygopalatine fossa. Note also the position of the infra-orbital groove in the floor of the posterior part of the orbit. This groove contains the maxillary division of the trigeminal nerve (V_2) [FIG. 18.02, p. 348].

The posterior wall of the fossa is most easily understood with the help of a disarticulated sphenoid [FIG. 17.05] The wings are so named, and the word rostrum [FIG. 17.05] means a beak. The pterygoid plates fuse together over a width of about 1 cm, and form the posterior wall. Note the position of the **foramen rotundum** which leads directly from the middle cranial fossa and transmits the maxillary division of V. Medial to the foramen rotundum, above the medial pterygoid plate and often nearly as large is the opening of the pterygoid canal. A wire pushed backwards into the canal emerges half way up the anterior margin of the foramen lacerum. In the intact skull this emergence is hidden by a small projecting ridge of bone.

The most important content is the maxillary division of V. This enters via the foramen rotundum and leaves by the inferior orbital fissure and infra orbital groove, on its way to the infra orbital canal and foramen. The foramen rotundum and the infra orbital groove are not in line, and it is obvious from the skull that the nerve must side-step laterally as it passes through the pterygopalatine fossa.

Dangling from the maxillary nerve in the fossa is the **pterygopalatine ganglion.** The function of this ganglion is just like that of the ciliary otic and submandibular ganglia—that is it is a parasympathetic relay station for fibres which are secretomotor to the lacrimal gland and to the mucous membrane of the nasal cavity and paranasal sinuses [p. 308]. The preganglionic fibres all come

† The old terms *sphenopalatine ganglion* and *sphenopalatine fossa* have been changed to pterygopalatine ganglion and fossa. The sphenopalatine foramen however retains its old name (see p. 277).

from the facial nerve (VII) via the corda tympani and supply the opening of the auditory tube. Their distribution is like that of other nerves which supply the nasal cavity only they are further back. Branches also pass backwards through the pterygoid canal and finally reach the dura. The main content of the canal however is the nerve of the pterygoid canal.

At its commencement posteriorly in the foramen lacerum, the parasympathetic fibres (preganglionic) from the facial nerve (VII) mingle with sympathetic fibres (postganglionic) which formed part of the plexus around the internal carotid artery. The parasympathetic fibres in the nerve of the pterygoid canal relay in the pterygopalatine ganglion. All the autonomic and other fibres then mingle with each other, and the nerves already described which reach the mouth, nasal cavity, and nasopharynx are mixed nerves.

Note that although the lacrimal gland lies above and lateral to the eye under the upper eye lid, in the territory of V_1 it receives its secretomotor supply via the zygomatic branch of V_2.

As soon as the maxillary nerve (V_2) enters the fossa it gives off branches which leave:

1. **Anteriorly** via the inferior orbital fissure. This is the continuation of the nerve itself now called the **infra-orbital nerve**.
2. **Medially** via the sphenopalatine foramen. One nerve is distributed to the lateral wall of the nasal cavity, the posterior superior lateral nasal nerve; the other passes across the roof of the nasal cavity, and then downwards and forwards in the nasal septum to the incisive foramen [FIG. 17.01], through which it passes to be distributed to the incisor teeth and hard palate. This is the **nasopalatine nerve.**
3. **Downwards** to the palate. Find the greater and lesser palatine foramina [FIG. 17.01]. A seeker pushed from below will lie in the pterygopalatine fossa. The greater and lesser palatine nerves pass downwards in these canals to both the hard and soft palates.
4. **Posteriorly** the nerve gives off branches which run to the roof of the nasopharynx.

The **maxillary artery** enters the fossa through the pterygomaxillary fissure, and breaks up into branches which accompany every one of the nerves which we have mentioned. The large size of the pterygoid canal is due to those blood vessels rather than the nerves.

Temporal bone

Look at the **temporal bone**. Notice the position of the **mastoid** and **styloid processes** and of the **stylomastoid foramen** that lies between them transmitting the facial nerve. On the medial side of the styloid process you will notice two foramina—the larger is the **jugular foramen** which we have already seen from above, and the other in front is the beginning of the **carotid canal**. The carotid canal transmits the internal carotid artery and the plexus of sympathetic nerves which surround it. Having entered the bone at the beginning of the carotid canal, the artery bends through a right-angle and then runs forwards towards the **foramen lacerum** which it enters from behind. It then turns upwards and enters the **cavernous sinus** and it is actually washed on its outer adventitial surface by the blood lying in the sinus. The artery is covered by a layer of intima on its outer surface which prevents the cavernous sinus blood from clotting. The artery bends again in the so-called *carotid siphon* so that its course through the base of the skull and the cavernous sinus is surprisingly tortuous [p. 254].

The upper attachment of the fibres of the carotid sheath will be to the base of the skull in such a way as to include the beginning of both the carotid canal and the jugular foramen.

Look at the **external acoustic meatus** and notice the size and shape of the tympanic plate which forms the floor and anterior and posterior parts of the wall of the bony part of it. You may be able to see the foramen ovale and foramen rotundum, the promentory, and sometimes the pyramid. Pass a bent wire through the auditory tube into the middle ear.

It is remarkable how much you can see if you look into the middle ear in the adult skull with an instrument such as an ophthalmoscope.

Go back to the **medial pterygoid plate**. In many skulls the upper part of the medial pterygoid plate has a shallow notch near the base of the skull. In this notch the cartilaginous part of the auditory tube opens and if you look immediately behind and slightly lateral to the posterior edge of the medial pterygoid plate you will see the groove in the base of the skull formed by the adjacent parts of the greater wing of the sphenoid and petrous part of the temporal bone. This groove is called the scaphoid fossa in which the cartilaginous part of the auditory tube lies.

THE MANDIBLE

Closing and opening the lower jaw

Consider the **mandible**. Put your finger on the head of your own mandible which lies immediately in front of the tragus of the ear. When you open your mouth wide you will find a large depression just in front of the tragus and you can actually feel the head of the mandible moving forwards as you do so. Put your finger now on the angle of the jaw and open your mouth once more. You will notice that the angle of the jaw moves backwards while the head of the mandible is moving forwards. It is obvious that there must be a point approximately half-way between the head and angle of the jaw where the mandible moves neither backwards nor forwards and it must be in this region that we find the axis around which the mandible apparently moves [FIG. 15.19, p. 303].

Look at the inner surface of the ramus of the mandible and find the **mandibular foramen**. It lies immediately behind the **lingula** of the mandible. The mandibular foramen transmits the inferior alveolar vessels and nerves and it is just at this point where the nerve enters the bone that the axis of the movement of the mandible is to be found. It is very fortunate that this is so as it means that the jaw can be opened without pulling on the inferior alveolar nerve producing presumably the most terrible tooth-ache.

If you look at the articular surface of the temporal bone you will notice that it is concave behind and convex in front. It would obviously be impossible for anything as rigid as the head of the mandible to fit both a concave and convex surface. This problem of the articulation of the head of the mandible with the concavoconvex surface of the temporal bone is solved by the presence of the flexible interarticular disc which adapts its shape appropriately whether the mouth is open or shut. The head of the mandible articulates with the same part of the inferior surface of the inter-articular disc but no matter what the position of the jaw may be the flexible disc itself can adapt its shape appropriately to the concave or convex parts of the articular surface of the temporal bone [FIG. 15.17, p. 302].

Identify the position of the **masseter muscle** in yourself. The fibres run downwards and backwards from the zygomatic arch towards the angle of the jaw. The muscle contracts strongly whenever the teeth are clenched. It follows from the somewhat oblique direction of most of the fibres of the masseter that it could be active in protruding the jaw.

Feel the side of your head above the zygomatic arch and you will

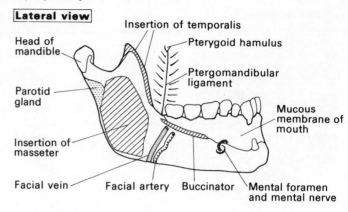

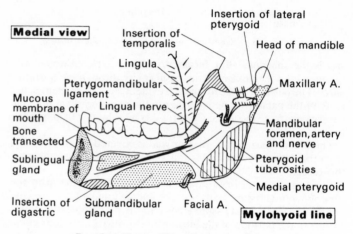

FIG. 17.14. The mandible and related structures.

find the fan-like belly of the **temporalis muscle** which again contracts every time you clench your jaw.

If you look at the angle of the mandible on its superficial surface you will see quite clearly the region where the masseter is inserted. Now if you look at the deep surface of the angle of the jaw you will see another very similar area marked by the so-called **pterygoid tuberosities** which are scallop-like ridges running downwards and backwards. These are produced by the insertion of the **medial pterygoid muscle**.

Now the medial pterygoid muscle takes origin from the medial surface of the lateral pterygoid plate. You will remember that the lateral pterygoid plate is simply an ossification of the membrane which would normally separate the medial and lateral pterygoid muscles from each other. If you look at the skull and mandible from the side, it will be seen that the direction of the fibres of the medial pterygoid are the same as those of the masseter. The only difference between them is that the masseter is inserted into the angle of the jaw on its superficial aspect and the medial pterygoid is inserted into the angle of the jaw on its deep aspect. If you push your thumb upwards and forwards and laterally just in front of the most prominent part of the angle of the jaw and press on the deep surface, and clench your teeth two or three times, you feel the contraction of the medial pterygoid muscle very easily. It coincides exactly with the contractions of the masseter.

From what you can demonstrate in yourself, the medial pterygoid muscle is principally concerned with clenching the jaw just like the masseter and temporalis.

What about opening the jaw? There are two ways in which this could be done. Either the chin could be pulled downwards or the head of the mandible could be pulled forwards. The body uses both methods.

A muscle attached to the chin will have very considerable leverage because of the distance which separates the chin from the axis of the movement of the mandible. The **digastric**, although it is rather small, produces a considerable effect for this reason.

You will notice that when the mouth is open the head of the mandible moves forwards more or less in the horizontal plane. It follows that any muscle which is attached anteriorly to the head or neck of the mandible and whose fibres run horizontally will be ideally situated to produce this forward movement. The **lateral pterygoid** muscle satisfies all the criteria just mentioned. Take the skull and the mandible and put them together. Think of the origin of the lateral pterygoid from the lateral side of the lateral pterygoid plate and from the adjacent part of the base of the skull in the infratemporal fossa and realize that the fibres do run horizontally, parallel with the base of the skull, and are inserted into the neck of the mandible and, as it says in the books, into the interarticular disc. You cannot feel this muscle in yourself, but it will be seen that it is in an ideal position to pull the head and neck of the mandible forwards.

Morphologically the interarticular disc is developed from the tendon of the lateral pterygoid muscle. They are a single structure. It is an ideal arrangement to have the principal mouth-opening muscle attached to both the neck of the mandible and to the interarticular disc. It cannot be attached to the head of the mandible itself because the head of the mandible is entirely enclosed inside the temperomandibular joint.

The two medial pterygoid muscles shut the mouth and the two lateral pterygoid muscles open it. If you contract the two ipsilateral pterygoid muscles, that is to say the medial and lateral pterygoid muscles on the same side, one of which is capable of opening your mouth and the other of shutting it, the chin swings to the opposite side. If the two pterygoids on the left contract then the chin will move to the right. If the pterygoids on the right contract the chin will move to the left. If all four pterygoid muscles contract simultaneously the lateral movements will cancel each other out and the jaw will be protracted so that the chin will stick out to the front. If you open your mouth wide you will find that you can no longer protrude your jaw because the jaw is already fully protracted.

The four pterygoid muscles between them can produce nearly all the movements of which the jaw is capable, including the most important one, and the one requiring the greatest degree of force, the clenching of the jaw. In this movement the medial pterygoid muscle has the assistance of the **masseter** and of the **temporalis**. Jaw clenching is very much more powerful than any of the other movements. The only movement which the pterygoids are not able to produce is the surprisingly powerful movement of retraction of the mandible. This is produced by the most posterior horizontal fibres of the temporalis muscle and also the digastric.

As has been seen in Chapter 15 these muscles are known as the muscles of mastication and are supplied by the mandibular division of the fifth cranial nerve (V_3), except for the posterior belly of the digastric which is supplied by the facial nerve (VII).

The **mylohyoid line** on the inner surface of the mandible [Fig. 17.14] is of the greatest importance in understanding the relations of the mandible. The mylohyoid muscle forms the diaphragm of the floor of the mouth and the mylohyoid line is the attachment of this muscle to the mandible. Anything which lies above this line is in the mouth—anything which is below this line is in the neck.

Look at the inner surface of the mandible just above the mylohyoid line in relationship with the premolar teeth and you will see a shallow depression called the **sublingual fovea**. This is produced by the sublingual salivary gland. In the floor of the mouth of most people the sublingual gland actually projects into the surface of the mouth just behind the incisor teeth and you can feel it with the tip of your tongue. Secretion of the sublingual gland is rich in mucus and it opens by means of a large number of separate ducts.

Below the middle of the mylohyoid line is another slightly depressed area on the bone called the **submandibular fovea**. You can identify the position of the submandibular gland in yourself very easily as it projects slightly beneath the skin just in front of the angle of the jaw. Also you can see in yourself the openings of the two submandibular ducts which lie on each side of the midline in front of the two sublingual glands.

The submandibular gland lies mainly in the neck below the diaphragm of the floor of the mouth. The direction of the superficial cervical part of the submandibular gland is backwards and hooks round the posterior free border of the mylohyoid muscle and then runs horizontally forwards inside the mouth to its destination near the midline.

Notice the relationship of the mandible to the mucous membrane inside the mouth and to the mucous membrane which lines the vestibule of the mouth. It follows that any fracture of the mandible is likely to tear through this mucous membrane so that bacteria inside the mouth have direct access to the bone.

The third salivary gland, the **parotid**, is intimately related to the jaw and, as its name tells you, also of course to the ear. The first thing to do is to identify its duct in yourself. This opens opposite the neck of the second upper molar tooth. Being opposite the neck, rather than the crown of the tooth, it is impossible to bite the end. The main gland lies between the ramus of the mandible and the anterior surface of the sternomastoid muscle and it extends deep down into the neck for a surprisingly great distance. It goes in front of the styloid process and may come in contact with the lateral recess of the pharynx.

Note the position of the **pterygomandibular ligament** which will run from the pterygoid hamulus downwards to be attached to the mandible just behind the last molar tooth. To the back of this ligament the superior constrictor muscle is attached and it spreads out in its origin a little bit on to the medial pterygoid plate and hamulus and below it will be attached to the mandible in close proximity to the attachment of the ligament itself.

To the front of this ligament the **buccinator muscle** is attached. This is the muscle whose fibres run forwards and which is attached not only to the ligament but also to the upper jaw just outside the line of the upper teeth and to the lower jaw just outside the line of the lower teeth. Its fibres will run forwards from the musculature of the cheek and contribute substantially to the fibres of the orbiculoris oris, the sphincter muscle which surrounds the lips. It is called the buccinator muscle because it is employed by people playing the bugle. It is supplied by the facial nerve (and not by V_3).

Just in front of the insertion of the masseter muscle identify the position of the facial artery. Make sure you can identify it quickly in yourself. Clenching your teeth will enable you to locate the anterior border of masseter.

Look at the medial surface of the ramus of the mandible and visualize the downward course of the inferior alveolar nerve with its accompanying artery which will supply the teeth of the lower jaw and eventually become the mental nerve at the mental foramen. Lying immediately in front of the nerve is the lingual nerve which supplies the anterior two-thirds of the tongue with its ordinary sensation from the trigeminal nerve and the special sense of taste from the facial nerve fibres which run in the chorda tympani. This nerve must run just above the mylohyoid line. It must also run below the attachment of the fibres of the superior constrictor to the mandible. There is a very narrow gap through which this nerve can run. Because of the close proximity of the inferior alveolar nerve and the lingual nerve to each other, it often happens when a dentist anaesthetizes the inferior alveolar nerve that he also paralyses the sensory nerve supplying the anterior two-thirds of the tongue. Both nerves are intimately related to the mandible and may be damaged following fractures of the mandible.

The **head of the mandible** is ossified in hyaline cartilage and not in membrane. It is an extremely interesting structure. You will remember that in an ordinary long bone the epiphysis, which lies on top of the diaphysis, consists entirely of cartilage to begin with. When ossification begins in the middle of the epiphysis the superficial part of the epiphysial cartilage becomes the articular cartilage whereas the deep part, which separates it from the shaft of the bone, becomes the epiphysial cartilage and is responsible for growth in length. The point about the cartilage of the head of the mandible is that the articular cartilage and epiphysial cartilage are one and the same. That part of it which participates in the temporomandibular joint is like articular cartilage, whereas the lower part of the same cartilage looks exactly like epiphysial cartilage during the time of growth. A disease of the temporomandibular joint which also destroys the articular cartilage at the same time destroys the growth cartilage and this can produce gross asymmetry of the face. Any kind of injury to the head of the growing mandible is of very serious consequence.

Futher information concerning the temporomandibular joint will be found on page 302.

18 The eye and ear

THE ORBIT

The bony cavity containing the eye, the orbit, has the shape of a square-based pyramid with its apex pointing backwards [FIG. 18.01]. The medial wall of both orbits are parallel to each other. The lateral walls are roughly at right angles to each other. The roof of the orbit is formed by the frontal bone and the lesser wing of the sphenoid. The lateral wall is formed by the greater wing of the sphenoid and the zygomatic bone. The floor is the orbital process of the maxilla bone and the orbital process of the palatine bone. The medial wall is the frontal process of the maxilla, the ethmoid, and the body of the sphenoid.

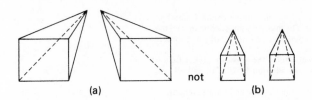

(a)　　　　not　　　　(b)

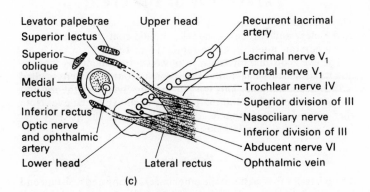

Levator palpebrae
Superior lectus
Superior oblique
Medial rectus
Inferior rectus
Optic nerve and ophthalmic artery
Lower head
Upper head
Lateral rectus
(c)

Recurrent lacrimal artery
Lacrimal nerve V₁
Frontal nerve V₁
Trochlear nerve IV
Superior division of III
Nasociliary nerve
Inferior division of III
Abducent nerve VI
Ophthalmic vein

FIG. 18.01. The orbit.
(*Upper*) The orbit is a square-based pyramid but with medial walls of both orbits parallel to one another as in (a) and not diverging as in (b).
(*Lower*) Arrangement of structures passing through superior orbital fissure on the left side.

The maxillary antrum [p. 308] lies below the floor of the orbit.
The orbital margin at the front protects the eye. It has the frontal bone above, the frontal and zygomatic bones laterally, the zygomatic and maxilla bones below and the frontal process of the maxilla bone medially.

ORBITAL FISSURES AND FORAMINA

1. Optic foramen.
2. Superior orbital fissure.
3. Inferior orbital fissure.
4. Posterior ethmoidal foramen.
5. Anterior ethmoidal foramen.

The ophthalmic artery and optic nerve enter the orbit together through the optic foramen. Many of the other important structures enter or leave the orbit through the superior orbital (sphenoidal) fissure.

Optic foramen

The optic foramen, through which the optic nerve and ophthalmic artery reach the eye, lies at the apex of the orbit. The optic foramen is found beneath the lesser wing of the sphenoid at the anterior termination of the optic groove. The **optic groove** (or chiasmatic groove) is the groove on the upper surface of the sphenoid bone for the optic chiasma anterior to the pituitary fossa.

Superior orbital fissure

The superior orbital fissure (or sphenoidal fissure) lies between the greater and lesser wings of the sphenoid. The two heads of the lateral rectus muscle to the eye divides the fissure into two compartments. The arrangement of structures passing through the superior orbital fissure [FIG. 18.01] from lateral to medial is as follows:
　Lateral compartment—lateral to the upper head of lateral rectus:
1. Recurrent lacrimal artery.
2. Lacrimal nerve (V_1).
3. Frontal nerve (V_1).
4. Trochlear nerve (IV).
　Medial compartment—between the two heads of lateral rectus:
1. Superior division of oculomotor nerve (III).
2. Nasociliary nerve (V_1).
3. Inferior division of oculomotor nerve (III).
4. Abducent nerve (VI).
5. Ophthalmic vein.

Inferior orbital fissure

The inferior orbital fissure or the sphenomaxillary fissure lies at the junction between the lateral wall and the floor of the orbit. Infraorbital blood vessels and ascending nerve branches from the pterygopalatine ganglion pass through this fissure.

Posterior ethmoidal foramen

The foramen lies on the medial wall of the orbit. The posterior ethmoidal artery and nerve pass through it to reach the nasal cavity [FIG. 18.02].

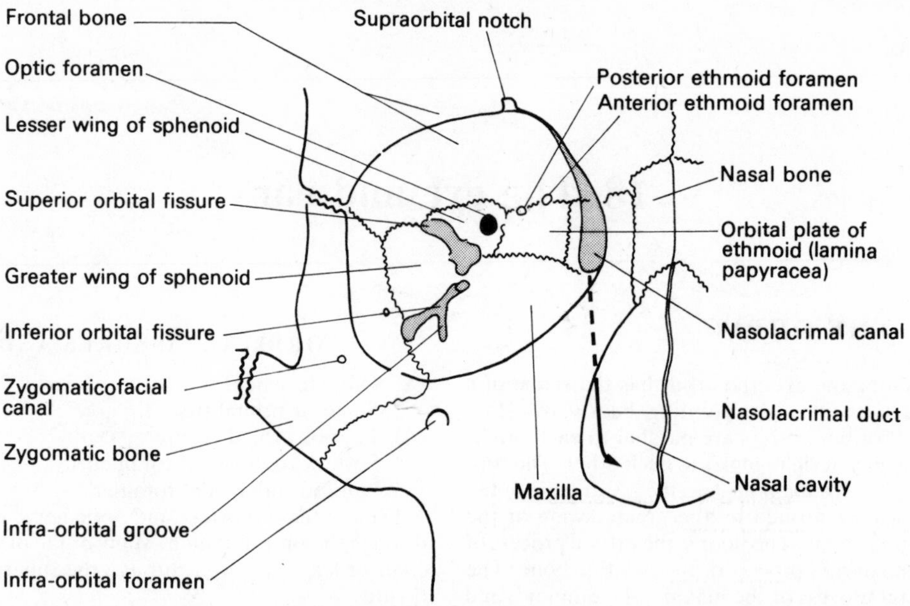

FIG. 18.02. Right orbit from the front.

The socket containing the eye is called the orbit. The bones forming the margins of the orbit are the frontal, the zygomatic, and the maxilla. The lacrimal bone, the lamina papyracea of the sphenoid, and both wings of the sphenoid also contribute to the depths of the orbit.

Note that the orbital plate of the frontal bone separates the cavity from the anterior cranial fossa while the greater wing of the sphenoid separates it from the middle fossa. The maxillary sinus lies below and the ethmoid air cells and nasal cavity lie medially. The orbit may become involved secondarily in disorders involving any of these neighbouring regions.

The optic foramen transmits the optic nerve and ophthalmic artery, the superior orbital fissure transmits the lacrimal, frontal, and nasociliary branches of V_1, in addition to III, IV, and VI and ophthalmic veins on their way to the cavernous sinus.

The inferior orbital fissure transmits V_2 which runs in the infra-orbital groove and canal emerging on the surface of the maxilla at the infra-orbital foramen.

Anterior ethmoidal foramen

The foramen lies on the medial wall of the orbit as a canal between the ethmoid and frontal bones anterior to the posterior ethmoidal foramen.

The nasociliary nerve (branch of ophthalmic division of the trigeminal nerve) and the anterior ethmoidal artery and vein pass through it to the nasal cavity [FIG. 18.02].

THE EYEBALL

The globe of the eye is spherical in shape with a diameter of approximately 2.5 cm. It does not lie centrally in the orbit, but is closer to the roof and the lateral wall. The front surface of the eye projects forwards to a plane joining the superior and inferior orbital margins which therefore protect the eye from injury from a frontal blow. Laterally the eye is not so well protected since about one-third of the globe lies in front of a line joining the lateral and medial orbital margins. On the medial side, however, the eye is protected by the nose. This leaves the lateral side as the least protected part.

On the other hand the field of view is greatest laterally. It is possible to detect movement taking place at the sides behind the head. Try this yourself. Look straight ahead and move a pencil at arm's length farther and farther round to the side. You will find that your lateral visual field extends through an angle of about 110°, that is 20° more than the plane running laterally through the cornea.

STRUCTURE OF THE EYE

The eye is usually considered to have three coats, a fibrous outer coat (sclera and cornea), a vascular middle coat (the uveal tract), and a nervous inner coat (the retina) [FIG. 18.03].

The eye is divided internally into two compartments by the lens and its suspensory ligaments. The space in front of the lens is filled with **aqueous humour**. This space is further divided by the iris into the **anterior chamber** and the **posterior chamber** of the eye. Note that the posterior chamber lies in front of the lens. The space behind the lens is filled with **vitreous humour**.

SCLERA

The posterior five-sixths of the outer coat is strong and resistant and is termed the **sclera** (Greek = hard). It is white in colour and opaque. The sclera extends as far forward as the anterior chamber of the eye where it joins the cornea at the corneo scleral junction.

The weakest point of the sclera is the optic disc where the optic nerve enters. This is termed the **lamina cribrosa** since the membrane has perforations here for the fibres of the optic nerve.

When the intra-ocular pressure is raised in **glaucoma** the optic disc is pushed backwards. It is then said to be cupped. In childhood, where the sclera is less strong, the entire eye may swell in this condition and in addition the cornea may protrude forwards. The condition is termed **hydrophthalmos** or **buphthalmos** since it then resembles an ox's eye (*bous* (Greek) = ox; *ophthalmos* (Greek) = eye).

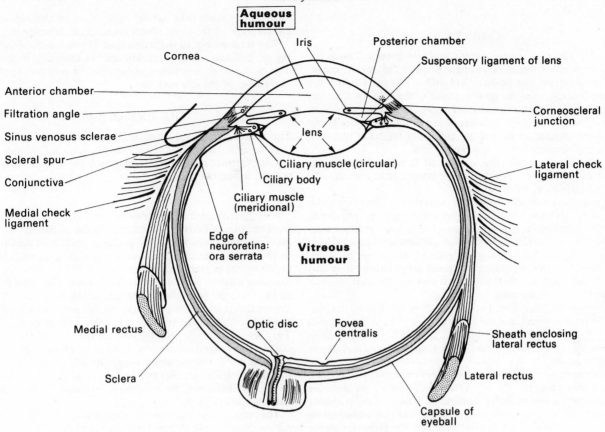

Fig. 18.03. The eyeball.

The diagram is a horizontal section through the right eyeball seen from above.

The optic axis passes in front through the middle of the cornea and strikes the fovea centralis behind. The optic nerve passes through the lamina cribrosa of the sclera and its head is called the optic disc which lies two disc widths on the nasal side of the fovea.

The vitreous humour is a clear jelly-like structure. It contains a delicate meshwork of fibres but no cells. If, after injury some escapes, it cannot be replaced. Its presence is essential to keep the neuroretina in contact with the pigmented layer and choroid.

Note that the lens is by nature more or less spherical in shape. Its actual shape depends on the constant pull of the suspensory ligaments which are attached all round the equator of the lens and to the ciliary bodies. The tension in the ligaments depends on the intra-ocular pressure, but adjustments are possible. Thus when the ciliary muscle contracts the circular muscle acts like a sphincter and reduces the diameter of the circle to which the rim of the suspensory ligament is attached. In addition the meridional fibres, taking origin from the so-called scleral spur also pull the ciliary bodies towards the lens and slacken off the suspensory ligament. The lens is thus able to become more spherical, the eye focusing for near vision.

When the ciliary muscle relaxes the intra-ocular tension reasserts itself and the lens is once more pulled back into the shape for distant vision. The reduction in focusing power (from about 16D to < 1D) is due to progressive reduction in the elasticity of the lens.

The ciliary bodies are remarkable also in that they secrete aqueous humour. The dynamic equilibrium established by the secretion and absorption of aqueous is the main factor in determining intra-ocular pressure. The fluid is secreted into the posterior chamber and then escapes by passing between the iris and lens into the anterior chamber. It finds its way into the sinus venosus sclerae through the spongy tissue in the cleft between iris and cornea called the filtration angle. It is important that this should be 'open' as in this diagram; a narrow angle impedes the absorption of aqueous.

The blood supply of the eyeball reaches it mainly from the long and short posterior ciliary arteries which pierce the sclera close to the optic nerve. The short vessels break up immediately into a dense capillary plexus supplying especially the macula lutea which is not supplied by the central artery of the retina. The long posterior ciliaries do not participate in this plexus but pass forwards to the sclerocorneal junction region and supply the ciliary bodies, and iris and the smooth muscle in this region. The eye also receives auxillary blood vessels which run along the extrinsic muscle and pierce the sclera close to their insertions. All these vessels anastomose in the region of the ciliary bodies. The retina gets its principal blood supply from the central artery of the retina.

Note that the copious blood supply in the anterior part of the uveal tract probably makes possible the survival of the lens which has no blood supply whatever of its own. The same is true of the cornea.

The fibrous capsule of the eyeball encloses it completely except in front. It is reflected away from the eye ball along each of the extrinsic muscles as a surprisingly tough sleeve of fascia. The lateral extensions near the insertions of the medial and lateral recti are attached to the side walls of the orbit, and this means that the socket in which the eye fits is itself anchored in position.

CORNEA

The anterior one-sixth of the outer coat is the **cornea**. This is the transparent window which allows light to enter the eye. Its curvature is greater than that of the sclera and as a result the two are separated by a slight furrow on the outside of the anterior chamber of the eye. This curvature is of great importance since, as will be seen later, most of the lens focusing power of the eye is due to the fact that the cornea has air in front and water (aqueous humour) behind and therefore acts as a powerful lens.

The curvature is usually the same in all directions, that is, the front surface is spherical. Any refraction errors can be corrected by the use of additional spherical lenses.

If the curvature is greater in one direction than another (say more curved vertically than horizontally) the cornea will be cylindrical. As a result, although it may still be possible to focus clearly either on vertical lines or on horizontal lines, it will not be possible to do so simultaneously as when viewing the letter E. As a result print will appear blurred. This condition is termed **astigmatism**. It is corrected by the use of a cylindrical lens to make up the deficiency of refraction in one plane only.

The cornea is a fixed lens. In addition there is a lens in the eye. Although this is not so powerful as the cornea, since it has aqueous humour in front and vitreous humour behind both with similar refractive indices as the lens itself, it is a variable lens and allows focusing to occur.

The **cornea** consists of a transparent layer of fibrous tissue termed the **substantia propria**. This is covered in front by the **anterior elastic membrane** and **stratified epithelium**. Behind is the **posterior elastic membrane** (Descemet's membrane) and the **endothelium** which lines the whole of the anterior chamber. The posterior elastic membrane may become involved in inflammation of the ciliary body (keratitis punctata or descemetitis) in leprosy and syphilis. White blood cells from the ciliary body then appear at the back of Descemet's membrane.

Blood supply to cornea

Nil. The cornea has no blood supply. The cells receive their nutrients from the aqueous humour behind and tears in front. They probably absorb oxygen from the air. Vascularization of the cornea leads to blindness.

The absence of a blood supply may account for the fact that corneal grafting may be carried out without the problem of rejection. Presumably the absence of a blood supply prevents the lymphocyte reaching the tissue and producing antibodies.

Nerve supply of cornea

The ciliary and conjunctival branches of the ophthalmic division of the trigeminal nerve (V_1) supply the cornea, which is very sensitive to touch. Without this sensory innervation, the cornea will be injured by foreign particles leading to corneal ulceration. The sclera is not nearly so sensitive.

Tenon's capsule

The capsule of the eyeball (Tenon's capsule) is a fibrous membrane which covers the posterior two-thirds of the eye. It is continuous behind with the optic nerve sheath. The capsule forms an articular socket which allows the eye to move in any direction. The extra-ocular muscles pass through the capsule which encloses each muscle in a tubular sheath. The action of the medial and lateral recti muscles are limited by lateral expansions of this capsule to the lateral walls of the orbit which act as check ligaments.

The relationship of the eyeball to Tenon's capsule is that of a perfect ball and socket joint. The *joint* is lubricated by tissue fluid. The centre of the eyeball remains fixed in position whatever the movement of the eye may be.

THE UVEAL TRACT

The middle pigmented vascular coat or uveal tract consists of the iris, the ciliary body, and the choroid behind. Inflammation of this region is termed **uveitis**. Inflammation of the iris and ciliary body is termed **iridocyclitis**.

Iris

The iris is a coloured circular disc suspended in the aqueous humour from the ciliary body. It is attached to the middle of the anterior surface of the ciliary body. The posterior surface of the iris tends to lie touching the lens but the aqueous humour can pass between the two from the posterior chamber where it is formed to the anterior chamber where it is absorbed. In iris bombé, where the iris is stuck to the lens, the iris bulges forwards at the periphery due to the accumulation of aqueous humour in the posterior chamber.

The aperture in the iris through which light passes is the pupil. Two sets of smooth muscles control the size of the pupil. The circularly arranged muscle fibres form the **sphincter pupillae** which close the pupil. The radially arranged muscle fibres form the **dilator pupillae** which open the pupil. The pupil is constricted by the parasympathetic nervous system. It is dilated by the sympathetic nervous system [p. 9].

The colour of the iris is determined by the number of melanophores (cells with melanin granules) in the stroma of the iris.

The iris consists of loose vascular connective tissue the stroma which contains the sphincter pupillae muscle fibres, nerves, pigment cells and blood vessels. It is covered in front by a thin anterior membrane and a flattened anterior epithelium. Behind, the stroma is covered by the posterior membrane which contains the dilator pupillae muscle fibres.

The anterior membrane is continuous with the posterior elastic membrane (Descemet's membrane) of the cornea.

Ciliary body

The ciliary body runs like an anulus around the eye. It is triangular in cross-section with the shortest side of the triangle pointing forwards. As has been seen, the iris is attached to the middle of the anterior surface of the ciliary body. The lens is attached by the suspensory ligament to the inner side of the triangle of the ciliary process.

Ciliary muscle

The outer part of the ciliary body constitutes the ciliary muscle. The inner fibres of the ciliary muscle run circularly round the eye. The outer fibres run backwards. Constriction of both sets of fibres reduces the tension in the suspensory ligaments and allows the lens to become more powerful so that near objects come into focus. They are supplied by the third cranial parasympathetic nerve fibres.

Ciliary process and aqueous humour

The aqueous humour (intra-ocular fluid) is formed by the ciliary process on the inner surface of the ciliary body in front of the attachment of the suspensory ligaments. The aqueous humour passes from the posterior chamber, where it is formed, through the pupil to the anterior chamber where it is absorbed at the **sclerocorneal junction** (filtration angle) into the **venous sinus of the sclera** which is better known as the canal of Schlemm. This is a circular space running round the iridial angle lined with endothelium which communicates with the anterior ciliary veins and with Fontana's spaces. These are a series of small spaces in the spongy peripheral processes of the iris at the angle of the anterior chamber. They are formed by the interlacing connective tissue fibres. They communicate on their inner side with the anterior chamber of the eye and on their outer side with the canal of Schlemm.

The circulation of aqueous humour is of critical importance to the eyesight. If the pressure within the eye falls (due to puncture of the eye) the retina is no longer held in position and may be detached. If the pressure rises the circulation of blood in the capillaries of the retina is reduced and blindness ensues. The eye in glaucoma feels much harder than normal and is painful.

RETINA

The innermost layer of the eye is the light receptive layer, the retina. It consists of nine layers from outside to inwards [FIG. 18.04]:

1. Pigment layer.
2. Neuro-epithelial layer with rods and cones.
3. Outer limiting membrane.
4. Outer nuclear layer.
5. Outer granular (reticular) layer.
6. Inner nuclear layer.
7. Inner granular (reticular) layer.
8. Ganglionic layer.
9. Nerve fibre layer.

The retina, which is derived embryologically from the optic cup, can be considered to be an extension of the brain joined to the main part by the optic nerve. The retina has three layers of nerve cells or neurones. These are:

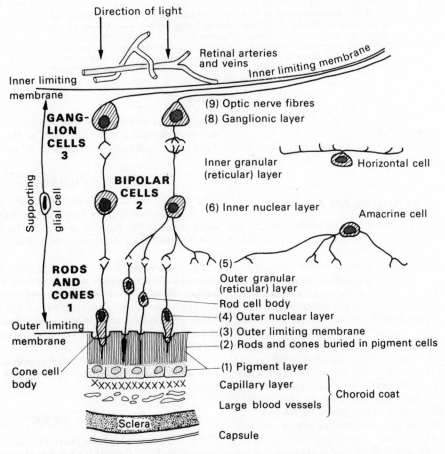

FIG. 18.04. Diagram showing the principal layers of the retina. The numbering corresponds to that in the text.

The three neurone sequence is shown. The neurone sequence to the right suggests that the idea of a chain of three neurones is an oversimplification, the bipolar cells as well as the rods and cones have complex multiple connections. In addition there are other cells, amacrine and horizontal cells which add another dimension of complexity to the retina which is essentially part of the brain.

The retina is supported by neuroglia like the brain. Most, but not all of it is enclosed within the inner and outer limiting membranes. They are held together by supporting glial cells, usually called Müller's cells.

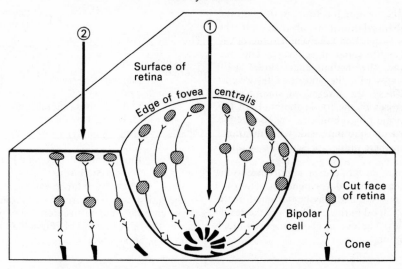

Fig. 18.05. This diagram of the central pit (fovea centralis) makes the point that light can fall on the light sensitive rods and cones, without having first to traverse the other more superficial layer. The size of a circular pit is limited by the fact that the cones (and rods) must remain in contact with the pigment retina. A larger pit is impossible, but a groove, with cross section similar to the profile of the fovea is made use of in many vertebrates. Light path 1 falls directly on the rods and cones. Light path 2 is subject to blurring as it passes through the full thickness of the retina.

1. Rods and cones.
2. Bipolar cells.
3. Ganglion cells.

The retinal blood vessels run with the nerve fibres which are usually unmyelinated.

RETINAL BLOOD SUPPLY

The arteria centralis retinae is a branch of the ophthalmic artery. It enters the substance of the optic nerve and appears at the optic disc. It divides into superior and inferior branches.

The retinal veins form the central vein of the retina which drains into the superior ophthalmic vein. Obstruction of these veins gives rise to **papilloedema** (oedema of the papilla of the eye or of the disc). When the oedema spreads to the macula colour vision is lost and the patient lives in a grey twilight world, and may become completely blind.

When viewed with an ophthalmoscope (p. 358) the back of the eye or fundus appears red. The most distinctive structure is the optic disc. This is a small circular area 1.5 mm in diameter which marks the site of entry of the optic nerve. The fundus appears pink in colour due to the numerous capillaries on its surface. The depression in the optic disc is termed the physiological cup. The ophthalmic artery and vein can be seen climbing out of this cup. Rather surprisingly, the veins pulsate. The arteries do not.

As has been seen the whole disc is depressed in **glaucoma** giving a glaucomic cup. On the other hand, in **papilloedema** the physiological cup may be replaced by a raised area.

The **macula**, or area of most distinct vision, is a small circular area two disc-widths lateral to the optic disc. It has no obvious blood vessels close to it. In the centre is a small pit the **fovea centralis** (*fovea* (Latin) = small pit) where cones predominate [Fig. 18.05]. It is in line with the visual axis. Hence very faint objects will only be seen by averting the gaze so that the light falls on to the more lateral but more sensitive rods. This may be the origin of ghosts since a faint object will seem to disappear when you look straight at it.

CHOROID

The choroid is the pigmented vascular layer or tunic of the eye which lies between the sclera and the retina. It extends from the optic nerve at the back of the eye to the **ora serrata** which is the serrated margin of the sensory part of the retina behind the ciliary body.

The choroid has four layers. The inner layer is the basal membrane (of Bruch) which is structureless. It is also known as the **lamina vitrea** (lamina of the vitreous humour). The inner blood vessel layer has small choriocapillaries. The outer blood vessel layer has large blood vessels. On the outer side is a lymph space which is in continuity with the extradural lymph space surrounding the optic nerve (lymph space of Schwalbe).

Four veins, having a whorled appearance (venae vorticosae), pass through the sclera from the choroid behind the equator of the eye.

LENS

The lens is convex with the posterior surface more curved than the anterior surface. It consists of layers of transparent elastic material enclosed in a capsule and suspended from the ciliary body immediately behind the pupil. The suspensory ligaments run from the circumference of the lens to the ciliary process of the ciliary body behind the zone for the production of aqueous humour.

The lens has a variable thickness, 'Thick for near vision, thin for distant vision'. It is the difference in the refractive index of the aqueous humour in front of the lens, the lens itself, and the vitreous humour behind the lens that allows the image of the object viewed to be brought sharply into focus on the retina.

The lens embryologically is derived from ectoderm in a vesicle and is laid down layer by layer with the oldest part in the centre. If the lens becomes opaque the condition is termed a cataract. The lens is removed, usually leaving the capsule, and the loss of refractive power is overcome by wearing +10D spectacles. If the capsule were also removed +16D spectacles would be needed.

The most frequent cause of a cataract is old age (senile cataract) but a cataract may be congenital (defect of the ectoderm whilst the lens was being laid down as in German measles). It may be caused by trauma or by exposure to intense radiation (ionizing radiation, X-rays, and even infrared heat rays as in glassblower's cataract).

In the embryo an artery (artery to the lens) runs through the vitreous humour to the lens. This becomes obliterated, changes into the transparent hyaloid canal, and virtually disappears. It may then act as a lymphatic canal for the vitreous humour. If a premature baby is exposed to too high an oxygen tension in an incubator, fibroblasts instead of disappearing may survive in the vitreous humour and lay down fibrous tissue. This retrolenticular (behind the lens) fibroplasia leads to blindness. An epidemic of blindness occurred in premature infants following the introduction of incubators with high oxygen tension.

EYELIDS

The eyelids (*palpebrae* (Latin) = eyelids) protect the eyes. When the eyes are open, the upper eyelids still cover the upper part of the cornea. Retraction of the upper lid above this point gives a scared staring expression seen typically in exophthalmic goitre. In old age the eyelids may not close completely during sleep and this may lead to corneal ulceration.

The space between the eyelids is termed the **palpebral fissure**. The junctions of the two eyelids are termed the palpebral **angles** of the eye. The lateral angle (lateral canthus) is more acute and the lids are in contact with the eyeball. The medial angle (medial canthus) is more rounded and contains the **plica semilunaris** (the remains of the third eyelid), the lacrimal papillae and puncta for the drainage of tears and the **lacrimal caruncle**. This is a small reddish mass between the two lacrimal papillae. Identify the inferior lacrimal punctum in your own eye.

In certain races the lacrimal caruncle and medial canthus are covered by a fold of skin termed the **epicanthus**.

Structure of the eyelid
Although when considering the eyelid, we are normally thinking of the upper eyelid, the lower eyelid has a similar structure. However, the upper eyelid is the more moveable of the two and is in more direct communication with the scalp [FIG. 18.06].

From within outwards the eyelid has six layers:
1. Conjuctiva.
2. Tarsus (tarsal plate).
3. Areolar layer.
4. Striated muscle.
5. Subcutaneous tissue.
6. Skin.

Tarsal plates
The framework of the eyelid is the tarsal plate (tarsus) formed of dense connective tissue. (Note that the Greek word tarsus also refers to the ankle area, as for example in the *tarsal* bones.) The

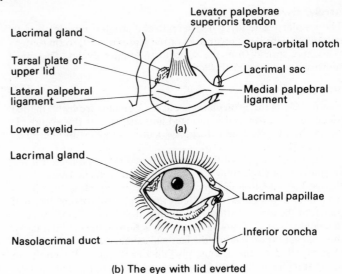

FIG. 18.06. The eyelids.
(a) Upper and lower eyelids.
(b) Tears formed by the lacrimal gland flow across the eye and drain into the nasolacrimal duct.
(c) Structure of the upper eyelid.

tarsal plates are in continuity with the palpebral fascia so that when the eyes are shut they are protected by the tarsal plates and the palpebral fascia.

The tarsal plates are attached to the zygomatic and lacrimal bones by the medial and lateral tarsal ligaments.

Glands
The **tarsal glands** or Meibomian glands (named after Meibom who first described them) open at the lid margin. Their secretions ensure the airtight closure of the lids. A swelling of the eyelid due to the retained secretions of these glands is termed a **chalazion**. They produce a lump beneath the skin about the size of a small pea (3 mm).

The **ciliary glands** (or Moll's glands) are modified sweat glands which open near to the follicles of the eyelashes (*cilium*, pleural *cilia* (Latin = eyelash). In addition there are **sebaceous glands** associated with the eyelashes. These are known as Zeis's glands.

Areolar tissue layer of eyelid

The areolar tissue layer lies deep to the striated muscle. It is in continuity with the danger area of the scalp [p. 000] so that infections may spread from the scalp to the upper eyelid. The main sensory nerves run in this layer.

Muscle layer

The muscle is the **orbicularis oculi**. It originates from the medial aspect of the orbit and is inserted into the skin of the eyelids. It closes the eyelids. The orbicularis oculi is supplied by the facial nerve (VII).

The palpebral part of the orbicularis oculi is responsible for the closing of the eyes in blinking. These fibres are attached medially to the medial palpebral ligament. When the eyelids close they do so from lateral to medial so that tears are swept towards the lacrimal puncta at the inner canthus. The pulling of the muscle of the medial ligament has a squeezing and releasing action on the lacrimal sac so that tears are sucked down the lacrimal canaliculae into the sac and then expelled via the nasolacrimal duct to the nose. Infection can easily spread via this duct from the nasal cavity to the conjunctival sac.

Subcutaneous tissue layer of eyelid

This layer consists of loose subcutaneous tissue which is easily distended with oedema fluid or blood, hence swollen eyelids and black eyes.

Skin

The skin of the eyelid is the thinnest in the body.

Conjunctiva

The conjunctiva (*conjugare* (Latin) = to join) is the mucous membrane which covers the anterior part of the eye (ocular conjunctiva) and which is reflected on to the posterior surface of both eyelids (palpebral conjunctiva). It thus joins the eyeball to the lids.

The conjunctiva is formly attached to the tarsal plates, but it is only attached loosely to the sclera. The reflexions under the upper and lower lids are termed the upper and lower **fornix**. It will be noted that because of this conjunctival reflexion, a foreign object such as a small fly which has unfortunately entered between the eyelid and the eyeball, will never get around to the back of the eye!

There is a shallow groove in the upper eyelid about 2 mm from its margin. Foreign bodies are liable to be caught here. They can sometimes be removed by pulling the upper eyelids over the lower eyelid thus enabling them to be displaced by the lower eyelashes. Clinically they are removed by everting the upper eyelid over the upper tarsal plate. This exposes the inner surface of the upper eyelid [FIG. 18.06]. The tarsal plate can be turned back-to-front to expose the deep surface of the upper eyelid. This is easier if the subject has long eyelashes.

THE LACRIMAL GLAND

The lacrimal gland produces tears. The larger orbital part of the gland lies in the fossa for the lacrimal gland in the outer part of the root of the orbit. The smaller palpebral part lies above the upper fornix. It can be seen by everting the upper lid.

Tears enter the upper fornix by 8–12 lacrimal ducts. They pass over the anterior surface of the globe to the puncta lacrimalia which are small holes at the inner end of the lid margins. They pass to the canaliculi and then to the lacrymal sac which lies in a fossa anteriorly on the medial wall of the orbit. The canaliculi, rather surprisingly, run first vertically, then horizontally inwards to reach the lacrimal sac.

The fundus of the lacrimal sac extends slightly above the medial tarsal ligament. If the skin at the outer side of the eye is pulled laterally, the medial tarsal ligament can be seen. Deep to this ligament lies the sac.

The **nasolacrimal duct** is 1.5 cm long. It extends from the lacrimal sac to the anterior part of the inferior meatus below the inferior concha [FIG. 18.06].

The normal wetness of the eye depends on the secretions of the conjunctiva and of the tarsal glands. The lacrimal gland can be removed without damage to the eye. The presence of a film of grease from the tarsal glands slows down the evaporation of tears.

THE EXTRA-OCULAR MUSCLES

The extra-ocular muscles are the levator palpebral superioris, the four rectus muscles and the two oblique muscles.

Levator palpebrae superioris

This muscle runs from the roof of the orbit to the superior tarsus and the skin of the upper eyelid. Its origin is in front of the optic foramen. It is supplied by the upper division of the oculomotor nerve (III) and its function is to open the eye by raising the upper lid.

Rectus and oblique muscles

The **rectus muscles** arise from a fibrous ring which encircles the optic foramen and the inner part of the superior orbital fissure [FIG. 18.07]. These muscles are inserted into the sclera in front of the equator of the eye. The **medial rectus** turns the eye inwards. The **lateral rectus** turns the eye outwards. The **superior rectus** turns the eye upwards. The **inferior rectus** turns the eye downwards. The lateral rectus is supplied by the abducent nerve (VI). The other three recti are supplied by the oculomotor nerve (III).

The **superior oblique** is a very unusual muscle. It arises from the orbit medial to the superior rectus and runs forwards to a pulley (*trochlea* (Latin) = pulley) formed by a ligamentous ring attached to the upper margin of the orbit. The tendon passes through this pulley and changes direction to run laterally and backwards, to the eye. It is inserted into the sclera behind the equator. As a result, contraction of the muscle turns the eye to look down and out (hence sometimes referred to rather facetiously as the tramp's muscle!). It is supplied by the trochlear (IV) nerve.

The **inferior oblique** is a much simpler muscle. It runs across the anterior part of the floor of the orbit. The origin is on the medial side. The muscle also runs laterally and backwards and is inserted into the sclera behind the equator. Its action is to turn the eye so that it looks upwards and outwards. It is supplied by the oculomotor nerve III.

Rotation of the eye

In addition to turning the eye so that we can look up, down, in, and out, the eye can be rotated slightly around its equator like a wheel. When looking at a patient's right eye the superior oblique will tend to rotate the eye clockwise, whilst the inferior oblique will tend to rotate it anticlockwise.

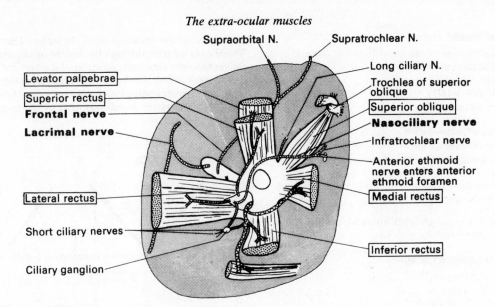

Fig. 18.07. The origins of the rectus muscles, superior and inferior oblique muscles, and levator palpebrae muscle.

Rather surprisingly the superior and inferior recti also tend to rotate the eye by their actions. The inferior rectus rotates the eye in the opposite way to the superior oblique, whilst the superior rectus rotates the eye in the opposite direction to the inferior oblique. The reason for this is as follows. When looking straight ahead the image of the object viewed falls on the fovea centralis. The optic disc where the optic nerve enters the eye lies some distance medial to

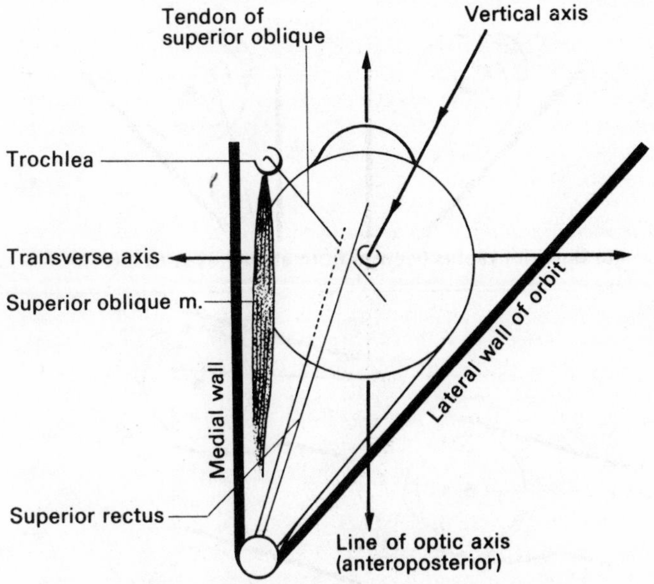

component of force in their action in a medial direction which will tend to rotate the eye. The direction is easy to remember since the two superior muscles (superior oblique and superior rectus) rotate the eye in the same direction.

SUMMARY OF EXTRA-OCULAR MUSCLES

Levator palpebrae superioris

Origin: Roof of orbit
Insertion: Skin of upper eyelid
                Superior tarsus
Nerve supply: Oculomotor (III) upper division
                Sympathetic from T1 and 2 via superior cervical
                ganglion
Action: Elevates upper lid

Fig. 18.08. Rotation of the eye by the superior oblique and the superior rectus.

Rectus oculi inferior (inferior rectus)

Origin: Lower border of the
                optic foramen
Insertion: Sclera
Nerve supply: Oculomotor (III)
Action: Turns the eye downwards

This muscle also turns the eye slightly inwards and rotates the eye in the same direction as the inferior oblique.

Rectus oculi lateralis (lateral rectus)

Origin: Lateral border of the
                optic foramen
Insertion: Sclera
Nerve supply: Abducens (VI)
Action: Turns eye laterally

this on the retina (hence the blind spot is always in the lateral field of view). Since the superior and inferior recti originate from the optic foramen above and below the optic nerve, it means that they are running to the eye at an angle from the medial side. This means that in addition to moving the eye up and down, there will be a

The muscle arises by two heads between which pass, from above downwards, the upper division of the oculomotor nerve, the nasociliary nerve, the lower division of the oculomotor nerve, the abducent nerve, and the ophthalmic veins.

Rectus oculi medialis (medial rectus)

Origin:	Medial border of the optic foramen
Insertion:	Sclera
Nerve supply:	Oculomotor (III)
Action:	Turns eye medially

Rectus oculi superior (superior rectus)

Origin:	Upper border of the optic foramen
Insertion:	Sclera
Nerve supply:	Oculomotor (III)
Action:	Turns eye upwards

This muscle also turns the eye slightly inwards and rotates the eye in the same direction as the superior oblique.

Obliquus oculi inferior (inferior oblique)

Origin:	Medial aspect of floor of orbit
Insertion:	Sclera behind equator
Nerve supply:	Oculomotor (III)
Action:	Turns eye upwards and outwards

This muscle also rotates the eye.

Obliquus oculi superior (superior oblique)

Origin:	Margin of optic foramen
Insertion:	Sclera behind equator
Nerve supply:	Trochlear nerve IV
Action:	Turns eye downwards and outwards

This muscle also rotates the eye. The muscle starts by running forwards, but its tendon changes direction to reach the eye by looping through the ligamentous ring to the upper margin of the orbit and then running laterally and backwards.

A FURTHER CONSIDERATION OF THE MOVEMENT OF THE EYE IN THE ORBIT

The eyeball lies in the orbit (or eye socket). It is protected by the overhanging orbital margin from the majority of injuries. Look at the skull and note that the *medial* walls of the orbits are parallel, that is, each lies in the parasagittal plane, whilst the *lateral* walls are about at right angles to each other, that is, each lies at an angle of about 45° to the parasagittal plane.

The optic nerve (with its surrounding meninges) and the ophthalmic artery enter via the optic foramen at the back of the orbit at the junction of the medial and lateral walls. There is a fibrous ring which surrounds the optic nerve at the optic foramen to which most of the external ocular muscles are attached.

Many people find the detailed action of these muscles difficult to understand because they do not take into account the above basic anatomical facts.

The orbit contains a considerable quantity of very soft non-fibrous fat tissue, and the eyeball is enclosed in a fibrous sac, within which it moves. This arrangement constitutes the most perfect ball and socket joint in existence in the body.

When the eye moves, in looking from side to side or up and down, the centre of the eyeball remains in exactly the same place. It is convenient to describe the movements of the eyeball as if they take place around three axes at right angles to each other [FIG. 18.09]. These axes all pass through the centre of the eyeball. The eyeball moves inwards and outwards around the vertical axis, upwards and downwards around the horizontal axis, and rotates around the optic axis. The corresponding movements of sailing boats and aeroplanes are described by the words: (i) to yaw; (ii) to pitch; (iii) to roll.

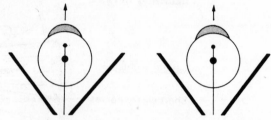

(a) Common misconception concerning action of superior rectus

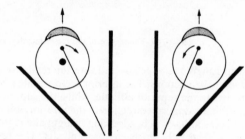

(b) Superior rectus (viewed from above) eyes straight

(c) Superior rectus (viewed from above) eyes to left

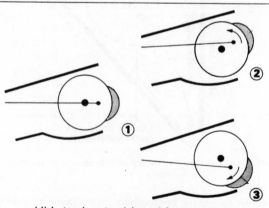

(d) Lateral rectus (viewed from side)

FIG. 18.09. The muscles acting on the eyeball produce movements of the eyeball which take place around three axes at right angles to one another producing yawing, pitching, and rolling movements.

There are six external ocular muscles (not including the levator palpebrae superioris). There are four rectus muscles: superior, inferior, medial, and lateral, and the two oblique muscles, superior and inferior.

Before you can work out what must happen when a given muscle contracts, it is essential to know:
  (i) the origin;
  (ii) the insertion of the muscle;
  (iii) the relationship of its line of pull to the axis around which the particular eyeball movement is taking place.

The lateral rectus turns the eyeball laterally, because of the relationship of the line of pull to the vertical axis around which the eyeball moves inwards or outwards. But what about movements around the other two axes? We consider only the mediolateral horizontal axis around which the eye rotates when you look upwards or downwards. To answer this question you must look at the eye from the side and note the position of the insertion and hence of the line of pull of the muscle with respect to this axis [FIG. 18.09(d)]. If the line of pull is above it or below, the muscle will turn the eye upwards or downwards as well as outwards. In fact [FIG. 18.09(d)①] starting in the standard position the insertion of both medial and lateral rectus muscles is exactly level with this axis and there is thus no tendency to turn the eye up or down.

Thus we can say that the action of the medial and lateral rectus muscles produces movement in one plane only (starting from the standard position).

The action of the other muscles is not so simple and they produce movement in three planes.

The problem in understanding the superior and inferior rectus muscles (which turn the eye ball not only upwards or downwards but also inwards) is because the position of the origin of the muscles from the fibrous ring at the apex of the orbit does not lie directly behind the eye but lies to the medial side. Beginners often have a very strong preconceived idea that the orbit seen from above looks like FIGURE 18.09(a). In such an arrangement the superior rectus would turn the eye upwards and nothing else. In fact the orbit is like FIGURE 18.09(b).

Consider the superior rectus. Obviously in the first place it must turn the eye upwards.

Consider FIGURE 18.08. The right eye is looked at from above. The superior rectus, because of the shape of the orbit, takes origin behind the medial edge of the eyeball. Notice that it is attached nearer the front than the back of the eyeball. In FIGURE 18.09(c) the right eye is already looking inwards and as a result the line of pull of the superior rectus muscle passes far over on the medial side of the vertical axis and it is bound to turn the eye further inwards, as well as turning it upwards. In FIGURE 18.09(c) the left eye is looking outwards. The relationship of the line of pull to the axis has now changed, the muscle will now turn the eye further outwards, as well as turning it upwards.

When the eye is looking straight forwards (the anatomical position) the line of pull [FIG. 81.09(b)] is just medial to the axis, so that the superior rectus is said to turn the eye upwards and inwards. In the same way the inferior rectus turns the eye downwards and inwards.

Work out the action of the superior and inferior oblique muscle by using similar arguments [FIG. 18.08].

If you wished the superior rectus to act without producing any inward movement, what position of the eyeball would you start from?

What would the action of the lateral rectus be if you first looked upwards (or downwards) and then allowed this muscle to contract?

What muscles would you use if you wanted to look straight upwards, or straight downwards starting in the anatomical position?

Note that the oblique muscles are capable of rotating the eye around the optic axis. They do this when the head rolls from side to side in such a way that the eyes remain on an even keel in spite of the head movements. They are assisted by superior and inferior recti.

When an object is moving rapidly towards you an indication of its speed and direction is obtained, not only by the activity of the extra-ocular muscle of the two eyes needed to follow the motion, but also from the rate of increase in size of the image of the object on each retina.

VISUAL PATHWAYS

RETINA

The visual pathway commences in the retina and terminates in the visual cortex of the occipital lobe of the brain. As it extends from the front to the back of the head, it is frequently involved secondarily with other disease processes inside the skull or brain.

Layers of the retina

The retina has two essentially different layers:
  (i) the neuroretina;
  (ii) the pigment retina.

These two layers are closely in apposition but they do not 'stick' together. Indeed they are joined only by a few frail connective tissue fibres, and as a result the neuroretina is rather easily torn and becomes separated from the pigment retina as the result usually of trauma such as a blow in a boxing match. A detached retina causes blindness. In such cases reattachment of the retina restores vision.

The layers of the retina are shown in FIGURE 18.04. Note that, except in the **macula lutea**, the light must pass through all the other neuroretina layers, before reaching the rods and cones. The **macula lutea** is the most important region of the retina. The arrangement [FIG. 18.05] in the **fovea centralis** means that the visual image can be focused by the media of the eye on these cones without the interposition of blood vessels or other layers of the neuroretina.

The optic nerve commences on the nasal side of the macula. This region is the most obvious feature of the **fundus of the retina** when it is examined with an ophthalmoscope. It is often called the disc or papilla. It has a pale greyish colour often with a pigmented rim, which contrasts markedly with the strong pink of the rest of the retina. The **central artery of the retina** emerges from it as two branches running superiorly and inferiorly, giving off branches as they go. The vessels stand out very obviously against the retina, the veins being much darker than the arteries. The disc has no rods or cones and consists entirely of optic nerve fibres, and is therefore 'blind'. The macula is separated by 'two disc widths' from the disc as estimated by eye. This fact as well as the manner in which the vessels approach but never reach the macula make its position easy to identify.

The absence of retinal vessels near the macula raises the question as to where it does get its blood supply from. It can only come from the choroid, and there is an especially rich capillary bed behind the macula. This is very easily seen when the fundus is examined in albino patients. In this disorder the outer layer of the retina is not pigmented so that the choroid vessels can be seen very easily.

The neuroretina [FIG. 18.04] is a complex structure containing three layers of neuronal cells, that of the rods and cones, the bipolar cell layer, and the molecular layer. The retina also contains horizontal cells and amacrine cells, which play an ancillary role in the visual process. Neuroglia cells are also found in the retina as well as the optic nerve.

In the retina the optic nerve fibres are unmyelinated and transparent. They pass through small holes in the sclera behind the optic disc and become myelinated forming the **optic nerve**. The nerve is enveloped in pia mater, arachnoid, and the inner layer of dura. The cerebrospinal fluid lies between the pia and arachnoid. The central artery of the retina (branch of the ophthalmic artery) and its companion veins pass through the theca of the optic nerve about halfway along the nerve, and then enter the nerve itself.

Raising the intracranial pressure may distend the theca here and interfere with the veins as they pierce the dura causing obstruction and oedema of the retina, which spreads from the disc peripherally. This is **papilloedema**. When it reaches the macula colour vision is lost, followed by blindness. This is one of the cardinal signs of a raised intracranial pressure. However, papilloedema does not always occur when the intracranial pressure is raised.

Ophthalmoscopy

The retina is examined with an **ophthalmoscope** [FIG. 18.10]. The whole region viewed is called 'the fundus'. The optic disc (the 'head' of the optic nerve) appears pale whitish-grey in colour often with a pigmented rim. The macula lies in the 'optic axis' on the temporal side of the disc. The central artery of the retina emerges at the centre of the disc. It branches superiorly and inferiorly giving off branches. The macula never has blood vessels in front of it.

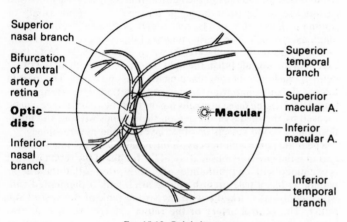

FIG. 18.10. Ophthalmoscopy.
The basic arrangements of the blood vessels of the retina as seen through an ophthalmoscope.

An examination of the circulatory system of the body is not complete without looking at the fundus. It is the only place in the body where blood vessels can actually be seen. The veins pass superficial or deep to their companion arteries. The veins pulsate whilst the arteries do not.

OPTIC TRACTS

The optic nerves pass backwards to the optic foramina and enter the cranium. They lose their dural and arachnoid coverings.

The nerves meet in the midline just in front of the pituitary region and give rise to the optic chiasma (= Greek letter chi χ; note 'ch' is hard as in Christ). Behind the chiasma the pathway is called the **optic tract**, which can be followed into the **lateral geniculate body**, a part of the thalamus. In addition a bundle can be seen entering the midbrain at the level of the **superior colliculi** (corpora quadrigemina) [FIG. 18.11].

In the lateral geniculate body the optic nerve fibres relay and project to the visual cortex as the **optic radiation** which is intimately related to the posterior horn of the lateral ventricle [p. 383].

In the optic chiasma the nerve fibres of the two optic nerves are reclassified. All fibres conveying visual information from one side of the midline pass into the optic tract of the opposite side. Thus you see to the right side with the nasal half of your right retina and the temporal half of the left and all these fibres pass to the left optic tract. These fields of the two eyes overlap to a considerable degree. The fibres from the nasal halves of both retinae cross the midline in the chiasma whilst those from the temporal halves do not.

It is essential always to be clear whether you are speaking of the visual field, or of the retina itself. A clinician maps out defects in the visual field and then works out which part of the retina or visual pathway could be affected.

In most domestic and other familiar animals the eyes are at the side of the head and the two visual fields hardly overlap. Binocular vision does not occur. In such cases the arrangement in the chiasma is that all the fibres cross over, so that the information from each eye all goes to the contralateral side of the brain. Taking the visual fields as a whole [FIG. 18.11] it is clear that what is seen to the right goes to the left side. This is still exactly what happens in man in spite of the considerable overlap of the two visual fields, because of the segregation at the chiasma of all the fibres (from whichever eye) which 'see' to the left (nasal retina L and temporal retina R) into the right

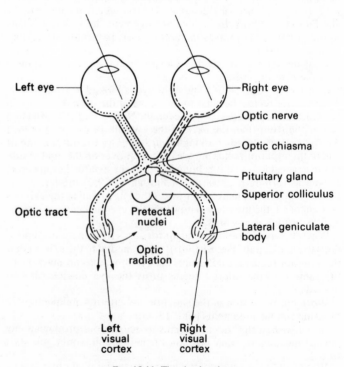

FIG. 18.11. The visual pathway.

optic tract and those which see to the right into the left optic tract. The difference is that two slightly different images are seen in binocular vision, and it is the function of the visual cortex to compare these images and to interpret the results in terms of 'distance' away from the observer.

Those fibres of the optic tract which do not go into the lateral geniculate body, pass instead into the midbrain and relay there in the pretectal nuclei on both sides [FIG. 18.11] at the level of the superior colliculi. From here fibres pass to the Edinger–Westphal nuclei of the oculomotor nerves (III). The parasympathetic efferents from these nuclei travel in the oculomotor nerves and relay in the orbit in the ciliary ganglia which lies on each side just lateral to the optic nerve a few millimetres from the optic foramen. Efferents from here pass in the short ciliary nerves to the eyeball, and supply the constrictor pupillae muscle. This is the pathway of the **light reflex**.

Note that because of the bilateral innervation of the pretectal nuclei, shining light into one eye causes pupillary constriction in both eyes. This is called the **consensual light reflex**.

Certain lesions in the tectum of the midbrain abolish the light reflex, but leave unaffected the rest of the visual pathway. The patient sees normally, and the constriction of the pupil which occurs with accommodation is also unaffected. This is the Argyll–Robertson pupil (two different people) and occurs characteristically in certain syphilitic diseases of the C.N.S.

It is possible to localize lesions of the visual pathway by studying the visual fields, the light reflex, and the convergence-accommodation reaction [FIG. 18.12].

DIAGNOSIS OF DEFECTS IN VISUAL PATHWAY

Site of lesion	Effect	Technical description of visual field defect	Likely cause	Visual field
1. Retina optic nerve	Blind in one eye		Trauma, atrophy, etc.	
2. Optic chiasma	Nasal halves both retinae affected—blind in both temporal fields	Bitemporal hemianopia— gets progressively worse. 'Tunnel' vision	Pituitary tumour	
3. Lateral edge of chiasma	Loss of temporal retina and blind in one nasal field	Nasal hemianopia	Tumour in cavernous sinus, e.g. aneurysm of internal carotid	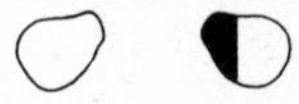
4. Optic tract	Loss of half visual field on side opposite to lesion	Bilateral homonymous hemianopia	Tumour	
5. Meyer's loop fibres stray into temporal lobe	Blind in upper half nasal field and in temporal field of other eye	Quadrantic homonymous hemianopia	Tumour, trauma to temporal lobe. Stroke	
6. Optic radiation	Blind in lower halves of nasal and temporal	Quadrantic homonymous hemianopia	Tumour, trauma to temporal lobe. Stroke	
7. Left visual cortex	Blind to the right	Bilateral homonymous hemianopia	Posterior cerebral artery thrombus	
8. Right visual cortex	Blind to the left			

FIG. 18.12.

THE EAR
EXTERNAL EAR

Auricle

The **auricle** (*auricula* (Latin) = external ear) or **pinna** (Latin = feather) is the projecting part of the external ear. It consists of fibrocartilage covered by skin which is firmly adherent [FIG. 18.13(a)].

The **tragus** is the small prominence of cartilage projecting over the external acoustic meatus. The term **tragi** (*tragos* (Greek) = goat) is used for the hairs of the external acoustic meatus. The projection of the pinna just opposite and posterior to the tragus is termed the **antitragus**. The **lobule** at the inferior aspect of the auricle has no cartilage.

Nerve supply

Auriculotemporal nerve	(Branch of mandibular nerve)
Great auricular nerve	(C2 and C3 cervical plexus)
Lesser occipital nerve	(C2 and C3 cervical plexus)

Lymphatic drainage

Drains to posterior auricular, pre-auricular, and external jugular lymph nodes.

External acoustic (auditory) meatus

The external acoustic canal or external auditory meatus (the term *meatus* refers to both the opening and the canal) extends from the concha to the tympanic membrane. It is about 2.5 cm long. The external 1.0 cm is cartilaginous which is incontinuity with the cartilage of the auricle. The internal 1.5 cm is bony.

The canal starts at the concha in an inwards and forwards direction hence a stethoscope's ear-pieces must point in this direction. The cartilaginous part runs inwards, downwards, and backwards. If the external ear is pulled upwards and backwards, the canal is straightened so that the tympanic membrane can be viewed. However, the floor of the bony canal may bulge upwards and obsure the vision.

In the child the external acoustic canal is short and the drum is closer to the surface than in the adult, as you would expect.

The external acoustic canal has the temporomandibular joint and the parotid salivary gland in front and the mastoid air cells behind.

Nerve supply of external acoustic meatus

1. Auriculotemporal nerve (V_3).
2. Auricular branch of vagus.

The auricle branch of the vagus (the alderman's nerves or nerve of Arnold) reaches the external acoustic meatus through a small canal in the petrous part of the temporal bone.

These nerves are associated with completely unexpected reflexes and referred pains. Thus molar tooth-ache may appear to be ear-ache. Wax in the ear may stimulate the vagus and cause coughing (ear-cough). Squeezing the concha slows the heart. Pressing with the finger on the inside of the canal causes vomiting. This was used following a heavy banquet so that another could be attended, hence the Alderman's nerve.

Tympanic membrane

The tympanic membrane or ear-drum lies obliquely across the far end of the external acoustic meatus with the upper and posterior edge nearer to the outside than the lower and anterior edge.

The tympanic membrane consists of fibrous tissue lined on its outer surface with stratified squamous epithelium, and on its inner surface with mucous membrane. It is pearly-grey in colour when examined with an auriscope [Fig. 18.13(c)]. The handle of the malleus, which is attached to the tympanic membrane, can be seen extending downwards and backwards. The lateral process of the malleus which is also attached to the tympanic membrane appears as a white spot. Above the malleus is a part of the tympanic membrane where the fibrous tissue is deficient. This is termed the membrana flaccida (Shrapnell's membrane).

This part of the tympanic membrane causes the notch at the upper border of the tympanic sulcus, the incisura tympanica (notch of Rivinus).

The chorda tympani nerve (which carries parasympathetic secretomotor fibres to the submandibular and sublingual salivary glands and taste fibres from the anterior two-thirds of the tongue) crosses the upper part of the tympanic membrane and the neck of the malleus. The nerve runs in the substance of the ear drum.

The long process of the incus lies behind and parallel to the handle of the malleus. It can sometimes be seen through the drum with an auriscope.

To drain the middle ear in otitis media the drum is incised in the postero-inferior quadrant.

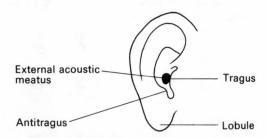

External acoustic meatus
Antitragus
Tragus
Lobule

(a) AURICLE OR PINNA

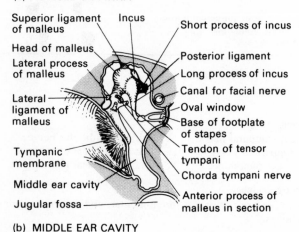

Superior ligament of malleus
Incus
Short process of incus
Head of malleus
Posterior ligament
Lateral process of malleus
Long process of incus
Canal for facial nerve
Lateral ligament of malleus
Oval window
Base of footplate of stapes
Tympanic membrane
Tendon of tensor tympani
Chorda tympani nerve
Middle ear cavity
Jugular fossa
Anterior process of malleus in section

(b) MIDDLE EAR CAVITY

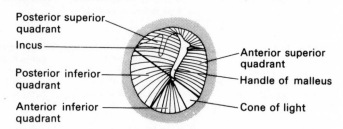

Posterior superior quadrant
Incus
Posterior inferior quadrant
Anterior inferior quadrant
Anterior superior quadrant
Handle of malleus
Cone of light

(c) TYMPANIC MEMBRANE (auriscope view)

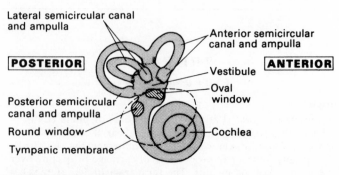

Lateral semicircular canal and ampulla
Anterior semicircular canal and ampulla
POSTERIOR
ANTERIOR
Vestibule
Oval window
Posterior semicircular canal and ampulla
Round window
Tympanic membrane
Cochlea

(d) RIGHT BONY LABYRINTH (lateral view)

Fig. 18.13. The ear.
(a) The external ear.
(b) The middle ear cavity from above.
(c) Tympanic membrane (ear drum) viewed through an auriscope.
(d) Lateral view of right bony labyrinth showing semicircular canals and cochlea.

Nerve supply to ear drum

Auricular branch of the auriculotemporal nerve and tympanic plexus branches of IX.

MIDDLE EAR

The middle ear (tympanic cavity or tympanum) is a very narrow matchbox-like cavity. Although it is about 1.25 cm long and 1.25 cm high, it is only 0.3 cm in width. Hence when puncturing the tympanic membrane care must be taken not to hit the medial wall.

Vibrations in the tympanic membrane which forms most of the lateral wall are transmitted via the three ossicles, the malleus, incus, and stapes, to the oval window on the medial wall [FIG. 18.13(b)].

The oval window (fenestra ovalis) receives the foot-plate of the stapes. Below and behind is the round window (fenestra rotunda) which by moving outwards relieves the pressure in the inner ear when the oval window moves inwards and vice versa [FIG. 18.13(d)].

The first turn of the cochlea forms a rounded elevation, the **promontory** in front of the windows.

The canal of the facial nerve (VII) lies between the roof and medial wall. It forms a slight ridge above the oval window and the promontory before turning down the posterior wall [FIG. 18.14].

The **posterior wall** has the pyramid from which the stapedius muscle emerges to run to the stapes, and the **aditus** (*ad antrum*) which leads to the epitympanic recess, the part of the middle ear above the tympanic membrane. It runs upwards, backwards, and outwards to the mastoid antrum.

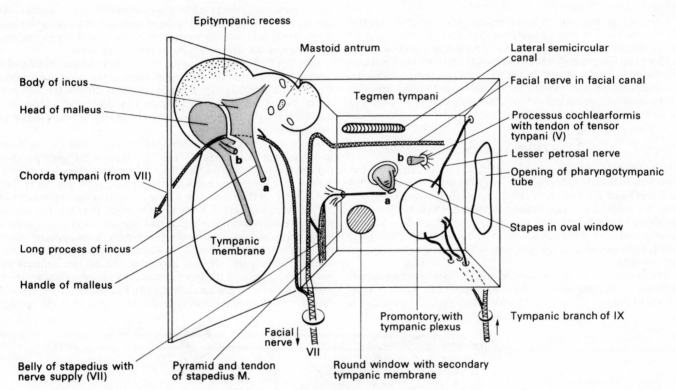

FIG. 18.14. Middle ear.

The diagram shows the left middle-ear cavity seen from the lateral side. The cavity is like a match box standing on edge. The lateral wall has been turned backwards and stands in the diagram like a door opened over 90°. The box has six sides:

1. Roof: the tegmen tympani. Thin plate of bone separating middle ear from the middle cranial fossa.
2. Floor: thin bone separating middle ear from the jugular bulb.
3. Anterior wall: largely occupied by the widening pharyngotympanic tube. The internal carotid canal is immediately in front of the anterior wall.
4. Lateral wall: largely occupied by the tympanic membrane. Above the membrane the head of the malleus and the body of the incus are housed in the epitympanic recess. Behind the recess is the tympanic antrum in which are the openings leading to the mastoid air cells.
5. Medial wall: the main features are the stapes whose footpiece is in the oval window; the round window with its secondary tympanic membrane. The promontory is produced by the first large turn of the cochlea. The tympanic plexus lies in its surface. This plexus has afferent branches from VII and IX, and gives rise to the lesser petrosal nerve.
   The processus cochlearformis encloses the tendon of the tensor tympani which turns laterally at this point heading for its insertion into the neck of the malleus (note: this tendon had to be cut in the diagram so as to open 'the door').
   The facial nerve VII runs along the medial wall and produces a longitudinal ridge. It may or may not be separated from the middle ear by a thin sliver of bone.
6. Posterior wall: VII runs down behind the posterior wall and emerges at the stylomastoid foramen. It is close to the belly of the stapedius muscle, whose tendon emerges at the apex of the pyramid passing horizontally to the stapes.
   Behind the posterior wall is the posterior cranial fossa and sigmoid sinus.

The **anterior wall** is close to the internal carotid artery. The auditory tube enters the anterior wall below the canal for the tensor tympani.

The **roof** of the middle ear or tegmen tympani (*tegmen* (Latin) = covering) is very thin. Above is the dura covering the temporal lobe of the brain. Below the **floor** of the middle ear is the bulb of the internal jugular vein.

The mucous membrane of the middle ear is closely adherent to periosteum forming a mucoperiosteum.

Mastoid antrum

The mastoid or tympanic antrum (*antron* (Greek) = cave) lies behind the external auditory canal and middle ear to which it is joined by the aditus. It has the mastoid air cells in its floor. Its posterior wall separates it from the cerebellum and the lateral or sigmoid venous sinus.

Since the mastoid antrum is in communication with the pharynx via the middle ear and auditory tube, it may become a site of infection especially in childhood. The problem of infection of the mastoid antrum is one of drainage. This may involve an incision into the mastoid air cells or mastoid antrum (mastoidotomy operation). In chronic infections which do not respond to antibiotics, removal of the mastoid antrum and air-cells (mastoidectomy) may have to be undertaken.

INTERNAL EAR

The internal ear consists of a series of tubes in the petrous temporal bone termed the labyrinth [FIG. 18.13(d)]. The **bony labyrinth** is filled with **perilymph** fluid which resembles extracellular fluid. Inside the bony labyrinth is the **membranous labyrinth** which is filled with **endolymph** fluid. Endolymph resembles intracellular fluid in composition and has potassium as its main positively charged ion. Thus endolymph contains potassium salts whilst perilymph contains sodium salts.

The labyrinth may be sub-divided into the **cochlea** which is concerned with hearing and the **vestibular apparatus** which is concerned with balance. Each has its own nerve supply. The cochlea is supplied by the cochlear division of the vestibulocochlear nerve (VIII). The cochlear ganglion is in the modiolus. The vestibular apparatus is supplied by the vestibular division of the vestibulocochlear nerve (VIII). The vestibular ganglion is in the internal acoustic meatus [FIG. 14.15, p. 285].

Cochlea

The cochlea consists of three tubes wound spirally for $2\frac{3}{4}$ turns round a central structure termed the **modiolus**. The upper tube, the scala vestibuli, and the lower tube, the scala tympani, are filled with perilymph. They communicate with one another through an aperture at the top termed the **helicotrema**. The middle tube, the scala media or cochlear duct, is filled with endolymph.

The auditory pathway

Sound waves are converted to action potentials in the cochlea. At low frequencies (20–500 Hz) there appears to be a one-to-one relationship between the frequency of the sound waves and the frequency of the nerve impulse in the eighth nerve.

Between 500–5000 Hz a group of nerves are phase linked to the sound frequency so that the overall effect is the same as for lower frequencies. For very high frequencies 5000–20 000 Hz only part of the basilar membrane is stimulated by the sound and the frequency of the sound wave is conveyed by the pattern of vibration of the basilar membrane.

The cells of origin of the nerve fibres which first convey these impulses from the cochlea (1) [FIG. 18.15] are in the spiral ganglion (2) which is within the cochlea itself. The cells are bipolar and their central axons form the acoustic part of the eighth nerve (3). The fibres relay on their own side in one of the several components of the acoustic ganglion (4). From here the main efferent pathways are to the inferior colliculi in the midbrain on both sides (7). The part of the auditory pathway which runs transversely to the opposite side is the trapezoid body (5). The fibres when they turn rostrally form the lateral lemniscus (6). Having arrived at the inferior colliculi the pathway continues in the brachium of the inferior colliculus which leads to the medial geniculate body (8). From here having relayed for the third time the pathway leads over the roof of the inferior

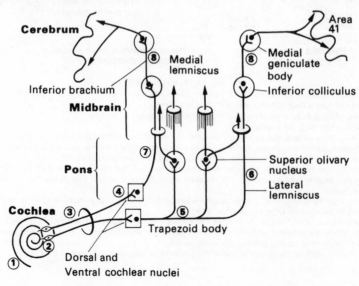

FIG. 18.15. Diagram of central connections of cochlear nerve.

horn of the lateral ventricle to reach the auditory area of the cortex which is in the middle of the surface of the superior temporal gyrus, which is mostly hidden in the depths of the lateral sulcus.

The trapezoid body and lateral lemniscus are both the site of many small scattered nuclei in which relays also take place including the superior olivary nucleus.

Note that the trapezoid body and lateral lemniscus contain fibres representing both ears. In addition the inferior colliculi are connected by commissural fibres. Both ears are represented in both auditory areas of the cortex. Damage to one side of the brainstem or cortex results in very little impairment of hearing.

VESTIBULAR APPARATUS

The vestibular apparatus consists of the saccule, the utricle, and the three **semicircular canals**. These are part of the membranous labyrinth and are filled with endolymph.

The saccule and utricle have **maculae**. These are areas of columnar cells with microscopic hair follicles. Chalky particles (otoliths) are embedded in these hairs making the nerve endings in the hair follicles sensitive to changes in gravity and linear acceleration.

The anterior, posterior, and lateral semicircular canals are arranged at right angles to one another. Each has a dilated end or **ampulla** in which is found the sensory area, the **crista**. The crista consists of hair cells which act as a barrier to the endolymph in the semicircular canal itself. The barrier is deflected when changes in angular acceleration occur.

Blood supply

Labyrinthine branch of the basilar artery. It is an 'end artery' like the central artery of the retina.

Venous drainage to superior and inferior petrosal sinuses.

19 The central nervous system

The central nervous system consists of the brain and spinal cord. The brain lies in the skull and the spinal cord in the vertebral canal down to (usually) the disc between L1 and L2.

The brain is divided into three parts which may each be further subdivided:

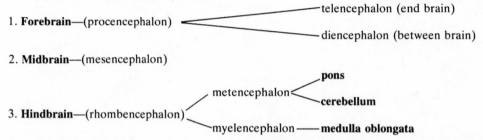

1. **Forebrain**—(procencephalon) ——————— telencephalon (end brain)
                                          diencephalon (between brain)

2. **Midbrain**—(mesencephalon)

                        metencephalon ——— **pons**
                                           cerebellum
3. **Hindbrain**—(rhombencephalon)
                        myelencephalon ——— **medulla oblongata**

The terms in common use are in bold type.

In the early stages of embryonic development the whole central nervous system is formed by two flat medullary plates, one on each side of the midline. The lateral borders of the medullary plates curl up and fuse with each other forming the **neural tube** [FIG. 19.01]. Failure of medullary tube closure results in anencephaly or spina bifida.

The neural tube soon develops three vesicular swellings, the primordia of the fore-, mid-, and hindbrain. The rostral† part of the hindbrain from the beginning has a thin rhomboid shaped roof (the future roof of the fourth ventricle) whence the name 'rhombencephalon'; its caudal part is indistinguishable from the spinal cord and is the myelencephalon.

From the telencephalic part of the forebrain a hollow outgrowth appears on each side of the midline. They give rise to the two **cerebral hemispheres** [FIG. 19.02]. From close to the rostral edge of the roof of the rhombencephalon another swelling appears which will become the **cerebellum**. From these simple beginnings the whole of the brain and spinal cord develops.

The mature central nervous system (CNS) is said to contain fourteen thousand million neurones (14×10^9). But the sophisticated functioning of the human brain is achieved not only by the number of neurones, but also by the complexity of their connections. Certain neurones are said to have a million 'connections'.

The CNS is lined on its inner surface by a specialized epithelium, the **ependyma**, and on its outer surface by the **pia mater**, a delicate membrane in which many blood vessels ramify before they enter

† Rostrum = beak: it is common practice to use the words 'cranial' (towards the head) and 'caudal' (towards the tail). When describing the brain itself 'cranial' is no longer a useful word. We use 'rostral' instead which means towards the beak or nose.

the brain. The pia is intimately related to the brain in a manner like the intimate relationship of periosteum to bone.

The cavity inside each cerebral hemisphere is the **lateral ventricle**. The cavity of the original forebrain (diencephalic cavity) becomes the **third ventricle**. The regions where the cerebral hemispheres first ballooned out are the **interventricular foramina**.

The **aqueduct** of the midbrain communicates above with the third ventricle and below with the cavity of the upper part of the hindbrain (rhombencephalon) which becomes the **fourth ventricle**. In the lower part (myelencephalon) the cavity narrows and leads to the **central canal of the spinal cord**.

Note that there are no first or second ventricles.

At birth all the neurone bodies of the adult CNS are already present. No further mitosis occurs in these cells. During childhood the increase in the size of the brain and spinal cord is brought about by the fibres which grow out from these cells as well as by the development of their myelin sheaths and the growth of connective tissue cells.

The connective tissue of the nervous system is called **neuroglia** (*glia* = glue). There are only three types of neuroglial cells in the CNS:

1. **Oligodendrocytes** (*Oligo* = few; *dendron* = process; *cyte* = cell). The oligodendrocytes are responsible for producing the myelin sheaths of fibres in the CNS.

2. **Astrocytes** (*Astro* = star-like, i.e. with many processes). Probably nourish many cells: they have flattened 'end-feet' on their processes which are closely applied to the outer surfaces of capillaries. These may have a supporting role in maintaining the 'blood–brain' barrier of the capillary endothelium. These cells also fill in the spaces left behind by damaged neurones.

3. **Microglial cells** are connective tissue cells whose origin is uncertain. They gain access to the CNS from outside along

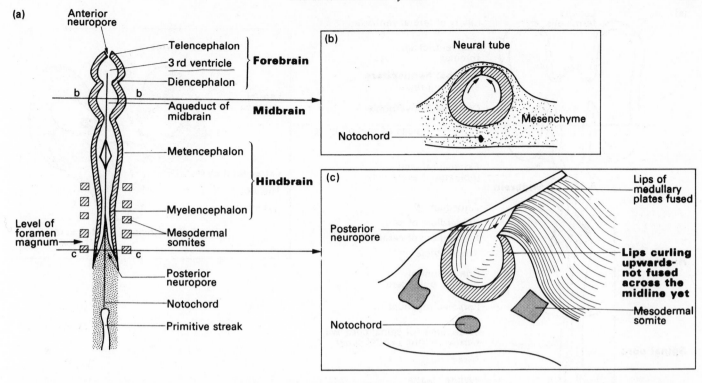

Fig. 19.01.

(a) Dorsal view of central nervous system at the stage when the medullary folds have united with each other forming the primordia of the brain. The openings at each end of the nerve tube are the anterior and posterior neuropores which unite and close off by the end of the fourth week of development.

The hindbrain joins the spinal cord (C1 segment) about the level of the fifth pair of mesodermal somites.

(b) Cross-section at level 'b' to show fusion of medullary plates completing neural tube at this level.

(c) Shows medullary plates fusing at level of the first cervical segment (somite 5). This process extends caudally until the spinal cord is completed.

with the invasion of blood vessels. Like phagocytes, they are mobile and engulf foreign material and debris.

DEVELOPMENT AND MATURATION OF NEURONES

The early cell bodies in the central nervous system have no fibres projecting from them. Similar neuronal cell bodies without fibres appear outside the CNS in the position of the spinal (posterior root) ganglia, cranial nerve ganglia and the sympathetic and parasympathetic ganglia, which form major parts of the peripheral nervous system. Most of them migrate from a special region at the lateral edges of the medullary plates, the neural crest.

Axons and dendrites grow out from their parent cell bodies getting longer and longer, like toothpaste squeezed out of a toothpaste tube. Bundles of fibres with similar origins and destinations are called **tracts**. Once the tracts have been established, the individual fibres are insulated functionally from each other by the development of the **myelin sheath**. Thus corticospinal (pyramidal) tract axons [p. 376] are present at birth, but it is not for another 12–18 months that their myelin sheaths develop and this tract becomes mature. Note that in normal infancy there is a Babinski response, as in an adult patient who has a pyramidal tract lesion following a stroke.

The myelin sheath is produced by Schwann cells in the periphery. They first of all enclose the fibres individually and then rotate around them leaving behind two layers of cell membrane. The ultimate thickness of the myelin sheath depends on the number of Schwann cell revolutions.

Make a model of this by wrapping a stocking (not a bandage) around your forearm. The toe of the stocking starts the deepest layer. Each circuit leaves two layers behind it. The Schwann cell body would be near the mouth of the sock.

Each Schwann cell surrounds only a short length (1 mm) of the fibre. There is a gap between one Schwann cell and the next where the fibres are exposed for a short length. This is the **node of Ranvier**. If a fibre branches this is where it will do so [FIG. 1.06, p. 4].

In the postganglionic sympathetic and parasympathetic fibres, the Schwann cells envelop but do not rotate around the fibres. These fibres are said to be 'unmyelinated'. The appearance of preganglionic and postganglionic sympathetic nerves connecting the spinal nerves to the sympathetic trunk, look different even to the naked eye. The former are the '**white rami communicantes**' (myelinated) and the latter '**grey rami communicantes**' (non-myelinated).

In the CNS there are no Schwann cells. Myelinization is the job of the oligodendrocytes. The mechanism is not clearly understood. One oligodendrocyte usually myelinates several fibres.

Myelin sheath development in both brain and spinal cord does not take place region by region, but tract by tract. This has been helpful in working out the anatomy of some nerve tracts.

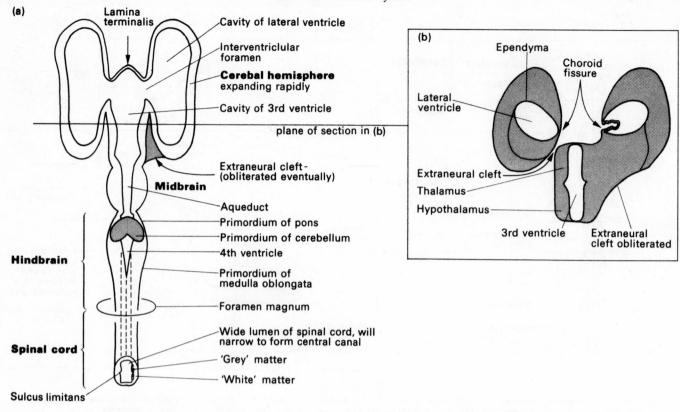

FIG. 19.02.

(a) The cerebral hemispheres are expanding rapidly. The right side of the figure is at a later stage than the left showing obliteration of the extraneural cleft. The ependymal lining secretes cerebrospinal fluid which is essential for the balloon-like expansion. The cerebellum develops from the rostral border of the 4th ventricle.

(b) Cross-section of the brain at level shown in (a). The obliteration of the cleft is shown on the right side. This allows the thalamus which originates in the diencephalon and the corpus striatum (giving rise to the caudate nucleus and lentiform nucleus) which originates in the cerebral hemisphere to lie in juxtaposition.

Note also the choroid fissure where ependyma and pia are in contact. The membranes together give rise to the choroid plexus.

There are important disorders called **demyelinating diseases**, in which the myelin sheaths degenerate. In the peripheral nerves this produces anaesthesia of skin and paralysis of muscle. In the CNS the commonest demyelinating disease is multiple sclerosis.

Tumours do not develop from mature neurones of the CNS because mitosis does not occur in them. But neurone tumours do occur in young children. They have their origins in immature embryonic (i.e. dividing) cells. Mitosis occurs however in the neuroglia cells of the adult brain and cerebral tumours in adults usually have their origin in astrocytes (astrocytoma) or oligodendrocytes (oligodendrocytoma) (in composite words '-*oma*' = tumour of).

LAYOUT OF WHITE AND GREY MATTER IN THE CNS

A cross-section of the brain or spinal cord during development shows two main layers. The superficial layer consists of nerve fibres and is '**white matter**' while the deep layer contains nerve cell bodies (as well as fibres) and is the '**grey matter**'. The names come from the naked eye appearances of corresponding areas in the adult CNS.

This pattern survives permanently in the spinal cord, medulla, and midbrain [FIG. 19.03]. But in the cerebral hemispheres grey matter is also found on the surface to a depth of 2–4 mm. This is the **cerebral cortex**. Nerve cells migrate from the deep layer and populate the surface which has six distinct layers of cells with intervening layers of white matter. Cells also migrate and colonize the **cerebellar cortex**.

In the cerebrum and cerebellum the cortical grey matter is in addition to, not instead of, the 'central' grey matter. The central grey matter is represented in the cerebrum by the **basal ganglia** and other ganglia and in the cerebellum by the **dentate nucleus** and its associated nuclei.

The grey matter of the neural tube is divided by the sulcus limitans into an alar laminar and a basal laminar. Sensory nerves develop from the alar laminar. Motor nerves develop from the basal laminar [FIG. 19.03].

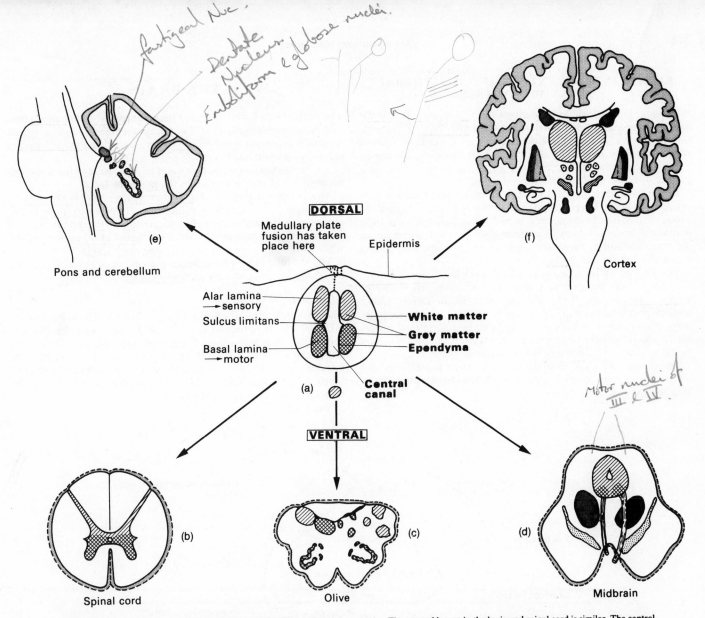

Handwritten annotations: Fastigeal Nuc. · Dentate Nucleus Emboliform & globose nuclei · Motor nuclei of III & IV.

DORSAL

Medullary plate fusion has taken place here — Epidermis

(e) Pons and cerebellum

Alar lamina → sensory
Sulcus limitans
Basal lamina → motor

White matter
Grey matter
Ependyma

(a)

Central canal

(f) Cortex

VENTRAL

(b) Spinal cord

(c) Olive

(d) Midbrain

FIG. 19.03. Diagram to show development of brain and spinal cord from neural tube. The general layout in the brain and spinal cord is similar. The central canal is always lined with ependyma, the surface with the pia mater. The neuronal tissue has white matter (fibres only) superficially and grey matter (fibres plus cell bodies) deep. The arrangement in the cerebral hemisphere is similar. The thalamus, hypothalamus, caudate and lentiform nuclei are, or have been during development, next to the ependyma. The grey matter of the cortex develops secondarily, as an extra layer, both in the cerebrum and cerebellum, and is a sign of their sophistication compared with the brainstem and spinal cord.

(a) Cross-section of the neural tube early in development. The grey matter is divided by the sulcus limitans into the alar laminar (shaded) and the basal laminar (cross-hatched). The same shading is used throughout the diagram.

(b) Cross-section of the thoracic spinal cord. The lateral column is motor in function and is derived from the basal lamina.

(c) Cross-section of the hindbrain at level of olive. The floor of the 4th ventricle represents the side walls of the early neural tube shown in (a). Note the sulcus limitans. The alar and basal laminae form only a small proportion of the whole medulla. This is because a great many fibres are passing through the medulla to or from the higher centres, and these occupy a considerable proportion of the cross-section area.

The grey matter derived from the basal and alar laminae does not form a continuous mass as in the spinal cord. Instead it breaks up into isolated nuclei which supply the cranial nerves. Not only is the grey matter discontinuous when looked at in cross-section but it also breaks up when looked at longitudinally. Thus the nuclei of the IIIrd, IVth, VIth, and XIIth cranial nerves which once formed a continuous longitudinal series, break up into four separate motor nuclei. Similar fragmentation takes place between other sets of cranial nerve nuclei.

(d) Cross-section of the midbrain. The grey matter surrounds the central canal (aqueduct). Its most anterior cells form the motor nuclei of III and IV. Behind them are the afferent cells of the mesencephalic nucleus of V.

(e) Sagittal section through the cerebellum to show the central grey matter represented by the dentate nuclues, the emboliform and globose nuclei (which in lower forms are a single mass) and the fastigial which lies close to the fastigium (Latin = roof tree) of the 4th ventricle.

(f) Coronal section through the cerebrum to show the central grey matter, which lies close to the lateral and third ventricles. It is also broken up into discontinuous masses. The main constituents are the thalamus, caudate nucleus and lentiform nucleus. Also included are the red nucleus and substantia nigra which both reach upwards from the midbrain into the subthalamic region. The subthalamic and hypothalamic regions contain nuclear masses of their own. The thalamus is derived from the alar lamina and is sensory in function, while the subthalamic and hypothalamic regions are derived from the basal lamina, and are basically motor and effector in their actions.

Handwritten note: Thalamus is sensory in function – from alar lam. Subthalamic & hypothalamic – motor – from basal lamina.

REPAIR IN THE CENTRAL AND PERIPHERAL NERVOUS SYSTEM

When neuronal cell bodies are damaged no replacement is possible because **neuronal cell division does not occur after birth**. Regeneration of nerve fibres does not occur either, because the damaged region is rapidly invaded by astrocytes which block any attempt by axons to bridge the gap. Destruction of any part of the CNS including the optic nerve is therefore permanent.

In the peripheral nervous system, however, regeneration may occur after **nerve fibre** section. Below the transected region the axons degenerate because they have been separated from their parent cell bodies but new fibres from above grow downward into the paths vacated by the now missing axons. The new **axoplasm** is synthesized in the cell bodies.

The degree of final recovery in the peripheral nerve system depends on the motor and sensory fibres finding their way to appropriate endings. If, by chance, many motor fibres replace sensory fibres, and sensory fibres are channelled into motor pathways, then recovery will be minimal. In certain peripheral nerves such as the radial nerve, recovery can be excellent but in others such as the sciatic nerve, recovery is very poor.

The changes which occur in the whole neurone including the cell body after nerve section is called **Wallerian degeneration** (after Waller who described them).

THE BRAIN

The brain [FIG. 19.04] consists of the **cerebrum**, the **midbrain, pons**, and **medulla oblongata** as well as the **cerebellum** which lies behind the brainstem and is connected to it on each side by three separate **cerebellar peduncles** [FIG. 19.38].

The **cerebrum** fills the **anterior** and **middle fossae** of the skull; the **hindbrain** and **cerebellum** lie in the **posterior fossa**.

The cerebrum contains the higher centres and is the seat of consciousness. The cerebellum the second largest part of the brain, is a lower centre and is normally concerned with the monitoring and co-ordination of movements usually initiated in other parts of the CNS. It functions entirely at an unconscious level.

The medulla oblongata contains the vital centres which regulate breathing, heart-beat, blood pressure, etc.

Grey matter of cerebrum

The two cerebral hemispheres [FIG. 19.05] are separated from each other by the longitudinal cerebral fissure in which lies the falx cerebri. In each hemisphere the convex superolateral surface fits under the vault of the skull; the medial surface is flat, and the inferior surface lies on top of the floor of the anterior fossa (frontal lobe), middle fossa (temporal lobe) and on the upper surface of the tentorium cerebelli (occipital lobe).

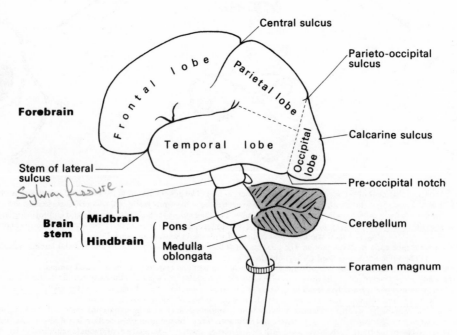

FIG. 19.04. Diagram showing the general layout of the major components of the brain: lateral view.

The brain consists of the cerebrum, the midbrain, the pons, the cerebellum, and the medulla oblongata. The cerebrum, because of its great size, is subdivided into four 'lobes'. The central and lateral sulci are made use of as shown, but in addition two arbitrary lines are also employed to demarcate the lobes. One runs from the pre-occipital notch to the parieto-occipital sulcus, the second is at right angles as shown, and joins the posterior extremity of the lateral sulcus.

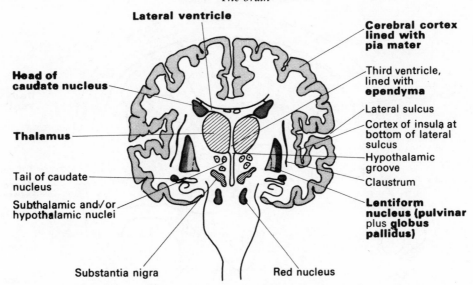

Head of
caudate nucleus

Lateral ventricle

Cerebral cortex
lined with
pia mater

Thalamus

Third ventricle,
lined with
ependyma

Lateral sulcus

Cortex of insula at
bottom of lateral
sulcus

Hypothalamic
groove

Tail of caudate
nucleus

Claustrum

Subthalamic and/or
hypothalamic nuclei

Lentiform
nucleus (pulvinar
plus globus
pallidus)

Substantia nigra

Red nucleus

FIG. 19.05. Coronal section through the cerebrum and brainstem to show the general relationships of the deep masses of grey matter.

The surface of both cerebral hemispheres instead of being smooth as in the embryo, consists of convex folds or **gyri** separated from each other by deep grooves or **sulci**. Unless the brain is pathologically shrunken it is not possible to see into the sulci without pulling the gyri apart. By means of these convolutions, the surface area of the cortex is increased about three times. The surface area is equivalent to about 45×50 cm².

The superolateral surface of each hemisphere is divided arbitrarily into the following so-called lobes:

1. Frontal.
2. Parietal.
3. Temporal.
4. Occipital.

The names are suggested by, but by no means coincide closely with, the principal overlying skull bones. The boundaries between the 'lobes' are shown in FIGURE 19.04.

The **central sulcus** (fissure of Rolando) separates the frontal and parietal lobes. It commences on the medial surface, and then runs downwards and forwards at about 70° to the sagittal plane. Its surface markings are as follows: take a point in the sagittal plane 1 cm behind the midpoint between nasion and inion [FIG. 19.06]. Place your hand, with the index finger in the midline, and extend your thumb on the top of the patient's head. The angle which the thumb makes under these conditons is about 70°, and its tip will just about reach the lower end of the sulcus [FIG. 19.07].

The parieto-occipital and calcarine sulci lie mainly on the medial surface of the brain and only appear in lateral view for a short distance.

The lateral sulcus (fissure of Sylvius) is fundamentally different from all the other sulci. While they are produced by the folding of the previously flat brain surface, the lateral sulcus results from the growth of the hemisphere as a whole. It is very helpful to understand what happens. The anatomy of the lateral ventricle is unintelligible unless the basic facts of development are known.

At first the cerebral hemispheres, distended by their lateral ventricles are shaped as in FIGURE 19.08(a). Note that the ventricle is eccentrically placed so that pia and ependyma are in apposition along a straight line from just behind the interventricular foramen almost to the posterior extremity (point Z) of the hemisphere [FIG. 19.08(b)]. This line of apposition is the choroidal fissure [FIG. 19.02(b)].

But the hemisphere then grows in a *curved* fashion so that the point Z [FIG. 19.08(c)] comes to be the tip of the temporal not the occipital lobe. Inevitably the lateral ventricle and the choroid fissure follow the same curved course. That part of the lateral ventricle, which lies in the temporal lobe, is the **inferior horn** of the lateral ventricle. The remainder is **the body of the lateral ventricle**.

The **anterior and posterior horns** are unimportant in understanding the basic plan of the lateral ventricle. They are simply extensions of the body of the ventricle drawn out by the frontal and occipital lobes when they develop later. Notice in FIGURE 19.08(d) that the lateral position of both the inferior horn and temporal lobe with respect to the body of the ventricle shows that the 'bend' must be a spiral or a helix. Note [FIG. 19.05] that the thalamus lies in the wall of the diencephalon, while the corpus striatum, which becomes both the caudate and lentiform nuclei, develops within the cerebral hemisphere. These structures, come together because of the obliteration of the extraneural cleft. They are major components of the central grey matter of the cerebrum [FIG. 19.02].

Up to the time when the cleft has disappeared, the only place where there is physical continuity between one hemisphere and the other, is via the tissue surrounding the interventricular foramina. This is why the corpus callosum first appears anteriorly, and then works backwards [p. 386].

The cerebral cortex is divided into distinct areas concerned with motor function, with sensation, with vision, hearing, speech, etc. These areas are sub-divided when appropriate on an anatomical basis so that each sub-division is then concerned with a particular region of the body [FIG. 19.09].

The function of the different areas of the cortex is reflected in histological differences and, in the visual area, even naked-eye differences. Maps of the cortex are based on histology. The histologically distinct areas often do not coincide very precisely with

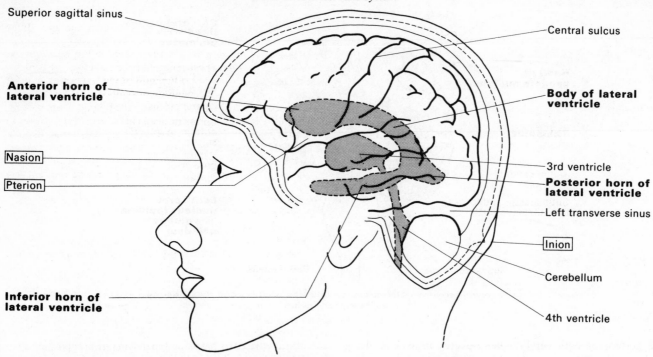

FIG. 19.06. The general relationships of the brain and ventricles (dotted lines) to the face and skull. The central sulcus commences 1 cm behind the midpoint between nasion and inion measured over the vertex.

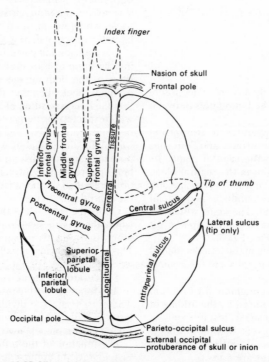

FIG. 19.07. View of the brain from above showing the way in which the hand and thumb may be used to indicate the angle and length of the central sulcus.

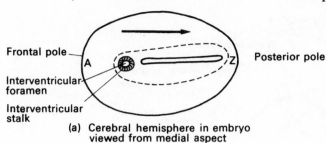

Frontal pole

Interventricular foramen

Interventricular stalk

A · · · · · Z · · · · · Posterior pole

(a) Cerebral hemisphere in embryo viewed from medial aspect

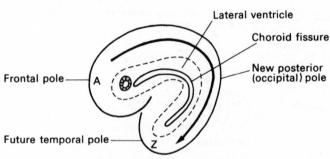

Lateral ventricle

Choroid fissure

New posterior (occipital) pole

Frontal pole — A

Future temporal pole — Z

(b) Cerebral hemisphere in embryo viewed from medial aspect at later stage

Body lat vent. [handwritten]

Later ventric foramen [handwritten]

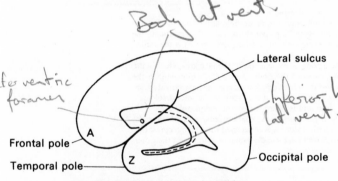

Lateral sulcus

Inferior horn Lat vent. [handwritten]

Frontal pole — A

Temporal pole — Z — Occipital pole

(c) Cerebral hemisphere in adult viewed from lateral aspect

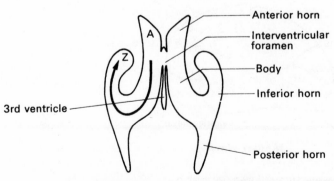

Anterior horn

Interventricular foramen

Body

Inferior horn

3rd ventricle

Posterior horn

(d) Lateral ventricles in mature brain viewed from above

the anatomical landmarks [Figs. 19.10 and 19.11] provided by the sulci and gyri. In any case the patterns of sulci and gyri differ considerably from brain to brain.

Besides the localization of functions seen within each cortex, there is also (in the human brain) specialization between the hemispheres. Thus speech is a prime function of the left cortex in most right-handed people. The recognition of symbols and the interpretation of shapes and textures is the function of the right hemisphere.

The terms 'left or right handedness' describes specialization in nature of the mature cerebrum. Thus a cerebral vascular accident (a stroke) in a right-handed person may completely paralyse the left half of the body (left hemiplegia) yet leave speech unaffected. Specialization between hemispheres develops during the first few years of childhood.

Medial surface of cerebral hemisphere

In Figure 19.12 note the position of:
1. Central sulcus.
2. Cingulate sulcus (*cingulum* = girdle). This sulcus gets its name from the cingulate gyrus).
3. Calcarine sulcus.
   (This sulcus produces a ridge in the posterior horn of the lateral ventricle, the calcar avis (*calcar* = a spur; *avis* = of a bird) which gives its name to the sulcus.)
4. Parieto-occipital sulcus.

Inferior surface

In Figure 19.13 note the position of:
1. The calcarine sulcus.
2. The collateral sulcus.

The dotted line shows the position of the attached edge of the tentorium cerebelli. In front of this line the brain lies on the floor of the middle cranial fossa; behind it rests on the tentorium.

BRAINSTEM

The brainstem lies between the cerebrum and the spinal cord. It consists of the midbrain, pons, and medulla oblongata. The cerebellum [Fig. 19.14] is joined to the three brainstem components by three pairs of cerebellar peduncles. The superior cerebellar peduncles join it to the midbrain, the middle peduncles to the pons, the inferior to the medulla. The medulla and pons are often called 'the bulb'. The term is useful to indicate certain tracts, or clinical disorders in this region, such as 'bulbar paralysis'.

The brainstem contains the nerve fibres connecting the cerebrum to the cranial nerve nuclei, to lower centres and to the spinal cord. It

---

Fig. 19.08. Development of lateral ventricles.
(a) Medial aspect of cerebral hemisphere which has been separated from the rest of the brain by cutting through the interventricular stalk. The choroid fissure is in a straight line. Cavity of ventricle shown by dots.
(b) Later in development the brain curves so that point Z instead of forming the occipital pole of the brain comes to form the tip of the temporal lobe. The ventricle and choroid fissures are necessarily also curved in shape.
(c) The lateral sulcus is produced on the inside of this curve, in the depth of which is the insula. The body and inferior horn of the lateral ventricle are present from stage (b). The anterior and posterior horns develop with the expansion of the frontal and occipital lobes.
(d) The mature lateral ventricles, seen from above to show the curve (indicated by arrow).

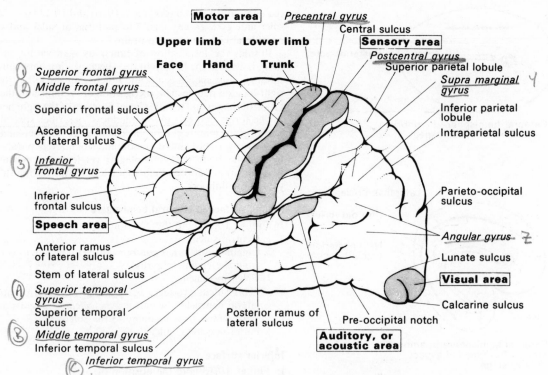

FIG. 19.09. Cerebral cortex: lateral veiw.

The stem of the lateral sulcus has three branches (*ramus* = branch) anterior, ascending, and posterior rami. In practice the term 'lateral sulcus' means the stem and posterior ramus. By pulling open the lips of the lateral sulcus, the cortex of the insula is revealed. The ascending and anterior rami enclose the pars triangularis which is the motor speech area (Broca's area).

In the voluntary motor area the body as a whole is represented upside-down, except for the face area which is the right way up. The areas devoted to sensation from particular regions of the body are roughly equivalent in the postcentral gyrus to the motor areas in the precentral gyrus. Sensation reaches consciousness in the postcentral gyrus.

The auditory area is much more extensive than it appears because much of it is hidden in the lateral sulcus.

The visual area terminates around the tip of the calcarine sulcus and is delimited by the lunate sulcus.

The auditory and visual areas are surrounded by cortex which give rise to association areas and which make possible the understanding of what is heard and seen. The inferior parietal lobule is concerned with the interpretation of symbols. Association areas for speech, heard or seen, motor or sensory are usually found on one side or the other, usually the left in right-handed people.

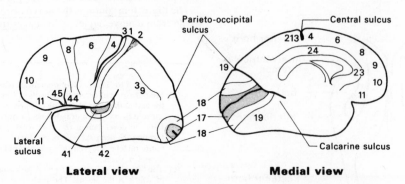

Lateral view   **Medial view**

FIG. 19.10.
Certain regions of the cortex are histologically distinct from each other. The numbers in the diagrams follow Brodman's description of cortex cytoarchitectonics. They are not in a logical order, although they may have appeared to be so when first described.

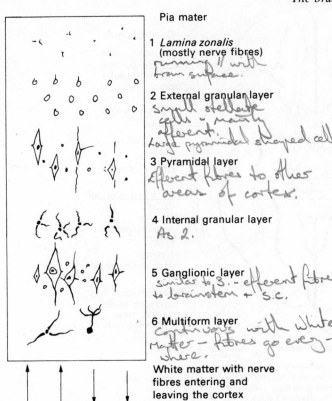

Pia mater

1 *Lamina zonalis*
(mostly nerve fibres)

[handwritten: running // with brain surface.]

2 External granular layer

[handwritten: small stellate cells - mainly afferent. Large pyramidal shaped cells]

3 Pyramidal layer

[handwritten: Efferent fibres to other areas of cortex.]

4 Internal granular layer

[handwritten: As 2.]

5 Ganglionic layer

[handwritten: similar to 3. - efferent fibres to brainstem + s.c.]

6 Multiform layer

[handwritten: continuous with white matter - fibres go everywhere.]

White matter with nerve
fibres entering and
leaving the cortex

FIG. 19.11. Layers of the cerebral cortex.

1. The lamina zonalis, or molecular layer, lies immediately beneath the pia mater. There are few neuronal cell bodies; it consists mainly of fine unmyelinated or poorly myelinated fibres mostly running parallel with the brain surface.
2. External granular layer: contains small stellate cells. Mainly afferent in function.
3. Pyramidal layer: also contains small stellate cells as in (2), but is characterized by large pyramidal shaped cells which are the source of efferent fibres to other parts of the cortex; association fibres, commissural etc.
4. Internal granular layer: like (2).
5. Ganglionic layer: similar to (3) but has a few giant pyramidal cells, such as the Betz cells of the motor cortex in area 4. Source of efferent fibres to brainstem and spinal cord.
6. Multiform layer: contains many different cell types, many of which send fibres into the other layers of the cortex. Is continuous on its deep surface with the white matter of the brain.

The pyramidal cells of the pyramidal (3) and ganglionic (5) layers are characteristic of 'motor cortex' (areas 4 and 6). The external (2) and internal (4) granular layers are characteristic of sensory cortex, in which they hypertrophy at the expense of layers 3 and 5. Examples are areas 41 and 17. In the visual area (17) the internal granular layer is reduplicated and the two parts are separated by a white streak easily seen with the naked eye. This region is called the striate cortex for this reason. At the bottom of the central sulcus the transition from motor to sensory cortex (area 4 to area 3) is also obvious to the naked eye.

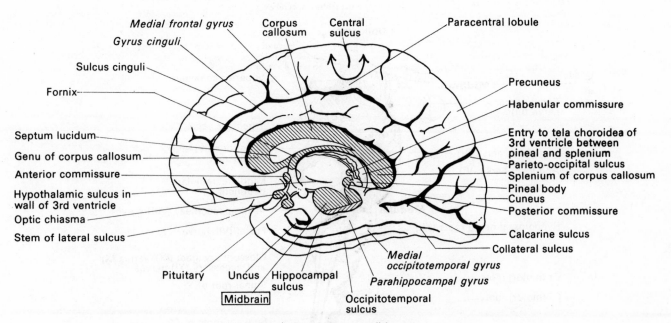

FIG. 19.12. Cerebral cortex: medial aspect.

The central sulcus reaches the medial surface and its termination is surrounded by the paracentral lobule, which is partly motor and partly sensory, concerned with the sphincters in the perineum.

The calcarine sulcus has the visual area on each side of it, and also in its depths.

The uncus of the parahippocampal gyrus is the area where smell reaches consciousness.

The gyrus cinguli is a major component of the limbic system.

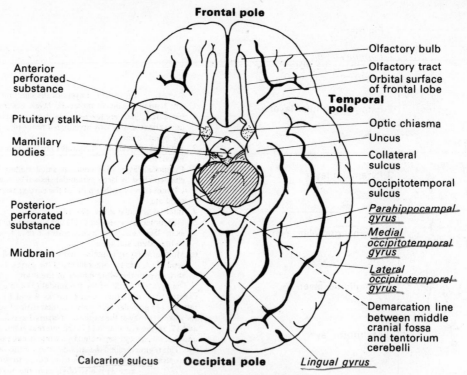

Frontal pole

Olfactory bulb

Olfactory tract

Orbital surface
of frontal lobe

Anterior
perforated
substance

**Temporal
pole**

Pituitary stalk — Optic chiasma

Mamillary
bodies

Uncus

Collateral
sulcus

Occipitotemporal
sulcus

Posterior
perforated
substance

*Parahippocampal
gyrus*

*Medial
occipitotemporal
gyrus*

Midbrain

*Lateral
occipitotemporal
gyrus*

Demarcation line
between middle
cranial fossa
and tentorium
cerebelli

Calcarine sulcus   **Occipital pole**   *Lingual gyrus*

Fig. 19.13.   Cerebral cortex: inferior view.
The midbrain has been cut across, and the cranial nerves sectioned in removing the brain from the
skull. The pituitary has been left behind in the pituitary fossa, and only the stalk is seen in the figure.
Note the position of the calcarine sulcus. The olfactory bulbs relay the impulses reaching it from
the olfactory nerves. The olfactory tract divides into the medial and lateral olfactory striae. The latter
is concerned with smell, the former with input to the limbic system.
The dotted line shows the posterior edge of the middle cranial fossa. Behind this line the occipital
lobe lies on the tentorium cerebelli.

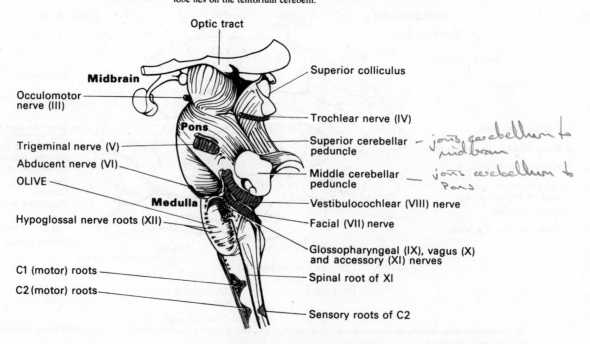

Optic tract

Midbrain

Superior colliculus

Occulomotor
nerve (III)

Trochlear nerve (IV)

Pons

Superior cerebellar
peduncle   — *joins cerebellum to
midbrain*

Trigeminal nerve (V)

Abducent nerve (VI)

Middle cerebellar
peduncle   — *joins cerebellum to
Pons*

OLIVE

Vestibulocochlear (VIII) nerve

Medulla

Facial (VII) nerve

Hypoglossal nerve roots (XII)

Glossopharyngeal (IX), vagus (X)
and accessory (XI) nerves

C1 (motor) roots

Spinal root of XI

C2 (motor) roots

Sensory roots of C2

Fig. 19.14. Brainstem from left side.
The cerebellum has been removed.
The hypoglossal roots are in series with the motor roots of C1.

also contains the cranial nerve nuclei and is involved with the special sense pathways except that of olfaction.

Much of the **reticular formation** lies in the brainstem, in particular those parts of this very diffuse system called the vital centres. Note that these 'centres', respiratory, cardiac, and vasomotor, are not anatomical entities, any more than is the 'centre' of gravity.

MIDBRAIN

Seen from the front, the midbrain runs from the upper border of the pons to the mamillary (breast-like) bodies [FIG. 19.15]. In cross-section [FIG. 19.16] it can be seen to contain the **aqueduct** (of Sylvius). This is surrounded by grey matter, the ventral part of which contains the motor nuclei of the oculomotor (IIIrd) and trochlear (IVth) nerves. In a more dorsal position is the sensory nucleus of the trigeminal nerve (V). Note the positions of the **tectum** (= roof) the **tegmentum** and the **cerebral peduncles**. Between the cerebral peduncles is the **interpeduncular fossa**. The

substantia nigra (black substance; it really does look black) separates the peduncles from the tegmentum, in which is the **red nucleus**, which is a pastel shade of pink in a fresh brain. From behind the tectum shows the superior and inferior colliculi (hillocks), also called the corpora quadrigemina (four twin bodies).

The reticular formation contribute substantially to the tegmentum.

PONS

The pons is the most prominent feature of the brainstem seen from the front [FIG. 19.15]. The trigeminal nerve (V) is the arbitrary landmark which shows where the pons joins the middle cerebellar peduncle. The pons (= bridge) physically joins one side of the cerebellum to the other but it does not do so functionally. The upper half of the fourth ventricle lies behind the pons, so that the tectum of the midbrain is not represented in any way in the pons.

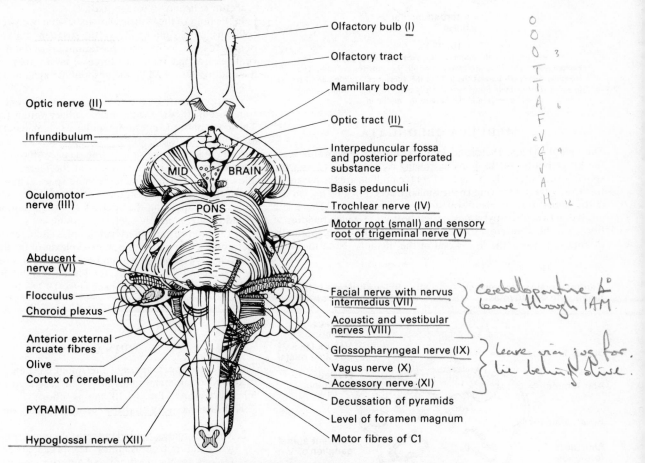

FIG. 19.15. Brainstem from the front.

The midbrain, and hence the brainstem, commences just caudal to the mamillary bodies. Note the points of emergence of the cranial nerves, which are numbered from before backwards according to their points of emergence. The olfactory tract does not reach the brainstem, but has been included for completeness.

The facial (VII) and vestibulocochlear nerves (VIII) emerge close together in the cerebellopontine angle. They leave the cranial cavity together in the internal auditory meatus. They lie close to the flocculus of the cerebellum and to the choroid plexus which reaches the subarachnoid space via the lateral foramen of the 4th ventricle.

Nerves IX, X, and XI lie close together behind the olive, and leave the cranium together in the jugular foramen.

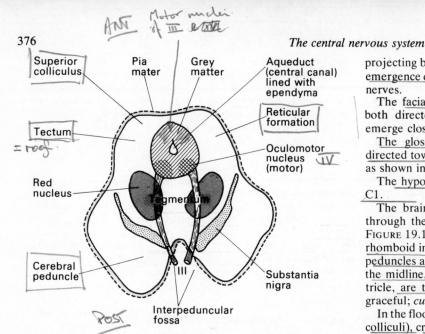

FIG. 19.16.

Cross-section through midbrain to show principal features. The oculomotor nuclei (III) are at the level of the superior colliculi. The trochlear nuclei (IV) are at the level of the inferior colliculi in line with those of III, but their fibres pass dorsally and cross the midline before emerging at the back of the midbrain.

MEDULLA OBLONGATA

The medulla [FIGS. 19.15 and 19.17] continues downwards from the lower border of the pons narrowing as it does so to join the spinal cord. Anteriorly on each side of the midline are the pyramids formed by the underlying **corticospinal tracts**. For this reason the corticospinal tract is also termed the *pyramidal tract*. There is a groove between the pyramids, at the bottom of which bundles of fibres can be seen crossing the midline. The pyramids gradually disappear towards the lower end of the medulla. Note the olive

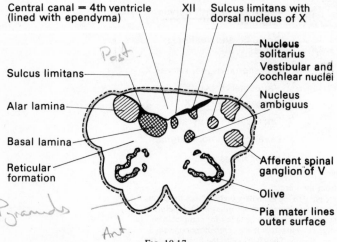

FIG. 19.17.

Cross-section through the upper part of the medulla. Compare the lumen of the 4th ventricle with the aqueduct of the midbrain.
The transparent roof of the 4th ventricle corresponds to the tectum of the midbrain.

projecting behind the upper part of the pyramid, and the points of emergence of the VI, VII, and VIII; IX, X, and XI; and XII cranial nerves.

The facial nerve VII and vestibulocochlear nerves (VIII) are both directed towards the internal acoustic meatus and they emerge close together in the cerebellopontine angle.

The glossopharyngeal, vagus, and accessory nerves are all directed towards the jugular foramen. They emerge along the line as shown in FIGURE 19.15.

The hypoglossal nerve fibres are exactly in series with those of C1.

The brainstem can only be examined posteriorly, by cutting through the cerebellar peduncles and removing the cerebellum. FIGURE 19.18 shows the principal features. The fourth ventricle is rhomboid in shape. It is bounded above by the superior cerebellar peduncles and below by the fibres of the inferior peduncles. Near the midline, on each side of the inferior angle of the fourth ventricle, are the nucleus gracilis and nucleus cuneatus (*gracilis* = graceful; *cuneatus* = wedge-like).

In the floor of the ventricle are to be seen two swellings (the facial colliculi), cross-running fibres (the striae medullares), and a series of grooves which divide the posterior part of the floor of the fourth ventricle so that there are three separate areas on each side of the midline [FIG. 19.18], the vestibular, vagal, and hypoglossal triangles.

The upper part of the roof of the fourth ventricle is thin and transparent. It is the superior medullary velum (*velum* = a veil). The lower part, the inferior medullary velum, has a **median opening** in it which allows cerebrospinal fluid (c.s.f.) to escape into the cisterna magna of the subarachnoid space [FIG. 19.19]. Two other **lateral foramina** are also present at the lateral angles which also allow c.s.f. to reach the subarachnoid space. These three foramina are of the greatest importance. They are the only means whereby c.s.f. can escape from the ventricles of the brain, which otherwise would be a completely enclosed system.

In the inferior medullary velum, pia, and ependyma are in contact with each other and the choroid plexus of the fourth ventricle is found here. The plexus is found not only in the ventricle itself but, by pushing out through the lateral foramina, the distal ends comes to lie in the subarachnoid space [FIG. 19.15] on each side of the midline.

SPINAL CORD

The external appearances of the spinal cord and its relationships to the covering membranes has been described on page 53. The internal features [FIG. 19.20] are as follows:

1. It has a very narrow central canal lined with ependyma.
2. The grey matter behind the central canal is sensory in function, while that in front is motor. The principal sensory and motor regions are the posterior and anterior 'columns' of grey matter. It is very common practice also to call them anterior and posterior 'horns'. Between T1 and L2, and S2 and S4, there is also a lateral column.
3. The white matter of the spinal cord surrounds the central grey matter. It is impossible to make out any special features in the gross anatomy of the white matter.

The blood supply to the spinal cord is shown in FIGURE 19.21.

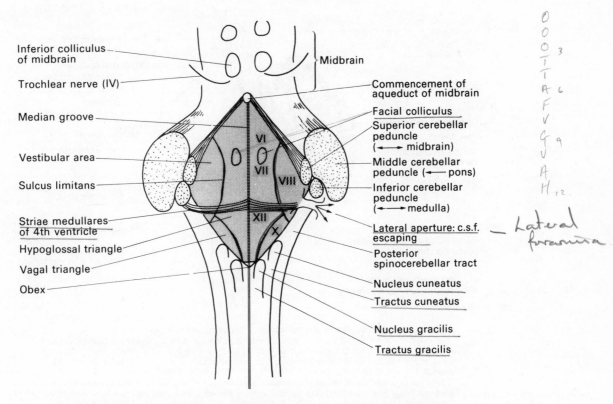

Inferior colliculus of midbrain

Trochlear nerve (IV)

Median groove

Vestibular area

Sulcus limitans

Striae medullares of 4th ventricle

Hypoglossal triangle

Vagal triangle

Obex

Midbrain

Commencement of aqueduct of midbrain

Facial colliculus

Superior cerebellar peduncle (←→ midbrain)

Middle cerebellar peduncle (←— pons)

Inferior cerebellar peduncle (←—→ medulla)

Lateral aperture: c.s.f. escaping

Posterior spinocerebellar tract

Nucleus cuneatus

Tractus cuneatus

Nucleus gracilis

Tractus gracilis

Cross-running fibres.

— Lateral foramina

0
0
0
T
A
F
V
G
V
A
H

3
6
9
12

FIG. 19.18.

The 4th ventricle has been exposed from behind by cutting through the three cerebellar peduncles and removing the cerebellum.

The floor of the 4th ventricle is divided into lateral halves by the median groove. Each half is subdivided by its own sulcus limitans and the horizontally running stria medullaris. The two striae indicate the lower level of the pons.

The facial colliculus is produced by the nucleus of the abducent nerve (VI) over which the fibres of the facial nerve (VII) are looped. The fibres intervene between the nucleus of VI and the floor of the ventricle.

The part of the floor lateral to the sulcus limitans is the vestibular area, which contains the vestibular as well as the cochlear nuclei. The vagal (X) and hypoglossal (XII) triangles agree only approximately with their underlying nuclei.

The superior medullary velum joins the two superior, and the inferior medullary velum joins the two inferior cerebellar peduncles to each other across the midline (not shown in Figure). The obex (= partition or barrier) is at the union of the two inferior peduncles. It is shaped like the web joining the base of the fingers. It is a useful anatomical reference point and landmark.

The position of the nucleus gracilis and nucleus cuneatus on each side of the midline is shown.

The lateral aperture (foramen of Luscha) of the ventricle is shown on the right side only. This is a direct communication between the 4th ventricle and subarachnoid space. The free outflow of c.s.f. is impeded by the presence in the aperture of the choroid plexus, whose tip lies free in the subarachnoid space.

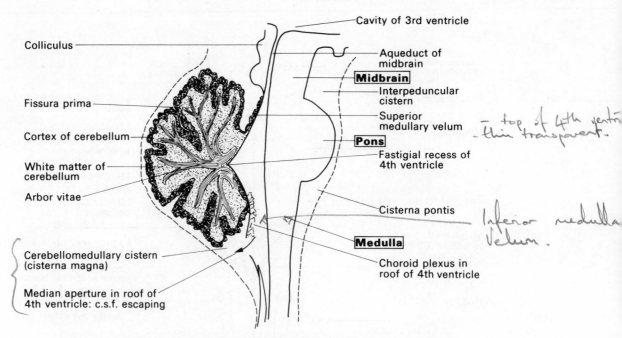

Cavity of 3rd ventricle

Colliculus

Fissura prima

Cortex of cerebellum

White matter of cerebellum

Arbor vitae

Cerebellomedullary cistern (cisterna magna)

Median aperture in roof of 4th ventricle: c.s.f. escaping

Aqueduct of midbrain

Midbrain

Interpeduncular cistern

Superior medullary velum

— top of 4th ventri *— thin transparent.*

Pons

Fastigial recess of 4th ventricle

Cisterna pontis

Medulla

Choroid plexus in roof of 4th ventricle

Inferior medulla Velum.

Fɪɢ. 19.19. Sagittal section of brainstem and cerebellum.

The shape of the 4th ventricle is seen from the side. Note the aperture in the inferior medullary velum (foramen of Magendie) which is the principal route for c.s.f. escaping from the ventricle into the cisterna magna, which is part of the subarachnoid space. The outflow of c.s.f., unlike that of the lateral aperture, is unimpeded.

The cerebellum is sectioned through the vermis. The white matter branches like a tree and is traditionally called arbor vitae (the tree of life). The fastigial recess of the 4th ventricle deeply indents the cerebellum: (fastigial here means like a gabled roof). The fastigial nuclei of central grey matter lie close by (not shown in the Figure). If the convolutions and irregularities of the cerebellar cortex are laid out flat, they are said to measure about 1 metre in length!

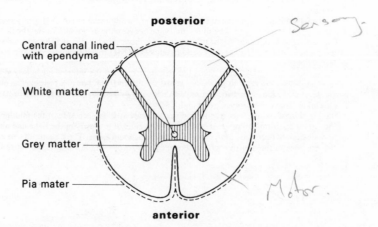

posterior

Sensory.

Central canal lined with ependyma

White matter

Grey matter

Pia mater

Motor.

anterior

Fɪɢ. 19.20. Cross-section of the spinal cord.
The central canal, lined with ependyma, is very narrow. The grey matter in front of the canal is motor. That behind the canal is sensory.

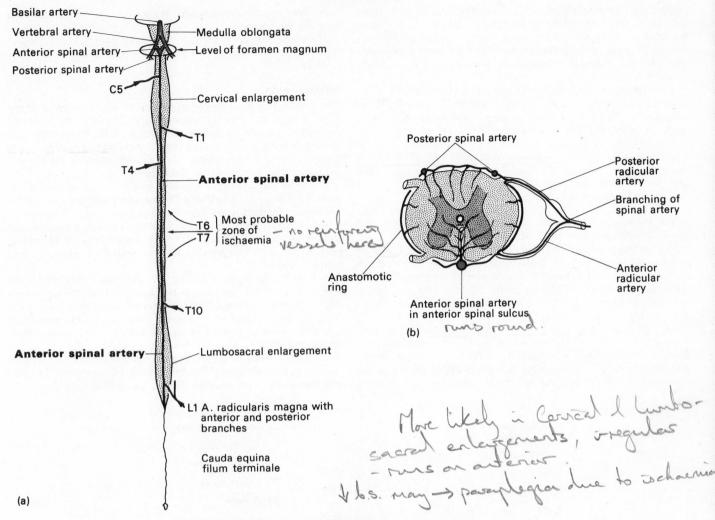

Basilar artery

Vertebral artery

Anterior spinal artery

Posterior spinal artery

C5

Medulla oblongata

Level of foramen magnum

Cervical enlargement

T1

T4

Anterior spinal artery

T6
T7 } Most probable zone of ischaemia — *no reinforcing vessels here*

T10

Anterior spinal artery

Lumbosacral enlargement

L1 A. radicularis magna with anterior and posterior branches

Cauda equina filum terminale

(a)

Posterior spinal artery

Posterior radicular artery

Branching of spinal artery

Anastomotic ring

Anterior spinal artery in anterior spinal sulcus *runs round.*

Anterior radicular artery

(b)

More likely in cervical & lumbo-sacral enlargements, irregular — runs on anterior

↓ b.s. may → paraplegia due to ischaemia

FIG. 19.21.
The blood supply of the spinal cord is shown from the front (a).

The anterior spinal artery is formed just above the foramen magnum by the union of branches from the vertebral arteries. It lies in the anterior spinal sulcus and runs the entire length of the spinal cord. Its branches reach the depth of the sulcus and supply the spinal cord on both sides.

The anterior spinal artery is reinforced by branches of the vertebral, intercostal, or lumbar arteries which enter the vertebral canal along the spinal roots.

The number of reinforcing vessels is not often more than 10. There may only be three. Five are shown in the diagram. They are asymmetrical and irregular in size, and are most frequent at the cervical and lumbosacral enlargements. The midthoracic region has no reinforcing vessels and the cord is most likely to suffer ischaemic changes in this region.

A small posterior spinal branch is given off by each vertebral artery setting up a longitudinal anastomotic channel on each side in close proximity to the posterior spinal nerve roots. They are also reinforced by branches from the segmental vessels.

Vessels pass beyond the termination of the spinal cord and supply the cauda equina.

In cross-section (b) the positions of the anterior and two posterior spinal arteries are shown. These vessels all communicate with each other by means of encircling vascular rings.

In certain disorders parts of the aorta may have to be replaced by a graft. Removing the diseased artery may involve cutting off the blood supply of part of the spinal cord. This may result in paraplegia.

CRANIAL NERVE NUCLEI

The olfactory and optic nerves join the cerebral cortex. The remaining ten cranial nerves are attached to the brainstem. The details of their peripheral distribution has been described in Chapter 12. Cranial nerves may be purely motor, purely sensory, or mixed. The plan of these nerves is an elaborate variation of that of the spinal nerves. The motor fibres always have their cells of origin within the central nervous system, while the sensory fibres (with one exception) have their cells of origin in peripheral ganglia which resemble spinal ganglia.

Efferent nuclei

The motor nuclei of the oculomotor (III), trochlear (IV), abducent (VI), and hypoglossal nerves (XII) are in line with each other near

∴ motor fibres have nuclei (cells of origin) in CNS
Sensory " " " " in peripheral ganglia.

the midline at different levels in the brainstem. They lie at the same level as the nerves to which they give origin. The nucleus of III is at the level of the superior colliculus, IV at that of the inferior colliculus, VI at the 'facial' colliculus, and XII deep to the hypoglossal triangle, which can both be seen in the floor of the fourth ventricle [FIG. 19.22].

The **trigeminal motor nucleus** (V) is level with the middle of the pons, while the **facial nucleus** (VII) is in line with it a few millimetres caudally, supplying the muscles of facial expression. Both these nuclei are somewhat lateral to the midline.

The facial nerve contains autonomic (parasympathetic) fibres which supply the lacrimal gland and submandibular and sublingual salivary glands. These fibres come from a separate nucleus, the **superior salivatory nucleus**. This nucleus lies close to the facial nucleus on its medial side.

Like the facial nerve, the glossopharyngeal nerve (IX) and vagus (X) also supply striated muscle as well as autonomic structures, and separate nuclei are also involved here.

The **nucleus ambiguus** runs from just caudal to the facial nucleus as far as the lower medulla. It gives rise to motor fibres which run not only into IX and X, but also into the accessory nerve (XI). The **spinal nucleus of XI** is lined up on the nucleus ambiguus but is separated from it by a gap through which the pyramidal tract fibres pass, having just decussated.

The corresponding autonomic (parasympathetic) fibres also come from the dorsal nucleus of the vagus which in spite of its name also supplies the glossopharyngeal nerve (IX). (The parasympathetic fibres to the heart come from the nucleus ambiguus.) Between the rostral end of this nucleus, and the superior salivatory nucleus,

is the **inferior salivatory nucleus**. It gives rise to the secretomotor fibres to the parotid.

The autonomic component of the oculomotor nerve (Edinger–Westphal nucleus) is part of the oculomotor nucleus previously described.

Afferent nuclei

The most extensive sensory nucleus is the sensory nucleus of the trigeminal nerve. This stretches [FIG. 19.23] from the midbrain throughout the whole length of the rest of the brainstem. It is ultimately continuous with the posterior column of grey matter in the spinal cord. It does not by any means only subserve the trigeminal nerve. This complex is in three parts.

The **principal nucleus** is adjacent to the motor nucleus in the pons. It is concerned with touch. The **mesencephalic nucleus** and the accompanying tract is concerned with proprioception. The cells of origin of proprioceptive fibres is actually in this midbrain nucleus and not as you would expect in the trigeminal ganglion. The **spinal nucleus** and its accompanying tract is concerned with pain and temperature.

Although trigeminal afferents are distributed throughout the whole length of this ganglion, it also receives afferents from other cranial nerves. Afferents from the external ocular muscles (III, IV, VI), the face (VII), and tongue (XII) travel with peripheral branches of the trigeminal nerve for part of their course, before terminating in its sensory nuclei.

The afferents corresponding to the nucleus ambiguus terminate in the **nucleus solitarius**. The fibres from IX and X are bundled together close to the nucleus to form the tractus solitarius.

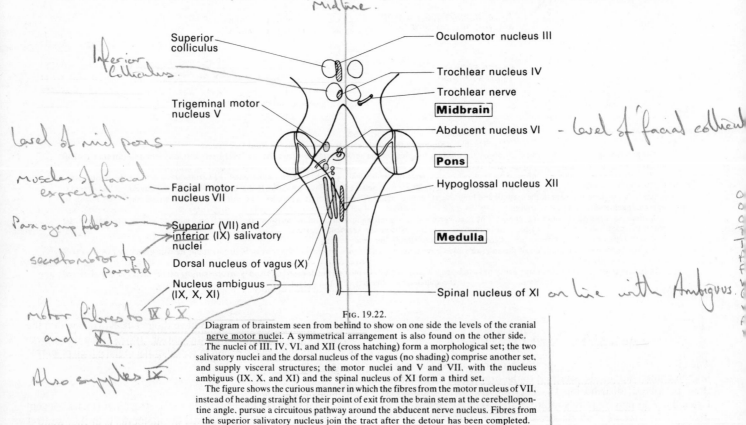

Midline.

Superior colliculus
Oculomotor nucleus III
Inferior Colliculus.
Trochlear nucleus IV
Trochlear nerve
Midbrain
Trigeminal motor nucleus V
Abducent nucleus VI — *level of facial colliculus*
level of mid pons.
Pons
muscles of facial expression.
Hypoglossal nucleus XII
Parasymp fibres
Facial motor nucleus VII
secretomotor to parotid
Superior (VII) and inferior (IX) salivatory nuclei
Medulla
Dorsal nucleus of vagus (X)
Nucleus ambiguus (IX, X, XI)
motor fibres to IX & X and XI.
Spinal nucleus of XI *on line with Ambiguus.*
Also supplies IX

FIG. 19.22.
Diagram of brainstem seen from behind to show on one side the levels of the cranial nerve motor nuclei. A symmetrical arrangement is also found on the other side.
The nuclei of III, IV, VI, and XII (cross hatching) form a morphological set; the two salivatory nuclei and the dorsal nucleus of the vagus (no shading) comprise another set, and supply visceral structures; the motor nuclei and V and VII, with the nucleus ambiguus (IX, X, and XI) and the spinal nucleus of XI form a third set.
The figure shows the curious manner in which the fibres from the motor nucleus of VII, instead of heading straight for their point of exit from the brain stem at the cerebellopontine angle, pursue a circuitous pathway around the abducent nerve nucleus. Fibres from the superior salivatory nucleus join the tract after the detour has been completed.

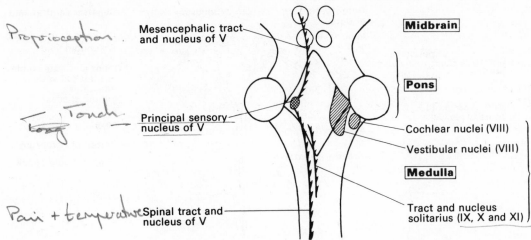

Proprioception

Touch

Pain + temperature

FIG. 19.23. Diagram of brainstem seen from behind to show the levels of the cranial nerve afferent nuclei. Certain nuclei are shown on one side; the others are shown on the other side.

The principal sensory nucleus of the trigeminal nerve, lies close to the motor nucleus but on its lateral side [FIG. 19.00] The trigeminal complex has two other components: one stretching from the principal nucleus to the upper midbrain, the other stretching downwards through the whole length of the medulla, until it eventually becomes continuous with the sensory grey matter of the spinal cord. Most of the fibres which relay in the complex enter the brainstem in the trigeminal nerve and then run upwards or downwards producing the association of nucleus and tract.

The nucleus solitarius lies close to the nucleus ambiguus receiving afferents from IX, X, and XI. The nucleus ambiguus and nucleus solitarius are the motor and sensory nuclei supplying the pharynx and larynx.

The vestibular nucleus, shown here as a single mass, is divided into four components, superior, inferior, medial, and lateral. The cochlear nucleus is also complex having two components, dorsal and ventral which in this view would be superimposed on each other.

Pressure A's → absorption/prod

Finally there are the nuclei of the VIIIth cranial nerve. The ganglia of the cochlear component are two in number, while the vestibular nerve has four nuclei. They are crowded together deep to the lateral angle of the floor of the fourth ventricle and occupy about 20 per cent of the total available area of the floor.

CEREBROSPINAL FLUID

The **central cavity** of the CNS which stretches from the lateral ventricles to the termination of the central canal of the spinal cord contains **cerebrospinal fluid**. This is a clear watery fluid which is continuously being secreted from the bloodstream by the **choroid plexuses** of the lateral ventricles, the third ventricle and the fourth ventricle.

The fluid escapes continuously from the ventricle system through the median and two lateral apertures in the roof and lateral angles of the fourth ventricle [p. 377], and now fills the subarachnoid space around the brain, spinal cord, and cauda equina. Altogether there is only about 150 cm³ of cerebrospinal fluid. The cerebrospinal fluid passes through the foramen magnum and fills the subarachnoid space right down to the level of S2, where the spinal theca terminates.

The removal of cerebrospinal fluid takes place partly in the spinal theca, especially in the diverticular sleeves surrounding the spinal nerve roots. In the skull there is a series of small villous-like projections into the bloodstream. They pass through the sustenta-

cular layer of dura into the superior sagittal sinus [FIG. 19.24]. It is not clear exactly how these projections function but when the cerebrospinal fluid pressure exceeds the venous pressure, cerebrospinal fluid is able to escape back into the bloodstream. Normally the venous pressure exceeds the cerebrospinal fluid pressure, otherwise the venous return would be obstructed. Very fine projections are called **arachnoid villi**. In later life they form agglomerations of villi, called **arachnoid granulations.**

SUBARACHNOID SPACE

The subarachnoid space, outside the CNS is situated between the pia mater, which follows the brain's most intimate surface irregularities, and the arachnoid mater which does not. During the early stages of development the arachnoid 'fits' the brain and spinal cord fairly closely. Its configuration becomes fixed about the time when the spinal cord is at the level of S2. When the spinal cord 'ascends' in the vertebral canal the arachnoid membrane remains at the lower level and does not follow the course of the spinal cord further upwards. Within the space so formed are found the nerves of the lumbosacral plexus, forming the cauda equina.

The growth of the cerebellum pushes the arachnoid away from the dorsum of the medulla and produces the **cisterna magna** or **Cerebellomedullary cistern**. The growth of the pons produces the **cisterna pontis**, and the bending of the brain forwards to about a right angle between the midbrain and procencephalon produces the interpeduncular cisterna.

The surface of the cerebral cortex in the foetus is at first quite

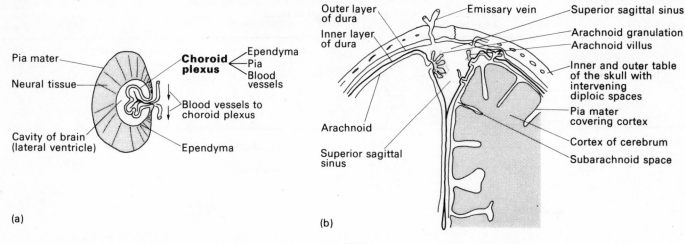

(a)

(b)

FIG. 19.24.

(a) The choroid plexuses are all basically similar, and form:
    (i) in the lateral ventricles along almost the whole length of the choroid fissure;
   (ii) in the roof of the 3rd ventricle;
  (iii) in the roof of the 4th ventricle.
    The central nervous system consists of a hollow tube lined on its outer surface by the highly vascular pia mater, and on its inner surface by the ependyma. As a rule the inner and outer membranes are separated by neuronal tissue. But in the above regions they are not, and when the two membranes come in contact, the choroid plexuses form. The plexuses are highly vascular.
(b) The arachnoid granulations are found in the skull from infancy onwards. They project into the superior sagittal sinus and become more numerous in the parietal and occipital regions compared with the frontal. They probably have a role in returning c.s.f. to the blood stream but the mechanism is not fully understood. Arachnoid granulations are matted agglomerations of villi and are more numerous in the aged. Occasionally granulations push their way through the inner table of the skull and make contact with the diploic venous spaces.

smooth and it fits the arachnoid well. But when the surface becomes folded, with the development of the gyri and sulci, the arachnoid does not follow into the sulci. The brain fits inside the arachnoid as well as a man fits inside a sack! [FIG. 19.19].

The subarachnoid space has an important relationship to the olfactory nerves. It surrounds these nerves as they pass through the cribriform plate and is continuous with the tissue spaces in the nasal mucosa. This provides a route whereby certain virus and bacterial infections can spread from the nasal cavity to the CNS. Epidemics of poliomyelitis and meningitis are spread in this way. Once the organisms reach the cerebrospinal fluid they can rapidly invade the whole nervous system.

The optic nerve is a tract of the brain, the retina being part of the grey matter. The nerve is lined on its outer surface by pia and surrounded by arachnoid as well as the inner layer of the dura. Between the pia and arachnoid is cerebrospinal fluid. The subarachnoid space reaches right to the back of the eyeball [p. 349].

The subarachnoid space extends for a short distance around the other cranial nerves as they pass through the foramina in the base of the skull. Their relationship is similar to that of the spinal nerves as they pass through the intervertebral foramina.

Finally the subarachnoid space extends for a short distance along the blood vessels which enter and leave the surface of the nervous system. A thin film of cerebrospinal fluid and a layer of pia intervene between these vessels and the central nervous system.

There are no lymphatics in the central nervous system. But infection can spread rapidly in the subarachnoid space, and certain malignant tumours initiated in the brainstem (medulloblastoma) give rise to secondary tumours in the cauda equina simply by drifting downwards in the cerebrospinal fluid.

The spinal subarachnoid space has been described on page 53.

Recall that the spinal cord finishes at the upper border of L2, and the spinal theca containing the subarachnoid space extends down to the lower border of S2. Cerebrospinal fluid can be removed by a **lumbar puncture** in which a needle is usually put into the spinal theca either above or below the spine of L4.

A needle can also be passed in the midline of the neck between the posterior arch of the atlas and the overhanging edge of the foramen magnum into the cisterna magna. This is a **cisternal puncture**. It is clearly important not to impale the medulla.

By the *circulation* of cerebrospinal fluid is meant the course taken by the fluid from its secretion in the lateral ventricles to its reabsorption back into the bloodstream [FIG. 19.25]. Clinical problems arise when this circulation is obstructed. If it is inside the brain or in the roof of the fourth ventricle, the ventricle system is distended and the brain is compressed against the skull. This is an **internal hydrocephalus**. Congenital abnormalities and tumours may be the cause.

The malabsorption of c.s.f. once it has reached the subarachnoid space gives rise to an **external hydrocephalus**. The space is distended somewhat at the expense of the brain. Meningitis (infection of the meninges) from any cause interferes with the free circulation in the subarachnoid space and with the mechanism of the arachnoid villi.

LATERAL VENTRICLE

The lateral ventricle [FIG. 19.26] is divided into the following regions:
    (i) the body;
   (ii) the inferior horn;
  (iii) the anterior horn;
   (iv) the posterior horn.

Origin, passage and absorption of cerebrospinal fluid

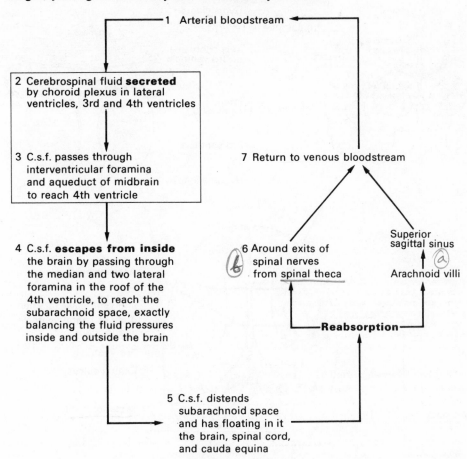

FIG. 19.25. Origin, passage, and absorption of cerebrospinal fluid.

We have already described how the body and inferior horn come to be C-shaped when looked at from the side.

1. The **body** is defined as starting anteriorly at the interventricular foramen, and as continuing backwards as far as the splenium of the corpus callosum. The **medial wall** contains the slot-like **choroid fissure**, where pia and ependyma are in contact and where the choroid plexus pushes into the ventricle. It is essential to understand the relationship of the choroid fissure to the fornix [FIG. 19.26]. The rest of the medial wall is the **septum pellucidum** and the **fornix** which is suspended from the inferior surface of the corpus callosum as it were by a mesentery.

The **roof** is the inferior surface of the corpus callosum, and the **floor** consists of the thalamus medially and the body of the caudate nucleus laterally.

2. The **inferior horn** is the true continuation of the body of the ventricle. The choroid fissure continues in its medial wall almost as far as the tip of the temporal lobe.

As the fornix is traced posteriorly into the temporal lobe it changes its name to the **fimbria** whose fibres eventually sink into the upper surface of the hippocampus, the gyrus lying in the floor of the inferior horn.

As the fissure is *below* the fornix in the body of the ventricle, it must lie above the fimbria in the inferior horn.

Similarly the body of the caudate nucleus lies *below* the fissure in the body of the ventricle, and when the tail of the nucleus (*cauda* = a tail) is traced round into the temporal lobe, it lies above the fissure, in the roof of the inferior horn. Note that continuous with the grey matter of the tip of the tail, is the amygdaloid nucleus (= almond like) which also lies in the roof.

Anything above the fissure in the body of the ventricle, must be below it in the inferior horn: anything below it in the body must be above it in the inferior horn.

3. The **anterior horn** is triangular in section, bounded medially by the septum pellucidum (the 'mesentery' of the fornix) and roofed by the under belly of the corpus callosum. The floor is the head of the caudate nucleus.

4. The **posterior horn** (may be absent) is covered on three sides by part of the corpus callosum called the **tapetum** (= a carpet), which lines the roof side wall and floor. The inferomedial aspect has two prominent ridges. One, produced by the **forceps major** of the corpus callosum, is the bulb of the posterior horn and the other, produced by the calcarine sulcus, is the calcar avis (= spur

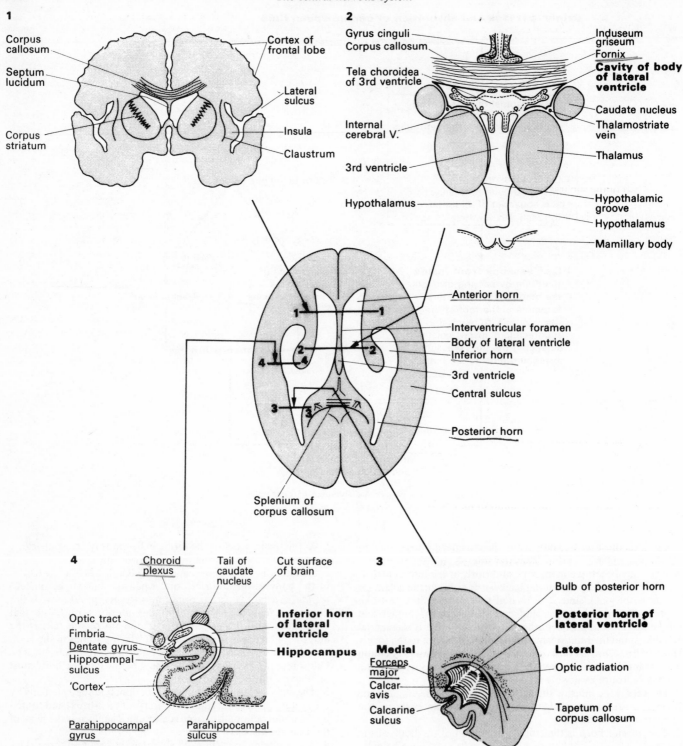

1

Corpus callosum
Septum lucidum
Corpus striatum
Cortex of frontal lobe
Lateral sulcus
Insula
Claustrum

2

Gyrus cinguli
Corpus callosum
Tela choroidea of 3rd ventricle
Internal cerebral V.
3rd ventricle
Hypothalamus
Induseum griseum
Fornix
Cavity of body of lateral ventricle
Caudate nucleus
Thalamostriate vein
Thalamus
Hypothalamic groove
Hypothalamus
Mamillary body

Anterior horn
Interventricular foramen
Body of lateral ventricle
Inferior horn
3rd ventricle
Central sulcus
Posterior horn
Splenium of corpus callosum

4

Choroid plexus
Tail of caudate nucleus
Cut surface of brain
Optic tract
Fimbria
Dentate gyrus
Hippocampal sulcus
'Cortex'
Parahippocampal gyrus
Parahippocampal sulcus
Inferior horn of lateral ventricle
Hippocampus

3

Bulb of posterior horn
Posterior horn of lateral ventricle
Medial
Forceps major
Calcar avis
Calcarine sulcus
Lateral
Optic radiation
Tapetum of corpus callosum

Fig. 19.26. The lateral ventricle seen from above. The transverse lines 1, 2, 3, 4 show the plane of section of the four peripheral diagrams looking in the direction of the arrows.

of a bird). The optic radiation is closely related to the posterior horn, as well as to the inferior horn [Fig. 19.26].

The details of the anatomy of the lateral ventricle, are now of great significance because of the increasing use of X-ray computerized tomographic scanning methods in the diagnosis of intracerebral lesions. Quite small variations in shape and size of the ventricle can now be recognized.

TELA CHOROIDEA OF FOURTH AND THIRD VENTRICLE

The tela choroidea of the **fourth** ventricle is the recess between the roof of the fourth ventricle and the inferior surface of the cerebellum. It is part of the subarachnoid space. The overhanging position of the cerebellum is entirely responsible for the presence of the tela, which would not exist but for the presence of the cerebellum. In the floor of the tela (or the roof of the fourth ventricle, they are the same thing) the choroid plexus of the fourth ventricle develops [Fig. 19.19, p. 378]. The open end of the tela leads straight into the cisterna magna.

The tela choroidea of the **third** ventricle is in principal formed in a similar manner. The corpus callosum overhangs the roof of the third ventricle in the same way as the cerebellum overhangs the roof of the fourth [Fig. 19.28]. When the corpus callosum fails to develop in man the tela choroidea of the third ventricle is completely absent. The corpus callosum begins development as a few fibres which run across from one side to the other via the hollow stalks which surround the interventricular foramina. There is no other place where physical continuity already exists between the two hemispheres at this time. The corpus callosum develops from before backwards.

The thin roof of the third ventricle lies just below the level of the choroid fissure of the cerebral hemisphere. When the corpus callosum grows back it must do so above the choroid fissure. The result is a space stretching almost the whole length of the third ventricle roofed over by the underbelly of the corpus callosum with the choroid fissures of the hemispheres in its side walls. The thin roof of the third ventricle is the floor of the tela.

In the mature brain the tela is entered by a narrow transverse cleft between the pineal and the splenium. It is also part of the subarachnoid space.

In both its floor and side walls pia and ependyma are in contact and the vascular pia hypertrophies to form the choroid plexuses of the third and lateral ventricles. It pushes the ependyma before it advancing downwards into the third ventricle and laterally into the body of both lateral ventricles. A similar invagination of choroid plexus takes place in the roof of the fourth ventricle.

As soon as the choroid fissure sweeps downwards to become the medial wall of the inferior horn of the lateral ventricle, it loses its contact with the tela. Also when the corpus callosum is absent choroid plexus formation and cerebrospinal fluid secretion are all quite normal. The association of the tela choroidea of the third and fourth ventricles with choroid plexus formation is developmentally speaking, a coincidence. There is no causal association.

WHITE MATTER OF THE CEREBRUM

The white matter of the brain is virtually devoid of nerve cell bodies, and consists of nerve fibres whose origins and destinations are in different parts of the grey matter of the central nervous system. But note that grey matter contains more nerve fibres than white matter because of the richness of connections between neighbouring nerve cells, which are in addition to the fibres of the white matter. Grey matter owes its appearance to the presence of nerve cell bodies not to the absence of nerve fibres.

The nerve fibres in the white matter are grouped into three divisions:

(i) association fibres;
(ii) commissural fibres;
(iii) projection fibres.

(i) **Association fibres** run:
  (a) From one gyrus to the next one. They are the **arcuate fibres** of the cerebrum (*arcus* = a bow).
  (b) From one lobe to another in the same hemisphere [Fig 19.27], without crossing the midline. If each of the four named 'lobes' in the cortex were jointed to each of the other three, then there should be six long association tracts. The **long association fibres tracts** are shown in Figure 19.27. Note that there are only four named tracts.

(ii) **Commissural fibres** join corresponding regions of the two hemispheres to each other across the midline [Figs. 19.28 and 19.29]. The greatest commissure is the corpus callosum (sometimes congenitally absent!) In addition there is the anterior commissure [Fig. 19.12] which connects frontal and temporal lobes with each other and is part of the limbic system. The posterior and habenular commissure (*habenulus* = a rein) are close to the stalk of the pineal gland.

The posterior commissure provides a somewhat circuitous route whereby the cephalic halves of the midbrain communicate

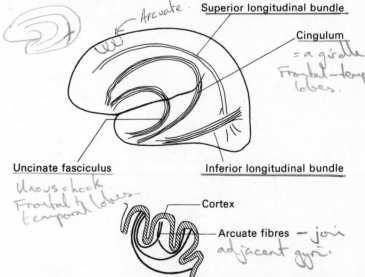

FIG. 19.27. Association fibres.
The main association fibre bundles which can easily be demonstrated to the naked eye in a fixed brain, are shown.
The uncinate fibres (*uncus* (L) = a hook) and the cingulum (L = a girdle) join the frontal and temporal lobes to each other, and between them allow two-way traffic. They are part of the olfactory and limbic systems.
The superior and inferior longitudinal bundles connect the occipital cortex, whose function concerns vision, with the frontal and temporal lobes. Besides these four groups, there are other association tracts in the brain.
Arcuate fibres join adjacent gyri to each other. There are also fibres which connect to more distant gyri.

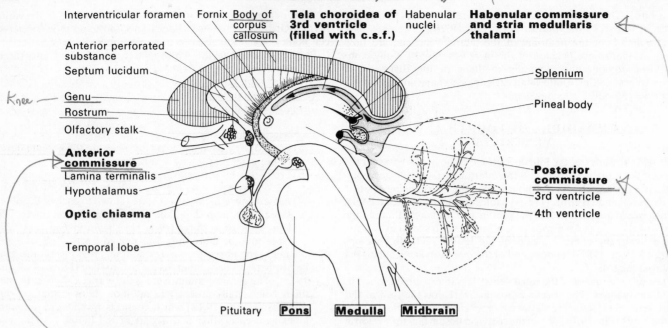

FIG. 19.28. Sagittal section of the brain to show commissural fibres and the tela choroidea between the roof of the 3rd ventricle and the inferior surface of the corpus callosum.

Commissural pathways join symmetrical parts of the cerebral cortex to each other. In addition, in the corpus callosum are fibres which join dissimilar parts of the left and right cortices to each other.

The corpus callosum is by far the largest collection of commissural fibres. The posterior extremity is the splenium; in front is the genu tapering down to the rostrum (bird's beak); between the genu and splenium is the body of the corpus callosum.

The anterior commissure conveys fibres from the gyri of one temporal lobe to the other. It also contains fibres (parts of the olfactory and limbic systems) which have their origins in the olfactory bulbs (running via the olfactory tracts) or in the anterior perforated substance or in the amygdaloid nuclei.

The habenular commissure lies just above the stalk of the pineal. It connects the habenular nuclei which lie in the wall of the third ventricle as shown. The habenular nuclei are relay stations between the grey matter of the anterior perforated substance and hypothalamus and amygdaloid nucleus. Their afferents are collected together into a bundle which makes a projection on the side wall of the third ventricle on the medial surface of the thalamus, the stria medullaris thalami. Efferents from the habenular nuclei reach the superior colliculi, the salivatory nuclei and reticular formation. Reflexes are conveyed by this route to produce eye movements, salivation and gastric section, masticatory movements, and deglutition.

The posterior commissure allows communication between nuclei situation in the midbrain; in particular it allows both pretectal nuclei, which are involved in the consensual light reflex, to influence both oculomotor nuclei. This commissure does not really agree well with the definition of a commissure.

Note: frontal lobes communicate via forceps minor
temporal lobes communicate via anterior commissure and body of corpus callosum
parietal lobes communicate via body of corpus callosum
occipital lobes communicate via forceps major
habenular nuclei in the wall of 3rd ventricle communicate via habenular commissure
certain brainstem nuclei communicate via posterior commissure

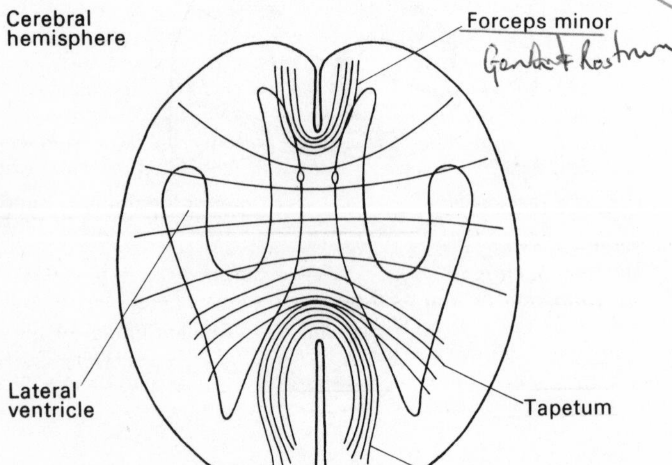

FIG. 19.29. Shows a diagrammatic view of the corpus callosum from above. The named parts are the forceps minor (whose fibres go through the genu and rostrum) which joins the two frontal lobes, and the forceps major (whose fibres go through the splenium) which joins the occipital lobes. The tapetum fibres are crowded together by their proximity to the lateral ventricle.

with each other. The habenular commissure connects the habenular nuclei which lie in the lateral wall of the third ventricle. They are part of the limbic system.

Fibres cross the midline in the two fornices and also join the optic nerves and cross to the opposite side in the optic chiasma. Both these commissures are associated with the limbic system.

(iii) **Projection fibres.** Finally there are cortical projection fibres which may be afferent or efferent. These are the fibres which allow either the cerebrum to communicate with other parts of the central nervous system or the rest of the central nervous system to communicate with the cerebrum.

CORTICOSPINAL TRACT (PYRAMIDAL TRACT)

In the cerebrum

Most tracts are named by taking their two end points and combining them into a composite name, putting the 'origin' of the tract first. '*Cortico-spinal*' means 'from the cortex to the spinal cord'. '*Spino-thalamic*' tract means 'from spinal cord to thalamus'.

Voluntary movements are initiated in 'the motor cortex'. This occupies the whole of the precentral gyrus (40 per cent of fibres) and extends anteriorly from here into the precentral zone and other parts of the cortex.

The spatial organization of the motor cortex is shown in FIGURE 19.09. The area devoted to the control of particular parts of the body depends not on body proportions but on the complexity of movement of which each member is capable. Compare the trunk motor area to that of the thumb. The motor area also extends on to the medial surface of the hemisphere. The paraterminal gyrus [FIG. 19.12] controls the body sphincters on both sides.

The fibres project downwards from the motor cortex in the **corona radiata** (*corona* = a crown) and converge on the **genu and posterior limb** of the **internal capsule** [FIG. 19.30]. The internal capsule lies on the internal aspect of the lentiform nucleus and is bounded on its own anteromedial aspect by the **head of the caudate nucleus** and on its posteromedial aspect by the **thalamus**. It has an anterior and posterior limb which join each other at an angle which is the genu (*genu* = a knee).

Note that the posterior limb of the internal capsule lies at about a right angle to the motor cortex. It is evident that the corona radiata must twist as it passes downwards.

Look at the cortex from the side. Note that the 'face area' is slightly anterior to the rest of the motor area. In the internal capsule the face fibres are still in front, only more so, because they lie in the genu, while the rest of the motor fibres in the posterior limb are strung out behind the genu in the same sequence as in the motor cortex.

In the midbrain

In the midbrain the fibres occupy the middle three-fifths of the cerebral peduncles [FIG. 19.30(c)]. The fibres are organized in the same sequence as in the motor cortex and internal capsule, but the 'face fibres' are now in the most medial position, while the lower limb and perineum are in the most lateral. The twist which was initiated between the motor cortex and internal capsule simply continues in the same direction lower down. This means that in the brainstem the 'face fibres' which synapse in cranial nerve nuclei of the opposite side or on both sides are strategically placed on the medial aspect of the pyramidal tract. Such fibres as terminate in

cranial nerve nuclei are the corticonuclear components of the pyramidal tract. The term corticobulbar is also used (*bulb* is an old term for *pons* and *medulla*).

In the hindbrain and spinal cord

The close grouping of these motor fibres in a single large bundle is broken up in the pons where the tract forms a large number of small bundles [FIG. 19.30(d)]. At the lower border of the pons the fibres reaggregate and form the **pyramid**. Just before they reach the level of the foramen magnum the fibres cross over to the other side by weaving through each other forming the **pyramidal decussation**. They come to lie in the spinal cord in the **lateral column** of white matter. The whole pathway is often called 'the pyramidal tract' (Galen invented the term 'pyramid' in the second century AD). A few fibres do not cross over at this point, but most of these do so later [FIG. 19.30(e) and (f)].

It is obvious that a pyramidal tract lesion in the central nervous system anywhere above the 'decussation of the pyramids' will produce paralysis on the contralateral side of the body. A lesion below the decussation will produce ipsilateral paralysis. An unlucky lesion at the decussation would produce bilateral paralysis.

The majority of impulses travelling in the pyramidal tracts reach the anterior column cells of the spinal cord directly, and cause these lower motor neurone cells to fire off, and thus bring about contractions of the appropriate motor units. However, with spinal reflexes there may be a series of short internuncial neurones (*nuncius* = a messenger, as in 'Papal nuncio') before the final common path is reached.

Other impulses transmitted with the pyramidal tract are associated with the reticular formation, and loss of these fibres when the tract is interrupted has a profound effect on muscle tone and on the muscle reflexes.

It is clinical practice to speak of 'upper' and 'lower' neurone lesions. This conveys the traditional idea that only two neurones are involved in the voluntary pathway, the first from the motor cortex to the anterior horn cell, the second from the anterior horn cell to the peripheral muscle. In view of the presence of internuncial neurones between the long fibres which have originated in the motor cortex and the anterior column cells, perhaps it is better to speak of a lesion of the upper motor *pathway*, an expression which leaves open the question as to the actual number of neurones which are involved.

Cranial nerve nuclei and anterior column cells are activated or inhibited by impulses which reach them not only from the pyramidal tract but also from the vestibular nuclei, the tectum, and reticular formation as well as from other sources. Whether or not an individual cell fires off or not according to the rules of the all-or-none law, depends on the balance of power, or algebraic sum, of all the excitatory and inhibitor stimuli which play simultaneously on the nerve-cell surface.

EXTRAPYRAMIDAL TRACTS

Vestibulospinal tract

The cells of origin lie especially in the lateral vestibular nucleus on each side in the lateral recess of the floor of the fourth ventricle [FIG. 19.23]. The tracts stay on their own side and run downwards in the anterior funiculi of the spinal cord [FIG. 19.31]. This tract also acts as the pathway for impulses whose ultimate source is not in the vestibular apparatus but in the cerebellum on the ipsilateral side. There is no direct cerebellospinal tract.

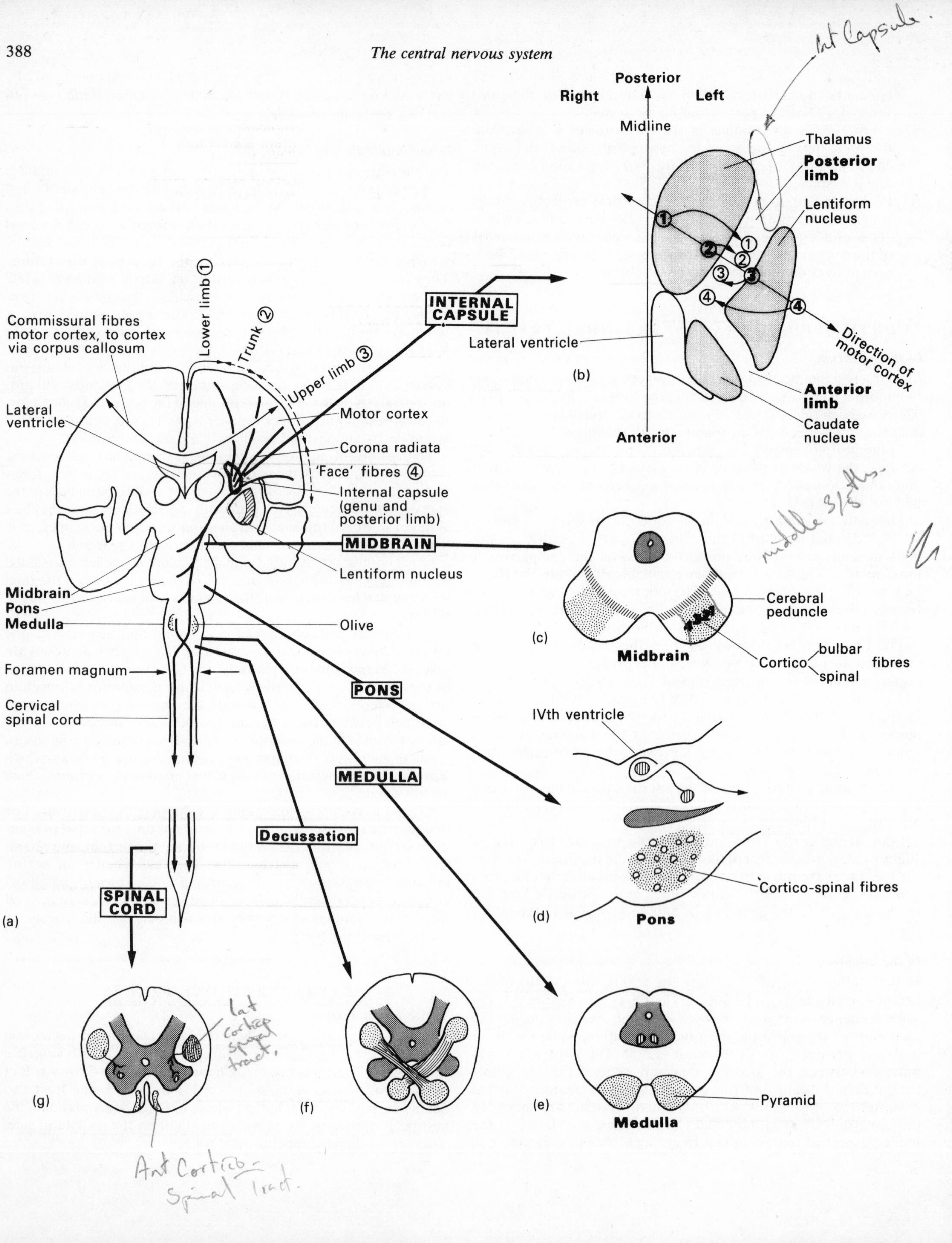

Ant Capsule.

(a)

Commissural fibres motor cortex, to cortex via corpus callosum

Lateral ventricle

Lower limb ①
Trunk ②
Upper limb ③

Motor cortex
Corona radiata
'Face' fibres ④
Internal capsule (genu and posterior limb)
Lentiform nucleus

Midbrain
Pons
Medulla

Olive

Foramen magnum

Cervical spinal cord

INTERNAL CAPSULE

MIDBRAIN

PONS

MEDULLA

Decussation

SPINAL CORD

(b)

Posterior
Right   **Left**
Midline

Thalamus
Posterior limb
Lentiform nucleus

Lateral ventricle

Anterior

Direction of motor cortex

Anterior limb
Caudate nucleus

(c)

middle 3/5 ths

Cerebral peduncle

Cortico bulbar fibres
spinal

Midbrain

(d)

IVth ventricle

Cortico-spinal fibres

Pons

(e)

Pyramid

Medulla

(f)

(g)

lat cortico spinal tract.

Ant Cortico-spinal Tract.

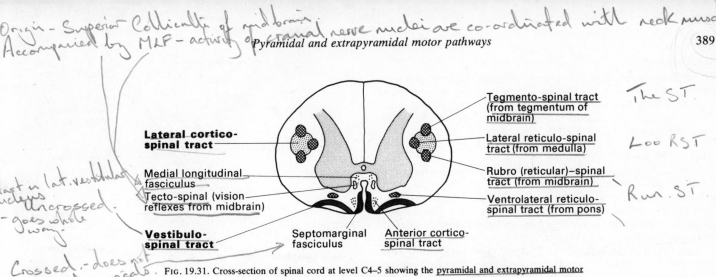

Handwritten top margin: Origin – Superior Colliculus of midbrain. Accompanied by MLF – activity of cranial nerve nuclei are co-ordinated with neck muscles.

Figure labels (left): Lateral cortico-spinal tract; Medial longitudinal fasciculus; Tecto-spinal (vision reflexes from midbrain); Vestibulo-spinal tract; Septomarginal fasciculus; Anterior cortico-spinal tract

Figure labels (right): Tegmento-spinal tract (from tegmentum of midbrain); Lateral reticulo-spinal tract (from medulla); Rubro (reticular)–spinal tract (from midbrain); Ventrolateral reticulo-spinal tract (from pons)

Handwritten left margin: Start in lat. vestibular nucleus. Uncrossed. – goes whole way. Crossed – does not go beyond cervicals.

Handwritten right margin: The ST; Lat. RST; Rub. ST

FIG. 19.31. Cross-section of spinal cord at level C4–5 showing the pyramidal and extrapyramidal motor pathways.

Note that on the right of the figure, the labels refer to the four tracts which have their origin in different parts of the reticular formation, in the midbrain, red nucleus, pons or medulla. These long tracts are augmented by polyneuronal pathways which run in the fasciculus proprius of the spinal cord. The fasciculus proprius is the white matter which lies in immediate contact with the grey matter of the spinal cord and is the main route whereby nearby segments of the spinal cord communicate with each other. Most reticulospinal fibres terminate in the cervical cord and have their origin in the reticular formation on both sides.

The vestibulospinal tract runs throughout the whole length of the spinal cord and is uncrossed. The septomarginal fasciculus however contains fibres from both sides, but does not extend beyond the cervical cord.

The tectospinal tract has its origin mainly in the superior colliculi of the midbrain. It is accompanied by the caudal extremity of the medial longitudinal bundle which is the principal pathway whereby the activity of the cranial nerve nuclei are co-ordinated with each other and with the neck muscles.

Finally, note the position of the anterior corticospinal pathway between the components of the vestibular pathway. Note also that the pyramidal tract itself is surrounded by three satellite pathways which all begin in parts of the reticular formation.

FIG. 19.30. The corticospinal (pyramidal) tract.

The pyramidal system commences in the cerebral cortex. Only about 40 per cent of pyramidal fibres are said to come from the motor area (area 4): the remainder come from the adjacent cortex, especially area 6.

The fibres run downwards from the motor cortex to the internal capsule in the corona radiata. This is an unfortunate term, because the fibres focus on the internal capsule, they do not radiate at all (a).

The motor cortex is approximately at right-angles to the posterior limb of the internal capsule. The fibres must therefore rotate through 90° as they traverse the corona radiata. The numbers in (a) and (b) correspond. Note that the 'face area' (4) [FIG. 19.09] is somewhat in front of the rest of the motor cortex, and this suggests that the direction of rotation should be anteromedial, that is, seen from above, clockwise on the left and anticlockwise on the right.

Having traversed the vertical dimension (2–3 cm) of the internal capsule, the pyramidal tract fibres come to lie in the middle three-fifths of the cerebral peduncle of the midbrain (c). The direction of rotation initiated in the corona radiata has continued through approximately another 90° so that the 'face fibres' now lie in the most medial position in the tract. They are strategically placed to supply cranial nerve nuclei either on their own side or on the contralateral side.

In the pons the pyramidal fibres are broken up into a number of separate bundles by the numerous pontine nuclei (d). These are intermediate relay stations in the cerebropontocerebellar pathway. Their afferent fibres come principally from the frontal and temporal lobes. The frontopontine fibres run in the anterior limb of the internal capsule and in the medial one-fifth of the cerebral peduncle. The temperopontine fibres run in the posterior part of the posterior limb and in the lateral one-fifth of the cerebral peduncle. They both take the obvious route.

Having 'lost' the cerebropontine fibres in the pons the pyramidal tracts now form the two pyramids on the anterior aspect of the medulla (e). They run between the two olives, and then cross over to the contralateral side giving rise to the decussation of the pyramids. The decussation is completed just as the level of the foramen magnum is reached (a, f). The result of the decussation is that the tract comes to lie in the lateral column of white matter of the spinal cord (g), becoming the lateral corticospinal tract.

A small proportion of fibres do not cross over in the medulla, and they give rise to the anterior corticospinal tract. Most of these fibres eventually cross in the spinal cord.

The pyramidal tract fibres mostly act on anterior horn cells via a small number of internuncial neurones.

The pyramidal tract fibres are organised in an orderly manner. Thus the fibres which are destined for the neck muscles (C1–4) or the brachial plexus (C5–T1) are most medial; next to them are the fibres controlling the trunk musculature and most superficially those controlling the lower limb. They will not be needed until the lower lumbar and sacral segments are reached.

Note that the fibres in the midbrain (c) would be in the logical sequence provided the tract stayed on its own side. The fact that it crosses over in the decussation means that the tract must again rotate on itself through 180°.

Handwritten left margin: Start in cerebral cortex twist with 90° as though int. capsule through Cerebral nucleus in midbrain. Splits in pons. Group in medulla through decussation. Lat. cortico-spinal tract.

Tectospinal tract — (vision reflexes)

This tract starts mainly in the superior colliculus and conveys reflex pathways to the cervical spinal cord from the visual system. It is concerned with the co-ordination of movements of head and eyes. The fibres are in the anterior funiculus of white matter behind the vestibulospinal tract [FIG. 19.31].

Rubrospinal tract

In spite of the large size of the red nucleus the rubrospinal tract is small and possibly unimportant in man. Owing to the problem of assessing the significance of very small (below the resolution of the light microscope) fibres, this question remains open. The red nucleus is a relay station for impulses from the contralateral cerebellar hemisphere. If the cerebellum is to control its own side of the body, it is obvious that the rubrospinal tract will cross back again. This happens in the midbrain, and the tract having descended through the hindbrain comes to lie in the spinal cord just in front of the (lateral) corticospinal tract [FIG. 19.31]. It is closely associated with the tegmentospinal tract from the midbrain tegmentum.

Reticulospinal tracts   Not really

The reticular formation is to be found throughout the whole length of the brainstem and in the subthalamic part of the diencephalon between the midbrain and the thalamus. There are no definite tracts or nuclei in the reticular formation, only what looks like a haphazard conglomeration of nerves cells and fibres. It is phylogenetically the oldest part of the brain. It acts on the anterior horn cells by two tracts [FIG. 19.31] which are both crossed and uncrossed. Fibres which are derived from the pons (ventrolateral reticulospinal) accompany the vestibulospinal tract, while those derived from the midbrain and medulla oblongata are in the lateral funiculus close to the pyramidal tract.

Afferent tracts corresponding to these motor tracts are present in the white matter of the spinal cord. They are either intermingled or very close to each other.

CEREBROPONTOCEREBELLAR PATHWAY

The cerebral cortex controls the opposite side of the body. At the same time information is sent to the cerebellum, the great co-ordination centre of the CNS also on the opposite side.

There is no direct cerebrocerebellar pathway. There is instead an intermediate relay station in the pons, and the pathway is the cerebropontocerebellar.

The cerebropontine fibres pass down closely associated with the pyramidal tract and relay in the pontine nuclei [FIGS. 19.32 and 19.33] on their own side. These nuclei are very numerous and break up the pyramidal tract into small bundles. The nuclei give rise to pontocerebellar fibres which run horizontally across the midline into the cortex of the opposite hemisphere. They comprise the whole of the middle cerebellar peduncle.

Two other small tracts, the anterior external arcuate fibres and the striae medullares are regarded as part of the cerebropontocerebellar pathway, whose nuclei have 'slipped' downwards from the pons into the upper part of the medulla. They both have their cells of origin in the two arcuate nuclei, one on each side in front of the pyramids. The nuclei are looked upon as ectopic pontine nuclei. The anterior external arcuate fibres cross the midline and sweep round the surface of the medulla in a lateral direction and reach the cerebellum in the inferior peduncle. The fibres destined for the striae cross the midline immediately and then run

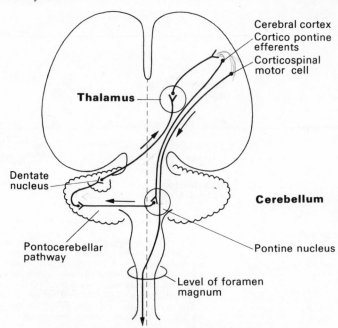

FIG. 19.32. Diagram of corticospinal and corticopontocerebellar and cerebellothalamocortical pathways.

These pathways run in parallel with the pyramidal tract, and has a very widespread origin being by no means confined to the 'motor area' of the cortex. Most of the fibres reach the internal capsule and travel in the cerebral peduncles. They relay in the numerous pontine nuclei (only one shown as a token in the Figure). The pontocerebellar fibres cross the midline and reach the cerebellum via the middle cerebellar peduncle. Note that the pyramidal fibres also cross the midline. The impulses are conveyed to the cerebellar cortex and relay in the dentate nucleus before passing via the superior peduncle to the midbrain on their way to the contralateral thalamus. They relay here and play back into the cerebral cortex. The cerebellum is therefore able to modify the behaviour of the cerebrum.

horizontally backwards next to the midline deep in the substance of the medulla. They emerge in the floor of the fourth ventricle, turn laterally forming landmarks which divide the fourth ventricle floor into two fairly equal areas. Finally they also run into the inferior peduncles. It is extraordinary that these fibres growing from their egregious parent nuclei should nevertheless, by different circuitous routes, eventually reach the same destinations. Note that they are concerned neither with the internal arcuate fibres [p. 393], nor with the posterior external arcuate fibres.

The cerebellum is a co-ordinating centre especially for movement. It must be informed from all available exteroceptive and proprioceptive sources; it must also receive information from those parts of the CNS which themselves are capable of initiating action. At the same time the cerebellum must also be able to 'interfere with' and 'block' or 'moderate' activity [FIG. 19.34].

It should be possible to make a theoretical list of the connections afferent and efferent which the cerebellum must have if it is to operate successfully. One could not say what pathway the connections will take, but that is a secondary matter.

You should expect the cerebellum to play back into the cerebrum of the opposite side and into the spinal cord on its own side.

Just as there is no direct cerebrocerebellar tract so there is no cerebellocerebral tract. The cerebellar impulses reach the opposite cerebrum after relaying in the thalamus (and other basal nuclei)

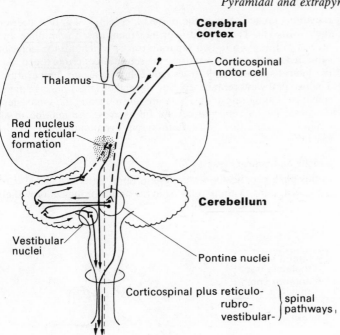

FIG. 19.33. The pyramidal tract acts on the anterior column cells of the spinal cord. The cerebellum also acts on the same cells. The principal route is the cerebellovestibulospinal, the relay taking place in the vestibular ganglion on the ipsilateral side.

An alternative route to the anterior column cells is via the red nucleus and reticular formation mostly on the opposite side. The fibres leave the cerebellum in the superior peduncle, cross the midline, relay and then cross back again immediately, and run downwards in the rubro and reticulospinal tracts.

If FIGURES 19.32 and 19.33 are taken together it can be seen that the cerebellum can modify the effect of corticospinal activity either (i) by playing back into the cortex where corticospinal activity is generated, or (ii) by playing forward on to the cells which are the spinal targets of the tract.

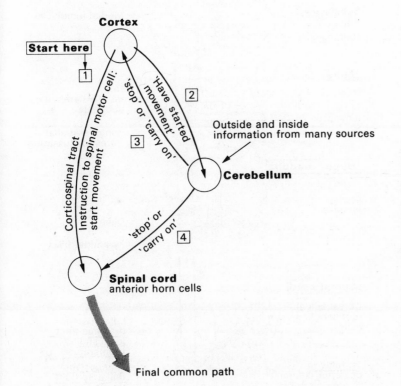

FIG. 19.34. The basic ideas of FIGURES 19.32 and 19.33 are shown. Consider the pathways in their numbered sequence, 1, 2, 3, 4. In the light of information arriving by tract 2, the cerebellum may suppress activity in the cortex (via 3) or render it ineffective when it reaches the brainstem or spinal cord (via 4).

These pathways have two or more neurones:
1. Corticospinal pathway;
2. Corticopontocerebellar pathway;
3. Cerebellothalamocortical pathway;
4. Cerebello {rubro- reticulo- vestibulo-} spinal pathway.

like other afferent impulses. The cerebellum can itself activate the ipsilateral anterior column cells in the spinal cord via the reticulo-spinal and vestibulospinal tracts, the efferents from the cerebellum having relayed in the reticular or vestibular nuclei.

AFFERENT PATHWAYS

Different modalities of sensation stimulate the peripheral nervous system in such a way that touch, hot, cold, pain, and vibration sense are all recognized as different sensations by the CNS. In addition **position sense** is made possible because of afferent information from joint receptors and muscle spindles. Each sensory pathway has at least three separate neurones before the information reaches the cerebral cortex. The first neurones have their cells of origin in the spinal ganglia; the second neurones are part of the central nervous system with their cells or origin either in the spinal cord or in the hindbrain; the central connections of the second neurones terminate in the thalamus where the cells of origin are found of the third neurones whose fibres project over most of the cerebral cortex, not merely to the postcentral gyrus. Like the motor pathways, the sensory pathway contains internuncial neurones and the thalamus is approached through the subthalamic part of the recticular formation.

Motor fibres to a particular region of the body travel together forming a particular part of the pyramidal tract. But sensory fibres from a particular region do not all travel together. As soon as the fibres reach the spinal cord they are segregated into separate pathways which differ from each other on a functional basis.

Spinothalamic tracts

The spinothalamic tracts are phylogenetically old and convey the sensations of **crude touch**, **pain**, and **temperature**. Their information reaches consciousness. The cells of origin of the peripheral neurones are in the spinal ganglia. When they enter the spinal cord

they relay immediately in the posterior columns of grey matter on their own side. The second neurones then cross the midline and ascend to the brain in two separate tracts, which are the **anterior spinothalamic tracts** transmitting the sensation of crude touch, and the **lateral spinothalamic tract** conveying pain and temperature. The two pathways remain separate throughout their passage in the spinal cord and brainstem. There are internuncial relays in or near the posterior column and certain fibres cross the midline in an oblique upward direction. Both tracts eventually relay in the thalamus [FIG. 19.35].

Gracile and cuneate tracts

In this phylogenetically newer system fibres enter the spinal cord but pass upwards without relaying in the posterior funiculus on their

---

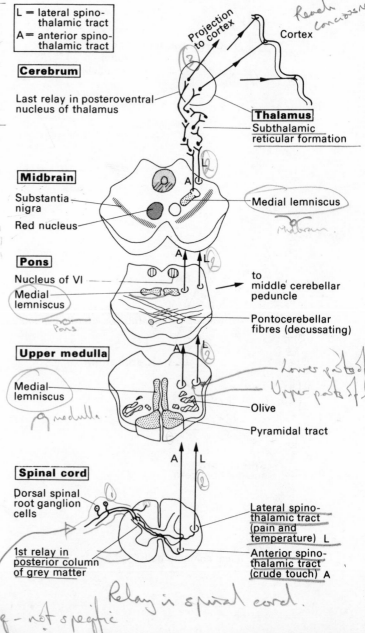

FIG. 19.35. Follow tract from below upwards.
    The spinothalamic pathways are named according to their positions in the anterior and lateral parts of the white matter of the spinal cord. These tracts do not form tightly packed bundles of fibres, but tend to be diffuse and to intermingle with neighbouring non-spinothalamic fibres.
    The first neurones in the sensory pathway have their cell of origin in the dorsal spinal root ganglia. The central connections run into the dorsal column of grey matter and relay there. The body of the second neurone is therefore in the dorsal column of grey matter and its axon then crosses to the contralateral side between the central canal of the spinal cord and the anterior median cleft. The crude touch fibres are segregated into the anterior spinothalamic tract, while the pain and temperature fibres collect in the lateral spinothalamic. Within each tract there is somatotopic localiza-tion with the fibres conveying sensation from the lowest parts of the body lying most superficial (this is the simplest possible arrangement).
    In the pons the anterior tract lies close to the medial lemniscus and by the midbrain they are in contact with each other. Light discriminating touch is conveyed by the medial lemniscus so that now crude touch and light touch are travelling in juxtaposition. Pain and temperature, conveyed in the lateral spinothalamic tract remains more or less isolated until the reticular formation in the subthalamic region conveys the impulses to the thalamus, where (ignoring by convention the multiple relays in the reticu-lar formation) the third and final neurone body is situated. Their axons project to many parts of the cortex, especially the sensory cortex where these, mostly unpleasant, sensatons reach consciousness.

The diagram labels (part of figure):

L = lateral spino-thalamic tract
A = anterior spino-thalamic tract

Cerebrum
Last relay in posteroventral nucleus of thalamus

Projection to cortex — Cortex — **Thalamus** — Subthalamic reticular formation

Midbrain
Substantia nigra
Red nucleus
Medial lemniscus

Pons
Nucleus of VI
Medial lemniscus
to middle cerebellar peduncle
Pontocerebellar fibres (decussating)

Upper medulla
Medial lemniscus
Olive
Pyramidal tract

Spinal cord
Dorsal spinal root ganglion cells
1st relay in posterior column of grey matter
Lateral spino-thalamic tract (pain and temperature) L
Anterior spino-thalamic tract (crude touch) A

own side. They do not relay until they reach the gracile and cuneate nuclei which are in the posterior aspect of the medulla, that is, in the brainstem not in the spinal cord. These are the gracile and cuneate tracts [FIG. 19.36].

The fibres, as they curve upwards on entering the spinal cord, give off collateral branches to posterior column cells and to the thoracic nuclei. Consequently their impulses will travel up simultaneously in the spinothalamic and spinocerebellar tracts, as well as in the gracile and cuneate tracts themselves.

The gracile and cuneate tracts convey the sensation of light or **discriminating touch**, as well as **proprioception**. The branches of some of these spinal (dorsal root) ganglion cells may stretch from the foot to the skull. They are without doubt the longest cells in the body. The fibres relay in the nucleus gracilis and nucleus cuneatus,

and their efferents run forwards and cross to the opposite side in the **internal arcuate fibres**. The crossed tract so formed is the **medial lemniscus**.

In the medulla the medial lemniscus lies in the parasagittal plane immediately behind the pyramidal tracts. As it passes upwards it rotates. In the tegmentum of the midbrain the anterior spinothalamic fibres join it, but the lateral tract does not.

All these fibres pass into the ventral nuclei of the thalamus, where they relay. Their impulses pass to the sensory cortex along the third neurones and also project widely to other parts of the cortex.

The decussation of the internal arcuate fibres take place at the level of the lower half of the fourth ventricle above the level of the decussation of the pyramids. Note that it is not equivalent to the great motor decussation because the spinothalamic pathways are not involved. The reason for this is that they have already crossed over in the spinal cord.

Any injury destroying the right or left half of the spinal cord, will produce complete **paralysis on one side** only (the injury side). But **sensation** is affected on **both sides**. There is **dissociated loss** of sensation. Thus on the injury side there is loss of proprioception and light touch due to the interruption of the gracile and cuneate tracts; but on the other (non injury) side there is loss of pain and temperature below the level of the lesion due to the interruption of the spinothalamic tracts. Such a lesion was described by Brown–Séquard in 1851 and is often referred to as the Brown–Séquard syndrome.

A rare disorder arises usually in the cervical enlargement of the spinal cord close to the central canal. It is **syringomyelia** (*syrinx* = a tube, as in syringa the plant from whose stems flutes were once made; *myelia* = spinal cord). A localized cyst appears which destroys only the fibres which are crossing the midline at that level, that is part of the spinothalamic pathway. There is loss of crude touch and pain and temperature but not proprioception or light touch. Patients can still play the piano but burn their fingers when smoking

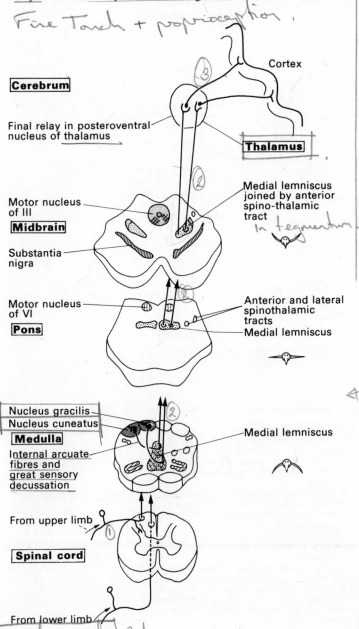

Fine Touch + proprioception.

Cerebrum

Cortex

Final relay in posteroventral nucleus of thalamus

Thalamus

Motor nucleus of III

Midbrain

Medial lemniscus joined by anterior spino-thalamic tract

In tegmentum

Substantia nigra

Motor nucleus of VI

Pons

Anterior and lateral spinothalamic tracts
Medial lemniscus

Nucleus gracilis
Nucleus cuneatus

Medulla

Medial lemniscus

Internal arcuate fibres and great sensory decussation

From upper limb

Spinal cord

From lower limb

eg gracilis muscle   *Relay in Gracilis & Cuneate nuclei*
in medulla
Cross in med lemniscus

FIG. 19.36. Gracile and cuneate tracts and medial lemniscus. Follow tract from below upwards.

Nearly all the fibres in the posterior white matter of the spinal cord belong to this system, which is responsible for 'fine touch' and proprioception. This system is entirely responsible for position sense.

The first neurones are shared with the spinothalamic system and the cell bodies lie in the spinal ganglia. Their afferent axons give off collateral branches which pass into the posterior white column on the same side. They travel upwards in the spinal cord and eventually reach the lower part of the medulla oblongata. The fibres entering the lower part of the spinal cord lie medial to those entering higher up. It is possible to see a division of the white matter into two columns, one medial, the gracile tract, and one lateral, the cuneate tract. The former conveys sensation from the lower half of the body, the latter from the upper half and the upper limb, but not of course the face which is supplied by cranial nerves.

The gracile tract relays in the nucleus gracilis while the cuneate tract relays in the nucleus cuneatus.

The second neurones have their cell bodies in these two nuclei.

In the medulla the fibres of the second neurones pass forwards and medially so as to cross the midline, forming the sensory decussation. As soon as they have crossed the midline they turn upwards forming a large tract, immediately on each side of the midline which is the medial lemniscus. The position and angle of the lemniscus is shown in the pons and midbrain. The medial lemniscus is joined by fibres from the trigeminal nuclei so that light touch and proprioception from the 'face' then join the fibres conveying these sensations from the rest of the body. All the fibres travel to the thalamus where they relay and are projected by the third neurone to parts of the cerebral cortex, especially the postcentral gyrus where the sensations reach consciousness.

[If you are wondering about the birds in the diagram, their wings are a memory aid to the shape of the tracts involved at different levels of brain-stem!]

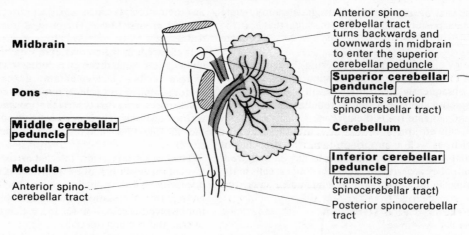

Midbrain

Anterior spino-
cerebellar tract
turns backwards and
downwards in midbrain
to enter the superior
cerebellar peduncle

**Superior cerebellar
peduncle**
(transmits anterior
spinocerebellar tract)

— for Ant tract

Pons

Cerebellum

**Middle cerebellar
peduncle**

Medulla

**Inferior cerebellar
peduncle**
(transmits posterior
spinocerebellar tract)

Anterior spino-
cerebellar tract

Posterior spinocerebellar
tract

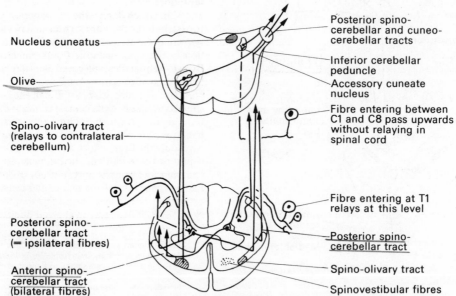

Nucleus cuneatus

Posterior spino-
cerebellar and cuneo-
cerebellar tracts

Inferior cerebellar
peduncle

*Post spino cere
& thro'
Inf cerebellar
peduncle.*

Olive

Accessory cuneate
nucleus

Fibre entering between
C1 and C8 pass upwards
without relaying in
spinal cord

Spino-olivary tract
(relays to contralateral
cerebellum)

Fibre entering at T1
relays at this level

Posterior spino-
cerebellar tract
(= ipsilateral fibres)

Posterior spino-
cerebellar tract

Spino-olivary tract

Anterior spino-
cerebellar tract
(bilateral fibres)

Spinovestibular fibres

Fig. 19.37. Spinocerebellar tracts.

There are two spinocerebellar tracts. They lie very superficially on the anterior lateral aspect of the spinal cord.

The cells of origin of these tracts lie at the base of the posterior column of grey matter. The posterior spinocerebellar tract comes from the dorsal nucleus (Clarke's column) which is only seen between T1 and L2 (mainly in the dorsal or thoracic segments of the spinal cord). The posterior tract is ipsilateral in origin: the fibres do not cross the midline. When they reach the brain stem they take the most direct route to the cerebellum by running into the inferior cerebellar peduncle. Most of the fibres reach the cerebellar cortex in or near the midline. They give rise to 'mossy' fibres.

The fibres entering the spinal cord above T1, from the upper limb and neck muscles travel upwards in the fasciculus cuneatus. There is no dorsal nucleus in the neck and they relay instead in the *accessory* or lateral cuneate nucleus. This lies close to the lowest part of the inferior cerebellar peduncle, and the fibres pass directly into it and join company with the posterior spinocerebellar tract.

The anterior spinocerebellar tract is bilateral in origin. It reaches the cerebellum in a curiously circuitous manner, by travelling upwards in the brain stem as far as the midbrain, when it turns backwards, downwards, and enters the superior cerebellar peduncle.

Two pathways are closely associated with the anterior spinocerebellar tract. One conveys information to the cerebellum not directly but via the olive. The spino-olivary tract enters the olive on its own side, and then having relayed, the pathway continues as the olivocerebellar pathway, which enters the inferior cerebellar peduncle of the opposite side. Some of the information in the spinovestibular pathways also reaches the cerebellum having relayed in the vestibular nuclei. The vestibulocerebellar fibres do not cross the midline.

a cigarette. As the cyst progresses the pyramidal and other long tracts are paralysed.

SPINOCEREBELLAR TRACTS

In FIGURE 19.37 the position of the **anterior and posterior spinocerebellar** tracts are shown. The cells of origin are the columns of **thoracic nuclei** which are at the base of the medial side of the posterior horn on each side of the midline, strategically placed next to the gracile and cuneate tracts. These cells are stimulated by collateral branches from ipsilateral gracile and cuneate tract fibres. The same impulses therefore are fed into these cerebellar pathways as are fed into the posterior column pathway.

The cerebellum principally controls its own side of the body. You would expect the feedback to be from the same side also. When the **posterior spinocerebellar tract** is traced upwards it passes into the **inferior cerebellar peduncle** of its own side. The **anterior spinocerebellar tract** runs up in the brainstem as far as the midbrain and then turns downwards in the **superior cerebellar peduncle**. It may contain some contralateral fibres.

The thoracic nuclei however are only found between T1 and L2. The afferent information entering above T1 is very important as it comes not only from the upper limb but especially from the neck muscles which control head movement. Above T1 the gracile and cuneate tract fibres do not give off collaterals until they have reached the medulla. Here the **accessory cuneate** nucleus provides the relay station just lateral to the nucleus cuneatus. The pathway to the cerebellum then continues in the **posterior external arcuate fibres**, which run a short distance to reach the **inferior cerebellar peduncle** where they join the posterior spinocerebellar tract.

CEREBELLUM

The cerebellum is the co-ordinator of movements initiated in other parts of the CNS, especially in the cerebral cortex. It is present in the most primitive vertebrates. Disorders of the cerebellum declare themselves not in actual paralysis of movement but in gross disorders of muscle co-ordination.

The cerebellum is attached to the dorsum of the brainstem by the superior, middle, and inferior cerebellar peduncles, which connect it to the midbrain, the pons, and medulla respectively. It occupies most of the posterior cranial fossa, and gives its name to the tentorium cerebelli [FIG. 19.37].

The cerebellum has **two hemispheres** joined to each other across the midline by the **vermis**. Viewed from the ventral aspect there is a deep groove separating the two halves at the bottom of which is the vermis. From the dorsal aspect the vermis smoothes over the gap and actually forms an elevated ridge in the midline.

Instead of the six layers of grey matter found in the cerebral cortex, the cerebellar cortex has only three. Although there is special localization so that different parts of the middle lobe, like the motor cortex of the cerebrum, are concerned with individual parts of the body, or the processing of different kinds of afferent information, there has been as yet no success in distinguishing histological differences between different parts of the cerebellar cortex.

The cortex appears to be fed by two histologically different afferent fibres (mossy and climbing). Its efferent fibres are all axons of Purkinje cells, which have large cell bodies and enormously elaborate dendritic processes. The individual cells have a very large number of synapses.

The cerebellum receives afferent information not only by the corticopontocerebellar and spinocerebellar pathways described above but it also receives a substantial input from the inferior olivary nucleus. This is a relay station for impulses from the thalamus and basal nuclei and reticular formation which will eventually reach the cerebellum on the opposite side. They relay in the olive on their own side. The fibres leaving the inferior olivary nucleus run straight across the midline into the inferior cerebellar peduncle.

The olive is the source of the climbing fibres which synapse directly with the Purkinje cells. The other pathways are the source of the mossy fibres which feed directly or indirectly into all three cell layers in the cerebellar cortex.

The oldest and smallest part of the cerebellum is the **flocculonodular lobe** [FIG. 19.38]. It is concerned with the **vestibular system** and not only receives most of its afferents from vestibular nuclei in the medulla but it also has direct (hot line) connections straight from the vestibular apparatus itself. On the efferent side it is connected to the brainstem and spinal cord via the reticular and vestibular nuclei.

The next phylogenetically younger part of the cerebellum is represented by the **anterior lobe** and by the **uvula** and **pyramids**. It is here that proprioceptive and tactile information is received. The whole of the rest of the cerebellum including all the **middle lobe** is the youngest part or **neocerebellum** (*neo* = new). It has developed in association with the **cerebral cortex**, with which it is connected by the two way traffic system, the **afferent** cerebropontocerebellar and **efferent** cerebellothalamocortical pathways.

The efferent Purkinje fibres collect in bundles directed towards four paired nuclei of grey matter which lie deep in the white matter of the cerebellum.

The principal nucleus is the **dentate nucleus**. The three others are the **emboliform, globose**, and the **fastigial**. The Purkinje cell axons relay in these nuclei and the efferent fibres then pass via the superior peduncle into the midbrain where they reach the opposite side. Some fibres relay in the red nucleus and reticular formation and cross back immediately to the first side and pass to the brainstem nuclei and spinal cord. Many of the efferent fibres however reach the ventral nuclei of the thalamus on their way to the cerebral cortex.

SUMMARY
CEREBELLAR PEDUNCLES

Superior cerebellar peduncle — to midbrain

This contains the following pathways:

To: cerebral cortex 1. Dentatothalamic—also from emboliform and globose nuclei—the main feedback pathway from neocerebellum to cerebral neocortex.

From: spinal cord 2. Anterior spinocerebellar.

From: brainstem 3. Proprioceptive information from cranial nerves nuclei in the midbrain, that is III and IV and especially the mesencephalic nucleus of V.

Middle cerebellar peduncle to pons.

Pontocerebellar fibres only. Part of the cerebropontocerebellar pathways, in which again neocortex communicates with neocortex, but in the opposite direction to (1) above.

Anterior lobe–intermediate 'age'
Receives afferents from spinal cord and brain stem

Fissura prima

Lingula

oldest part of cerebellum
concerned with vestibular
nuclei

Nodule

Posterior lobe:new cerebellum
cortico-ponto-cerebellar
system

(a)

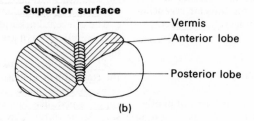

Superior surface

Vermis
Anterior lobe

Posterior lobe

(b)

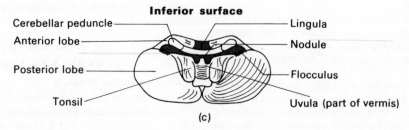

Inferior surface

Cerebellar peduncle ——————————— Lingula

Anterior lobe ——————————— Nodule

Posterior lobe ——————————— Flocculus

Tonsil ——————————— Uvula (part of vermis)

(c)

Fig. 19.38.

The cerebellum has two 'hemispheres' joined across the midline by the vermis.

The nodular floccular 'lobe' (a and c) is relatively quite small; it is phylogenetically the oldest part of the cerebellum and is connected to the vestibular nuclei and also directly to the vestibular apparatus. Efferents from this lobe relay in the fastigial nuclei, one on each side of the midline.

The anterior lobe (a, b, and c) receives information from the spinal cord via the spinocerebellar pathways, as well as corresponding afferents from the brainstem.

The posterior lobe is only found in mammals: it is the newest part of the cerebellum. It is associated with the evolutionary development of the pyramidal tracts and provides these tracts with the essential back up of the necessary co-ordinating system. This lobe is often called the neocerebellum.

The white matter of the cerebellum (arbor vitae) is shown in sagittal section in (a). On each side of the midline, and not shown in the figure, are four masses of grey matter (eight altogether) where efferent pathways synapse on their way to the superior peduncle. Apart from the fastigial nucleus; there are two small masses, the emboliform and globose nuclei and the large dentate nucleus. The dentate nucleus and olive are about the same size.

Inferior cerebellar peduncle

Action in the central nervous system can be initiated in the cerebrum, in the midbrain tectum, in the reticular formation and especially in the vestibular nuclei. Except for the midbrain, all these regions communicate with the cerebellum, via the inferior cerebellar peduncle, so that the cerebellum 'knows what is going on'.

The following **afferent tracts** are present:
1. From vestibular nuclei: vestibulocerebellar.
2. From reticular formation: reticulocerebellar.
3. From cerebrum: (a) olivocerebellar: from the lower centres in contralateral cerebrum relaying in the contralateral olive. This pathway is the main source of the climbing fibres which enter the cerebellar cortex.
    (b) anterior external arcuate fibres: striae medullares (misplaced parts of cerebropontocerebellar pathway).
4. From spinal cord: (a) the posterior spinocerebellar tract and the posterior external arcuate fibres.
    (b) information similar to that reaching the cerebellum from the spinal cord must also reach it from the peripheral components of cranial nerves.

You may have noticed that a tectocerebellar tract has not been mentioned. Perhaps it passes through or relays in the reticular formation. It is certain that such a pathway exists, and that there is also a pathway in the reverse direction. It is a basic tenet of the logic of communication theory that a message cannot be said to have gone from A to B, when you are at A, until B acknowledges it. It is very helpful to bear this fact in mind when considering the communication systems of the CNS.

Efferent tracts

To the vestibular nuclei: (a) fastigiovestibular tract. These fibres have their cells of origin in the nucleus fastigius [Fig. 19.03(e)]. This nucleus is close to the inferior peduncle.
    (b) noduloflocculovestibular. The nodule is the oldest part of the cerebellum which arose in close association with the vestibular system.

Both these tracts represent the first step in the cerebellovestibular-spinal pathways and are the principal effector mechanisms of the cerebellum. There are also cerebellovestibulobulbar pathways which allow the cerebellum to regulate movements involving the cranial nerves, in the same way as it regulates movements involving the spinal nerves.

OLFACTORY AREAS AND THE 'LIMBIC SYSTEM'

The sense of smell in man is poorly developed. Yet the olfactory parts of the brain and certain associated regions called the 'limbic system' are the largest of any mammal in both absolute and relative terms. It is widely believed that the limbic system is no longer much concerned with olfaction but with the determination and expression of mood and affect. Major psychotic illnesses, including schizo-

phrenia, may be disorders of the limbic system. Memory of recent events is abolished by damage to parts of the limbic system. The anatomy of this region is complex and poorly understood. The possibility of its importance was suggested by Le Gros Clark in the 1930s.

Olfaction

The olfactory mucosa lies in the upper part of the nasal cavity and has an area of only 3–4 cm². The chemical substances in the inspired air excite the hairs on the olfactory mucosa cells which give rise to action potentials. There seems no end to the number of different smells which the nose can distinguish, and the brain remember.

The central axons of the olfactory cells pass through the cribriform plate in bundles and relay in the olfactory bulb. From here fibres pass backwards in the olfactory tract, which divides into the medial and lateral olfactory striae. The latter passes into the floor of the inferior horn of the lateral ventricle and to the uncus, which is probably the region where smell reaches consciousness. Tumours in this region give rise to olfactory hallucinations. Complete loss of smell occurs after fractures of the base of the skull in which the cribriform plates are often severely involved.

LIMBIC SYSTEM

The medial stria of the olfactory tract terminates in grey matter around the anterior commissure in the **medial olfactory** or **septal area**. This region probably has nothing to do with smell, but is part of the limbic system. Also included in this system are the **hippocampus** and **parahippocampal gyrus**, the **cingulate gyrus** and the surface of the **insula**, the grey matter of the **amygdeloid nucleus** and the **dentate gyrus** which continues over the upper surface of the corpus callosum as the **induseum gresium** (*induseum* = a shirt). The **fimbria** and **fornix** join the hippocampal region to the **mamillary bodies**. All these structures, with their connecting pathways, constitute the very extensive and elaborate limbic system [Fig. 19.39].

These areas are connected to each other by tracts, of which the **fornix** is important connecting the hippocampal region to the mamillary bodies. The stria terminalis runs from the medial olfactory region to the floor of the fourth ventricle. There are many other tracts allowing intercommunication within the system.

There are other tracts which allow intercommunication between the limbic system and the other, non-limbic, regions of the brain.

Thus the mamillary bodies are connected to the thalamus, by the **mamillothalamic tract,** and to the reticular formation in the tegmentum of the brainstem by the mamillotegmental tract.

RETICULAR FORMATION

The reticular formation is found in all levels of the brainstem and it reaches upwards into the **subthalamic** region.

The formation has three major component groups of cell and fibres:
1. The **paramedian**: near the midline. Like those of the inferior olive, the efferents go to the **cerebellum**.
2. The **central**: further from the midline. This unfortunate term is only meaningful when only half the brainstem is thought of. This part of the reticular formation seems to have an effector function and is inhibitory to the lower centres of the spinal cord, and excitatory to the higher centres.
3. The **lateral**: from its position it would be expected to be *largely sensory*. It receives collateral branches from neurones of the sensory pathways.

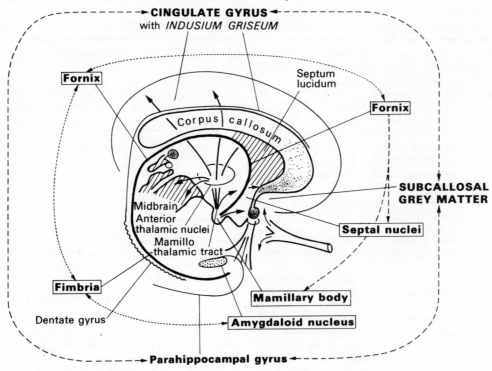

FIG. 19.39. The limbic system.

The limbic system is complex to look at because of the curved shape of the cerebral hemisphere as seen from the side, and also the presence of the corpus callosum.

It is possible to regard it as consisting of an outer circle (capitals in the figure) and an inner circle (small letters in the figure). The components of the two circles are joined here by two-headed arrows to imply that the circuitry of this system is not yet understood.

The outer circle consists mainly of cortex, in particular the parahippocampal gyrus, the cingulate gyrus and the subcallosal grey matter.

The inner circle consists of grey matter around the inferior horn of the lateral ventricle which reaches the mamillary bodies on both sides. From here the large mamillothalamic tract reaches the anterior nucleus of the thalamus, and also fibres spray out into the hypothalamus, the septal region and the brainstem, especially the reticular formation in the tegmentum of the midbrain. The anterior commissure allows connections to be established between the temporal lobes on each side. The habenular commissure is described in FIGURE 19.28.

The reticular formation is the oldest part of the brain. It is probably connected to all the newer systems of the brain and spinal cord. Extero- and proprioceptive information therefore passes through the reticular formation on the way to the thalamus and cortex.

The reticular formation has a profound effect on the state of 'arousal' of the cortex, and this component is known as the ascending reticular activating system.

Certain anaesthetics and tranquillizers work indirectly on the cortex by influencing the activity of the reticular formation.

The reticular formation contains all the vital centres, and in addition regulates wakefulness and sleep. Damage results in coma and death. Like the limbic system the reticular formation influences behaviour, and is of interest to psychologists and psychiatrists.

THALAMUS

The thalamus (= a bed!) is the final relay station for sensory pathways from the brainstem and spinal cord whose information is destined for the cortex.

The **medial** and **lateral geniculate bodies** are part of the thalamus. The former is the final relay in the auditory pathway before it reaches the auditory cortex, while the lateral geniculate body is the corresponding final relay station in the visual pathway. These have already been described. They are very small in comparison with the rest of the thalamus.

Cut in horizontal section [FIG. 19.40] the posterior half of the thalamus is seen to be divided by the internal medullary lamina in the parasagittal plane into **medial** and **lateral thalamic nuclei**. The **internal medullary lamina** (of white matter) splits to envelope the central thalamic nuclei. It is very obvious to the naked eye.

Further forward the central nuclei disappear and the lamina forks to enclose the **anterior nuclei**. At the same time the lateral nucleus to a large extent comes to be replaced by the **ventral nucleus**.

Besides these five sets of nuclei the thalamus has the thalamic reticular nucleus on its lateral side, separating it from the internal capsule.

The various nuclei of the thalamus between them project to specific regions covering most of the surface of the cortex [FIG. 19.41]. Commuting traffic between the cortex and the thalamus is responsible for the electrical changes of the EEG (electroencephalogram Berger rhythm, etc.).

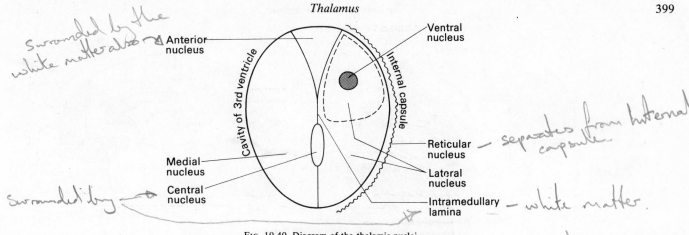

Fig. 19.40. Diagram of the thalamic nuclei.
The 'nuclei' shown in the diagram are subdivisions of the thalamus
which are all <u>obvious to the naked eye.</u> In this view the ventral nuclei are
hidden beneath the lateral nucleus. The reticular nucleus is separated from
the rest of the thalamus by <u>thin layers of grey and white matter.</u>

[handwritten annotations: "Surrounded by the white matter also" pointing to Anterior nucleus; "Surrounded by" pointing to Central nucleus; "separates from Internal capsule" near Reticular nucleus; "white matter." near Intramedullary lamina; "horizontal section."]

THALAMIC NUCLEI

Nucleus	Connections	Comment
Anterior	from: mamillothalamic tract to: cyngulate gyrus	<u>Limbic system</u>
Ventral	from: medial lemniscus spinothalamic tracts brainstem to: corpus striatum frontal parietal cortex	<u>Pathway to consciousness</u>
Central	from: thalamic nuclei reticular formation to: cortex	<u>Cortical arousal system</u>
Reticular	from: thalamus, brainstem, reticular formation cortex to: cortex	
Medial	from: thalamus basal ganglia frontal cortex to: prefrontal region Thalamofrontal and front-thalamic fibres run in the anterior limb of the internal capsule	The prefrontal cortex has profound importance in determining '<u>personality</u>' and 'character'. Section of these efferent fibres turns an ambitious, clever, foresighted person into a purposeless cabbage
Lateral including the pulvinar	from: parietal lobe occipital lobes to: parietal lobe occipital lobes	
Medial geniculate body	from: inferior brachium to: auditory cortex	<u>Conscious auditory pathway</u>
Lateral geniculate body	from: optic tract **to: calcarine cortex**	<u>Conscious visual pathway</u>

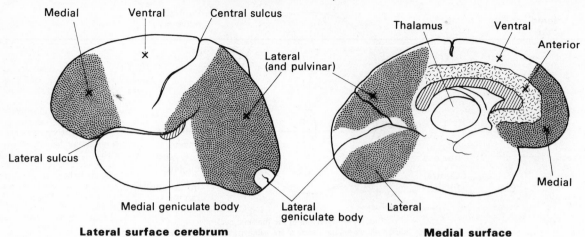

Medial   Ventral   Central sulcus

Lateral
(and pulvinar)

Lateral sulcus

Thalamus   Ventral

Anterior

Medial

Medial geniculate body   Lateral
geniculate body   Lateral

Lateral surface cerebrum

Medial surface

FIG. 19.41. Diagram of thalamocortical projections.
The figure shows those parts of the cerebral cortex to which the thalamic nuclei project.
Note that in certain areas the boundaries between projection areas (visual, auditory, limbic system) coincide with the boundaries of
certain of the cytoarchitectonic areas in FIGURE 19.12.

BASAL NUCLEI

The basal nuclei (basal ganglia) comprise those masses of grey matter which develop deep in the cerebral hemisphere on each side, and are therefore part of the **telencephalon**. They are closely associated functionally (as well as anatomically) with the thalamus, which develops in the side wall of the third ventricle, in the **diencephalon**.

These nuclei are the **caudate** and **amygdaloid** nuclei and the **lentiform nucleus**. The lentiform nucleus is in two parts, the **putamen** (= a shell or husk) and the **globus pallidus**. The former is dark grey, the latter pale. In addition a thin layer of grey matter separates the outer surface of the putamen from the cortex of the insula [FIG. 19.43] called the **claustrum** (= a bar, not a good name). The anterior putamen and head of the caudate nucleus are continuous with each other with strips of grey and white matter alternating. For this reason they are referred to together as the **corpus striatum**.

The basal nuclei are related to each other and to the internal capsule, third ventricle and insula as shown in FIGURE 19.42. The lateral ventricle circumnavigates the caudate nucleus and its tail.

INTERNAL CAPSULE

The internal capsule has an anterior and posterior limb connected by the genu [FIG. 19.43]. The anterior limb contains thalamofrontal and corticopontine nerve fibres. Further back in the genu are cortical fibres destined for the medulla and for the reticular formation in the brainstem.

Also in the genu and in the anterior two-thirds of the posterior limb are the pyramidal tract fibres.

The 'face' fibres are in the genu, and the lower limb and sphincter areas are the most posterior [see p. 387]. In the rest of the posterior limb are the afferent thalamocortical fibres destined for the post-central gyrus.

Further back in the retrolenticular and sublenticular regions are the optic and auditory radiations. Sweeping over the basal nuclei

and lateral ventricle are the fibres of the corpus callosum which weave through the cortical projection fibres at right angles.

Thrombosis and haemorrhage in the central branches of the middle cerebral artery which supply the internal capsule may give rise to hemiplegia in which the whole of one side of the body is paralysed because of the crowding together of so many motor fibres into a small area.

SUMMARY
EXTRAPYRAMIDAL SYSTEM

The pyramidal system is found only in mammals. In other vertebrates the basal nuclei control movement exclusively. Evolutionary improvements have taken place by adding to the earlier systems rather than replacing them. So Man's pyramidal system is in addition to his still active and important 'extrapyramidal' system.

By the extrapyramidal system is meant those motor systems of the CNS other than the pyramidal system itself. Too much should not be made of the concept that the pyramidal and extrapyramidal systems are mutually exclusive.

The main effector components are:
1. Corpus striatum.
2. Reticular formation including subthalamus, red nucleus, and substantia nigra.
3. Cerebellum and vestibular nuclei.

The corpus striatum has many connections from the cortex, that is many projection fibres terminate in the head of the caudate lobe or in the putamen. From here impulses are relayed to the globus pallidus. From the pallidum impulses are transmitted in short relays to the subthalamus, the reticular formation and to the reticulospinal tracts. The red nucleus is regarded as part of the reticular formation and conveys impulses to the rubrospinal and rubroreticulospinal tracts, and also to the olive.

The cerebellum and vestibular ganglia have already been described.

Diseases which damage the extrapyramidal system produce characteristic disabilities such as paralysis agitans (Parkinsonism).

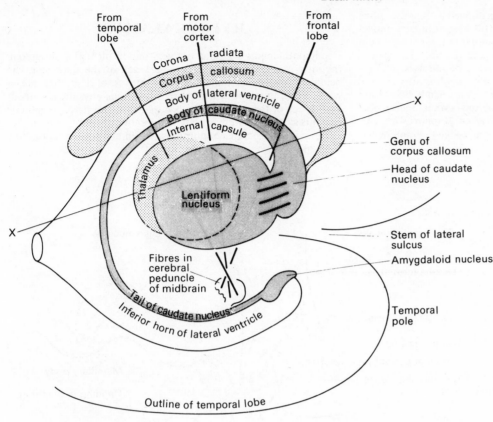

From temporal lobe

From motor cortex

From frontal lobe

Corona radiata
Corpus callosum
Body of lateral ventricle
Body of caudate nucleus
Internal capsule

Thalamus

Lentiform nucleus

X

X

Genu of corpus callosum

Head of caudate nucleus

Stem of lateral sulcus

Amygdaloid nucleus

Fibres in cerebral peduncle of midbrain

Tail of caudate nucleus

Inferior horn of lateral ventricle

Temporal pole

Outline of temporal lobe

FIG. 19.42. Basal nuclei.

The basal nuclei are shown from the right side after removal of all the neuronal tissues which normally hide the nuclei.

The lateral part of the lentiform nucleus. the putamen and the head of the caudate nucleus are continuous with each other and are referred to as the corpus striatum. The separation of the upper part of the lentiform nucleus from the body of the caudate nucleus is brought about by pyramidal corticobulbar and corticospinal fibres. The corona radiata fibres pass at right angles to the corpus callosum fibres.

The relationship of parts of the caudate nucleus to the several regions of the lateral ventricle are shown. Note the amygdaloid nucleus at the end of the tail of the caudate nucleus. and the anticlockwise twist on the corticofugal fibres as they pass down to the brainstem.

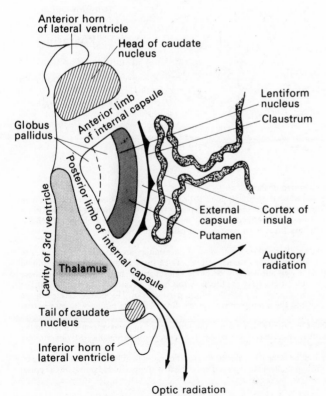

Anterior horn of lateral ventricle

Head of caudate nucleus

Globus pallidus

Anterior limb of internal capsule

Posterior limb of internal capsule

Cavity of 3rd ventricle

Thalamus

Tail of caudate nucleus

Inferior horn of lateral ventricle

Optic radiation

Lentiform nucleus

Claustrum

External capsule

Putamen

Cortex of insula

Auditory radiation

FIG. 19.43. Section through X–X FIG. 19.42.

The motor and corresponding sensory fibres from the pre- and post-central gyri lie in the posterior genu and the adjacent part (only) of the posterior limb.

The auditory and optic radiations lie behind the posterior limb. In hemiplegia the vascular disorder usually occurs in small blood vessels which pass through the holes in the anterior perforated substance and supply the region of the genu. Almost the whole of one side of the body may be affected. the 'face' and upper limb being most severely affected with the lower limbs and sphincters more or less escaping. sight and hearing being unaffected. Obviously the disability in individual cases depends on the exact location of the thrombosis or haemorrhage.

It is important to understand the paradox implied in this descriptive term. The patient has difficulty in performing voluntary movements, yet the head, forearm, and hands are in constant movement (pill rolling movements of the fingers and thumb). The gait is characteristically shuffling and the posture bent forwards, face mask-like and unexpressive. The patient's speech is affected and other communication systems; movements of the hands, bodily and facial expression, the ability to look interested in what other people are saying and so on, are all severely affected. Rapport with such patients may be difficult for these reasons although their intellect is usually unimpaired.

HYPOTHALAMUS

The hypothalamus is a small but very important region situated in the antero-inferior part of the lateral wall of the third ventricle below the level of the thalamus. It contains several different nuclei with diverse functions which control in various ways the activity of both lobes of the pituitary, the sympathetic and parasympathetic nervous system, as well as body temperature, appetite, and wakefulness. It also secretes endorphins, which are potent pain-killing substances.

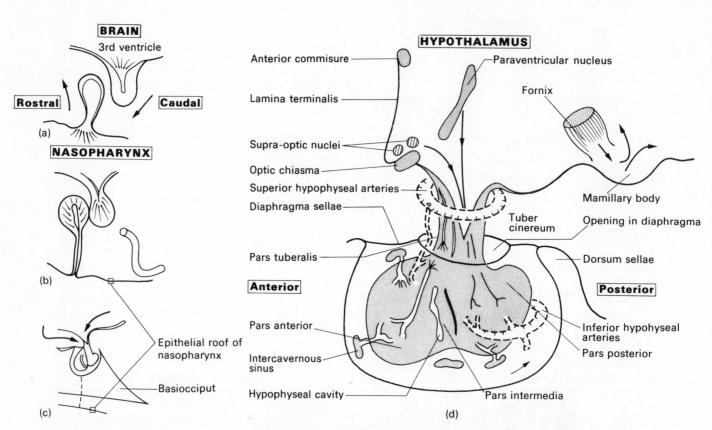

FIG. 19.44. The pituitary gland and hypothalamus.

(a), (b), (c) The early stages of development of the pituitary are shown. From the epithelium, which is eventually going to lie in the roof of the nasopharynx, there is an upgrowth called Rathke's pouch. A downgrowth from the floor of the procencephalon (floor of 3rd ventricle) also takes place. The two outgrowths meet. The former loses its connection with the nasopharynx, as the neck of the diverticulum atrophies and most of the cells disappear (b and c). It is usually possible to find cellular remnants of the neck of Rathke's pouch embedded in the sphenoid bone below the pituitary fossa. Such remnants are called 'cell rests' (from the German word rest = remnant). Cell rests give rise to tumours.

(d) The pituitary is shown from the left, cut in sagittal section. The blood supply of the anterior and posterior lobes of the pituitary are largely independent. The former receives nearly all its blood supply from the superior hypophyseal arteries, branches of the internal carotid, which form an anastomotic ring around the upper part of the pituitary stalk. Most of the branches pass into the stalk and discharge into fairly wide blood sinuses in the pars tuberalis. From here the sinuses empty into veins which run down within the stalk and reach the anterior pituitary itself. Here the blood vessels break up once more into sinuses. The hormones picked up by the first group of sinuses are carried downwards in the veins and regulate the activity of the pars anterior. Hormones secreted in the pars anterior pass into the second set of veins which drain into the cavernous sinus, or into the intercavernous vessels. From here the pituitary hormones pass into the general circulation. This is the pituitary portal system.

Posterior pituitary activity is controlled more directly by the hypothalmus, in particular by fibres which begin in the supraoptic and paraventricular nuclei in the lateral wall of the third ventricle and travel down in the infundibulum (= a funnel) and stalk to reach the posterior lobe. The venous return from the posterior pituitary also discharges into the cavernous sinus.

Note that the hypophyseal cavity is a persistent part of the cavity of Rathke's pouch. Behind it the pars intermedia is part of the wall of Rathke's pouch. The pars tuberalis is very important as it is here that the hormones secreted here or elsewhere in the hypothalamus are picked up by the first group of blood sinuses in the portal system.

In many instances the activities of the hypothalamus have diurnal rhythms which may take many weeks to re-adjust to a new time-table, as jet-lagged travellers can testify. Other functions, such as the menstrual cycle, are regulated on a monthly basis. The hypothalamus can also work to a much longer time scale, controlling the length of pregnancy, the onset of puberty, of maturity, and even senility. Its importance to the body as a whole is out of all proportion to its size.

The hypothalamus extends from the mamillary bodies to as far forward as the lamina terminalis [FIG. 19.44]. Laterally it extends to the internal capsule. The infundibulum connects the hypothalamus to the pituitary gland. Between the mamillary bodies and infundibulum is the **tuber cinereum** (hump of ashes). It is greyish-blue in colour. It is also called the median eminence.

The hypothalamus contains many nuclei, of which about six have been identified in the supra-optic region. Between them they control the secretions of the anterior (adenohypophysis) and posterior pituitary (neurohypophysis). The tuber cinereum is also concerned here.

The supra-optic and paraventricular nuclei have fibres which terminate in the posterior pituitary [FIG. 19.44]. They pass through the pituitary stalk. Their nerve terminals do not release a chemical transmitter such as acetylcholine or adrenaline like 'ordinary' nerve terminals, but instead they secrete peptide hormones such as vasopressin (the antidiuretic hormone) and oxytocin (the child-birth and milk ejection hormone). Osmoreceptors in the region of the hypothalamus sample osmotic pressure exerted by the bloodstream, and increase or decrease appropriately the synthesis and release of the antidiuretic hormone by the posterior pituitary into the cavernous sinus which in turn regulates the urinary volume. The neurohypophysis is truly part of the brain: in development it is a downgrowth from the floor of the third ventricle. The presence of nerve fibres running from the hypothalamus to the posterior pituitary is therefore not surprising. They constitute a neurosecretory system.

The posterior pituitary is supplied by the **inferior hypophyseal arteries**, which come off the carotids in the cavernous sinus on each side of the midline. Their branches break up into sinusoids in the neurohypophysis, and forming venules, finally drain into the cavernous sinus. Branches of the inferior hypophyseal artery do not reach the anterior pituitary.

The **anterior pituitary**, however, develops in the embryo in quite a different manner: it is at first an epithelial upgrowth called Rathke's pouch from the roof of the future nasal cavity. The blind end of this pouch comes to lie immediately in front of the neurohypophysis. The neck of Rathke's pouch soon atrophies and disappears [FIG. 19.44].

The method of control of the hormones secreted by the anterior pituitary (adenohypophysis) reflects its different mode of development. There are no nerve fibres running from the brain to the adenohypophysis: instead it is complex vascular connections which establish communications from the hypothalamus.

The two **superior hypophyseal arteries** are branches of the internal carotids. They encircle the pituitary stalk and anastomose with each other. They give off branches which break up into capillaries in the infundibulum and median eminence. The capillaries then rejoin each other forming small veins which run down the pituitary stalk. When they reach the anterior pituitary they break up for the *second time* into capillaries and blood sinusoids which lie between the secretory cells. These sinusoids drain via small veins into the cavernous and intercavernous sinuses and thence into the general bloodstream. This complex of vessels is the pituitary portal system (hypothalamohypophyseal portal system).

Anterior pituitary hormones are synthesized and stored in the cells. Entry to the bloodstream is controlled by other releasing and release-inhibiting hormones. These substances are synthesized in the hypothalamic cells including those of the median eminence, and pass down their axons to the nerve terminals whereby they are secreted into the first group of capillaries of the pituitary portal system. They then reach the anterior pituitary blood sinusoids causing the release of the stored hormones into the blood vessels. From here the hormones reach the cavernous sinus and enter the general circulation.

Besides its control of the pituitary, the hypothalamus is concerned with body temperature (temperature regulating centre): the anterior part lowers body temperature, causing vasodilatation and sweating, while the posterior part raises body temperature via the subthalamus and vasomotor centre in the reticular formation. The tuber cinereum contains a 'hunger centre'.

The hypothalamus is closely related to the limbic system, with the mammillary bodies, the fornix, and various limbic system tracts passing next to or through it. In addition the reticular formation is represented by the adjacent subthalamus, and the basal nuclei are in the immediate superolateral vicinity. The parts of the brain concerned with the endocrines, with arousal, mood and affect are all closely associated in these non cortical parts of the cerebrum.

BLOOD–BRAIN BARRIER

When a dye such as trypan blue is injected into the bloodstream, it passes through the capillary walls and stains the tissues blue—but not the brain. The capillary endothelial cells provide a barrier which keeps out certain substances such as bile pigments (usually the brain is not yellow like other body tissues in jaundice) while letting others through. Only in severe cases of jaundice do the basal nuclei become stained (kernicterus)

Drugs which do not cross the barrier are without effect on the CNS. When the CNS is damaged, say by a stroke, the blood–brain barrier breaks down locally. It is possible to demonstrate this radiographically by injecting suitable radiopaque material into the bloodstream. This is of great diagnostic value.

Index

Index